Arctic One Health

Morten Tryland
Editor

Arctic One Health

Challenges for Northern Animals and People

Springer

Editor
Morten Tryland
Department of Arctic and Marine Biology
UiT The Arctic University of Tromsø
Tromsø, Norway

Department of Forestry and Wildlife Management
Inland Norway University of Applied Sciences
Koppang, Norway

ISBN 978-3-030-87852-8 ISBN 978-3-030-87853-5 (eBook)
https://doi.org/10.1007/978-3-030-87853-5

This Springer imprint is published by the registered company Springer Nature Switzerland AG.
The registered company address is: Gewerbestrasse 11, 6330 Cham, Switzerland

Preface

As a researcher, we may like to think that the findings we published ten years ago are still valid, and I keep referring to them as if they are. However, life conditions and ecosystems are constantly changing. The Intergovernmental Panel on Climate Change (IPCC) just recently (August 2021) published their new report, "Climate Change 2021," with a summary of the current state of the climate and scenarios for the future. According to this report, we will probably see increasing shifts in climate conditions which will cause more extreme weather events and their consequences (drought, wildfires, floods, erosion, and landslides) than ever before, affecting the life of animals and many people around the globe.

What about the Arctic region? The frozen north—where people are still harvesting from what nature may offer and relying on traditional knowledge gathered through many generations to survive? Unfortunately, the prospective climate change and its impact are expected to be even more severe in northern regions than in other parts of the globe.

This map shows the March (dark blue) and September (light blue) median extent of the ice in the Northern Hemisphere during 1981–2010, a period that has been used as an important reference for climate. In comparison, the ice extent for August 2020, which is demarcated by a red line, indicates that the amount of ice is rapidly declining in this region. This will likely have a severe impact on the life conditions for many northern species, such as birds, seals, polar bears, and the marine organisms in general—and for people.

Compared to other regions, very few people live in the Arctic today. However, as the temperatures are increasing in many places, glaciers are retracting revealing barren land that may have been covered by ice for thousands of years. The findings of stone arrows, leather clothes, and bones from hunters and gatherers who perished long ago remind us that people have actually been living off this land for thousands of years—shaping stones for tools and learning how to make shelter and clothing to combat the harsh and relentless natural forces of the Arctic.

As a veterinarian, I have enjoyed the opportunity to work with infectious diseases and zoonoses in wildlife species and populations in the northern regions for the most of my career. However, as the work with this book developed, I realized that this book project was not just about veterinary medicine. In an effort to increase the understanding of the complexity of the biology, health, and diseases in these northern ecosystems as well as its animal populations and its peoples, it instead evolved into an attempt to introduce multidisciplinarity.

Thus, although a core part of this book, the aim was not just to present a review on relevant pathogens and their epidemiology in northern regions. The group of authors have all strived to reveal what is unique about the northern ecosystems and its inhabitants. We have tried to portray the Arctic region as a whole, describing how it is affected by different environmental, biological, and anthropological processes, including climate change, seasonal animal migrations, loss of original wildlife habitats, and pollution. We also present chapters on toxic hazards by hunting with lead, potential hazards associated with traditional ways of conservation and preparation of arctic food, wildlife health surveillance, and challenges by providing good veterinary services in small and local communities, as well as a short introduction to reindeer herding and health and disease challenges.

As I am writing this, a small virus with just a few genes is ruling the world! After nearly two years of the COVID-19 pandemic, we are still struggling to find the best way forward for our families and local communities, as well as on national and global scales. The pandemic has changed our lives in so many ways. It is a serious reminder that a virus hosted by animals may sometimes overcome the host restrictions and become a threat to people´s lives and our vulnerable societies and economies.

This experience contributed to a strong tailwind, encouraging us all to have this book published. It is more clear than ever that the well-being of the environment and its living components are all connected. One Health is more than zoonotic diseases—it is a multidisciplinary approach to broaden our views and knowledge. It allows us to make better and more holistic decisions and to find sustainable solutions on local, regional, and global levels. We hope this book will inspire students and professionals alike to become engaged in the Arctic region and to learn about the many ways this vulnerable region of the planet is challenged.

Tromsø, Norway
Koppang, Norway

Morten Tryland

Acknowledgments

I would like to thank all the students that, brick by brick and over many years, have brought their culture, language, curiosity, energy, inspiration, and interest for wildlife to our rather small research group, Arctic Infection Biology, at UiT, The Arctic University of Norway located in Tromsø, Northern Norway. You have contributed to so many interesting projects, discussions, and publications and have also continuously inspired your fellow students and colleagues to work on infection biology and One Health issues in the northern regions.

Fifty-four authors have contributed to one or more chapters of this book. They represent the most excellent and updated knowledge in their fields and are acknowledged for their time and engagement in compiling the chapters of this book.

A review process always improves a text, and all the chapters have been reviewed by experts in their respective fields. I would like to thank all the reviewers who have contributed with their valuable time and knowledge: Drs. Bart Nolet, Grete K. Hovelsrud, Mia Landauer, Heli Routti, Mandy Lindeberg, Brenda Howard, Simon Weli, Jack C. Rhyan, Antti Oksanen, Lucy Robertson, Rebecca Davidson, Kjersti Wik Larssen, Pierre-Yves Daoust, Ingebjørg Helena Nymo, Niels Kanstrup, Heather Fenton, Marianne Lian, Jan Åge Riseth, and Andrea Miller.

A special thank you to Dr. Andrey Gogin (Deputy Director for International Cooperation), to Dr. Denis Kolbasov (Director), and to Dr. Nataly Pavelko (Head of International Coordination) at Federal Research Center for Virology and Microbiology, Pokrov, Russia, for their extraordinary help with organizing and sharing data on wildlife diseases in the northern regions of the Russian Federation. The Governors of Svalbard (Sysselmannen) and Ragnar Sønstebø (Norwegian Meteorological Institute) are acknowledged for sharing information about disease outbreaks in the region of Svalbard, and Dr. Vilhjalmur Svansson (University of Iceland) is acknowledged for sharing historical information on the present and previous outbreaks of infectious diseases in Iceland.

I would also like to thank Editor Silvia Herold and Project Coordinator Srinivasan Manavalan at Springer Nature for taking the initiative to compile a book about One Health in the Arctic region and for taking care of the production process.

Contents

Editor and Contributors

About the Editor

Morten Tryland is a professor in veterinary medicine—infection biology and One Health. After a period of clinical veterinary practice, he transitioned to research in virology and wildlife diseases. He has worked for 30 years in northern and arctic regions with infectious diseases, zoonoses, wildlife reservoirs, and disease transmission in populations. He has been a member of the Norwegian Scientific Committee for Food Safety (Panel on Biological Hazards) and the European Food Safety Authority (EFSA; Panel on Biological Hazards), and he is currently the Norwegian editor of the scientific journal *Acta Veterinaria Scandinavica*. Although Dr. Tryland has been involved in several research projects in South-East African countries, he has spent most of his research career investigating infectious diseases and zoonoses in arctic wildlife and semi-domesticated reindeer. He has led or participated in projects addressing how infectious agents impact humans and wildlife in the vulnerable northern ecosystems in Alaska, Canada, Iceland, and Fennoscandia, including the Svalbard Archipelago. He has spent a sabbatical period at the University of Fairbanks, Alaska, USA, and at East Iceland Nature Research Centre, Egilsstadir, Iceland. Dr. Tryland recently started in a new position as a professor in One Health at Inland Norway University of Applied Sciences, Koppang, Norway.

Contributors

Xavier Fernández Aguilar, MS, DVM, PhD University of Calgary, Calgary, AB, Canada

O. Alejandro Aleuy, DVM, PhD Department of Ecosystem and Public Health, University of Calgary, Calgary, AB, Canada

Sarah Allan, PhD Assessment and Restoration Division, NOAA Office of Response and Restoration/U.S. Department of Commerce, Homer, AK, USA

Jon M. Arnemo, DVM, PhD Inland Norway University of Applied Sciences, Koppang, Norway
Swedish University of Agricultural Sciences, Umeå, Sweden

Tessa Baker, DVM, PhD Department of Ecosystem and Public Health, University of Calgary, Calgary, AB, Canada

Silke Bauer Department of Bird Migration, Swiss Ornithological Institute, Sempach, Switzerland
Federal Institute of Technology ETH Zürich, Zürich, Switzerland

Kimberlee Beckmen, MS, DVM, PhD Alaska Department of Fish and Game, Fairbanks, AK, USA

Émilie Bouchard, DVM, MSc Western College of Veterinary Medicine, University of Saskatchewan, Saskatoon, SK, Canada

Anouk Brisebois, MSc CICERO Shades of Green, CICERO Center for International Climate Research, Oslo, Norway

Nigel Caulkett, DVM, MVetSc, DACVAA University of Calgary, Calgary, AB, Canada

Sylvia L. Checkley, DVM, PhD University of Calgary, Calgary, AB, Canada

Rebecca Davidson, DVM, PhD Norwegian Veterinary Institute, Tromsø, Norway

John J. Debenham, BVSc, CertAVP, PhD Norwegian University of Life Sciences, Ås, Norway

Rune Dietz, MSc, DSc Faculty of Science and Technology, Department of Ecoscience, Arctic Research Centre (ARC), Aarhus University, Aarhus, Denmark

Svetlana Dresvyanikova, DVM, PhD Federal Research Center for Virology and Microbiology, CVM Pokrov, Vladimir, Russia

Frauke Ecke, PhD Department of Wildlife, Fish, and Environmental Studies, Swedish University of Agricultural Sciences (SLU), Umeå, Sweden

Irina Egorova Federal Research Center for Virology and Microbiology, CVM Pokrov, Vladimir, Russia

Anastasia Emelyanova, PhD Thule Institute, University of Oulu, Oulu, Finland
University of the Arctic, Thematic Networks and Research Liaison Office, Oulu, Finland

Magnus Evander, PhD Department of Clinical Microbiology, Arcum, Umeå University, Umeå, Sweden

Heather Fenton, DVM, MVSC, DACVP Ross University School of Veterinary Medicine, Basseterre, St. Kitts, West Indies

Taya Forde, DVM, PhD Institute of Biodiversity Animal Health & Comparative Medicine, University of Glasgow, Glasgow, UK

Boris Fuchs, MSc Inland Norway University of Applied Sciences, Koppang, Norway

Barbara A. Han, PhD Cary Institute of Ecosystem Studies, Millbrook, NY, USA

Cristina Hansen, DVM, PhD Department of Veterinary Medicine, University of Alaska Fairbanks, Fairbanks, AK, USA

Karsten Hueffer, DVM, PhD FHEA Department of Veterinary Medicine, University of Alaska Fairbanks, Fairbanks, AK, USA

Emily J. Jenkins, BSc Hon, DVM, PhD Department of Veterinary Microbiology, WCVM, University of Saskatchewan, Saskatoon, SK, Canada

Bjørn Munro Jenssen, MSc, PhD Norwegian University of Science and Technology, Trondheim, Norway

Pikka Jokelainen, DVM, PhD Statens Serum Institut, Copenhagen, Denmark

Temitope U. Kolapo, DVM, MEc Department of Veterinary Microbiology, WCVM, University of Saskatchewan, Saskatoon, SK, Canada

Susan J. Kutz, DVM, PhD Department of Ecosystem and Public Health, University of Calgary, Calgary, AB, Canada

Robert J. Letcher, MSc, PhD Environment and Climate Change Canada, Ottawa, ON, Canada

Marianne Tronstad Lund, PhD CICERO Center for International Climate Research, Oslo, Norway

Magnus Magnusson, PhD Swedish Forest Agency, Umeå, Sweden
Department of Wildlife, Fish, and Environmental Studies, Swedish University of Agricultural Sciences (SLU), Sweden

Fabien Mavrot, DVM, PhD Department of Ecosystem and Public Health, University of Calgary, Calgary, AB, Canada

Laurie Meythaler-Mullins, DVM, MSc Colorado State University, Fort Collins, CO, USA
Center for One Health Research, University of Alaska Fairbanks, Fairbanks, AK, USA

Ingebjørg Helena Nymo, DVM, PhD Norwegian Veterinary Institute, Tromsø, Norway

Antti Oksanen, DVM, PhD Finnish Food Authority (FINPAR), Oulu, Finland

Roland Pape, PhD Faculty of Technology, Natural Sciences and Maritime Sciences, Department of Natural Sciences and Environmental Health, University of South-Eastern Norway, Bø, Norway

Edoardo Pozio, DBS, PhD Istituto Superiore di Sanità, Rome, Italy

Arleigh Reynolds, DVM, PhD, DACVN Department of Veterinary Medicine, University of Alaska Fairbanks, Fairbanks, AK, USA

Lucy J. Robertson, PhD Norwegian University of Life Sciences, Ås, Norway

Rajnish Sharma, BVSc & AH, MVSc, PhD Department of Veterinary Microbiology, Western College of Veterinary Medicine, University of Saskatchewan, Saskatoon, SK, Canada
Centre for One Health, College of Veterinary Science, Guru Angad Dev Veterinary and Animal Sciences University, Ludhiana, Punjab, India

Gay Sheffield, MSc University of Alaska Fairbanks, Nome, AK, USA

Christian Sonne, DVM, PhD, DscVetMed Faculty of Science and Technology, Department of Ecoscience, Arctic Research Centre (ARC), Aarhus University, Aarhus, Denmark

Raphaela Stimmelmayr, DVM, MSc PhD North Slope Department of Wildlife Management, Utqiagvik, AK, USA
Institute of Arctic Biology, University of Alaska, Fairbanks, AK, USA

Sigbjørn Stokke, MSc, PhD Norwegian Institute for Nature Research, Trondheim, Norway

Matilde Tomaselli, DVM, PhD Canadian High Arctic Research Station - Polar Knowledge Canada, Cambridge Bay, NU, Canada

Morten Tryland, DVM, PhD Department of Arctic and Marine Biology, UiT The Arctic University of Tromsø, Tromsø, Norway
Department of Forestry and Wildlife Management, Inland Norway University of Applied Sciences, Koppang, Norway

Bob van Oort, PhD CICERO Center for International Climate Research, Oslo, Norway

Øystein Varpe, PhD Department of Biological Sciences, University of Bergen, Bergen, Norway
Norwegian Institute for Nature Research, Bergen, Norway

Sarah Wilkin, MSc Coordinator Marine Mammal Health and Stranding Response Program, National Marine Mammal Stranding and Emergency Response Coordinator, Office of Protected Resources, NOAA Fisheries/U.S. Department of Commerce, Silver Spring, MD, USA

Sadie Wright, MSc Alaska Region Oil Spill Response Coordinator, Protected Resources Division, NOAA Fisheries/U.S. Department of Commerce, Juneau, AK, USA

Michael Ziccardi, DVM, PhD Executive Director-One Health Institute, School of Veterinary Medicine, University of California, Davis, CA, USA

Birgitta Åhman, PhD Swedish University of Agricultural Sciences, Uppsala, Sweden

Part I

Introduction

The Arctic Region and Its Inhabitants

Anastasia Emelyanova

1 Borders of the Arctic

There is no unified border of "The Arctic." Eight countries, different in many ways, have land and waters in the arctic region—the Russian Federation, the United States of America, Canada, Norway, Denmark (Greenland and the Faroe Islands), Sweden, Finland, and Iceland. Russia has the longest border to the Arctic, with a coastline spanning 22,600 km of the total 38,700 km (58.4%). Natural and social sciences employ different definitions of "the Arctic," and there needs to be awareness and appreciation of these differences. The first Arctic Human Development Report (AHDR) published in 2004 (Einarsson et al. 2004) and the second one in 2015 (Larsen and Fondahl 2015) employ the same delimitation of the Arctic, discussing the various ways of defining it. They provide a commonly used definition for the social sciences, although this definition has certain drawbacks. In particular, it underscores the great subregional differentiation across the arctic territory, although "Arctic residents do face many of the same challenges across the entire region" (Larsen and Fondahl 2015). The detailed definition of the Arctic in the AHDR (Einarsson et al. 2004) takes onboard the approach, where the location of jurisdictional or administrative boundaries and the availability of data are prioritized for practical reasons. For the specific focus of this chapter, we will follow the same approach, utilizing the available population and health data (Fig. 1). As can be noted from Fig. 1, large regions defined as Arctic administrative areas are in fact south of the Arctic circle, which sometimes is used as a strict boundary of the Arctic.

The Arctic Council is a high-level intergovernmental organization that works with the issues faced by the Arctic countries. It consists of several working groups,

A. Emelyanova (✉)
Thule Institute, University of Oulu, Oulu, Finland

University of the Arctic, Thematic Networks and Research Liaison Office, Oulu, Finland
e-mail: anastasia.emelyanova@oulu.fi

M. Tryland (ed.), *Arctic One Health*, https://doi.org/10.1007/978-3-030-87853-5_1

Fig. 1 Arctic administrative areas are denoted by a purple line. Map compiled by Winfried K. Dallmann, Norwegian Polar Institute

and they use different definitions of the Arctic, known as the Arctic Monitoring and Assessment Programme (AMAP) working group's borders, Emergency Prevention, Preparedness, and Response (EPPR) borders, and the ones from the Conservation of Arctic Flora and Fauna biodiversity working group of the Arctic Council (CAFF). These definitions of the Arctic consider various bioclimatic subzones, vegetation, glaciers, oceans, governance, laws and regulations, tourism, husbandry, and many other key elements of the Arctic. These various borders show that "Arctic regionalism is territorially erratic and potential boundaries are set (and change) along principally functional landmarks and towards problem-oriented governance" (see the figure, e.g., in Knecht 2013).

Several interconnected interdisciplinary approaches provide an exhaustive basis for determining the boundaries of the Arctic. One approach, recently published in the largest Arctic Encyclopedia (Lykin 2017), mentioned some distinctive features of the Arctic border, e.g., the Arctic Circle (66° North latitude). This definition is particularly different from the AHDR definition that includes the territory north of the 60° latitude, especially in the Western Northern hemisphere. Others include physical features, e.g., differentiation of landscapes, zoning, tundra, forest tundra, and taiga, as well as societal features such as discomfort for human life in high latitudes, a rise in labor costs, production costs, depreciation of fixed assets, and quality of life of the population (Lykin 2017). Zaikov et al. adds that the growth season in the Arctic is much shorter than in lower latitudes, the periods with colder air temperatures are longer, there are large territories of permafrost, and that also other physical features of the Arctic are economically challenging. Some examples include higher costs of industry and infrastructure development and maintenance, higher costs of energy supply and transportation, special requirements for the communal services in the arctic settlements, and in many places a limited diversity of the local economy (Zaikov et al. 2019).

All the above-stated approaches to defining the borders of the Arctic are evolving due to the processes of globalization and climate change, which represent new challenges and opportunities for modern arctic people. Perhaps these definitions do not so much focus on, e.g., nomadic Arctic people's livelihoods. One example is reindeer herding: When Sámi herders ruled northern Fennoscandia, they travelled with reindeer between the coast and inland, utilizing different pastures. Then, the borders between Norway, Sweden, Finland, and Russia were drawn between the eighteenth and twentieth centuries, and "their country" was divided between four nations, with perhaps little interest in the Sámi or reindeer herding (Evengård et al. 2015). Hence, the borders are important structures when talking about inhabitants, both human and wildlife, of the Arctic.

2 Prehistoric Human Settling of the Arctic

For thousands of years, the Arctic was a home to many diverse groups of Indigenous people, "surviving in at times unforgiving conditions while developing vibrant cultures" (Evengård et al. 2015).

A number of sources discuss various aspects and dates of human colonization of the Arctic, going back 30,000 to 45,000 years ago (e.g., Pitulko et al. 2016). The times of colonization vary across the different Arctic regions. Proto-Eskimo culture (ancestors of all modern Eskimo) in the Far Eastern North appeared about 12,000 years ago, when the first signs of humans were found in Alaska (Gusev and Shumkin 2011; Guseinov 2012). The first groups of hunters entered the high latitudes in America already 24,000 years ago (Diakonov 2019). Gusev and Shumkin stated that people may have been attracted to the Far North to hunt animals of the large Pleistocene fauna for food and valuable raw materials (e.g., mammoth, woolly rhinoceros, musk buffalo, bison) and reindeer. Later, about 8000–7800 years ago, the main objects of hunting were reindeer, polar bear, smaller prey, and birds, according to the early hunter settlements of the Arctic. Domestication of dogs was one unique feature of human activity at that time (Diakonov 2019). Humans peopled the North American Arctic (northern Alaska and Canada) and Greenland around 6000 years ago, leaving behind a complex archaeological record that consisted of different cultural units and distinct ways of life (Raghavan et al. 2014). According to Diakonov, Gusev, and Shumkin, Greenland was settled from the East about 4500 years ago when larger groups of proto-Eskimo hunters for marine mammals moved into the area (Gusev and Shumkin 2011; Diakonov 2019). After that, the Bering Sea culture began to develop about 3500 years ago (Guseinov 2012). The inhabitants of the Bering Strait and the Arctic coastal area of Fennoscandia started forming stationary settlements and community support systems, improving housing, and establishing specialized marine mammal hunting with the use of a harpoon (Gusev and Shumkin 2011).

Numerous detailed descriptions of how the Arctic was settled have been published (Anderson et al. 2014; Raghavan et al. 2014; e.g., Filatov 2017; Kotlyakov et al. 2017; Diakonov 2019). In Russia, the inclusion of the territories of the Russian European North (tenth to twelfth until seventeenth century) and Siberia and Far East (sixteenth to twentieth century) made this country the largest of the Arctic nations. The colonization or resettlement of the Russian arctic regions was done by many different peoples living in Russia and started almost a century later than the colonization processes run by some of the European countries. It continued until the late twentieth century, when many people moved up north from the middle latitudes of Russia, partly forcefully (prisoners) but mainly voluntarily (labor migration). The essence and peculiarities of the colonization of the northernmost regions of Russia, especially the Asian regions, were compared to similar processes in America and other parts of the world (Rybakovsky 2018).

The population of northern Norway grew three to four times bigger after acquiring independence from Denmark (1814) and building infrastructure; for example, in 1902, a railway was built that connected the north of the country to its south and neighboring countries. For a very long time, the Canadian North was a colony and not free to make decisions, including those regarding settling policy and processes. It was visited by professional workers in a so-called shift method (several weeks/months in the Arctic, several weeks/months out) until the main cities were established in the twentieth century (Filatov 2017).

Many scientists and adventurous explorers from outside of the Arctic entered the northernmost territories in the eighteenth century (Evengård et al. 2015). Those far away Arctic lands were already well settled with Indigenous peoples and a small number of settlers from the outside. The new visitors were also the first people to introduce the Arctic to the global world, to describe its landscape, the wildlife fauna, humans, and the various cultures met there (Evengård et al. 2015). While curiosity was one factor driving people up North, commercial interests over the Arctic resources started to prevail in the last three centuries, bringing more people to the region. The economic reasoning behind the Arctic settlement was changing massively over time. In the period from the seventeenth to nineteenth centuries to the beginning of the twentieth century, biological resources (fur and food) were traded, and the main industries were hunting and fishing. From the end of the nineteenth century until now, transportation development became important to satisfy not only national but also world interests. Since the 1930s, military defense activities in the Arctic were strengthened, flourishing especially in the 1950s and 1970s, during the Cold War. Today, the minerals and raw material, oil, and gas business of the Arctic economy are the main driving economic forces, which started to take shape from the end of the nineteenth century (the beginning of gold mining in Alaska—1880) (Fauzer and Smirnov 2018).

3 Ongoing Population and Health Developments in the Arctic

When assessing the health changes in the Arctic, it is crucial to understand the demographic elements. In the recent past, population dynamics in the whole arctic region have been shaped in many ways by the Russian depopulation processes. This is because the Russian population comprises the largest share of the total Arctic population.

From 1989 to 2019, the Russian Arctic lost about 30% of its population (Danilova 2020; Fauzer et al. 2020). In contrast, population growth has increased by 16.3% in the non-Russian Arctic during the same period, due to several concurrent population processes. In the North American and North Atlantic Arctic, the arctic communities experienced a marked population growth in the twentieth century due to internal and external factors (Emelyanova 2018). In 1945, the population of Alaska (USA) was 100,000 and grew sevenfold by 2015. In Greenland, the increase has been more than fivefold in the same period, and a fourfold increase occurred in Iceland since 1945. At present, only Alaska, Iceland, and the Canadian Arctic have continued to experience population growth due to positive net migration and natural population increases (more births than deaths) (Larsen and Fondahl 2015). Figure 2 depicts the regions of the Arctic that underwent population growth in the last decade (colored blue) and regions with a declining population (colored red). The numbers of inhabitants of northern parts of Sweden and Finland have declined up to 10% as well as in many regions of Russia due to accelerated out-migration and natural population decrease; the latter happens when there are more deaths than births. Starting in the 1990s, the profound growth seen in Greenland and the Faroe Islands reversed to a declining population trend. By 2018, the populations in these two countries had returned back to the levels of 1990 (Gløersen et al. 2006).

In the Russian Arctic and Far Eastern regions,[1] one recognized challenge for the Russian government today is to cope with significant population decline (The President Executive Order of 9.10.2007 №1351 2007). These include the Russian regions with the highest levels of population decline (Republic of Komi, Arkhangelsk, Murmansk, Magadan, and Sakhalin oblasts) as well as the Republic

[1]Eleven areas of Russia with a population of about seven million people total in 2018: republics of Karelia, Komi, Sakha Yakutia, autonomous okrugs of Nenets, Khanty-Mansi, Yamalo-Nenets and Chukotka, as well as Arkhangelsk, Murmansk and Magadan oblasts, and Kamchatka kray.

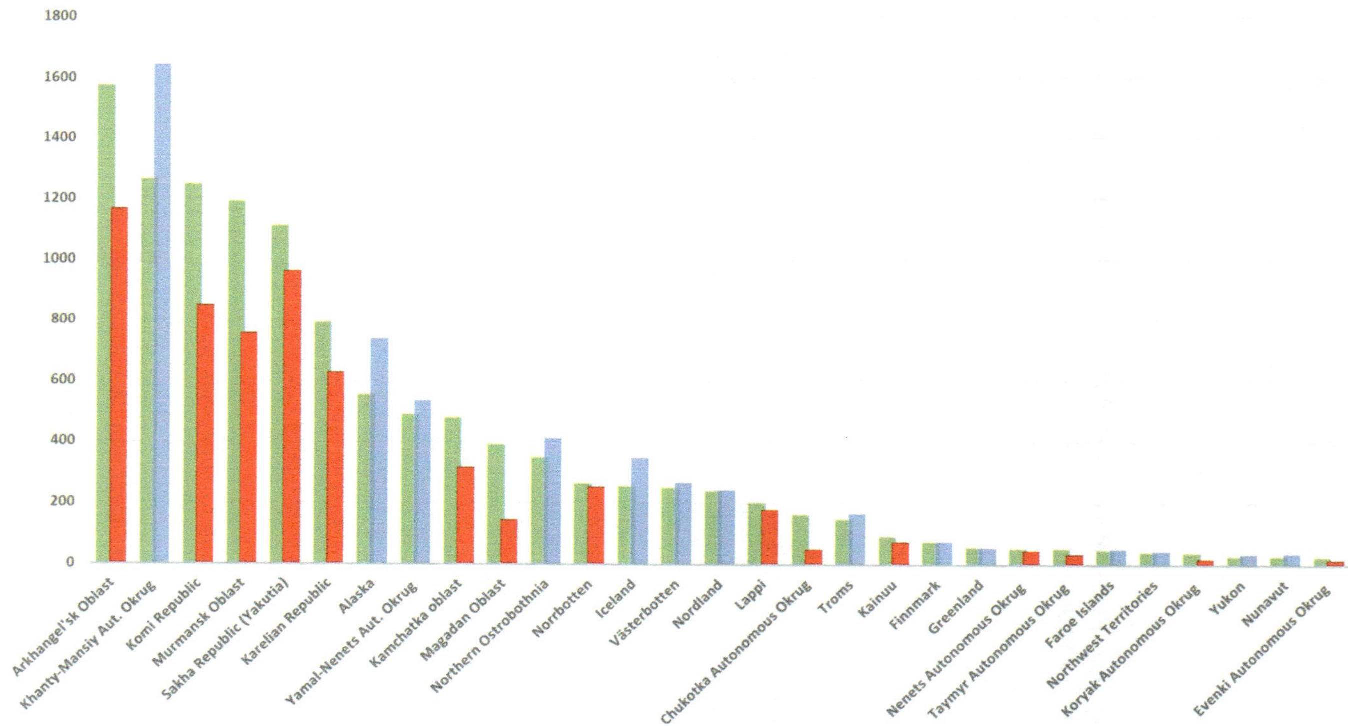

Fig. 2 Population changes (number of people) in the Arctic from 1990 (first bar of a region) to 2018 (second bar of a region), based on data from the national statistical database. (For Nunavut, the population change is from 2000 to 2017, for Russian Arctic regions from 1990 to 2017, and for the Taymyr, Evenki, and Koryak okrugs, from 1990 to 2011)

of Karelia and Sakha (Yakutia), which lose their population but at a lower scale. The continued shrinking of the Russian Arctic population is projected to decline from seven million people to only 5.9 million during 2018–2050. The fastest relative population loss is expected in the Barents or North-West of the Russian Arctic (Table 1). The populations of the Arkhangelsk, Murmansk, and Magadan regions keep declining even under the optimistic Arctic Boost scenario in one of the projection exercises (Emelyanova 2017).

The main reasons for the decline in the Russian Arctic population during different periods were related to the booms and busts in exploration of natural resources, lower quality of life in comparison with the central regions (Smirnov 2020), and the policy of the state (Danilova 2020). Population density in the Far East is the lowest in Russia, less than one person per 1 km^2, which was addressed in the policy for the development of the region (Golubeva and Emelyanova 2020).

In 2017, the Russian government developed the Concept of Demographic Policy in Russia's Far East for the period up to 2025 (The Government Decree of 20.06.2017 №1298-р 2017). One priority is to reverse population decline caused by large out-migration and natural decrease. Several related governmental orders are to be implemented in two stages, during 2017–2020 and 2021–2025. In 2020, the President signed the Decree to update "The Strategy for the Development of the Arctic Zone of the Russian Federation and Ensuring National Security for the Period until 2035," where it was reiterated that out-migration, population loss, lower quality of life and lower health indicators in the Arctic compared to the rest of the nation are threats to national security (The President Executive Order of 26.10.2020 №645 2020).

Selin states that in order to avoid economically driven demographic losses, financial mechanisms must be strongly considered in addition to the demographic policy framework; a boost of development must be encouraged, rather than the current stabilization approach (Selin 2016). People from the arctic regions of Russia are much poorer than nation-wide, even though they transfer more money into the federal budget than they receive back after federal reallocation (Selin and Bashmakova 2010). Compared to other arctic regions where larger companies tend to involve and hire the local arctic population, Russian companies prefer to hire shift workers as well as experts and machinery from abroad (Kryukov and Kryukov 2017).

Fauzer and Smirnov stated that the global Arctic has 415 settlements with a population of over one thousand people, 135 of them located in its Russian part (Fauzer and Smirnov 2018). The process of urbanization has done its job of enlarging urban centers and regional capitals, e.g., 9 out of 15 larger cities in the Arctic with a population above 50,000 are located in Russia (Fauzer et al. 2020) (Fig. 3). However, in Russia, the largest northernmost cities have also lost their population to other southern cities in Russia or abroad. The current Russian settlement policy focuses on developing larger community centers using already developed infrastructure with more shift work and transportation opportunities to the remote locations of the Arctic, rather than building new settlements for temporary projects (The Government Decree of 15.12.1994 №31 1994; Blagodeteleva 2017).

Table 1 The indicators related to the demographic and health profile of the Russian Arctic (RA) regions

	Population size as of Jan 1st (thous.)	Projected population size as of Jan 1st (thous.)	Projected population size as of Jan 1st. Zero migration scenario (thous.)	Net migration (total)	Relative population change	Number of women per 100 men at ages 60+	Total fertility rate. Number of children per woman	Female life expectancy at birth (years)	Male life expectancy at birth (years)	Difference between female and male life expectancy at birth (years)	Population median age (years)	Projected population median age (years)	Human life indicator. Both sexes (years)[a]
Reference period	2018	2050	2050	2017	2018–2050	2018	2017	2017	2017	2017	2018	2050	2017
The Russian Federation	146,880	137,360	131,218	211,878	−6.5	181	1.62	77.6	67.5	10.1	39.2	43.9	67.7
Republic of Karelia	622	470	530	−1916	−24.4	202	1.56	76.2	64.9	11.3	40.9	46.3	65.4
Republic of Komi	841	491	830	−9470	−41.6	193	1.78	76.6	65.3	11.3	38.7	43.6	66.6
Nenets AO	44	49	59	−231	−11.0	164	2.35	77.2	65.9	11.2	35.4	37.0	65.4
Arkhangelsk oblast	1111	668	994	−7814	−39.8	190	1.65	77.2	66.1	11.6	40.4	43.3	66.8
Murmansk oblast	754	588	666	−3503	−22.0	210	1.56	76.3	66.5	9.8	38.5	40.9	67.0
Khanty-Mansi AO[b]—Yugra	1655	1756	1757	−4067	6.1	149	1.88	78.3	69.3	9.1	35.2	38.3	69.4
Yamal-Nenets AO	539	578	601	−2418	7.4	137	1.95	77.9	69.0	8.9	34.8	36.1	68.2
Republic of Sakha (Yakutia)	964	885	1105	−4649	−8.2	159	1.93	77.1	66.4	10.7	33.1	37.3	66.6
Kamchatka kray	316	272	280	544	−13.7	167	1.78	75.3	65.2	10.0	37.9	40.5	64.9

Magadan oblast	144	99	120	−1398	−31.6	164	1.60	75.5	63.4	12.1	39.0	41.4	64.8
Chukotka AO	49	41	67	−656	−17.5	126	2.08	71.7	60.3	11.3	36.1	36.1	58.0
RA total	7039	5897	7009	−35,578	−19.6	169.2	1.83	76.3	65.7	10.7	37.3	40.1	65.7

Source: Scherbov et al. (2019)

[a]The Human Life Indicator expresses well-being in terms of years of life, similar to life expectancy at birth, and takes the inequality in longevity into account

[b]*AO* autonomous okrug

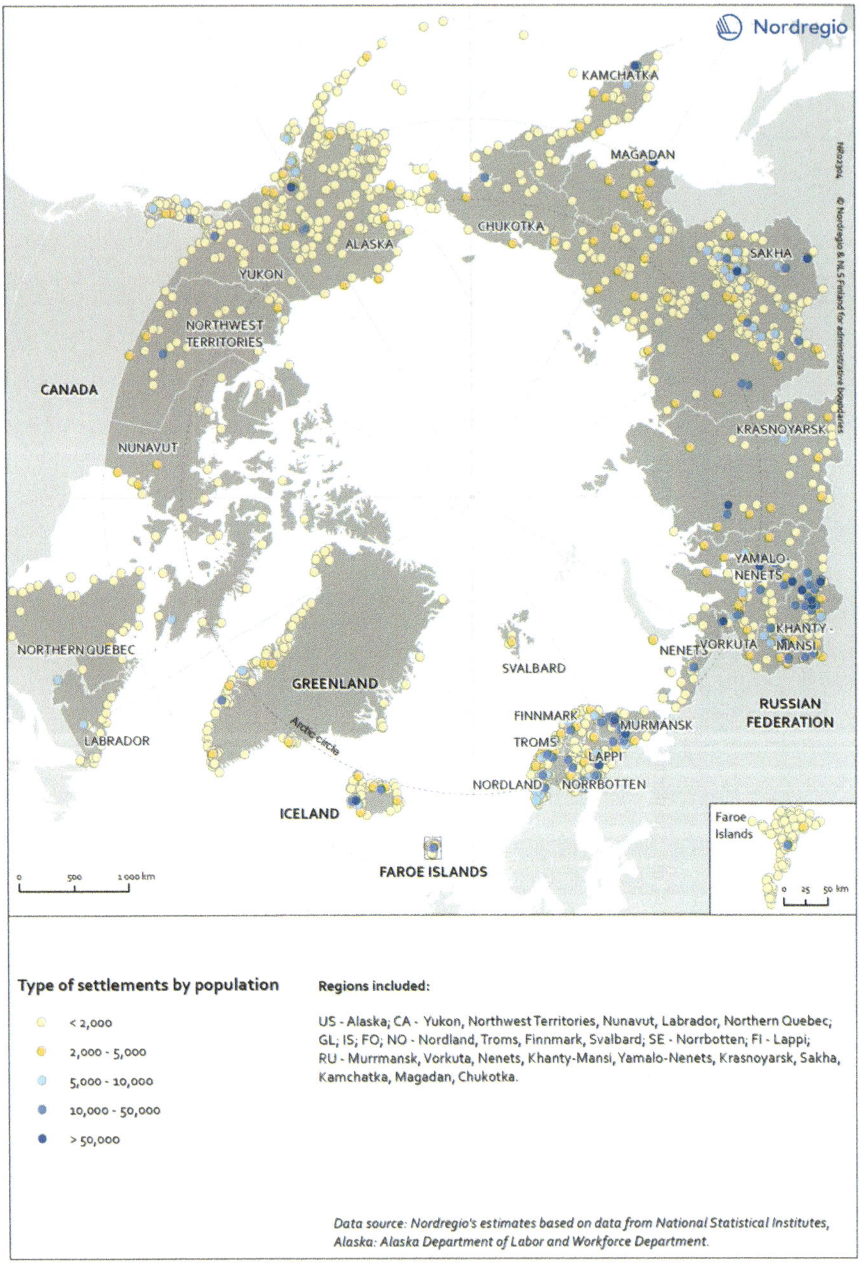

Fig. 3 Settlements by population size in the Arctic by Nordregio (Wang 2019)

The larger settlements of the Arctic are mainly regional or administrative centers especially in the Canadian Arctic, Greenland, Iceland, and the Faroe Islands (Fig. 3). In Alaska (USA), Arctic Fennoscandia, and in a few areas of Russia, the settlement density is in general high, including many large settlements (Jungsberg et al. 2019; Wang 2019). In Nunavut and Northern Quebec (Canada), Greenland, the Faroe Islands, and Finnmark (Norway), most of the population live in small settlements. The Yukon, bordering Alaska in the west, differs from other Canadian Arctic regions due to its regional center Whitehorse, which comprises 65% of the total population of Yukon. In the Northwest Territories and Labrador (Canada), Iceland, Troms and Nordland (Norway), Norrbotten (Sweden) and Lapland (Finland), there is not such a remarkable difference in the share of the population in small and large settlements (Jungsberg et al. 2019).

The population residing in the Arctic is in fact a very small share of the overall population of eight Arctic countries and keeps declining. In Russia, each of the regions Chukotka, Kamchatka, Magadan, Karelia, Nenets, and Yamalo-Nenets autonomous areas located in the Arctic has <0.05% of the total Russian population, similar to the Canadian Arctic and Alaska (Emelyanova 2018). This small percentage contrasts sharply with the colossal part of the country's land mass these arctic provinces occupy. In many cases, this characterizes the region as a place of pristine wilderness with low anthropogenic activity. The settlements are divided into highly urbanized cities and, on the other hand, highly dispersed small communities and villages situated across the region (Emelyanova 2017).

Fertility processes changed toward less children born in many regions (Fig. 4). In 2016, the total fertility rate (TFR) has been around the replacement level at which a population exactly replaces itself from one generation to the next, without migration (2.1 children per woman) in Nunavut, the Faroe Islands, Greenland, Nenets autonomous okrug (AO), Chukotka, and other areas next to the red line in Fig. 4. The remaining areas have fertility rates below the replacement level, with the lowest level found in Magadan and Murmansk oblasts, Troms in Norway, and Lapland in Finland.

Even though the general population in Arctic areas is considered rather young compared to other central and southern areas of Arctic countries, the process of population aging is underway. We can see it in declining fertility rates, changing population structure toward a declining proportion of people of working age and an increasing number of older persons. The old-age dependency ratio (OADR; % of working-age population based on pension age $\times$ 100) is as high in the Russian Karelia and Arkhangelsk regions as in their Fennoscandian neighbors, whereas the Siberian regions reached OADR levels similar to those in Arctic areas mostly inhabited by Indigenous people, e.g., Greenland, and the Canadian Arctic. As Scherbov et al. further details for the Russian Arctic, the OADR index will grow in all Russian Arctic regions during the period 2018–2050 with an average growth of 26.8 index points, increasing at the fastest speed in the republic of Karelia (40.7) (Scherbov et al. 2019).

Prospective old-age dependency ratio (POADR) ($\times$ 100) is another important indicator of population aging. Unlike the OADR, POADR is adjusted according to

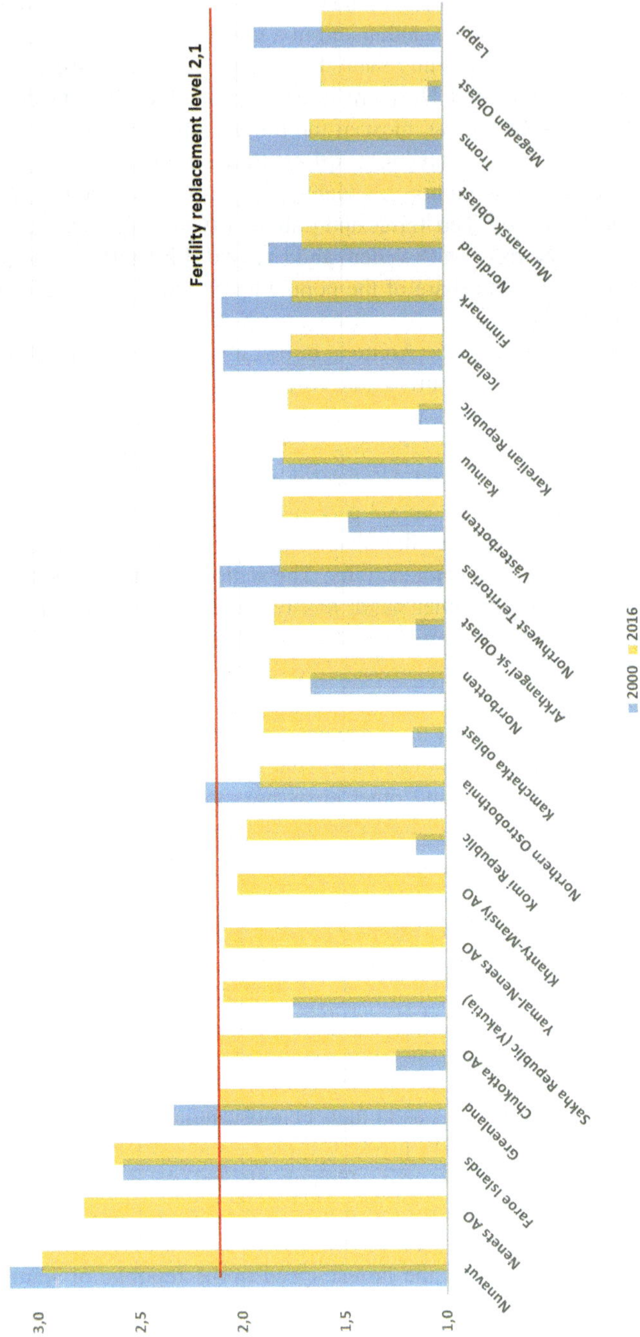

Fig. 4 Total fertility rate in the Arctic region in 2000 and 2016

the health gains and increasing life expectancy. This ratio is based on a flexible threshold of who is considered old. It is calculated as a ratio of the number of people older than the old-age threshold to the number of people between age 20 and the old-age threshold. The ratio is multiplied by 100 (Scherbov et al. 2019). POADR will grow less rapidly than OADR; the average change is 6.0 points, with the fastest aging registered in Arkhangelsk oblast (10.4). It is interesting to note that according to POADR, people in Chukotka will not age at all as a group, with only a 0.2 index value change over three decades. Hence, societal challenges associated with population aging might not yet apply there (index values come from Scherbov et al. 2019).

There is a significantly varying pattern of longevity in the Arctic. The average life expectancy (LE) at birth in the Russian Arctic is shorter than in many other Arctic places and was 65.7 years for males and 76.3 years for females in 2017. The male LE is especially low compared to that of men in, e.g., Iceland, the Faroe Islands, and the Swedish north (>80 years). When comparing the Russian northern areas, there is a particularly large gap between the areas of around 7 years in female and 9 years in male LEs with the lowest indication in Chukotka, and the highest indication in the Khanty-Mansi AO—Yugra. The difference between female and male life expectancy at birth has been up to 12 years longer for females in the Magadan oblast, with the lowest difference between sexes found in Yamal-Nenets AO (9 years)—another unique and sad feature of Russian longevity. The gender gap is significantly lower in other parts of the Arctic.

The ethnic composition greatly affects processes in health and population development. The United States classify people by race, Canada by ethnicity of Aboriginal peoples, Greenland—based on place of birth. Iceland, and the Faroe Islands do not have Indigenous populations since they were settled relatively recently by outsiders. Fennoscandia has only one Indigenous group in the north—the Sámi. The only areas where Indigenous people reside in the Russian Arctic are Khanty-Mansi, Sakha (Yakutia), and Nenets AO. Arctic Indigenous populations often show different "results" in demographic and health indicators compared to those of the non-Indigenous population (Coates and Holroyd 2020). The Indigenous population often has a much younger age structure, higher fertility and mortality, and faster growth and is more likely to be located in predominantly rural regions (Emelyanova 2018; OECD 2020). Indigenous groups in Russia are facing a lower quality of life, lower quality of social and health services as well as poorer housing, educational, and communication opportunities in the remote areas of their residence compared to the general population of Russia (The President Executive Order of 26.10.2020 №645 2020). This is often true based on the analysis of the wealth and the livelihoods and living conditions of the Inuit, Sámi, and other Indigenous populations in arctic settlements (Poppel et al. 2015; OECD 2020).

The health of people living in rural, remote Indigenous communities in the Arctic is persistently poorer than that of their urban and non-Indigenous counterparts. According to Bjerregaard et al. (2004), the incidence of infectious diseases in the Inuit and other Indigenous peoples of the circumpolar north has declined considerably in the second half of the twentieth century but is still high compared to many European countries. Dramatic rises in mortality have been registered every winter for most developed countries (Keatinge 2002), with a reliably higher share of pathological weather sensitivity among the northernmost European population compared to the regions with a warmer temperate climate and high levels of emotional stress in arctic residents (Hasnulin and Hasnulina 2012). Chronic diseases such as diabetes and cardiovascular diseases are on the increase, while accidents, suicides, violence, and substance abuse are of major importance for the pattern of ill health in most Inuit communities in their homelands stretching from the easternmost tip of Russia in the west to Greenland in the east (Bjerregaard et al. 2004). Lifestyle changes, social change, changes in society, and the environment and circulating contaminants are major determinants of health among the Inuit in many countries of the arctic region (see, e.g., Bjerregaard et al. 2004; Curtis et al. 2005; Singh et al. 2014).

Many global and local processes have changed the community well-being in the Arctic (Ribova 2000), including climate change and other environmental changes. These processes have posed more longstanding and also emerging health challenges to the inhabitants related to e.g., food and water security, changes in disease patterns, and impacts on healthcare infrastructure (Parkinson and Evengård 2009; Larsen and Fondahl 2015; Ruscio et al. 2015; Evengård and Thierfelder 2021). Climate change has been predicted to be the most influential factor in the emergence of infectious diseases (Sonne et al. 2017; Waits et al. 2018). Climate change can promote multiplication rates of pathogens and introduce new infections to previously isolated areas (Dudley et al. 2015; Waits et al. 2018). Climate change can also indirectly affect health by changing human behavior, e.g., lead to more people using public bathing facilities, providing increased risks for waterborne disease outbreaks (Eze et al. 2014). In the case of Arctic warming, people can spend more time in nature and public places, subsequently increasing chances to contract more infectious diseases (Chashchin et al. 2017). Climate change can also affect contaminant levels, which can lead to human health effects (Pacyna et al. 2015; Abass et al. 2018).

Warming of the climate might slightly prolong commercial anthropogenic activity in the Arctic; however, there will still be unfavorable conditions for shipping, fishing, and resource extraction during a major part of the year. There will also be lots of obstacles (e.g., lack of appropriate infrastructure, machinery and equipment) to overcome with considerable financial resources and time before the very much discussed "economic boom" connected to climate change is realized (Voronkov 2015). In addition to affecting socioeconomic activities like transportation, marine sea food production, resource exploitation, governance issues and many others (Crépin et al. 2017), the dominant drivers of arctic societal changes such as climate change and globalization will change the traditional livelihoods in the North, e.g., bringing diversity to the cultures and languages, with a transition from traditional foods based on hunting and fishing (e.g., "country food") to the Western diet (Larsen and Fondahl 2015). Overall, the manifestations and future predictions of climate change vary across the Arctic and between and within communities, and consequently, the need for adaptive responses varies (Larsen and Fondahl 2015).

The abovementioned increasing globalization is another megatrend affecting the inhabitants of the Arctic in many ways. For example, health and disease patterns (increased obesity, diabetes, and cardiovascular diseases) are changing, population out-migration is promoted as people seek opportunities and alternatives outside the Arctic, and energy security is compromised by shifting energy prices. Increased globalization also has many other both positive and negative effects on the daily lives of inhabitants in the Arctic (Larsen and Fondahl 2015). Globalization in the circumpolar world and its various effects on the economies of the North, traditional livelihoods, security, well-being, and other systems have been described in detail in the book by Heininen and Southcott (Heininen and Southcott 2010; Jabour 2011). Some chapters in the book examine the impacts of globalization on groups of Indigenous people, e.g., the Sea Sámi in northern Norway and the Nenets reindeer herders in Russia. Both groups have been able to find new ways to deal with the challenges they face (Heininen and Southcott 2010).

More research is required on innovative, economically effective ways to improve human and wildlife well-being (Larsen and Fondahl 2015). In the view of Stephen (2018), challenges for future research on societal impacts of a rapidly changing Arctic include achieving greater clarity and critical reflection around key concepts. Arctic sustainable development, climate change, and globalization as the dominant drivers of societal impacts in the Arctic, the Global Arctic, and the Arctic stakeholder concept are some concepts to be considered. The use of comparative methods to investigate societal impacts of a changing Arctic is essential for cases both within and also beyond the Arctic, as the Arctic is certainly unique but not the only place affected by global change processes (Stephen 2018). The use of best practices and reinforcing all possible assets is essential. One Health is also a prominent approach to consider these connections between the environment, plant, animal, and human health. Understanding this is increasingly critical in assessing the impact of global climate change on the health and well-being of Arctic inhabitants (Ruscio et al. 2015).

4 Conclusions

The approaches to defining the borders of the Arctic stated in the chapter are numerous and currently evolving due to the processes of globalization and climate change. These definitions have not yet shown much focus on nomadic Arctic people's livelihoods.

Based on historical data, human colonization of the Arctic by Indigenous people goes back 30,000 to 45,000 years ago, depending on the area of the Arctic. The population of the northernmost areas grew much faster in the nineteenth to twentieth centuries due to commercial interests over the Arctic natural and biological resources.

Currently, most territories in the Russian Arctic experience a natural decrease (30% population lost from 1989 to 2019) due to severe out-migration and prevailing count of deaths over births, while a population growth of 16% was found in many other non-Russian Arctic territories during the same period. Special policies and governmental strategies are nowadays in place to stimulate population growth in the

Arctic. The ethnic composition greatly affects processes in health and population development. The Indigenous population often has a much younger age structure, higher fertility and mortality, and faster growth and is more likely to be located in predominantly rural regions. They are facing a lower quality of life, lower quality of social and health services, as well as poorer housing and educational and communication opportunities, which requires continuous effort from all decision makers to improve the situation.

Climate change and environmental changes also affect community well-being to various degrees across the Arctic and between and within communities. In addition, as documented above in this chapter, there is a clear trend toward the emergence of new human infectious diseases in the Arctic and a strong impact on contaminant levels. There is a consequential need for adaptive responses. One Health is one of the prominent approaches to consider the connections between the environment, plant, animal, and human health in a rapidly changing Arctic.

Acknowledgements The author thanks Timothy Heleniak and Linda Kivi from Nordregio for providing some population data and recommending literature sources useful for the chapter. The author also thanks the Editor and reviewers of the chapter.

References

(2012) Circumpolar Health Atlas. University of Toronto Press

Abass K, Emelyanova A, Rautio A (2018) Temporal trends of contaminants in Arctic human populations. Environ Sci Pollut Res 25:28834–28850. https://doi.org/10.1007/s11356-018-2936-8

Anderson PM, Balanovsky OP, Besprozvanny EM et al (2014) The initial settlement of the Arctic by humans in a changing natural environment. ГЕОС, Moscow

Bjerregaard P, Young TK, Dewailly E, Ebbesson SOE (2004) Indigenous health in the Arctic: an overview of the circumpolar Inuit population. Scand J Public Health 32:390–395. https://doi.org/10.1080/14034940410028398

Blagodeteleva OM (2017) Evolution and current trends of development of settlement systems in the north areas (on the example of the USA, Canada and Russia). Bull Perm Natl Res Polytech Univ Appl Ecol Urban 3:5–25. https://doi.org/10.15593/2409-5125/2017.03.01

Chashchin VP, Popova ON, Byzinov RB, et al (2017) Epidemiological character of tick-borne viral encephalitis extension in the Arkhangelsk region. Hum Ecol 12–19

Coates K, Holroyd C (eds) (2020) The Palgrave handbook of Arctic policy and politics. Palgrave Macmillan

Crépin A-S, Karcher M, Gascard J-C (2017) Arctic climate change, economy and society (ACCESS): integrated perspectives. Ambio 46:341–354. https://doi.org/10.1007/s13280-017-0953-3

Curtis T, Kvernmo S, Bjerregaard P (2005) Changing living conditions, life style and health. Int J Circumpolar Health 64:442–450. https://doi.org/10.3402/ijch.v64i5.18025

Danilova EV (2020) Population of the Russian Arctic: the number of processes, forecasts. Innov Invest 260–265

Diakonov VM (2019) Adaptation or survival? Strategies of life support an ancient people in the extreme cold arctic conditions. Sci Electron J Merid 9–11

Dudley JP, Hoberg EP, Jenkins EJ, Parkinson AJ (2015) Climate change in the north American Arctic: a one health perspective. EcoHealth 12:713–725. https://doi.org/10.1007/s10393-015-1036-1

Einarsson N, Nymand Larsen J, Nilsson A, Young OR (2004) Arctic human development report. Stefansson Arctic Institute, under the auspices of the Icelandic Chairmanship of the Arctic Council 2002-2004, Akureyri
Emelyanova A (2017) Population projections of the Arctic by levels of education. http://pure.iiasa.ac.at/14981/. Accessed 4 Jan 2018
Emelyanova A (2018) Exploring the future population and educational dynamics in the Arctic: 2015 to 2050. Finn Yearbook Popul Res 53:1–24. https://doi.org/10.23979/fypr.70159
Evengård B, Thierfelder T (2021) CLINF: climate-change effects on the epidemiology of infectious diseases, and the associated impacts on northern societies. In: Nord DC (ed) Nordic perspectives on the responsible development of the Arctic: pathways to action. Springer International Publishing, Cham, pp 49–70
Evengård B, Larsen JN, Paasche Ø (eds) (2015) The New Arctic. Springer International Publishing
Eze JI, Scott EM, Pollock KG et al (2014) The association of weather and bathing water quality on the incidence of gastrointestinal illness in the west of Scotland. Epidemiol Infect 142:1289–1299. https://doi.org/10.1017/S0950268813002148
Fauzer VV, Smirnov AV (2018) The world's Arctic: natural resources, population distribution, economics. The Arctic: Ecol Econ 6–22. https://doi.org/10.25283/2223-4594-2018-3-6-22
Fauzer VV, Lytkina TS, Smirnov AV (2020) Population of the world Arctic: Russian and foreign approaches to studying demographic problems and settlement of territories. Econ Soc Changes: Facts, Trends, Forecast 13:158–174. https://doi.org/10.15838/esc.2020.3.69.11
Filatov II (2017) Exploration of the Arctic before and after the 1917 revolution. State Regulation of Public Relations, Gosreg, pp 58–66
Gløersen E, Dubois A, Copus A, Schürmann C (2006) Northern peripheral, sparsely populated regions in the European Union and in Norway
Golubeva E, Emelyanova A (2020) Policy initiatives on healthy ageing in Russia from 2010-2020. Eur J Mental Health 15:93–110. https://doi.org/10.5708/EJMH.15.2020.2.2
Guseinov V (2012) The "New Arctic" and interests of Russia. Krasnaya Zvezda
Gusev SV, Shumkin VY (2011) Initial settlement of the Eurasian Arctic: time, routes, common and special features. Federal State Budgetary Institution of Science Institute of the History of Material Culture of the Russian Academy of Sciences, pp 133–134
Hasnulin VI, Hasnulina AV (2012) Psycho-emotional stress and meteoreacton as systemic manifestations of human disadaptation under changing climatic conditions in the north of Russia. Hum Ecol 3–7
Heininen L, Southcott C (eds) (2010) Globalization and the circumpolar north. University of Alaska Press, Fairbanks
Jabour J (2011) Globalization and the circumpolar north. Polar J 1:140–142. https://doi.org/10.1080/2154896X.2011.568797
Jungsberg L, Turunen E, Heleniak T et al (2019) Atlas of population, society and economy in the Arctic. Nordregio, Stockholm
Keatinge WR (2002) Winter mortality and its causes. Int J Circumpolar Health 61:292–299. https://doi.org/10.3402/ijch.v61i4.17477
Knecht S (2013) Arctic regionalism in theory and practice: from cooperation to integration? pp 164–183
Kotlyakov VM, Velichko AA, Vasil'ev SA (eds) (2017) Human colonization of the Arctic: the interaction between early migration and the paleoenvironment, 1st edition. Academic, London
Kryukov VA, Kryukov YaV (2017) How to expand frames of the Arctic projects. Eco 5–32
Larsen JN, Fondahl G (2015) Arctic human development report: regional processes and global linkages. Nordic Council of Ministers
Lykin YF (2017) The Arctic encyclopedia in two volumes. Paulsen Publishing House, Moscow
OECD (2020) Linking indigenous communities with regional development in Canada. OECD Publishing, Paris

Pacyna JM, Cousins IT, Halsall C et al (2015) Impacts on human health in the Arctic owing to climate-induced changes in contaminant cycling – the EU ArcRisk project policy outcome. Environ Sci Pol 50:200–213. https://doi.org/10.1016/j.envsci.2015.02.010

Parkinson AJ, Evengård B (2009) Climate change, its impact on human health in the Arctic and the public health response to threats of emerging infectious diseases. Glob Health Action 2. https://doi.org/10.3402/gha.v2i0.2075

Pitulko VV, Tikhonov AN, Pavlova EY et al (2016) Early human presence in the Arctic: evidence from 45,000-year-old mammoth remains. Science 351:260–263. https://doi.org/10.1126/science.aad0554

Poppel B, Andersen T, Beach H et al (2015) SLiCA: Arctic living conditions: living conditions and quality of life among Inuit, Saami and indigenous peoples of Chukotka and the Kola Peninsula. Nordisk Ministerråd

Raghavan M, DeGiorgio M, Albrechtsen A et al (2014) The genetic prehistory of the New World Arctic. Science 345. https://doi.org/10.1126/science.1255832

Ribova L (2000) Individual and community well-being. http://www.thearctic.is/articles/topics/wellbeing/enska/index.htm. Accessed 19 Mar 2021

Ruscio BA, Brubaker M, Glasser J et al (2015) One health – a strategy for resilience in a changing arctic. Int J Circumpolar Health 74:27913. https://doi.org/10.3402/ijch.v74.27913

Rybakovsky LL (2018) Colonization of the Asian part of Russia: the features of implementation and geopolitical consequences. Sociol Res:38–47. https://doi.org/10.31857/S013216250000760-9

Scherbov S, Shulgin S, Andruchowitz S et al (2019) Russian demographic data sheet 2019. http://www.populationrussia.ru. Accessed 4 Mar 2019

Selin VS (2016) Economic policy in the Arctic: comparative analysis. In: The north and the Arctic in a new paradigm of world development: Luzin readings. ИЭП КНЦ РАН, Apatity, p 126–130

Selin VS, Bashmakova ER (2010) Northern and Arctic regions: what role they play in Russian development under new geo-economic conditions. Region: Econ Sociol:23–29

Singh K, Bjerregaard P, Man Chan H (2014) Association between environmental contaminants and health outcomes in indigenous populations of the circumpolar north. Int J Circumpolar Health 73. https://doi.org/10.3402/ijch.v73.25808

Smirnov AV (2020) The Arctic population: dynamics and centers of the settlement system. Arctic North 270–290. https://doi.org/10.37482/issn2221-2698.2020.40.270

Sonne C, Letcher RJ, Jenssen BM et al (2017) A veterinary perspective on one health in the Arctic. Acta Vet Scand 59:84. https://doi.org/10.1186/s13028-017-0353-5

Stephen K (2018) Societal impacts of a rapidly changing Arctic. Curr Clim Change Rep 4:223–237. https://doi.org/10.1007/s40641-018-0106-1

The Government Decree of 15.12.1994 №31 (1994) General scheme of settlement on the territory of the Russian Federation (basic provisions)

The Government Decree of 20.06.2017 №1298-p (2017) Concept of Demographic policy in the Russia's Far East for the period up to 2025

The President Executive Order of 9.10.2007 №1351 (2007) Concept of demographic policy of the Russian Federation for the period up to 2025

The President Executive Order of 26.10.2020 №645 (2020) Strategy for the development of the Arctic zone of Russia and ensuring national security until 2035

Voronkov LS (2015) Climate changes and the imperatives of all round-a-year human activity in the Arctic. Yearbook of the Institute for International Studies of the Moscow State Institute of International Relations of the Ministry of Foreign Affairs of the Russian Federation 1:9–18

Waits A, Emelyanova A, Oksanen A et al (2018) Human infectious diseases and the changing climate in the Arctic. Environ Int 121:703–713. https://doi.org/10.1016/j.envint.2018.09.042

Wang S (2019) Settlements by size in the Arctic. In: Nordregio. https://nordregio.org/maps/settlements-by-size-in-the-arctic/. Accessed 17 Mar 2021

Zaikov KS, Kondratov NA, Kudryashova EV, et al (2019) Scenarios for the development of the Arctic region (2020-2035). The Arctic and the North 5–24

A Holistic Approach to One Health in the Arctic

Arleigh Reynolds, Susan Kutz, and Tessa Baker

1 What Is One Health?

The concept of One Health recognizes the interdependence of human, animal, and environmental health, and that a holistic approach to the wellbeing of all will lead to improved health outcomes and enhanced resilience. As the discipline is evolving, our understanding of the interdependence of animal, human and environmental health has broadened with the realization that none of these can be truly healthy unless they are all simultaneously healthy (Hueffer et al. 2019). At the time of this writing, the world is engulfed in a pandemic that has globally affected every aspect of life. The causes and impacts of this pandemic are a powerful example of a One Health issue. As we look to understand the causes of such problems it becomes immediately apparent that such understanding will require expertise from many disciplines and the ability to share that knowledge not just across academic disciplines, industries, and government sectors, but across cultures as well.

The term "One Health" was adopted by the veterinary and human medical professions to identify the relationship between human and animal health, and the influence the environment exerts on this relationship (Gibbs 2014; Zinsstag et al. 2010). Between 65 and 70% of emerging diseases in humans are of zoonotic origin (Wendt et al. 2015). The way we impact our environment and how that influences human–animal interactions play significant roles in how these diseases develop and spread. Human sourced drivers such as loss of biodiversity, repurposing of wildlife habitat, the expansion of large intensive livestock enterprises, and rapid

A. Reynolds (✉)
Center for One Health Research, University of Alaska Fairbanks, Fairbanks, AK, USA
e-mail: ajreynolds@alaska.edu

S. Kutz · T. Baker
Faculty of Veterinary Medicine, Department of Ecosystem and Public Health, University of Calgary, Calgary, AB, Canada
e-mail: skutz@ucalgary.ca; tessa.baker1@ucalgary.ca

M. Tryland (ed.), *Arctic One Health*, https://doi.org/10.1007/978-3-030-87853-5_2

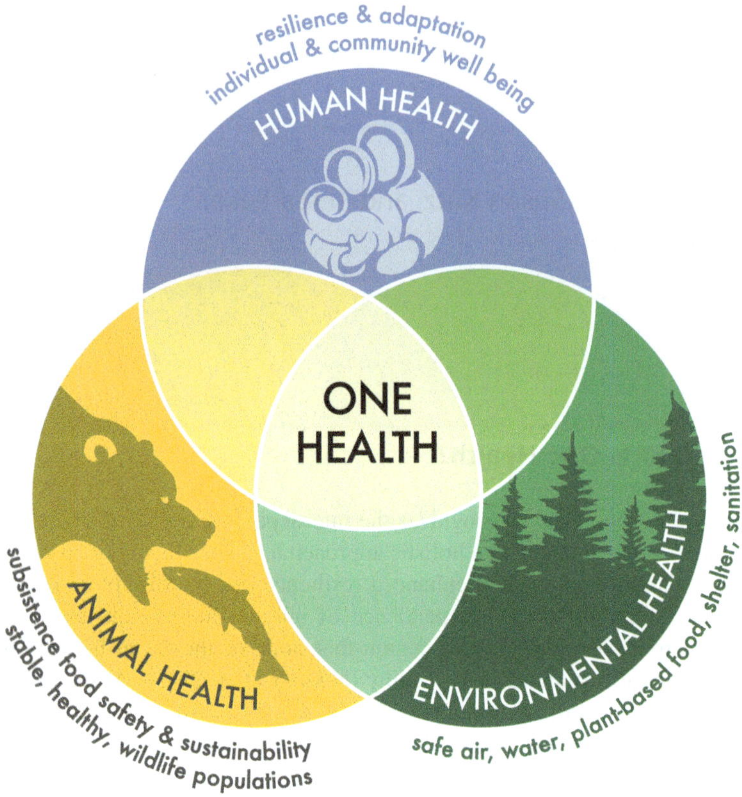

Fig. 1 A holistic approach to One Health in the circumpolar North

anthropogenic driven climate and environmental change, all impact the potential for endemic wildlife pathogens to become zoonotic disease threats (Zinsstag et al. 2010; Gibbs 2014; Wendt et al. 2015; Hueffer et al. 2013).

While human and animal health professions have only relatively recently developed and adopted the term One Health, concepts and ideas, recognizing the interconnectedness of all living beings and their environment, have been at the core of Indigenous worldviews for millennia (Kutz and Tomaselli 2019; Jack et al. 2020). Such an inclusive and holistic approach views health as more than the absence of disease, but rather as a state of individual and community well-being with a focus not only on physical health, but on behavioral, emotional, cultural, and spiritual health as well. Taking this holistic approach to health and applying it to the One Health paradigm, as presented in Fig. 1, allows us to bring in expertise across natural and social sciences and synergize western science with traditional Indigenous Ways of Knowing. Such a broad and, at the same time, deep integration of knowledge and experience provides opportunities to understand large issues like food safety, security, and sovereignty, zoonotic disease threats, and environmental contamination at their roots and engage diverse stakeholders to build effective solutions (Ruscio et al. 2015).

Two-eyed seeing, or Etuaptmunk, as stated by Mi'kmaq elder Elder Albert Marshall is an Indigenous concept that truly encompasses the spirit of One Health. It means "learning to see from one eye with the strengths of Indigenous knowledges…and from the other eye with the strengths of Western knowledges…and learning to use both these eyes together, for the benefit of all" (Denny and Fanning 2016). This concept explicitly acknowledges and values the views of different participants, recognizing the value of incorporating different worldviews. The two-eyed seeing approach has been increasingly applied to wildlife co-management where Indigenous rightsholders, government wildlife managers, and academics are coming together to better understand wildlife health in a more holistic and inclusive manner (Box 1). This approach leads to greater depths of understanding of complex issues and better informed decision-making. It responds to the call and requirements of many governments and conservation agencies to include Indigenous knowledge in decision-making, and importantly, it also responds to the calls of the UN Declaration on the Rights of Indigenous Peoples as well as the Truth and Reconciliation Commission, Canada.

2 The Role of the Veterinarian in One Health

For many reasons, veterinarians are uniquely suited to facilitate the transfer and application of this knowledge between disciplines, sectors, and across cultures at the interface of human, animal, and environmental health. Broadly, veterinarians are trained in comparative medicine, understanding health, and the vast array of determinants of health, across numerous species. Veterinarians have an intricate knowledge of physiology, anatomy, and pathology at an individual level, yet at the same time, are trained in animal welfare, herd health, public health and population medicine, understanding the epidemiology and control of disease at a population level, as well as the socio-economic and environmental factors that will influence the implementation and efficacy of health interventions. Veterinarians are adept at communicating with clients across a very broad socio-economic spectrum and adjusting their communication and treatment offerings to meet the needs and capacity of their clients.

Working within the public health domain, veterinarians are trained to identify zoonotic disease threats and frequently have a deeper understanding of the occurrence and prevention of the common domestic animal-derived zoonoses than their human health counterparts. Veterinarians also routinely work with (or as) wildlife and infectious disease researchers in the surveillance for zoonotic diseases and their vectors of transmission.

Veterinarians also play a critical role in ensuring food and water safety. In urban and non-remote areas veterinarians inspect animal sources of food for safety concerns. They play an important role in food safety in subsistence areas, where climate change, contaminant exposure, and emerging zoonotic diseases are threatening food safety and security in Northern communities in new rapidly changing ways. These challenges require adjustments in the application of both Traditional and Western ways of knowing to effectively monitor and manage. Due to their training and the natural connections they develop with people around animals, veterinarians can also serve as liaisons between community members and research and government agencies, including health and social welfare, as well as facilitators of knowledge transfer and best-practice implementation from these sources back to the communities involved. The breadth of people and organizations that veterinarians work with around individual and population health of wild and domestic animals results in working relationships that span a multitude of stakeholders in local, regional and national sectors.

3 One Health Concerns in the Circumpolar North

The Arctic has unique, sensitive ecosystems that are undergoing rapid change and this is profoundly influencing socio-ecological systems. The rate of Arctic climate warming is occurring at twice the rate of that experienced at lower latitudes (USGCRP 2018) (see also chapter "Climate Change in Northern Regions"). Simultaneously, the region is increasingly stressed by amplifying anthropogenic disturbance in the way of landscape change, shipping, and accelerating economic development. The flora and fauna of the Arctic are adapted to a highly seasonal environment with extremes in temperature and humidity and as this landscape changes the stressors on the endemic flora and fauna increase and invasive species become more common. At the same time, across many Arctic taxa, species diversity is low and there is little redundancy, thus challenging the capacity of the Arctic ecosystem as we know it to cope with these increasing pressures. For the people of the Arctic, these changes are superimposed over a population where poverty, marginalization, and food security are common. The complex interacting factors and rapidly changing socio-ecological system in the Arctic leads to many complex challenges that are ideally suited for a One Health approach.

Healthy domestic and wild animals are central to ecosystem health as well as to the physical, mental, and economic health of people (Fig. 1). In the following sections, we explore the One Health issues around these relationships (Fig. 2).

3.1 Zoonotic Diseases

Many Northern communities are at least partially dependent upon subsistence activities for their dietary needs and cultural activities. This engages people and wildlife in an intimate relationship that may pose risks for emerging and endemic zoonoses. For example, tularemia from muskrats, anisakis and tapeworm from fish, echinococcosis and rabies from wild or domestic canids, and brucellosis and anthrax from caribou, reindeer or bison, are all recognized, and relatively common, zoonoses found around the Arctic (see also related chapters in this book). Less well understood

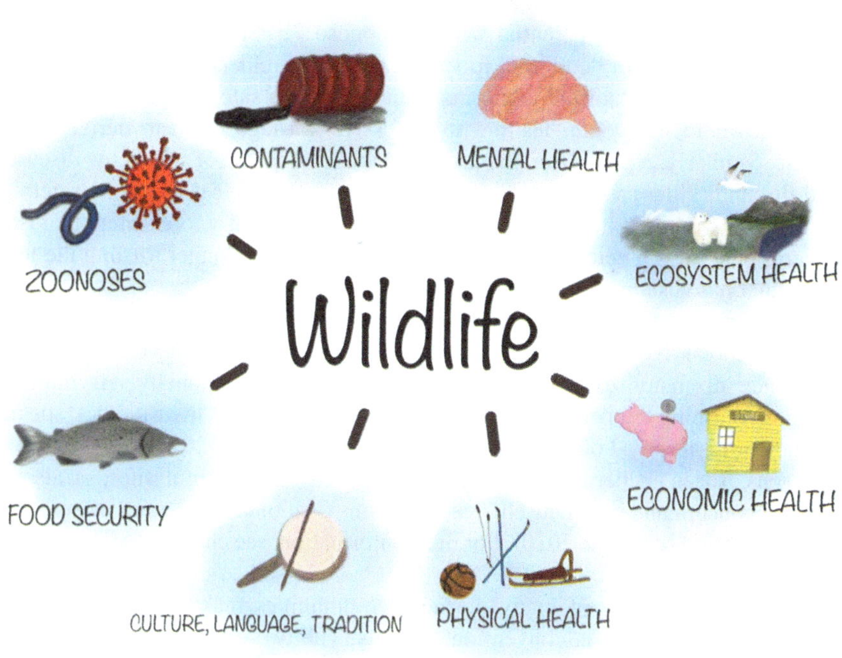

Fig. 2 Wildlife and One Health. Wildlife is central to One Health relationships in the circumpolar North. The history, culture, health and livelihoods of northern Indigenous peoples are intricately woven with that of the wildlife with which they co-exist. Around the Arctic, a diversity of wildlife species have served as a source of food, clothing, and tools, played a central role in cultural activities and transgenerational learning, and provided trade and economic opportunities. These fundamental contributions of wildlife to the health of Arctic peoples continues today. Figure designed by Renate Schlaht

potential zoonoses in the Arctic include pathogens such as *Erysipelothrix, Leptospira, Chlamydia*, Q-fever (*Coxiella burnetii*), Orf virus, a variety of arboviruses (arthropod-borne; viruses transmitted by blood-sucking insects), tick-borne pathogens and others. While we most often focus on zoonotic disease in people, pathogen transmission can occur in the opposite direction, as is thought to have occurred for *Giardia* in muskoxen where the human genotype is found circulating in muskoxen on Banks Island (Kutz et al. 2008). Other proposed/potentially emerging risks include COVID-19, where spill-over from people has caused widespread outbreaks in farmed mink (Munnink et al. 2020).

Many endemic zoonotic diseases have long been known and recognized by Indigenous peoples. In some cases, the knowledge of how to prepare food in a way to prevent transmission has been passed down through generations, and for others, public messaging efforts to reduce transmission have been implemented broadly for over a century. However, the decline of intergenerational knowledge sharing, in Canada largely an outcome of children being removed from their homes to attend residential school, together with overly zealous news reports and public health messaging around potential wildlife zoonoses, has led to a decline in confidence in subsistence or "country" foods. For example, elders are frequently heard saying that what is a 'normal' abnormality and 'what is safe to eat' have not been passed down to the youth, leading to excessive wastage of meat derived from wildlife. Similarly, reports from the press and in social media of 'mad cow disease', 'bird flu' and 'killer cat parasites' can lead to inappropriate fear about the safety of country foods. Thus, in this case, it is not the reality of country food safety that is of concern, rather it is a perception that may lead to people no longer trusting the food source that has sustained them for generations.

However, zoonotic pathogens can pose significant health risks to communities in the circumpolar North. Climate change is one of the main drivers behind the emergence of many zoonotic diseases and their vectors globally. At northern latitudes, warming temperatures support enhanced survival of invasive tick species and the northern spread of the diseases they carry (Waits et al. 2018). The release of pathogens frozen in permafrost, including from historical burial sites, is also of potential concern under climate change conditions (National Academies of Sciences, Engineering and Medicine 2020). For more information, see chapter "Anthrax in the North".

The remote location of many northern communities makes it challenging to monitor zoonotic risks via conventional means. The two-eyed approach described above provides a platform for integrating traditional knowledge with Western science to create a synergetic knowledge base that is more comprehensive than either would be separately. Recent development of a network for local citizens to report anomalies has improved data collection and potential early recognition of emerging zoonotic threats across the Circumpolar North. The Local Environmental Observer or LEO network, (www.leonetwork.org), sponsored by the Alaska Native Tribal Health Consortium connects local observers with scientists, government agencies and health care providers. This network has been used to alert experts at research, government agency, and health care hubs of marine mammal die-offs, unexpected post-mortem observations by hunters, emerging disease vectors, and unusual environmental events that occur hundreds of kilometers away in remote areas from which they would otherwise have very limited access to information. Recent advances in convenient, minimally invasive surveillance techniques such as filter paper whole blood sampling (see Box 1) can be used by hunters to monitor harvested wildlife for endemic and emerging diseases and give researchers and communities early warning for potential zoonotic threats. This has also been used in reindeer herding, with field necropsies conducted by herders. These new technologies help tie rural communities with urban research centers and greatly broaden information gathering to the benefit of all parties involved.

4 Food Safety, Security, and Sovereignty

Rapid environmental changes have made food safety increasingly difficult to achieve in Northern communities over the past 30 years. As the Arctic warms and weather patterns change, traditional means of food storage have been challenged severely. Ice cellars that have been used to preserve food for generations are failing across the circumpolar north (Brubaker et al. 2009). Unusually wet summers have at times made it difficult to make dried fish (see Fig. 3). In lower latitudes veterinarians inspect animals used for food. Many northern communities are under-served in veterinary services due to their small population and remote location, and so do not have access to veterinary inspectors to ensure that the animals they consume are safe to eat.

Climate change has also threatened the safety of marine-based foods. Warming ocean temperatures have increased the duration and severity of harmful algal blooms resulting in dangerously high levels of paralytic and amnesic toxins in filter-feeding shellfish. These changes have also supported the growth of the offending organisms further north than has been previously observed. In 2019, hazardous levels of these toxins were measured in shellfish on the northwestern coast of Alaska. Simultaneously, significant levels of these toxins were also measured in walrus and Bowhead whales on the northwestern and northern coastal areas of Alaska (Lefebvre et al. 2016).

Fig. 3 Fish drying in Emmonak Alaska. Photo courtesy of Dr. Walkie Charles

As described below, the Arctic has become a sink for many persistent toxins produced in the industrial centers of lower latitudes. Bioaccumulation of mercury and persistent organic pollutants have led to harmful contaminant levels in apex predators such as seals, polar bears, and northern pike (Atwell et al. 1998; Fisk et al. 2001; Braune et al. 2005) (see also chapters "Arctic Ecosystems, Wildlife and Man: Threats from Persistent Organic Pollutants and Mercury", "Oil Spills in the Arctic", "Nuclear Radiation", and "Rabies in the Arctic"). As these problems have evolved, concerns regarding the safety of traditional or country foods have led to confusion on the safety of these dietary choices. In the Yukon-Kuskokwim area of Alaska, an area where fish make up the majority of calories and protein of a primarily subsistence-based diet, reports of high mercury levels in northern pike and other key species has led to a syndrome referred to as "fish fear" and resulted in families moving away from traditional foods and towards a more "western" diet. A four-decade-long retrospective study of women in this area showed a progressive decline in plasma vitamin D levels which were concomitantly associated with a significant increase in pediatric rickets in the region (O'Brien et al. 2017; Singleton et al. 2015). Collaborative work between researchers, health care providers, and community members has concluded that although it is important to monitor contaminant levels in subsistence species, people were healthier eating these traditional foods (Mehruba et al. 2016). Rather than switching diets, food safety could be attained by regulating the way in which these foods were selected stored, and prepared. One example of such a recommendation aimed at reducing mercury exposure is to continue eating Northern Pike, but to avoid the larger fish, and focus on eating more small fish (Berner 2019).

Northern communities have a high rate of food insecurity (Huet et al. 2017). Socio-economic, infrastructural, regulatory, and environmental changes have negatively impacted food security in the North (Hueffer et al. 2019). These changes may also require hunters to cover greater distances to access game resources. Shifts to a cash economy and reliance upon using mechanized transportation may make harvesting more efficient but also puts time restraints on those that have to work to pay for these conveniences (Hueffer et al. 2019). The resulting challenges in access to subsistence food sources have negative impacts on food security, cultural practices, knowledge transfer, and mental and behavioral health. For communities off the road system, conventional foods often must be transported by air or barge at considerable cost. Household incomes in these areas are often below national averages impairing the ability to purchase high-priced food items (Huet et al. 2017). Many of these communities still rely heavily on subsistence foods for the majority of their caloric intake (Johnson et al. 2019).

Unprecedentedly rapid environmental changes have challenged long-standing traditional knowledge on game movements, salmon returns, berry ripening times, and most hazardously, travel on ice. Severe and widespread population declines of caribou have left 'caribou people', those Indigenous groups that rely heavily on caribou for subsistence, without one of their main sources of food. In the fall of 2020, the Yukon river chum and silver salmon runs experienced an unprecedented collapse. Commercial harvests reported for this period were 97% lower than the 5-year average. Traditionally, millions of both species return at this time of year when it is easy to preserve them for use as winter food for humans and their dog teams. The catastrophic failure of this run has put both people and their dogs in a position of severe food insecurity.

No discussion of food security in the North would be complete without including the concept of food sovereignty. Access to habitat for reindeer grazing and plant and berry harvesting and to fish and game resources that are central to subsistence living is being challenged by rapid environmental change, socioeconomic shifts, and competition from both commercial and expanding urban personal use harvesting. The Inuit Circumpolar Council has addressed this issue very thoroughly in two documents relating to food security and food sovereignty, and the readers are directed to these resources for a more in-depth coverage of this issue (ICC 2015, 2020). Government regulation of these resources often does not consider traditional knowledge of the resource and traditional harvesting practices. This often results in a conflict when population assessments differ between traditional harvesters and western scientists. Traditional knowledge applied in this sense is often more adept at predicting and detecting population changes by evaluation of harvested animal body condition and overall health, than the technologically driven modelling methods often used by government agencies which set regulations (Kutz and Tomaselli 2019). The assessments used to make these regulations are frequently based on measurements made over a few places and a few days due to cost and time restrictions. In contrast, subsistence hunters are constantly on the land and observing the movement and state of the animals they rely upon and often have a more complete temporal and spatial understanding of these populations than the biologists formulating harvest regulations. Kutz and Tomaselli (2019) describe a "two-eyed approach" to wildlife management that integrates Traditional and Western knowledge in a way that combines the information bases and cooperatively generates solutions that may be superior to those reached by either alone. Under this approach, traditional knowledge holders can combine their knowledge with scientists and develop a more comprehensive model for understanding and predicting the state of fish and game populations (Box 1). For a more detailed description of the harvesting and storage of traditional foods please see chapter "Traditional Conservation Methods and Food Habits in the Arctic".

> In northern Canada, community members, academics, and government wildlife agencies have come together to implement a collaborative wildlife health surveillance program. The communities of Ulukhaktok, Northwest Territories, Kugluktuk and Ekaluktutiak, Nunavut, rely heavily on local caribou and muskox populations (Tomaselli et al. 2018a, b; Hanke et al. 2021; Di Francesco et al. 2021). In response to community concerns about the health and population trajectories of these species, community-based caribou and muskox health surveillance programs were established with the hunters and trappers organizations in all three communities. These programs are multi-

(continued)

pronged, bringing traditional knowledge and scientific knowledge together to understand wildlife population health, disease, and zoonoses. They consist of: (1) baseline wildlife health interviews h (2) hunter-based sampling and (3) ongoing annual interviews. Baseline interviews on the past and current health and population status of caribou and/or muskoxen are done using a combination of individual and group interviews and participatory epidemiology methods (e.g., Tomaselli et al. 2017). This process documents important information on the ecology, health, and trajectory of the populations, identifies community concerns, and forms the basis for further monitoring and investigations. Harvesters are provided with standardized field-friendly sampling kits that they use to collect samples and data from caribou and muskoxen that they harvest for subsistence or through local guides/outfitting operations. Kits consist of data sheets, pre-labeled sampling bags, and Nobuto filter paper strips for blood collection (Fig. 4).

Samples are initially processed in the community by a hired monitor with the hunters and trappers organization and/or government wildlife employees and then sent for further laboratory analyses. Various health indicators, such as infectious disease, stress, nutritional status, genetics, and condition, are determined and the results are brought back to the community in the forms of presentations and reports with key community partners as co-authors on final publications. Ongoing annual interviews are used to document the Indigenous knowledge on population health and trends. These interviews serve to track populations from year to year in real time and identify changes and concerns on a much more rapid time scale than may be detected by the infrequent population surveys. Together, these three steps bring local, traditional and scientific knowledge together to establish historical baselines and trends, document the current status of the populations, and detect any new/emerging conditions, diseases, or concerns. Extensive co-learning is manifested through training of hunters on sampling, monitors on sample processing, scientists/graduate students on traditional harvest methods, animal uses, and knowledge of the land, and the general public on wildlife health and disease. Through this enhanced interaction among community, government and academic partners there is ongoing knowledge sharing, trust building, and vastly improved communication networks which leads to more effective co-management.

Conventional western approaches to the management of these resources may also impose time and individual harvest limits which may not fit the new migration patterns of the animals or the traditional cultural practices of the local Indigenous people. Indigenous culture has developed practices over millennia that harvest fish and game in a manner that takes only what is needed, shares with those in need, and leaves behind sufficient animals to maintain a healthy population (Fig. 5). Traditional harvests take place at the time the animals are available and when the conditions are most favorable to preserve them. Conflict often arises when harvest

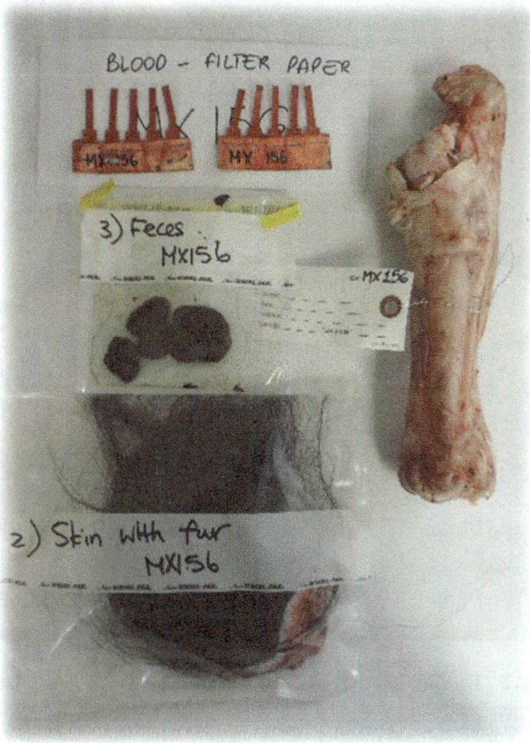

Fig. 4 Muskox samples collected through community-based wildlife health surveillance program. Blood on filter paper allows the easy collection of blood that can then be frozen immediately (e.g., at ambient winter temperatures). Blood collected on filter paper can be used to do a variety of serological assays, as well as DNA isolation. The ease of sampling in the field, which does not require test tubes or any technical or time sensitivity makes it a simple, yet elegant tool for hunter-based wildlife health surveillance

windows are set that do not incorporate traditional practices. An example of this conflict can be seen when a salmon harvest opening occurs during a rainy period when the fish cannot be dried (Fig. 3). Traditional practices would not support harvesting and potentially wasting the fish but would allow people to fish when the conditions are correct for preserving the catch. Governmental regulations often place limits on game harvest to protect over-harvesting, particularly under circumstances when species may be susceptible to this problem. These regulations are often based on singe person allotments for hunters who are in the field only a few days each year and are appropriate for urban households. Indigenous harvesting is often focused on providing food for the whole community. Indigenous hunters usually share their harvest with others outside their household and particularly with elders who may be physically limited. These conflicts are another place where co-production of knowledge may be engaged to support regulations that work optimally for all involved. The prioritization of subsistence resource use brings us back to the concept of food sovereignty as an integral part of food security in areas where Indigenous people have lived for millennia but now may not have say over their access to traditional foods. As described below, the operationalization of One Health as a bridge between Indigenous worldview and Western Science may provide a platform for this type of policy development.

Fig. 5 The Yup'ik Men's Dance Fan. This fan is used in ceremonial dance. The fan represents the human hand with each feather representing a finger. The space underneath the feathers represents the Yup'ik cultural practice of taking only what you need and leaving the rest behind for others that come after and also to maintain sustainable wildlife and plant populations for generations to come. Photo courtesy of Dr. Walkie Charles

5 Contaminant Monitoring

The Arctic Council Arctic Monitoring and Assessment Programme working group (AMAP) has demonstrated that the Arctic is a sink for Anthropogenic pollution and climate change (Gibson et al. 2016). Ocean and atmospheric currents bring organic and heavy metal contaminants from lower latitudes, where they are generated by industrial societies, to the Arctic, where they accumulate in the physical environment and bioaccumulate in the food web (Fisk et al. 2001; Braune et al. 2005; Atwell et al. 1998). Climate change has exacerbated the movement and impact of these contaminants (Braune et al. 2005). The recent and rapid accumulation of these toxins has resulted in many new stressors upon Arctic ecosystems threatening the survival of several species and endangering the safety of subsistence food sources.

For over 20 years the AMAP working group has monitored contaminant levels in humans and sentinel species across the Arctic (Gibson et al. 2016). These studies have found significant and potentially health-threatening concentrations of industrially produced mercury and persistent organic pollutants (POPs) in humans and apex predators such as polar bears, seals, narwhals, and northern pike. Mercury accumulation can impair central nervous system functions and therefore affect cognition and locomotion which may decrease hunting efficiency and result in aberrant behavior (Black et al. 2016). POPs have multi-systemic effects. They can alter hormone transportation and receptor activity resulting in decreased fertility, enhanced rates of miscarriage, low birthweight, and enhanced neonatal mortality (Black et al. 2016). They also impair immune function and enhance the risk of developing certain forms of neoplasia. Climate change has enhanced not only the transport of these contaminants to the Arctic but also their impact. Shrinking sea ice has forced marine mammals to swim further than normally required to obtain food and shelter, causing an enhanced utilization of lipid depots and resulting in mobilization of lipophilic compounds during periods of high stress.

Species such as seals, narwhals, and even polar bears have traditionally been staples in the subsistence diets of coastal inhabitants across the Circumpolar North. This puts these people at the highest position in the food chain and therefore at the greatest risk of the bioaccumulatory impacts of these toxins. POP concentrations in Inuit living in Eastern Greenland are among the highest measured anywhere and have been associated with an increased incidence of cancer and immune-related issues in this population (Gibson et al. 2016).

Contaminant accumulation in the Arctic is a clear example of an issue that can be addressed well through a One Health lens. AMAP, CAFF, and ACAP have incorporated a One Health approach by combining environmental monitoring with the monitoring of humans and sentinel animal species. Programs that monitor sentinel species in the food web such as seals, narwhals, and polar bears, provide an understanding of the trends and severity of contaminant bioaccumulation in the food web. Monitoring companion sentinel species, such as sled dogs, may also provide useful information in developing dietary recommendations for people living a subsistence lifestyle in these areas. Studies of sled dogs have been useful in determining mercury and POP bioaccumulation (Sonne et al. 2017; Dunlap et al. 2011) as these dogs often eat similar diets to the humans they live and work with. The relatively higher metabolic rate of these dogs in comparison to their human counterparts may also permit scientists to see health issues in the dogs before they become problems in people.

6 Mental and Behavioral Health and Well-Being

In Indigenous populations, rapid social and economic change associated with colonization and assimilation practices has been associated with pervasive social issues including suicide, substance abuse, and domestic violence (Hueffer et al. 2019). Loss of language (Krauss 1980; Gone 2013), cultural practices, and cultural knowledge have been associated with a severe increase in the incidence of these problems. Added to these stressors are rapid environmental changes which have impacted, traditional activities such as travel on ice, hunting, fishing, and gathering of plant resources. When traditional knowledge struggles to accommodate rapid rates and previously unexperienced types of environmental change, it can adversely affect the self-esteem of resource providers and the self-efficacy of a society. Indigenous people have lived in and stewarded their traditional lands for millennia. Indigenous worldview has historically seen the terrestrial and marine environments and the flora and fauna they encompass as benefactors which people are not only dependent upon but also inseparable and indivisible from. When such cultural foundations change from being benefactors to becoming threats in the form of contaminated foods, unreliable fish and game populations, and unsafe or unreliable ice conditions, this challenges belief systems and can result in severe and negative impacts on mental and behavioral health and well-being.

Suicide is now the leading cause of death for Alaska Native people between the ages of 15 and 25 (Berman 2014; Hicks 2007) and in Canada, the suicide rate of Inuit is approximately 9 times that of non-Indigenous Canadians (Kumar and Tjepkema 2019). Prior to 1950, suicide was rare and most common among aged men that no longer felt capable of contributing to the needs of their community. Conventional approaches, which isolate and treat individuals perceived to be at high risk, have done little to prevent the continuation of these issues (Hicks 2007). Recent approaches focusing on building strengths rather than managing outcomes have begun to show promise (Rasmus et al. 2019; Rivkin et al. 2019). Reintroduction or retention of cultural practices, transfer of traditional ways of knowing, and fluency in the original language are all traits common to Indigenous communities that have shown resilience to suicide in the Circumpolar North (Rivkin et al. 2019). Recently, programs, using a One Health approach, have shown real promise in preventing negative mental and behavioral health outcomes in Northern Indigenous communities. These community-based programs incorporate the relationship and inseparable nature of the human, animal, and environmental health as foundations of strengths that can be used to build resilience to these problems. One prominent example of this is the Alaska Native Cultural Hub for Resilience Research (ANCHRR). This NIH-funded community-based program partners University (UAF) researchers with community members and elders to study and define best practices in resilient communities and share them with communities that are struggling. Instead of identifying and isolating at-risk individuals, which may exacerbate the issues by focusing on these individuals as being "different," this program emphasizes building strengths through cultural activities, sharing of personal stories, and transfer of traditional knowledge. This work is often done out on the land during hunting, fishing, or berry picking activities. In this way, potentially susceptible youth are "wrapped in a blanket of community support and strength" that builds self-efficacy and self-esteem through gaining proficiency in skills and acknowledgement of accomplishments from respected community members.

Other examples of such programs are the Frank Attla Youth and Sled Dog Care Program (FAYSDCP) and the Alaska Care and Husbandry Instruction for Lifelong Living (ACHILL) described in the chapter "Dogs and People: Providing Veterinary Services to Remote Arctic Communities" (Veterinary medicine in remote arctic communities). These community-based programs holistically address mental and behavioral issues by simultaneously addressing environmental and animal health.

The interdependence of the One health triad is central to the culture and knowledge base of these communities and so this approach addresses the human health issues at their root causes rather than treating their outcomes. Veterinarians, who by the nature of their education are trained in preventive health care and understand the value of the human–animal bond, are essential stakeholders in the development and implementation of these One Health processes.

7 Operationalizing One Health

The One Health approach to describing, understanding, and managing large issues that span the interface between human, animal, and environmental health is gaining support from community members, health care professionals, academics, governmental agencies, and NGOs across the globe (Ruscio et al. 2015; Arctic Council SDWG 2017). This approach is particularly relevant in the Circumpolar North where environmental changes are happening at a rate that has been unprecedented and making it difficult for social and ecological systems to adapt in a healthy manner. Although these changes pose a tremendous challenge to northern communities, they simultaneously present an opportunity to understand and address related changes that are happening to a lesser degree at lower latitudes.

While One Health is being embraced as the way to work on these "Wicked Problems," it is often easier to conceptualize this approach than to operationalize it (Vesterinen et al. 2019). Putting One Health into action requires stakeholders to work across disciplines and cultures and work in a constructionist approach that addresses issues starting at the community perspective and working outwards. Each part of this strategy requires a paradigm shift from conventional academic and scientific approaches to problem-solving. This paradigm shift is simultaneously the greatest potential strength and the greatest potential challenge encountered in operationalizing One Health.

For centuries, western scientists have used a reductionist approach to study and solve problems. This method entails breaking down or reducing problems to a single underlying cause and has been used to identify individual pathogens as the cause of a disease, or a single gene mutation as the source of an error in metabolism. Many of the hallmark successes of modern science have resulted from this approach which has become the default method of scientific problem-solving. While reductionist reasoning has worked well for single-issue problems, this approach falls short when addressing issues which may have multiple causes and interactions.

"Wicked problems" require knowledge that not only penetrates deeply into a single discipline but also spans across all of the disciplines involved. At this writing (May 2021), vaccines against the COVID-19 have arrived and are being used to control the pandemic which has spanned the globe and affected nearly every aspect of life. These vaccines may well stem the spread and effects of the virus but their development and implementation will not help us understand why this catastrophe occurred in the first place and what we might be able to do to prevent or mitigate the next pandemic from gaining a foothold. For this, we must understand how anthropogenic environmental changes have impacted wildlife populations that serve as potential reservoirs for emerging zoonotic threats and how these are influenced by current livestock husbandry, cultural, social, and economic practices. This understanding requires a constructionist approach that integrates information across disciplines and cultures and approaches the issue from the perspective of the communities involved in a bottom-up, rather than top-down, prescriptive approach. One Health applied in this manner supports the gathering of knowledge both broadly and deeply, and the solutions acquired are likely to be effective because they have arisen with the input of the communities where they will be implemented.

Working across disciplines challenges the current paradigm under which research and problem management are conducted. Instead of working within a single discipline and communicating findings to others with similar education and training, those working in a One Health approach must be capable of both giving and receiving information to and from those with different backgrounds than their own. This can be challenging even across conventional western disciplines. Natural scientists use different methodology and terminology in their work than that used by social scientists, and these differences require significant adjustments when studies are designed to incorporate both approaches. Bridging the gap across cultural knowledge systems in a "Two eyed approach" is another example of the advantages and challenges associated with working outside of conventional western scientific methods (Kutz and Tomaselli 2019). Scientists are often uncomfortable transferring the implications of their work outside of their own narrow fields of study, however, science communication to non-scientists, such as community members and policy makers, is central to the success of a One Health approach. The shortfall in science literacy and the resulting negative impacts from non-adherence to CDC recommendations led to the world's greatest per capita case and fatality rates in the United States during the first 9 months of the COVID-19 pandemic. This example serves as a warning to all involved for the need to improve science communication between researchers, medical professionals, and the general public (Eysenbach 2020).

While there is no handbook for operationalizing One Health, several systems have been developed for use as a platform to begin the process. The US CDC has developed a One Health Zoonotic Disease Prioritization process (CDC-OHZDP) for emerging zoonotic disease threats (Salyer et al. 2017). This process uses a workshop format and engages stakeholders including community members, academic researchers, health care providers, and those working for government agencies. The process and its application in regional, national, and international situations have been published in peer-reviewed journals (Salyer et al. 2017). The CDC co-sponsored such a workshop with the UAF Center for One Health Research (COHR) in March of 2019 to prioritize emerging zoonotic disease threats in Alaska. The top seven threats identified in this workshop are listed in Table 1. This was the first time this process had been implemented in the Circumpolar North and serves as an example of what could be developed in other Arctic countries. The resulting report can be used to support surveillance and research efforts focused on these

Table 1 Priority zoonotic diseases selected in Alaska by participants in the One Health Zoonotic Disease Prioritization workshop conducted May 20–21, 2019. Reproduced from Goroyka et al. (2020)

Zoonotic disease	Human disease burden	Animal disease burden	Diagnostics, treatment and prevention
Amnesic shellfish poisoning/ paralytic shellfish poisoning	Between 1973 and 1996 over 200 cases of paralytic shellfish poisoning were reported in Alaska and were attributed to more than 70 outbreaks across the state	Nearly all molluscan shellfish in Alaska are affected by paralytic shellfish poisoning and the Alaska Department of Environmental Conservation regularly tests commercially harvested shellfish In a recent study by the University of Alaska SE, PSP measurements in mussels at sites around Juneau reached 4500 micrograms per 100 grams of shellfish. This level is fatal to a person after only consuming a few mussels	Clinical diagnosis is based on recent shellfish ingestion and the presence of clinical manifestations of toxicity such as nausea, vomiting, paresthesia, dysarthria, dysphagia, and weakness. The toxin can also be confirmed in a clinical specimen such as blood or urine To stay safe, clean shellfish thoroughly, removing all butter and discarding the gut. Also only consume shellfish sold commercially and routinely tested as cooking and freezing will not destroy the toxin Treatment for severe cases is the use of a mechanical respirator and oxygen
Zoonotic influenza	There have been no human infections with Asian HPAI H5N1 virus reported in the United States. However, sporadic human infections with avian influenza A (H7) viruses have been identified in the United States Since 2010, 466 cases of swine flu have been reported in the United States	H1N1 and H3N2 swine flu viruses are endemic among pig populations in the US with outbreaks normally occurring in colder weather months As part of a large-scale Avian influenza surveillance study from 2007–2011, researchers reported a mean apparent prevalence of avian influenza virus of 11.4% within wild birds. Prevalence was highest in dabbling	As a general precaution, people should avoid wild birds, contact with domestic birds that appear ill or have died, and avoid contact with surfaces that appear to be contaminated with feces from wild birds One mode of prevention is via the seasonal flu vaccine which can be given to humans and animals The best way to prevent infection is to avoid sources of

(continued)

Table 1 (continued)

Zoonotic disease	Human disease burden	Animal disease burden	Diagnostics, treatment and prevention
		ducks whose mean prevalence was 15.8% As part of the USDA ongoing surveillance for swine, over 120,000 samples have been tested between 2010 and 2016 resulting in over 10,000 positive cases for influenza	exposure specifically contact with infected poultry Treatment includes antiviral drugs and continued monitoring Diagnosis for influenza and novel types of zoonotic influenza includes respiratory specimens for laboratory testing using PCR
Rabies	Three human cases have been reported in Alaska since 1914 but none have been reported since 1942	Between 15 and 50 cases of wildlife cases are reported each year in Alaska. Rabies is enzootic among the fox populations in the North and West regions in Alaska. There have been periodic epizootics documented every 3 to 5 years	Rabies is diagnosed in animals using direct fluorescent antibody tests. Several rapid laboratory tests are required for diagnosis in humans There is a vaccine available to both animals and humans. Following any contact or bite from a rabid animal, medical attention is immediately necessary Prophylaxis is the immediate treatment; however, following the onset of clinical symptoms, there is no treatment, and the disease is fatal
Cryptosporidiosis/ Giardiasis	A recent study reported a 28.8% seroprevalence of cryptosporidium in people with or without wild bird contact in Alaska. The same study reported an 18.9% seroprevalence of Giardia intestinalis in the same population From 2001–2010, there were 1042	One study looking at the prevalence of cryptosporidium and giardia subspecies found that prevalence was highest among ring seals (22.6% cryptosporidium, 64.5% giardia) and right whales (24.5% cryptosporidium, 71.4% giardia)	Both cryptosporidiosis and giardiasis are diagnosed through microscopic analysis of stool samples. In both cases, PCR can be used to determine species. Those with competent immune systems will recover from cryptosporidiosis without treatment,

(continued)

Table 1 (continued)

Zoonotic disease	Human disease burden	Animal disease burden	Diagnostics, treatment and prevention
	human cases of giardiasis reported. Annual rates of giardiasis in Alaska have repeatedly been higher than in the rest of the United States Another study looking at the prevalence among Alaska residents found the prevalence of giardia antibody was highest among subsistence hunters and their families at 30%		fluid replacement and nitazoxanide may be recommended. For giardiasis metronidazole, tinidazole, and nitazoxanide are recommended. Prevention for both are primarily good hygiene practices and avoiding contaminated food and water
Toxoplasmosis	A 2019 study reported a 2.9% seroprevalence for Toxoplasma gondii in people with or without wild bird contact in Alaska	A recent study looking at seroprevalence among sea otters reported 32% of sea otters tested positive for *T. gondii* Another study looking at serum antibody prevalence for T. gondii within Alaska wildlife reported 23% positive among moose, 43% for black bears, 9% for wolves, and 7% for Dall sheep	Toxoplasmosis is primarily diagnosed through serologic testing. Healthy individuals typically do not require treatment to recover. However, pyrimethamine and sulfadiazine, plus folinic acid can be administered. Prevention includes cooking foods to proper temperatures and avoiding contact with cat feces
Brucellosis	A 2019 study reported a 0.1% seroprevalence for Brucella spp. in people with or without wild bird contact in Alaska	There are 10 species of *Brucella* recognized in animals One recent study looking at seroprevalence of Brucella in Alaskan harbor seals found that overall, 52% of adult seals tested positive for antibody seroprevalence A study looking at serum antibody prevalences for Brucella among	Diagnosing brucellosis is done through bacterial isolation in blood cultures and serologic testing There is no standardized diagnostic tests for different species of animals Antibiotics, generally doxycycline and rifampin, are given to treat the infection. Brucellosis can be

(continued)

Table 1 (continued)

Zoonotic disease	Human disease burden	Animal disease burden	Diagnostics, treatment and prevention
		caribou, wolves, and bears reported the highest prevalence in the northwest region of Alaska	prevented by avoiding the consumption of undercooked meat and unpasteurized dairy products. Additionally, those handling animal tissues should wear protective clothing Prevention includes vaccination of domestic livestock
Q Fever	A recent study reported an 8.3% seroprevalence of *Coxiella burnetii* in people with or without wild bird contact in Alaska	A 2015 study reported a 17% seroprevalenve of *Coxiella burnetii* in live seemingly healthy northern sea otters Enhydra lutris kenyoni of Alaska. Another study in 2013 found an 80% seroprevalence in northern fur seals of Alaska	Q fever is diagnosed through a blood test. The majority of those infected are able to recover without treatment. But, a 2-week course of doxycycline may be recommended. Prevention methods for Q fever include avoiding contact with animals and refrain from consuming raw milk products

diseases and the policy and funding necessary for this work to proceed (https://www.cdc.gov/onehealth/pdfs/Alaska-508.pdf).

The CD-OHZDPP is the first step in operationalizing One Health as it can play a key role in prioritizing One Health issues. Once an issue, such as a zoonotic disease threat has been identified, the next step is to use a One Health approach to analyze and manage it. The One Health Systems Mapping and Analysis Toolkit Process (OH-SMART) uses systems mapping and analysis to achieve these goals (Vesterinen et al. 2019). OH-SMART was developed in a joint effort between the University of Minnesota and the USDA to analyze and facilitate communication and collaboration across government agencies and other stakeholders as presented in Fig. 6.

Beginning with an identified One Health challenge, the first step is to identify the network of stakeholders that will be involved in the process. These stakeholders are then interviewed to determine their approach to the issue and which other stakeholders they collaborate with and the depth of these collaborations. These interviews are then placed on a swim lane map so that the flow of resources, data, and lines of communication can be followed. The map is analyzed for best practices

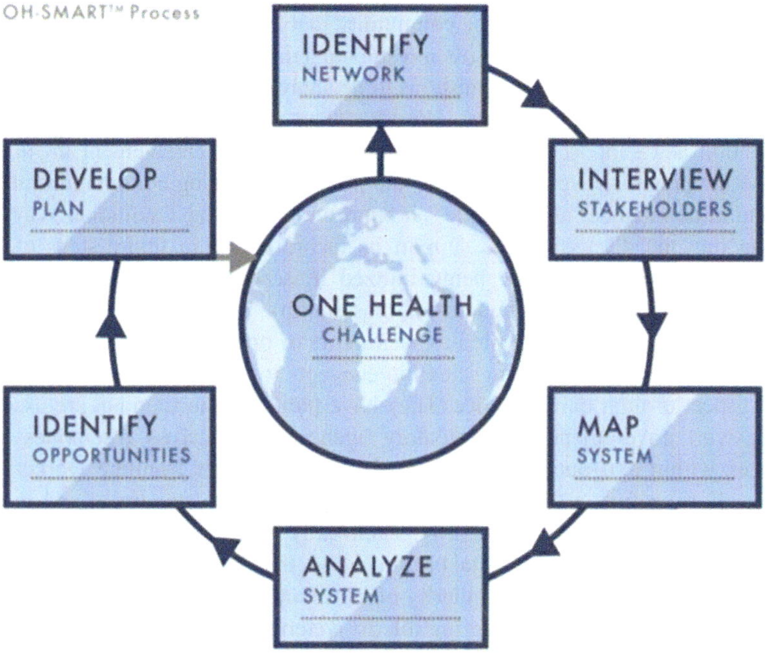

Fig. 6 The OH-SMART model. Used with permission of the authors (Vesterinen et al. 2019)

and discrepancies. Opportunities are identified for improving and strengthening the system. These are best practices which can be institutionalization or otherwise made stronger and discrepancies which can be resolved. These opportunities are prioritized based upon the impact, resource requirement, feasibility, and sustainability of doing so. In the final step, action plans are developed to implement the highest priority opportunities as identified by the group. This is an iterative process at all levels, and each progressive step will often uncover new components of previous steps which need to be considered to improve the effectiveness of the outcomes.

This technique can be applied in a several-day, in-place workshop, or over a longer period of time by distance. It can address problems retrospectively, prospectively, or while they are occurring. Retrospective analysis seeks to understand what went well and what can be improved for future situations. Prospective analysis helps prepare One Health workers for potential future problems by analyzing the current system for efficiencies and improvements. OH-SMART analysis during a situation helps workers analyze how well the system in place is working and to make needed adjustments in real time. For a detailed description of the OH-SMART process and its implementation, the readers are referred to Vesterinen et al. (2019).

While the OH-SMART technique was developed to assist in the sectoral analysis of a system focused on government agencies, it can be applied to community-based problems as well. In this case, the first step is to develop an understanding of the problem from the perspective of the community that is experiencing it. This entails time and relationship building and often results in a different focus than would have been the case if the problem and questions to be analyzed were determined by those working outside of the community. How well the first step engages community collaboration and support will determine the community's engagement in further steps and the potential overall success of the process. Engaging communities at the onset and in this manner also provides an opportunity to bring Traditional Ways of Knowing into the discussion and support its incorporation in future steps of the process. This approach is now being utilized in academic programs in Alaska (https://www.uaf.edu/onehealth/education/master.php).

Veterinarians are uniquely suited as facilitators of operationalizing One Health. They are trained to communicate scientific concepts to lay clients and do so as an integral aspect of their daily practice. They have public health training, are used to working with government and regulatory agencies, and across disciplines and specialties within and outside of their profession. They are frontline workers in animal welfare, zoonotic disease surveillance, reporting, and treatment and in the maintenance of food safety and security. Their daily job encompasses the interface of human, animal, and environmental health, and so, they have a working knowledge in all areas of One Health and familiarity of communicating that knowledge across disciplines and cultures. This emerging role for veterinarians is demonstrated in the inclusion of One Health in the mission statement of veterinary colleges across the globe. As One Health becomes a more conventional approach to understanding and managing large, complicated issues at the interface of human, animal, and environmental health, veterinarians will be increasingly called upon to facilitate and implement the operationalization of One Health. This will be especially true in the North where people still have close ties to the land and the animals they live with and depend upon, and, during a time when these relationships continue to change rapidly.

The Arctic is experiencing environmental, social, and economic change at a historically unprecedentedly rapid rate. This poses great challenges and, simultaneously, great opportunities to operationalize paradigm shifts supporting adaptation and resilience to these changes and which can then serve as a management model for similar changes that are occurring more gradually on a global scale. Addressing these issues effectively requires a One Health approach that integrates knowledge across disciplines and cultures, recognizes the interdependence of human, animal, and environmental health, and begins the process from a community-based perspective.

References

Arctic Council Sustainable Development Working Group (2017) One Health: Operationalizing One Health in the Arctic. DWG, Tromsø

Atwell L, Hobson KA, Welch HE (1998) Biomagnification and bioaccumulation of mercury in arctic marine food web: insights from stable nitrogen isotope analysis. Can J Fish Aquat Sci 55:1114–1121

Berman M (2014) Suicide among young Alaska native men: community risk factors and alcohol control. Am J Public Health 104(Suppl 3):S329–335. https://doi.org/10.2105/AJPH.2013.301503

Berner J (2019) One Health One Future conference presentation. Fairbanks, Alaska November 5, 2019

Black S, Duigan P, Akeegok J et al (2016) Marine animal health in a changing environment. In: Cook S, Hall, D, Liljebelke K (eds) One Health case studies. 5M Publishing, Sheffield, pp 164–165

Braune BM, Outridge PM, Fisk AT et al (2005) Persistent organic pollutants and mercury in marine biota of the Canadian Arctic: an overview of spatial and temporal trends. Sci Total Environ 351–352:4–56. https://doi.org/10.1016/j.scitotenv.2004.10.034

Brubaker M, Bell J, Rollin A (2009) Climate change effects on traditional Inupiak food cellars. Center for Climate Change and Health, Bulletin No.1, October 19. Alaska Native Medical Center, Anchorage Alaska

Denny SK, Fanning LM (2016) A Mi'kmaw perspective on advancing Salmon governance in Nova Scotia, Canada: setting the stage for collaborative co-existence. Int Indig Policy J 7(3). https://doi.org/10.18584/iipj.2016.7.3.4

Di Francesco J, Hanke A, Milton T et al (2021) Documenting indigenous knowledge to identify and understand the stressors of muskoxen (*Ovibos moschatus*) in Nunavut, Canada. Arctic (in press)

Dunlap KL, Reynolds AJ, Gerlach SC et al (2011) Mercury interferes with endogenous antioxidant levels in Yukon River subsistence-fed sled dogs. Environ Res Lett 6(4):044015. https://doi.org/10.1088/1748-9326/6/4/044015

Eysenbach G (2020) How to fight an infodemic: the four pillars of infodemic management. J Med Internet Res 22(6). https://doi.org/10.2196/21820

Fisk AT, Hobson KA, Norstrom RJ (2001) Influence of chemical and biological factors on trophic transfer of persistent organic pollutants in the northwater polynya marine food web. Environ Sci Technol 35(4):732–738. https://doi.org/10.1021/es001459w (Erratum in: Environ Sci Technol 35(8):1700)

Gibbs EPJ (2014) The evolution of One Health: a decade of progress and challenges for the future. Vet Rec 174(4):85–91

Gibson J, Adlard B, Olafsdottir K et al (2016) Levels and trends of contaminants in humans of the Arctic. Int J Circumpolar Health 75:33804. https://doi.org/10.3402/ijch.v75.33804

Gone JP (2013) Redressing first nations historical trauma: theorizing mechanisms for indigenous culture as mental health treatment. Transcult Psychiatry 50(5):683–706. https://doi.org/10.1177/1363461513487669

Hanke AH, Angohiatok M, Leclerc L-M et al (2021) A caribou decline foreshadowed by Inuit in the Central Canadian Arctic: a retrospective analysis. Arctic (in press)

Hicks J (2007) The social determinants of elevated rates of suicide among Inuit youth. Indig Affairs 4:30–37

Hueffer K, Parkinson AJ, Gerlac R et al (2013) Zoonotic infections in Alaska: disease prevalence, potential impact of climate change, and recommended actions for earlier disease detection, research, prevention, and control. Int J Circumpolar Health 72:1–11

Hueffer K, Ehrlander M, Eyz K et al (2019) One Health in the circumpolar North. Int J Circumpolar Health 78:1. https://doi.org/10.1080/22423982.2019.1607502

Huet C, Ford JD, Edge VL et al (2017) Food insecurity and food consumption by season in households with children in an Arctic city: a cross-sectional study. BMC Public Health 17(1):578. https://doi.org/10.1186/s12889-017-4393-6

Inuit Circumpolar Council (2015) Alaskan Inuit food security conceptual framework: how to assess the Arctic from and Inuit perspective; summary and recommendations report. Inuit Circumpolar

Council – Alaska. Available via https://iccalaska.org/wp-icc/wp-content/uploads/2016/03/Food-Security-Summary-and-Recommendations-Report.pdf. Accessed 01 April 2020

Inuit Circumpolar Council (2020) Food sovereignty and self governance: Inuit role in managing Arctic marine resources. Inuit Circumpolar Council – Alaska. Available via https://iccalaska.org/wp-icc/wp-content/uploads/2020/09/FSSG-Report_-LR.pdf. Accessed 01 Nov 2020

Jack JC, Gonet J, Mease A et al (2020) Traditional knowledge underlies One Health. Science 369 (6511):1576. https://doi.org/10.1126/science

Johnson JS, Nobmann ED, Asay E et al (2019) Dietary intake of Alaska native people in two regions and implications for health: the Alaska native dietary and subsistence food assessment project. Int J Circumpolar Health 68(2):109–122. https://doi.org/10.3402/ijch.v68i2.18320

Krauss ME (1980) Alaska native languages, past, present, and future. Alaska Native Lang Center 4:1–121

Kumar MB, Tjepkema M (2019) Suicide among First Nations people, Métis and Inuit (2011-2016): Findings from the 2011 Canadian Census Health and Environment Cohort (CanCHEC). Statistics Canada. 99-011-X2019001. ISBN 978-0-660-31402-0. https://www150.statcan.gc.ca/n1/en/catalogue/99-011-X2019001

Kutz S, Tomaselli M (2019) "Two-eyed seeing" supports wildlife health – bridging indigenous and scientific knowledge improves wildlife surveillance and fosters reconciliation. Science 364 (6446):1135–1137

Kutz SJ, Thompson RA, Polley L et al (2008) *Giardia* assemblage A: human genotype in muskoxen in the Canadian Arctic. Parasites Vect 1(32). https://doi.org/10.1186/1756-3305-1-32

Lefebvre KA, Quakenbush L, Frame E et al (2016) Prevalence of algal toxins in Alaskan marine mammals foraging in a changing arctic and subartic environment. Harmful Algae 55:13–24. https://doi.org/10.1016/j.hal.2016.01.007

Mehruba A, Ridpath A, Berner J et al (2016) Medical toxicology and public health – update on research and activities at the Centers for Disease Control and Prevention and the Agency for Toxic Substances and Disease Registry. J Med Toxicol 12(3):315–317

Munnink BBO, Sikkema RS, Nieuwenhuijse DF et al (2020) Jumping back and forth: anthropozoonotic and zoonotic transmission of SARS-CoV-2 on mink farms. bioRxiv. https://doi.org/10.1101/2020.09.01.277152

National Academies of Sciences, Engineering, and Medicine (2020) Understanding and responding to global health security risks from microbial threats in the Arctic: proceedings of a workshop. The National Academies Press, Washington, DC. https://doi.org/10.17226/25887

O'Brien DM, Thummel KE, Bulkow LR et al (2017) Declines in traditional marine food intake and vitamin D levels from the 1960s to present in young Alaska native women. Public Health Nutr 20(10):1738–1745. https://doi.org/10.1017/S1368980016001853

Rasmus SM, Trickett E, Charles B et al (2019) The qasgiq model as an indigenous intervention: using the cultural logic of contexts to build protective factors for Alaska native suicide and alcohol misuse prevention 25(1)

Rivkin I, Lopez EDS, Trimble JE et al (2019) Cultural values, coping, and hope in Yup'ik communities facing rapid cultural change 47(3)

Ruscio BA, Brubaker M, Glasser J et al (2015) One health – a strategy for resilience in a changing Arctic. Int J Circumpolar Health 74:27913

Salyer SJ, Silver R, Simone K et al (2017) Prioritizing zoonoses for global health capacity building-themes from one health zoonotic disease workshops in 7 countries, 2014-2016. Emerg Infect Dis (13):S55–64. https://doi.org/10.3201/eid2313.170418

Singleton R, Lescher R, Gessner BD et al (2015) Rickets and vitamin D deficiency in Alaska native children. J Pediatr Endocrinol Metab 28(7–8):815–823. https://doi.org/10.1515/jpem-2014-0446

Sonne C, Letcher RJ, Jenssen BM et al (2017) A veterinary perspective on One Health in the Arctic. Acta Vet Scand 59(1):84. https://doi.org/10.1186/s13028-017-0353-5

Tomaselli M, Kutz SJ, Gerlach C et al (2017) Local knowledge to enhance wildlife population health surveillance: conserving muskoxen and caribou in the Canadian Arctic. Biol Conserv 217 (1):337–348. https://doi.org/10.1016/j.biocon.2017.11.010

Tomaselli M, Gerlach SC, Kutz SJ et al. (2018a) Community of Iqaluktutiaq. Iqaluktutiaq voices: local perspectives about the importance of muskoxen, contemporary and traditional use and practices. Arctic 71(1):1–4

Tomaselli M, Kutz S, Gerlach C et al (2018b) Local knowledge to enhance wildlife population health surveillance: conserving muskoxen and caribou in the Canadian Arctic. Biol Conserv 217:337–348

USGCRP (U.S. Global Change Research Program) (2018) Fourth National Climate Assess II:1–470

Vesterinen HM, Dutcher TV, Errecaborde KM et al (2019) Strengthening multi-sectoral collaboration on critical health issues: one health systems mapping and analysis resource toolkit (OH-SMART) for operationalizing one health. PLoS One 14(7). https://doi.org/10.1371/journal.pone.0219197

Waits A, Emelyanova A, Oksanen A et al (2018) Human infectious diseases and the changing climate in the Arctic. Environ Int 121:703–713

Wendt A, Kreienbrock L, Campe A (2015) Zoonotic disease surveillance – inventory of systems integrating human and animal disease. Zoonoses Public Health 62(1):61–74

Zinsstag J, Schekking E, Waltner-Toews D et al (2010) From "one medicine" to "one health" and systemic approaches to health and well-being. Prev Vet Med 101(3–4):148–156

Seasonal Animal Migrations and the Arctic: Ecology, Diversity, and Spread of Infectious Agents

Øystein Varpe and Silke Bauer

1 Introduction

Long before humans started travelling long distances, many animals have been on the move. Through long-distance migrations, animals move year after year and in relatively predictable manners, between regions, continents, and hemispheres. Migrations come in many forms and fascinate people far beyond the biologists that study them. Many migrants and migrations are iconic, such as the European eels (*Anguilla anguilla*) that after born in the Sargasso Sea find their way to lakes and waterways in Europe, grow up there, and return as adults, for the eels a once in a lifetime trip (Righton et al. 2016). Another example is the common cuckoo (*Cuculus canorus*), which spends the non-breeding season south of Sahara and whose return to Europe is one of the distinguished signatures of spring at northern latitudes (Willemoes et al. 2014). Prominent examples also include migrations to Arctic regions, such as the Arctic tern's (*Sterna paradisaea*) annual move between the Arctic and the Southern Ocean (Hromádková et al. 2020). Well-known are also the abundant and highly visible goose migrations, from temperate wintering grounds to Arctic breeding grounds—a phenomenon seen around the entire northern hemisphere (CAFF 2018). Clearly, migrants are actors in multiple ecosystems through the year and through their life, and they connect very different parts of the planet (Bauer and Hoye 2014).

Ø. Varpe (✉)
Department of Biological Sciences, University of Bergen, Bergen, Norway

Norwegian Institute for Nature Research (NINA), Bergen, Norway
e-mail: oystein.varpe@uib.no

S. Bauer
Department of Bird Migration, Swiss Ornithological Institute, Sempach, Switzerland

M. Tryland (ed.), *Arctic One Health*, https://doi.org/10.1007/978-3-030-87853-5_3

Why have seasonal migrations evolved? Migrants exploit spatially and temporally varying resources by inhabiting multiple habitats, often far apart, at different times of the year, and movements between them have in many cases clearly been favoured over stationary (or resident) strategies. That is, migrations have allowed growth, survival, and reproduction to be combined in ways that lead to higher fitness than for the resident alternative (Fokkema et al. 2020). Seasonal food availability is a major selection pressure favouring seasonal migrations (Alerstam et al. 2003; Varpe 2017), but similar variations in predation risk, infection risks, or abiotic environmental conditions can be underlying drivers too. The main costs of migrations are high energy requirements of movement and the need to undergo physiological or morphological changes of operating in different environments (e.g. anadromous fish inhabiting fresh and salt water). Costs also include the need for sophisticated navigation and orientation capacities as well as mortality risk from encountering predators along the migration routes. For broader and more complete coverage of migrations and their evolutionary ecology, see (Alerstam et al. 2003; Dingle 2014; Hansson and Åkesson 2014; Milner-Gulland et al. 2011).

Migrations come in many forms (Chapman et al. 2014) and have been categorized, e.g. according to the distances, timescales or habitats covered, the proportion of a population migrating, or the existence and use of intermediate feeding locations. In this chapter, we focus on *seasonal migrations*, where individuals move between areas on an annual basis, with the movement part usually happening during well-defined and relatively brief time windows and hence with distinct phenology. For species that are long lived and reproduce multiple times through life (iteroparity), an individual performs such seasonal migrations repeatedly. Some migrations take place over particularly long distances, either in the absolute sense or relative to the body size of the migrant, and can be termed *long-distance migrations* (Alerstam et al. 2003), in contrast to *short-distance migrations*. Many migrations cross continents, and in the case of birds, major corridors and routes have emerged that are referred to as flyways (e.g. Wauchope et al. 2017). Clearly, there are also shorter migrations, where some individuals just move a little, typically to lower latitudes. It is a gradient, from the longest migrations to permanent residency. The variety of environments visited during migration is of particular relevance in relation to infections with parasites and pathogens. Some migrations take place as one long move (*non-stop migrations*) with limited interactions with the interjacent environments, whereas most migrants stop over to rest and refuel (*stop-over migrations*) (Alerstam and Bäckman 2018; Hedenström 2010). Finally, there may be variation in the proportion of a population that migrates, i.e. some individuals may be year round residents, whereas others migrate (*partial migration*) (Chapman et al. 2011).

Migrations have many ecological consequences and influence ecosystem functions and the structures of ecological communities. Importantly, migrants are living links, or vectors, between ecosystems (Bauer and Hoye 2014); hence, a meta-ecosystem perspective is required (Loreau et al. 2003; Varpe et al. 2005). Migrants transport energy and nutrients, primarily through defecation, offspring (such as eggs), and the death of individuals. The offspring produced by migrants may be placed in a different system than where the resources for their production were obtained, such as fish feeding in the open ocean and spawning in coastal ecosystems (Varpe et al. 2005; Willson and Womble 2006). Some migrants such as several salmon species have a strategy where death follows reproduction (semelparity), and individuals hence become a resource for scavengers and decomposers (Finney et al. 2000; Naiman et al. 2002). Part of the matter transported can be pollutants that have accumulated in the organism, particularly so for lipid rich migrants high in the food chain (Krümmel et al. 2003). An individual is also a whole ecosystem with many other smaller species inhabiting its interior as well as its exterior. Some of these are, for instance, symbionts in the gut, but many are also parasites and infectious agents that travel with the migrants and may jump on or off along the route (Viana et al. 2016). Consequently, parasites and infectious agents are dispersed between systems through migrants, a central element of this chapter and the whole book. When migrations change, these changes will have ramifications for the spread of infectious agents.

This chapter focuses on seasonal migrations to and from, or within, the Arctic and sub-Arctic regions, to form an ecological and evolutionary basis about migrations that a One Health perspective on Arctic ecosystems can build on. We cover the particular aspects of migrations to these northernmost areas of the planet, such as the pronounced seasonality of the environment, the oftentimes highly synchronized migrations, and the large aggregations of a diverse range of animals in the Arctic summer, arriving there from many different parts of the world. This melting pot effect seems particularly interesting from the perspective of infectious agents and the diseases they may cause. We give ecological background on the linkages between migrants and disease, both more generally, and with respect to the particular migrations we see in the Arctic. Overall, we advocate a meta-ecosystem perspective where places and environments are connected through organisms whose migrations and other behaviours may change, sometimes rapidly through phenotypic plasticity and other times more slowly through evolution. As a consequence, a dynamic perspective is needed when interpreting ecological consequences of migrations and the interplay between infectious agents, migratory hosts, and changing environments.

2 Arctic Migrations

2.1 A Seasonal Environment

The fundamental cause of seasonality is the tilted axis of the earth and the resulting seasonal pattern of irradiance experienced at all places on the planet. This seasonality, however, increases with latitude and is therefore particularly pronounced in high-latitude Arctic and sub-Arctic regions (and similarly so at equivalent southern hemisphere latitudes). The seasonal light and heat fluxes affect a range of derived physical variables, such as air and water temperatures, snow cover, sea ice cover, freshwater availability and run-off from land to sea, and so forth. These direct and derived abiotic variables have many consequences for biological processes and, in particular, for primary producers dependent on light for photosynthesis. The primary production in the Arctic is highly seasonal, both on land (Zeng et al. 2011) and in water (Ji et al. 2013), and basically divides a year into a productive and an unproductive period. A range of adaptations to this seasonality follows among grazers and for consumers higher up in the food chain (Varpe 2017). One such adaptation is the capacity to move to the Artic when the productivity is high and leave when it is low, through seasonal long-distance migrations (Alerstam et al. 2003).

Importantly, food availability is about more than the seasonal timing of photosynthesis. For plant material fed on by herbivores, some of the resources produced during one productive window remain in the environment, such as in roots or seeds, and can be fed on through the winter as well as next spring. Arctic year-round residents such as reindeer (*Rangifer tarandus*) and the rock ptarmigan (*Lagopus muta*) would do so. Physical conditions may also constrain access to food. For instance, anadromous arctic char cannot leave the lake for the ocean before the connecting river or creek is open and with water flowing (Gulseth and Nilssen 2000). Similarly, migrating birds arriving early to a snow covered and frozen Arctic have little or no food (Ebbinge and Spaans 1995; Grabowski et al. 2013; Prop and de Vries 1993), and the feeding season in the Arctic can be short, impacting breeding, moulting, and migration schedules (Holmes 1972). In the marine environment, many species cannot access their feeding environment if it is covered by sea ice, such as in seabirds where the extent of sea ice near the breeding colony may impact breeding phenology and success (Chaulk and Mahoney 2012; Gaston et al. 2005). Also, because of sea ice, most estuaries are only accessible to shorebirds well into the summer and during the autumn (Churchwell et al. 2016; Taylor et al. 2010). Furthermore, marine mammals need to surface for air and cannot swim far into ice covered areas (Brierley et al. 2002) unless there are open leads. As sea ice also blocks most of the surface light from entering the water column, it makes it harder for visually searching predators, such as fish and birds, to detect prey (Langbehn and Varpe 2017). Little light during the winter will also make visual search difficult on land. Finally, food availability is further impacted by seasonal trophic interactions, such as competition (DuBowy 1988) and kleptoparasitism (Varpe 2010).

Seasonality also impacts food independent processes such as environmental temperature and access to breeding habitat. The ground must typically be snow-free and relatively dry before birds establish their nest and lay eggs (Hendricks 2003; Holmes 1966; Prop and de Vries 1993). For instance, many species in Greenland recently experienced reproductive failures due to unusually much snow during winter and a subsequently delayed melt during the summer (Schmidt et al. 2019). Furthermore, low temperatures may prevent egg laying and lead to a skipped breeding season, as suggested for Brant geese (*Branta bernicla*) in the Canadian Arctic (O'Briain et al. 1998), or add costs that may have most severe consequences in species where only one of the parents is incubating, such as in some sandpipers (Meyer et al. 2021).

2.2 Why Migrate to the Arctic?

The cost-benefit analyses for Arctic migrations typically involve the same variables as for other regions, but their seasonality aspect is more distinct. Historically, high food availability and long days (24 h daylight for long parts of the summer) have been argued to be the main benefit of migrating to the Arctic to breed. This is a key selection pressure in many taxa (Alerstam et al. 2003; Corkeron and Connor 1999; Schekkerman et al. 2003), but additional benefits have also been highlighted. Notably, an experiment with artificial eggs across a latitudinal gradient documented lower predation risk for ground nesting birds the higher north (McKinnon et al. 2010), suggesting that predator avoidance could also be key to why breeding in the Arctic is beneficial. Similarly, the Arctic tundra ecosystem may be regarded parasite-poor, which would make low parasite pressure a potential benefit to spending parts of the year in the region (Piersma 1997).

From a cost perspective, arriving too early, or not being ready to leave in time in the autumn, may lead to costs and mortality. There is a rapid transition both from winter to spring and from autumn to more winter-like conditions. Low temperatures, frozen ground and seas, and little food are all among the challenges involved if arriving too early or staying too long. Similarly, arriving too late may have costs such as too little time to successfully raise young or failure to compete for and occupy the best breeding sites. Consequently, selection pressures on the timing of migrations and subsequent breeding seem particularly strong for breeding migrations to the Arctic (Burr et al. 2016; Holmes 1966; Love et al. 2010; van Gils et al. 2016).

2.3 Diversity of Arctic Migrations

Long-distance seasonal migrations to breed in the Arctic are particularly common in birds, both in seabirds and terrestrial birds. Some of these species are also extremely abundant and therefore constitute considerable biomasses (Fox et al. 2019; Gonzalez-Bergonzoni et al. 2017). The migration routes and modes of bird migrations to the Arctic are also very diverse (Fig. 1, Table 1). Some cover long non-stop flights and cross large oceans, whereas others stop frequently and have complex routes with intermittent resting and refueling. Furthermore, other organisms such as some fish and marine mammals primarily migrate to the Arctic to feed while they reproduce further south. These marine migrants often combine seasonal migrations with energy storage which in turn allow for reproduction at times of the year and at places where food availability is not necessarily high, but where it is beneficial to spawn or give birth (Fokkema et al. 2020; Varpe 2017). Breeding based on such reserves is referred to as capital breeding (Jönsson 1997; Stephens et al. 2009; Varpe et al. 2009).

Below, we provide four exemplary cases to illustrate the diverse migration patterns and selection pressures. Our cases are chosen to include some of the particularly abundant groups of migrants that breed on the arctic tundra, notably

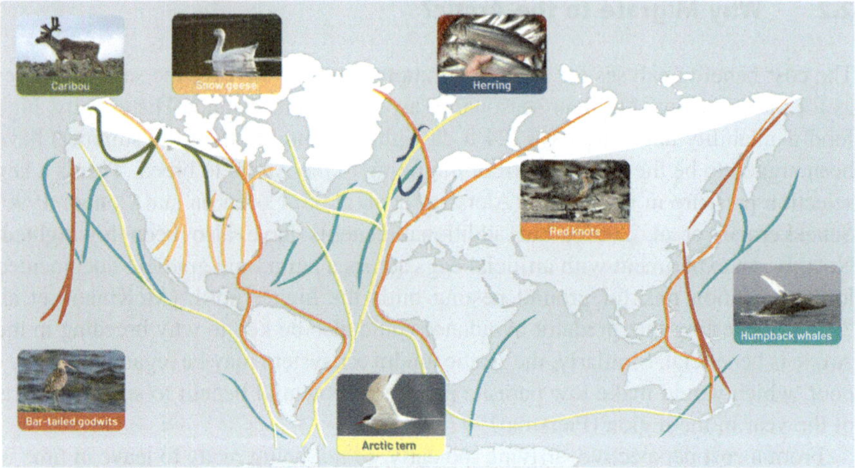

Fig. 1 Long-distance migrations and the Arctic, schematically illustrated through examples covered in the chapter. Migrants connect the Arctic to the rest of the world and make it a seasonal melting pot during the productive summer, when large numbers of organism move in from all over the world, some to breed, others to just feed. For photo credit, see the Acknowledgment

shorebirds and geese. These two groups also differ in a range of attributes related to their biology and migration strategies and thus serve as interesting contrasts. We have also included a case for within-Arctic migrations of the dominant terrestrial ungulate of the Arctic, the reindeer or caribou (*Rangifer tarandus*). Their notable interaction with humans through reindeer hunting and herding, is covered briefly. Finally, we discuss migrations in marine species. We focus these cases on the migrations and interactions with food, predators, and the physical environment. Interactions with infectious agents are covered in Sect. 3.

2.3.1 Shorebirds: Global Distributions and Major Flyways

Shorebirds are small to medium-sized birds that are commonly found along shorelines and mudflats, where they forage on invertebrates in the mud or sand. Of the >200 species worldwide, many shorebird species feature high among the most impressive migrants on earth—the distances they cover annually are simply astounding. For instance, bar-tailed godwits (*Limosa lapponica baueri*) migrate from non-breeding sites in New Zealand and Australia to breeding grounds in Alaska (Battley et al. 2012; Gill et al. 2009), red knots (*Calidris canutus*) migrate from South Africa or Mauretania to the Russian Arctic (van Gils et al. 2016), and Western sandpipers (*Calidris mauri*) migrate from the Pacific coast in North and South America to Alaska or Eastern Siberia (Lank et al. 2003) (Fig. 1, Table 1). However, some shorebird species breeding in the Arctic travel moderate distances, e.g. purple sandpipers (*Calidris maritima*) that winter in temperate areas and migrate to the coasts around the North Sea during their non-breeding season (Summers et al. 2001).

Table 1 Examples of migrations to and from or within the Arctic. Species are selected to illustrate a diversity of migration routes and types

Arctic migrants	Migration route characteristics	Time in the Arctic	Literature
Birds Dominant group of seasonal migrants to the Arctic. Geese, shorebirds, gulls, and auks are particularly numerous			
Arctic tern	Migrate between the Southern and Northern Atlantic ocean. Use prevailing winds and the route is therefore meandering. One-way distance >18,000 km	May–July	Egevang et al. (2010), Hromádková et al. (2020)
Brent goose	Migrate to the Arctic from non-breeding areas in temperate regions of Western Europe via several intermittent stops for resting and refueling to breeding areas in the Arctic	May/June–September	Clausen et al. (2003), Green et al. (2002)
Long-tailed skua	Migrate between the Arctic and the South Atlantic and Indian Ocean. The Grand Banks of Newfoundland appears to be important marine staging areas. Individuals estimated to travel more than 40,000 km over 1 year	June–August	Gilg et al. (2013)
Bar-tailed godwits	Migrate from non-breeding areas in New Zealand and Australia to breeding areas in Alaska and Russian Arctic. Extraordinary long non-stop flights, up to ~11,000 km. Migration routes differ between the way north and south	June–September	Battley et al. (2012), Gill et al. (2009)
Terrestrial mammals One dominant migrant, the ungulate *Rangifer tarandus*. It is an abundant herbivore with circumpolar distribution			
Reindeer or caribou	Migrate within the Arctic. Some herds have among the longest migrations of any terrestrial mammal (annual distance >5000 km for some herds), whereas other herds (for instance, inhabiting islands) are mostly stationary. Distinct calving areas often form key parts of the migration structure	Year round	Fancy et al. (1989), Gunn et al. (2011), Gurarie et al. (2019)

(continued)

Table 1 (continued)

Arctic migrants	Migration route characteristics	Time in the Arctic	Literature
Marine mammals			
Several species have summer feeding migrations to the Arctic and breed in low-latitude waters, often oligotrophic waters with low food availability			
Grey whale	Feed on Arctic benthos in Bering, Chukchi, and Beaufort Seas. Migrating along the coasts of western North America. Calving areas near Mexico. Migrations segregated with sex, age, and reproductive-stage differences in migration timing	~ half a year during the productive summer season	Guazzo et al. (2019), Rice and Wolman (1971), Swartz et al. (2006)
Humpback whale	High-latitude (e.g. sub-Arctic) feeding grounds and low-latitude (often tropical) breeding and calving grounds. Capital breeding combined with seasonal migrations	~ half a year during the productive summer season	Calambokidis et al. (2001), Clapham and Mead (1999)
Beluga whale	Migrate within the Arctic. As shown for the Pacific Arctic beluga, the timing and spatial extent of the migrations are linked to seasonality in sea ice. Offspring migrate together with their mother, and the migrations are matrilineally maintained	Year round	Colbeck et al. (2013), Hauser et al. (2017)
Fish			
Several species have seasonal feeding migrations to Arctic and sub-Arctic seas while spawning grounds are located further south. Others have a life and annual cycle that combines life in lakes with feeding migration into the ocean during the productive summer			
Northeast Arctic cod	Feed in the Barents Sea as young. Migrates south to spawn, with areas near Lofoten in Northern Norway being a key spawning area, but a wide latitudinal range of spawning grounds exist, particularly historically. Mature fish live long and will migrate many times between feeding and spawning grounds. Larvae drift northwards from spawning grounds to the nursery area	May–December (for the migratory adult part of the population)	Jørgensen et al. (2008), Ottersen et al. (2014)
Atlantic herring (Norwegian	Planktivorous fish with oceanic feeding migrations that reach sub-Arctic seas	August–October (for the migratory	Claireaux et al. (2020), Dragesund

(continued)

Table 1 (continued)

Arctic migrants	Migration route characteristics	Time in the Arctic	Literature
spring-spawning)	during the later parts of the seasonal feeding migration. Spawning along the coast of Norway and responsible for huge fluxes of biomass from ocean to coast as eggs are placed in distinct areas and at the seabed. Wintering areas have been highly variable. Larvae drift northwards from spawning grounds to the nursery area in the Barents Sea	adult part of the population)	et al. (1997), Varpe et al. (2005)
Arctic char	Anadromous fish with seasonal feeding migrations from lake ecosystems into the ocean and back. Time at sea relatively brief, from weeks to a few months. For some populations, some individuals migrate, while others are stationary and spend the whole year in the lake. The lakes are sometimes very close to the seashore, and the actual migration distance between the systems can be short	Year round	Gulseth and Nilssen (2000), Klemetsen et al. (2003)

With return journeys exceeding tens of thousands of kilometres, shorebirds are truly connecting continents and even hemispheres. As many species rely on specific coastal habitats, e.g. tidal mudflats, for fueling up before and during migration, the availability of these habitats determines overall migration routes and culminates in major flyways (Mathot et al. 2018). Along these flyways, some stopover sites are of particular importance as they are hotspots of productivity (Butler et al. 2001). There, shorebirds find their preferred food in excess, such as bivalves, snails, crustaceans, and polychaete worms, and consequently achieve unprecedented fueling rates (Kvist and Lindström 2003).

Shorebirds species differ in how they move along these flyways depending on how frequently they stop and how much fuel they accumulate: "*hop*" migrants cover short distances between successive sites and accumulate only little extra fuel stores, "*skip*" migrants are intermediate, and "*jump*" migrants accumulate large fuel stores and make long non-stop migrations (Piersma 1987). Although this distinction may seem marginal, how many and which sites migrants use determines migratory connectivity, which has far-reaching implications for individual fitness and population dynamics, gene flow and genetic mixing, the transmission of parasites and other infectious agents, and, ultimately, the conservation of migrants (Taylor et al. 2016; Webster et al. 2002).

Shorebird populations have severely declined over the past decades, such as the species breeding in the North-American Arctic (Smith et al. 2020). In the East-Asian Australasian flyway, even more species are threatened (Clemens et al. 2016; Piersma et al. 2016). Although we still need to fully understand which threats and disturbances impact shorebirds, the rapidly changing climate in the Arctic (see below), human impacts on coastal systems, and hunting and agricultural shifts are among the prime culprits (Colwell 2010; Mu and Wilcove 2020).

One major human-made threat to shorebirds is habitat destruction and alteration. Especially along the East-Asian Australasian flyway, coastal developments over the past decades have resulted in a >50% loss of the coastal wetlands (MacKinnon et al. 2012), notably in the Yellow Sea where 28% of the tidal flats existing in the 1980s have disappeared (Murray et al. 2014). This trend will continue through future reclamation projects such as landfills in estuaries and result in further declines of shorebirds (Studds et al. 2017). Similarly, in North American Delaware Bay, shorebirds feast on energy-rich eggs of horseshoe crabs (*Limulus polyphemus*), but overharvesting of crabs has deprived the shorebirds of much-needed fuel and resulted in strong population declines (Baker et al. 2004).

Also on the list of threats to shorebirds are recent and rapid increases in many goose populations (see below). For instance, snow geese have degraded wetlands along the coast of Hudson bay, a key refueling area (Jefferies et al. 2006), 'mowed' down the grass in shorebird breeding habitats, and thus exposed eggs to predators and attracted predators such as Arctic foxes into breeding areas (Flemming et al. 2019; Lamarre et al. 2017).

2.3.2 Arctic Geese: Large Birds and Populations, with Major Ecosystem Signatures

Migratory geese are long-distance migrants that typically spend the non-breeding season in temperate regions. On the way to their Arctic breeding grounds, they use intermittent stopover sites for resting and refueling. Although migration routes are diverse, broad patterns exist with European geese migrating to the Russian Arctic, e.g. Taymyr peninsula, or to Svalbard and Greenland, North-American geese migrating to Canadian Arctic and Alaska, and Asian geese migrating to Siberia or Alaska (CAFF 2018).

Prior to migration and between migratory bouts, geese accumulate body reserves. As body reserves are mainly stored as body fat deposits, they can be relatively easily scored using abdominal profiles (Madsen and Klaassen 2006). Consequently, changes in abdominal profiles over repeated sightings of individually marked (neck- or leg-banded) birds can indicate fundamental energetics, e.g. fueling rates or energy required for a migratory flight bout (Chudzinska et al. 2016). The timing of migration of geese is determined by a combination of cues—with a general cue such as photoperiod indicating time of the year and a more specific cue such as vegetation development for fine-tuning migratory progression to local phenology (Bauer et al. 2008a, 2011). Also, depending on the spatial scales involved, the conditions at one stopover site may be correlated with conditions at the next and hence indicate the conditions the geese will meet on next location (Bauer et al. 2020; Kölzsch et al. 2015; Tombre et al. 2008).

In contrast to most other migratory species, the populations of many migratory goose species have increased tremendously over the past decades: in the Western Palaearctic alone, there are 4.7 million geese from nine species (Fox et al. 2019; Madsen et al. 1999). This upsurge can probably be attributed to two major factors: agricultural change and policy and legislation. In the regions where migratory geese typically spend the non-breeding season, agricultural practices have undergone a major transformation, and many former natural habitats have been converted to farmland. Consequently, food during winter has become a superabundant high-quality resource, sustaining larger populations than ever before (Fox et al. 2005). Increasing goose populations are also a result of effective legislation: International treaties such as the Agreement on the Conservation of African-Eurasian Migratory Waterbirds (AEWA) or the Ramsar convention protect migratory waterbirds and their habitats across Africa, Europe, the Middle East, Central Asia, Greenland, and the Canadian Archipelago (https://www.unep-aewa.org/en and https://www.ramsar.org/). Hunting bans and restrictions as well as the fading popularity of hunting have also reduced mortality.

Another factor contributing to the geese' success story is their social transmission of migration behaviour, i.e. young learn where and when to migrate by accompanying their parents, and therefore, they can respond to environmental changes much faster (than under genetic transmission) (Sutherland 1998). Indeed, over the past few decades, many migratory goose species have changed migration behaviour, e.g. pink-footed geese adjusted migration timing to changes in phenology of vegetation growth (Bauer et al. 2008b) and barnacle geese shifted the spring-staging area northward (Tombre et al. 2019). Barnacle geese have also established new breeding areas, some of them far south of the Arctic, e.g. at the island Gotland in Sweden, and thereby drastically reduced migration distances (Larsson and Forslund 1994; Larsson et al. 1988; Van Der Jeugd et al. 2009). Some barnacle geese have even ceased migrating altogether and live year-round in the Netherlands (Jonker et al. 2012).

Although migrants are generally thought to affect ecosystem functioning and community structure (Bauer and Hoye 2014), this all the more applies to the immense numbers of migratory geese. Their intense and extended herbivory may turn the Arctic from a carbon sink into a carbon source (van der Wal et al. 2007), increase nutrient loading of Arctic freshwaters (Hessen et al. 2017), or lead to near-irreversible shifts in coastal marsh ecosystems (Jefferies et al. 2006). Conflicts with agriculture in temperate non-breeding and stopover areas have also intensified (Bauer et al. 2018; Fox et al. 2017). To alleviate these conflicts while still sustaining migratory wildlife populations as an internationally shared objective, adaptive flyway management plans have been developed for several goose populations over the past years, e.g. (Madsen et al. 2017).

2.3.3 Caribou and Reindeer: Seasonal Movements Within the Arctic

The reindeer (or caribou) inhabit tundra, alpine, and forest regions in Arctic and sub-Arctic ecosystems, with several sub-species identified (Gunn et al. 2011), and with distinct herds within sub-species. Some herds are very abundant, such as the Taymyr reindeer herd in Russia, estimated to around one million individuals (Kolpashikov et al. 2015). Reindeer can be mostly sedentary, such as the Svalbard reindeer (Tyler and Øritsland 1989) and the Peary caribou (Gunn et al. 2011), whereas many perform seasonal migrations, including several North American sub-species and herds (Gunn et al. 2011). The migrations are truly astonishing and among the longest of any terrestrial mammal (Berger 2004). Individuals in the Porcupine and Central Arctic herds cover more than 5000 km per year (Fancy et al. 1989). Reindeer and caribou migrations are typically between distinct calving areas, sometimes in coastal areas, and larger and more interior feeding areas during winter, on the tundra or in forest environments (Gunn et al. 2011; Nicholson et al. 2016). Some herds also cross long stretches of sea ice both during their spring and autumn migration, such as the herds calving and spending the summer on Victoria Island in the Canadian Arctic (Poole et al. 2010). Reindeer rarely swim long distances and, if possible, circumvent open water along their tundra routes (Leblond et al. 2016).

Migration patterns of caribou may vary between years, such as the large variability recently analyzed for the Riviere-George (RGH) and the Riviere-aux-Feuilles (RFH) herds (Le Corre et al. 2020), where wintering areas in different years may be situated several hundred kilometres apart. Individuals were more likely to change migration route and wintering area when the population size was large, suggesting competition for food, also between the herds, to be the underlying reason for the plasticity in migration movements (Le Corre et al. 2020).

Many Arctic and sub-Arctic people interact closely with reindeer and their annual cycle and migrations. Some are primarily hunting them (Kolpashikov et al. 2015), whereas others, such as many Saami in northern Fennoscandia, keep reindeer as their property and are involved in reindeer husbandry (Weladji and Holand 2006). Reindeer kept by herding communities are considered semi-domesticated and this reindeer husbandry exists among several people throughout northern Eurasia, for instance, by the Nenets. A nomadic lifestyle and seasonal migrations are often integrated in this interaction, where reindeer and people move together through the landscape. For instance, in Finnmark (Northern-Norway), it is common practice to move the reindeer between coastal calving and summer pastures and winter feeding grounds in the interior where the climate is colder and with less or drier snow (Weladji and Holand 2006). Migration distances in Finnmark, Norway, range from 100 to 350 km one way from summer to winter pastures (Kelman and Næss 2019; Weladji and Holand 2006). The owners earmark their individuals, but individuals from different owners typically mix and move together and are looked after by cooperative herding units (Næss 2012).

2.3.4 Marine Migrants: Travelling to the Arctic to Feed

Despite the mainly terrestrial focus of this chapter, many marine species also use Arctic and sub-Arctic waters for parts of the year or migrate within the Arctic. A notable feature is that instead of migrating to the Arctic to *breed* which is typical for birds (Drent et al. 2006), many marine species migrate to the Arctic to *feed*. They then use the highly but only seasonally productive northern seas for growth and storage and for preparation for breeding elsewhere. For instance, several species of large whales feed in high-latitude systems during the northern summer and move south to low latitudes, often tropical locations, for breeding and calving (Table 1). The eastern pacific population of grey whale (*Eschrichtius robustus*), which migrates south and north along the western North American coast, is one such example (Guazzo et al. 2019; Rice and Wolman 1971; Swartz et al. 2006). Its core feeding grounds are in the Bering and Chukchi seas where it primarily feeds on benthic prey, in particular tube-dwelling amphipods (Rice and Wolman 1971). The grey whales are observed on their southward migration from November onwards, with near-term pregnant females migrating first (Rice and Wolman 1971). The lagoons in Baja California, Mexico, are key calving and nursing areas. The return migration to the Arctic takes place in spring, with arrival to the feeding grounds from May onwards (Swartz et al. 2006). Humpback whales (*Megaptera novaeangliae*) follow a similar scheme (Darling and McSweeney 1985), but with major calving and breeding areas near oceanic islands, such as Hawaii for the Pacific population (Calambokidis et al. 2001) and Capp Verde and Caribbean islands for the North Atlantic population (Stevick et al. 2016). For humpback whales, genetic evidence suggests that the North Pacific, North Atlantic, and Southern Hemisphere populations are different and could be regarded sub-species (Jackson et al. 2014). The humpback and the grey whale illustrate how several large baleen whales rely on Arctic feeding grounds and combine that with capital breeding, long distance migrations, and none or limited feeding during the period away from the Arctic. Lower predation risk from killer whales at low latitudes have been proposed as one potential ultimate driver of these breeding migrations into biologically unproductive waters (Corkeron and Connor 1999). As for most migratory species, understanding the complex set of cues and navigational aids used by the individuals is challenging, with whales being no exception (Burnham 2020).

Several fish populations in the North Atlantic are also examples of the travelling to feed strategy. They migrate northward into foraging areas during summer and return south to the spawning grounds. Some of the most northern feeding grounds are within the Arctic, and several species are feeding in sub-Arctic seas. Atlantic cod (*Gadus morhua*) and herring (*Clupea harengus*) are two examples of large populations with such seasonal migrations. The migratory Northeast Arctic cod feeds in the Barents Sea, including the waters around Svalbard, and migrate south to Lofoten and other areas along the Norwegian coast to spawn (Jørgensen et al. 2008; Ottersen et al. 2014). The Norwegian spring-spawning herring has a more oceanic feeding migration into the Norwegian Sea (Claireaux et al. 2020; Varpe et al. 2005) and typically reach sub-Arctic waters towards the end of the summer. The extent that planktivourus fish, such as herring but also capelin (*Mallotus villosus*) and Atlantic mackerel (*Scomber scombrus*), move north during a given year is dependent on population size, as captured by the concept of density dependent migratory waves (Fauchald et al. 2006). After its seasonal feeding migration, the herring returns to the coast of Norway for wintering followed by spawning in March and April. For herring as well as cod, this annual routine has evolved in close connection with a seasonal food source (Varpe and Fiksen 2010) and with the ocean currents. Larvae of both species drift northward and into the Barents Sea where the individuals grow and develop for several years until maturation (Ottersen et al. 2014; Skagseth et al. 2015).

Marine species also display long seasonal migrations within the Arctic. Seals, whales, the polar bear (*Ursus maritimus*), and the Arctic char (*Salvelinus alpinus*), all have distinct movement patterns through the year. Ringed seals (*Phoca hispida*) breed in fjords and in relation to fjord ice during spring, but some individuals have offshore migrations during their non-breeding and post-molting period (Freitas et al. 2008). Similarly, some polar bears tend to be more local year round, whereas others are oceanic and operate in the open ocean sea ice ecosystems (Mauritzen et al. 2001). Pregnant polar bears will return to land during autumn and to suitable denning areas. Beluga whales (*Delphinapterus leucas*) offer another example. In the waters near Alaska and Russia, these whales perform foraging migrations into the seasonally sea-ice covered Eastern Chukchi Sea and Eastern Beaufort Sea (Hauser et al. 2017), and prior to freeze up in the late autumn, they return to wintering areas in the Bering Sea. Offspring are thought to learn these migrations from their mothers (Hauser et al. 2017), a phenomenon also highlighted for other Beluga whale populations (Colbeck et al. 2013).

The anadromous Arctic char also illustrates seasonal migrations within the Arctic. Arctic char has a circumpolar distribution in Arctic and sub-Arctic coastal waters and can be found both as landlocked populations spending all their life in the lake or as anadromous populations that spend part of the summer in the ocean (Klemetsen et al. 2003). In some lakes, some individuals are anadromous, whereas others stay in the lake all year round, with permanent residents typically reaching much smaller body size. For high-latitude populations, the melt-up of the river connecting lake and sea usually determines the timing of the migration to the sea, which can be highly synchronous (Gulseth and Nilssen 2000). The high productivity of the ocean and the good marine feeding conditions is the central benefit of this migration, but the time at sea comes with higher risk as well as energetic costs related to osmoregulation in salt water (Klemetsen et al. 2003; Nilssen et al. 1997). During the winter, all individuals are in the lake. Feeding within the lake also displays pronounced seasonality, partly driven by the seasonal light conditions (Svenning et al. 2007).

Finally, a variety of zooplankton species in northern waters have evolved prominent seasonal vertical migrations in the water column (Bandara et al. 2021). These migrations form key parts of annual routines, such as for herbivorous species that feed in surface waters and migrate to deeper diapause habitats (Conover 1988; Varpe 2012), and result in seasonal patterns of food availability for higher trophic levels (Varpe and Fiksen 2010).

3 Migratory Animals and Their Role as Long-Distance Dispersers of Infectious Agents

Migratory animals are commonly assumed to act as long-distance dispersers of infectious agents (Altizer et al. 2011; McKay and Hoye 2016; Westerdahl et al. 2014). As many of the infectious agents of wildlife have the potential to also infect humans and livestock, understanding the role of migratory animals in the spread of zoonotic agents has important implications for human health and economy. Still, and as argued in a recent review, much remains to be investigated to integrate migrations within disease ecology (McKay and Hoye 2016).

Migrants travel astonishing distances, establish unique geographic links between otherwise separated communities and thus encounter a greater diversity of infectious agents than resident species (Gutiérrez et al. 2019). For instance, migratory birds harboured a higher diversity of intestinal (Koprivnikar and Leung 2015; Leung and Koprivnikar 2016) or blood parasites (Emmenegger et al. 2018) and migratory ungulates generally had a higher parasite diversity than resident or nomadic species (Teitelbaum et al. 2018).

Furthermore, many migrants aggregate during migration—either for the migratory movement itself or on intermittent stopover sites where they refuel for the next migratory bout. Individuals are then in close contact, which constitutes an ideal transmission scenario. Prominent examples for aggregations on highly profitable stopover sites are the Yellow sea or Hudson Bay, where millions of shorebirds gather during a short time period in spring (Krauss et al. 2010). The high densities of multiple species on these sites facilitate the transmission of infectious agents such that, e.g., the prevalence of avian influenza in Delaware Bay is higher than anywhere in the world (Krauss et al. 2010).

Moreover, the metabolically costly migrations usually involve physiological and morphological adjustments that might be made at the expense of immune function (Buehler et al. 2010). If migrants suppress their immune system, latent infections may be reactivated and passed on to vectors or new hosts (Becker et al. 2020). This mechanism has been suggested for migrating birds that carry ixodid ticks which are vectors of *Borrelia burgdorferi* sensu *lato* (Gylfe et al. 2000), a bacteria that can cause Lyme disease in humans.

Despite the long distances that migratory animals travel, the diversity of infectious agents they encounter and the dense aggregations they form, migrations can also provide effective mechanism for hampering the transmission of infectious agents, through migratory escape, migratory culling, and migratory separation (Fig. 2) (Hall et al. 2014; McElroy and de Buron 2014; Satterfield et al. 2015; Shaw and Binning 2016).

Migratory Escape For infectious agents that are transmitted via the environment or via local vectors, migration may lead animals away from pathogen- or vector-rich places. If conditions during the absence of migratory hosts become unfavourable for the persistence of pathogens or vectors, migrants will then return to an almost pathogen-free environment.

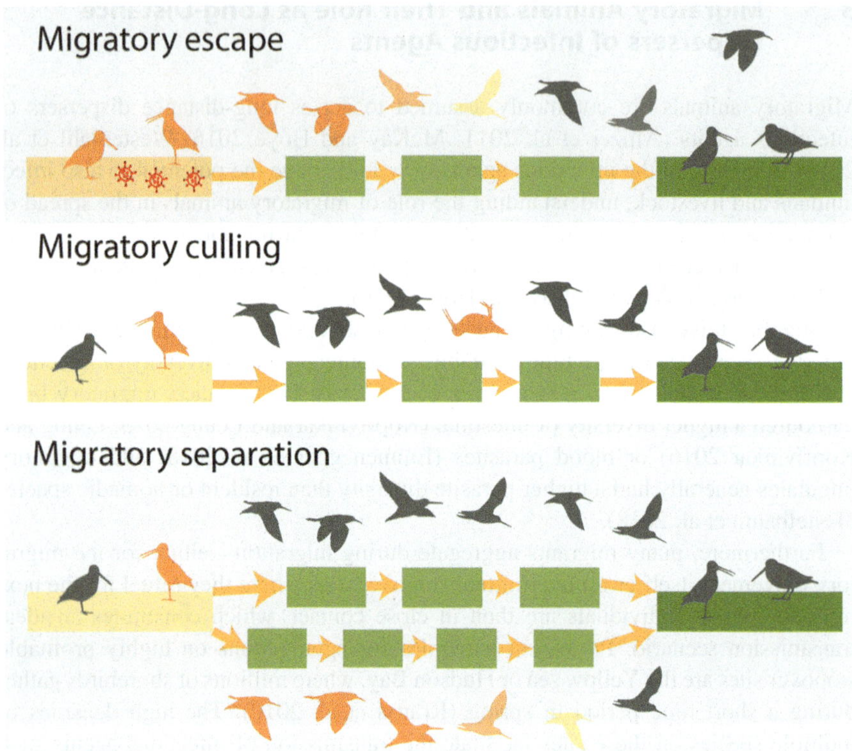

Fig. 2 Migration from a non-breeding area (left) to a breeding area (right) via several intermittent stopover sites can effectively interrupt the transmission-cycle of infectious agents through migratory escape, culling, or separation. In migratory escape, animals leave areas with high infection pressure, clear infections on the way, as indicated by color change from infected (red) to uninfected (grey), and reach a relatively pathogen-free breeding location. Similarly, for migratory culling, infected individuals die at a higher rate such that mostly uninfected individuals reach the breeding area. Finally, under migratory separation, infected individuals migrate along a different route or at different times than uninfected individuals, and this separation of infected from uninfected individuals hampers the spread of parasites in the population

Migratory culling If infectious agents are relatively virulent and/or hosts relatively susceptible, infected individuals may die at a higher rate than uninfected individuals. Consequently, infected individuals are "weeded" out from the population, and the risk of transmission to uninfected individuals decreases overall.

Migratory Separation Perhaps a less severe form of migratory culling, for migratory separation, migratory hosts are also affected by parasites but sublethally and lead to infected individuals taking other migration routes, using different stopover sites or migrating at other times than uninfected individuals. For instance, many infections and diseases reduce their hosts' appetite, induce fatigue, or shift daily activity rhythms towards more resting. Pathogens can also change their hosts' gut surface area or passage time, effectively reducing the uptake of nutrients and energy, which, in turn, prolongs fueling and delays migratory departures, e.g. in Bewicks' swans (*Cygnus columbianus bewickii*) infected with Avian influenza (van Gils et al. 2007).

Pathogens can also increase their hosts' daily energy expenditure, as they evoke immune responses and/or increase metabolic rates. Both immune responses and higher metabolic rates require nutrients and energy, which cannot be devoted to other processes such as migration. Indeed, higher resting metabolic rates have been shown for birds with parasites, e.g. blood-sucking ectoparasites in tree swallows (Sun et al. 2020), but these effects are not consistent across species or life-history stages (Hahn et al. 2018; Robar et al. 2011). Similarly, ectoparasites such as feather mites or sea lice may invoke structural changes that alter aerodynamics or hydrodynamics and increase costs for locomotion, e.g. through more energy required for a distance flown or swum, and/or lower flight or swim speed (Binning et al. 2013, 2017).

Any of these changes may result in infected individuals not replenishing resources at their typical rate or using them up faster. Consequently, infected individuals may require more frequent fueling during migration or more time to complete migration, and although this temporal or spatial separation usually lasts over a relatively short period only, this might be sufficient for separating infected and uninfected individuals and interrupting the transmission cycle and ultimately lower prevalence (Altizer et al. 2011; Bauer et al. 2016; Hall et al. 2014).

4 Outlook

4.1 Changing Migrations and Threats to Migrants

Our era has been coined the Anthropocene as human impacts on nature are substantial, and as a consequence, biodiversity is lost at an unprecedented rate (IPBES 2019). Many migratory populations have declined as a result of changing landscapes and habitat alterations, new obstacles with rapid expansions of human structures and activities, and climate change effects such as altered phenology and abundance of both food and predators along their migration routes (e.g. Wilcove and Wikelski 2008).

Climate Change Over the past decades, global average temperatures have risen with 0.2 °C per decade and are projected to continue to rise (IPCC 2013). However, climate change is highly uneven across regions and times of the year with, e.g. temperatures increasing more rapidly in the Arctic (IPCC 2013) and extreme events occurring more frequently, which will have a broad range of effects in nature and societies (Hansen et al. 2014; Langbehn and Varpe 2017; Schmidt et al. 2019; Vincent et al. 2011; Wassmann et al. 2011). One consequence is a rapidly changing phenology in the Arctic with concomitant consequences for the optimal time window for many life-history processes of Arctic migrants (Fjelldal et al. 2020; Lameris et al. 2018). For instance, in migratory birds, reproductive success often depends on arriving early enough to the Arctic to synchronize breeding (or feeding) with the spring onset (Doiron et al. 2015; Lameris et al. 2018). If migrants cannot respond to an advancing spring, the periods of high demand will not match those of high productivity with negative effects on individual fitness and eventually on population trends (Cohen et al. 2018; Samplonius et al. 2021). For instance, red knots that breed in the Russian Arctic produce smaller offspring during years when spring starts early (van Gils et al. 2016), likely as a result of parents arriving (and starting to breed) too late. Consequently, their chicks miss the peak of insect abundance and grow slower or to a smaller adult size. When those smaller individuals migrate to non-breeding grounds in West Africa, they face the problem that their short bills cannot reach their preferred food, which is deeply buried clams. Although the size differences may seem small, they culminate in lower survival and shorter life expectancy for smaller compared to larger individuals (van Gils et al. 2016).

A warming Arctic is impacting the breeding environment of birds and may reduce the areas climatically suitable for breeding. Arctic organisms have few opportunities to shift their distributions further north (Lehikoinen and Virkkala 2016; Maclean et al. 2008) and therefore suffer from contracted breeding ranges. Also, for those able to shift their distributions to higher elevation or to other areas, such shifts may interact with migration routes and potentially lead to restructuring of entire flyways (Wauchope et al. 2017).

In the Arctic marine environment, the reduced spatial extent and thickness of sea ice and its shorter seasonal occurrence are among the most pronounced climate-induced changes (Stroeve et al. 2012). Consequently, more light enters the water column, impacting both pelagic and benthic organisms through possibilities for increased photosynthesis and primary production and through improved conditions for visually searching predators (Clark et al. 2013; Varpe et al. 2015). Fish, for instance, often rely on their vision when hunting, and sea ice loss leads to large areas becoming more profitable for detecting prey. Northward shifts in their distributions may result, particularly likely for seasonal migrants that can make use of the abundant light during summer, but leave the Arctic during the dark winter (Langbehn and Varpe 2017). Sea ice is also a physical hindrance for fish filling their swim bladder by gulping air at the surface (physostomous fish) such as herring (Kaartvedt and Titelman 2018). This constraint has similarities to how marine mammals are prevented from breathing during under-ice excursions (Brierley et al. 2002) or risk entrapment in the sea ice landscape (Ferguson et al. 2010; Stafford 2019).

Human Structures, Overexploitation, Habitat Loss, and Alterations Collisions with man-made structures such as power lines, wind turbines, and buildings kill large numbers of birds and bats annually, but estimates of the resulting fatalities and their ecological significance vary (Cryan et al. 2014; Lambertucci et al. 2015). For instance, mortality from wind turbines appears to represent a substantial population-level threat in bat populations in North America (Frick et al. 2017), but further research is urgently needed.

Similarly, fences, roads, pipelines, and other structures can hinder ungulates during their migrations, and there is concern regarding loss of migration routes and an inability to secure the management measures that could conserve them (Berger 2004). This is also the case for reindeer in the Arctic (Kuemmerle et al. 2014; Taillon et al. 2012). Landscapes used for reindeer husbandry are becoming increasingly fragmented, for instance, through fencing practices (including at national borders), sometimes preventing traditional movement patterns and seasonal migrations by Saami and their reindeer (Kelman and Næss 2019).

Human structures and activities that may prevent or modify migrations are not unique to terrestrial environments. Marine mammals use sounds to communicate and navigate and are likely sensitive to artificial noise (Duarte et al. 2021). There are therefore concerns related to the interaction between marine mammals and the increasing industrial activities, including shipping, in the Arctic (Halliday et al. 2017).

Other species have seen new and increased challenges related to parasites and infections along their migration routes. Atlantic salmon (*Salmo salar*) is one example. Born in rivers and spending much of their life in the ocean, they pass coastal regions twice before reproducing (Klemetsen et al. 2003). In some countries, such as Norway, salmon farms are abundant in fjords and coastal environments, and wild salmon suffer from exposure to salmon lice and other parasites and infectious agents as they proliferate among farmed fish and spread back to the wild (Forseth et al. 2017). With warming waters and less sea ice, aquaculture may be extended to even more northern communities and expand the areas of potential parasite and pathogen transmission.

4.2 What Do These Changes Mean for Migrants as Dispersers of Parasites and Pathogens?

Overall, much remains to be understood about the role of migratory animals in the spread of infectious agents. Only a handful of infectious agents and diseases have been investigated in detail, and these usually include those with zoonotic potential or severe economic implications such as avian influenza or brucellosis.

If global and climatic changes alter migration routes, migrants may reach novel breeding, non-breeding, or stopover places and introduce infectious agents into possibly "naïve" communities. Furthermore, a warming and changing Arctic may impact which pathogens can survive year round in the Arctic and hence pose new interactions with seasonal migrants. Increased attention to infectious agents seems both important and fruitful as additions to migration studies that focus on the roles of the changing physical environment or the changing abundance and phenology of food or predators.

4.3 Conservation of Migrants and Migrations

With the multiple threats that migratory species face and their declining populations, conserving migrants and migrations has become an internationally shared objective. However, conserving mobile species is a challenging endeavour as migrants rely on multiple locations to complete their annual or life cycles (Runge et al. 2014). The use of these sites is interdependent such that the conditions on one site will have consequences for the use of others, and therefore, management actions on single sites may be futile if they do not take into account actions elsewhere (Bauer et al. 2018).

For many species, elements of their migrations may also change considerably between years, such as variability in the wintering area of caribou (Le Corre et al. 2020) or in location and size of calving grounds (Taillon et al. 2012). Such dynamics are clearly challenging to account for when most management and conservation measures are built around static approaches such as reserves with constant boundaries. Dynamic approaches are called for (Taillon et al. 2012). Similar challenges exist for managing moving fish populations where seasonal migrations include waters managed by multiple countries, as well as international waters, for which international negotiations are needed to implement sustainable management. For instance, the wintering area of the Norwegian spring-spawning herring has varied considerably during the last century, including areas many hundreds of kilometres apart, such as near Iceland during the 1950s and early 1960s but wintering primarily in near-coast areas of Northern Norway from about 1980 and onwards (Claireaux et al. 2020; Dragesund et al. 1997). Clearly, such dynamics present challenges not only for fisheries management but also for predicting impacts of climate change and other stressors on the population and its large-scale movement.

Management practices may lead to changes in movements and migrations. The presence and extent of goose hunting impact the timing of migrations and how long geese stay on a particular stopover site (Bauer et al. 2018). Furthermore, management may sometimes include the eradication of entire populations, e.g. to halt the spread of the chronic wasting disease from an infected reindeer population (Mysterud et al. 2020). However, we need to be aware that any management measure, but particularly massive culling, may have side- or carryover effects, including increased movement and geographic spread (Mysterud et al. 2020).

Moreover, the multiple sites used by migrants and the movements in between form migratory networks that range from linear networks in which sites are used in a sequential manner to diffuse networks where many sites are used interchangeably. The structure of these networks plays an important role in how vulnerable migrant populations are to the degradation or loss of sites (Betini et al. 2015; Xu et al. 2019).

In addition to the structure of migratory networks, the vulnerability of species depends on the capacity of migrants to adapt to changing conditions. Although some species seem to adapt rapidly through altered migration routes or timing, for many other species, it is largely unknown whether and how fast they can respond. Furthermore, threats differ in their consequences for different demographic rates, and likewise, management actions may influence survival or reproductive success differently.

Thus, efficient conservation needs an international approach that entails the entire migration range, spatial prioritization by identifying and conserving crucial sites and crucial connectivity, and the identification of actions that are most effective.

5 Concluding Remarks

Migrations are unique movements that connect the world, and in this network of species and migration routes, the Arctic stands out as a melting pot where migrants from all over the world meet during the Arctic summer, some to breed, others to just feed. Travelling animals connect ecosystems and serve as spatial vectors, of energy and nutrients and of other organisms that follow, sometimes as active hitchhikers, for parts or the entire route. Understanding drivers and patterns of migrations is essential for understanding the dynamics of diseases and must therefore be considered in veterinary and human medicine and the One Health perspective. Current global changes are particularly pronounced in the Arctic with many ongoing environmental changes impacting migrants in a diversity of ways. Interactions with infectious agents are also likely to change, and the abundance of pathogens may increase, for instance, with episodic outbreaks during particularly warm weather (Hueffer et al. 2020). The spatial configuration of many migrations are also rapidly changing, potentially exposing migrants to new infectious agents and creating new connections between previously separated areas.

Acknowledgment We thank all the photographers who captured the animals that made it to Fig. 1. The caribou (Victor Adam), humpback whale (shadowfaxone), bar-tailed godwit (psubraty), red knot (PublicDomainImages), and the snow geese (Astrid Zellmann) are from pixabay; https://pixabay.com/photos, with the name of the photographer or profile name on pixabay in parenthesis. Arctic tern is photographed by Martins Briedis and the herring by Leif Nøttestad/Havforskningsinstituttet.

References

Alerstam T, Bäckman J (2018) Ecology of animal migration. Curr Biol 28:R968–R972

Alerstam T, Hedenström A, Åkesson S (2003) Long-distance migration: evolution and determinants. Oikos 103:247–260

Altizer S, Bartel R, Han BA (2011) Animal migration and infectious disease risk. Science 331:296–302

Baker AJ, Gonzalez PM, Piersma T, Niles LJ, do Nascimento IDS, Atkinson PW, Clark NA, CDT M, Peck MK, Aarts G (2004) Rapid population decline in red knots: fitness consequences of decreased refuelling rates and late arrival in Delaware Bay. Proc R Soc B 271:875–882

Bandara K, Varpe Ø, Wijewardene L, Tverberg V, Eiane K (2021) Two hundred years of zooplankton vertical migration research. Biol Rev 96:1547–1589

Battley PF, Warnock N, Tibbitts TL, Gill RE, Piersma T, Hassell CJ, Douglas DC, Mulcahy DM, Gartrell BD, Schuckard R, Melville DS, Riegen AC (2012) Contrasting extreme long-distance migration patterns in bar-tailed godwits *Limosa lapponica*. J Avian Biol 43:21–32

Bauer S, Hoye BJ (2014) Migratory animals couple biodiversity and ecosystem functioning worldwide. Science 344:54-+
Bauer S, Gienapp P, Madsen J (2008a) The relevance of environmental conditions for departure decision changes en route in migrating geese. Ecology 89:1953–1960
Bauer S, Van Dinther M, Hogda K-A, Klaassen M, Madsen J (2008b) The consequences of climate-driven stop-over sites changes on migration schedules and fitness of Arctic geese. J Anim Ecol 77:654–660
Bauer S, Nolet BA, Giske J, Chapman JW, Åkesson S, Hedenström A, Fryxell JM (2011) Cues and decision rules in animal migration. In: Milner-Gulland EJ, Fryxell JM, Sinclair ARE (eds) Animal migration: a synthesis, vol 68. Oxford University Press, pp 68–87
Bauer S, Lisovski S, Hahn S (2016) Timing is crucial for consequences of migratory connectivity. Oikos 125:605–612
Bauer S, Lisovski S, Eikelenboom-Kil RJ, Shariati M, Nolet BA (2018) Shooting may aggravate rather than alleviate conflicts between migratory geese and agriculture. J Appl Ecol
Bauer S, McNamara JM, Barta Z (2020) Environmental variability, reliability of information and the timing of migration. Proc R Soc B 287
Becker DJ, Ketterson ED, Hall RJ (2020) Reactivation of latent infections with migration shapes population-level disease dynamics. Proc R Soc B Biol Sci 287:20201829
Berger J (2004) The last mile: how to sustain long-distance migration in mammals. Conserv Biol 18:320–331
Betini GS, Fitzpatrick MJ, Norris DR (2015) Experimental evidence for the effect of habitat loss on the dynamics of migratory networks. Ecol Lett 18:526–534
Binning SA, Roche DG, Layton C (2013) Ectoparasites increase swimming costs in a coral reef fish. Biol Lett 9
Binning SA, Shaw AK, Roche DG (2017) Parasites and host performance: incorporating infection into our understanding of animal movement. Integr Comp Biol 57:267–280
Brierley AS, Fernandes PG, Brandon MA, Armstromg F, Millard NW, McPhail SD, Stevenson P, Pebody M, Perrett J, Squires M, Bone DG, Griffiths G (2002) Antarctic krill under sea ice: elevated abundance in a narrow band just south of ice edge. Science 295:1890–1892
Buehler DM, Tieleman BI, Piersma T (2010) How do migratory species stay healthy over the annual cycle? A conceptual model for immune function and for resistance to disease. Integr Comp Biol 50:346–357
Burnham RE (2020) Whale geography: a species-centric approach applied to migration. Progr Phys Geogr-Earth Environ 44:419–434
Burr ZM, Varpe Ø, Anker-Nilssen T, Erikstad KE, Descamps S, Barrett RT, Bech C, Christensen-Dalsgaard S, Lorentsen S-H, Moe B, Reiertsen TK, Strøm H (2016) Later at higher latitudes: large-scale variability in seabird breeding timing and synchronicity. Ecosphere 7:1–12
Butler RW, Davidson NC, Morrison RIG (2001) Global-scale shorebird distribution in relation to productivity of near-shore ocean waters. Waterbirds 24:224–232
CAFF (2018) A global audit of the status and trends of Arctic and northern hemisphere goose populations. In: Fox AD, Leafloor JO (eds) Conservation of Arctic flora and fauna. International Secretariat, Akureyri
Calambokidis J, Steiger GH, Straley JM, Herman LM, Cerchio S, Salden DR, Urban J, Jacobsen JK, von Ziegesar O, Balcomb KC, Gabriele CM, Dahlheim ME, Uchida S, Ellis G, Miyamura Y, de Guevara PL, Yamaguchi M, Sato F, Mizroch SA, Schlender L, Rasmussen K, Barlow J, Quinn TJ (2001) Movements and population structure of humpback whales in the North Pacific. Mar Mamm Sci 17:769–794
Chapman BB, Brönmark C, Nilsson J-Å, Hansson L-A (2011) The ecology and evolution of partial migration. Oikos 120:1764–1775
Chapman BB, Hulthén K, Wellenreuther M, Hansson L-A, Nilsson J-Å, Brönmark C (2014) Patterns of animal migration. Animal movement across scales. Oxford Universtiy Press, Oxford, pp 11–35

Chaulk KG, Mahoney ML (2012) Does spring ice cover influence nest initiation date and clutch size in common eiders? Polar Biol 35:645–653

Chudzinska ME, Nabe-Nielsen J, Nolet BA, Madsen J (2016) Foraging behaviour and fuel accumulation of capital breeders during spring migration as derived from a combination of satellite- and ground-based observations. J Avian Biol 47:563–574

Churchwell RT, Kendall SJ, Blanchard AL, Dunton KH, Powell AN (2016) Natural disturbance shapes benthic intertidal macroinvertebrate communities of high Latitude River deltas. Estuar Coasts 39:798–814

Claireaux M, dos Santos Schmidt TC, Olsen EM, Slotte A, Varpe Ø, Heino M, Enberg K (2020) Eight decades of adaptive changes in herring reproductive investment: the joint effect of environment and exploitation. ICES J Mar Sci

Clapham PJ, Mead JG (1999) Megaptera novaeangliae. Mamm Species 1–9

Clark GF, Stark JS, Johnston EL, Runcie JW, Goldsworthy PM, Raymond B, Riddle MJ (2013) Light-driven tipping points in polar ecosystems. Glob Change Biol 19:3749–3761

Clausen P, Green M, Alerstam T (2003) Energy limitations for spring migration and breeding: the case of brent geese *Branta bernicla* tracked by satellite telemetry to Svalbard and Greenland. Oikos 103:426–445

Clemens RS, Rogers DI, Hansen BD, Gosbell K, Minton CDT, Straw P, Bamford M, Woehler EJ, Milton DA, Weston MA, Venables B, Weller D, Hassell C, Rutherford B, Onton K, Herrod A, Studds CE, Choi CY, Dhanjal-Adams KL, Murray NJ, Skilleter GA, Fuller RA (2016) Continental-scale decreases in shorebird populations in Australia. Emu-Austral Ornithol 116: 119–135

Cohen JM, Lajeunesse MJ, Rohr JR (2018) A global synthesis of animal phenological responses to climate change. Nat Clim Chang 8:224–228

Colbeck GJ, Duchesne P, Postma LD, Lesage V, Hammill MO, Turgeon J (2013) Groups of related belugas (*Delphinapterus leucas*) travel together during their seasonal migrations in and around Hudson Bay. Proc R Soc B 280

Colwell MA (2010) Shorebird ecology, conservation, and management. University of California Press

Conover RJ (1988) Comparative life histories in the genera *Calanus* and *Neocalanus* in high-latitudes of the northern hemisphere. Hydrobiologia 167:127–142

Corkeron PJ, Connor RC (1999) Why do baleen whales migrate? Mar Mamm Sci 15:1228–1245

Cryan PM, Gorresen PM, Hein CD, Schirmacher MR, Diehl RH, Huso MM, Hayman DTS, Fricker PD, Bonaccorso FJ, Johnson DH, Heist K, Dalton DC (2014) Behavior of bats at wind turbines. Proc Natl Acad Sci USA 111:15126–15131

Darling JD, McSweeney DJ (1985) Observations on the migrations of North Pacific humpback whales (*Megaptera novaeangliae*). Can J Zool 63:308–314

Dingle H (2014) Migration – the biology of life on the move. Oxford University Press

Doiron M, Gauthier G, Lévesque E (2015) Trophic mismatch and its effects on the growth of young in an Arctic herbivore. Glob Change Biol 21:4364–4376

Dragesund O, Johannessen A, Ulltang O (1997) Variation in migration and abundance of Norwegian spring spawning herring (*Clupea harengus* L.). Sarsia 82:97–105

Drent RJ, Fox AD, Stahl J (2006) Travelling to breed. J Ornithol 147:122–134

Duarte CM, Chapuis L, Collin SP, Costa DP, Devassy RP, Eguiluz VM, Erbe C, Gordon TAC, Halpern BS, Harding HR, Havlik MN, Meekan M, Merchant ND, Miksis-Olds JL, Parsons M, Predragovic M, Radford AN, Radford CA, Simpson SD, Slabbekoorn H, Staaterman E, Van Opzeeland IC, Winderen J, Zhang X, Juanes F (2021) The soundscape of the Anthropocene Ocean. Science 371:eaba4658

DuBowy PJ (1988) Waterfowl communities and seasonal environments: temporal variability in interspecific competition. Ecology 69:1439–1453

Ebbinge BS, Spaans B (1995) The importance of body reserves accumulated in spring staging areas in the temperate zone for breeding in dark-bellied brent geese *Branta b. bernicla* in the high arctic. J Avian Biol 26:105–113

Egevang C, Stenhouse IJ, Phillips RA, Petersen A, Fox JW, Silk JRD (2010) Tracking of Arctic terns *Sterna paradisaea* reveals longest animal migration. Proc Natl Acad Sci USA 107:2078–2081

Emmenegger T et al (2018) Blood parasites prevalence of migrating passerines increases over the spring passage period. J Zool 306:23–27

Fancy SG, Pank LF, Whitten KR, Regelin WL (1989) Seasonal movements of caribou in arctic Alaska as determined by satellite. Can J Zool 67:644–650

Fauchald P, Mauritzen M, Gjøsæter H (2006) Density-dependent migratory waves in the marine pelagic ecosystem. Ecology 87:2915–2924

Ferguson SH, Higdon JW, Chmelnitsky EG (2010) The rise of killer whales as a major Arctic predator. In: Ferguson SH, Loseto LL, Mallory ML (eds) A little less Arctic: top predators in the world's largest northern Inland Sea, Hudson Bay. Springer, Dordrecht, pp 117–136

Finney BP, Gregory-Eaves I, Sweetman J, Dougas MSV, Smol JP (2000) Impacts of climatic change and fishing on Pacific salmon abundance over the past 300 years. Science 290:795–799

Fjelldal MA, Layton-Matthews K, Lee AM, Grotan V, Loonen M, Hansen BB (2020) High-Arctic family planning: earlier spring onset advances age at first reproduction in barnacle geese. Biol Lett 16

Flemming SA, Nol E, Kennedy LV, Smith PA (2019) Hyperabundant herbivores limit habitat availability and influence nest site selection of Arctic-breeding birds. J Appl Ecol 56:976–987

Fokkema W, van der Jeugd HP, Lameris TK, Dokter AM, Ebbinge BS, de Roos AM, Nolet BA, Piersma T, Olff H (2020) Ontogenetic niche shifts as a driver of seasonal migration. Oecologia 193:285–297

Forseth T, Barlaup BT, Finstad B, Fiske P, Gjoaester H, Falkegard M, Hindar A, Mo TA, Rikardsen AH, Thorstad EB, Vollestad LA, Wennevik V (2017) The major threats to Atlantic salmon in Norway. ICES J Mar Sci 74:1496–1513

Fox AD, Madsen J, Boyd H, Kuijken E, Norriss DW, Tombre IM, Stroud DA (2005) Effects of agricultural change on abundance, fitness components and distribution of two arctic-nesting goose populations. Glob Change Biol 11:881–893

Fox AD, Elmberg J, Tombre IM, Hessel R (2017) Agriculture and herbivorous waterfowl: a review of the scientific basis for improved management. Biol Rev 92:854–877

Fox AD, Ebbinge BS, Mitchell C, Heinicke T, Aarvak T, Colhoun K, Clausen P, Dereliev S, Faragó S, Koffijberg K, Kruckenberg H, Loonen MJJE, Madsen J, Mooij J, Musil P, Nilsson L, Pihl S, van der Jeugd H (2019) Current estimates of goose population sizes in western Europe, a gap analysis and an assessment of trends. Ornis Svecica 20:115–127

Freitas C, Kovacs KM, Ims RA, Fedak MA, Lydersen C (2008) Ringed seal post-moulting movement tactics and habitat selection. Oecologia 155:193–204

Frick WF, Baerwald EF, Pollock JF, Barclay RMR, Szymanski JA, Weller TJ, Russell AL, Loeb SC, Medellin RA, McGuire LP (2017) Fatalities at wind turbines may threaten population viability of a migratory bat. Biol Conserv 209:172–177

Gaston AJ, Gilchrist HG, Mallory ML (2005) Variation in ice conditions has strong effects on the breeding of marine birds at Prince Leopold Island, Nunavut. Ecography 28:331–344

Gilg O, Moe B, Hanssen SA, Schmidt NM, Sittler B, Hansen J, Reneerkens J, Sabard B, Chastel O, Moreau J, Phillips RA, Oudman T, Biersma EM, Fenstad AA, Lang J, Bollache L (2013) Trans-equatorial migration routes, staging sites and wintering areas of a high-Arctic avian predator: the long-tailed Skua (*Stercorarius longicaudus*). PLoS One 8

Gill RE, Tibbitts TL, Douglas DC, Handel CM, Mulcahy DM, Gottschalck JC, Warnock N, McCaffery BJ, Battley PF, Piersma T (2009) Extreme endurance flights by landbirds crossing the Pacific Ocean: ecological corridor rather than barrier? Proc R Soc B 276:447–458

Gonzalez-Bergonzoni I, Johansen KL, Mosbech A, Landkildehus F, Jeppesen E, Davidson TA (2017) Small birds, big effects: the little auk (*Alle alle*) transforms high Arctic ecosystems. Proc R Soc B 284

Grabowski MM, Doyle FI, Reid DG, Mossop D, Talarico D (2013) Do Arctic-nesting birds respond to earlier snowmelt? A multi-species study in north Yukon, Canada. Polar Biol 36:1097–1105

Green M, Alerstam T, Clausen P, Drent R, Ebbinge RS (2002) Dark-bellied Brent Geese *Branta bernicla bernicla*, as recorded by satellite telemetry, do not minimize flight distance during spring migration. Ibis 144:106–121

Guazzo RA, Schulman-Janiger A, Smith MH, Barlow J, D'Spain GL, Rimington DB, Hildebrand JA (2019) Gray whale migration patterns through the Southern California Bight from multi-year visual and acoustic monitoring. Mar Ecol Prog Ser 625:181–203

Gulseth OA, Nilssen KJ (2000) The brief period of spring migration, short marine residence, and high return rate of a northern Svalbard population of Arctic char. Trans Am Fish Soc 129:782–796

Gunn A, Russell D, Eamer J (2011) Northern caribou population trends in Canada. Canadian Councils of Resource Ministers Technical Thematic Report no 10

Gurarie E, Hebblewhite M, Joly K, Kelly AP, Adamczewski J, Davidson SC, Davison T, Gunn A, Suitor MJ, Fagan WF, Boelman N (2019) Tactical departures and strategic arrivals: divergent effects of climate and weather on caribou spring migrations. Ecosphere 10

Gutiérrez JS, Piersma T, Thieltges DW (2019) Micro- and macroparasite species richness in birds: the role of host life history and ecology. J Anim Ecol 88:1226–1239

Gylfe A, Bergström S, Lunström J, Olsen B (2000) Reactivation of Borrelia infection in birds. Nature 403:724–725

Hahn S, Bauer S, Dimitrov D, Emmenegger T, Ivanova K, Zehtindjiev P, Buttemer WA (2018) Low intensity blood parasite infections do not reduce the aerobic performance of migratory birds. Proc R Soc B 285

Hall RJ, Altizer S, Bartel RA (2014) Greater migratory propensity in hosts lowers pathogen transmission and impacts. J Anim Ecol 83:1068–1077

Halliday WD, Insley SJ, Hilliard RC, de Jong T, Pine MK (2017) Potential impacts of shipping noise on marine mammals in the western Canadian Arctic. Mar Pollut Bull 123:73–82

Hansen BB, Isaksen K, Benestad RE, Kohler J, Pedersen ÅØ, Loe LE, Coulson SJ, Larsen JO, Varpe Ø (2014) Warmer and wetter winters: characteristics and implications of an extreme weather event in the high Arctic. Environ Res Lett 9

Hansson LA, Åkesson S (eds) (2014) Animal movement across scales. Oxford Univeristy Press

Hauser DDW, Laidre KL, Stafford KM, Stern HL, Suydam RS, Richard PR (2017) Decadal shifts in autumn migration timing by Pacific Arctic beluga whales are related to delayed annual sea ice formation. Glob Change Biol 23:2206–2217

Hedenström A (2010) Extreme endurance migration: what is the limit to non-stop flight? PLoS Biol 8:e1000362

Hendricks P (2003) Spring snow conditions, laying date, and clutch size in an alpine population of American Pipits. J Field Ornithol 74:423–429, 427

Hessen DO, Tombre IM, van Geest G, Alfsnes K (2017) Global change and ecosystem connectivity: how geese link fields of Central Europe to eutrophication of Arctic freshwaters. Ambio 46: 40–47

Holmes RT (1966) Breeding ecology and annual cycle adaptations of the red-backed sandpiper (*Calidris alpina*) in northern Alaska. Condor 68:3–46

Holmes RT (1972) Ecological factors influencing the breeding season schedule of Western Sandpipers (*Calidris mauri*) in subarctic Alaska. Am Midl Nat 472–491

Hromádková T, Pavel V, Flousek J, Briedis M (2020) Seasonally specific responses to wind patterns and ocean productivity facilitate the longest animal migration on earth. Mar Ecol Prog Ser 638:1–12

Hueffer K, Drown D, Romanovsky V, Hennessy T (2020) Factors contributing to anthrax outbreaks in the circumpolar north. EcoHealth 17:174–180

IPBES (2019) Global assessment report on biodiversity and ecosystem services of the intergovernmental science-policy platform on biodiversity and ecosystem service. In: Brondizio ES, Settele J, Díaz S, Ngo (eds) IPBES secretariat. Bonn

IPCC (2013) Climate change 2013: the physical science basis. Contribution of working group I to the fifth assessment report of the intergovernmental panel on climate change. In: Stocker TF,

Qin D, Plattner G-K, Tignor M, Allen SK, Boschung J, Nauels A, Xia Y, Bex V, Midgley PM (eds)

Jackson JA, Steel DJ, Beerli P, Congdon BC, Olavarria C, Leslie MS, Pomilla C, Rosenbaum H, Baker CS (2014) Global diversity and oceanic divergence of humpback whales (*Megaptera novaeangliae*). Proc R Soc B 281

Jefferies RL, Jano AP, Abraham KF (2006) A biotic agent promotes large-scale catastrophic change in the coastal marshes of Hudson Bay. J Ecol 94:234–242

Ji R, Jin M, Varpe Ø (2013) Sea ice phenology and timing of primary production pulses in the Arctic Ocean. Glob Change Biol 19:734–741

Jonker R, Kurvers RHJM, Bilt A, Faber M, Wieren SE, Prins HHT, Ydenberg RC (2012) Rapid adaptive adjustment of parental care coincident with altered migratory behaviour. Evol Ecol 26: 657–667

Jönsson KI (1997) Capital and income breeding as alternative tactics of resource use in reproduction. Oikos 78:57–66

Jørgensen C, Dunlop ES, Opdal AF, Fiksen Ø (2008) The evolution of spawning migrations: state dependence and fishing-induced changes. Ecology 89:3436–3448

Kaartvedt S, Titelman J (2018) Planktivorous fish in a future Arctic Ocean of changing ice and unchanged photoperiod. ICES J Mar Sci 75:2312–2318

Kelman I, Næss MW (2019) Climate change and migration for Scandinavian Saami: a review of possible impacts. Climate 7

Klemetsen A, Amundsen PA, Dempson J, Jonsson B, Jonsson N, O'connell M, Mortensen E (2003) Atlantic salmon *Salmo salar* L., brown trout *Salmo trutta* L. and Arctic charr *Salvelinus alpinus* (L.): a review of aspects of their life histories. Ecol Freshw Fish 12:1–59

Kolpashikov L, Makhailov V, Russell DE (2015) The role of harvest, predators, and socio-political environment in the dynamics of the Taimyr wild reindeer herd with some lessons for North America. Ecol Soc 20

Koprivnikar J, Leung TLF (2015) Flying with diverse passengers: greater richness of parasitic nematodes in migratory birds. Oikos 124:399–405

Kölzsch A, Bauer S, de Boer R, Griffin L, Cabot D, Exo KM, van der Jeugd HP, Nolet BA (2015) Forecasting spring from afar? Timing of migration and predictability of phenology along different migration routes of an avian herbivore. J Anim Ecol 84:272–283

Krauss S, Stallknecht DE, Negovetich NJ, Niles LJ, Webby RJ, Webster RG (2010) Coincident ruddy turnstone migration and horseshoe crab spawning creates an ecological 'hot spot' for influenza viruses. Proc R Soc B 277:3373–3379

Krümmel EM, Macdonald RW, Kimpe LE, Gregory-Eaves I, Demers MJ, Smol JP, Finney B, Blais JM (2003) Delivery of pollutants by spawning salmon: fish dump toxic industrial compounds in Alaskan lakes on their return from the ocean. Nature 425:255–256

Kuemmerle T, Baskin L, Leitao PJ, Prishchepov AV, Thonicke K, Radeloff VC (2014) Potential impacts of oil and gas development and climate change on migratory reindeer calving grounds across the Russian Arctic. Divers Distrib 20:416–429

Kvist A, Lindström Å (2003) Gluttony in migratory waders – unprecedented energy assimilation rates in vertebrates. Oikos 103:397–402

Lamarre J-F, Legagneux P, Gauthier G, Reed ET, Bêty J (2017) Predator-mediated negative effects of overabundant snow geese on arctic-nesting shorebirds. Ecosphere 8:e01788

Lambertucci SA, Shepard ELC, Wilson RP (2015) Human-wildlife conflicts in a crowded airspace. Science 348:502–504

Lameris TK, van der Jeugd HP, Eichhorn G, Dokter AM, Bouten W, Boom MP, Litvin KE, Ens BJ, Nolet BA (2018) Arctic geese tune migration to a warming climate but still suffer from a phenological mismatch. Curr Biol 28:2467–2473.e2464

Langbehn TJ, Varpe Ø (2017) Sea-ice loss boosts visual search: fish foraging and changing pelagic interactions in polar oceans. Glob Change Biol 23:5318–5330

Lank DB, Butler RW, Ireland J, Ydenberg RC (2003) Effects of predation danger on migration strategies of sandpipers. Oikos 103:303–319

Larsson K, Forslund P (1994) Population dynamics of the barnacle goose *Branta leucopsis* in the Baltic area: density-dependent effects on reproduction. J Anim Ecol 63:954–962
Larsson K, Forslund P, Gustafsson L, Ebbinge BS (1988) From the high Arctic to the Baltic - the successful establishment of a barnacle goose *Branta leucopsis* population on Gotland, Sweden. Ornis Scand 19:182–189
Le Corre M, Dussault C, Cote SD (2020) Where to spend the winter? The role of intraspecific competition and climate in determining the selection of wintering areas by migratory caribou. Oikos 129:512–525
Leblond M, St-Laurent MH, Cote SD (2016) Caribou, water, and ice: fine-scale movements of a migratory arctic ungulate in the context of climate change. Movement Ecol 4
Lehikoinen A, Virkkala R (2016) North by north-west: climate change and directions of density shifts in birds. Glob Change Biol 22:1121–1129
Leung TLF, Koprivnikar J (2016) Nematode parasite diversity in birds: the role of host ecology, life history and migration. J Anim Ecol 85:1471–1480
Loreau M, Mouquet N, Holt RD (2003) Meta-ecosystems: a theoretical framework for a spatial ecosystem ecology. Ecol Lett 6:673–679
Love OP, Gilchrist HG, Descamps S, Semeniuk CA, Bêty J (2010) Pre-laying climatic cues can time reproduction to optimally match offspring hatching and ice conditions in an Arctic marine bird. Oecologia 164:277–286
MacKinnon J, Verkuil YI, Murray N (2012) IUCN situation analysis on east and southeast Asian intertidal habitats, with particular reference to the Yellow Sea (including the Bohai Sea). Occasional paper of the IUCN species survival commission 47
Maclean IMD, Austin GE, Rehfisch MM, Blew J, Crowe O, Delany S, Devos K, Deceuninck B, Gunther K, Laursen K, Van Roomen M, Wahl J (2008) Climate change causes rapid changes in the distribution and site abundance of birds in winter. Glob Change Biol 14:2489–2500
Madsen J, Klaassen M (2006) Assessing body condition and energy budget components by scoring abdominal profiles in free-ranging pink-footed geese *Anser brachyrhynchus*. J Avian Biol 37: 283–287
Madsen J, Cracknell G, Fox T (eds) (1999) Goose populations of the Western Palaearctic: a review of the status and distribution. Wetlands International Publication
Madsen J, Williams JH, Johnson FA, Tombre IM, Dereliev S, Kuijken E (2017) Implementation of the first adaptive management plan for a European migratory waterbird population: the case of the Svalbard pink-footed goose *Anser brachyrhynchus*. Ambio 46:275–289
Mathot KJ, Piersma T, Elner RW (2018) Shorebirds as integrators and indicators of mudflat ecology. In: Beninger PG (ed) Mudflat ecology. Springer International, Cham, pp 309–338
Mauritzen M, Derocher AE, Wiig O (2001) Space-use strategies of female polar bears in a dynamic sea ice habitat. Can J Zool 79:1704–1713
McElroy EJ, de Buron I (2014) Host performance as a target of manipulation by parasites: a meta-analysis. J Parasitol 100:399–410
McKay AF, Hoye BJ (2016) Are migratory animals superspreaders of infection? Integr Comp Biol 56:260–267
McKinnon L, Smith PA, Nol E, Martin JL, Doyle FI, Abraham KF, Gilchrist HG, Morrison RIG, Bety J (2010) Lower predation risk for migratory birds at high latitudes. Science 327:326–327
Meyer N, Bollache L, Galipaud M, Moreau J, Dechaume-Moncharmont F-X, Afonso E, Angerbjörn A, Bêty J, Brown G, Ehrich D, Gilg V, Giroux M-A, Hansen J, Lanctot R, Lang J, Latty C, Lecomte N, McKinnon L, Kennedy L, Reneerkens J, Saalfeld S, Sabard B, Schmidt NM, Sittler B, Smith P, Sokolov A, Sokolov V, Sokolova N, van Bemmelen R, Varpe Ø, Gilg O (2021) Behavioural responses of breeding arctic sandpipers to ground-surface temperature and primary productivity. Sci Total Environ 755:142485
Milner-Gulland EJ, Fryxell JM, Sinclair ARE (eds) (2011) Animal migration: a synthesis. Oxford University Press, New York
Mu T, Wilcove DS (2020) Upper tidal flats are disproportionately important for the conservation of migratory shorebirds. Proc R Soc B Biol Sci 287:20200278

Murray NJ, Clemens RS, Phinn SR, Possingham HP, Fuller RA (2014) Tracking the rapid loss of tidal wetlands in the Yellow Sea. Front Ecol Environ 12:267–272

Mysterud A, Rauset GR, Van Moorter B, Andersen R, Strand O, Rivrud IM (2020) The last moves: the effect of hunting and culling on the risk of disease spread from a population of reindeer. J Appl Ecol 57:2509–2518

Næss MW (2012) Cooperative pastoral production: reconceptualizing the relationship between pastoral labor and production. Am Anthropol 114:309–321

Naiman RJ, Bilby RE, Schindler DE, Helfield JM (2002) Pacific salmon, nutrients, and the dynamics of freshwater and riparian ecosystems. Ecosystems 5:399–417

Nicholson KL, Arthur SM, Horne JS, Garton EO, Del Vecchio PA (2016) Modeling Caribou movements: seasonal ranges and migration routes of the Central Arctic Herd. PLoS One 11

Nilssen KJ, Gulseth OA, Iversen M, Kjol R (1997) Summer osmoregulatory capacity of the world's northernmost living salmonid. Am J Phys Regul Integr Comp Phys 272:R743–R749

O'Briain M, Reed A, Macdonald SD (1998) Breeding, moulting, and site fidelity of Brant (*Branta bernicla*) on Bathurst and Seymour Islands in the Canadian high Arctic. Arctic:350–360

Ottersen G, Bogstad B, Yaragina NA, Stige LC, Vikebo FB, Dalpadado P (2014) A review of early life history dynamics of Barents Sea cod (*Gadus morhua*). ICES J Mar Sci 71:2064–2087

Piersma T (1987) Hop, skip or jump? Constraints on migration of arctic waders by feeding, fattening and flight speed. Limsoa Dutch Engl Summ 60:185–194

Piersma T (1997) Do global patterns of habitat use and migration strategics co-evolve with relative investments in immunocompetence due to spatial variation in parasite pressure? Oikos 80:623–631

Piersma T, Lok T, Chen Y, Hassell CJ, Yang HY, Boyle A, Slaymaker M, Chan YC, Melville DS, Zhang ZW, Ma ZJ (2016) Simultaneous declines in summer survival of three shorebird species signals a flyway at risk. J Appl Ecol 53:479–490

Poole KG, Gunn A, Patterson BR, Dumond M (2010) Sea ice and migration of the dolphin and union caribou herd in the Canadian Arctic: an uncertain future. Arctic 414–428

Prop J, de Vries J (1993) Impact of snow and food conditions on the reproductive performance of barnacle geese *Branta leucopsis*. Ornis Scand 24:110–121

Rice DW, Wolman AA (1971) The life history and ecology of the gray whale (*Eschrichtius robustus*). The American Society of Mammalogists

Righton D, Westerberg H, Feunteun E, Økland F, Gargan P, Amilhat E, Metcalfe J, Lobon-Cervia J, Sjöberg N, Simon J, Acou A, Vedor M, Walker A, Trancart T, Brämick U, Aarestrup K (2016) Empirical observations of the spawning migration of European eels: the long and dangerous road to the Sargasso Sea. Sci Adv 2

Robar N, Murray DL, Burness G (2011) Effects of parasites on host energy expenditure: the resting metabolic rate stalemate. Can J Zool 89:1146–1155

Runge CA, Martini TG, Possingham HP, Willis SG, Fuller RA (2014) Conserving mobile species. Front Ecol Environ 12:395–402

Samplonius JM, Atkinson A, Hassall C, Keogan K, Thackeray SJ, Assmann JJ, Burgess MD, Johansson J, Macphie KH, Pearce-Higgins JW, Simmonds EG, Varpe Ø, Weir JC, Childs DZ, Cole EF, Daunt F, Hart T, Lewis OT, Pettorelli N, Sheldon BC, Phillimore AB (2021) Strengthening the evidence base for temperature-mediated phenological asynchrony and its impacts. Nat Ecol Evol 5:155–164

Satterfield DA, Maerz JC, Altizer S (2015) Loss of migratory behaviour increases infection risk for a butterfly host. Proc R Soc B Biol Sci 282:20141734

Schekkerman H, Tulp I, Piersma T, Visser GH (2003) Mechanisms promoting higher growth rate in arctic than in temperate shorebirds. Oecologia 134:332–342

Schmidt NM, Reneerkens J, Christensen JH, Olesen M, Roslin T (2019) An ecosystem-wide reproductive failure with more snow in the Arctic. PLoS Biol 17

Shaw AK, Binning SA (2016) Migratory recovery from infection as a selective pressure for the evolution of migration. Am Nat 187:491–501

Skagseth Ø, Slotte A, Stenevik EK, Nash RDM (2015) Characteristics of the Norwegian coastal current during years with high recruitment of Norwegian spring spawning herring (*Clupea harengus* L.). PLoS One 10:e0144117

Smith PA, McKinnon L, Meltofte H, Lanctot RB, Fox AD, Leafloor JO, Soloviev M, Franke A, Falk K, Golovatin M, Sokolov V, Sokolov A, Smith AC (2020) Status and trends of tundra birds across the circumpolar Arctic. Ambio 49:732–748

Stafford KM (2019) Increasing detections of killer whales (*Orcinus orca*), in the Pacific Arctic. Mar Mamm Sci 35:696–706

Stephens PA, Boyd IL, McNamara JM, Houston AI (2009) Capital breeding and income breeding: their meaning, measurement, and worth. Ecology 90:2057–2067

Stevick PT, Berrow SD, Berube M, Bouveret L, Broms F, Jann B, Kennedy A, Suarez PL, Meunier M, Ryan C, Wenzel F (2016) There and back again: multiple and return exchange of humpback whales between breeding habitats separated by an ocean basin. J Mar Biol Assoc UK 96:885–890

Stroeve JC, Serreze MC, Holland MM, Kay JE, Malanik J, Barrett AP (2012) The Arctic's rapidly shrinking sea ice cover: a research synthesis. Clim Chang 110:1005–1027

Studds CE, Kendall BE, Murray NJ, Wilson HB, Rogers DI, Clemens RS, Gosbell K, Hassell CJ, Jessop R, Melville DS, Milton DA, Minton CDT, Possingham HP, Riegen AC, Straw P, Woehler EJ, Fuller RA (2017) Rapid population decline in migratory shorebirds relying on Yellow Sea tidal mudflats as stopover sites. Nat Commun 8:14895

Summers RW, Nicoll M, Peach W (2001) Numbers, migration phenology and survival of purple sandpipers *Calidris maritima* at Gourdon, eastern Scotland. Bird Study 48:139–146

Sun NW, Goodwin SE, Griego MS, Gerson AR, Clotfelter ED (2020) Does blood loss explain higher resting metabolic rates in nestling birds with hematophagous ectoparasites? J Avian Biol 51

Sutherland WJ (1998) Evidence for flexibility and constraint in migration systems. J Avian Biol 29: 441–446

Svenning MA, Klemetsen A, Olsen T (2007) Habitat and food choice of Arctic charr in Linnevatn on Spitsbergen, Svalbard: the first year-round investigation in a high Arctic lake. Ecol Freshw Fish 16:70–77

Swartz SL, Taylor BL, Rugh DJ (2006) Gray whale *Eschrichtius robustus* population and stock identity. Mammal Rev 36:66–84

Taillon J, Festa-Bianchet M, Cote SD (2012) Shifting targets in the tundra: protection of migratory caribou calving grounds must account for spatial changes over time. Biol Conserv 147:163–173

Taylor AR, Lanctot RB, Powell AN, Huettmann F, Nigro DA, Kendall SJ (2010) Distribution and community characteristics of staging shorebirds on the northern coast of Alaska. Arctic 63:451–467

Taylor CM, Laughlin AJ, Hall RJ (2016) The response of migratory populations to phenological change: a migratory flow network modelling approach. J Anim Ecol 85:648–659

Teitelbaum CS, Huang S, Hall RJ, Altizer S (2018) Migratory behaviour predicts greater parasite diversity in ungulates. Proc R Soc B Biol Sci 285:20180089

Tombre IM, Høgda KA, Madsen J, Griffin LR, Kuijken E, Shimmings P, Rees E, Verscheure C (2008) The onset of spring and timing of migration in two arctic nesting goose populations: the pink-footed goose *Anser bachyrhynchus* and the barnacle goose *Branta leucopsis*. J Avian Biol 39:691–703

Tombre IM, Oudman T, Shimmings P, Griffin L, Prop J (2019) Northward range expansion in spring-staging barnacle geese is a response to climate change and population growth, mediated by individual experience. Glob Change Biol 25:3680–3693

Tyler NJC, Øritsland NA (1989) Why don't Svalbard reindeer migrate? Holarct Ecol 12:369–376

Van Der Jeugd HP, Eichhorn G, Litvin KE, Stahl J, Larsson K, Van Der Graaf AJ, Drent RH (2009) Keeping up with early springs: rapid range expansion in an avian herbivore incurs a mismatch between reproductive timing and food supply. Glob Change Biol 15:1057–1071 van der Wal R, Sjogersten S, Woodin SJ, Cooper EJ, Jonsdottir IS, Kuijper D, Fox TAD, Huiskes AD (2007) Spring feeding by pink-footed geese reduces carbon stocks and sink strength in tundra ecosystems. Glob Change Biol 13:539–545

van Gils JA, Munster VJ, Radersma R, Liefhebber D, Fouchier RAM, Klaassen M (2007) Hampered foraging and migratory performance in swans infected with low-pathogenic avian influenza a virus. PLoS One 2

van Gils JA, Lisovski S, Lok T, Meissner W, Ozarowska A, de Fouw J, Rakhimberdiev E, Soloviev MY, Piersma T, Klaassen M (2016) Body shrinkage due to Arctic warming reduces red knot fitness in tropical wintering range. Science 352:819–821

Varpe Ø (2010) Stealing bivalves from common eiders: kleptoparasitism by glaucous gulls in spring. Polar Biol 33:359–365

Varpe Ø (2012) Fitness and phenology: annual routines and zooplankton adaptations to seasonal cycles. J Plankton Res 34:267–276

Varpe Ø (2017) Life history adaptations to seasonality. Integr Comp Biol 57:943–960

Varpe Ø, Fiksen Ø (2010) Seasonal plankton-fish interactions: light regime, prey phenology, and herring foraging. Ecology 91:311–318

Varpe Ø, Fiksen Ø, Slotte A (2005) Meta-ecosystems and biological energy transport from ocean to coast: the ecological importance of herring migration. Oecologia 146:443–451

Varpe Ø, Jørgensen C, Tarling GA, Fiksen Ø (2009) The adaptive value of energy storage and capital breeding in seasonal environments. Oikos 118:363–370

Varpe Ø, Daase M, Kristiansen T (2015) A fish-eye view on the new Arctic lightscape. ICES J Mar Sci 72:2532–2538

Viana DS, Santamaria L, Figuerola J (2016) Migratory birds as global dispersal vectors. Trends Ecol Evol 31:763–775

Vincent WF, Callaghan TV, Dahl-Jensen D, Johansson M, Kovacs KM, Michel C, Prowse T, Reist JD, Sharp M (2011) Ecological implications of changes in the Arctic cryosphere. Ambio 40:87–99

Wassmann P, Duarte CM, Agusti S, Sejr MK (2011) Footprints of climate change in the Arctic marine ecosystem. Glob Change Biol 17:1235–1249

Wauchope HS, Shaw JD, Varpe Ø, Lappo EG, Boertmann D, Lanctot RB, Fuller RA (2017) Rapid climate-driven loss of breeding habitat for Arctic migratory birds. Glob Change Biol 23:1085–1094

Webster MS, Marra PP, Haig SM, Bensch S, Holmes RT (2002) Links between worlds: unraveling migratory connectivity. Trends Ecol Evol 17:76–83

Weladji RB, Holand O (2006) Influences of large-scale climatic variability on reindeer population dynamics: implications for reindeer husbandry in Norway. Clim Res 32:119–127

Westerdahl H, Bensch S, Nilsson JÅ, O'Connor E, Sehgal R, Tesson S, Hasselquist D (2014) Pathogens and hosts on the move. In: Hansson LA, Åkesson S (eds) Animal movement across scales, pp 126–148

Wilcove DS, Wikelski M (2008) Going, going, gone: is animal migration disappearing. PLoS Biol 6:e188

Willemoes M, Strandberg R, Klaassen RHG, Tottrup AP, Vardanis Y, Howey PW, Thorup K, Wikelski M, Alerstam T (2014) Narrow-front loop migration in a population of the common cuckoo *Cuculus canorus*, as revealed by satellite telemetry. PLoS One 9

Willson MF, Womble JN (2006) Vertebrate exploitation of pulsed marine prey: a review and the example of spawning herring. Rev Fish Biol Fish 16:183–200

Xu Y, Si Y, Wang Y, Zhang Y, Prins HHT, Cao L, de Boer WF (2019) Loss of functional connectivity in migration networks induces population decline in migratory birds. Ecol Appl 29:e01960

Zeng HQ, Jia GS, Epstein H (2011) Recent changes in phenology over the northern high latitudes detected from multi-satellite data. Environ Res Lett 6

Part II

Major Health Threats to Arctic Animals and People

Climate Change in Northern Regions

Bob van Oort, Marianne Tronstad Lund, and Anouk Brisebois

1 Scope

This chapter describes the past, present, and projected changes in global- and arctic climate and environment, the underlying drivers, and the associated hazards and impacts on natural and human systems. For the purpose of this book, we focus on the time period from 1750 (defined as preindustrial) to 2100 (when most projections of future emissions and socioeconomic development end). This chapter presents a general picture of some of the key climate changes and related challenges for arctic nature and society. However, the Arctic is a large area with diverse climates, landscapes, settlements, and livelihoods, and there are a variety of limits defining "the Arctic" (Fig. 1). "The Arctic" can hence be a very different place dependent on definitions and geographical scope. The characteristics of these different arctic regions, their inhabitants, and their ecosystems have been presented in the preceding chapters. Thus, when describing impacts on nature and society, it is necessary to distinguish between regions. It is also important to acknowledge and realize that people and institutions have agency and that they are not passive actors when exposed to change. To describe all details and nuances for each arctic region and challenge is beyond the scope of this chapter. While we will give some examples of impacts for specific geographical areas within the Arctic, many region-specific examples related to contaminants, oil spills, traditional foods, and zoonoses will be given in the following chapters.

B. van Oort (✉) · M. T. Lund · A. Brisebois
CICERO Center for International Climate Research, Oslo, Norway
e-mail: oort@cicero.oslo.no

M. Tryland (ed.), *Arctic One Health*, https://doi.org/10.1007/978-3-030-87853-5_4

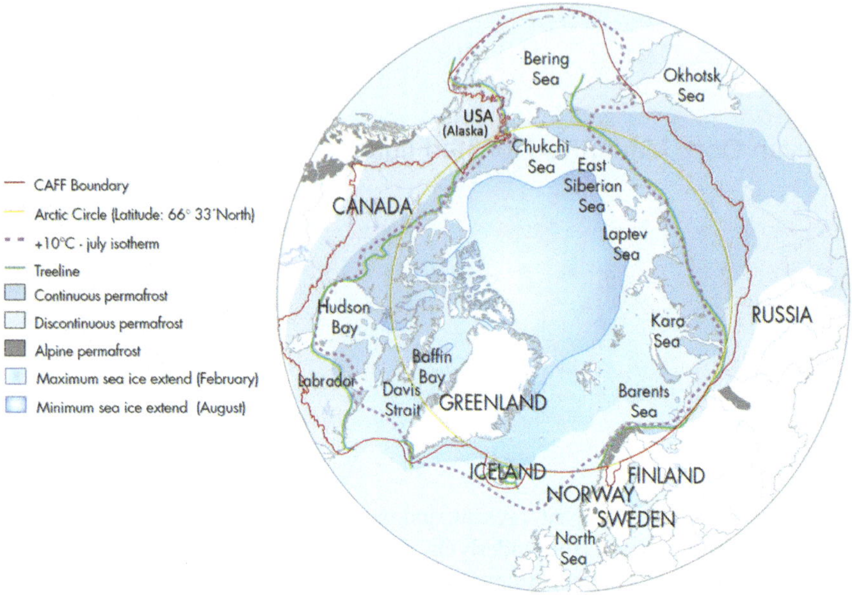

Fig. 1 The northern polar region showing the different delineations of the Arctic, areas with permafrost, minimum and maximum sea ice cover, and the eight Arctic states. Source: Huntington (2001)

2 Climate Change: The Global Perspective

Following the definition of the Intergovernmental Panel on Climate Change (IPCC), *climate* is "in a narrow sense usually defined as the 'average weather,' or more rigorously, as the statistical description in terms of the mean and variability of relevant quantities over a period ranging from months to thousands or millions of years" (IPCC 2013). The averaging period most commonly used is 30 years (World Meteorological Organization 2017). Climate is thus clearly distinguished from *weather*, which describes the conditions of the atmosphere—pressure, temperature, wind, clouds, and precipitation—at a certain place and time. While weather varies widely from day to day or season to season, *climate change* describes the change in the state of the climate that persists for an extended period, typically decades or longer. In a broader sense, climate is the state of the components of the Earth system: the atmosphere, ocean, biosphere, and cryosphere.

2.1 Consistent Indicators of a Warming Climate

While global surface air temperature change is the most commonly used and communicated measure of climate change, numerous other essential indicators

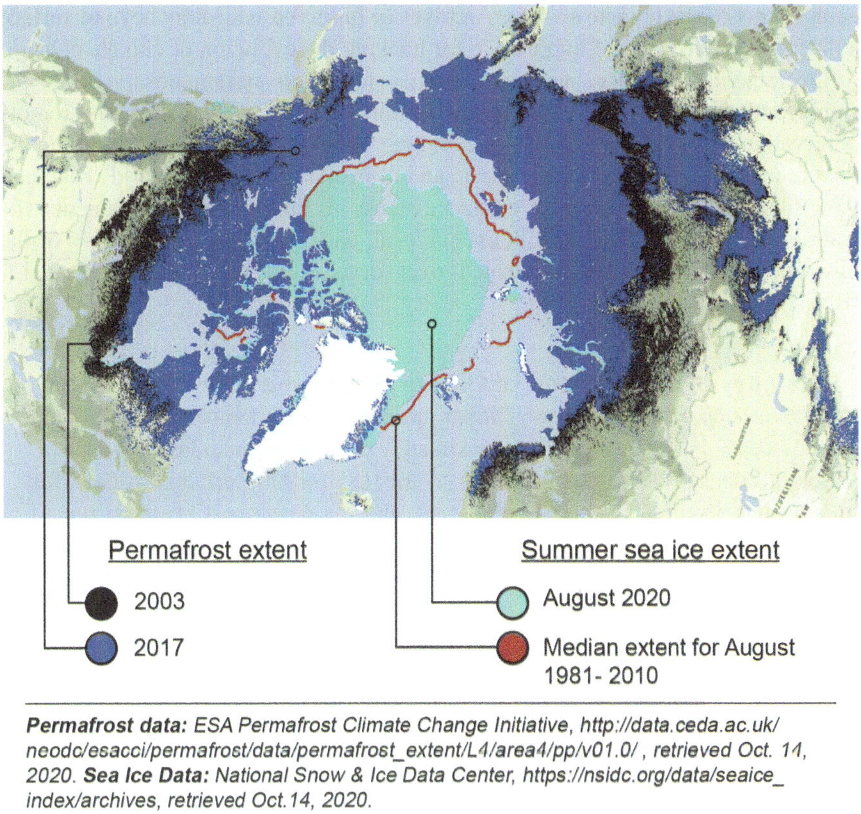

Fig. 2 Changes in permafrost and summer sea ice extent from 2003 to 2017 (permafrost extent, including both continuous and discontinuous permafrost) and 1981–2010 to 2020 (summer sea ice extent)

exist. These include changes in physical quantities and associated environmental parameters, such as levels of greenhouse gases in the atmosphere, energy balance, precipitation, ocean heat content, sea level, and sea ice extent. All these are now being monitored, although the geographical coverage and lengths of observational time series vary.

Many of the most direct and rapidly responding indicators of climate change pertain to the atmosphere and land surface. For instance, changes to the cryosphere, comprising snow, river and lake ice, sea ice, glaciers, ice shelves and ice sheets, and frozen ground, are among the most visible physical indicators of a changing climate. Although the timescales differ, all parts of the cryosphere are inherently sensitive to changes in air temperature and precipitation and thus to a changing climate, and the cryosphere is sometimes referred to as a "natural thermometer" (IPCC 2013). Due to its inertia, the oceans often provide a clearer signal of longer-term change than other components of the climate system. Archives of observed indicators beyond surface temperature are of critical importance for tracking the evolution of climate change.

Over the last century, new technologies for Earth system observation have significantly increased our capacity to monitor the climate. Stationary and remote observation systems now provide us with large amounts of measurements, from the top of the atmosphere to the deep ocean, and from all regions of the world. Together, these observations provide clear evidence across all Earth system components: the warming of the climate system is unequivocal, and, since the 1950s, many of the observed changes are unprecedented over decades to millennia (IPCC 2013) (Box 1).

Box 1 Key Global Indicators of Climate Change and Their Recent Trend

Global temperature increase: Global mean surface air temperature over land and oceans has increased by approximately 1 °C since preindustrial (0.87 °C for the decade 2006–2015 relative to the 1850–1900 period. Most of this warming occurred in the past 35 years, and the 6 warmest years on record have all taken place since 2014. Since the mid-twentieth century, there has also been pronounced warming of the troposphere and cooling of the stratosphere globally.

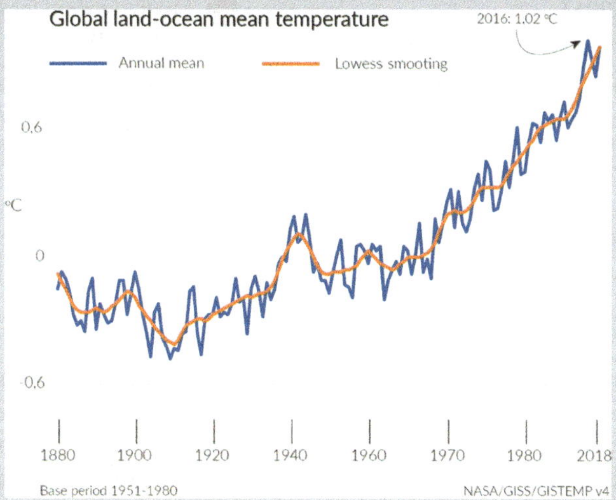

(continued)

Box 1 (continued)

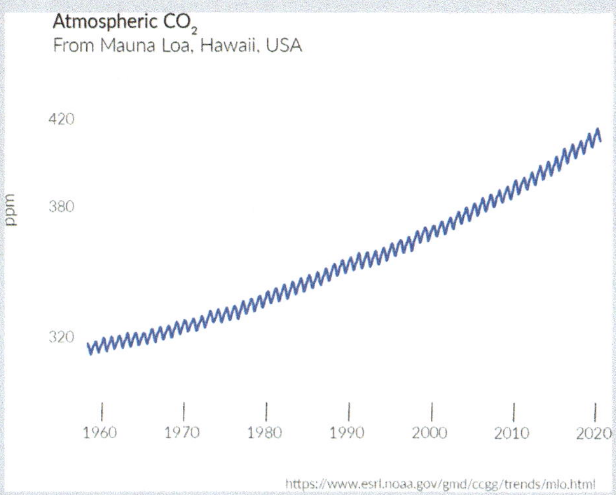

Increasing greenhouse gas concentrations: The global average atmospheric CO_2 concentration was 409 parts per million (ppm) in 2019. This is a 47% increase from 1750, and higher than any level recorded in the past 800,000 years. During the same time interval, CH_4 increased by 150% from 722 parts per billion (ppb) to 1803 ppb, and N_2O by 20% from 271 ppb to 324.2 ppb in 2011. These current concentrations exceed any level measured during at least the past 800,000 years; the period covered by ice cores.

Warming oceans: More than 90% of the excess heat energy stored by the Earth over the last 50 years has been absorbed by the global oceans. This has resulted in significant ocean warming, most strongly in the upper 700 m, but in the past three decades extending all the way down to 3000 m. The graph shows the increase in global ocean heat content in the top 0 700 m since 1955. For context, an increase of 1 unit on this graph (1×10^{22} joules) approximately equals 18 times the total amount of energy used by all the people on Earth in a year.

(continued)

Box 1 (continued)

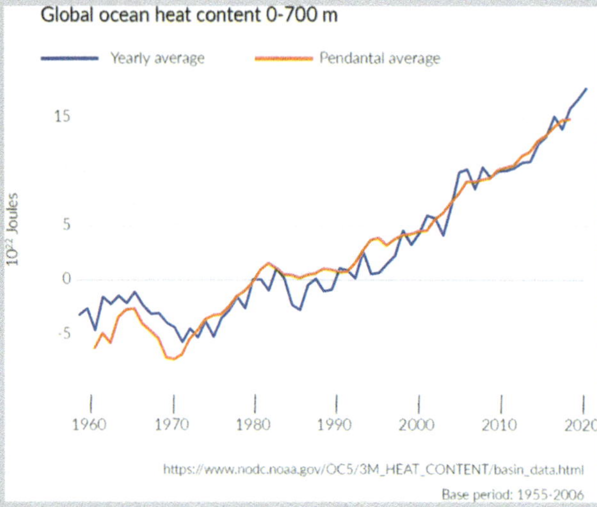

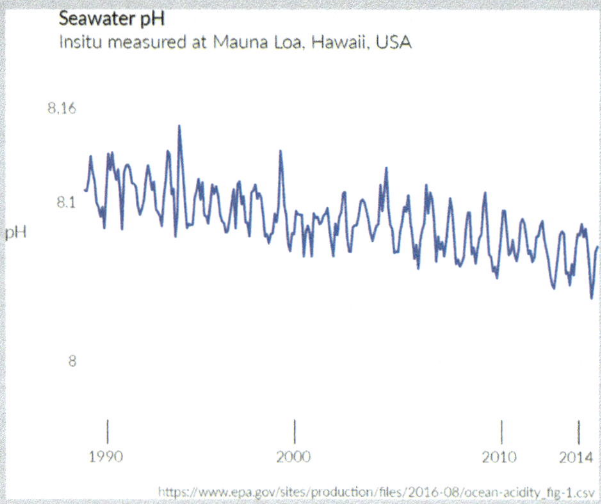

Ocean acidification: The global oceans are presently slowing the rate of global warming by absorbing about 30% of carbon dioxide (CO_2) from human emissions. Concurrently, there is an increasing ocean acidification. The acidity of surface ocean waters has increased by 30% since the Industrial Revolution.

Sea level rise: Global sea level rose about 0.2 m in the last century. The increase accelerated in the past two decades. Sea level rise is caused by both

(continued)

Box 1 (continued)
thermal expansion, as the oceans warm, and by influx of freshwater from melting of snow and ice on land.

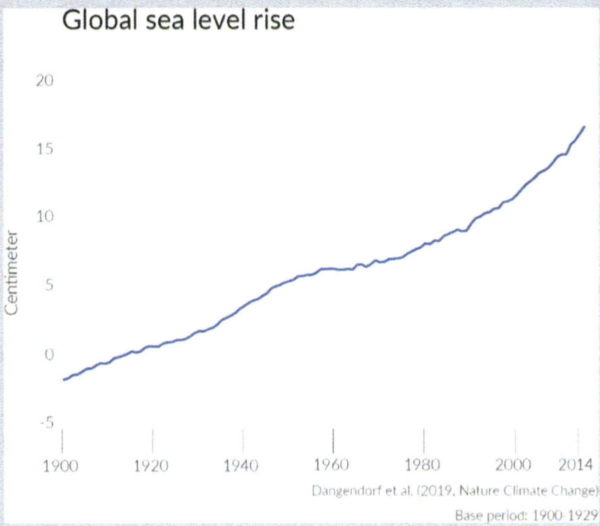

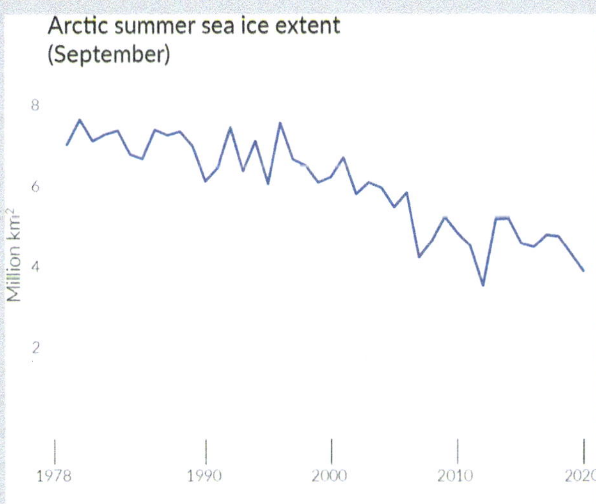

Snow and ice: Satellite observations show that snow is melting earlier, and the amount of snow in the spring has decreased in the Northern Hemisphere in the past five decades. Glaciers are retreating in almost all regions, and both the

(continued)

Box 1 (continued)
extent and thickness of the arctic sea ice have declined rapidly since the beginning of the instrumental record.

Extreme events: An increase in the number of heat waves, droughts, and heavy precipitation events have been observed in many regions. There is also increasing evidence of more intense Atlantic hurricane activity. Attribution studies show that human-induced climate change has made several of the severe extreme events over the past years much more likely to occur than without underlying global warming.

2.2 Lines of Evidence of Human-Made Climate Change

Just as the warming of the climate system is clear, so is the human influence on this warming. As stated by the 2013 fifth assessment report by the IPCC, "it is extremely likely that human influence has been the dominant cause of the observed warming since the mid-twentieth century" (IPCC 2013). This is clear from multiple lines of evidence, from observations, detection, and attribution using climate models and statistical techniques, and the fundamental understanding of the climate system.

To understand what drives the rapid observed warming, we investigate all possible natural and human-made factors known to influence our climate. When we do this, it becomes clear that natural factors alone, such as solar activity or volcanic eruptions, cannot explain contemporary climate change. The observed global warming over the past century can only be reproduced when also human-induced increases in levels of greenhouse gases and other human activities are accounted for. This is true for global mean warming but also for observed warming on all continents. Of the human-made impact, CO_2 increase from fossil fuel combustion constitutes the largest individual driver, followed by an increase in methane and other greenhouse gases. Air pollution in the form of particles has caused a net cooling contribution that has masked some of the greenhouse gas-induced warming to date.

Further evidence of the human-made contribution is derived from observations of energy exchanges at the top of the atmosphere and at the surface. These measurements reveal that over the past 40 years, less energy is escaping to space at the wavelengths associated with CO_2, while more heat is received by the planet's surface. This provides a direct, empirical causal link between CO_2 and global warming. Measurements also show that the pattern of warming is consistent with an increased greenhouse effect: The largest warming is taking place during nighttime and during boreal winter, and the lower 8–17 km of the atmosphere (the so-called troposphere) is warming while levels higher up are cooling. Finally, not only do we measure an increased amount of CO_2 in the atmosphere, but measurements confirm that this added CO_2 comes from fossil fuel combustion. Variations in the Earth's climate are also caused by internal forcing (see definitions Box 2), so-called natural variability. However, while able to significantly affect global and regional climate on annual and decadal scales, natural variability cannot explain the long-term warming trend.

Box 2 Definitions of Some Key Concepts Used in This Chapter

Albedo: The fraction of solar radiation reflected by a surface or object, often expressed as a percentage. Snow-covered surfaces have a high albedo [reflecting radiation and heat], while the albedo of soils ranges from high to low, and vegetation-covered surfaces and oceans have a low albedo [absorbing heat].

Climate forcing: Climate forcing "forces" the climate to change, disturbing the energy balance between incoming radiation from the Sun and outgoing terrestrial radiation from the Earth (the global energy balance).

Climate forcing can be separated into *internal and external types*, which operate from within and outside the Earth's climate system. External forcing includes variations in the Earth's orbit around the Sun, and changes in the amount of energy coming from the Sun, and internal forcing includes changes in the composition of the atmosphere, or volcanic activity.

RCP (Representative Concentration Pathways): Scenarios that include time series of emissions and concentrations of the full suite of greenhouse gases (GHGs), aerosols and chemically active gases, as well as land use/land cover. The word representative signifies that each RCP provides only one of many possible scenarios that would lead to the specific radiative forcing characteristics. The term pathway emphasizes that not only the long-term concentration levels are of interest but also the trajectory taken over time to reach that outcome.

Hazard: The potential occurrence of a natural or human-induced physical event or trend or physical impact that may cause loss of life, injury, or other health impacts, as well as damage and loss to property, infrastructure, livelihoods, service provision, ecosystems, and environmental resources.

Exposure: The presence of people, livelihoods, species or ecosystems, environmental functions, -services, and -resources, infrastructure, or economic, social, or cultural assets in places and settings that could be adversely affected.

Impacts (consequences, outcomes): Effects on lives, livelihoods, health, ecosystems, economies, societies, cultures, services, and infrastructure due to the interaction of climate changes or hazardous climate events occurring within a specific time period and the vulnerability of an exposed society or system. Impacts are also referred to as consequences and outcomes.

Contaminants are defined as inputs of alien and potentially toxic substances into the environment; not all contaminants cause pollution, as their concentrations may be too low.

(continued)

Box 2 (continued)

Pollutants are defined as anthropogenically introduced substances that have harmful effects on the environment.

Vulnerability: The propensity or predisposition to be adversely affected. Vulnerability encompasses a variety of concepts and elements including sensitivity or susceptibility to harm and lack of capacity to cope and adapt.

Risk: Risk results from the interaction of vulnerability, exposure, and hazard, and is the potential for consequences where something of value is at stake and where the outcome is uncertain.

Resilience: The capacity of a social-ecological system to cope with disturbance, responding or reorganizing in ways that maintain its essential function, identity, and structure while also maintaining the capacity for adaptation, learning, and transformation.

3 Climate Change at High Northern Latitudes

The polar regions encompass a vast share of the world's ocean and cryosphere, including more than 90% of the total permafrost area, 69% of the world's glacier area and almost all of the world's sea ice (IPCC 2019a). The polar regions are also an integral part of the global climate system, acting as the cooling chambers of the planet, and influencing lower-latitude weather. The polar regions receive limited solar radiation, and have a net loss of heat over the course of the year. In contrast, lower latitudes receive more heat from the Sun than they can radiate back to space. This difference causes warm air and ocean currents from lower latitudes to move into the Arctic and Antarctic, and cold air and water from the polar regions to move to the lower latitudes. Through this process, the polar regions regulate the distribution of heat on the Earth over the course of the year. The unique geography of the Arctic also leads to recurring weather patterns that influence weather far beyond the region. Examples include the Arctic oscillation, an atmospheric circulation pattern that occurs over the mid-to-high latitudes of the Northern Hemisphere, and the semipermanent pressure centers such as the Aleutian and Icelandic low. Currently, the Arctic climate and environment are undergoing dramatic changes. The consequences of these changes affect people and ecosystems in a multitude of ways with implications extending to the whole planet.

3.1 Arctic Climate Change

From 2011 to 2015, the average Arctic surface temperature was higher than at any other time since instrumental records began in around 1900. In January 2016, the Arctic was 5 °C warmer than the 1981–2010 average for the region, a full 2 °C higher than the previous record set in 2008, and monthly mean temperatures in October through December 2016 were 6 °C higher than average for these months (AMAP 2017a). Recent events indicate that new *extremes* are occurring in the Arctic climate system. In northern Alaska and northeastern Russia, recent observations suggest a widespread decline in periods of extreme cold during both winter and summer, and increases in extreme warm periods during autumn and spring (AMAP 2017a). In the Central Arctic, surface air temperatures up to +6 °C higher than normal (i.e., the average over 1981–2010) were recorded during winter (January to March) in 2016 and 2018. This is nearly twice the magnitude of such previously observed temperature records (Overland et al. 2019b). According to the World Weather Attribution, the recent long and record-breaking heat waves in the Arctic are rare, even with climate change, but would be virtually impossible without human-induced climate change (World Weather Attribution 2020).

The oceans store and transport heat, freshwater, and carbon, and exchange these with the atmosphere, and therefore strongly influence climate. Consistent with the global mean trend, the polar ocean has continued to warm and acidify over the recent years. In large areas of the seasonally ice-free Arctic, summer temperatures in the upper ocean layers increased at around 0.5 °C per decade during 1982–2017. This warming is primarily associated with increased heating due to an educed sea ice cover, and an increased inflow of ocean heat from lower latitudes. The global oceans have taken up 90% of the additional heat in the climate system over the past 50 years. This means that only a fraction of warming is taking place in the atmosphere for now. However, heat taken up by the ocean is only moved from one place to another but does not disappear. Eventually heat stored in the ocean is released back to the atmosphere, committing Earth to additional surface warming in the future.

The pan-Arctic loss of sea ice is one of the most dramatic changes in the Arctic environment, and a prominent indicator of the ongoing climate change. The Arctic sea ice extent, defined as the total area of the Arctic with at least 15% sea ice concentration, has declined in all months of the year since the beginning of the instrumental record in 1979. The smallest decline is seen during winter, and the largest during summer, with the strongest trend in September when the ice extent tends to be at its lowest. The summer ice loss is most prominent in seas above Alaska, Canada, and Russia (see Fig. 1). The observed change in Arctic summer sea ice loss is driven partly by human-made warming increased greenhouse gases and warming and partly by natural climate variability (Kay et al. 2011; Notz and Marotzke 2012).

Box 3 Why Is the Arctic Warming Faster than the Global Average?

Over the last two decades, the surface air temperature in the Arctic has increased at more than twice the global average. Warming has been especially pronounced since the year 2000, with Arctic annual surface temperatures exceeding previous records for the past 6 years (2014–2019) (Overland et al. 2019a, b). Termed Arctic amplification, this higher-than-average warming is a

(continued)

Box 3 (continued)
robust feature in both observations and climate model projections. The most important and well-known driving mechanism is the surface albedo feedback, whereby reduced sea ice and snow cover reveals the darker, lower reflectivity ocean and land surfaces which absorb more solar radiation. This in turn results in further warming and sea ice loss. However, several additional mechanisms have been proposed and been found to contribute, including increased Arctic atmospheric water vapor abundance, changes in summer-time cloudiness, additional heat generated as new sea ice forms across more extensive open water areas in the autumn, and northward transport of heat and moisture.

Climate change in the Arctic is not limited to the ocean, and the terrestrial components of the cryosphere are also undergoing dramatic changes at high northern latitudes. On land, *permafrost* temperatures are increasing. The permafrost region represents a large reservoir of organic carbon, almost twice as much carbon as is currently in the atmosphere (Schuur et al. 2018). This reservoir is also highly climate sensitive. Melting permafrost can therefore influence global climate through emissions of carbon dioxide and methane released from an increase in microbial breakdown of organic carbon, and the release of trapped methane. It is not fully clear whether northern permafrost region is currently a source of additional methane and CO_2 to the atmosphere (IPCC 2019a), but recent observations and projections suggest that the positive feedback of increased greenhouse gas release due to thawing permafrost and subsequent additional climate warming may already be underway (Anthony et al. 2018; Schuur 2018).

Glaciers are another critical indicator of climate change. Summer melting of the *Greenland Ice Sheet* (GIS) has increased since the 1990s, driven by both oceanic and atmospheric warming, to a level unprecedented over at least the last 350 years. The GIS is estimated to have been close to a state of balance in the 1990s, seasonally growing and melting but maintaining a stable average, but annual losses have risen since then, peaking at 345 ± 66 billion tons per year in 2011 (Shepherd et al. 2020). In the summer of 2019, the melting rate was record high due to a very persistent high-pressure system over the region bringing warm weather. During this period alone, it was calculated that the GIS lost ice equivalent to around 80 million Olympic swimming pools. In contrast to melting sea ice, melting of glaciers and ice sheets contribute to sea level rise. Currently, Greenland is losing ice seven times faster than in the 1990s (The IMBIE Team 2020), and could contribute 5–33 cm to sea level by 2100, depending on emission scenario (Aschwanden et al. 2019). During 2006–2015, mass loss from *Arctic glaciers* of around 212 Gt per year contributed to sea level rise at a similar rate (0.6 ± 0.1 mm per year) as the melting of the GIS. The increasing surface melt on Arctic glaciers also leads to a positive climate feedback through lower surface *albedo* (see definitions in Box 2), where darker, ice-free surfaces revealed due to melting causes increased heat absorption, and hence further melting and more dark surfaces (Box et al. 2012).

Reductions in Arctic (land areas north of 60°N) spring *snow cover* extent has been recorded since satellite charting began in 1967. Changes in terrestrial snow cover are critical, as they influence the surface energy budget, influencing how much heat is absorbed and released, in addition to the freshwater availability.

3.2 Future Projections

The ongoing changes in the Arctic climate system are expected to continue into the twenty-first century, with a further decrease of sea ice, permafrost and snow on land, and reductions in the mass of glaciers. How fast these changes take place is dependent on global emission scenarios and climate change mitigation measures (IPCC 2019a). However, if continuing along the recent observed trend, a largely ice-free summer Arctic ocean could already be a reality by the late 2030s, which is earlier than projected by most climate models (AMAP 2017a). If global warming is stabilized at 1.5 °C, the chance of a sea ice-free September at the end of century is estimated at approximately 1%; for a 2 °C increase, this chance rises to 10–35%. Projected mass reductions for polar glaciers between 2015 and 2100 range from 16 ± 7% for a scenario corresponding to low emissions (RCP2.6 with low greenhouse gas concentrations—see Box 2 for explanation of "RCP") to 33 ± 11% in a case with very high human-made emissions (RCP8.5 with high greenhouse gas concentrations. Note: This scenario currently seems unrealistically high.) Near-surface permafrost area is projected to decrease by 2–66% for RCP2.6 and by 30–99% for RCP8.5 over the same period. About 20% of Arctic land permafrost is vulnerable to abrupt permafrost thaw and subsequent ground subsidence, which is expected to increase small lake areas by over 50% by 2100 for RCP8.5. The duration of snow cover is projected to decrease by an additional 10–20% from current levels over most of the Arctic by mid-century under a high emission scenario (AMAP 2017a). Even as the overall regional water cycle intensifies, including increased precipitation, evapotranspiration, and river discharge to the Arctic Ocean, decreases in snow and permafrost may lead to soil drying.

It is important to keep in mind that with a *global* warming of 1.5 °C or 2 °C, the projected warming in the *Arctic* is significantly higher due to Arctic amplification (Box 3). Projections suggest that under an emission scenario that is roughly similar to that required to meet the 2 °C goal of the Paris Agreement, the wintertime surface temperature over the Arctic Ocean could stabilize at 5–9 °C above the 1986–2005 average (AMAP 2017a). It could further stabilize the duration of snow cover at about 10% below current levels and near-surface permafrost extent at around 45% below current levels. While successful implementation of the Paris Agreement would limit the climate and environmental changes in the Arctic, the region will still be a very different place in 2100 than today.

Box 4 "What Happens in the Arctic Does Not Stay in the Arctic"

A typical saying is that "what happens in the Arctic does not stay in the Arctic." Changes in the Arctic atmosphere, cryosphere, and ocean can have wide-reaching implications. Arctic change has been proposed to affect weather patterns in lower latitudes, even influencing Southeast Asian monsoons (AMAP, 2017e). The linkages between Arctic change and mid-latitude weather are an active research area and a societally important one because hundreds of millions of people can potentially be impacted (Jung et al. 2015). Part of the scientific disagreement is due to irregular connections in the Arctic to mid-latitude linkage pathways, both within and between years (Overland et al. 2019a). Considerable literature exists on the potential for sea ice loss in the Barents and Kara Seas to drive cold episodes in eastern Asia (Kim et al. 2014; Kretschmer et al. 2016), while sea ice anomalies in the Chukchi Sea and areas west of Greenland are associated with cold events in eastern North America (Kug et al. 2015; Overland et al. 2019a; Ballinger et al. 2019). Recent research suggests that one of the major ocean currents in the Arctic, called the Beaufort Gyre, is becoming faster and more turbulent as a result of rapid sea ice melt (Armitage et al. 2020). The Beaufort Gyre keeps the Arctic polar environment in equilibrium by storing freshwater near the surface of the ocean. If this excess fresh water were to be released into the Atlantic Ocean, it could potentially slow down its circulation and have hemisphere-wide climate implications. Future projections are however still unclear.

4 Consequences and Responses of Natural and Societal Systems to Climate Change

Climate changes globally and in the Arctic are already underway, and they are affecting ecosystems, lives, and livelihoods, including health, throughout the different arctic regions (AMAP 2017a). Climate change will have direct impacts, but will also interact with other drivers of change, including globalization, geopolitical shifts, increased industrial, shipping and tourism activities, and other socioeconomic, cultural, and political conditions (e.g., Nilsson et al., 2017). Such combined changes create complex and cumulative impacts on all Arctic communities—indigenous, nonindigenous, rural and urban—where climate change will amplify existing climate-related risks and create new risks for natural and human systems (IPCC 2014). Arctic societies and activities are in various ways connected to the natural systems, and climate-driven changes in the cryosphere, flora, or fauna may have secondary climate impacts on societies (Crépin et al. 2017). These cumulative, direct, and indirect effects act simultaneously to reshape nature and society and will have consequences leading both to challenges and opportunities in the Arctic (Hovelsrud et al. 2010, 2011). Consequences shaped by multiple drivers will need adaptation actions not to climate change alone, but to this complex set of drivers together (Hovelsrud et al. 2010; AMAP 2017b, c, d; Nilsson and Larsen 2020). Climate change is not always the strongest or main driver, and its impacts may sometimes be difficult to separate from impacts driven by other factors (Hovelsrud et al. 2010; Arbo et al. 2013; Nilsson et al. 2015, 2017). Due to these interlinkages, some of these indirect and multiple driver issues will therefore be discussed together with the direct climate change effects which are the key focus in this section. It is further important to understand that the Arctic is not a homogenous region but consists of different ecosystems, countries, people, policies, activities, and interests (see Chap. 1). Risks and consequences thus often depend heavily on the specific regional contexts. We will address some of these through examples in the following sections.

4.1 Climate Risk

Climate risk is a result of the interaction between physical change, or hazards, and the vulnerability and exposure of human and natural systems (see Box 2 for definitions). The severity of climate risk thus depends on the magnitude and rate of warming but also on geographic location, levels of development and vulnerability, and capacity for adaptive action. Here we must underline that people are not passive victims of climate impacts, but that they have agency, the ability to act and react, and that the combined use of different kinds of knowledge and institutions can greatly contribute to reduce climate risk (Dannevig et al. 2013; Hovelsrud et al. 2018; Wheeler et al. 2020).

From a global perspective, populations and regions which are at a disproportionately high risk of negative consequences from global warming include especially vulnerable or natural resource-dependent populations, arctic ecosystems, and dryland regions (Hoegh-Guldberg et al. 2018). Five key concerns about climate change-related risk have been visualized in Fig. 3, which summarizes how rising global temperatures may affect different systems under different degrees of warming (Mahony and Hulme 2012). This figure conveys clear messages.

While the figure takes a global perspective, each of the reasons for concern applies for the Arctic. The figure shows that climate risk is already heightened at the present level of global warming and that climate risk increase with increasing warming, slowly and gradually for some reasons for concern, rapidly and abruptly for others. A global warming of 2 °C or 3 °C is in many cases enough to push the selected human and natural systems into a future with significant and widespread impacts, risks, and even irreversible changes. Future levels of climate risk will thus depend on the rate, peak, and duration of warming (IPCC 2014), which in turn are largely determined by the future evolution of global anthropogenic emissions of greenhouse gases and other gases and particles.

Table 1 further illustrates how climate-driven changes in land-based processes can cascade further into increased risks for people and nature. In many cases, such climate risk exacerbate already existing challenges and will increase differences in vulnerability within and between populations.

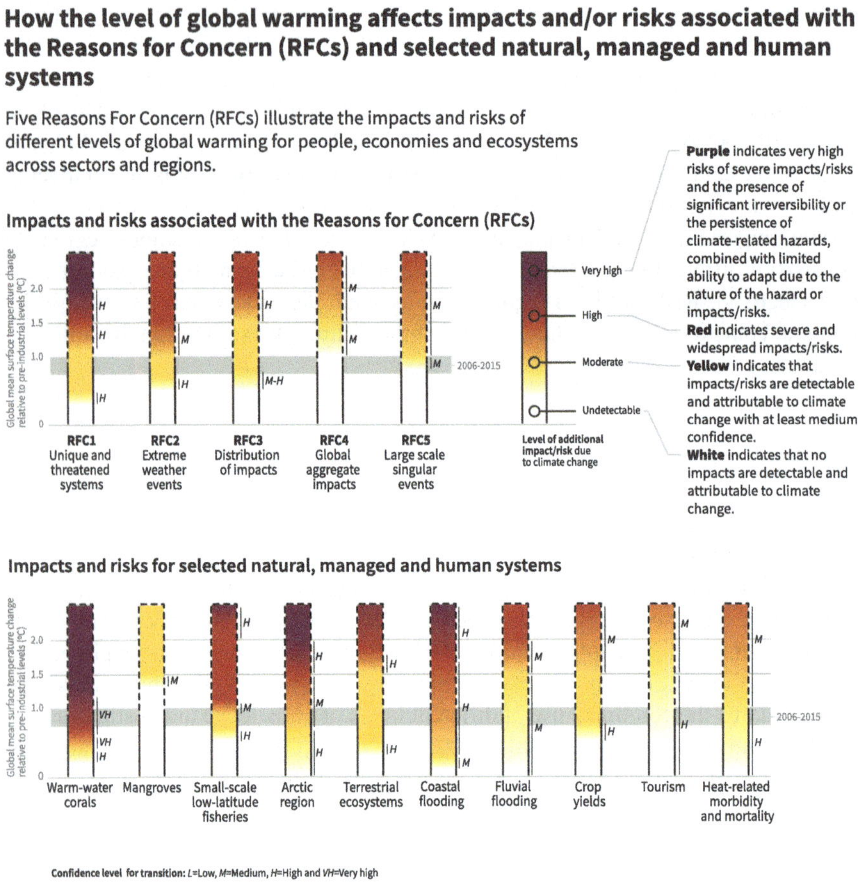

Fig. 3 Five integrative reasons for concern (RFCs) provide a framework for summarizing key impacts and risks across sectors and regions and were introduced in the IPCC Third Assessment Report. RFCs illustrate the implications of global warming for people, economies, and ecosystems. Impacts and/or risks for each RFC are based on assessment of the new literature that has appeared. As in AR5, this literature was used to make expert judgments to assess the levels of global warming at which levels of impact and/or risk are undetectable, moderate, high, or very high. The selection of impacts and risks to natural, managed, and human systems in the lower panel is illustrative and is not intended to be fully comprehensive (referring to IPCC report, sections 3.4, 3.5, 3.5.2.1, 3.5.2.2, 3.5.2.3, 3.5.2.4, 3.5.2.5, 5.4.1, 5.5.3, 5.6.1, Box 3.4). RFC1 Unique and threatened systems: ecological and human systems that have restricted geographic ranges constrained by climate-related conditions and have high endemism or other distinctive properties. Examples include coral reefs, the Arctic and its Indigenous people, mountain glaciers, and biodiversity hotspots. RFC2 Extreme weather events: risks/impacts to human health, livelihoods, assets, and ecosystems from extreme weather events such as heat waves, heavy rain, drought and associated wildfires, and coastal flooding. RFC3 Distribution of impacts: risks/impacts that disproportionately affect particular groups due to uneven distribution of physical climate change hazards, exposure, or vulnerability. RFC4 Global aggregate impacts: global monetary damage, global-scale degradation, and loss of ecosystems and biodiversity. RFC5 Large-scale singular events: are relatively large, abrupt, and sometimes irreversible changes in systems that are caused by global warming. Examples include disintegration of the Greenland and Antarctic ice sheets. Source: Figure SPM.2 from (IPCC 2018)

Fig. 4 Arctic greening or Arctic browning? Permafrost polygons show up in stark contrast against the tundra vegetation. Variations in moisture on the perimeter of the polygons encourage plant growth of a different color. Source: Western Arctic National Parkland photostream, Aerial view of polygons on tundra, flickr.com, public domain

Continuing at the current trend, global warming is likely to reach 1.5 °C between 2030 and 2052 (IPCC 2018). Current international mitigation commitments are insufficient to meet the Paris Agreement ambitions of limiting global temperature rise this century to well below 2 °C above preindustrial levels[1] and to pursue efforts to limit the temperature increase even further to 1.5 °C. Without further action, current commitments likely result in global warming of between 2.6 and 3.2 °C (Rogelj et al. 2016; Olhoff and Christensen 2018).

4.2 Natural Systems in the Arctic

4.2.1 Terrestrial and Freshwater Ecosystems

The Arctic and sub-Arctic consist of various types of ecosystems, including forests, tundra, wetlands, freshwater ecosystems, grasslands, and rocky- and snow-and-ice habitats on the arctic land areas, in addition to the marine ecosystems. Climate change impacts in these systems include both gradual changes and extreme events or pulse disturbances. Many of these changes will in turn feedback to enhance further climate change, also discussed in Sect. 3.

[1] (https://climateactiontracker.org/)

Table 1 Systems at risks due to climate change and cascading effects in land-based processes This table is based on (Hurlbert et al. 2019) and expanded with additional risks based on discussion and literature mentioned in the text of this chapter. Closed circles refer to effects discussed in this chapter, most with an Arctic relevance. Open circles are not discussed specifically in this chapter, most with a relevance not specific to the Arctic

Effects on land based processes from climate change:

Systems at risk:	Dry land water scarcity	Soil erosion	Vegetation loss	Wildfire damage	Permafrost degradatation	Tropical crop yield decline	Food supply instabilities
Food					●	○	●
Livelihoods	○	●	○		●	○	●
Value of land	○	●	○	○			●
Human health	○	●	○	●	●		●
Ecosystem health	○		●	●	●	○	
Infrastructure	○	●		○	●		

● Effects discussed in the text ○ Effects *not* discussed in the text

The terrestrial ecosystems already experience large vegetational shifts as a result of changes in permafrost, snow cover, temperature, precipitation, and hydrological changes but also as a consequence of human land use changes (Barry et al. 2013).

The overall trend for tundra vegetation across the 36-year satellite period (1982–2017) shows an increasing aboveground biomass known as "greening" throughout most of the circumpolar Arctic. This greening process is driven by several factors, including increases in summer, spring, and winter temperatures and length of the growing season and soil moisture. Within the overall trend of such greening, some tundra areas show "browning" or declines in aboveground biomass (Bhatt et al. 2017). Research on tundra browning is more limited, but suggests causal mechanisms that include increased tundra freezing and drying due to reduced snow cover and permafrost thawing and subsidence that increases surface water. Also, insect activity and increased herbivore grazing play a role in tundra changes. A warming climate allows geometrid moths to appear more frequently and defoliate shrubs and bushes, while reindeer or other grazing animals can manage tundra vegetation and mitigate the transition from grassland to shrubland in a warming Arctic (Rybråten and Hovelsrud 2010; Olofsson et al. 2013). Changes in tundra vegetation can have important ecosystem effects, in particular on hydrology, carbon, and nutrient cycling and surface energy balance. Vegetation also has a stabilizing effect on permafrost, which can collapse and increase methane emissions in the absence of vegetation (Myers-Smith and Hik 2013; Frost and Epstein 2014; Nauta et al. 2015). Changing vegetation also influences the diversity and abundance of herbivores, such as caribou declines connected to Arctic greening (Fauchald et al. 2017; Horstkotte et al. 2017).

Studies on boreal forests similarly identify climate change-related changes. In some well-studied regions such as Alaska and western Canada, wildfires are more frequent and affect larger areas now than in the last 100 to 10,000 years ago (AMAP 2017c). Abrupt thawing permafrost is also increasingly disturbing boreal forests. Such changes to boreal forests also impact its role in the carbon cycle in absorbing CO_2 from the atmosphere via vegetation growth and thus contribute to further climate change (Glomsrød and Aslaksen 2006).

Arctic freshwater ecosystems are influenced by the duration of snow and ice cover, changes in water temperature and nutrient concentrations, and changing inputs from the catchments and surrounding terrestrial ecosystem (Wrona et al. 2013). Perhaps even more dramatic are the appearance and disappearance of entire freshwater ecosystems themselves. For example, while Arctic lakes are rapidly draining and disappearing following the loss of permafrost and the increased evaporation due to higher air temperatures, increased snow and ice melt and thawing permafrost may also increase the formation of swamps and new lakes. Also human development has its imprint on these freshwater changes, such as flow regulation on major northerly flowing rivers, which adds complexity to the determination of climate-driven changes. All these changes have implications for hydrology, the surrounding vegetation, and the animals depending on this (Post et al. 2009), as well as for livelihoods (Arctic Council 2013).

4.2.2 Marine Ecosystems

Climate change impacts on marine ecosystems, including sea ice in the Arctic, are affecting or expected to affect all marine mammal species through its effects on the productivity of plankton in the water column, of bottom flora and fauna, and of fish (AMAP 2017d). The composition of arctic marine species and their activity is changing due to gradual warming, acidification, and other changes in water layers linked to melting glaciers and sea ice. Similar to the changes on land, the natural habitat of arctic marine species is becoming smaller, while sub-Arctic species are expanding their range northward, further increasing pressure on high-arctic species (IPCC 2019b).

Reduced sea ice extent has increased the total annual primary production for the Barents Sea. As the ice edge migrates northward, phytoplankton blooms and its food chain linked zooplankton and krill communities follow this increasingly earlier migration ever further north (see also Sect. 1.4 on animal migrations). The IPCC (2019a) stressed with high confidence that the cascading effects of multiple climate-related drivers on polar zooplankton and phytoplankton have affected food web structure and function. Changes in species distribution in response to temperature and sea ice changes and such food web migrations are already being documented regularly in the Barents Sea region for invertebrates, fish, and animals at the top of the food chain including marine mammal species. In addition, ever more boreal species are moving into northern waters. Examples of this are the increased abundance of mackerel at the northern limit of its distribution and the decrease of boreal cod on its southern limit, but increase at its northern limit (Rijnsdorp et al. 2009). As a result of earlier and longer availability of phytoplankton, zooplankton, and krill, commercial fish stocks are larger than they have been in decades. Also, the distribution of sea bottom living organisms (benthos), such as algae, worms, bivalves, sponges, and sea stars, is changing. In Arctic waters with sea ice, their abundance is much lower than in areas further south. These bottom living species are especially affected by changes in surface salinity due to melting of sea ice and freshwater input from rivers. Further, the recent and sudden increases in macroalgal presence and the abundance of benthos in Arctic fjords are thought to be linked to climatic factors (Kortsch et al. 2012).

Coastal breeding and seasonal migrating marine mammals such as gray and harbor seals and several whale species are likewise positively affected, as these species are shifting their distributions northward following the retreat of the sea ice, as their food web moves northward and the open water season gets longer.

However, there are also many examples of how sea ice reductions are negatively impacting marine life, especially for sea-ice- or pack-ice-dependent species which rely on sea ice for survival and reproduction (AMAP 2017a). For example, shifts in the temporal and spatial distribution and availability of suitable areas of sea ice for ice-breeding seals have led to increased stranding and pup mortality in years with little ice (IPCC 2019b). Also ice edge and marginal ice zone-dependent bearded seals are negatively affected by increased migration distances and possible changes in prey composition and availability (AMAP 2017e). Pacific walrus females are no longer able to haul out on ice over the shelf in summer due to the retraction of the southern ice edge into the deep Arctic Ocean. This has led to unusually large herds and onshore stampedes, leading to increased mortality of walrus calves.

Many seabird species are also experiencing negative consequences due to changes in prey availability linked to ocean climate change, and for some populations also changes in predators play a role. Figure 5 gives an overview of key drivers that are causing, or are projected to cause, direct effects in arctic marine ecosystem.

However, patterns of species change in the marine ecosystem are complex, and the same species can be affected differently by warming waters and decreasing ice cover, while also non-climatic factors, such as the fishing industry, interfere with the population patterns (AMAP 2017e). Impacts to several of these species, such as walrus or polar bears are probably masked, as population trends are currently stable or positive, with the species still recovering from excessive hunting practices in the past.

4.3 Societal Systems in the Arctic

Changes in global and arctic climate, and linked ecosystem changes as described in this chapter, result in a quickly changing environment and how both local and

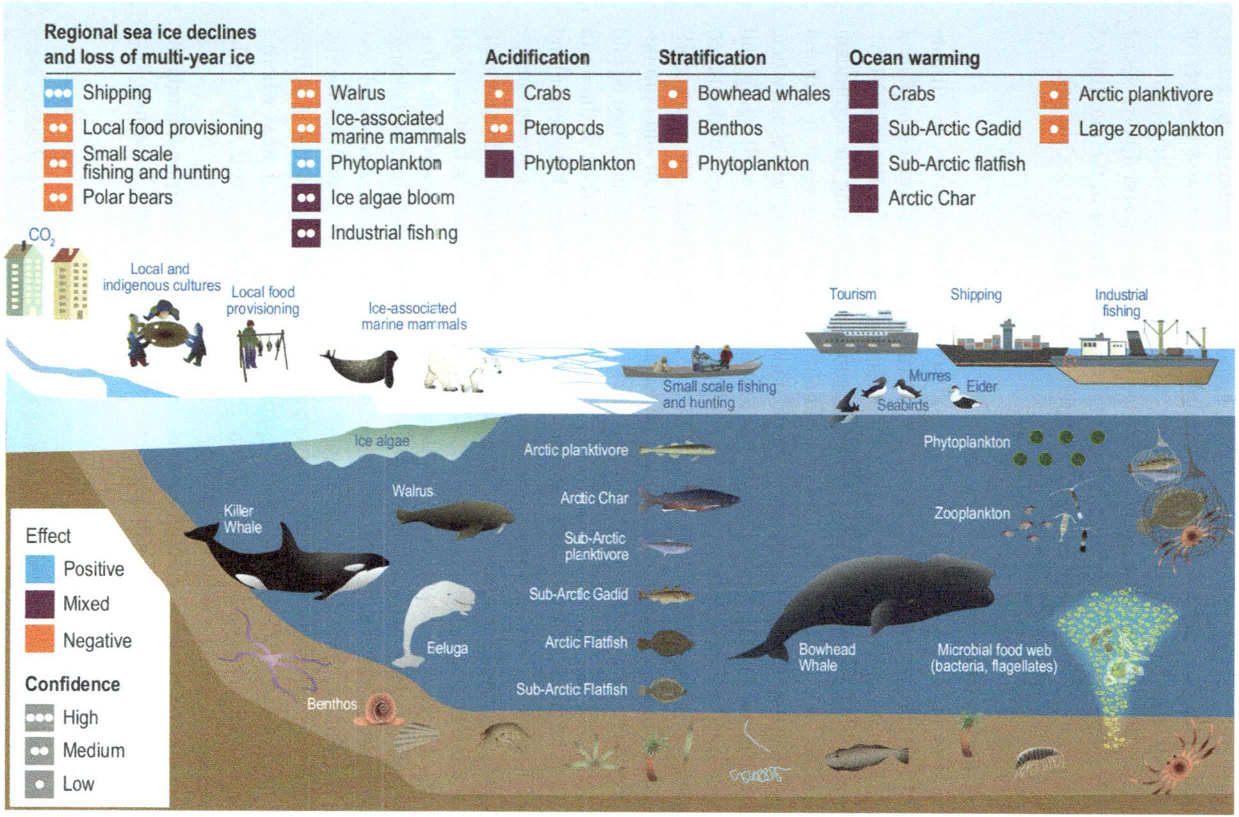

Fig. 5 Schematic summary of key drivers that are causing, or are projected to cause, direct effects on arctic marine ecosystems (Sect. 3.2.1.2). Effects presented here are described in the main text (referring to Meredith et al. 2019: Sections 3.2.3.1; 3.2.4.1.1; 3.2.4.2; 3.2.4.3) with associated confidence levels and citations. For mixed effects, no confidence level is given (see main text for details on how multiple drivers cause interacting positive and negative effects). Projected

seasonal migratory populations are connected to this environment. These changes may negatively affect ecosystem services, such as food provision, climate or flood regulation, and their cultural importance (IPCC 2019b). Such changes have implications for people's livelihoods and activities, their cultural practices, economies, education, and also health (Meredith et al. 2019). Current changes in Arctic societal systems are interlinked with globalization and many other socioeconomic drivers of change. While we acknowledge the complexity and diverse set of drivers, the following sections aim to highlight the role of climate in the interlinked issues of cultural identity, local economy, migration and relocation, and health.

4.3.1 Cultural Identity

The cultural identity of Arctic communities is affected in multiple ways by climate change. "Living in connection with nature" is identified as an important component of cultural identity and well-being in Arctic populations (Larsen and Fondahl 2015). Traditional activities such as hunting, herding, fishing, and gathering are significant as recreational activities for many Arctic inhabitants and also have a cultural importance to some of the indigenous Arctic populations. Subsistence activities, and the sharing of food, are considered vital to many households' well-being and cultural identity. While climatic variation and weather events have undermined opportunities to harvest and process such culturally important foods, the vulnerability to changes in such traditional activities and foods differs by region and community and is a function of geographic location, nature of climate change impacts, and human factors (Ford et al. 2018). These issues have been much studied in the north American and Canadian Arctic, and these studies found differences between inland and coastal locations, with the latter being the uniquely sensitive to climate change impacts due to permafrost and sea ice dynamics. Local innovation and leadership provide important adaptations to minimize climate impacts on these activities (Ford et al. 2018). Nevertheless, landslides, flooding, and other events compromise infrastructure, and changing snow and ice regimes, less predictable weather, and changing wind patterns make travel by semipermanent trails more dangerous and less dependable (Ford et al. 2018). These changes affect community viability, identity, and together with globalization trends lead to partial outmigration from communities (Larsen and Fondahl 2015). This has consequences especially in rural areas where a reduction in population makes the maintenance of infrastructure, health, and social services more challenging (AMAP 2017a).

Also, in the absence of adaptation options, reduced access to traditional foods can have a negative effect on the nutritional status of Indigenous people for whom local

---

Fig. 5 (continued) effects are conceptual representations based on high emission scenarios (Sect. 3.2.1.2). The cross-sectional view of the Arctic ecosystem shows the association of key functional groups (marine mammals, birds, fish, zooplankton, phytoplankton, and benthic assemblages) with arctic marine habitats. Species depicted in the fishing net are not a comprehensive depiction of all target species. Source: Figure 3.5 from (Meredith et al. 2019)

Fig. 6 Arctic sea ice. This picture was taken in August 2016, a year when the summer sea ice extent was at a record low. Source: L'Oriol, flickr.com, public domain

food sources can make up important contributions to a nutritious diet (Larsen and Fondahl 2015; IPCC 2019b).

Especially in the North American and Russian contexts, large-scale industrial resource development facilitated by an increasingly accessible Arctic also threatens the preservation of the local cultural identity of many communities, mainly due to environmental pollution or the limitations these industrial developments create in terms of access to hunting, fishing, and gathering areas (Larsen and Fondahl 2015). Increased mining and petroleum extraction activities in Arctic Russia amplify the pressure on land use and reduce the territory of the Indigenous peoples with consequences, for example, for reindeer herding across the tundra. Changing environmental conditions and increasing competition over resources lead to loss of social identity and an increased risk for social alienation and destructive behaviors (Revich et al. 2012; Larsen and Fondahl 2015).

Local Economy

Climate change impacts the economies of the North mostly through an increased accessibility to the Arctic. Its resources become more accessible as sea ice melts, permafrost thaws, and ice-free seasons are getting longer, both to local communities and to companies from beyond the region. The economic benefits may thus be spread, contested, and may come at the cost of loss of these areas to other activities, such as reindeer herding in the example given above.

Also the example of migrating fish stocks as described in Sect. 4.2 indicates both positive and negative economic consequences. New species, including for consumption, will migrate north from southern waters, with new fishing opportunities in Arctic waters. However, migrating species may also make coastal fisheries more challenging as they need to travel further and longer to reach migrating stocks. Warmer and wetter weather conditions may further challenge industries such as the dried cod industry through increased microbial spoilage (Raheem and Crosta 2019) and increased risk for infectious pancreatic necrosis (IPN) virus and nodavirus causing viral neural necrosis (VNN) (Raheem and Crosta 2019).

These changes affect local communities, but impacts are not limited to Arctic inhabitants alone. For instance, sea ice reduction has facilitated an increase in cruise tourism opportunities across the Arctic. Alaska attracts the highest number of cruise passengers annually at just over 1 million; Svalbard attracts 40,000–50,000; Greenland 20,000–30,000; and Arctic Canada 3500–5000 (Johnston et al. 2017). Compared to a decade ago, there are more cruises on offer, ships travel further, larger vessels with more passengers are in operation, more purpose-built polar cruise vessels are being constructed, and private pleasure crafts are appearing in the Arctic more frequently (IPCC 2019b). The Arctic Ocean's open water season has already increased by 1–3 months over much of the ocean since the late 1970s, creating an increase in marine shipping (Meredith et al. 2019) and thus jobs, though not necessarily for the Arctic communities (AMAP 2017a). Most cruise ship companies are based outside the Arctic, and most of the revenue will consequently leave or even enter the Arctic.

In the Norwegian and Russian Arctic, most of this increase is associated with oil and gas service vessels and tankers, marine cruises, while traffic in the Canadian Arctic is dominated by cargo vessels, government icebreakers, and research ships, while pleasure craft presents the fastest growing vessel type. The impacts of different types of shipping are likely to vary in scale and scope between different communities in the same arctic region, but a general concern is the increased impacts on the environment such as noise, pollution, invasive species, and garbage disposal (Olsen et al. 2019). There are variations across the Arctic in how local populations (can) deal with these challenges. Environment protective regulations can limit residents' use of the environment, but regulations and communication can also improve local preparedness (Olsen et al. 2019).

In Fennoscandia, there is a relatively large local control over local resources. Exceptions to this are hydropower, minerals, and forest, which are under national, large state-owned companies, and environmental law control. However, these state-owned companies often include plans for local employment or local redistribution of part of the revenues (Lim 2020). In Russia a history of central state control over Arctic lands limits the amount of local control, but in recent years, the Russian government has passed several strategic planning programs designed to invest billions of dollars in northern infrastructure and economy (Lim 2020). The effectiveness of these measures is unclear, but some parts of the Russian Barents area have seen as influx of capital (e.g., Nenets Autonomous Area, Murmansk). In the Canadian and American Arctic, there are similar discussions or models in place around local employment, taxation, and local redistribution of revenues.

Finally, climate change incidences and events pose threats to the infrastructure on which economic activities depend (AMAP 2017e). For example, the bearing capacity of building foundations has declined by 40–50% in some Siberian settlements since the 1960s, and the vast Bovanenkovo gas field in western Siberia has seen a recent increase in landslides related to thawing permafrost (AMAP 2017d). Table 2 gives an overview of the current and expected impacts of climate change across different economic sectors in the Arctic.

4.3.2 Migration and Relocation of People

Climate change-related environmental and societal changes also impact movements in Arctic and sub-Arctic communities at various geographic scales, including migration and relocation patterns. As described in the previous sections, seasonal tourism can cause large influxes of people, and seasonal migration related to natural resource development activities is common, e.g., in the Russian Arctic. A loss of cultural identity mixed with other factors can as discussed also lead to outmigration from rural Arctic communities. People move for many different reasons, and it is often difficult to isolate just one cause. However, there are settlements in the Arctic that are already impacted by climate change, such as coastal villages in Alaska threatened by relocation due to thawing permafrost related to coastal erosion and failing support for houses, water systems, and civil infrastructure (AMAP 2017c). Also, large urban settlements impacted by thawing permafrost are already an important concern in Russia (Larsen and Fondahl 2015). These combined factors may lead to outmigration and relocation as people seek opportunities and alternative housing and lives. In 2009, for example, 22% of the population of Alaska moved, which is well above the 15% of all Americans who moved that year, reflecting higher mobility of the Alaskan population in context of climate change in the Arctic (Larsen and Fondahl 2015). Furthermore, there is a general trend for women to leave and men to stay, creating a gender imbalance in the arctic regions (AMAP 2017e). Seasonal work migration related to natural resource development is likely to increase due to the increased accessibility of resources, possibly adding to the gender imbalance phenomenon in the Arctic.

4.4 Human Health

People in Arctic communities are particularly at risk of facing health issues (AMAP 2015). A rapid warming has gone hand in hand with increased globalization and accessibility, industrial activities, resource extraction, and an increased flux of people and new animal species (and their potential pathogens), exposing the Arctic population and communities to multiple new health challenges. Health impacts of climate change are both direct, resulting from changes in temperatures and/or extreme climate events, and indirect, resulting from how climate change affects livelihoods and foods, infrastructure, wildlife, the viability of parasites, bacteria and viruses, and thus health (Larsen and Fondahl 2015; IPCC 2019b). Especially some mental health risks linkaged with changes in social and natural systems have already been briefly mentioned in the preceding sections (Sect. 4.2.1). Here, we will especially highlight some of the direct climate links with physical health, local food security, and damaged healthcare infrastructure. We also highlight some

Table 2 Expected impacts of climate change on sectors in the Arctic

Sector/system	Consequence of climate change	Other forces affecting outcomes
Fishing	– Changes in stock and species abundance and distribution	Changes in: markets and demand, gear, policies, offshore development, transportation
	– Changes in stock productivity	
	– Differences for species and per region	
	– Coastal and marine fisheries, fish farming and drying will be differently affected	
Agriculture, livestock, forestry	– Vegetation shifts	Change in market value of meat; overgrazing; land use policies affecting access to pasture and migration routes; property rights; climate change; policy regarding carbon capture and storage in soil and above ground biomass
	– Expansion of forests	
	– Altered husbandry practices	
	– Probable expansion of agricultural opportunities	
	– Increased growing season and extreme weather events	
	– Wetland development	
	– Increased forest fires	
Renewables: Hydroelectric power	– Change in precipitation affecting the reservoirs and spring run-off	Changes in energy markets; environmental policies
	– Increased maintenance costs caused by thawing permafrost	
Non-Renewables: Mining and Petroleum	– Reduced sea ice and glaciers offering some new opportunities for development	Changes in policies affecting extent of sea and land use area; new extraction technologies; changes in markets and demand
	– Changes in hydrology, thawing permafrost and temperature affect production levels, ice roads, flooding events and infrastructure	
	– Shorter season for mining	
	– Offshore production likely to benefit from less sea ice	
	– Increased design and operational costs	
Tourism	– Warmer conditions, more open water, public perception of "last chance" opportunities	Travels costs; shifting tourism market; more enterprises
	– Reduced access costs	
	– Snow dependent activities more limited	
	– Longer season	
	– More cruise tourism	
	– Weather events, like storms, have unknown trends	

(continued)

Table 2 (continued)

Sector/system	Consequence of climate change	Other forces affecting outcomes
Maritime transportation	– Open seas allow for more vessels	Policies for safety requirements; timing and movements; changing insurance premiums
	– Reduced access costs	
	– Increase dependent on prevailing ice conditions	
	– Weather events, like storms, have unknown trends	
	– Greater constraints in using ice-roads	
Infrastructure on land	– Thawing permafrost affecting stability of ground and infrastructure	Weak regional and national economies; other disasters that divert resources; disinterest by southern-based law makers
	– Coastal erosion	
	– Increased costs from impacts of natural disasters	
	– Increased building costs	
	– Construction season extended	
Governmental services	– Increased demands on the federal budget associated with climate change	Other disasters/priorities that divert resources
	– Possible increased federal economic support	
Subsistence activities	– Change in harvest patterns	Changes in cost of fuel; land use affecting access; food preferences; harvesting rights; protection regulations
	– Changes in species distribution and abundance	
	– Impediments to access of harvesting areas	
	– Changes in seasonality	
	– Threats to culture and food security	

Sources: Glomsrød and Aslaksen (2006), Meredith et al. (2019)

indirect health risks linked to climate change, including new vector-borne diseases, the appearance of diseases that are not currently a problem in the Arctic, and the declining quality of surface water, largely affected by contaminants, causing problems for clean drinking water provision in some parts of the Arctic (AMAP 2017f). Pollutants, oil spills, and especially zoonotic diseases are discussed more in depth in other chapters in this book.

4.4.1 Direct Impacts of Climate on Health and Well-Being in Arctic Populations

The Arctic is becoming increasingly accessible, and this trend will continue in the future as climate changes. The pace of natural resource development such as oil and gas and mining has accelerated. The scale of other human activities, such as shipping and tourism, has likewise increased, and impacts of increased anthropogenic activity and related emissions are very likely to intensify toward 2030 and beyond (AMAP 2017e). Current impacts associated with the observed increase in Arctic shipping by boats include a higher rate of reported accidents per km traveled compared to southern waters, increases in vessel noise propagation, and air pollution, though the rate and seriousness of impacts differ per region (Meredith et al. 2019; Olsen et al. 2019). Factors that can potentially lead to a shipping accident are several and include inadequacy of navigation aids and port infrastructure, ice, and harsh weather conditions. The long distances to search and rescue facilities or dedicated spill response organization can increase the potential for impacts (Meredith et al. 2019). Disruptions to cultural and subsistence hunting activities from increased shipping compound climate-related impacts to local people (Olsen et al. 2019).

Climate change impacting infrastructure can also lead to risks for human health in the Arctic. For example, in many Arctic communities, physical infrastructure is built on permafrost. The weakening of this permafrost foundation will likely damage the structure build on top. It will also likely damage water intake systems and pipes, thus resulting in contamination of community water supplies. Moreover, the failure of the foundation of access roads, boardwalks, water storage tanks, and wastewater treatment facilities can turn water distribution and wastewater treatment systems inoperable. Changes in permafrost, erosion, precipitation, and extreme weather events may also affect infrastructure related to nuclear activities as well as key transportation routes (AMAP 2015). Failing infrastructure is also an issue for ice roads due to earlier melt of rivers and lakes, which may lead to direct health risks and in addition lead to changes in accessibility of remote locations. Thus, climate-related impacts on various infrastructures can pose risks for injuries, accidents, and death (Meredith et al. 2019).

Many Arctic residents rely on a stable climate for food storage and preservation methods which include aboveground air drying and smoking of fish and meat at ambient temperature, belowground cold storage on or near the permafrost, as well as fermentation. Loss of permafrost may result in spoilage of food stored belowground. The loss of these traditional food storage methods may also contribute to a shift from traditional foods based on hunting and fishing to a "western" type of diet (Parkinson et al. 2014), with people becoming increasingly dependent on often expensive, unhealthy choices, further affecting food security, nutritional status, and overall health (Revich et al. 2012; Larsen and Fondahl 2015; AMAP 2017e); some of the associated health risks are further discussed in the next section on diseases.

Mental health effects are already mentioned. The combined changes are leading to risk of marginalization, depression, stress, and distress (Glomsrød and Aslaksen 2006; Larsen and Fondahl 2015; Meredith et al. 2019). For example, due to the combined impacts of climate change and socioeconomic change on their traditional livelihood, Indigenous reindeer herders have a higher prevalence of suicide and mental health disorders than the national average over recent decades, particularly for young males in Fennoscandia and in Russia (Furberg 2016; AMAP 2017d).

4.4.2 Diseases and Infections

A warming Arctic is associated with a northward expansion of plants and animals, including their bacterial, viral, and parasitic flora. These associations will create favorable conditions for the emergence of infectious diseases in regions that were previously free of these pathogens (Revich et al. 2012). Changes in precipitation and temperature will also affect migration and survival of vector populations and influence the spread of vector-borne infections and zoonotic diseases, including to new hosts (Tryland et al. 2019; Omazic et al. 2019; van Oort et al. 2020). For instance, warmer temperatures may allow a parasite to survive in the environment for longer periods, increase an insect replication cycle, allow an infected host animal species to survive winters in larger numbers, increase in population, and expand their range of habitation, thus increasing the opportunity for transmission of the infection to humans (Hueffer et al. 2013). Thawing permafrost may also expose dormant diseases carried by bacteria or viruses (Legendre et al. 2014; Perron et al. 2015). An example of this includes an anthrax outbreak in Yamal—the first since 1941—which led to the culling of more than 200,000 reindeer and killed 1 human in 2016. This event was originally attributed to abnormally hot temperatures and thawing tundra where the bacterial spores had been dormant in permafrost (Luhn 2016). However, another study (Hueffer et al. 2013) identified a complex of factors at work, including extended periods of thawing in the active layer exposing contaminated soil and carcasses for longer durations, together with increased reindeer numbers and discontinued vaccination. Together, however, these studies suggest that the thawing of arctic soils could potentially lead to negative health impacts.

Studies on the potential of climate-related increase of vector-borne diseases include studies on tick-borne disease (TBD), while other vectors such as mosquitos and in-migration of host species from southern latitudes may similarly contribute to increased occurrence of zoonoses. TBDs such as tularemia, anaplasmosis, and tick-borne encephalitis or *Borrelia* are already occurring but are likely to increase and emerge in new places, as climate change in arctic regions is projected to lead to both wetter and warmer seasons, associated with climate- and management change-related vegetation changes and increased animal migration (van Oort et al. 2020). As several Nordic studies on ticks and TBD have shown, hard tick species (*Ixodes* spp.) have changed their distribution from coast to inland, and several hundreds of kilometers northward in the last decennia across the Arctic. Such vector migrations impose an increased TBD risk for potentially new hosts such as reindeer (van Oort et al. 2020).

Box 5 Tick-Borne Disease Spread

Cases of both anaplasmosis (tick-born fever) and babesiosis (piroplasmosis) have been reported in domesticated livestock in northern Norway, and fatal cases of anaplasmosis have been reported in reindeer in Norway. Ticks have also been found on moose, deer, and roe deer in coastal areas in northern

(continued)

Box 5 (continued)
Norway, and a crossover to reindeer is likely. Even if not infected, reindeer can be carriers of ticks enabling tick and TBD migration, as there are anecdotal reports of a single reindeer being infested with over 400,000 individuals of a tick species. The potential for reindeer as host would add to the challenges in reindeer herding, but it also increases the risk of tick and TBD dispersion to inland habitats when and where reindeer migrate (van Oort et al. 2020).

Other studies identified an increasing trend in the Arctic regions in the incidence of non-tick-related zoonotic diseases, which are linked to arthropod vectors, water, soil, wild, or domestic animals (Parkinson et al. 2014). Similar to TBDs, expansion of these diseases will likely follow the northward migration of the corresponding host animal- or bird populations, as climate induces changes in habitats and migration routes of wild animals, birds, and insects (Hueffer et al. 2013). An increase in winter temperatures improves the chances of animal or bird host survival, as well as the survival and replication rate of the many insect vectors, which can transmit infectious agents (Revich et al. 2012). A recent paper identified a list of 37 potential climate-sensitive infections. Of these, 28 are zoonotic and have a risk of expanding their distribution northward, in part due to a changing climate (Omazic et al. 2019).

Arctic animals can also host many microbial agents that can cause zoonotic infectious diseases, for example, rabies (foxes), brucellosis (especially ungulates), echinococcosis/cystic and alveolar hydatid disease (dogs/other carnivores and rodents), as well as *Trichinella spiralis* (bears, walrus, certain types of seal, and foxes) (AMAP 2015). All of these zoonoses were identified to be climate sensitive (Omazic et al. 2019). Small mammals across the Arctic have also been documented to be reservoirs for a range of zoonoses including tick-borne encephalitis virus, hemorrhagic fever with kidney syndrome, Q fever, tularemia, toxoplasmosis, pseudotuberculosis, leptospirosis, and other infectious diseases (Revich et al. 2012; Hueffer et al. 2013). Not all, but some of these diseases may be climate sensitive either through their vectors or through their hosts, and increased risk of exposure may thus be linked to climate change.

Ironically, management of climate-driven events may also contribute to pathogen spreading. An example is mitigation action to feed shortages due to increased icing events in semidomesticated reindeer, such as corralling and giving supplementary feed, which may increase the spread of pathogens (Tryland et al. 2019; Omazic et al. 2019). Here, climate is an indirect driver of increased disease or infections.

As discussed in the section on cultural identity, some Arctic communities can also be at risk of acquiring food-borne diseases, due to cultural traditions regarding subsistence hunting, gathering, food preparation, and preservation methods (Hueffer et al. 2013). Climate change could affect the incidence of food-borne diseases as higher temperatures may result in increased activity of temperature-sensitive food-borne intoxication. An example of this is botulism, for which the toxin production by the bacterium *Clostridium botulinum* is temperature dependent. In Arctic areas of

Canada, Alaska, and Greenland, a high incidence of botulism was attributed to the preparation of traditional fermented (or in fact "putrefied") foods, such as fish, seal, and whale (AMAP 2015). Another theory is the introduction of plastic wrapping, i.e., creating anaerobic conditions for *Clostridium botulinum*, which will be mentioned in a later chapter (DHSS 2017).

While the potential links with climate change in many food-borne diseases are not well studied, climate change may increase the risk of water-borne diseases in the Arctic via warming water temperatures, or via debris and bacteria being washed by surface water into drinking or affecting water-sourced foods. After periods of rapid snowmelt, bacteria can increase in untreated drinking water, with associated increases in acute gastrointestinal illness, and food-borne gastroenteritis is associated with shellfish harvested from warming waters (Meredith et al. 2019). Further, decreased access to water could be associated with a development of infections such as gastroenteritis, respiratory infections, and vector-borne diseases (Larsen and Fondahl 2015).

Probably as important as the introduction of new pathogens, is that climate change, globalization, and increased industrial activities induce shifts from traditional foods based on hunting and fishing to a "western" type of diet in many indigenous arctic communities. These communities are experiencing increased rates of malnutrition, obesity, diabetes, and cardiovascular diseases related to the overconsumption of these new foods (Larsen and Fondahl 2015).

4.4.3 Contaminants and Pollutants

Environmental toxins, contaminants, and pollutants are discussed in a separate chapter in this book. Here, we will focus on the climate linkages related to exposure and risks.

Climate change may affect the transport and processing of contaminants and pollutants into the Arctic and create new sources of local pollution. Presently, the largest source of contaminants and pollutants is emissions in the industrialized world following long-range transport. Transport into the Arctic varies from year to year depending on weather conditions and large-scale dynamics, such as the phase of the Arctic oscillation. Changes in atmospheric and oceanic circulation and precipitation patterns from global warming can affect the transport pathways from lower latitudes and hence the abundance in the Arctic. Climate-induced changes in the environmental levels of contaminants and pollutants can also be caused by change in the extent of ice-free ocean, which in turn affects the deposition and cycling of contaminants and toxins between the atmosphere and ocean.

Furthermore, current waste sites (chemical and mixed domestic wastes) build on permafrost may begin to leak their contents into water systems and onto land masses as increased permafrost thaw may lead to land subsidence. These events could increase human exposure to POPs (e.g., PCBs and some pesticides leaking from waste sites and dioxins and furans from waste burning) and could influence health outcomes (AMAP 2015).

4.5 Projections for Natural and Societal Systems

How climate change progresses depends on the pathways of greenhouse gas emissions and carbon uptake over time. Some potential changes in climate are described in Sect. 3.2. These different pathways of emissions and warming have different consequences for natural and societal systems. Even if emissions and consequent warming would be limited to 1.5 or 2 °C by 2100, emissions and warming may overshoot those targets to come down again *on their way* to that target. This is important, as by doing so, certain irreversible changes may occur.

There are several tipping points in arctic ecosystems (CarbonBrief 2020), thresholds where a tiny change could push the system into a completely new state. Exactly when these tipping points are reached is uncertain, and in some cases, the change is gradual and not abrupt. The IPCC (Meredith et al. 2019) notes that the cryosphere is already experiencing a number of globally consequential tipping elements and climate feedback loops that could permanently alter regional and global climate. In the next few decades, these will include mass loss from glaciers, permafrost thaw, and decline in Arctic snow cover and sea ice. A northward shift in boreal forests can also be added to this, seeing expansion into tundra to the north and dieback to the south.

Each of these changes constitutes a large ecosystem change, with consequences for its related species or human communities with their activities, traditions, livelihoods, etc. Several of these consequences have been discussed in the preceding sections, and many of the consequences are dependent on the direction of other drivers and on mitigation and adaptation actions.

Projected risks for ecosystems include continued climate-induced changes in the polar oceans, sea ice, snow, and permafrost. These will drive habitat and biome shifts, with associated changes in the ranges and abundance of ecologically important species. Projected shifts will include further habitat contraction and changes in abundance for polar species, including marine mammals, birds, fish, and antarctic krill. On arctic land, projections indicate a loss of globally unique biodiversity as some high arctic species will be outcompeted by more temperate species and very limited refugia exist (Meredith et al. 2019).

Projected permafrost thaw and a decrease in snow will affect Arctic hydrology and lead to generally warmer and drier conditions for the rest of this century. This will increase the risk of wildfire across most tundra and boreal regions, while interactions between climate and shifting vegetation will influence future fire intensity and frequency with impacts on vegetation and wildlife (Meredith et al. 2019). This may accelerate further ecological shifts, such as the expansion of tall shrubs and trees to cover 24–52% of Arctic tundra by 2050, and the changing composition of boreal forests as coniferous trees are increasingly being replaced by deciduous species normally found further south (AMAP 2017a; Meredith et al. 2019).

An increase and shifts in the growing season will have consequences for herbivores and further throughout the food web, including small-rodent cycles. For example, while plant growth starts earlier in response to warming, the timing of calving in caribou/reindeer has not—this means that the most nutritious early plants are arriving before calves and milk production need these. This mismatch in timing between resource need- and availability already now contributes to reduced production and survival of calves (Post et al. 2009).

Climatic change will continue to reduce seasonal ice cover in Arctic rivers, lakes, and ponds and thus increase water temperature and both shift and increase the length of the growing season (Wrona et al. 2013; Prowse et al. 2015). These changes will impact freshwater ecosystem productivity, from increased algal growth to earlier invertebrate emergence, to unknown cascading effects of these for fish and fishing activities (Wrona et al. 2013; Prowse et al. 2015). Climate-induced changes will probably also cause reductions in the populations of cold-water fish, especially salmonids (Wrona et al. 2013). Many warmwater fish on the other hand will expand their current range into northern habitats (Wrona et al. 2006).

Warming, ocean acidification, reduced seasonal sea ice extent, and continued loss of multi-year sea ice are projected to impact polar marine ecosystems through direct and indirect effects on habitats, populations, and their viability. Marine net primary production in the Arctic is projected to increase, modified by changing nutrient supply due to shifts in upwelling and stratification. As a result, 95% or more of the deep sea (3000–6000 m depth) seafloor area and cold-water coral ecosystems are projected to experience declines in benthic biomass under the high warming RCP8.5 scenario. Projected range expansion of subarctic marine species will increase pressure for high-Arctic species, with regionally variable impacts (IPCC 2019b).

Disaster risks for people, livelihoods, and ecosystem services in the Arctic are expected to increase due to projected changes in floods, fires, landslides, avalanches, unreliable ice and snow conditions, and permafrost changes. Risks can also be mitigated: for example, much of the arctic infrastructure is located in regions where permafrost thaw is projected to intensify by mid-century, and costs arising from permafrost thaw and related climate change impacts by 2100 can be halved by redesigning existing infrastructure. The distribution or abundance of harvested species will change, and this will also challenge fisheries governance, especially under high emission scenarios. While the local capacity to adapt to such changes is high, long-term societal changes and limited resources can undermine this capacity. At the same time, there are different visions on adaptation: in Inuit culture, long-term adaptive planning is seen as arrogance to assume you can predict the future, reducing the ability to react flexibly on situations as they arise (Ford et al. 2018).

Few studies in the region have examined how health outcomes will be affected by climate change, although it is generally expected that existing risks will increase in magnitude and frequency (AMAP 2017f). As the Arctic becomes more ice free, projections suggest that the level of industrial activity will increase. Increased oil and gas exploration and extraction, mining, shipping, and other human activities can contribute to larger local sources of pollution and contaminants to the air. Additional risks to ecosystems and health arise from oil spills. An opening Arctic with increased shipping and industrial activities will increase the risk for a continued increase of bioaccumulation of persistent organic pollutants (POPs) and mercury in marine plants and animals. Consequently, nutritional health will be increasingly at risk due to shifts in shifts in diets and food systems caused by social and economic changes and climate change as discussed in Sect. 4.3.

4.6 Way Forward: Resilience, Adaptation, and Governance

The consequences of climate change in the Arctic depend on global emission pathways, the likelihood of the projected impacts, the vulnerability and resilience of the ecosystems and communities, and on the (limits to the) adaptive actions that can be taken either preventively or after an event has taken place.

The Arctic and its indigenous cultures have evolved in a highly variable environment. Nomadic lifestyles and ways of making decisions that include attention to diversity in food sources and subsistence practices have been important sources of resilience when environmental conditions vary. Forced settlement, loss of land, and management strategies that do not allow for diversity have eroded some of this flexibility (Arctic Council 2013). Other policies have also eroded traditional institutions, practices, languages, while local knowledge is undermined by a drastically changing environment. The current Arctic is undergoing many changes, and the perception of highly adaptive northerners may no longer be valid (Arctic Council 2013).

Nordic regions have a generally high resilience with respect to economy, technology, access to information, resources, knowledge and skills, infrastructure, enabling institutions, equity, and attitudes and perceptions of change. Also, in other arctic regions, there are examples of adaptive actions at different governance levels. In Canada, the government of Nunavut developed a strategic adaptation plan in 2011, the Canadian government targeted funding for vulnerability assessments and adaptation planning, and Inuit organizations are lobbying on climate change action domestically and internationally. The Greenland government has a focus on the integration of climate change into sector plans for fishing and hunting, shipping, and agriculture and in 2015 decided to mainstream climate change adaptation into sectoral planning (AMAP 2017f). In the Bering/Chukchi/Beaufort Region, adaptation is underway both informally and as a result of proactive planning (AMAP 2017g).

Some actions may have an explicit focus on climate change, but in most cases, responses to climate change are incorporated into other goals related to improving community health, housing, self-reliance, and sustainability, illustrative for the larger context of change and adaptation. Some adaptation actions are immediate, such as changes to when, what, and where people hunt, fish, and gather food. Others involve the development and use of better technologies to help in these activities. Some communities are engaging in formal and informal monitoring, to better detect changes and understand their trends in order to project future conditions (AMAP 2017g).

Challenges to implementation of responses to long-term changes, such as shifts in the frequency and intensity of extreme events and their risks to important species and ecosystems in the Arctic, depend on a great number of factors. One such challenge is the different time horizons for climate change and governance. Governance is, in many contexts, too fragmented across administrative boundaries and sectors to provide integrated responses to the increasing and cascading risks from climate-related changes in the ocean and/or cryosphere. The current strengthening of capacities is not sufficiently rapid or robust to adequately address the scale of increasing projected risks. Financial, technological, institutional, and other barriers exist for implementing responses to current and projected negative impacts of climate-related changes, and ecosystem recovery rates are slower than the recurrence of climate impacts. Adaptive capacities differ between as well as within communities and societies. Recent reviews discuss adaptation progress in the Arctic requires new governance mechanisms and institutional frameworks. This includes inclusion consideration of adaptation across different levels of government reducing institutional fragmentation, inclusion of traditional knowledge and cultural values in adaptation planning, and cross-departmental mandates for integrating climate considerations into planning (Ford et al. 2018).

Health adaptation strategies to climate change are generally underrepresented in policies, planning, and programming. For instance, all Annex I members of the UNFCCC report their progress toward agreed climate goals through national communications every 4 years. While climate change is strongly linked to health, of the Fifth National Communications of Annex I parties to the UNFCCC, only 15% of the initiatives has an explicit human health component described. The Arctic is no exception to this global trend. Despite the substantial health risks associated with climate change in the Arctic, health adaptation responses remain sparse, with the health sector substantially underrepresented in adaptation initiatives compared to other sectors (Meredith et al. 2019).

However, some Arctic health adaptation efforts and actions have been undertaken at the local level, focusing on increasing awareness of the health impacts of climate change and conducting vulnerability assessments. Depending on the region and risks, adaptation at the local scale is broad. In the Canadian (Nunavut) context, for example, adaptations range from community freezers to increase food security, to community-based monitoring programs to detect and respond to health impacting climate events, and to Elders mentoring youth in cultural activities to promote mental health (Meredith et al. 2019). To reduce ecosystem change-related health impacts, response options could be strengthened through maintaining networks of protected areas, e.g., as a barrier to range shifts in species. Also, precautionary approaches, such as rebuilding overexploited or depleted fisheries, regular monitoring of stocks and risks, and updating measures over time, could function as a buffer to negative climate change impacts for both species, fisheries, livelihoods, and regional economies.

Adaptation to climate change takes place in the context of other changes and challenges, and there is no "one-size-fits-all" solution for adaptation within the Arctic regions (AMAP 2017c). It requires case-specific consideration of each community within the context of the unfolding local interactions between environmental, cultural, social, and economic conditions (AMAP 2017c). Where adaptation actions are limited, profound economic and institutional transformative change is needed to enable climate-resilient development. Intensified cooperation and coordination among governing authorities across scales, jurisdictions, sectors, policy domains, and planning horizons can enable such effective responses. Adaptation and transformation will benefit from institutional arrangements that provide strong multiscale linkages with local and Indigenous communities and combined scientific and indigenous knowledge. Governance systems in the Arctic regions vary substantially from one country to another and are changing: the decentralization underway in Canada and Russia and self-government agreements with Indigenous peoples will be important for adaptation and the development of more resilient communities, because they vest more authority in local and regional governments, which is where key decisions about effective adaptation actions are likely to be made and enacted (AMAP 2017g).

Sustained monitoring is important to avoid adverse impacts on human health, food security, agriculture, wildfires, tourism, conservation, droughts, and floods. Prioritizing measures which address social vulnerability and equity will promote fair and just climate resilience and sustainable development and reduce negative consequences from climate impacts. There is high confidence that ambitious adaptation, including governance for transformative change, has the potential to reduce risks in many locations. However, adaptation limits (e.g., biophysical, geographical, financial, technical, social, political, and institutional) depend on the emission scenario and context-specific risk tolerance and are projected to be reached in more areas beyond 2100.

5 Concluding Remarks

The Earth is warming at an unprecedented rate, and the Arctic is warming about twice as fast as the global average. Climate change is affecting all regions, with consequences posing risks to society, health, infrastructure, and ecosystems. Also, in the Arctic, changes to the cryosphere and ecosystems cascade into consequences for flora and fauna and society. Climate is not the only driver of change in the region, and many climate impacts compound on other drivers such as increased accessibility and globalization, land use changes, and human impacts on ecosystems. These impacts and consequences are expected to intensify in the coming decades. Exactly how large the impacts will depend on the rate, peak, and duration of warming, while the risk is determined by local and regional variable impacts, vulnerability and resilience, adaptive capacity, and management actions of the exposed natural and societal systems.

The Arctic will see large changes which are beyond historical variations, and risks to health for both ecosystem and society are increasing. For climate and ecosystems, tipping points are critical and will determine the speed and reversibility of change, while for society, food and health are red threads connecting many of the ongoing and projected changes in climate and ecosystems.

Even with ambitious mitigation, many ocean- and cryosphere-dependent communities are projected to face challenges to adaptation limits (e.g., biophysical, geographical, financial, technical, social, political, and institutional) during the second half of the twenty-first century under high emission scenarios. Low-emission pathways limit the risks from ocean and cryosphere changes in this century and beyond and enable more effective responses while also creating co-benefits.

Acknowledgments We thank Eilif Ursin Reed at CICERO for his help in producing figures (Box 1, Fig. 2, Table 1). We also thank the photographers of Figs. 4, and 6 for making their photos available for use via the public domain. We like to thank the producers of Figs. 1, 3, and 5 for their permissions for using the figures in this chapter. Finally, we like to thank the reviewers for their important and valuable feedbacks, which put more details and nuances in the broad and complex topic of this chapter.

References

AMAP (2015) AMAP assessment 2015: human health in the Arctic. Arctic Monitoring and Assessment Programme (AMAP), Oslo, Norway

AMAP (2017a) Snow, water, ice and permafrost in the Arctic (SWIPA) 2017. AMAP

AMAP (2017b) Baffin Bay/Davis Strait region – overview report. AMAP

AMAP (2017c) Bering/Chukchi/Beaufort region – overview report. AMAP

AMAP (2017d) Adaptation actions for a changing Arctic (AACA)-Barents area overview report

AMAP (2017e) Adaptation actions for a changing Arctic. Perspectives from the Barents area. Arctic Monitoring and Assessment Programme (AMAP)

AMAP (2017f) Adaptation actions for a changing Arctic perspectives from the Baffin Bay/Davis Strait region

AMAP (2017g) Adaptation actions for a changing Arctic perspectives from the Bering-Chukchi-Beaufort region

Anthony KW, von Deimling TS, Nitze I, Frolking S, Emond A, Daanen R, Anthony P, Lindgren P, Jones B, Grosse G (2018) 21st-century modeled permafrost carbon emissions accelerated by abrupt thaw beneath lakes. Nat Commun 9(1):1–11

Arbo P, Iversen A, Knol M, Ringholm T, Sander G (2013) Arctic futures: conceptualizations and images of a changing Arctic. Polar Geogr 36(3):163–182. https://doi.org/10.1080/1088937X.2012.724462

Arctic Council (2013) Arctic resilience: interim report 2013. Stockholm Environment Institute and Stockholm Resilience Centre, Stockholm

Aschwanden A, Fahnestock MA, Truffer M, Brinkerhoff DJ, Hock R, Khroulev C, Mottram R, Khan SA (2019) Contribution of the Greenland ice sheet to sea level over the next millennium. Sci Adv 5(6):eaav9396. https://doi.org/10.1126/sciadv.aav9396

Ballinger TJ, Lee CC, Sheridan SC, Crawford AD, Overland JE, Wang M (2019) Subseasonal atmospheric regimes and ocean background forcing of Pacific Arctic Sea ice melt onset. Clim Dyn 52(9–10):5657–5672. https://doi.org/10.1007/s00382-018-4467-x

Barry T, Berteaux D, Bültmann H (eds) (2013) Arctic biodiversity assessment: status and trends in Arctic biodiversity. The Conservation of Arctic Flora and Fauna, Akureyri

Bhatt US, Walker, Donald A, Raynolds MK, Bieniek PA, Epstein HE, Comiso JC, Pinzon JE, Tucker CJ, Steele M, Ermold W (2017) Changing seasonality of panarctic tundra vegetation in relationship to climatic variables – IOPscience. Environ Res Lett 12(5)

Box JE, Fettweis X, Stroeve JC, Tedesco M, Hall DK, Steffen K (2012) Greenland ice sheet albedo feedback: thermodynamics and atmospheric drivers. Cryosphere 6(4):821–839. https://doi.org/10.5194/tc-6-821-2012

CarbonBrief (2020) Explainer: nine 'tipping points' that could be triggered by climate change. In: Carbon brief. https://www.carbonbrief.org/explainer-nine-tipping-points-that-could-be-triggered-by-climate-change. Accessed 27 Oct 2020

Crépin A-S, Karcher M, Gascard J-C (2017) Arctic climate change, economy and society (ACCESS): integrated perspectives. Ambio 46(S3):341–354. https://doi.org/10.1007/s13280-017-0953-3

Dannevig H, Hovelsrud GK, Husabø IA (2013) Driving the agenda for climate change adaptation in Norwegian municipalities. Environ Plan C: Gov Policy 31(3):490–505. https://doi.org/10.1068/c1152

DHSS (2017) Botulism in Alaska. Department of Health and Social Services (DHSS), Division of Public Health, Section of Epidemiology, State of Alaska

Fauchald P, Park T, Tømmervik H, Myneni R, Hausner VH (2017) Arctic greening from warming promotes declines in caribou populations. Sci Adv 3(4):e1601365

Ford JD, Couture N, Bell T, Clark DG (2018) Climate change and Canada's north coast: research trends, progress, and future directions. Environ Rev 26(1):82–92. https://doi.org/10.1139/er-2017-0027

Frost GV, Epstein HE (2014) Tall shrub and tree expansion in Siberian tundra ecotones since the 1960s. Glob Chang Biol 20(4):1264–1277. https://doi.org/10.1111/gcb.12406

Furberg M (2016) Towards the limits – climate change aspects of life and health in northern Sweden studies of tularemia and regional experiences of changes in the environment. Master thesis, Umeå University

Glomsrød S, Aslaksen I (2006) The economy of the north. Statistics Norway, Oslo

Hoegh-Guldberg O, Jacob D, Taylor M et al (2018) Impacts of 1.5°C global warming on natural and human systems. IPCC

Horstkotte T, Utsi TA, Larsson-Blind Å, Burgess P, Johansen B, Käyhkö J, Oksanen L, Forbes BC (2017) Human-animal agency in reindeer management: Sámi herders' perspectives on vegetation dynamics under climate change. Ecosphere 8(9):e01931–e01931. https://doi.org/10.1002/ecs2.1931

Hovelsrud GK, White JL, Andrachuk M, Smit B (2010) Community adaptation and vulnerability in Arctic regions. Community adaptation and vulnerability in Arctic regions 335–348. https://doi.org/10.1007/978-90-481-9174-1_14

Hovelsrud GK, Poppel B, van Oort B, Reist JD (2011) Arctic societies, cultures, and peoples in a changing cryosphere. AMBIO 40(Supp. 1):100–110

Hovelsrud GK, Karlsson M, Olsen J (2018) Prepared and flexible: local adaptation strategies for avalanche risk. Cogent Soc Sci 4(1):1460899. https://doi.org/10.1080/23311886.2018.1460899

Hueffer K, Parkinson AJ, Gerlach R, Berner J (2013) Zoonotic infections in Alaska: disease prevalence, potential impact of climate change and recommended actions for earlier disease detection, research, prevention and control. Int J Circumpolar Health 72(0). https://doi.org/10.3402/ijch.v72i0.19562

Huntington HP (2001) Arctic flora and fauna: status and conservation. Conserv Arctic Flora Fauna (CAFF)

Hurlbert M, Krishnaswamy J, Davin E, Johnson FX, Fernando C, Morton J, Viner D, Bailis R, Byers E, Calvin K, Diaz-Chavez R, Evans J, Fletcher A, Ford J, Manialawy Y, McElwee P, Quan J, Renwick A, Thiery W, Warner A, Rodrigues R, Ii BLT, Zikhali T (2019) Risk management and decision-making in relation to sustainable development. In: Climate change and land: an IPCC special report on climate change, desertification, land degradation, sustainable land management, food security, and greenhouse gas fluxes in terrestrial ecosystems, pp 673–800

IPCC (2013) Climate change 2013: the physical science basis. Contribution of working group I to the fifth assessment report of the intergovernmental panel on climate change. Cambridge University Press, Cambridge

IPCC (2014) Climate change 2014: impacts, adaptation, and vulnerability. Part B: regional aspects. Contribution of working group II to the fifth assessment report of the intergovernmental panel on climate change. Cambridge University Press, Cambridge

IPCC (2018) Summary for policymakers. In: Masson-Delmotte V, Zhai P, Pörtner H-O, Roberts D, Skea J, Shukla PR, Pirani A, Moufouma-Okia W, Péan C, Pidcock R, Connors S, Matthews

JBR, Chen Y, Zhou X, Gomis MI, Lonnoy E, Maycock T, Tignor M, Waterfield T (eds) Global warming of 1.5 °C. An IPCC special report on the impacts of global warming of 1.5 °C above pre-industrial levels and related global greenhouse gas emission pathways, in the context of strengthening the global response to the threat of climate change, sustainable development, and efforts to eradicate poverty. IPCC

IPCC (2019a) Summary for policymakers. In: IPCC Special Report on the Ocean and Cryosphere in a Changing Climate

IPCC (2019b) IPCC special report on the ocean and cryosphere in a changing climate. Intergovernmental Panel on Climate Change

Johnston M, Dawson J, De Souza E, Stewart EJ (2017) Management challenges for the fastest growing marine shipping sector in Arctic Canada: pleasure crafts. Polar Record 53(1):67–78. https://doi.org/10.1017/S0032247416000565

Kay JE, Holland MM, Jahn A (2011) Inter-annual to multi-decadal Arctic Sea ice extent trends in a warming world. Geophys Res Lett 38(15). https://doi.org/10.1029/2011GL048008

Kortsch S, Primicerio R, Beuchel F, Renaud PE, Rodrigues J, Lønne OJ, Gulliksen B (2012) Climate-driven regime shifts in Arctic marine benthos. Proc Natl Acad Sci USA 109(35): 14052–14057. https://doi.org/10.1073/pnas.1207509109

Kug JS, Jeong JH, Jang YS, Kim BM, Folland CK, Min SK, Son SW (2015) Two distinct influences of Arctic warming on cold winters over North America and East Asia. Nat Geosci 8(10):759–762. https://doi.org/10.1038/ngeo2517

Larsen JN, Fondahl G (2015) Arctic human development report. Nordic Council of Ministers

Legendre M, Bartoli J, Shmakova L, Jeudy S, Labadie K, Adrait A, Lescot M, Poirot O, Bertaux L, Bruley C, Couté Y, Rivkina E, Abergel C, Claverie JM (2014) Thirty-thousand-year-old distant relative of giant icosahedral DNA viruses with a pandoravirus morphology. Proc Natl Acad Sci U S A 111(11):4274–4279. https://doi.org/10.1073/pnas.1320670111

Lim KS (2020) Soft law instruments on Arctic investment and sustainable development. Polar Record

Luhn A (2016) Reindeer to be culled in Russia's far north due to anthrax outbreak | Russia | The Guardian. The Guardian

Mahony M, Hulme M (2012) The colour of risk: an exploration of the IPCC's "burning embers" diagram. Generations 6(1):75–89. https://doi.org/10.4245/sponge.v6i1.16075

Meredith M, Sommerkorn M, Casotta S, Derksen C, Ekaykin A, Melbourne-Thomas J, Muelbert MCC, Ottersen G, Pritchard H, Schuur EAG (2019) Polar regions. In: Pörtner H-O, Roberts DC, Masson-Delmotte V, Zhai P, Tignor M, Poloczanska E, Mintenbeck K, Alegría A, Nicolai M, Okem A, Petzold J, Rama, Wyer NM (eds) IPCC special report on the Ocean and cryosphere in a changing climate

Myers-Smith IH, Hik DS (2013) Shrub canopies influence soil temperatures but not nutrient dynamics: an experimental test of tundra snow-shrub interactions. Ecol Evol 3(11): 3683–3700. https://doi.org/10.1002/ece3.710

Nauta AL, Heijmans MMPD, Blok D, Limpens J, Elberling B, Gallagher A, Li B, Petrov RE, Maximov TC, Van Huissteden J, Berendse F (2015) Permafrost collapse after shrub removal shifts tundra ecosystem to a methane source. Nat Clim Chang 5(1):67–70. https://doi.org/10.1038/nclimate2446

Nilsson AE, Larsen JN (2020) Making regional sense of global sustainable development indicators for the Arctic. Sustainability 12(3):1027. https://doi.org/10.3390/su12031027

Nilsson AE, Carlsen H, van der Watt L-M (2015) Uncertain futures: the changing global context of the European Arctic. Report of a scenario-building workshop in Pajala, Sweden. SEI

Nilsson AE, Bay-Larsen I, Carlsen H, van Oort B, Bjørkan M, Jylhä K, Klyuchnikova E, Masloboev V, van de Watt L-M (2017) Towards extended shared socioeconomic pathways: a combined participatory bottom-up and top-down methodology with results from the Barents region. Glob Environ Chang 45:124–132. https://doi.org/10.1016/j.gloenvcha.2017.06.001

Notz D, Marotzke J (2012) Observations reveal external driver for Arctic Sea-ice retreat. Geophys Res Lett 39(8). https://doi.org/10.1029/2012GL051094

Olhoff A, Christensen JM (2018) Emissions gap report 2018. UNEP DTU Partnership

Olofsson J, te Beest M, Ericson L (2013) Complex biotic interactions drive long-term vegetation dynamics in a subarctic ecosystem. Philos Trans R Soc B 368(1624):20120486. https://doi.org/10.1098/rstb.2012.0486

Olsen J, Carter NA, Dawson J (2019) Community perspectives on the environmental impacts of Arctic shipping: case studies from Russia, Norway and Canada. Cogent Soc Sci 5(1):1609189. https://doi.org/10.1080/23311886.2019.1609189

Omazic A, Bylund H, Boqvist S, Högberg A, Björkman C, Tryland M, Evengård B, Koch A, Berggren C, Malogolovkin A, Kolbasov D, Pavelko N, Thierfelder T, Albihn A (2019) Identifying climate-sensitive infectious diseases in animals and humans in northern regions. Acta Vet Scand 61(1):53. https://doi.org/10.1186/s13028-019-0490-0

Overland JE, Hanna E, Hanssen-Bauer I, Kim SJ, Walsh JE, Wang M, Bhatt US, Thoman RL, Ballinger TJ (2019a) Surface air temperature. NOAA Arctic Report Card: Update for 2019

Overland JE, Wang M, Box JE (2019b) An integrated index of recent pan-Arctic climate change. Environ Res Lett 14(3):035006. https://doi.org/10.1088/1748-9326/aaf665

Parkinson AJ, Evengard B, Semenza JC, Ogden N, Børresen ML, Berner J, Brubaker M, Sjöstedt A, Evander M, Hondula DM, Menne B, Pshenichnaya N, Gounder P, Larose T, Revich B, Hueffer K, Albihn A (2014) Climate change and infectious diseases in the Arctic: establishment of a circumpolar working group. Int J Circumpolar Health 73:1–7. https://doi.org/10.3402/ijch.v73.25163

Perron GG, Whyte L, Turnbaugh PJ, Goordial J, Hanage WP, Dantas G, Desai MM (2015) Functional characterization of bacteria isolated from ancient Arctic soil exposes diverse resistance mechanisms to modern antibiotics. PLoS One 10(3):e0069533–e0069533. https://doi.org/10.1371/journal.pone.0069533

Post E, Forchhammer MC, Bret-Harte MS, Callaghan TV, Christensen TR, Elberling B, Fox AD, Gilg O, Hik DS, Høye TT, Ims RA, Jeppesen E, Klein DR, Madsen J, McGuire AD, Rysgaard S, Schindler DE, Stirling I, Tamstorf MP, Tyler NJC, van der Wal R, Welker J, Wookey PA, Schmidt NM, Aastrup P (2009) Ecological dynamics across the Arctic associated with recent climate change. Science 325(5946):1355–1358. https://doi.org/10.1126/science.1173113

Prowse T, Bring A, Mård J, Carmack E, Holland M, Instanes A, Vihma T, Wrona FJ (2015) Arctic freshwater synthesis: summary of key emerging issues. J Geophys Res Biogeo 120(10):1887–1893. https://doi.org/10.1002/2015JG003128

Raheem D, Crosta C (2019) Climate change, cod production and consumption. Curr Dev Arctic Law 7:15–22

Revich B, Tokarevich N, Parkinson AJ (2012) Climate change and zoonotic infections in the Russian Arctic. Int J Circumpolar Health 71(0). https://doi.org/10.3402/ijch.v71i0.18792

Rijnsdorp AD, Peck MA, Engelhard GH, Möllmann C, Pinnegar JK (2009) Resolving the effect of climate change on fish populations. ICES J Mar Sci 66(7):1570–1583

Rogelj J, Den Elzen M, Höhne N, Fransen T, Fekete H, Winkler H, Schaeffer R, Sha F, Riahi K, Meinshausen M (2016) Paris agreement climate proposals need a boost to keep warming well below 2 °c. Nature 534(7609):631–639. https://doi.org/10.1038/nature18307

Rybråten S, Hovelsrud GK (2010) Local effects of global climate change: differential experiences of sheep farmers and reindeer herders in Unjárga/Nesseby, a coastal Sámi Community in Northern Norway. In: Hovelsrud GK, Smit B (eds) Community adaptation and vulnerability in Arctic regions. Springer, Dordrecht, pp 313–333

Schuur T (2018) Permafrost and the global carbon cycle. In: second state of the carbon cycle report (SOCCR2): a sustained assessment report

Schuur EAG, McGuire AD, Romanovsky V, Schädel C, Mack M (2018) Arctic and boreal carbon. Second state of the carbon cycle report (SOCCR2): a sustained assessment report, pp 428–468

Shepherd A, Ivins E, Rignot E, Smith B, van den Broeke M, Velicogna I, Whitehouse P, Briggs K, Joughin I, Krinner G, Nowicki S, Payne T, Scambos T, Schlegel N, Agosta C, Ahlstrøm A, Babonis G, Barletta VR, Bjørk AA, Blazquez A, Bonin J, Colgan W, Csatho B, Cullather R, Engdahl ME, Felikson D, Fettweis X, Forsberg R, Hogg AE, Gallee H, Gardner A, Gilbert L,

Gourmelen N, Groh A, Gunter B, Hanna E, Harig C, Helm V, Horvath A, Horwath M, Khan S, Kjeldsen KK, Konrad H, Langen PL, Lecavalier B, Loomis B, Luthcke S, McMillan M, Melini D, Mernild S, Mohajerani Y, Moore P, Mottram R, Mouginot J, Moyano G, Muir A, Nagler T, Nield G, Nilsson J, Noël B, Otosaka I, Pattle ME, Peltier WR, Pie N, Rietbroek R, Rott H, Sandberg Sørensen L, Sasgen I, Save H, Scheuchl B, Schrama E, Schröder L, Seo KW, Simonsen SB, Slater T, Spada G, Sutterley T, Talpe M, Tarasov L, van de Berg WJ, van der Wal W, van Wessem M, Vishwakarma BD, Wiese D, Wilton D, Wagner T, Wouters B, Wuite J (2020) Mass balance of the Greenland ice sheet from 1992 to 2018. Nature 579(7798):233–239. https://doi.org/10.1038/s41586-019-1855-2

The IMBIE Team (2020) Mass balance of the Greenland ice sheet from 1992 to 2018. Nature 579(7798):233–239. https://doi.org/10.1038/s41586-019-1855-2

Tryland M, Ravolainen V, Pedersen ÅØ (2019) Climate change: potential impacts on pasture resources, health and diseases of reindeer and caribou. In: Tryland M, Kutz SJ (eds) Reindeer and Caribou: health and disease. CRC Press, pp 493–514

van Oort BEH, Hovelsrud GK, Risvoll C, Mohr CW, Jore S (2020) A mini-review of Ixodes ticks climate sensitive infection dispersion risk in the Nordic region. Int J Environ Res Public Health 17(15):5387. https://doi.org/10.3390/ijerph17155387

Wheeler HC, Danielsen F, Fidel M, Hausner V, Horstkotte T, Johnson N, Lee O, Mukherjee N, Amos A, Ashthorn H, Ballari Ø, Behe C, Breton-Honeyman K, Retter G, Buschman V, Jakobsen P, Johnson F, Lyberth B, Parrott JA, Pogodaev M, Sulyandziga R, Vronski N (2020) The need for transformative changes in the use of indigenous knowledge along with science for environmental decision-making in the Arctic. People Nat 2(3):544–556. https://doi.org/10.1002/pan3.10131

World Meteorological Organization (2017) WMO Guidelines on the Calculation of Climate Normals

World Weather Attribution (2020) Siberian heatwave of 2020 almost impossible without climate change

Wrona FJ, Prowse TD, Reist JD, Hobbie JE, Lévesque LMJ, Vincent WF (2006) Climate change effects on aquatic biota, ecosystem structure and function. Ambio 35(7):359–369. https://doi.org/10.1579/0044-7447(2006)35[359:CCEOAB]2.0.CO;2

Wrona FJ, Reist JD, Lehtonen H, Kahilainen K, Forsström L (2013) Freshwater ecosystems. Narayana Press

Loss of Untouched Land

Roland Pape

1 Introduction

Cold and remote, thus often considered less hospitable by us humans, the Arctic region has—so far—largely escaped the extensive direct human modifications that are experienced by the rest of the planet. However, how much 'untouched' land is left within the Arctic and why is it actually important to have? What are the looming threats to these 'wild' areas, and how could they affect the integrity of ecosystems, thereby undermining the entire socio-ecological system within the Arctic and, potentially, beyond? This chapter will provide answers to these questions, shedding light on the current development around the loss of 'untouched' land within the Arctic, which is triggered by the interaction of primarily external, but also of internal forces. Three different 'cases' are portrayed to exemplify the various human-induced impacts and their ecosystems' consequences throughout the Arctic region, primarily from a natural science perspective.

1.1 Coupled Socio-Ecological Systems and the Role of 'Untouched' Land

Society and nature form a coupled, socio-ecological system (Liu et al. 2007), within which both parts are interdependent and constantly coevolving. Such a coupling is argued to be especially strong within the Arctic, where many indigenous and non-indigenous people depend, to a greater or lesser extent, on the natural resources for some combination of food, clothing, shelter and spiritual fulfilment, as they have been for millennia (Forbes 2008). Also beyond the Arctic, our social and economic

R. Pape (✉)
Faculty of Technology, Natural Sciences and Maritime Sciences, Department of Natural Sciences and Environmental Health, University of South-Eastern Norway, Bø, Norway
e-mail: Roland.Pape@usn.no

M. Tryland (ed.), *Arctic One Health*, https://doi.org/10.1007/978-3-030-87853-5_5

systems heavily rely on the ecological capital and ecosystem services provided by nature. At the same time, our utilization and 'enhancement' of this ecological capital and these ecosystem services exert cumulative pressure on the natural system (Venter et al. 2016). Consequently, we are about to convert the natural system into a novel, (semi-)artificial state as soon as the inherent ability of this system to cope with human-induced pressures is exceeded. It is, however, especially the less-impacted, 'untouched' land, that contributes vitally to our well-being (Díaz et al. 2018), provides important ecosystem services (e.g. provisioning and control of food and water, carbon storage, and pollination; Watson et al. 2018), buffers against climate change (Martin and Watson 2016), and houses biodiversity (Di Marco et al. 2019). These so-called 'wild' areas are also important for human health, as being places of spiritual and mental renewal, exploration and wonder (Ewert et al. 2011). Moreover, they serve many local communities by sustaining traditions and long-term cultural connections with these places (e.g. Riggio et al. 2020).

Meanwhile, signs of human impact have been detected for 95% of the Earth's land surface, about 50% of it being even characterized as moderately to highly modified (Kennedy et al. 2019). As such, we have dramatically transformed the natural system and forced it into novel states. Now, knowing degradation, fragmentation and loss of habitat (all of which occur in the wake of this transformation) to be key drivers of biodiversity loss (e.g. Jacobson et al. 2019), accelerating detrimental effects on global biodiversity (e.g. Newbold et al. 2015) come not as a surprise. At the same time, however, biodiversity, by ensuring the complementarity of traits, acts as one major insurance against failures within natural systems (cf. Körner 2004). Once biodiversity is diminished, functionality and stability of the Earth's ecosystems are weakened (Steffen et al. 2015), which affect also the capital and services they provide and, in turn, the entire socio-ecological system (Millennium Ecosystem Assessment 2005). As such, the pace and the scale of human-induced environmental change, resulting in the loss of 'untouched' land, are nowadays undermining the conditions for environmental health and, subsequently, human health. The interlinkages between environmental and human health, however, remain poorly integrated within research, policy and practice (Ford et al. 2015).

1.2 The Arctic: Last of the 'Wild'?

It was already in 2002 that Sanderson and colleagues took advantage of the availability of global data sets to map human stressors across the world, facilitated by remote sensing ('Human Footprint': Sanderson et al. 2002). Globally consistent digital data (on, e.g. human population, infrastructure, and cropland) were combined to generate a high-resolution map (1 km) of summed human pressure, leading to a global 'last of the wild' map. Building off this earlier attempt, four up-to-date approaches to map human influence have recently been compared by Riggio et al. (2020). From the global perspective of this comparison, it is equivocally the Arctic that is least impacted by direct human activity, followed by the Boreal Forest. About 90% of the Arctic region show very low and 8.7% show low human influence

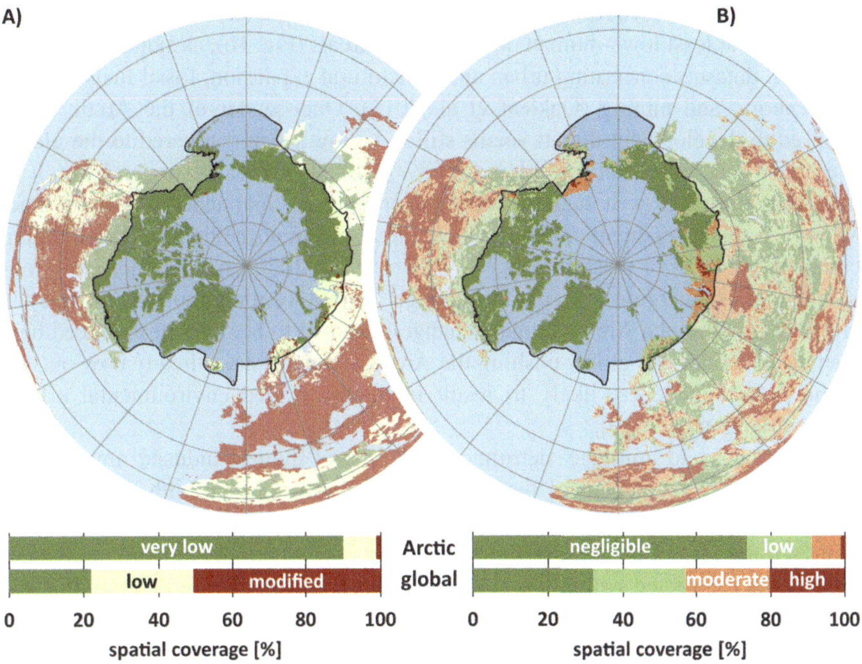

Fig. 1 Global maps of human modification (**a**) and development threats (**b**), revealing most of the Arctic region (as delimited by the Conservation of Arctic Flora and Fauna (CAFF) working group of the Arctic Council) to be 'untouched' by direct human impacts, but partly exposed to multiple development threats. Data: Kennedy et al. (2018), Oakleaf et al. (2015)

(Fig. 1a), based on the 'Global Human Modification' dataset of Kennedy et al. (2018) and a geographic delimitation of the Arctic region in accordance with the Conservation of Arctic Flora and Fauna (CAFF) working group. Compared to a global average of just 21.9% of very low impacted areas, the Arctic region is, thus, characterized by some of the largest continuous intact, 'untouched' ecosystems on Earth (Johnsen et al. 2010), ascribing the Arctic region indeed the role as the last of the 'wild'.

1.3 Human Impacts and Ecosystem Vulnerability in the Arctic: A General Appraisal

Even though major parts of the Arctic remained comparably untouched by direct human activity, some parts are facing increasingly larger human-induced environmental threats, including the full range of human stressors known from other parts of the world. Habitat loss and fragmentation from industrial development and infrastructure, unsustainable harvesting, pollution, and invading exotic (southern) species: they all exert their often cumulative and interacting impacts—also within the

Arctic (Johnsen et al. 2010). Consequently, almost one quarter of the Arctic region is subject to—at least low—human development threats (Fig. 1b), as synthesized from 50 global datasets concerning urban and agricultural expansion, fossil fuels, renewable energy, and mining (Oakleaf et al. 2015). One quarter of the Arctic being exposed to development threats seems still to be low when compared to the global average, where more than two thirds (or 68%) of the land area are under threat. It is, however, alarming as it mainly concerns previously 'untouched', pristine land. Moreover, this assessment of development threats accounts neither for the emerging cumulative effects of these threats (cf. Bowler et al. 2020), nor for those environmental threats emerging from climate change. With a warming more than twice the global rate within the Arctic, climate change is known to be more pronounced here than elsewhere on the globe, pushing the Arctic's climate state already now outside of previous experience, likely to result in unprecedented environmental effects (cf. Overland et al. 2019).

To be able to mitigate detrimental effects of human-induced impacts on ecosystems, an assessment of the ecosystems' ability to cope with such impacts ahead of these impacts would be needed. Consequently, Weißhuhn et al. (2018) proposed the idea of an ecosystem vulnerability assessment. Originally, the idea of vulnerability is rooted within social sciences, based on research on natural hazards that affect human structures and communities. Vulnerability is commonly described as a function of exposure, sensitivity, and adaptive capacity—expressing, generally speaking, a potential for loss. Within this equation of vulnerability, exposure refers to the probability of a hazard (or, more generally, disturbance) to occur, sensitivity is a measure of susceptibility to this disturbance, and the adaptive capacity characterizes the ability to cope with the disturbance and its consequences. When such a vulnerability assessment is applied to ecosystems, an environmental system like the Arctic moves from the traditional anthropocentric view as a source of hazard that influences human systems to a biocentric view of being a responding system that is affected by both natural and human drivers (Weißhuhn et al. 2018). Obviously, the increasing development threats, intermingled with those environmental threats that arise from the changing climate, result in an increased exposure of the Arctic—but what about the two remaining components within that equation describing ecosystem vulnerability: sensitivity and adaptive capacity?

Generally, the harsh Arctic environment, characterized by low temperatures, short growing seasons, low availability of nutrients, and—if at all—thin humus layers (e.g. Billings 1987), puts major constraints on plant growth and recovery after disturbance. As such, Arctic environments are considered particularly sensitive to human disturbance and slow to recover (e.g. CAFF 2013). Trampling effects might serve as an illustrative example for the general sensitivity of Arctic environments, as they remain not restricted to the plant cover but have cascading effects on ecosystem structure and functioning (Ravolainen et al. 2020). Not more than 25 trampling passes by humans are needed to cause visible signs within the vegetation (Tolvanen et al. 2001), while 500 trampling passes even led to cover losses in vegetation of approximately 50% in mesic tundra and 70% in moist to wet tussock tundra (Monz 2002). Moreover, trampling effects were still visible more than 60 years later, indicative of—if any—a slow recovery of the original vegetation after disturbance (e.g. Becker and Pollard 2016, Ylisirniö and Allén 2016). After trampling, the vegetation is rather trapped within an alternative stable state, where grasses, rushes, and sedges have replaced the original plant species and persist even for centuries after the initial impact (e.g. Egelkraut et al. 2018).

The third and last variable within the vulnerability equation, adaptive capacity, is constrained by the fact that terrestrial ecosystems within the Arctic are assumed to be less complex than their counterparts at lower latitudes (Callaghan 2005). Consisting of relatively few species in the food chains, with even fewer keystone species, implies that population changes in a single (keystone) species may have strong cascading effects across the entire ecosystem (Gilg et al. 2012; CAFF 2013). Many Arctic wildlife species have wide distributions with most habitats still intact, which actually imply adaptive capacity to some habitat loss from conversion, degradation and infrastructure. These species, however, respond to habitat patchiness and seasonality with a significant selection for certain localized key habitats across the landscape during certain times of the year or across years. Thus, their likely intersection with encroaching human activities puts them at a high risk of losing key habitats that are important to maintain a species' adaptive capacity within the highly stochastic Arctic environment (CAFF 2013).

Generally, the Arctic is affected by human stressors and resulting impacts that originate from a multitude of sources: some of which are indigenous to the Arctic, while most, however, have their origin partially or fully outside the Arctic region. Across these different origins, the following three sections will shed light on exemplary human-induced stressors that challenge the adaptive capacity of the Arctic ecosystems and, consequently, environmental and human health.

1.3.1 Resource Exploitation: Destructive by Nature?

The warming of the Arctic facilitates access to large portions of land that have previously been isolated or logistically unfeasible for development (IPCC 2013; Tolvanen et al. 2018). Within this area, which is considered a 'storehouse of resources' (e.g. Boyd et al. 2016), growing global demands, high market prices and improved technology have triggered extractive activities, pushed by new regulatory changes and policies that encourage further exploration and increased land use (e.g. Haley et al. 2011). According to earlier estimates of the US Geological Survey (USGS), 90 billion barrels of oil, 1669 trillion cubic feet of natural gas, and 44 billion barrels of natural gas liquids may remain to be found in the Arctic, accounting for about 22% of the undiscovered, technically recoverable resources in the world (Bird et al. 2008). Compared to hydrocarbon production, however, mining has the longest history across the Arctic, starting in the late nineteenth and early twentieth centuries (Emmerson and Lahn 2012). For 2012, the total value of nonferrous and noble metals mined from the Arctic amounted for 22.6 billion USD (Bortnikov et al. 2015), not to mention commodities from coal mining. As of 2015, there were some 373 mineral mines across Alaska, Canada, Greenland, Iceland, The Faroes, Norway (including Svalbard), Sweden, Finland and Russia, with the top five minerals being gold, iron, copper, nickel and zinc (Haddaway et al. 2019). As indicated by this brief outline, mining and other extractive industries are nowadays considered the key economic drivers within the Arctic (Tolvanen et al. 2018). An unprecedented exposure of the Arctic to human development threats is the consequence, rooted in the resource availability within the Arctic, but driven by economic interests that originate primarily outside the Arctic.

Extractive activities, including prospecting, exploration, construction, operation and maintenance, but also deconstruction and remains, are likely to affect social and environmental systems in a range of positive and negative ways, both directly and indirectly (Haddaway et al. 2019). As a source of employment, contributing to local and regional economies, they can yield a range of benefits to society. They are, however, also likely to cause conflicts—not least by having the potential to disrupt traditional ways of life in nearby communities, with adverse effects on human well-being and health. Moreover, being a disruptive industry by nature, its ecological footprint with negative impacts on the environment is large—originating from, among other factors, deforestation, erosion, contamination and alteration of soil profiles, contamination of local streams and wetlands, as well as an increase in noise level, dust and emissions and their cascading effects on the ecosystems (e.g. CAFF 2019).

Focusing solely at the effects of extractive activities per se, however, falls short to reckon the total dimension of associated environmental impacts. These impacts extend far beyond the area being 'consumed' by extractive activities themselves, i.e. exceeding the actual physical footprint of these activities. The infrastructure needed to prepare and support extractive activities, such as seismic lines, roads, housings, ports, railway tracks, airstrips and power lines not just adds to the overall physical footprint by their sheer presence (Haddaway et al. 2019) but also exerts a multitude of various indirect effects. These indirect ramifications, despite often being less apparent, may take effect across vast spatial expanses. Seismic lines, for instance, which are narrow corridors used to transport and deploy geophysical survey equipment during the delineation of oil and gas reservoirs, have been reported to be a major feature of the ecological footprint of hydrocarbon exploration. Although they are just temporary used, the heavy machinery used along these lines causes significant and persistent environmental changes regarding, among others, soil conditions, carbon storage and fluxes, as well as plant and animal species (Dabros et al. 2018; Raynolds et al. 2020). Such direct impacts remain confined to a rather small, local area: the line itself. Zoomed out, to the larger setting of the entire surrounding landscape, however, the spatial configuration of such lines takes additional effect. By forming tightly spaced grids at 50–100 m distance, the length of seismic lines per area unit may exceed by more than twice that of all other, manmade linear features (like roads, rail lines, power lines, and pipelines) taken together. As such, within the boreal forest of Western Canada, seismic lines accounted for 80% of all edge effects and associated fragmentation of the landscape, outmatching roads as the otherwise major agent of habitat fragmentation (Pattison et al. 2016). Roads—and especially their spatial network—have a pervasive multifaceted influence on ecosystems, including pronounced impacts on wildlife movements (Bischof et al. 2017). For small, relatively slow-moving species, they pose barriers, while, for larger animals, they act either as corridors that facilitate movement (for large predators like the wolf; Dickie et al. 2020) or as semi-permeable areas of risk that are avoided (e.g. for caribou/reindeer; Wilson et al. 2016).

The increasing net of artificial linear features, like seismic lines and roads, divides the habitat of species into smaller and more isolated pieces of land, separated by a matrix of human-transformed land cover. The loss of undisturbed area, increase in isolation and greater exposure to direct human impacts along fragment edges initiate long-term changes to the structure and function of the remaining fragments (Haddad et al. 2015). It often starts with seemingly small and harmless impacts, but the cumulative effects of a piecemeal development will, in the end, lead to devastating impacts on native species, accompanied by a loss of species richness, a total change to the landscape and the destruction of the area's wilderness heritage (CAFF 2001; EEA 2004).

Caribou and reindeer (*Rangifer tarandus*) are particularly well known to react sensitive to human infrastructure and activity, avoiding relatively large areas in the vicinity of human impacts—see Wolfe et al. (2000) for a general review and Eftestøl et al. (2019) especially for mining activities. At the same time, caribou migrate up to 1200 km a year round trip, clocking in at the world's longest terrestrial migration, driven by the low annual productivity of the vegetation in conjunction with the highly seasonal Arctic environment (Joly et al. 2019). Their dependence on suitable grazing grounds across those vast spatial expanses makes caribou and reindeer highly susceptible to the observed encroachment of human development within the Arctic region. The sustainable management of this keystone species of Arctic and sub-Arctic ecosystems, however, has important ecological, cultural, and societal implications (Pape and Löffler 2012). It forms the socio-ecological cornerstone of circumpolar indigenous cultures, from subsistence hunting of caribou by native peoples in Canada, Greenland and Alaska to reindeer husbandry by Sámi in northern Fennoscandia and numerous herding cultures across Siberia (Vors and Boyce 2009; Holand et al. 2021). Within Canada, nowadays 28 of 57 caribou populations are declining: a process driven by forestry, oil, gas and mining—and of major concern to Indigenous Nations (Collard et al. 2020). In Eurasia, reindeer husbandry relies, like the reindeer itself, on large areas of available land, as opportunities for a flexible use of land across space and time are at the foundation of its adaptive capacity (e.g. Kuoljok 2019). Like in North America, however, also in Eurasia, large spatial extents of grazing areas have been physically or functionally lost due to human impact.

Human impacts, however, are unevenly distributed across the circumpolar Arctic, with a few regions in North America and Russia experiencing most extensive threats due to extractive activities related to oil and gas development (Forbes et al. 2009). Among these regions is the Yamal-Nenets Autonomous Okrug, where approximately 90% of Russia's and 20% of the world's gas production are located. At the same time, it represents the world's most productive reindeer herding region with 631,000 domestic reindeer (Degteva and Nellemann 2013). Here, migratory Nenets and their large reindeer herds are nowadays forced to move between their summer pastures near the Kara Sea and their winter pastures south of the Bay of Ob back and

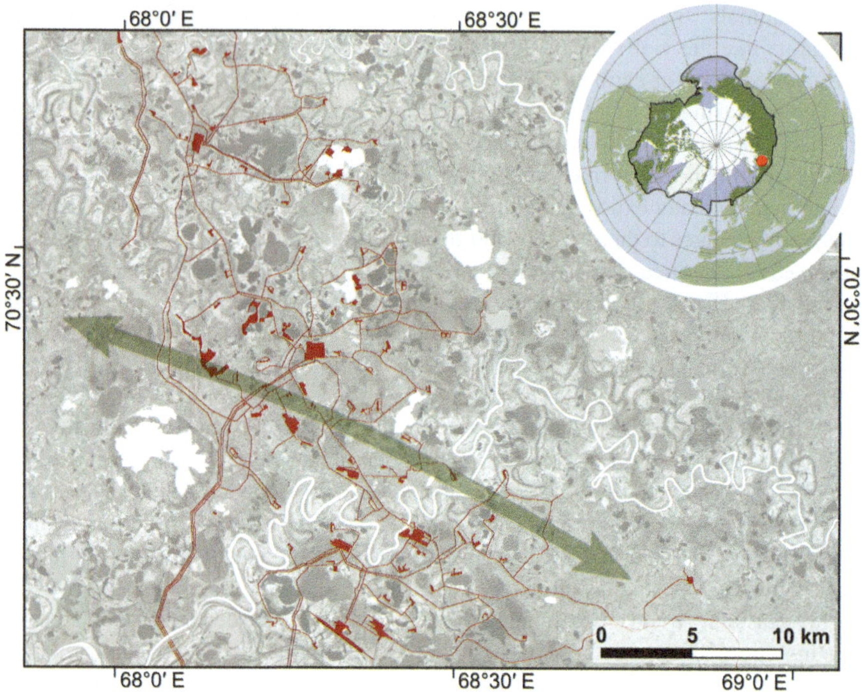

Fig. 2 Human-induced land cover change at the area of Bovanenkovo gas field within the Yamal-Nenets Autonomous Okrug, as detected based on satellite data (Landsat TM and Landsat 8). Detected change between 1985 and 2018 is marked in red, whereas the green arrow illustrates the migration route of the Nenets reindeer herders

forth across actively exploited fields, like the Bovanenkovo gas field (Fig. 2). Extensive development since the 1980s has led to blocked migration routes, degradation and loss of pasture area and loss of campsites. Reduced migration opportunities, quality and access to natural resources have affected the herding cycle, and, thus, the lives and well-being of indigenous herders in Yamal (Degteva and Nellemann 2013). Nevertheless, Nenets have proven adept in responding to a broad range of intensifying industrial impacts, primarily based upon the movements of people and animals in space and time being still unfettered (e.g. Kumpula et al. 2012). Further development, however, is likely to limit the proven adaptive capacity by constraining the opportunities of the reindeer herders for a flexible land use. In this context, it is not only crucial how much territory is affected by future development but what kind of territory and which migrations routes. Within an area dominated by mires and wetlands, for campsites, sacred sites and migration routes, the ongoing loss of high ground (where competition for land with industrial developers is particularly high) is particularly critical (Forbes et al. 2009, Skarin et al. 2020).

1.3.2 Reindeer and Reindeer Husbandry: Victim, Culprit, or Both?

Across terrestrial ecosystems, herbivores play generally a key role in shaping vegetation, soil nutrient dynamics, and the carbon cycle through complex herbivore–plant–soil interactions (Tuomi et al. 2021). Within the Boreal and Arctic region, reindeer and caribou have shaped ecosystem development since the last glacial epoch, ascribing them the role as an ecological keystone (e.g. Pape and Löffler 2012). Through grazing, trampling and excretion, reindeer and caribou affect vegetation abundance and cover, plant biomass, light limitation, plant nutrient levels, photosynthetic capacity and productivity. Apart from these effects on the vegetation, they influence soil properties and nutrient cycling by producing faeces and urine, altering the composition of the litter pool and soil temperatures, and affecting pore space and moisture of soils (Roy et al. 2020). The variety of effects that arises from interactions between herbivores and Arctic ecosystem functioning reveals a huge potential of both, exposure and susceptibility of the Arctic ecosystems to grazing by caribou and reindeer.

According to the keystone herbivore hypothesis of Zimov et al. (1995), grazers are manipulating their own food supply by forcing vegetation change from one state to another (i.e. between alternative stable states; cf. van der Wal 2006), thereby enhancing ecosystem processes like nutrient turnover and primary productivity. The change can be attributed to a number of drivers, such as reduced insulation of soils following a decrease in plant cover that results in higher soil temperatures and, subsequently, a higher nutrient turnover. Increased faecal nutrient input, and tolerance to repeated defoliation and trampling, allows graminoids (i.e. grasses, rushes and sedges) to outcompete shrub vegetation (Egelkraut et al. 2018). Evidence for this hypothesis arises from the large-scale transition of Beringian ecosystems from grass-dominated steppe to moss-dominated tundra following the extinction of mega-herbivores at the end of the Pleistocene (Bakker et al. 2016). Also the recently observed opposite trend in parts of the Arctic, from a shrub-dominated state to a more palatable, graminoid-dominated state, following an increase in reindeer grazing pressure (e.g. Forbes et al. 2009), fits well with this hypothesis.

Such shifts between alternative states do not necessarily refer to overgrazing (sensu Wilson and MacLeod 1991), at least as long as the productivity of the system is not negatively affected. For caribou/reindeer and reindeer husbandry, winter pastures and their lichen resources are often considered the bottleneck. Growing in snow-poor habitats—thus being easily accessible—ground and arboreal lichens constitute an important part of reindeer's winter diet if present (Gaare and Danell 1999). Their slow growth, however, makes lichens susceptible to overgrazing. Apart from their depletion due to grazing, ground lichens are especially prone to trampling: during a normal week of foraging, moving across dry 6- to 8-cm-thick continuous lichen mats, one single adult reindeer may trample a lichen volume corresponding to about a year's supply of lichen for that animal (Heggenes et al. 2017). Adverse effects of grazing on these sensitive resources are commonly mitigated by the reindeer's natural migratory behaviour between winter and summer pastures, which helps to avoid trampling of ground lichen resources during summer, and the implementation of additional pasture rotation methods and further measures into the management of herded animals.

This adaptive capacity of reindeer and reindeer husbandry to the sensitive pasture resources, which results from an unconstrained movement according to the prevailing conditions of pastures and climate, gets imperiled by the increasing loss of pastures areas (see also the previous section "Resource Exploitation: Destructive by Nature?"; Brännlund and Axelsson 2011). Originally a sustainable form of land use, reindeer husbandry, forced by the social-economic and political environment in which it is embedded, takes meanwhile a bivalent role regarding the loss of 'untouched' land. Being dependent on the availability of suitable pastures, it is clearly a victim of pasture loss. Despite the loss of suitable land, negative effects on reindeer husbandry in Sweden have, so far, largely been circumvented based upon increased supplementary feeding, changes in herd structure, more efficient herding practices and more intense use of the remaining pastures (Uboni et al. 2020). It is, however, especially this higher grazing pressure on the remaining pastures that challenges the role of reindeer husbandry as sustainable form of land use (e.g. Moen et al. 2010), ascribing it rather the role of being a potential culprit when it comes to the loss of 'untouched' land. It has been shown that overabundance of reindeer may lead to a substantial transformation of vegetation (e.g. Golovatin et al. 2012) that is even visible from space (see Fig. 3). With the closure of the Finnish-Norwegian border for reindeer migration in 1852, the herders on the Finnish side lost access to their former summer pastures, thus being forced to keep their reindeer year-round within areas that formerly constituted lichen-rich winter pastures (Brännlund and Axelsson 2011). Subsequently, these vegetation changes may result in both cascading ecosystem effects (e.g. Ims et al. 2007) and, as shown for parts of Norway, a loss of productivity within reindeer husbandry (Bråthen et al. 2007). Supplementary winter feeding, advocated as a potential solution, is practiced in Finland and spreading rapidly in Sweden and Norway. However, this will push reindeer husbandry into more static and confined herding systems, resembling farming more than pastoralism in the future (Holand et al. 2021).

1.3.3 Greening: Out of Arctic Control?

Tundra vegetation change, the so-called Arctic greening, has been identified as one of the clearest examples of the terrestrial impact of climate change (IPCC 2014). Time series analysis of satellite imagery since the early 1980s provides compelling evidence for an increase of the vegetation across the Arctic (Fig. 4). Technically, a measure for the relative amount of vegetation is provided by comparing the amount of visible versus infrared light being reflected from a surface and recorded by the satellite. Plants strongly absorb visible light while reflecting near-infrared light. As such, if there is much more radiation reflected in near-infrared wavelengths rather than in visible wavelengths, the surface is likely to be covered by dense vegetation. The resultant metric is known as the normalized difference vegetation index (NDVI), which serves as a proxy for both productivity and biomass of the vegetation (Pettorelli et al. 2005).

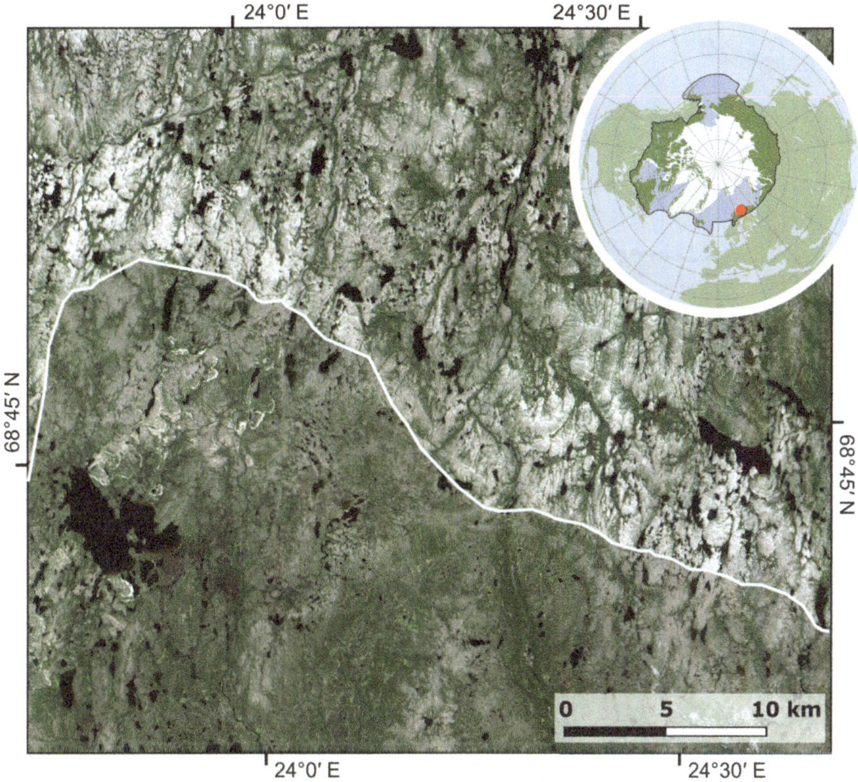

Fig. 3 Due to different grazing regimes, the Finnish-Norwegian border is clearly visible from space (data: Sentinel 2 from 2020). Continuous grazing and associated trampling on the Finnish side of the border (in the lower, southern part of the map) have led to a strong decline in lichens, while lichens are still abundant on the only seasonally grazed Norwegian side, as indicated by the whitish colours

In tundra vegetation, the NDVI is known to be strongly related to cover, leaf area and biomass of shrubs (Raynolds et al. 2012). Hence, increased biomass and coverage of dwarf shrubs, the so-called shrubification, has been identified as a key factor behind the observed greening of the Arctic (e.g. Myers-Smith et al. 2011). However, given the spatial heterogeneity of this trend, which is slowing and even showing signs of browning in many regions (cf. Fig. 4; Lara et al. 2018), experts cannot yet agree on the direction of change (Abbott et al. 2016).

Such changes in the shrub cover of the Arctic have important repercussions on ecosystems and their functioning—not least with global implications. The ecological consequences of an increase in shrub cover, both of deciduous and evergreen species, in tundra areas are manifold, although a differentiated view is needed (Vowles and Björk 2019). Taller and denser shrub patches generally reduce the reflection of sunlight (albedo), especially when encroaching lichen heaths (Aartsma et al. 2020). The latter have a higher albedo due to their whitish-yellowish colour

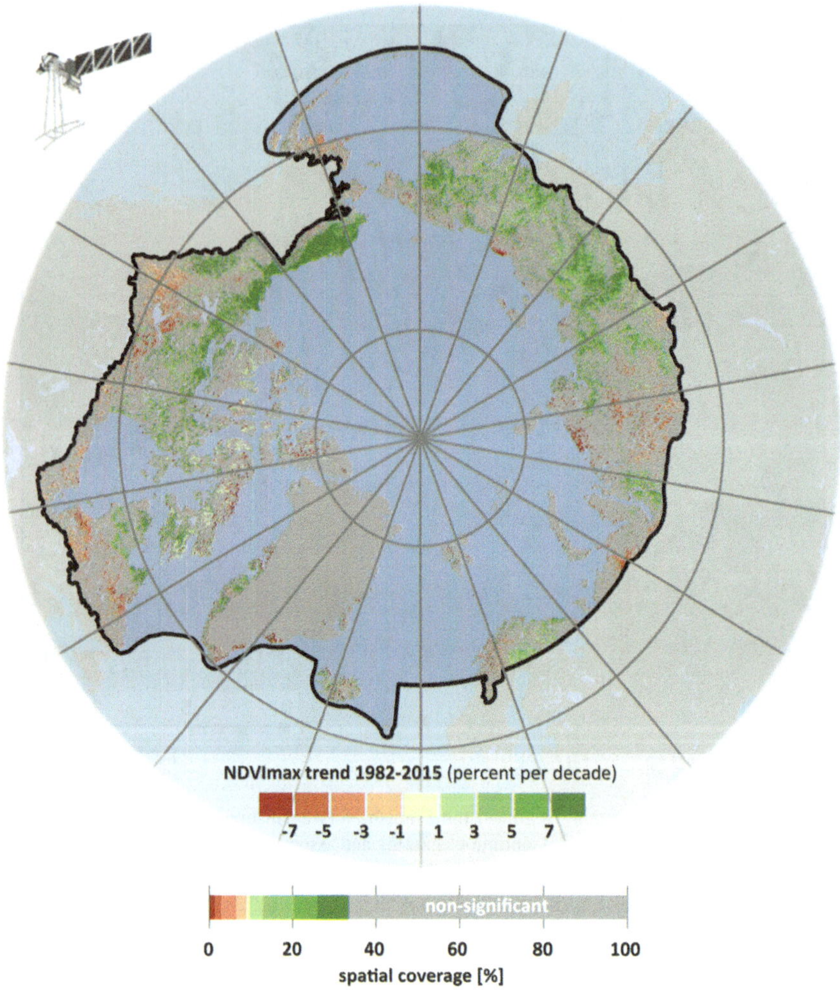

Fig. 4 Spatial patterns of greening and browning across the Arctic as derived from satellite data for the period 1982–2015 and visualized using greenish and reddish colours, respectively. About one third of the Arctic shows significant changes in vegetation, where greening, corresponding to an increase in coverage and biomass of shrubs ('shrubification'), is the prevalent trend (data: GIMMSv3, Tucker et al. 2005)

(cf. Fig. 3), so that their replacement by shrub heaths is likely to reinforce the warming of the Arctic. For erect deciduous species, such as birch (*Betula* spp.), willow (*Salix* spp.) and alder (*Alnus* spp.), their taller canopy traps more snow, which acts as insulation and raises winter soil temperatures (Sturm et al. 2005). In turn, these higher soil temperatures increase litter decomposition (Baptist et al. 2010), nitrogen mineralization (DeMarco et al. 2011) and winter respiration rates (Nobrega and Grogan 2007). The turnover of carbon gets further promoted by the increased input of more easily decomposed deciduous plant litter, which is even likely to be more important for nutrient cycling than higher soil temperatures (Vankoughnett and Grogan 2016) and contributing to changes in the (global) carbon balance. Expanding evergreen dwarf shrubs, like crowberry (*Empetrum nigrum*), have a markedly different effect on nutrient cycling and turnover of carbon. Owing to their low stature, they are unlikely to affect snow cover, or, in turn, soil temperatures. Moreover, different from deciduous species, many evergreen species produce organic compounds that are slow to decompose. As these compounds also contribute to the acidification of the soil, thus lowering microbial activity, nutrient cycling and the turnover of carbon gets further impeded (Adamczyk et al. 2016). As such, unlike for deciduous species, an expansion of evergreen species might facilitate carbon storage (Vowles and Björk 2019).

Apart from affecting nutrient cycling and carbon balance, shrubification is likely to have also strong effects on both, plant–plant and plant–animal interactions—hence affecting, at the end, the entire Arctic ecosystem. The increased coverage of shrubs is accompanied by a decrease in the diversity of understory vegetation (Pajunen et al. 2011), as plant species intolerant to shading, like many lichen species, are likely to disappear (Chagnon and Boudreau 2019). Moreover, shrubification has been linked to a higher risk of nest predation of several arctic-alpine bird species in tundra ecosystems, contributing to the observed decline of their populations (Ims et al. 2019). Large herbivores, like reindeer, have the potential to mediate or even halt the observed greening by selective browsing of deciduous shrubs (e.g. Verma et al. 2020); consequently, they are even proposed as 'a natural climate solution' (Macias-Fauria et al. 2020). However, a deterioration of pasture quality that may accompany the greening may also put these species at risk—as argued in the case of the declining caribou populations in North America (Fauchald et al. 2017).

Due to its cascading effect on the entire Arctic ecosystem, the observed greening of the Arctic, driven by increased shrub growth and abundance, is likely to be the first stage towards the disappearance of large tracts of what today is recognized as Arctic ecosystems (CAFF 2013). The actual driver behind this development, climate change, originates fully outside the Arctic and is, thus, out of Arctic control. Affecting, in contrast to the other stressors, the entire Arctic, it is also climate change that takes the lead over all stressors, which, cumulatively, contribute to the loss of 'untouched' land within the Arctic.

1.4 Loss of Untouched Land in the Arctic: A Synthesis

To constitute the last of the wild, where still about 90% of the land are characterized by very low direct human impacts, does not prevent the Arctic from being increasingly exposed to the full range of human stressors known from other parts of the world. Nowadays, extractive activities like mining and hydrocarbon production are the key economic drivers within the Arctic, facilitated not least by the warming of the Arctic—the latter of which actually forms a problem of its own. Native and originally sustainable forms of land use, like reindeer husbandry, are caught in between conflicting land uses and climate change and, due to that, sometimes forced beyond the limits of ecological sustainability. Hence, industrial development, land use and climate change—though to different proportions and, in the case of reindeer husbandry, involuntarily—act in concert to drive the loss of 'untouched' land in the Arctic. Thereby, human activities often take effect far beyond the actual physical footprint of those activities, due to various indirect and cascading environmental consequences. Within the Arctic, the increasing exposure to human disturbance in combination with a high susceptibility but low coping capacity of the natural system will inevitably lead to detrimental effects on environmental health. Moreover, originating from the rather tight coupling of nature and society in the Arctic, repercussions of this loss of 'untouched' land on the societal system seem inescapable, likewise with adverse effects also for human health.

References

Aartsma P, Asplund J, Odland A et al (2020) Surface albedo of alpine lichen heaths and shrub vegetation. Arct Antarct Alp Res 52:312–322

Abbott BW, Jones JB, Schuur EAG et al (2016) Biomass offsets little or none of permafrost carbon release from soils, streams, and wildfire: an expert assessment. Environ Res Lett 11:34014

Adamczyk B, Ahvenainen A, Sietio OM et al (2016) The contribution of ericoid plants to soil nitrogen chemistry and organic matter decomposition in boreal forest soil. Soil Biol Biochem 103:394–404

Bakker ES, Gill JL, Johnson CN et al (2016) Combining paleo-data and modern exclosure experiments to assess the impact of megafauna extinctions on woody vegetation. PNAS 113: 847–855

Baptist F, Yoccoz NG, Choler P (2010) Direct and indirect control by snow cover over decomposition in alpine tundra along a snowmelt gradient. Plant Soil 328:397–410

Becker MS, Pollard WH (2016) Sixty-years legacy of human impacts on a high Arctic ecosystem. J Appl Ecol 53:876–884

Billings WD (1987) Constraints to plant growth, reproduction, and establishment in Arctic environments. Arct Alp Res 19:357–365

Bird KJ, Charpentier RR, Gautier DL et al (2008) Circum-Arctic resource appraisal: estimates of undiscovered oil and gas north of the Arctic circle. U.S. Geological Survey Fact Sheet 2008-3049. USGS, Denver

Bischof R, Steyaert SMJG, Kindberg J (2017) Caught in the mesh: roads and their network-scale impediment to animal movement. Ecography 40:1369–1380

Bortnikov NS, Lobanov KV, Volkov AV et al (2015) Strategic metal deposits of the Arctic zone. Geol Ore Deposits 57:433–453

Bowler DE, Bjorkman AD, Dornelas M et al (2020) Mapping human pressures on biodiversity across the planet uncovers anthropogenic threat complexes. People Nat 2:380–394

Boyd R, Bjerkgård T, Nordahl B et al (2016) Mineral resources in the Arctic. Geological Survey of Norway, Trondheim

Brännlund I, Axelsson P (2011) Reindeer management during the colonization of Sami lands: a long term perspective of vulnerability and adaptation strategies. Glob Environ Change 21:1095–1105

Bråthen KA, Ims RA, Yoccoz NG et al (2007) Induced shift in ecosystem productivity? Extensive scale effects of abundant large herbivores. Ecosystems 10:773–789

CAFF (2001) Arctic flora and fauna: status and conservation. Conservation of Arctic Flora and Fauna International Secretariat, Akureyri

CAFF (2013) Arctic biodiversity assessment. Status and trends in Arctic biodiversity. Conservation of Arctic Flora and Fauna International Secretariat, Akureyri

CAFF (2019) Mainstreaming biodiversity in Arctic mining: challenges and proposed solutions. Conservation of Arctic Flora and Fauna International Secretariat, Akureyri

Callaghan TV (2005) Arctic tundra and polar desert ecosystems. ACIA. Arctic Climate Impact Assessment, Cambridge University Press, New York, pp 243–352

Chagnon C, Boudreau S (2019) Shrub canopy induces a decline in lichen abundance and diversity in Nunavik (Québec, Canada). Arct Antarct Alp Res 51:521–532

Collard R-C, Dempsey J, Holmberg M (2020) Extirpation despite regulation? Environmental assessment and caribou. Conserv Sci Pract 2:e166

Dabros A, Pyper M, Castilla G (2018) Seismic lines in the boreal and arctic ecosystems of North America: environmental impacts, challenges, and opportunities. Environ Rev 26:214–229

Degteva A, Nellemann C (2013) Nenets migration in the landscape: impacts of industrial development in Yamal peninsula, Russia. Pastoralism 3:15

DeMarco J, Mack MC, Bret-Harte MS (2011) The effects of snow, soil microenvironment, and soil organic matter quality on N availability in three Alaskan Arctic plant communities. Ecosystems 14:804–817

Di Marco M, Ferrier S, Harwood TD et al (2019) Wilderness areas halve the extinction risk of terrestrial biodiversity. Nature 573:582–585

Díaz S, Pascual U, Stenseke M et al (2018) Assessing nature's contributions to people. Science 359: 270–272

Dickie M, McNay SR, Sutherland GD et al (2020) Corridors or risk? Movement along, and use of, linear features varies predictably among large mammal predator and prey species. J Anim Ecol 89:623–634

EEA (2004) Arctic environment: European perspectives. Why should Europe care? Environmental issue report vol 38. EEA, Copenhagen

Eftestøl S, Flydal K, Tsegaye D et al (2019) Mining activity disturbs habitat use of reindeer in Finnmark, northern Norway. Polar Biol 42:1849–1858

Egelkraut D, Aronsson K-Å, Allard A et al (2018) Multiple feedbacks contribute to a centennial legacy of reindeer on tundra vegetation. Ecosystems 21:1545–1563

Emmerson C, Lahn G (2012) Arctic opening: opportunity and risk in the high north Lloyd's and Chatham House

Ewert A, Overholt J, Voight A et al (2011) Understanding the transformative aspects of the wilderness and protected lands experience upon human health. In: Watson A, Murrieta-Saldivar J, McBride B (eds) Science and stewardship to protect and sustain wilderness values: ninth world wilderness congress symposium, 6–13 Nov 2009, pp 140–146

Fauchald P, Park T, Tømmervik H et al (2017) Arctic greening from warming promotes declines in caribou populations. Sci Advanc 3:e1601365

Forbes BC (2008) Equity, vulnerability and resilience in social–ecological systems: a contemporary example from the Russian Arctic. Equity and the environment. Res Soc Probl Public Policy 15: 203–236

Forbes BC, Stammler F, Kumpula T et al (2009) High resilience in the Yamal-Nenets social–ecological system, west Siberian Arctic, Russia. Proc Natl Acad Sci 106:22041–22048

Ford AES, Graham H, White PCL (2015) Integrating human and ecosystem health through ecosystem services frameworks. EcoHealth 12:660–671

Gaare E, Danell Ö (1999) Use of pastures and area. In: Dahle HK, Danell Ö, Gaare E et al (eds) Reindeer management in Northwest-Europe 1998—biological opportunities and limitations. Nordic Council of Ministers, Copenhagen, pp 31–46

Gilg O, Kovacs KM, Aars J et al (2012) Climate change and the ecology and evolution of Arctic vertebrates. Ann N Y Acad Sci 1249:166–190

Golovatin MG, Morozova LM, Ektova SN (2012) Effect of reindeer overgrazing on vegetation and animals of tundra ecosystems of the Yamal peninsula. Czech Polar Rep 2:80–91

Haddad NM, Brudvig LA, Clobert J et al (2015) Habitat fragmentation and its lasting impact on Earth's ecosystems. Sci Advanc 1:e1500052

Haddaway NR, Cooke SJ, Lesser P et al (2019) Evidence of the impacts of metal mining and the effectiveness of mining mitigation measures on social–ecological systems in Arctic and boreal regions: a systematic map protocol. Environ Evid 8:9

Haley S, Klick M, Szymoniak N et al (2011) Observing trends and assessing data for Arctic mining. Polar Geogr 34:37–61

Heggenes J, Odland A, Chevalier T et al (2017) Herbivore grazing – or trampling? Trampling effects by a large ungulate in cold high-latitude ecosystems. Ecol Evol 7:6423–6431

Holand Ø, Moen J, Kumpula J et al (2021) Project ReiGN: reindeer husbandry in a globalizing north–resilience, adaptations and pathways for actions. In: Nord DC (ed) Nordic perspectives on the responsible development of the Arctic: pathways to action, pp 227–248

Ims RA, Yoccoz NG, Bråthen KA et al (2007) Can reindeer overabundance cause a trophic cascade? Ecosystems 10:607–622

Ims RA, Henden J-A, Strømmeng MA et al (2019) Arctic greening and bird nest predation risk across tundra ecotones. Nat Clim Chang 9:607–610

IPCC (2013) Climate change 2013: the physical science basis. Contribution of working group I to the fifth assessment report of the intergovernmental panel on climate change. Cambridge University Press, Cambridge

IPCC (2014) Climate change 2014: synthesis report. Contribution of working groups I, II and III to the Fifth Assessment Report of the Intergovernmental Panel on Climate Change, IPCC, Geneva

Jacobson AP, Riggio J, Tait AM et al (2019) Global areas of low human impact ('Low Impact Areas') and fragmentation of the natural world. Sci Rep 9:14179

Johnsen KI, Alfthan B, Hislop L et al (eds) (2010) Protecting Arctic biodiversity. United Nations Environment Programme, GRID-Arendal

Joly K, Gurarie E, Sorum MS et al (2019) Longest terrestrial migrations and movements around the world. Sci Rep 9:15333

Kennedy CM, Oakleaf JR, Theobald DM et al (2018) Global human modification. https://doi.org/10.6084/m9.figshare.7283087. Accessed 9 Mar 2021

Kennedy CM, Oakleaf JR, Theobald DM et al (2019) Managing the middle: a shift in conservation priorities based on the global human modification gradient. Glob Change Biol 25:811–826

Körner C (2004) Mountain biodiversity, its causes and function. Ambio 33:11–17

Kumpula T, Forbes BC, Stammler F et al (2012) Dynamics of a coupled system: multi-resolution remote sensing in assessing social-ecological responses during 25 years of gas field development in Arctic Russia. Remote Sens 4:1046–1068

Kuoljok K (2019) Without land we are lost: traditional knowledge, digital technology and power relations. AlterNative 15:349–358

Lara MJ, Nitze I, Grosse G et al (2018) Reduced arctic tundra productivity linked with landform and climate change interactions. Sci Rep 8:2345

Liu J, Dietz T, Carpenter SR et al (2007) Complexity of coupled human and natural systems. Science 317:1513–1516

Macias-Fauria M, Jepson P, Zimov N et al (2020) Pleistocene Arctic megafaunal ecological engineering as a natural climate solution? Philos Trans R Soc B 375:20190122

Martin TG, Watson JEM (2016) Intact ecosystems provide best defense against climate change. Nat Clim Chang 6:122–124

Millennium Ecosystem Assessment (2005) Ecosystems and human well-being. Island Press, Washington, Current state and trends

Moen J, Carina E, Keskitalo H (2010) Interlocking panarchies in multi-use boreal forests in Sweden. Ecol Soc 15:17

Monz CA (2002) The response of two arctic tundra plant communities to human trampling disturbance. J Environ Manag 64:207–217

Myers-Smith I, Forbes BC, Wilmking M et al (2011) Shrub expansion in tundra ecosystems: dynamics, impacts and research priorities. Environ Res Lett 6:45509

Newbold T, Hudson LN, Hill SLL et al (2015) Global effects of land use on local terrestrial biodiversity. Nature 520:45–50
Nobrega S, Grogan P (2007) Deeper snow enhances winter respiration from both plant-associated and bulk soil carbon pools in birch hummock tundra. Ecosystems 10:419–431
Oakleaf JR, Kennedy CM, Baruch-Mordo S et al (2015) A world at risk: aggregating development trends to forecast global habitat conversion. PLoS One 10:e0138334
Overland J, Dunlea E, Box JE et al (2019) The urgency of Arctic change. Pol Sci 21:6–13
Pajunen AM, Oksanen J, Virtanen R (2011) Impact of shrub canopies on understory vegetation in western Eurasian tundra. J Veg Sci 22:837–846
Pape R, Löffler J (2012) Climate change, land use conflicts, predation and ecological degradation as challenges for reindeer husbandry in northern Europe – what do we really know after half a century of research? Ambio 41:421–434
Pattison CA, Quinn MS, Dale P et al (2016) The landscape impact of linear seismic clearings for oil and gas development in boreal forest. Northwest Sci 90:340–354
Pettorelli N, Vik JO, Mysterud A et al (2005) Using the satellite-derived NDVI to assess ecological responses to environmental change. Trends Ecol Evol 20:503–510
Ravolainen V, Soininen EM, Jónsdóttir IS et al (2020) High Arctic ecosystem states: conceptual models of vegetation change to guide long-term monitoring and research. Ambio 49:666–677
Raynolds MK, Walker DA, Epstein HE et al (2012) A new estimate of tundra-biome phytomass from transarctic field data and AVHRR NDVI. Remote Sens Lett 3:403–411
Raynolds MK, Jorgenson JC, Jorgenson MT et al (2020) Landscape impacts of 3D-seismic surveys in the Arctic National Wildlife Refuge, Alaska. Ecol Appl 30:e02143
Riggio J, Baillie JEM, Brumby S et al (2020) Global human influence maps reveal clear opportunities in conserving Earth's remaining intact terrestrial ecosystems. Glob Change Biol 26:4344–4356
Roy A, Suchocki M, Gough L et al (2020) Above- and belowground responses to long-term herbivore exclusion. Arct Antarct Alp Res 52:109–119
Sanderson EW, Jaiteh M, Levy MA et al (2002) The human footprint and the last of the wild: the human footprint is a global map of human influence on the land surface, which suggests that human beings are stewards of nature, whether we like it or not. Bioscience 52:891–904
Skarin A, Verdonen M, Kumpula T et al (2020) Reindeer use of low Arctic tundra correlates with landscape structure. Environ Res Lett 15:115012
Steffen W, Richardson K, Rockström J et al (2015) Planetary boundaries: guiding human development on a changing planet. Science 347:1259855
Sturm M, Douglas T, Racine C et al (2005) Changing snow and shrub conditions affect albedo with global implications. J Geophys Res–Biogeo 110:G01004
Tolvanen A, Forbes BC, Rytkönen KE et al (2001) Regeneration of dominant plants after short-term pedestrian trampling in sub-arctic plant communities. In: Wielgolaski FE (ed) Nordic mountain birch ecosystems, Man and the biosphere series, vol 27. Parthenon, Paris, pp 359–368
Tolvanen A, Eilu P, Juutinen A et al (2018) Mining in the Arctic environment—a review from ecological, socioeconomic and legal perspectives. J Environ Manag 233:832–844
Tucker CJ, Pinzon JE, Brown ME et al (2005) An extended AVHRR-8km NDVI dataset compatible with MODIS and SPOT vegetation NDVI data. Int J Remote Sens 26:4485–4498
Tuomi M, Väisänen M, Ylänne H et al (2021) Stomping in silence: conceptualizing trampling effects on soils in polar tundra. Funct Ecol 35:306–317
Uboni A, Åhman B, Moen J (2020) Can management buffer pasture loss and fragmentation for Sami reindeer herding in Sweden? Pastoralism 10:23
van der Wal R (2006) Do herbivores cause habitat degradation or vegetation state transition? Evidence from the tundra. Oikos 114:177–186
Vankoughnett MR, Grogan P (2016) Plant production and nitrogen accumulation above- and belowground in low and tall birch tundra communities: the influence of snow and litter. Plant Soil 408:195–210

Venter O, Sanderson EW, Magrach A et al (2016) Sixteen years of change in the global terrestrial human footprint and implications for biodiversity conservation. Nat Commun 7:12558

Verma M, Schulte to Bühne H, Lopes M et al (2020) Can reindeer husbandry management slow down the shrubification of the Arctic? J Environ Manag 267:110636

Vors LS, Boyce MS (2009) Global declines of caribou and reindeer. Glob Change Biol 15:2626–2633

Vowles T, Björk RG (2019) Implications of evergreen shrub expansion in the Arctic. J Ecol 107: 650–655

Watson JEM, Venter O, Lee J et al (2018) Protect the last of the wild. Nature 563:27–30

Weißhuhn P, Müller F, Wiggereing H (2018) Ecosystem vulnerability review: proposal of an interdisciplinary ecosystem assessment approach. Environ Manag 61:904–915

Wilson AD, MacLeod ND (1991) Overgrazing: present or absent? J Range Manag 44:475–482

Wilson RR, Parrett LS, Joly K et al (2016) Effects of roads on individual caribou movements during migration. Biol Conserv 195:2–8

Wolfe SA, Griffith B, Gray Wolfe CA (2000) Response of reindeer and caribou to human activities. Polar Res 19:63–73

Ylisirniö A-L, Allén A (2016) Plant communities of Fennoscandian subarctic mountain ecosystems 60 years after human disturbance. Arct Antarct Alp Res 48:469–483

Zimov SA, Chuprynin VI, Oreshko AP et al (1995) Steppe-tundra transition: a herbivore driven biome shift at the end of the Pleistocene. Am Nat 146:765–794

Arctic Ecosystems, Wildlife and Man: Threats from Persistent Organic Pollutants and Mercury

Christian Sonne, Robert James Letcher, Bjørn Munro Jenssen, and Rune Dietz

1 Introduction to Persistent Organic Pollutants (POPs) and Mercury

The harmful effects of persistent organic pollutants (POPs) and mercury on wildlife and local human populations in the Arctic have been an issue that was recognised already back in the 1970s starting with the Arctic Monitoring and Assessment Program (AMAP) (AMAP 1998, 2004, 2005). It is a paradox that despite the Arctic is located far away from the use and emission sources of POPs and mercury, levels of these anthropogenic pollutants are reported to be high in both wildlife and humans (Muir et al. 1992; Ayotte et al. 1996). These high levels were, and still are, mainly due to long-range transport of these pollutants through the atmosphere, sea currents and rivers into the Arctic (AMAP 1998, 2009, 2011, 2015, 2018). Despite international regulations and collaborations through AMAP and the Stockholm (POPs) and Minamata (mercury) conventions, high concentrations still persist in many wildlife species, for example, in odontocetes such as killer whales (*Orcinus orca*) and long-finned pilot whales (*Globicephala melas*), but also in polar bears (*Ursus maritimus*)

C. Sonne (✉) · R. Dietz
Faculty of Science and Technology, Department of Ecoscience, Arctic Research Centre (ARC), Aarhus University, Aarhus, Denmark
e-mail: cs@bios.au.dk

R. J. Letcher
Ecotoxicology and Wildlife Health Division, Environment and Climate Change Canada, National Wildlife Research Centre, Carleton University, Ottawa, ON, Canada

B. M. Jenssen
Faculty of Science and Technology, Department of Ecoscience, Arctic Research Centre (ARC), Aarhus University, Aarhus, Denmark

Department of Biology, Norwegian University of Science and Technology, Trondheim, Norway

Department of Arctic Technology, University Centre in Svalbard, Longyearbyen, Norway

M. Tryland (ed.), *Arctic One Health*, https://doi.org/10.1007/978-3-030-87853-5_6

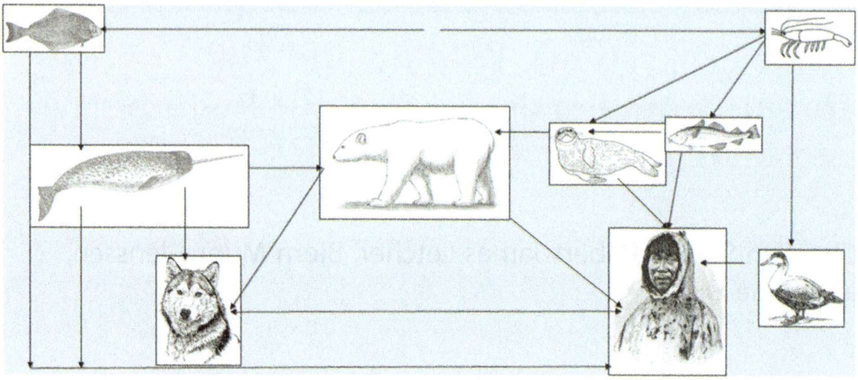

Fig. 1 Typical species assessed in the Arctic Monitoring and Assessment Program for the exposure and health studies of contaminants in the Arctic food webs

due to the biomagnification and high trophic position of these species in the Arctic ecosystems as well as the persistence of these compounds (e.g. Desforges et al. 2018; Dietz et al. 2018, 2019; McKinney et al. 2013; Sonne et al. 2018). Recently, it has been reported that also climatic fluctuations and global warming play an important role in retaining high levels of these anthropogenic compounds in Arctic wildlife (McKinney et al. 2013; Dietz et al. 2019).

The AMAP program has monitored the health of Arctic wildlife and humans for almost three decades. For the purpose of studying contaminant (POPs and mercury) concentrations, spatial and temporal trends and human exposure, these programs have used ringed seals (*Pusa hispida*) and polar bears as key monitoring species (Fig. 1). From the East Greenland region, a large number of organ tissue samples from polar bears have been obtained since the 1980s as part of the traditional hunt. Likewise, from Svalbard, Canada and Alaska, polar bear adipose tissue and blood have been archived and analysed, facilitating the spatial and temporal trends, which are further supported by similar samples from ringed seals. With respect to East Greenland polar bears, histopathological, gross morphological and bone composition and morphological investigations have been carried out since 1999 (Dietz et al. 2013a, b, c, 2018, 2019; Sonne 2010). These studies, in addition to many other, have provided a unique opportunity to investigate the potential organ-specific effects of POP and mercury exposure. Likewise, analyses of bone density and histopathology have been carried out on West Greenland ringed seal populations and on Alaska polar bears, ringed seals and toothed whale spp. as well as on polar bears and beluga whales and glaucous gulls in Alaska, Canada and Svalbard (Dietz et al. 2013a, b, c, 2019; Letcher et al. 2010).

In addition to the biological insights in wildlife, these studies have also provided information on transport of these pollutants to the Arctic and in Arctic food webs, geographical (spatial) and temporal trends, consequences on human health due to consumption of contaminated wildlife and the confounding effects of climate change on levels and effects of these pollutants on wildlife and human health (AMAP 2009,

2011, 2015, 2018). Furthermore, controlled studies have been conducted on Greenland sled dogs (*Canis familiaris*) and farmed Arctic foxes (*Vulpes lagopus*) to support and further understand toxic and endocrine mechanisms behind cause and effects in wildlife from exposure to these toxic compounds (Dietz et al. 2019; Letcher et al. 2010; Sonne 2010). In addition, as in vitro studies using, for example, cells from Arctic wildlife provided information modes of action of pollutants and causality including those of immune toxicity and endocrine disruption (Dietz et al. 2019). In the following sections, we will give examples of biological effects from contaminant exposure in Arctic wildlife and inform about other stressors in the Arctic environment including the cumulative effects from the multiple stressors. We also describe the One Health issues in the Arctic with respect to human exposure and come up with recommendations and future perspectives. It should, however, be noted that these are only highlights and that the primary sources should be consulted to get a thorough and more detailed overview of the information.

1.1 POPs and Mercury Biomagnification and Toxicity

Due to the lipid-rich Arctic marine food webs, the lipophilic or "fat-loving" substances transported to the Arctic biomagnify and exert adverse effects on exposed wildlife as well as humans (AMAP 2009, 2015; Letcher et al. 2010; Dietz et al. 2013a, b, c, 2019). Mercury is the only common metal which is liquid at ordinary temperatures and can evaporate into the atmosphere, and in its organic form, which is methylmercury, it is highly toxic and can biomagnify and exert adverse effects on exposed Arctic wildlife (Dietz et al. 2013a, b, c, 2019). With respect to mercury, the concentrations are highest in west Greenland and northeastern Canada (Dietz et al. 2019) (Fig. 2). Because these long-range transported toxic industrial substances biomagnify in food chains, most studies on the effects of these have therefore been carried out on top predators including marine mammals and seabirds in Greenland and Svalbard (Norway) where the highest POP and mercury concentrations are found (Dietz et al. 2019; Letcher et al. 2010; Sonne 2010). Multiple projects involving large international teams have provided the possibility of conducting correlative studies between concentrations of contaminants and biomarker endpoints. These correlations provide a weight of evidence for adverse biological effects of these pollutants (Dietz et al. 2019; Letcher et al. 2010; Sonne 2010).

Long-range transported pollutants have been extensively monitored in the Arctic to monitor exposure and threat to wildlife species and also to the Inuit populations that consume these country foods. Human exposure is especially relevant and occurs when there is consumption of a marine diet consisting especially of marine predators including polar bears and toothed whales high in contaminants (AMAP 2009, 2015; Dietz et al. 2019). Exposure to POPs such as polychlorinated biphenyls (PCBs) and mercury continue to be the most problematic and of high concern due to their toxicity, despite their ban decades ago (Dietz et al. 2019; Letcher et al. 2010; Sonne 2010). However, other POPs including several poly- and per-fluoroalkyl

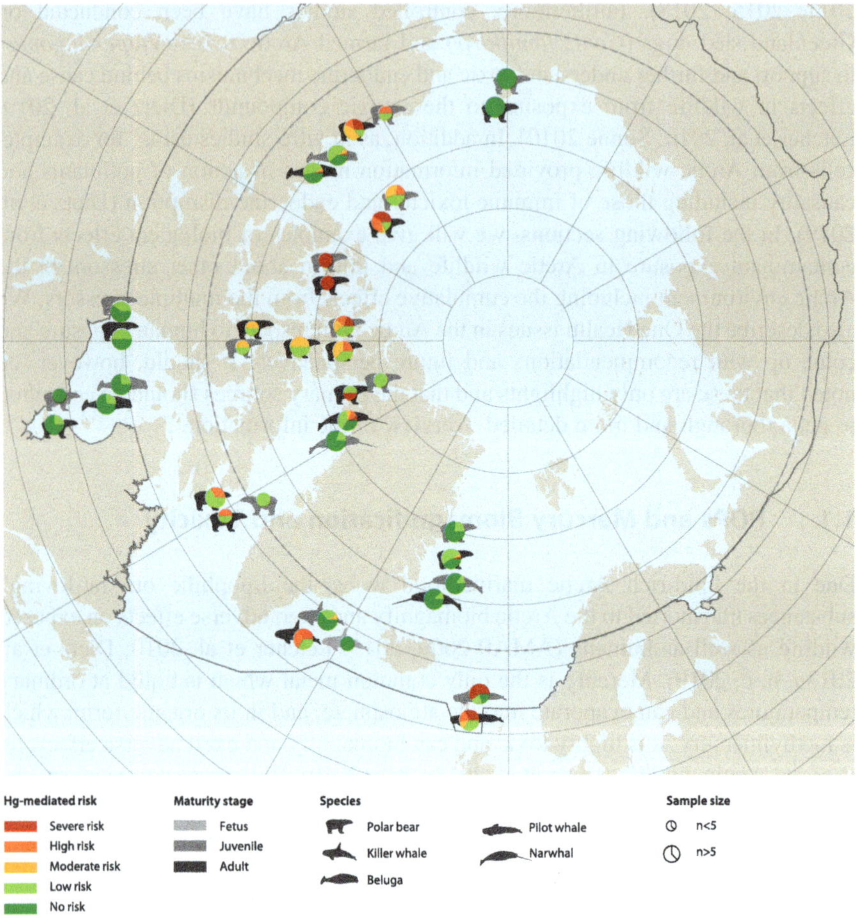

Fig. 2 Mercury exposure risk groups (top) and PCB PBPK risk group (bottom) for polar bears and toothed whales (Modified from Dietz et al. 2019)

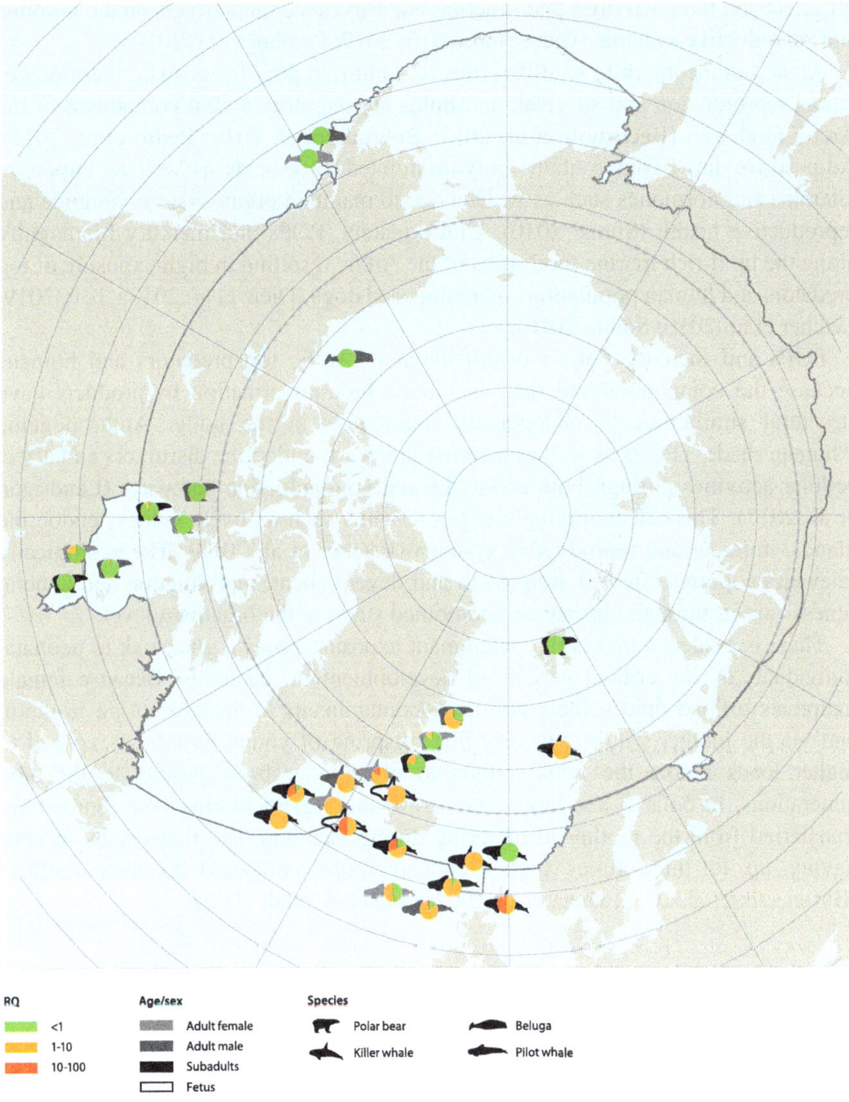

Fig. 2 (continued)

substances (PFAS), which are not lipid soluble but associated with proteins, and lipid soluble organochlorine pesticides (OCPs) are also found in concentrations that for some species in some areas lead to concern for health effects on top predators and humans (Dietz et al. 2009, 2013c, 2015, 2018). Perfluorooctane sulfonate (PFOS) is a highly bioaccumulative PFAS that belongs to the group known as perfluoro sulfonic acids (PFSAs). Along with longer-chain perfluorocarboxylic acids (PFCAs), PFOS is found at high concentrations in the liver, which lead to concerns for effects on liver activities and function but also cancer and effects on the immune and reproductive systems (Dietz et al. 2015, 2019; Letcher et al. 2010).

Arctic homoeothermic wildlife relies on lipid-rich prey for growth, thermoregulation, reproduction and survival, and lipids are therefore a vital component of the Arctic food web (Bechshøft et al. 2011; Bourque et al. 2018; Pedro et al. 2017). Adipose (or fat) tissue contains polyunsaturated fatty acids as well as important vitamins and hormones such as A, D and E to maintain organ-system, immune and reproductive health (Sonne 2010). Unfortunately, POPs and mercury biomagnify along the lipid-rich marine food webs of the Arctic resulting in high exposure of top predators and human populations including sled dogs (Dietz et al. 2013a, b, c, 2019; Letcher et al. 2010; Sonne 2010).

POPs and mercury pose a health threat to Arctic top predators and humans because the compounds and their metabolic biotransformation by-products have structural similarities to endogenous compounds in the body. Anthropogenic ("human-made") POPs as well as mercury are neuro-endocrine disruptors and affect cellular activities through both endocrine and non-endocrine pathways (Landrigan et al. 2017). This affects multiple organ systems such as liver, kidney, endocrine glands, immune and reproductive systems (Letcher et al. 2010). The ecologically relevant endpoints affected are growth and development, reproduction and general fitness caused through chronic and combined stress at the organism level.

There is evidence that high contaminant exposure poses a great risk to neonatal individuals during critical periods of development in mammals. Because female mammals offload lipid-soluble pollutants/contaminants to their offspring in utero, and via the mother's milk, the very first offspring of young female is exposed to higher loads before the female starts to offload her body burden to the next generations. In polar bears, large amounts of the organochlorine body burden are transferred from the mother to offspring during lactation, and that results in cubs having up to three times higher concentrations compared to their mothers (Bytingsvik et al. 2012; Dietz et al. 2004; Polischuk et al. 2002).

1.2 Cause and Effects

Interpretations and conclusions in wildlife health studies linked to contaminant exposure are typically based on correlations and descriptive interpretations, unavoidably confounded by a plethora of factors affecting the physiological state of a free-ranging animal. To improve the understanding of the effects of exposure to real-world contaminant cocktails, experimental exposure studies have been performed on sled dogs and domesticated Arctic foxes, being possible surrogate model species for *Canidae* spp. In captive sled dogs and Arctic fox studies, where both included a cohort fed a naturally POP contaminated diet of minke whale (*Balaenoptera acutorostrata*) blubber, it has been possible to define and compare POP exposed and unexposed reference groups in direct relation to an array of effects such as on reproductive organs and other internal organs, the skeletal system, immune and endocrine systems and POP dietary accumulation, biotransformation and toxicokinetics (Letcher et al. 2010; Sonne 2010, Sonne et al. 2017). These studies have shown that exposure to POPs leads to effects on the reproductive, endocrine and immune systems and also vitamin homeostasis and lesions in internal organs and thyroid glands. In addition, in vitro studies including those using live immune cells and cell lines have been used to further understand and improve the cause-and-effect understanding of pollution exposure and health effects in Arctic wildlife (Letcher et al. 2010; Dietz et al. 2013a, b, c, 2019). This is important as the establishment of cause and effects in wildlife toxicology studies is very difficult and therefore is the use of relevant surrogate species, in vitro studies and population level modelling a necessity (Dietz et al. 2019; Sonne 2010).

2 POPs and Mercury Biological Effects

An overview of the effects from contaminant exposure on the most threatened species is given in Fig. 2, and in the following sections, a few selected examples are given.

2.1 Hormones and Reproductive Organs

Endocrine disruption from POP exposure is likely a challenge in Arctic wildlife (Jenssen et al. 2015; Letcher et al. 2010). Both steroid hormones and thyroid hormones seem to be impacted by POP parent compounds and their metabolites, mimicking hormone activity in both the transport pathways and receptor mechanisms and affecting overall health and survival (Dietz et al. 2019; Letcher et al. 2010). These investigations have been reproduced in controlled experiments on sled dogs and Arctic foxes (Dietz et al. 2019; Sonne et al. 2014; Sonne 2010; Sonne et al. 2017). Studies on polar bears from Svalbard indicate that plasma steroid levels may be affected by POP exposure, particularly PCBs and their hydroxylated metabolites (OH-PCBs). In females, pregnenolone and androstenedione in blood appeared to be significantly negatively correlated to several OH-PCBs, indicating that these POPs and metabolites may affect steroidogenesis (Gustavson et al. 2015). PCBs also appear to affect male plasma steroids, in particular androgen levels in Svalbard polar bears (Ciesielski et al. 2017). A study by Dietz et al. (2015) showed that these PCB levels may affect fecundity based on and thereby the populations of the highest exposed polar bear subpopulations using pharmacological modelling (PBPK model) (Dietz et al. 2015, 2018).

Studying reproductive organs in wildlife is often based on necropsy samples as clinical investigations are rather difficult. Investigation of endocrinology and reproductive organs in wildlife and domesticated species (sled dogs and Arctic fox) can give some important information. In polar bears, exposure to POPs may cause effects related to the larger testicular dysgenesis syndrome (TDS) that affects their reproductive organs including the morphology, function and size of testes and also the penile bone mineral density (Sonne 2010; Sonne et al. 2006a, 2007a). Such TDS effects include, for example, the presence of orchitis (testes inflammation), and changes in size and function of the testes. In fact, it has been observed that about 50% of the polar bears in East Greenland show dysfunction of their testes during mating season (Spörndly-Nees et al. 2019). Similar correlations have been found between POP concentrations and size of female reproductive organs and that may affect the productivity of East Greenland polar bears (Dietz et al. 2015, 2018; Sonne et al. 2006a). Suboptimal testosterone and semen production have also been reported in Arctic foxes exposed to high concentrations of POPs as compared to that in a control group (Sonne et al. 2017). In addition, inverse correlations between POP exposure and size of female sexual organs were also found (Sonne et al. 2006a). The effects of POPs on female and male reproductive function and resultant population demographics are therefore some of the most important hypotheses to investigate further.

2.2 The Skeletal System

Studies of polar bears suggest that osteoporosis may be a problem and especially for male bears that do not have the evolutionary physiological mechanisms that females have. In contrast to polar bear males, female bears are due to their denning physiology and exposure to immobilisation, starvation and loss in body weight, less susceptible to bone deterioration from POP exposure (Daugaard-Petersen et al. 2018a, b; Sonne et al. 2004). This makes them resistant towards osteoporosis, presumably because an efficient bone metabolism is needed for supplementing their offspring with calcium for bone development (Sonne 2010; Sonne et al. 2004). This is suggested to put males at high risk of endocrine disruptions from POP exposure that lead to changes in their bone mineral density (Sonne 2010). The studies of male polar bears have shown modulation of bone mineral density by PCB exposure, and this is supported by studies on seals in the Baltic Sea and alligators (*Alligator mississippiensis*) in Florida, USA (Daugaard-Petersen et al. 2018a, b; Lind et al. 2003, 2004; Sonne et al. 2004, 2015). In polar bears, the effects from this are a decline in overall health and homeostasis of calcium but also a risk for effects on the population level given the risk of penile bone fractures (Sonne et al. 2015). Female polar bears are likely more robust to bone stress from, e.g. fluctuations in food availability and composition as well as endocrine-related effects from POPs.

2.3 The Immune System

Environmental contaminants have been shown to modulate, among other effects, all measurable aspects of cellular and humoral immunity in marine mammals that affect their resistance against infectious diseases through lower lymphocyte proliferation (Desforges et al. 2016). Several studies have reported immune effects in Arctic wildlife, including polar bears, sled dogs and Arctic foxes (Sonne 2010), demonstrating that contaminants are reaching levels that can cause significant changes in physiology and immune fitness. Polar bear immunity was assessed in a comprehensive study in Svalbard and Hudson Bay, and negative associations were found between PCB exposure and serum immunoglobulin G (IgG) levels, antibody titres against influenza virus and reovirus following immunisations and lymphocyte proliferation (Lie et al. 2004, 2005). These findings are supported by a controlled study on Greenland sled dogs showing reduced lymphocyte proliferation and antibody production (Sonne et al. 2006b, 2007b). Likewise, hepatic mRNA expression of interleukin-1β (IL-1β), an important pro-inflammatory cytokine, in ringed seals from Svalbard was positively correlated with hepatic POP levels hepatic IL-1 mRNA, and expression was positively correlated with blubber PCB burdens in 41 ringed seals in northern Labrador, Canada (Brown et al. 2014; Routti et al. 2010). These studies all illustrate that in vivo real-life exposure to contaminants may cause measurable changes in immune function in Arctic wildlife, while controlled studies are needed to further understand and conclude on that (Dietz et al. 2019; Sonne 2010).

In vitro experiments, where immune cells are exposed to contaminants under laboratory conditions in order to characterise effect levels, have also been carried out in Arctic wildlife species. East Greenland ringed seal leukocytes were exposed to four PCB congeners (CB-138, CB-153, CB-169 and CB-180) and two PFAS, i.e. PFOS and perfluorooctanoic acid (PFOA), and it was found that the coplanar CB congeners 138, 153 and 180 but not 169 and PFAS caused significant suppression of lymphocyte proliferation at relevant environmental concentrations for seals (Levin et al. 2016). Beluga whale (*Delphinapterus leucas*) leukocytes were also used to show that low levels of mercury exposure can cause significant reductions in lymphocyte proliferation and intracellular thiol production, and significant induction of metallothionein (Frouin et al. 2012). Altogether, studies of free-ranging animals and in vitro experiments suggest high contaminant loads in the Arctic can induce immune suppression which affects the ability to respond to intruding infectious pathogens in marine mammals.

2.4 Neuro-Endocrinology

Multiple environmental contaminants can cross the blood-brain barrier and exert effects on the brain functioning thus causing neurobehavioral effects. Through various mechanisms, contaminants can influence mating and other reproductive behaviours, motivation, communication, aggression, dominance and other social behaviours, as well as learning and other cognitive abilities (Grandjean and Landrigan 2006, 2014). Although several POPs are reported in brain tissue of polar bears (Pedersen et al. 2015, 2016), it is extremely difficult to conduct field studies that would provide information on behavioural effects of POPs in polar bears. However, investigations of correlations between contaminant levels and brain neurotransmitters and receptors have been conducted in polar bears from the indigenous hunt in Greenland. Despite the relatively low concentrations of mercury in the brains of these polar bears, significant negative correlations were found between both mercury concentrations and *N*-methyl-D-aspartic acid (NMDA) glutamate receptors (citation). NMDA glutamate and genomic methylation are important for animal health, behaviour, reproduction and survival, and their reduction may therefore have population-level effects for polar bears (Basu et al. 2009; Krey et al. 2014; Pilsner et al. 2010). Mercury loads are even higher in beluga (*Delphinapterus leucas*) brains as they do not have the excretion route for mercury through the hair formation (Dietz et al. 2013a; Ostertag et al. 2013). Another proposed mechanism of developmental neurotoxicity is through thyroid hormone disruption; thyroid hormones are essential for proper neurodevelopment of the foetus and early neonate (Ahmed et al. 2008; Zoeller and Crofton 2005). Therefore, high levels of POPs and their associated thyroid disrupting properties raise concern for neurodevelopmental effects in polar bears which may alter their behaviour and reproduction among others (Jenssen et al. 2015). Therefore, more studies are needed to understand the effects of pollution on behaviour and how that may alter biodiversity and ecosystem functioning.

2.5 Modelling Combined Effects

Circumpolar polar bear subpopulations are under influence of immunological, reproductive and carcinogenic consequences from POP exposure (Dietz et al. 2015, 2018, 2019). From a population conservation point of view, contaminants that affect reproduction and survival in both males and females are among the most important to monitor in different Arctic subpopulations of polar bears, toothed whales and seals. Consideration of combined effects from the "cocktail" of contaminant exposure is important when understanding and evaluating exposure of Arctic wildlife. Applying an individual and agent-based modelling predict collapses of certain killer whale populations from effects of PCB exposure due to effects on reproduction (Desforges et al. 2018). The model predicted ca. 50% (10 out of 19 populations worldwide) of the global populations to be significantly reduced or even go extinct within 30–100 years including the one in East Greenland. Although most of the populations were predicted to go extinct due to the reproductive effects of PCBs inhabit regions outside of the Arctic, it is important to note that Arctic ecotype populations that consume marine mammals, such as the East Greenland population, also were predicted to go extinct. Risk assessments of the effects from pollution have also been conducted using PBPK modelling, which assesses the concentrations of contaminants in tissues in relation to the critical concentration for immune and reproductive effects across the Arctic and over time in East Greenland (Dietz et al. 2015, 2018; Sonne et al. 2009). These studies show that although PCBs have been regulated for decades, the relatively high environmental levels of PCBs still represent the pollutant class that poses by far the highest risk for reproductive effects in polar bears. Other major contaminant classes of concern were PCBs and PFOS. The PBPK modelling of polar bears and other Arctic marine mammals for PCB exposure show that those at risk aside from polar bears are pilot whales and East Greenland polar bears (Dietz et al. 2015, 2019). For mercury, the risk was highest for killer whales and pilot whales in the North Atlantic and polar bears and seals in Canada and narwhals (*Monodon monoceros*) in Northwest Greenland (Dietz et al. 2019). In a meta-study on 14 polar bear subpopulations, an exploratory regression analysis identified clear negative relationships between polar bear subpopulation densities and the contaminant concentrations in their adipose tissue, such as PCBs, DDTs and polybrominated diphenyl ethers (Nuijten et al. 2016).

2.6 Metabolomics

Metabolomics, lipidomics and proteomics together with transcript levels of genes, ratios between fatty acids and plasma parameters, may also be used to estimate the risk of combined effects from climate change, i.e. fasting and emaciation and POP exposure on energy metabolism and endocrine disruption in Svalbard and Hudson Bay polar bears (Morris et al. 2019; Routti et al. 2019; Tartu et al. 2017). Energy requirements for fasting, breeding, lactation and migration lead to increased metabolism which release contaminants from adipose tissue that then become bioavailable and may exert endocrine and metabolic disrupting effects (Jenssen et al. 2015). Such studies of multiple stressors and combined effects, including those of infectious diseases, are important, and new omics technologies may help elucidating the complicated endocrine and physiological feedback mechanisms (Dietz et al. 2019).

3 Climate Change and Multiple Stressors

Arctic ecosystems are also exposed to other multiple stressors than contaminants, which lead to cumulate stress in wildlife and humans. For example, Arctic wildlife such as polar bears have received considerable focus as they, depending on the regional subpopulation, are threatened most dramatically by climate change and loss of sea ice (Derocher et al. 2013; Durner et al. 2009; Molnár et al. 2011). Modelling has shown that the southernmost polar bear subpopulations in the Hudson Bay are at greatest risk due to climate change and will struggle to exist throughout this century (Hamilton et al. 2014). In fact, models have predicted two-thirds of the world's polar bears could disappear if greenhouse gas emissions continue to increase as predicted (Amstrup et al. 2018; Molnár et al. 2020). This has been linked to the occupation of large home range sizes and the requirement of higher energetic costs and thus higher feeding rates, which again can lead to increasing blood PCB concentrations (Olsen et al. 2003; Pavlova et al. 2016a, b; van Beest et al. 2016). In polar bears, multiple types of anthropogenic stressors, or factors, such as adipose tissue concentrations of POPs, September ice cover (being proxy of climate change), human population density and polar bear harvest rates may influence population dynamics. A meta-study on 14 polar bear subpopulations indicated that the September ice cover, as a proxy for climate variability, was the best predictor for adult female survival, whereas the study also identified a clear negative correlation between polar bear population density and subpopulation specific contaminant concentrations in adipose tissue (Nuijten et al. 2016). This particular study indicates that multiple anthropogenic stressors may inflict population dynamics in a complex manner that may be difficult to entangle.

Furthermore, climate warming induced migration of warm water adapted fish species may act as bio-vectors increasing contaminant levels in marine Arctic ecosystems (Atwood et al. 2017; MacKenzie et al. 2014; VanWormer et al. 2019). Climate change has been deemed the most important factor in the emergence of infectious diseases, and nowhere else in the world is climate change occurring as fast as in the polar regions (Parkinson and Butler 2005). A warming climate may profoundly affect disease dynamics in the Arctic by changing the species composition and northward invasion of disease vectors and transport of zoonotic pathogens (Tryland et al. 2013). In addition, increased survival of infected animals during milder winters may further increase the risk of pathogen reservoir in marine mammals including that of zoonoses. Moreover and as discussed above, immunotoxic contaminants may increase disease-related mortality and morbidity of Arctic marine mammals (Dudley et al. 2015).

Climate change has a duel impact, acting through alteration of food web pathways for contaminants and spreading the zoonotic diseases. Pollution-induced increase in disease rates due to immunotoxic effects of POPs and mercury can increase the likelihood and risk of disease transfer from animals to humans (zoonoses) as the proportion of infected Arctic wildlife increases. A large volume of marine and terrestrial wildlife is consumed by humans in the Arctic, often raw and inadequately frozen, and this likely increases the risk of zoonotic diseases (Tryland et al. 2013). There has been a particular focus on indigenous people who consume large amounts of marine food, including the blubber of marine mammals with high concentrations of POPs, which affect their immune system (Dewailly et al. 2000; Donaldson et al. 2010). With a decreased function of the immune system comes an increased risk of zoonotic infectious, i.e. infections transferred between animals and humans. Few studies have examined the incidence of zoonotic transmitted diseases in humans in the Arctic. The majority of such studies have been performed in Alaska and Canada, with just a small number of studies in Greenland and Russia. This is despite a significant proportion of the indigenous Arctic residents living in close contact with wildlife. Indeed, traditional food preparation often avoids heat treatment. The process of heat treatment of food can be extremely effective against zoonotic diseases as shown by the case of *Trichinella* and *Toxoplasma gondii* (which causes lifelong parasitic cysts in the human brain): 80% of Inuit consuming dried marine mammal meat tested seropositive for the disease, while prevalence was only 10% in Cree who prefer cooked meat (Kotula et al. 1983; Messier et al. 2009). It is clear that zoonoses are a very real threat to Arctic residents, and this highlights the need for an overview of the most serious zoonotic diseases, especially for those Arctic regions where information is particularly sparse. The need for an overview is strengthened by the context of changing disease patterns, as well as ongoing bioaccumulation and exposure to toxic chemical contaminants, some of which have been proven to be immune-suppressive. This would render both animal and human hosts more vulnerable to diseases, especially hosts that are already immunologically vulnerable to novel diseases.

4 One Health: Linking Humans, Wildlife and Environment

Despite implementations of international regulation for industrial production and the use of POPs (and mercury) over the last two decades, the concentrations of the highly toxic PCBs, chlordane pesticides and mercury, have remained essentially unchanged in Arctic marine mammal (Dietz et al. 2019). This is likely due to climate change effects on food web interactions, generational transfer and continued secondary and unintentional emissions. For Hg, the body burdens even appear to be continuously increasing in most top predators in the central Arctic reaching up to 20-fold baseline levels of pre-industrialisation (Dietz et al. 2009, 2013a, b, c, 2019). One Health aims to improve health and well-being, including mental health, through the prevention of risks and the mitigation of effects of crises that originate at the interface between humans, animals and their various environments (http://www.onehealthglobal.net).

Since Greenlanders in this region traditionally ingest significant quantities of adipose tissue from wildlife including seals, whales and polar bears, they are among the people carrying the highest POP and mercury burdens (AMAP 2009, 2015). Such high exposure is likely to pose a health risk based on available literature on dioxin toxic equivalent (TEQ) concentrations, which are based on toxic equivalency factors (TEFs) for individual dioxins, furans and dioxin-like PCBs, and tolerable daily intake (TDI) guidelines (Sonne et al. 2013; Wielsøe et al. 2017, 2018). Human exposure to contaminants in Greenland has been evaluated from chemical analyses of prey species and food intake showing that the TDI was exceeded for chlordane (CHL) by a factor of 3–6, while PCB exposure did not (Deutch et al. 2006; Johansen et al. 2004). However, none of these studies reported on polar bear and ringed seal tissues which are important food items in East Greenland. Since POP loads in polar bears and ringed seals are known to be four times higher in East Greenland than in West Greenland, human exposure is subsequently higher in East Greenland (Dietz et al. 2019; Letcher et al. 2010). According to Nielsen et al. (2006), it is recommended that Greenlanders reduce their exposure to PCBs and CHLs by reducing their blubber intake. The AMAP is therefore concerned about Arctic human health within a contaminant exposure context, and their studies do support observations that Greenland hunters are particularly exposed to high PCB concentrations due to frequent ingestion of polar bear, killer whale, narwhal and seal tissues. Studies from the Russian Arctic have shown that dioxin, furans and PCB exposure of neo- and prenatal children exceeded TEQ TDI levels by up to 33 times in the year 2000 (AMAP 2009, 2015).

Chronic diseases, including diabetes, upper respiratory and recurrent middle ear infections, cancer, and osteoporosis, are becoming an epidemic in the Arctic and may be partially caused by chronic POP exposure and associated endocrine disruption (AMAP 2009, 2015). Likewise, disruption by POPs of the thyroid and steroid hormone endocrine axis is influencing physiological endpoints with effects on reproduction, cancer and immunity and may even affect the physical growth of Inuit children (Tahir et al. 2020). It is apparent that subsistence hunters in East Greenland are exposed to mixtures of POPs that include both toxic parent compounds and also their derived metabolites (Sonne et al. 2013). Some POPs can induce (hepatic) cytochrome (CYP) 450 enzymes in Arctic people that may interfere with normal metabolic processes and homeostasis of various intrinsic hormones and vitamins influencing immunity and reproduction. For example, prenatal exposure to PCBs has been suspected to play a role in the relatively high incidence of acute respiratory infections and middle ear inflammation in Inuit children. Likewise, Sandau et al. (2002) found that metabolism of PCBs was significantly negatively correlated with thyroid hormones, namely, free thyroxine, in northern peoples in the Ungava region in northern Québec, Canada. These findings of associations between POP exposure, induction of cytochrome P450 (CYP450) monooxygenases and changes in thyroid hormone concentrations may play a role in the development of chronic diseases. A single study has been published on osteoporosis in relation to POP exposure in Greenlanders (Côté et al. 2006). The study focused on quantitative ultrasound parameters (QUS) at the right calcaneus of 153 people from Southwest Greenland and found blood plasma CB-153 concentrations strongly and negatively associated with the three QUS parameters. While the relationship was no longer significant when normalising for age and weight, people in Southwest Greenland belong to some of the lowest POP-exposed Greenlanders, and effects on bone composition are therefore not likely to occur. Another study of Cree women from Eastern James Bay in Canada showed that bone stiffness index was negatively associated to blood concentrations of CB-105 and CB-118 (Paunescu et al. 2013).

It has been shown that blood concentrations of PFAS and especially PFOS in male Inuit from East Greenland can be two to three times higher as compared to the Faroese population where local exposure has already been attributed with effects on the immune system (Grandjean et al. 2012). Although certain PFAS have been associated with developmental and hormonal effects, immunotoxicity and tumour growth in rodents, the impact of these compounds on human health appear to be inconclusive. Of the PCBs found in Greenlanders, the highly dioxin-like congeners CB-77, CB-126 and CB-169 attain a coplanar configuration similar to the very toxic dioxins and furans and are in fact commonly found in Arctic wildlife (AMAP 2004, 2009). These coplanar PCB congeners are characteristically highly potent inducers via the aryl hydrocarbon receptor (AhR) and subsequent expression of CYP1A, CYP2A and CYPB1 monooxygenase activity. Furthermore, for Greenlanders, significant correlations were found between blood contaminant concentrations and calculated daily intake of POPs (Sonne et al. 2014). Mercury exposure of Inuit people is also of great health concern and has been recognised as a neuro-endocrine and immune health problem in the societies of Faroe Islands, West Greenland (Thule) and Canada.

5 Future Perspectives and Recommendations

Further efforts are required to understand the toxicokinetics and toxicodynamics of POPs and mercury in Greenland wildlife and peoples in this changing Arctic in order to better predict the individual-level health risks associated with contaminant exposure. Arctic top predators are sentinels for humans as they consume the same diet and act as potential vectors for zoonotic transfer to humans due to harvesting. Combining correlational studies on wildlife health with experimental work on surrogate species, such as the sled dog, will allow better understanding of the proximate toxic pathways behind exposure to contaminants and infectious diseases, their interactions and the driving role of a rapidly changing climate. Doing this offers a promising One Health approach to survey and pinpoint environmental change and multiple stressors that may have effects on wildlife and human health. The warming Arctic climate is suspected to influence abiotic and biotic long-range transport and exposure pathways of contaminants to the Arctic. As a result, there will be likely increases in POP exposure of Arctic wildlife and human populations, while exposure to vector-borne diseases and zoonoses may increase as well through range expansion and introduction from invasive species. Broad and nevertheless in-depth studies on the occurrence and human health risk of Arctic zoonoses and their interactive effects with climate change and contaminant exposure are pending, as well as an increased effort to educate the relevant groups of the public regarding safe handling of wildlife.

References

Ahmed EI, Zehr JL, Schulz KM, Lorenz BH, Don Carlos LL, Sisk CL (2008) Pubertal hormones modulate the addition of new cells to sexually dimorphic brain regions. Nat Neurosci 11:995–997

AMAP (1998) Arctic monitoring and assessment programme: AMAP Assessment Report – Arctic Pollution Issues 1998; Oslo. www.amap.no

AMAP (2004) Arctic monitoring and assessment programme: AMAP Assessment 2002 – Persistent Organic Pollutants in the Arctic 2004; Oslo. www.amap.no

AMAP (2005) Arctic monitoring and assessment programme: AMAP Assessment 2002 - Heavy Metals in the Arctic 2005; Oslo. www.amap.no

AMAP (2009) Arctic monitoring and assessment programme: AMAP Assessment 2009 – Human health in the Arctic 2009; Oslo. www.amap.no

AMAP (2015) Arctic monitoring and assessment programme: AMAP Assessment 2013 – Human Health in the Arctic 2013; Oslo. www.amap.no

AMAP (2018) Arctic monitoring and assessment programme: AMAP Assessment 2018 – Persistent Organic Pollutants in the Arctic 2018; Oslo. www.amap.no

AMAP (2011) Arctic monitoring and assessment programme: AMAP Assessment 2011 – Mercury in the Arctic 2011; Oslo. www.amap.no

Amstrup SC, Marcot BG, Douglas DC (2018) A Bayesian network modeling approach to forecasting the 21st century worldwide status of polar bears. In: DeWeaver ET, Bitz CM, Tremblay L-B (eds) Arctic Sea Ice decline: observations, projections, mechanisms, and implications, vol 180. American Geophysical Union, Washington, DC, pp 213–268

Atwood TC, Duncan C, Patyk KA, Nol P, Rhyan J, McCollum M, McKinney MA, Ramey AM, Cerqueira-Cézar CK, Kwok OCH, Dubey JP, Hennager S (2017) Environmental and behavioral changes may influence the exposure of an Arctic apex predator to pathogens and contaminants. Sci Rep 7:13193

Ayotte P, Carrier G, Dewailly E (1996) Health risk assessment for Inuit newborns exposed to dioxin-like compounds through breast feeding. Chemosphere 32:531–542

Basu N, Scheuhammer AM, Sonne C, Dietz R, Letcher RJ (2009) Is mercury in the environment of neurotoxic concern to polar bears? Environ Toxicol Chem 28:133–140

Bechshøft TØ, Jakobsen J, Sonne C, Dietz R (2011) Distribution of vitamin A (retinol) and E (α-tocopherol) in polar bear kidney: implications for biomarker studies. Sci Total Environ 409: 3508–3511

Bourque J, Dietz R, Sonne C, St. Leger J, Iverson S, Rosing-Asvid A, Hansen M, McKinney MA (2018) Feeding habits of new Arctic predator: insight from full-depth blubber fatty acid signatures of Greenland, Faroe Islands, Denmark, and managed-care killer whales Orcinus orca. Mar Ecol Prog Ser 603:1–12

Brown TM, Ross PS, Reimer KJ, Veldhoen N, Dangerfield NJ, Fisk AT, Helbing CC (2014) PCB related effects thresholds as derived through gene transcript profiles in locally contaminated ringed seals (Pusa hispida). Environ Sci Technol 48(21):12952–12961. https://doi.org/10.1021/es5032294

Bytingsvik J, Lie E, Aars J, Derocher AE, Wiig Ø, Jenssen BM (2012) PCBs and OH-PCBs in polar bear mother-cub pairs: a comparative plasma levels in 1998 and 2008. Sci Total Environ 417: 117–128

Ciesielski TM, Hansen IT, Bytingsvik J, Hansen M, Lie E, Aars J et al (2017) Relationships between POPs, biometrics and circulating steroids in male polar bears (*Ursus maritimus*) from Svalbard. Environ Pollut 230:598–608

Côté S, Ayotte P, Dodin S, Blanchet C, Mulvad G, Petersen HS et al (2006) Plasma organochlorine concentrations and bone ultrasound measurements: a cross-sectional study in peri-and postmenopausal Inuit women from Greenland. Environ Health 5:33

Daugaard-Petersen R, Langebæk R, Rigét FF, Letcher RJ, Hydstrup L, Bech Jensen JE, Bechshøft T, Wiig Ø, Jenssen BM, Pertoldi C, Dietz R, Sonne C (2018a) Persistent organic pollutants, skull size and bone density of polar bears (*Ursus maritimus*) from East Greenland and Svalbard 1892-2015. Environ Res 162:74–80

Daugaard-Petersen T, Langebæk R, Rigét FF, Dyck M, Letcher RJ, Hyldstrup L, Dietz R, Sonne C (2018b) Persistent organic pollutants and penile bone mineral density in East Greenland and Canadian polar bears (Ursus maritimus) during 1996–2015. Environ Int 114:212–218

Derocher AE, Aars J, Amstrup SC, Cutting A, Lunn NJ, Molnár PK et al (2013) Rapid ecosystem change and polar bear conservation. Conserv Lett 6:368–375

Desforges JPW, Sonne C, Levin M, Siebert U, De Guise S, Dietz R (2016) Immunotoxic effects of environmental pollutants in marine mammals. Environ Int 86:126–139

Desforges JP, Hall A, McConnell B, Rosing Asvid A, Barber JL, Brownlow A, De Guise S, Eulaers I, Jepson PD, Letcher RJ, Levin M, Ross PS, Samarra F, Víkingson G, Sonne C, Dietz R (2018) Predicting global killer whale population collapse from PCB pollution. Science 361:1373–1376

Deutch B, Dyerberg J, Pedersen HS, Asmund G, Møller P, Hansen JC (2006) Dietary composition and contaminants in North Greenland, in the 1970s and 2004. Sci Total Environ 370:372–381

Dewailly É, Ayotte P, Bruneau S, Laliberté C, Gingras S, Belles-Isles M et al (2000) Susceptibility to infections and immune status in Inuit infants exposed to organochlorines. Environ Health Perspect 108:205–211

Dietz R, Rigét FF, Sonne C, Letcher RJ, Born EW, Muir DCG (2004) Seasonal and temporal trends in polychlorinated biphenyls and organochlorine pesticides in East Greenland polar bears (*Ursus maritimus*), 1990-2001. Sci Total Environ 331:107–124

Dietz R, Outridge PM, Hobson KA (2009) Anthropogenic contribution to mercury levels in present-day Arctic animals – a review. Sci Total Environ 407:6120–6131

Dietz R, Sonne C, Basu N, Birgit Braune C, Todd O'Hara D, Letcher RJ, Scheuhammer T, Andersen M, Andreasen C, Andriashek D, Asmund G, Aubail A, Baagøe H, EW EB, Chan HM, Derocher AE, Grandjean P, Knott K, Kirkegaard M, Krey A, Lunn N, Messier F, Obbard M, Olsen MT, Ostertag S, Peacock E, Renzoni A, Rigét F, Skaare JU, Stern G, Stirling I, Taylor M, Wiig Ø, Wilson S, Aars J (2013a) What are the toxicological effects of mercury in Arctic biota? Sci Total Environ 443:775–790

Dietz R, Rigét FF, Sonne C, Born EW, Bechshøft T, McKinney MA, Letcher RJ (2013b) Part 1: Three decades (1984-2010) of legacy contaminant trends in East Greenland polar bears (*Ursus maritimus*). Environ Int 59:485–493

Dietz R, Rigét FF, Sonne C, Born EW, Bechshøft T, McKinney MA, Drimmei R, Muir DCG, Letcher RJ (2013c) Part 2: Three decades (1984-2010) of flame retardant trends in East Greenland polar bears (*Ursus maritimus*). Environ Int 59:494–500

Dietz R, Gustavson K, Sonne C, Desforges JP, Rigét FF, McKinney MA et al (2015) Physiologically-based pharmacokinetic modelling of immune, reproductive and carcinogenic effects from contaminant exposure in polar bears (*Ursus maritimus*). Environ Res 140:45–55

Dietz R, Desforges J-P, Gustavson K, Sonne C, Rigét FF, Born EW, Letcher RJ (2018) Immunologic, reproductive, and carcinogenic risk assessment in East Greenland polar bears (Ursus maritimus) during the period 1983-2013. Environ Int 118:169–178

Dietz R, Letcher R, Desforges J-P, Eulaers I, Sonne C, Wilson S, Andersen-Ranberg E, Basu N, Barst BD, Bustnes JO, Bytingsvik J, Ciesielski TM, Drevnick PE, Gabrielsen GW, Haarr A, Hylland K, Jenssen BM, Levin M, McKinney MA, Nørregaard RD, Pedersen KE, Provencher J, Styrishave B, Tartu S, Aars J, Ackerman JT, Rosing-Asvid A, Barrett R, Bignert A, Born EW, Branigan M, Braune B, Bryant CE, Dam M, Eagles-Smith CA, Evans M, Evans TJ, Fisk AT, Gamberg M, Gustavson K, Hartman CA, Helander B, Herzog MP, Hoekstra PF, Houde M, Hoydal K, Jackson AK, Kucklickt J, Lie E, Loseto L, Mallory ML, Miljeteig C, Mosbech A, Muir DCG, Nielsen ST, Peacock E, Pedro S, Peterson SH, Polder A, Rigét FF, Roach P, Saunes H, Sinding M, Skaare JU, Søndergaard J, Stenson G, Stern G, Treu G, Schuurtt SS, Víkingsson G (2019) Current state of knowledge on biological effects from contaminants on Arctic wildlife and fish. Sci Total Environ 696:133792

Donaldson SG, Van Oostdam J, Tikhonov C, Feeley M, Armstrong B, Ayotte P et al (2010) Environmental contaminants and human health in the Canadian Arctic. Sci Total Environ 408: 5165–5234

Dudley JP, Hoberg EP, Jenkins EJ, Parkinson AJ (2015) Climate change in the north American Arctic: a one health perspective. EcoHealth 12:713–725

Durner GM, Douglas DC, Nielson RM, Amstrup SC, McDonald TL, Stirling I et al (2009) Predicting 21st-century polar bear habitat distribution from global climate models. Ecol Monogr 79:25–58

Frouin H, Loseto LL, Stern G, Haulena M, Ross PS (2012) Mercury toxicity in beluga whale lymphocytes: limited effects of selenium protection. Aquat Toxicol 109:185–193

Grandjean P, Andersen EW, Budtz-Jørgensen E, Nielsen F, Mølbak K, Weihe P et al (2012) Serum vaccine antibody concentrations in children exposed to perfluorinated compounds. JAMA 307: 391–397

Grandjean P, Landrigan PJ (2014) Neurobehavioural effects of developmental toxicity. Lancet 13: 330–338

Grandjean P, Landrigan PJ (2006) Review: developmental neurotoxicity of industrial chemicals. Lancet 16:2167–2178

Gustavson L, Ciesielski TM, Bytingsvik J, Styrishave B, Hansen M, Lie E et al (2015) Hydroxylated polychlorinated biphenyls decrease circulating steroids in female polar bears (*Ursus maritimus*). Environ Res 138:191–201

Hamilton SG, de la Guardia LC, Derocher AE, Sahanatien V, Tremblay B, Huard D (2014) Projected polar bear sea ice habitat in the Canadian Arctic archipelago. PLoS One 9:e113746

Jenssen BM, Dehli Villanger G, Gabrielsen KM, Bytingsvik J, Ciesielski TM, Sonne C et al (2015) Anthropogenic flank attack on polar bears: interacting consequences of climate warming and pollutant exposure. Front Ecol 3:1–7

Johansen P, Muir DCG, Asmund G, Rigét FF (2004) Human exposure to contaminants in the traditional Greenland diet. Sci Total Environ 331:189–206

Kotula AW, Murrell KD, Lacosta-Stein L, Lamb L, Douglass L (1983) Destruction of Trichinella spiralis during cooking. J Food Sci 48(3):765–768. https://doi.org/10.1111/j.1365-2621.1983.tb14894.x

Krey A, Kwan M, Chan HM (2014) In vivo and in vitro changes in neurochemical parameters related to mercury concentrations from specific brain regions of polar bears (Ursus maritimus). Environ Toxicol Chem 33:2463–2471

Landrigan PJ, Fuller R, Acosta NJR, Adeyi O, Arnold R, Basu N et al (2017) The Lancet Commission on Pollution and Health. pii:S0140-6736(17)32345-0

Letcher RJ, Bustnes JO, Dietz R, Jenssen BM, Jørgensen EH, Sonne C et al (2010) Effects assessment of persistent organohalogen contaminants in Arctic wildlife and fish. Sci Total Environ 408:2995–3043

Levin M, Gebhard E, Jasperse L, Desforges JP, Dietz R, Sonne C et al (2016) Immunomodulatory effects of exposure to polychlorinated biphenyls and perfluoroalkyl acids in East Greenland ringed seals (*Pusa hispida*). Environ Res 151:244–250

Lie E, Larsen HJS, Larsen S, Johansen GM, Derocher AE, Lunn NJ et al (2004) Does high organochlorine (OC) exposure impair the resistance to infection in polar bears (*Ursus maritimus*)? Part I: Effect of OCs on the humoral immunity. J Toxicol Environ Health A 67: 555–582

Lie E, Larsen HJS, Larsen S, Johansen GM, Derocher AE, Lunn NJ et al (2005) Does high organochlorine (OC) exposure impair the resistance to infection in polar bears (*Ursus maritimus*)? Part II: possible effect of OCs on mitogen- and antigen-induced lymphocyte proliferation. J Toxicol Environ Health A 68:457–484

Lind PM, Bergman A, Olsson M, Örberg J (2003) Bone mineral density in male Baltic Grey seal (Halichoerus grypus). Ambio 32:385–388

Lind PM, Milnes MR, Lundberg R, Bermudez D, Örberg J, Guilette LJ (2004) Abnormal bone composition in female juvenile American alligators from a pesticide-polluted Lake (Lake Apopka, Florida). Environ Health Perspect 112:359–362

MacKenzie BR, Payne MR, Boje J, Hoyer JL, Siegstad H (2014) A cascade of warming impacts brings bluefin tuna to Greenland waters. Glob Chang Biol 20:2484–2491

McKinney MA, Iverson SJ, Fisk AT, Sonne C, Rigét FF, Letcher RJ et al (2013) Global change effects on the long-term feeding ecology and contaminant exposures of East Greenland polar bears. Glob Chang Biol 19:2360–2372

Messier V, Lévesque B, Proulx J-F, Rochette L, Libman MD, Ward BJ, Serhir B, Couillard M, Ogden NH, Dewailly É, Hubert B, Déry S, Barthe C, Murphy D, Dixon B (2009) Seroprevalence of Toxoplasma gondii Among Nunavik Inuit (Canada). Zoonoses and Public Health 56:188–197. https://doi.org/10.1111/j.1863-2378.2008.01177.x

Molnár PK, Derocher AE, Klanjscek T, Lewis MA (2011) Predicting climate change impacts on polar bear litter size. Nat Commun 2:186

Molnár PM, Bitz CM, Holland MM, Kay JE, Penk SR, Amstrup SC (2020) Fasting season length sets temporal limits for global polar bear persistence. Nat Clim Change 10:732–738

Morris AD, Letcher RJ, Dyck M, Chandramouli B, Cosgrove J (2019) Concentrations of legacy and new contaminants are related to metabolomic profiles in Hudson Bay polar bears. Environ Res 168:364–374

Muir DCG, Wagemann R, Hargrave BT, Thomas DJ, Peakall DB, Norstrom RJ (1992) Arctic marine ecosystem contamination. Sci Total Environ 122:75–134

Nielsen E, Larsen JC, Ladefoged O (2006) Risk assessment of contaminant intake from traditional Greenland food items. Danish Veterinary and Food Administration, 180 pp. www.dfvf.dk

Nuijten RJM, Hendriks AJ, Jenssen BM, Schipper AM (2016) Circumpolar contaminant concentrations in polar bears (Ursus maritimus) and potential population-level effects. Environ Res 151:50–57

Olsen GH, Mauritzen M, Derocher AE, Sørmo EG, Skaare JU, Wiig O et al (2003) Space-use strategy is an important determinant of PCB concentrations in female polar bears in the Barents Sea. Environ Sci Technol 37:4919–4924

Ostertag SK, Stern GA, Wang F, Lemes M, Chan HM (2013) Mercury distribution and speciation in different brain regions of beluga whales (Delphinapterus leucas). Sci Total Environ 456–457: 278–286

Parkinson AJ, Butler JC (2005) Potential impacts of climate change on infectious diseases in the Arctic. Int J Circumpolar Health 64:478–486

Paunescu AC, Dewailly E, Dodin S, Nieboer E, Ayotte P (2013) Dioxin-like compounds and bone quality in Cree women of Eastern James Bay (Canada): a cross-sectional study. Environ Health 12:54

Pavlova V, Grimm V, Dietz R, Sonne C, Vorkamp K, Rigét FF et al (2016a) Modelling population level effects of PCB contamination in East Greenland polar bears. Arch Environ Contam Toxicol 70:143–154

Pavlova V, Nabe-Nielsen J, Dietz R, Sonne C, Grimm V (2016b) PCB contamination as a potential driver of mate–finding associated Allee effect in Svalbard polar bears: the implications from an individual based model. Proc Biol Sci 283:1843

Pedersen KE, Basu N, Letcher RJ, Greaves AK, Sonne C, Dietz R et al (2015) Brain region-specific perfluoroalkylated sulfonate (PFSA) and carboxylic acid (PFCA) accumulation and neurochemical biomarker responses in East Greenland polar bears (*Ursus maritimus*). Environ Res 138:22–31

Pedersen KE, Basu N, Letcher RJ, Sonne C, Dietz R, Styrishave B (2016) Per- and polyfluoroalkyl substances (PFASs) – new endocrine disruptors in polar bears (*Ursus maritimus*)? Environ Int 138:22–31

Pedro S, Boba C, Dietz R, Sonne C, Rosing-Asvid A, Hansen M et al (2017) Blubber-depth distribution and bioaccumulation of PCBs and organochlorine pesticides in Arctic-invading killer whales. Sci Total Environ 601–602:237–246

Pilsner JR, Lazarus AL, Nam D, Letcher RJ, Scheuhammer T, Sonne C et al (2010) Mercury-associated DNA hypomethylation in polar bear brains via the LUminometric methylation assay (LUMA): a sensitive method to study epigenetics in wildlife. Mol Ecol 19:307–314

Polischuk SC, Norstrom RJ, Ramsay MA (2002) Body burdens and tissue concentrations of organochlorines in polar bears (*Ursus maritimus*) vary during seasonal fasts. Environ Pollut 118:29–39

Routti H, Arukwe A, Jenssen BM, Letcher RJ, Nyman M, Bäckman C et al (2010) Comparative endocrine disruptive effects of contaminants in ringed seals (*Phoca hispida*) from Svalbard and the Baltic Sea. Comp Biochem Physiol C Toxicol Pharmacol 152:306–312

Routti H, Atwood T, Bechshøft T, Boltunov A, Ciesielski T, Desforges JP, Dietz R, Gabrielsen GW, Jenssen BM, Letcher RJ, McKinney M, Morris A, Rigét FF, Sonne C, Styrishave B, Tartu S (2019) State of knowledge on current exposure, fate and potential health effects of contaminants in polar bears from the circumpolar Arctic. Sci Total Environ 664:1063–1083

Sandau CD, Ayotte P, Dewailly E, Duffe J, Norstrom RJ (2002) Pentachlorophenol and hydroxylated polychlorinated biphenyl metabolites in umbilical cord plasma of neonates from coastal populations in Québec. Environ Health Perspect 110:411–417

Sonne C (2010) Health effects from long-range transported contaminants in Arctic top predators: an integrated review based on studies of polar bears and relevant model species. Environ Int 36:461–491

Sonne C, Dietz R, Born EW, Rigét FF, Kirkegaard M, Hyldstrup L et al (2004) Is bone mineral composition disrupted by organochlorines in East Greenland polar bears (*Ursus maritimus*)? Environ Health Perspect 112:1711–1716

Sonne C, Leifsson PS, Dietz R, Born EW, Letcher RJ, Hyldstrup L (2006a) Xenoendocrine pollutants may reduce size of sexual organs in East Greenland polar bears (*Ursus maritimus*). Environ Sci Technol 40:5668–5674

Sonne C, Dietz R, Larsen HJS, Loft KE, Kirkegaard M, Letcher RJ, Shahmiri S, Møller P (2006b) Impairment of cellular immunity in West Greenland sledge dogs (*Canis familiaris*) dietary exposed to polluted minke whale (*Balaenoptera acutorostrata*) blubber. Environ Sci Technol 40:2056–2062

Sonne C, Dietz R, Born EW, Rigét FF, Leifsson PS, Bechshøft TØ, Kirkegaard M (2007a) Spatial and temporal variation in size of polar bear (*Ursus maritimus*) sexual organs and its use in pollution and climate change studies. Sci Total Environ 387:237–246

Sonne C, Fonfara S, Dietz R, Kirkegaard M, Letcher RJ, Shahmiri S, Andersen S, Møller P (2007b) Multiple cytokine and acute phase protein gene transcription in West Greenland sledge dogs

(*Canis familiaris*) dietary exposed to organic environmental pollutants. Arch Environ Contam Toxicol 53:110–118

Sonne C, Gustavson K, Rigét FF, Dietz R, Birkved M, Letcher RJ, Muir DCG, Bossi R, Vorkamp K, Born EW, Petersen G (2009) Reproductive performance in East Greenland polar bears (Ursus maritimus) may be affected by organohalogen contaminants as shown by critical body residue modelling and risk quotients estimation. Chemosphere 77:1558–1568

Sonne C, Dietz R, Letcher RJ (2013) Chemical cocktail party in East Greenland: a first time evaluation of human organohalogen exposure from consumption of ringed seal and polar bear tissues and possible health implications. Toxicol Environ Chem 95:853–859

Sonne C, Gustavson K, Rigét FF, Krüger T, Dietz R, Bonefeld-Jørgensen E (2014) Physiologically based pharmacokinetic modeling of contaminants in Greenlanders: a new tool for monitoring levels and possible health effects? Environ Int 64:91–97

Sonne C, Dyck M, Rigét FF, Bech-Jensen JE, Hyldstrup L, Letcher RJ et al (2015) Penile density and globally used chemicals in Canadian and Greenland polar bears. Environ Res 137:287–291

Sonne C, Torjesen PA, Berg KA, Fuglei E, Muir DCG, Jenssen BM et al (2017) Exposure to persistent organic pollutants reduces testosterone concentrations and affects sperm viability and morphology during the mating peak-period in a controlled experiment on farmed Arctic foxes (*Vulpes lagopus*). Environ Sci Technol 51:4673–4680

Sonne C, Jepson PD, Desforges JP, Alstrup AKO, Olsen MT, Eulaers I, Hansen M, Letcher RJ, Mckinney MA, Dietz R (2018) Pollution threatens toothed whales. Science 361:1208

Spörndly-Nees E, Holm L, van Beest FM, Fakhrzadeh A, Ekstedt E, Letcher RJ, Magnusson U, Desforges JP, Dietz R (2019) Sonne C age and seasonal variation in testis and baculum morphology in East Greenland polar bears (Ursus maritimus) in relation to high concentrations of persistent organic pollutants. Environ Res 173:246–254

Tahir E, Cordier S, Courtemanche Y, Forget-Dubois N, Desrochers-Couture M, Bélanger RE, Ayotte P, Jacobson JL, Jacobson SW, Muckle G (2020) Effects of polychlorinated biphenyls exposure on physical growth from birth to childhood and adolescence: a prospective cohort study. Environ Res 189:109924

Tartu S, Bourgeon S, Aars J, Andersen M, Polder A, Thiemann GW, Welker JM, Routti H (2017) Sea ice-associated decline in body condition leads to increased concentrations of lipophilic pollutants in polar bears (*Ursus maritimus*) from Svalbard. Norway Sci Total Environ 576:409–419

Tryland M, Nesbakken T, Robertson L, Grahek-Ogden D, Lunestad BT (2013) Human pathogens in marine mammal meat – a northern perspective. Zoonoses Publ Health 61:377–394

van Beest FM, Aars J, Routti H, Lie E, Andersen M, Pavlova V, Sonne C et al (2016) Spatiotemporal variation in home range size of female polar bears and correlations with individual contaminant load. Polar Biol 39:1479–1489

VanWormer E et al (2019) Viral emergence in marine mammals in the North Pacific may be linked to Arctic Sea ice reduction. Sci Rep 9:1–11

Wielsøe M, Kern P, Bonefeld-Jorgensen EC (2017) Serum levels of environmental pollutants is a risk factor for breast cancer in Inuit: a case control study. Environ Health 16:16

Wielsøe M, Bjerregaard-Olesen C, Kern P, Bonefeld-Jørgensen EC (2018) Receptor activities of persistent pollutant serum mixtures and breast cancer risk. Endocr Relat Cancer 25:201–215

Zoeller RT, Crofton KM (2005) Mode of action: developmental thyroid hormone insufficiency – neurological abnormalities resulting from exposure to propylthiouracil. Crit Rev Toxicol 35: 771–781

Oil Spills in the Arctic

Sadie K. Wright, Sarah Allan, Sarah M. Wilkin, and Michael Ziccardi

1 Introduction to Oil Spills

1.1 What Is Oil?

Oil, including crude oil extracted from the ground and products distilled and refined to isolate and extract different fractions for specific uses, are complex mixtures that can contain thousands of different organic and inorganic compounds. Petroleum products are predominantly (50–98%) composed of various hydrocarbons, divided into four main categories: saturates, aromatics, resins, and asphaltenes. The specific proportions of the chemical components in the product determine oil's physical and chemical properties, behavior in the environment, and toxicity.

Different types of oil vary in their physical, chemical, and toxicological characteristics, and impact the environment in different ways when spilled. In the United States, spill responders classify most oils into five basic groups according to their properties and behavior in the environment: (1) nonpersistent light oils, such as gasoline, are highly volatile and difficult to clean up due to high flammability. They have a high concentration of water-soluble toxic compounds and can cause severe, but generally localized, impacts in the environment. (2) Persistent light oils, such as

S. K. Wright (✉)
NOAA Fisheries, Alaska Region, Protected Resources Division, Juneau, AK, USA
e-mail: sadie.wright@noaa.gov

S. Allan
NOAA Office of Response and Restoration, Assessment and Restoration Division, Anchorage, AK, USA

S. M. Wilkin
NOAA Fisheries, Office of Protected Resources, Silver Spring, MD, USA

M. Ziccardi
Oiled Wildlife Care Network, One Health Institute, UC Davis School of Veterinary Medicine, Davis, CA, USA

M. Tryland (ed.), *Arctic One Health*, https://doi.org/10.1007/978-3-030-87853-5_7

diesel and light crude oils, are less volatile with a lower concentration of water-soluble toxic compounds. They can often be cleaned up effectively but also have the potential to leave residues in the environment and on organisms, with longer-term contamination potential. (3) Medium oils include most crude oils, which can be cleaned up most effectively directly after the product is spilled. These oils can cause severe and long-term impacts to shorelines and feather- and fur-bearing animals. (4) Heavy oils, including heavy crude oils and Bunker C, have a very low concentration of volatile or water-soluble compounds and are difficult to remove from shorelines. They weather very slowly and can cause heavy contamination of shorelines, long-term contamination of sediments, and severe impacts to waterfowl and fur-bearing mammals when fresh. (5) Sinking (or non-floating) oils, such as residual and slurry oils, which do not float on water and show no evaporation or dissolution when submerged. These oils can be removed from water bodies by dredging but can cause long-term contamination of sediments and severe impacts to benthic organisms. There are also oil products that do not fit nicely into these classifications, including diluted bitumen and non-petroleum oils, such as vegetable oil and animal fats.

1.2 Oil Transport and Fate

When oil spills, it begins to change physically and chemically in a process called weathering. Oil spilled on water spreads to form surface slicks and sheens and be transported by wind and currents. It undergoes evaporation, photolysis, oxidation, dispersion, emulsification, and dissolution to varying degrees depending on the oil type and environmental conditions. Oil that is entrained (i.e., incorporated), dispersed, or dissolved in the water is transported by horizontal and vertical diffusion, and may adsorb to particles, undergo sedimentation, or interact with biota (Fig. 1) (NRC 2003). Ultimately, most of the oil is degraded, though the more persistent constituents and oil protected from weathering may remain in sediments or on shorelines for decades (Prince et al. 2002; Lindeberg et al. 2018). The transport and fate of oil from subsurface releases in the ocean (e.g., drilling incidents or well blowouts) differs from surface releases in a number of ways, primarily related to subsurface movement with currents, subsurface entrainment of droplets and dissolved constituents, and weathering before reaching the surface—especially through enhanced dissolution and possibly emulsification. However, in both shallow and deep-water releases, the majority of the oil rises to the surface, forms slicks and sheens, and weathers and degrades (NRC 2003). Arctic conditions, in particular the formation and presence of sea ice, affect oil transport, weathering, and fate (see Regional Considerations for Spills in the Arctic, Sect. 1.5).

Oil spilled in terrestrial environments, including freshwater habitats, is generally more contained and easier to recover than oil spilled in marine waters. Oil spilled on land weathers through evaporation, photo-oxidation, biodegradation, and dissolution, to varying degrees depending on the oil type and environmental conditions. Oil may coat vegetation, can soak into soil, and may contaminate groundwater. Oil flows

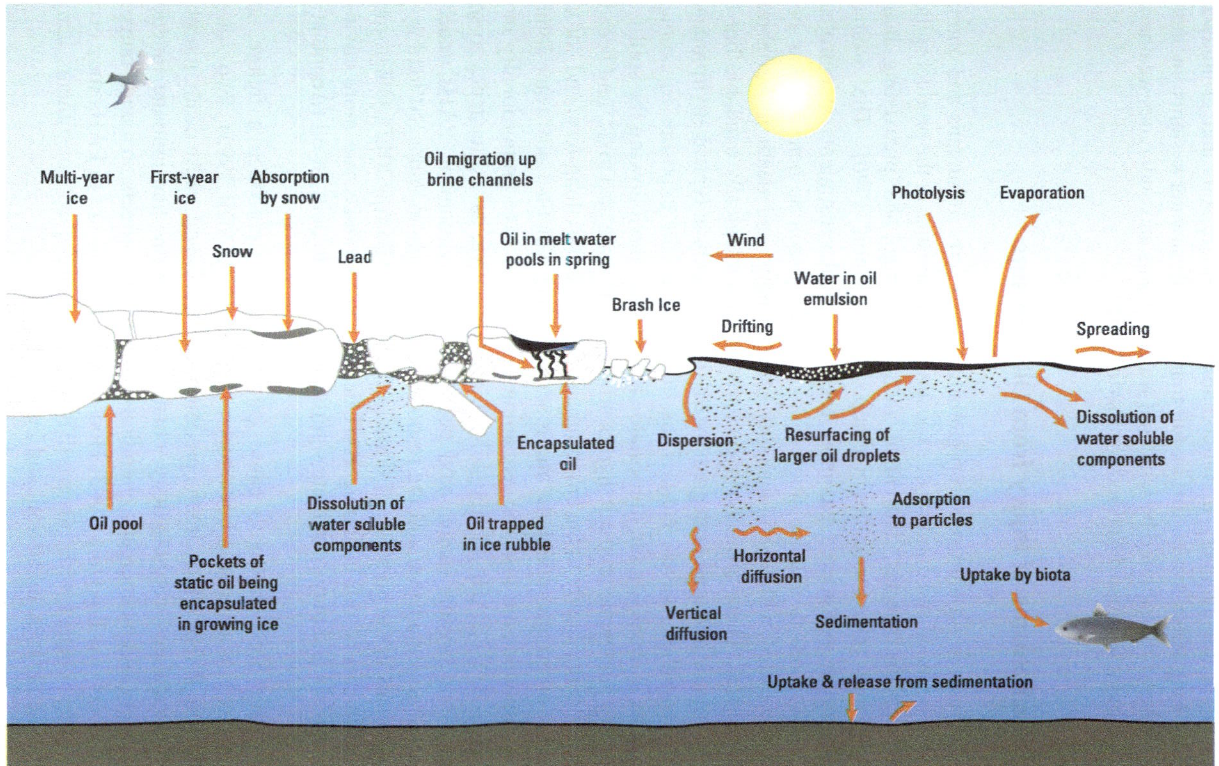

Fig. 1 Transport and fate of oil in open water and ice, including weathering processes and interaction with ice. Reprinted with permission (NRC 2014): modified from (Daling et al. 1990)

over land, following the natural topography. The subsurface movement of oil in soils depends on soil structure, water content, and other factors. Most of the oil ultimately is degraded, though heavier oils can form extremely long-lasting asphalts, and timelines for natural recovery of oil-contaminated anoxic soils and groundwater can be very long (Essaid et al. 2011). Oil that enters freshwater systems, including rivers and lakes, is transported by wind and water currents and undergoes weathering processes similar to those described for marine spills.

1.3 Mechanisms of Injury from Oil Exposure

Many chemicals in oil are toxic to organisms. Oil toxicity to humans and animals depends on the mode and duration of exposure, characteristics of the product spilled, sensitivity of the species, individual factors like age and preexisting health conditions, and other variables (Jessup and Leighton 1996). The aromatic hydrocarbons are considered the most toxic chemical constituents in petroleum products. Exposure to smaller single-ring aromatics, such as benzene, has been strongly associated with carcinogenicity, organopathy, and death in vertebrates (ATSDR 2007a). These aromatic hydrocarbons are biologically available but are extremely volatile and tend to degrade or dissipate quickly in typical environmental conditions (weather, waves, etc.). Thus, the concern for exposure to aromatic hydrocarbons is usually in the immediate vicinity and time following the spill.

Multi-ring polycyclic aromatic hydrocarbons (PAHs) are also toxic, and because they degrade more slowly in the environment, they can cause long-term and chronic impacts. As oil weathers and the more acutely toxic, volatile, and water-soluble organic compounds are lost, the remaining oil tends to have proportionally more PAHs (NRC 2003). PAHs are lipophilic, bioaccumulative, have a tendency to sorb particulates and organic matter, and persist in the environment. PAHs undergo photooxidative degradation and are responsible for the photo-enhanced toxicity of oil to organisms in the presence of ultraviolet radiation (Barron 2017). Other, more polar constituents of oil, including PAH degradation products, are more environmentally mobile and may contribute significantly to the toxicity of oil (Aeppli et al. 2012).

PAHs have been shown to cause carcinogenicity, reproductive failure, and immunotoxicity in animals in laboratory experiments (ATSDR 1995). Two-, three-, and four-ring PAHs, which are the predominant PAHs in oil, cause death, morphological abnormalities, cardiotoxicity, changes to endocrine and immune function, and circulatory and bioenergetic alterations, among other effects, in a wide range of species (Albers and Loughlin 2003; Hodson 2017). Exposure to PAHs may occur through direct contact with oil, passive uptake of dissolved constituents across biological membranes, inhalation, and ingestion.

In addition to the direct injuries identified, some species may experience other injuries as a result of becoming oiled. For birds and heavily furred aquatic mammals (e.g., sea otters, fur seals), the microscopic interlocking of barbules and barbicels in feathers (Albers 1995; Jessup and Leighton 1996) or the density and alignment of interlocking hair bundles (Tarasoff et al. 1972; Williams et al. 1992) creates a waterproof barrier that traps air next to the skin, providing critical insulation and buoyancy in the water. Oil exposure causes the collapse of this microstructure (Hartung 1967; Jenssen and Ekker 1988) and, with its comparatively lower surface tension, can allow water to penetrate deeply into this insulative air layer (Stephenson and Andrews 1997; Newman et al. 2000; O'Hara and Morandin 2010). The result is increased heat loss from the skin and, particularly in an arctic environment, a tendency to become hypothermic. The decreased insulation results in a significantly greater basal metabolic rate (to maintain core body temperature) (Jenssen and Ekker 1991; Jenssen 1994) and increases the animal's vulnerability to starvation due to the use of stored body fat (Hartung 1967; Fry and Lowenstine 1985). In addition to body temperature issues, the removal of this air layer can also cause birds and mammals to lose the capacity to swim or float in the water (McEwan and Koelink 1973; Vermeer and Vermeer 1975), leading to an inability to forage or to escape predators.

1.4 History of Spills in the Arctic

Units of Measurement

In this chapter, we use "gallons" as units of volume. In oil spill response and planning, barrels, or "bbls" is a typical unit of measurement. 42 gallons = 1 barrel (bbl). 1 Mbbl = 1000 bbl. As a reminder, 1 gallon = 3.785 l.

1.4.1 The United States

Alaska, the only state in the United States that extends north of the Arctic Circle, has a long history of oil and gas development and associated spills. Small crude and refined oil spills less than 2100 gallons are fairly common in the Northwest Arctic and North Slope regions, averaging over 85 spills annually (NOAA 2014). From 1995 to 2012, only 0.1% of spill incidents that impacted the ocean in Alaska were greater than 21,000 gallons in volume (NOAA 2014). Although they are rare in Alaska, large oil spills in the marine environment pose a high risk to humans, wildlife, and the environment. The largest spill in Alaska occurred in 1989 in Prince William Sound when the oil tanker *Exxon Valdez* ran aground, spilling over 11 million gallons of crude oil. Although Prince William Sound is south of the Arctic, what we learned from that event regarding cause, response, effects, and restoration is instructive for spill preparedness in the Arctic and subarctic (see EVOS case study, Sect. 3.3). The largest spill within the US Arctic occurred when the bulk carrier *Selendang Ayu* grounded near Unalaska in the eastern Aleutian Islands in December 2004, spilling over 335,000 gallons of fuel in addition to its cargo of soybeans (ADEC 2020).

Several large terrestrial crude oil spills from pipeline incidents have occurred in Alaska since 2000, including a 286,000 gallon spill from the Trans-Alaska Pipeline (TAPS) in 2001, a 212,000 gallon spill from a transit line in 2006, and a 108,000 gallon TAPS pump station release to secondary containment in 2010 (ADEC 2020). The two TAPS spills occurred inland in central Alaska just below the Arctic Circle, while the third (2006, resulting from a corroded pipeline) occurred on the North Slope near Prudhoe Bay.

1.4.2 Canada

Canada also has a long history of oil spills, including major well blowouts, pipeline spills, and spills from vessel and rail accidents. Most of these spills occurred at lower latitudes, and spills in the Canadian Arctic have been infrequent. Oil exploration and development has occurred in the Canadian Arctic since the early 1900s, and offshore drilling in the Beaufort Sea began in the 1970s. Numerous on-land spills originated from the CANOL crude oil pipeline No. 1 in the Northwest Territories and Yukon Territory before it was abandoned in 1945 (Kershaw and Kershaw 1986). A small crude oil spill occurred from the Romulus C-42 well on Ellesmere Island, the northernmost onshore oil producing well in the world, before it was decommissioned in 1972 (Ferguson et al. 2020). A series of small, experimental oil spills were conducted on the northern tip of Baffin Island from 1981 to 1983, providing valuable field data about the natural attenuation of oil under arctic conditions (Prince et al. 2002). In 1970, the tanker *Arrow* ran aground in Chedabucto Bay, Nova Scotia, releasing an estimated 7.5 million gallons of Bunker C that fouled hundreds of miles of shorelines and impacted the fishing industry. This spill occurred in subarctic waters and was the catalyst for major changes to the legal framework related to oil spills in Canada.

1.4.3 Russia

Identifying detailed information regarding oil spills in Russia is challenging due to shifting political climates and reporting procedures. Many significant spills have occurred at the main production and transport areas in or near Russian territorial waters in the Black and Caspian Seas, including a substantial leak from an oil pipeline near Tuapse in 2014 and a tanker release from *Volganeft 139* of at least 748,000 gallons into the Kerch Strait between Ukraine and Russia in 2007 (Fashchuk 2011). Reported spills in the Arctic region of Russia include a series of large crude oil spills from the Komi oil pipeline (74.8 million–187 million gallons) near Usinsk in 1994 (Sagers 1994). The oil was initially contained within a dike, but a collapse in October 1994 sent the oil into the taiga forests inhabited by the indigenous Komi people. ITOPF reports that no major oil spills occurred in the Russian Federation from 1999 to 2009, but numerous smaller spills (up to 150,000 gallons) occurred (ITOPF 2012). More recently, a significant release of diesel fuel (estimated at 5.55 million gallons: 1.59 million on land and 3.96 million into the Daldykan and Ambarnaya rivers) was released in May 2020 near Norlisk from a holding tank supplying a power plant (Nechepurenko 2020). The spill contaminated an area of 350 km^2 and directly affected an area of about 0.18 km^2 near the Daldykan River, causing the Russian government to declare a regional state of emergency. In the company's press release, it was acknowledged that "(t)he accident was caused by a sudden sinking of supporting posts in the basement of the storage tank," as a result of thawing of the permafrost on which it was built (BBC News 2020).

1.4.4 Europe

Significant oil spills have also occurred in the arctic waters of European nations, primarily Norway. Oil exploration off Norway began in the North Sea in the late 1960s with the discovery of the Ekofisk offshore oil field in 1969. Production began at Ekofisk in 1971, and in 1977, the *Bravo* platform experienced an oil and natural gas blowout, resulting in the first major oil release in the North Sea with over 8.4 million gallons of oil estimated spilled before the well was capped (NOAA 1977). Oil production in Norwegian waters has moved further north, beginning in the Norwegian Sea in 1993 and in the Barents Sea in 2007 (Norwegian Petroleum 2020). Vessel-based spills have also occurred in Norway, including the bulk carrier *Server* in 2007, which spilled an estimated 280,500 gallons of international fuel oil (IFO) 180, and the bulk carrier *Full City* which grounded in 2009 and contaminated approximately 100 km of coastline with IFO 180, although neither of these spills occurred north of the Arctic Circle (ITOPF 2018). The former USSR collaborated in response to two spills in the Baltic Sea under bilateral/regional agreements: the tanker *Antonio Gramsci* in 1987, which ran aground off the south coast of Finland, spilling 3.74 million–4.49 million gallons of crude oil (Bonsdorff 1981), and the Soviet tanker *Volgoneft 263*, which collided with a West German cargo ship *Betty* off of the Swedish coast, releasing 264,200 gallons of waste oil in 1990 (Fagoe 1991).

1.5 Regional Considerations for Spills in the Arctic

1.5.1 Response

The remoteness and extreme weather in the Arctic are key factors that make oil spill response more challenging than at lower latitudes. Transportation and infrastructure generally are less available and less consistent the further north you go. Winter snow and ice conditions can slow or halt ground transportation, as can summer permafrost thaw-induced flooding. Vessels or aircraft can provide fairly consistent transportation to arctic communities with established ports and landing strips during the summer, but access becomes more difficult during the dark winter months when ice can limit or preclude vessel access and blizzards can ground flights for lengthy periods. Infrastructure, personnel, commercial services, communications, and equipment are either unavailable or much more limited than in temperate climates, which can limit or significantly delay response in the Arctic.

Interaction with large, dangerous, or unpredictable oiled and unoiled wildlife is possible in arctic responses, including predators such as polar bears, and other large species, such as ice-adapted pinnipeds (e.g., bearded seals, walrus) and whales. Responders navigating through ice leads may encounter higher numbers of wildlife aggregated in the available open water.

1.5.2 Oil Transport and Fate

The transport and fate of oil that is spilled into waters with sea ice depends on the timing of the release relative to sea ice formation or breakup and the type of ice present. If oil is spilled when sea ice is forming, oil may become encapsulated in ice within hours, where it will cease to weather and be transported with the ice until the spring thaw releases the oil (NRC 2014). If oil is released below already formed sea ice, it will spread beneath the ice, pool in the rough surfaces under the sea ice, and be transported with the ice as it moves, unless the water current is strong enough to move the oil (Fig. 1) (Dickins and Buist 1999; Wilkinson et al. 2017). For subsea releases under ice, most weathering will not be significant until the oil reaches the surface of the ice or open water (Sørstrøm et al. 2010). Sea ice may also limit the spread of oil on water and concentrate it in leads, polynyas, and other open water areas. Evaporation will be reduced in areas where surface slicks are thicker, and wave-damping will reduce dispersion and emulsification, leading to overall slower weathering rates (Evers et al. 2004; Singsaas et al. 2020). Oil will drift separately from sea ice when ice coverage is low (less than approximately 30%), but will move with the ice when there is greater coverage (more than approximately 60–80%) (NRC 2014; French-McCay et al. 2017). Oil-degrading bacteria are present in arctic waters and degradation is not inhibited by cold water (McFarlin et al. 2018), though cold temperatures may limit biodegradation through other mechanisms and degradation in or under ice may be limited by lack of oxygen or nutrients (NRC 2014). Degradation rates for oil stranded on arctic shorelines are lower than in temperate regions, primarily due to seasonal snow and ice coverage (Prince et al. 2002).

The transport and fate of oil spilled in terrestrial environments in the Arctic depends on environmental conditions and timing. Oil spilled in the summer will undergo weathering and transport processes similar to those described for non-arctic environments, although tundra soils are often saturated with water, which may prevent oil penetration, and underlying permafrost and ice wedges limit subsurface vertical and horizontal spread. Though a limiting factor, oil can move in the seasonally thawed layer above the permafrost, especially in drier soils (Kershaw and Kershaw 1986), and there are documented cases of limited vertical migration into the permafrost layer (Biggar et al. 1998). Most of the spilled oil is likely to spread and pool on the surface. Degradation rates will be slower than at lower latitudes due to lower temperatures and seasonal freezing and snow cover.

Oil spilled on land when snow and ice are present will weather much more slowly. Oils with low viscosity, such as diesel, will flow downslope, under snow and over frozen ground or ice, accumulating in natural low points. Under these conditions, the oil can persist with minimal weathering until it is recovered or ice breakup occurs. Oils with higher viscosities will spread less or not at all. Snow will absorb large volumes of spilled oil. Oil-contaminated snow can be transported by wind, and can be collected and contained for spill cleanup. At spring breakup, any oil that was not recovered will still be relatively fresh and may impact aquatic systems or terrestrial plants and animals as it weathers and degrades. Because of the short summer season in the Arctic, it may take decades for oil to degrade (Prince et al. 2002).

1.5.3 Differences in Exposure Pathways

Differences in oil transport and fate can change oil exposure pathways for wildlife in arctic environments. In particular, the presence of sea ice affects how wildlife are exposed to oil. In the presence of sea ice, oil will concentrate at the ice-water interface, either under the ice or in leads and open water between ice (Fig. 1). In both cases, the ice will limit the spread of the oil, concentrating it in thicker water-surface or below-ice slicks. The underside of sea ice is a highly productive habitat where algae grow and zooplankton and fish congregate to feed (Fig. 2). This convergence can lead to longer exposures to higher concentrations of oil for organisms living on or right below the underside of the sea ice. Leads in sea ice are where marine mammals and birds surface to breathe, enhancing the likelihood, duration, and concentration of exposure to oil from direct contact and through respiration.

If oil is entrained (Petrich et al. 2013) or encapsulated in ice and transported with the ice pack, exposure may be delayed until thaw and fresh oil could impact water and wildlife far from where it was originally released. Oil can penetrate brine channels and migrate vertically, ultimately pooling on the surface of the ice (Oggier et al. 2020) and creating an exposure risk for wildlife that spend time there, including seals and polar bears. These same processes may also prolong the window of

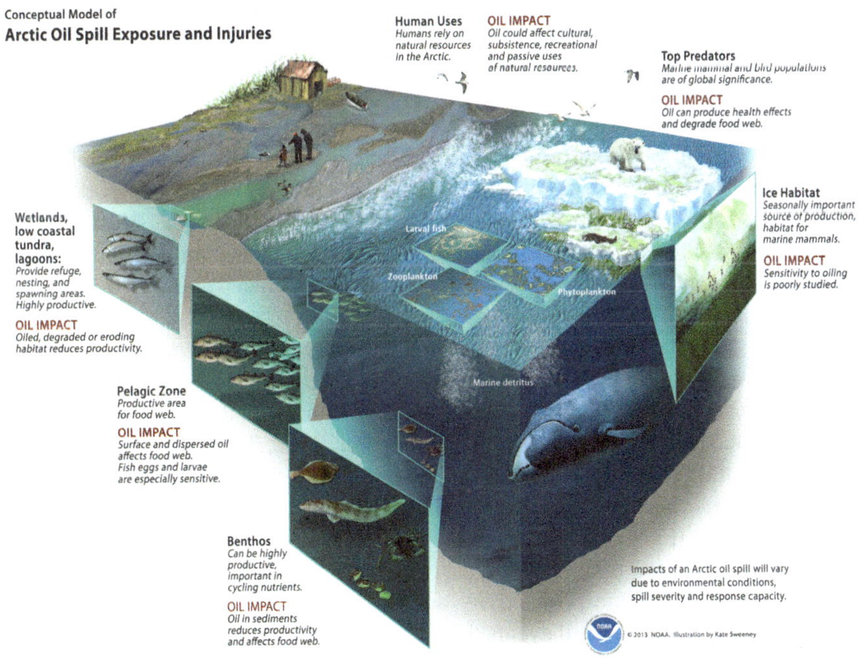

Fig. 2 Conceptual model showing anticipated exposure and injuries associated with oil spills in the Arctic. NOAA illustration by Kate Sweeney

opportunity for certain spill response tactics that are more effective with more concentrated and less weathered product.

2 Impacts of Oil Spills in the Arctic

2.1 Environment

2.1.1 Terrestrial, Freshwater, Wetlands, and Low Coastal Lagoons

Oil spills on land can cause a range of impacts in terrestrial systems. Tundra plants can be killed, even in lightly oiled areas, by smothering, chemical toxicity, and temperature changes (McKendrick and Mitchell 1978). Long-term impacts on plant cover and plant community composition are expected in the absence of cleanup and restoration (Kershaw and Kershaw 1986). The effects of oil on bacterial community profiles in arctic soils may persist for many decades at spill sites (Ferguson et al. 2020). Oil pollution of soil surfaces increases absorption of solar energy and may increase the depth of the seasonally thawed layer (Everett 1978; Chelombitko 2019), leading to the development of thermokarst terrain (i.e., irregular hummocks and marshy ponds) and subsequent degradation of the area (Chelombitko 2019). In rivers, oil will form sheens and may be rapidly dispersed, diluted, and moved downstream with the current, depending on the oil type, water volume, river morphology, flow rates, and other factors. Spilled oil can contaminate bottom and bank sediments, which then become a longer-term source of contamination to aquatic and riparian vegetation, benthic organisms, and the water. If oil enters a river or lake that is frozen, the ice cover will reduce weathering and evaporative loss and enhance dissolution, leading to greater toxicity in the water. Oil spilled in lakes and ponds will impact both primary production by phytoplankton and zooplankton abundance and community composition (Miller et al. 1978). Microbial populations in arctic freshwater ecosystems will respond to the presence of petroleum hydrocarbons, shift their composition, and effectively degrade many oil constituents. However, microbial degradation will slow significantly when waters are ice-covered (Horowitz and Atlas 1977).

Coastal lagoons, dynamic transition zones between freshwater and marine ecosystems, play a key role in arctic habitats, providing rich feeding and breeding grounds for arctic fish, migratory birds, marine mammals, and critical resources for local subsistence hunters and fishers (Dunton et al. 2012; Tibbles 2018). Wetlands make up as much as 60% of all arctic ecosystems and approximately 37% of the Alaskan Arctic coastline north of the Bering Strait (Gundlach and Hayes 1978; Haynes et al. 2017). Wetlands include a complex mix of bogs, fens, swamps, marshes, and shallow open water areas, providing crucial ecosystems functions. Because of their extreme productivity and sensitivity, these environments are at significant risk during oil spill events, leading responders to classify them among the most vulnerable to oil spill damage (Gundlach and Hayes 1978) (Fig. 2). In coastal wetlands and lagoons, oil type appears to be one of the most significant factors for predicting effects, with lighter oils more acutely toxic than heavier oils (Michel and

Rutherford 2013). Heavier oils in more protected areas will persist longer. Protection from oil entering these habitats, coupled with effective containment and cleanup of a spreading slick, are key functions in reducing effects. Some techniques to remove oil from coastal wetlands and lagoons may actually cause more harm to the habitat than allowing natural recovery to occur.

2.1.2 Marine

Oil spilled in the arctic marine environment can degrade the quality of pelagic, benthic, ice, nearshore, and shoreline environments (Fig. 2). Surface and dispersed oil and dissolved oil constituents will impact the productive pelagic environment, potentially disrupting food webs. Phytoplankton, zooplankton, and fish eggs and larvae that inhabit the upper part of the water column are critical components of the food web and especially vulnerable to oil exposure and resulting death or other toxic effects. The presence of oil in the pelagic environment can cause injury to fish, birds, and marine mammals, both by exposing them directly to oil and by altering or contaminating their prey base.

Some portion of the spilled oil can be transported to the benthos through sedimentation processes, including as marine snow (Daly et al. 2016; Ross et al. 2021). The benthic environment is highly productive, and in some parts of the Arctic, especially where there are shallower shelves or stable polynyas, there is tight benthic-pelagic coupling (Hobson et al. 1995; Dunton et al. 2005). Oil in the benthos may reduce productivity and nutrient cycling by altering the benthic community and could kill or contaminate important prey species, such as mollusks and isopods (Percy 1978). Oil degrades slowly in benthic sediments, and arctic benthic environments are expected to recover slowly following an impact (Conlan et al. 1998; Konar 2013).

2.2 Animals

2.2.1 Birds

For animals to survive in harsh, cold, arctic conditions, the primary challenge is to balance heat loss versus metabolic heat production; this is primarily done through control of heat loss via behavioral, physical, and physiological adaptations. Birds in the Arctic are particularly reliant on the interlocking mechanical structure of their feathers to trap a blanket of air next to their skin, and oil can negatively impact that protection. The majority of wildlife observed to be affected by most oil spills are seabirds, waterfowl, and shorebirds. In the Arctic, alcids (such as murres or guillemots) and diving ducks are species most prone to being found acutely oiled in large numbers (Piatt et al. 1990), but many other species may become oiled and are never found, or are more chronically affected, leading to population-level effects. Both crude and refined petroleum products affect birds externally in a similar manner. As described above, oiled feathers quickly lose their waterproofing and insulating properties, resulting in loss of buoyancy and subsequent hypothermia (Jenssen and Ekker 1988; Albers 1995). Affected birds may drown at sea or, if they are able to make it to shore, are vulnerable to dehydration, starvation, and/or predation. In addition, oiled birds spend more time preening, have difficulty foraging for prey, have decreased flight capacity, have an increased metabolic rate due to heat loss, and may have skin and corneal burns (Jessup and Leighton 1996). These difficulties often lead to dehydration and malnutrition exhibited first through loss of body fat and then, subsequently, to muscle wasting (Leighton 1993; Bursian et al. 2017).

Exposure of seabirds and eggs to oil in the breeding season, through both physical coating of the eggs and effects of ingested or inhaled components, may result in reproductive failures due to stress, changes in reproductive behaviors, failed laying, embryo mortality, teratogenesis, failed hatching, and increased chick abandonment (Leighton 1993). Eggs can become oiled from feathers of brooding birds or contaminated nesting material. External oiling of eggs causes reduced embryonic viability at higher concentrations and protracted embryonic development and cardiotoxicity at lower concentrations (Goodchild et al. 2020).

Ingested or inhaled petroleum products can result in a variety of acute and long-term health effects in birds. Ingested oils can result in irritation and erosion of the lining of the gastrointestinal tract, resulting in bleeding, decreased nutrient uptake, and dehydration from ensuing diarrhea (Fry and Lowenstine 1985; Tseng and Ziccardi 2019). Biotransformation of oil metabolites by the liver can lead to cellular damage, altering liver metabolic function and affecting immune function and production of cholesterol (Peakall et al. 1989; Harvey 1991; Briggs et al. 1996). Oxidative damage from exposure to PAH metabolites can lead to hemolytic anemia (Leighton et al. 1985; Harr et al. 2017) and further lead to liver impairment via hemochromatosis (Yamato et al. 1996; Balseiro et al. 2005). The ingestion of oil (via preening or through food items) can cause significant damage to the renal system, manifested by lesions and alterations in metabolic function (Fry and Lowenstine 1985; Harr et al. 2017). Inhalation of volatile fumes may lead to lesions within the upper or lower portions of the respiratory tract in birds.

2.2.2 Marine Mammals

Marine mammal exposure to spilled oil may occur through multiple pathways (Ziccardi and Wilkin 2018). Mammals breathing at or near the surface of the water may be exposed to the volatiles and aerosols related to the spilled product, which can be inhaled or aspirated (Fig. 3), potentially leading to neurological issues (Spraker et al. 1994) and/or lung damage (Schwacke et al. 2014). Marine mammals may directly encounter surface oil, causing impacts to the skin, eyes, or mucous membranes. Baleen whales may experience fouling as they skim feed at the surface or within the water column, as oil or oil/water mix coats the baleen (Geraci 1990), although recent laboratory experiments suggest this may not be as great a concern as previously feared (Werth et al. 2019). Oil in the water column may also be bioaccumulated by prey species (zooplankton or fish), with marine mammals subsequently exposed by ingestion of oiled prey or oil/water mixtures (Fig. 4). Oil that sinks to the benthos may likewise contaminate benthic prey that are consumed by some marine mammals. Shoreline or ice surface oiling provides an additional

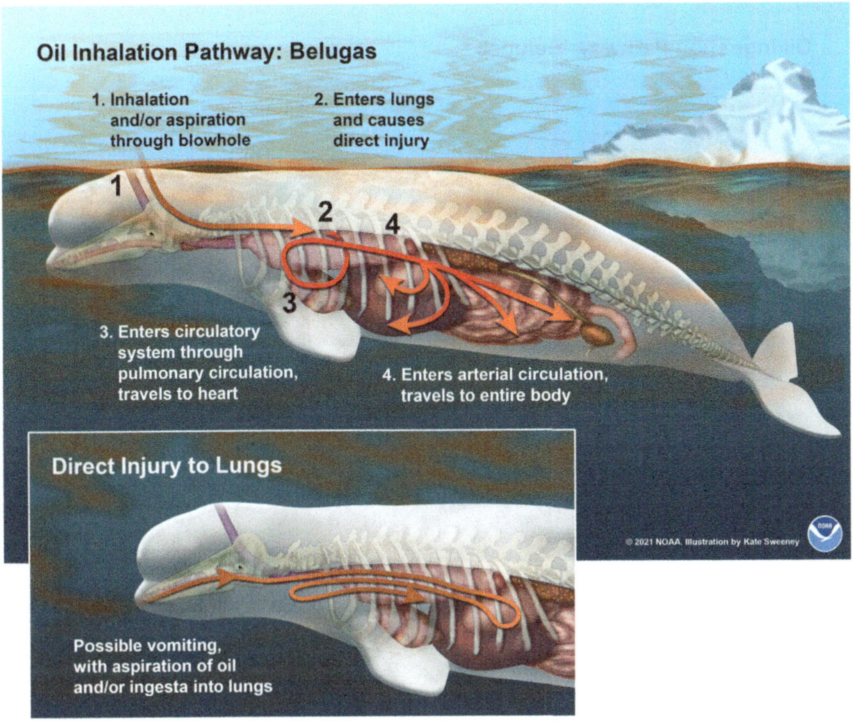

Fig. 3 Conceptual model of inhalation pathways of a marine mammal (beluga whale) exposed to spilled oil. NOAA illustration by Kate Sweeney

exposure pathway for pinnipeds that may haul out in those areas, and may result in fouling of fur, dermal exposure, and ingestion. Oil concentrated under ice may likewise contaminate animals transiting, foraging, or creating breathing holes.

In addition to these direct impacts from a spill, indirect effects may result from the marine mammal losing access to prey species or preferred habitat. Pinnipeds may choose not to haul out on a preferred beach or ice flow due to oil spill response activities and may have to swim longer distances, which increases energy expenditure. Animals may be forced to switch prey or forego feeding, resulting in nutritional stress.

Depending on the duration and severity of both direct and indirect impacts, animals may lose energy reserves, experience organ damage, be more susceptible to disease, and may experience reproductive loss or decreased fecundity and ultimately decreased survival. In contrast to birds, there may not be much acute mortality observed of marine mammals in the immediate aftermath of an oil spill, but sublethal impacts are likely to occur that may result in mortalities or population reduction and failure to recover over long periods of time, as was observed following the *Exxon Valdez* oil spill in 1989 (Matkin et al. 2008) and *Deepwater Horizon* oil spill in 2010 (Schwacke et al. 2014; Takeshita et al. 2017).

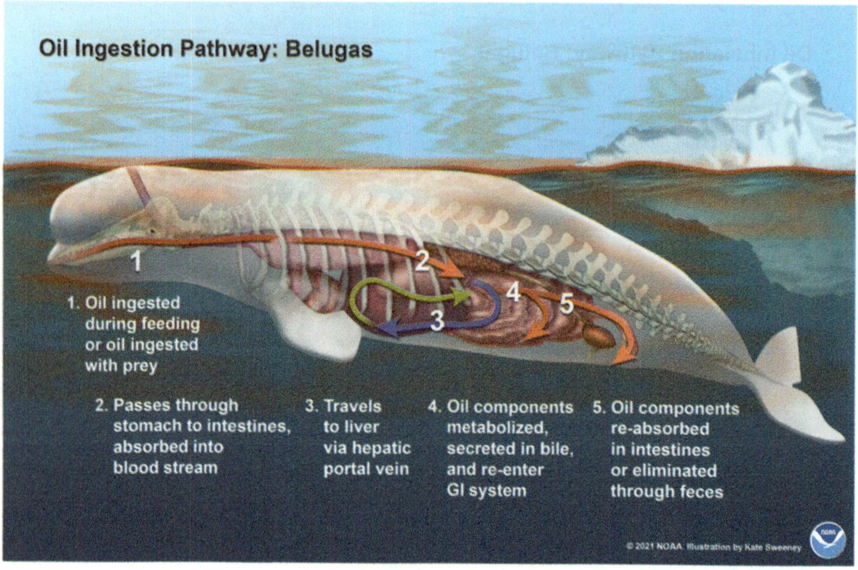

Fig. 4 Conceptual model of ingestion and aspiration pathways of a marine mammal (beluga whale) exposed to spilled oil. NOAA illustration by Kate Sweeney

Marine mammals in the Arctic include species that occur north of the Arctic Circle for most of the year and depend on the arctic ecosystem for all aspects of life, such as bowhead whales (*Balaena mysticetus*), narwhals (*Monodon monoceros*), ringed seals (*Phoca hispida*), bearded seals (*Erignathus barbatus*), walrus (*Odobenus rosmarus*), and polar bears (*Ursus maritimus*) (Laidre et al. 2008). Other marine mammal species migrate to the Arctic seasonally for essential parts of their life history (e.g., pupping in ice lairs, foraging), such as gray whales (*Eschrichtius robustus*), hooded seals (*Cystophora cristata*), and harp seals (*Pagophilus groenlandicus*). Finally, there are species that have a broader geographic distribution, but stocks or populations may remain within arctic or subarctic waters year-round, such as beluga whales (*Delphinapteras leucas*). Marine mammals are critical to arctic ecosystems and a fundamental cultural and subsistence resource for arctic peoples (Moore and Hauser 2019).

2.2.3 Fish

Fish may be exposed to spilled oil by spending time on or in contaminated sediments, taking up contaminants through body surfaces (including respiratory structures such as gills), and ingesting oil droplets, particulates contaminated with oil, and contaminated prey. The impacts of oil exposure on fish depend on the oil type, exposure conditions, species, and life stage, among other factors. Juvenile and adult fish that are exposed to oil may suffer acute mortality or develop other injuries that lead to delayed mortality. Early life stages of fish are highly sensitive to oil exposure, and even minimal exposure during embryonic development can lead to acute or delayed mortality (Incardona et al. 2015; Laurel et al. 2019). The presence of ultraviolet radiation enhances the toxicity of oil in early life stages of fish (Barron 2017), though ice, when present, may provide some protection from photoenhanced toxicity.

In the Arctic, fish are prey for many other fish, birds, and marine mammals, as well as an important subsistence resource for humans. Polar (or Arctic) cod (*Boreogadus saida*) is a keystone species in ice-associated arctic marine food webs because it provides a critical linkage between lower and higher trophic levels (Huserbråten et al. 2019). Polar cod is one of the most abundant and important fish species in circumpolar arctic ecosystems, and their life history makes them especially vulnerable to oil exposure. They spawn under ice, produce buoyant eggs, and grow into larvae that feed on ice-associated zooplankton, all at the ice-water interface where the highest concentrations of spilled oil would likely be present (Fig. 2). Oil exposure can cause acute mortality or a range of morphological and bioenergetic effects that lead to delayed mortality (Laurel et al. 2019). If an oil spill led to changes in the distribution and abundance of polar cod, it would likely have consequences for the entire arctic food web (Bluhm and Gradinger 2008; Choy et al. 2017).

2.2.4 Invertebrates

Aquatic invertebrates may be exposed to spilled oil in much the same way as fish, through contact with contaminated water and sediment and ingestion of oil droplets, particulates contaminated with oil, and contaminated food. Unlike fish, invertebrates do not readily metabolize many of the toxic chemicals in oil and can bioaccumulate PAHs and other lipophilic constituents (Agersted et al. 2018; Szczybelski et al. 2019). Lipid-rich arctic invertebrates, such as pelagic copepods and euphausiids (krill) and benthic mollusks, are important food sources for fish, birds, and marine mammals. Their high lipid content and slow elimination of oil chemicals (Nørregaard et al. 2015; Agersted et al. 2018) may create a persistent risk of trophic transfer of these constituents from invertebrate prey to higher-level consumers following an oil spill.

Arctic invertebrates can also suffer toxic effects from oil exposure (Hannam et al. 2010; Gardiner et al. 2013). Similar to fish, invertebrates that live at the ice-water interface, in the upper part of the water column, or in nearshore areas are most at risk for exposure. Deeper-water benthic invertebrates can also be exposed in situations where oil reaches the seafloor (Fig. 2). Though limited toxicity data are available for arctic species, there is not currently evidence that arctic invertebrates are inherently more sensitive to oil than temperate species (Camus et al. 2015; Bejarano et al. 2017). However, the impact of exposures that are above toxic thresholds on populations and communities in the Arctic is not well understood. Because of their importance to arctic food webs, changes in the abundance of invertebrates could affect energy flow throughout the ecosystem, and recovery of these populations is likely to be slower in the Arctic than at lower latitudes.

2.3 Humans

Humans are known to be at risk of adverse health effects from direct exposure to oil and its constituents. Of greatest concern are the chronic effects of benzene and some PAHs and their association with cancer. Risks of occupational PAH exposure were first recognized by Sir Percival Pott in 1775 through increased risk of scrotal cancer in chimney sweeps in England (Waldron 1983). Since that time, strong experimental and epidemiological evidence has linked numerous volatile congeners produced by petroleum to health impacts. Benzene has been shown to disrupt neuronal membranes and also causes hematological effects leading to aplastic anemia and acute myelogenous leukemia (ATSDR 2005). Other compounds in the BTEX group (benzene, toluene, ethylbenzene, and xylene) can also have acute effects on respiratory and central nervous systems (ATSDR 2007b, 2010). Further, 17 different PAHs have been identified as a great concern for inducing cancer, primarily skin and lung cancer, but also bladder and gastrointestinal cancers (ATSDR 1995), and have been shown to both cross the placenta and cause male idiopathic infertility (Madeen and Williams 2017). These facts are concerning not only for the direct environmental exposures of humans to oil and its byproducts but also the potential of bioaccumulation of these compounds in food items that people may consume. Much information has come out of the *Deepwater Horizon* oil spill related to risks associated with eating seafood from a spill location. While data showed that individual PAHs were found in low concentrations and, when detected, were at least two orders of magnitude lower than the level of concern for human health risk (Ylitalo et al. 2012), others argue that analyses were not extensive enough to completely interpret human risk (Farrington 2020).

While most epidemiologic studies have focused on the acute effects of crude oil exposure during spill cleanup, effects have been documented years after a spill response. Responders to the 2002 *Prestige* oil spill self-reported having respiratory symptoms 5 years after the spill (Zock et al. 2012), and similar findings occurred 1–3 years after the *Deepwater Horizon* response among decontamination workers and workers with high exposure to burning oil/gas compared with unexposed workers (Gam et al. 2018), though effects appear to have resolved after 4–6 years (Lawrence et al. 2020). There is also increasing evidence that exposure to total hydrocarbons and volatile organic compounds from air pollution during spills (especially spill responses using in-situ burns as a clean-up method) is associated with increased risk of coronary heart disease (Strelitz et al. 2019).

In addition to the direct affects to human health, oil spills impact regional community function. Indigenous peoples of the Arctic have hunted and gathered the local resources they need to survive and flourish in the extreme environment for thousands of years. Many subsistence communities are reliant on the plants and wildlife sustained by the unique arctic ecosystem to provide essential traditional food and other needed resources for their survival and cultural identity. A very large oil spill or a series of smaller spills in the same area have the potential to impact the subsistence way of life in the Arctic. Direct oil exposure in wildlife species (marine mammals, birds, fish, shellfish, etc.) that provide food to subsistence communities poses a health risk. Avoiding harvesting animals that may have been exposed to oil due to uncertainty about toxicity would reduce the available traditional food resources for subsistence communities.

Components of economies tied to healthy wildlife populations and a clean environment would suffer as a result of a large oil spill or chronic smaller spills. Arctic aquaculture is a growing commercial enterprise, particularly for salmon, with Norway having a sizeable aquaculture industry. Aquaculture sites would be especially impacted by an oil spill, as they would be impractical to relocate. Commercial fishing activities could be impacted in both the short and long term; while some dynamic shifting of fishing practices might be possible depending upon the parameters of the spill (i.e., relocating to a different geographic region for the season), it would increase costs and may not result in the same yield. Finally, tourism activities could be impacted, including businesses that rely upon tourists such as hotels, restaurants, native handicraft shops, whale and wildlife viewing companies, and cruise lines with arctic routes.

Additionally, the co-stressors that exposure to PAHs can have during spills with other physiological and psychosocial stressors are becoming more well understood following studies of recent spills. Beyond the direct effects described above, disasters affecting the health, economy, and quality of life in communities may lead to psychosocial impacts including anxiety, stress, and depression associated with being impacted by a large-scale disaster (Lowe et al. 2019; Ritchie and Long 2021). The effects of this stressor are just now becoming more widely understood and could have a tremendous impact in arctic communities where culture and livelihood are tied so closely to the natural environment.

3 Preparedness, Response, Restoration, and Prevention

3.1 Preparedness

Formal planning for oil spill response in the Arctic happens at many levels. For example, in Alaska, US federal laws regarding oil spill preparedness apply throughout state and federal waters, State of Alaska laws apply to state lands and waters, some communities have worked with industry to establish agreements to mitigate impacts to subsistence harvest should a spill occur, and non-governmental organizations have developed plans and outreach materials to aid preparedness and communication between community members and government agencies before and during spills. Wildlife trustee agencies have developed regional plans in collaboration with communities and wildlife response organizations (NMFS 2017; ARRT 2020), and taxon-specific plans (USFWS 2015; Ziccardi et al. 2015) in an effort to prepare for and streamline oiled wildlife response. Planning ahead is essential for caching spill response equipment in accessible locations in advance of spills to mitigate the risk of transportation limitations during a large response.

Oil exploration and development operations throughout the Arctic are largely required to have some level of spill contingency planning, but the criteria and oversight vary by country and region. History has shown that new laws and regulations have been established following major oil spills in an effort to fill gaps identified by the accident and attempted spill response (e.g., US Oil Pollution Act of 1990 [OPA 90] created in the aftermath of the 1989 Exxon Valdez Oil Spill, and recent amendments to the Law on Environmental Protection by the Government of the Russian Federation in the aftermath of the 2020 Norilsk diesel spill).

The challenges of responding to oil spills in the Arctic, including the dearth of response infrastructure and resources in the region and the potential for cross-border or multinational impacts, necessitate international cooperation on oil spill response. There are national and international laws and conventions for establishing mutual aid or mutual assistance agreements. The 1990 International Maritime Organization Convention on Oil Pollution Preparedness, Response, and Cooperation provides a global framework for cooperation in the event of major spills and provides the foundation for specific agreements between nations. For example, the United States has bilateral agreements with Russia and Canada related to oil spill preparedness and response. Additionally, the Agreement on Cooperation on Marine Oil Pollution Preparedness and Response in the Arctic (MOPSA) was signed by all eight Arctic States in 2013 and entered into force in 2016 (Arctic Council 2013). The objective of MOPSA is to strengthen cooperation, coordination, and mutual assistance among the Parties to prepare for responses to protect the marine environment from oil pollution.

3.2 Response Activities

3.2.1 Primary Response

A number of tactics exist to remove oil from the environment following a spill. These are generally referred to as "primary" response tactics and can be categorized as mechanical, physical, or nonmechanical (which includes in-situ burning and chemical/biological response).

Mechanical recovery of oil includes the use of booming, barriers, skimmers, and sorbent materials to capture and remove oil from the environment (Fig. 5). Booms are typically used for in-water spills and are floating devices akin to long snakes that contain surface oil on one side and prevent it from spreading further. Some booms are fireproof and can contain on-water in-situ burns of oil (Fig. 5). Skimmers are mechanical devices that collect oil on the water's surface and can be used in conjunction with booms. Sorbent materials can be used in conjunction with booms or on land to recover oil. Physical recovery of oil generally includes all shore-based or terrestrial cleaning and recovery of oil. This includes pressure washing, flushing and flooding with clean water (water can be heated to enhance removal), wiping with sorbents, steam cleaning and sandblasting, and removal of contaminated soil, sediment, vegetation, animal carcasses, or other natural debris (NMFS 2015).

Chemical or biological response tactics include the use of dispersants or herders. Dispersants are applied directly to oil and are designed to break the oil into smaller droplets to enhance natural weathering and removal processes (Fig. 5, and see Oil Transport and Fate, Sect. 1.5.2). Dispersants do not reduce the total amount of oil in

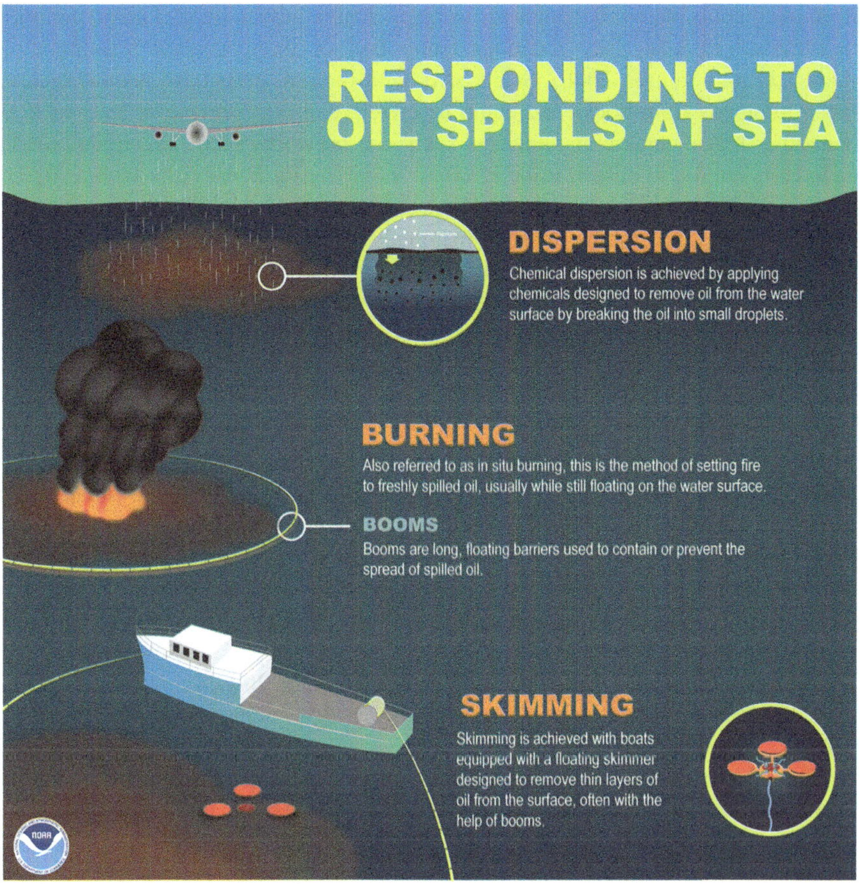

Fig. 5 Illustration showing several methods for on-water oil spill response designed to remove oil from the environment. Provided by NOAA

the environment but, instead, may change the characteristics of the oil, thereby changing the transport, fate, and potential effects of the oil (NMFS 2015). Other nonmechanical countermeasures include application of other chemical agents (e.g., solidifiers and fire foam), and application of biodegrading organisms or nutrient stimulants used to enhance biodegradation of oil.

Implementation of primary response tactics will be particularly challenging in the Arctic due to environmental conditions and paucity of infrastructure, equipment, and personnel. The same extreme weather conditions and lack of daylight in the winter that increases the risk of spills will hinder primary response.

3.2.2 Secondary Response

Secondary oil spill response tactics includes efforts to keep wildlife away from oil. Animals can be hazed using visual or auditory devices designed to scare them away from oiled areas, deterred from areas using exclusion devices (e.g., fences, netting), or attracted to less risky locations using bait or environmental manipulation (Gorenzel and Salmon 2008). Scarecrows with moving parts have been shown to deter pinnipeds from docks and could be effective at oiled terrestrial sites such as haulouts. Noise-generating devices such as fireworks and electronic sound generators have also shown to be effective in hazing birds. Devices such as Breco buoys and Phoenix Wailers playing high intensity noise have been successful in deterring birds and some marine mammals from on-water oil slicks. Audio recordings of predator calls (e.g., Bigg's killer whales) broadcast into the water near spills may be effective in deterring marine mammal prey species such as pinnipeds, whales, and porpoise. Oikomi pipes, a technique using metal pipes suspended over the side of a boat into the marine environment and struck with a hammer, are an approved method in the United States to deter killer whales (EPA and USCG 2020).

Secondary response also includes the preemptive capture, short-term holding, and release of unoiled wildlife. For example, in 2015, three endangered Hawaiian monk seals were captured by the NOAA Marine Mammal Health and Stranding Response Program and brought into captive care to avoid the threat of exposure to a fuel spill resulting from a sunken tug on Oahu. The seals were returned to the wild after the threat of oil exposure was over. Penguins and dotterels have also been preemptively captured to avoid oil exposure in response to large oil spills in New Zealand and South Africa, and subsequently released (Wolfaardt et al. 2008; Gartrell et al. 2014).

3.2.3 Tertiary Response

Tertiary response efforts are defined as the capture, handling, transport, stabilization, rehabilitation, and release of oiled wildlife. In the Arctic, these tactics will be particularly challenging and dangerous due to remoteness and extreme weather conditions. The conditions in the Arctic may limit the number of animals that trained wildlife responders can handle, and it requires advanced caching of equipment and training of personnel. Birds, terrestrial mammals, and marine mammals in the Arctic are capable of seriously injuring responders, and only trained responders with proper personal protective equipment and wildlife response equipment should be authorized to handle live animals. Considerations for holding and rehabilitating Arctic species will be very different than for species in warmer climates.

3.2.4 Potential Negative Impacts of Response Activities

Containing and preventing landfall of spilled oil is a top priority in a spill response to reduce long-term impacts of shoreline oiling. However, nearly all response tactics carry a risk of negatively impacting wildlife and people, and those risks should be considered and minimized when possible. Mechanical response equipment has the potential to interact with wildlife by entangling wildlife in lines, pipes, booms, or sorbents; crushing small animals or nests on land; or striking animals while equipment is in motion (e.g., ship strike of a whale). The noise from response activities may startle wildlife, leading to unusual or harmful behavior (e.g., stampeding pinnipeds crushing smaller animals, birds abandoning nests or chicks), or chasing

Fig. 6 Photo of killer whales approaching skimming operations and spilled oil in Prince William Sound, Alaska, in 1989 during the Exxon Valdez oil spill response. Photo taken by Dan Lawn. Used with permission from the Alaska State Library

wildlife inadvertently into the spilled oil. Response activities, including wildlife deterrence, can also attract curious marine mammals and other wildlife toward the activity and the spilled oil. It is a common misconception that response activities will deter marine mammals from a spill zone or that these animals can detect and avoid oil (St. Aubin et al. 1985), but repeatedly, curious, intelligent marine mammals have been observed to approach response equipment and spilled oil (Fig. 6). Deterrence and other noisy activities should be closely monitored to determine efficacy and the potential to negatively impact wildlife in the area. Wildlife observers can aid in minimizing potential stressors by notifying equipment operators of animals in the area or alerting wildlife response personnel if interactions occur.

Planning for in-situ burns should take into consideration the presence of people and wildlife in the vicinity of the operation. In addition to the potential for disturbance of wildlife, terrestrial or on-water burns result in significant smoke plumes that introduce particulates into the air which may be inhaled and embedded in lung tissue. Solid particulates and pyrogenic PAHs (which may have a higher mutagenicity than the original PAH components in oil) are emitted during in-situ burning (Sheppard et al. 1983; USCG and EPA 2014). Proximity to human communities and animal aggregations (e.g., pinniped haulouts and rookeries, bird colonies, caribou herds), wind direction, burn duration, and potential toxic components of the smoke plume should be considered in order to minimize unintended consequences.

Chemical dispersants have the potential to make the toxic components of oil more bioavailable to marine foragers by dispersing oil droplets throughout the water column. Dispersants reduce the potential for oil to contact wildlife and the shoreline but increase the potential exposure of pelagic and benthic biota to dispersed oil (NRC 2005, 2013). Though modern-use dispersants are generally much less toxic than oil or oil with dispersant, some aquatic species and life stages are sensitive to toxic effects from dispersant exposure. Exposure of marine mammals to oil, either through ingestion of oil or indirectly through prey with bioaccumulated PAHs, may cause digestive system distress, narcosis, lesions, developmental deformities, decreased growth, and mortality (USCG and EPA 2014). Dispersants do not remove oil from the environment, so a thorough evaluation of environmental tradeoffs, including less visible impacts to pelagic environments, should be conducted before deciding whether and how to apply dispersants.

3.3 Restoration and Recovery

The 1989 Exxon Valdez Oil Spill (EVOS) occurred in the subarctic waters of Prince William Sound, but many of the lessons learned from EVOS are applicable to spills in the Arctic and show that spill impacts in cold regions can be significant and long lasting (Fig. 7). Contamination of nearshore habitats can persist for many years, affecting multiple generations of early life stage fish, such as Pacific herring (*Clupea pallasi*) (Peterson et al. 2003; Incardona et al. 2015). After EVOS, clams and other invertebrates in some areas remained contaminated for more than a decade, affecting the animals that fed on them, including harlequin ducks and sea otters (Peterson et al. 2003). Some wildlife, such as the AT1 transient killer whale pod, may never recover from the impacts (Peterson et al. 2003; Matkin et al. 2008). Over three decades later, lingering oil from EVOS remains on shorelines in areas where it is protected from physical weathering and degradation (Lindeberg et al. 2018). The deleterious impacts of the spill on human communities were also substantial and long-term (Gill et al. 2016).

Human-led restoration of natural resources that are injured by an oil spill can take many different forms, such as removing lingering contamination sources, replanting vegetation, and improving fish habitat. Restoration may also address impacts to human uses of resources, such as fishing or recreation. The goal of restoration is to decrease the amount of time it takes for natural resources, and the ecological services that they provide, to return to pre-spill conditions. In some cases, natural attenuation and recovery are preferable to active restoration, but in other cases, restoration is essential to habitat or resource recovery. For example, tundra plant communities can take many decades to recover and may be permanently altered by an oil spill; therefore, soil remediation followed by transplanting or seeding are frequently used to restore that habitat. The Exxon Valdez Oil Spill Restoration Plan published in 1994 is another example and provides detailed analyses and descriptions of restoration of biological resources and lost services resulting from the 1989 spill (EVOSTC 1994).

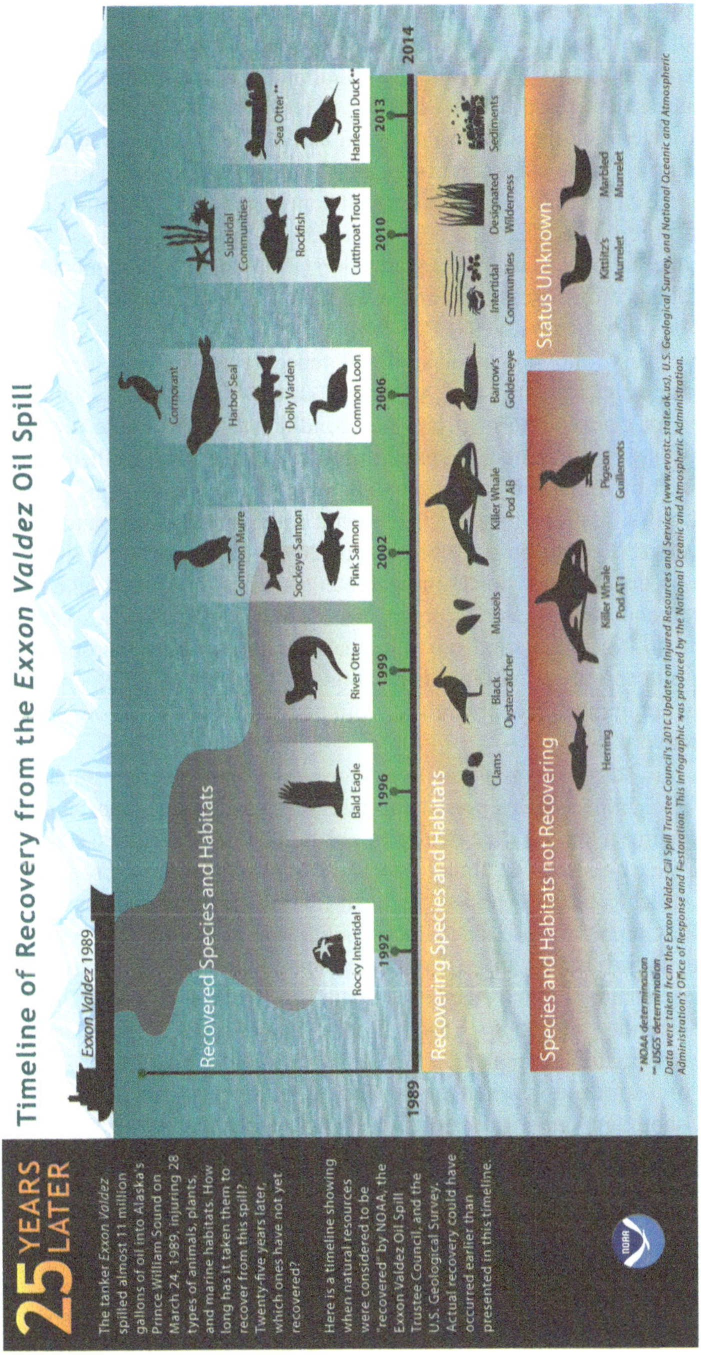

Fig. 7 Timeline of recovery of some of the species and environmental indicators impacted by the Exxon Valdez oil spill in Prince William Sound Alaska (1989–2014). Provided by NOAA

Many Arctic nations have laws that establish liability for public and private losses caused by oil spills. In some cases, those laws require the polluter to pay for restoration of public natural resources that were injured by the spill. One of the most robust examples of this is the US Natural Resource Damage Assessment (NRDA) process, described in the OPA 90, which is used to evaluate the impacts of spills and determine the appropriate types and amounts of restoration required. Another example is Canada's Environmental Damage Assessment (EDA) process and ship-source oil pollution fund, though its application to restoration of natural resources is more limited than the NRDA process in the United States (Mac Innis 2005). The International Maritime Organization (IMO) also has a regulatory framework and guidance for assessment and restoration of environmental damages following marine oil spills (IMO/UNEP 2009).

Natural recovery following an oil spill will generally be slower in the Arctic than in more temperate regions, and in some cases, habitats, populations, and communities may never return to pre-spill conditions. For example, tundra plant communities may be permanently altered if disturbed areas become thermokarsts or are recolonized by opportunistic species that were not dominant prior to the spill. Recolonization of seafloor areas in the Arctic following a disturbance can take decades or more (Conlan et al. 1998; Konar 2013). Long recovery times in the Arctic are a product of both slower natural attenuation of the spilled oil and slower recolonization and growth rates in organisms. Low temperatures, lack of sunlight, and ice cover slow oil weathering, as well as bacterial degradation. Chronic exposure to lingering oil can impact multiple generations of fish and wildlife, sometimes in ways that are not immediately obvious (e.g., impaired immune function), slowing or impeding recovery.

If early life stages or reproductive success of animals are impacted, the population-level impacts may persist for multiple generations. The tendency for some Arctic animals to aggregate in large numbers in confined spaces, such as open water leads, increases the likelihood of a spill impacting a large proportion of a population, which could also impede recovery. Many arctic animals have long life spans and slow reproductive rates, which could prolong population-level impacts. Furthermore, arctic systems are experiencing ongoing and increasing stresses from global climate change, which may exacerbate the impacts of a spill and hinder recovery. Scientific information about many arctic species, populations, and communities is largely incomplete, which could make it impossible to fully assess the impacts of a spill and if, when, or to what extent recovery has occurred.

An important component of restoration and recovery is to consider the human uses of the environment and resources that must be suspended or moved to a different location due to the impacts of the oil spill. Liabilities and requirements for addressing lost human uses vary between Arctic nations. Commercial losses, such as lost income due to fisheries closures, may be compensated through financial settlements with affected parties. Additionally, the Arctic environment and its resources are a critical part of the culture and subsistence of indigenous arctic peoples. Damage assessment and restoration efforts should consider food security and cultural uses of the resources including their importance to local communities.

Restoration is important for preserving ecological integrity, ecosystem services, and human uses of natural resources. However, some injuries caused by oil spills cannot be restored, replaced, or compensated for, particularly in the Arctic where unique habitats are relatively pristine and human communities are largely reliant on subsistence resources.

3.4 Prevention

Whereas restoration of damaged resources carries considerable uncertainty of success, prevention of oil spills is a sure way to avoid environmental impacts, and some effective strategies and tools are available. Some requirements exist for marine vessels to implement automatic identification systems (AIS) onboard while underway, depending on vessel size and governing organization. AIS tracking information can be a useful tool in preventing oil spills by enabling detection of vessels deviating from shipping routes and moving toward navigation hazards. Early detection of these deviations aids communication between response organizations and captains and crew of disabled vessels, reducing the likelihood of a grounding that could lead to a significant spill. Increased application of vessel tracking technology in the Arctic is a step in the right direction toward prevention. The necessary infrastructure and equipment for a response to a disabled vessel is needed in order to take action.

Double-hulled tankers are a measure that can be taken to reduce the likelihood of a spill occurring from a tanker grounding incident. MARPOL, originally adopted by the IMO in 1973 and coming into force in 1983, was amended in 1992 to closely match the requirements contained in OPA 90, requiring all newly built tankers to have double hulls. Further, under revised regulations passed in 2007, Category 1 (pre-MARPOL) tankers were required to be phased out by 2005, with Category 2 and 3 (MARPOL and smaller) tankers phased out by 2010. The United States, Canada, and the Russian Federation are part to this Convention, along with approximately 150 other nations, representing 99% of global shipping.

Pipelines of various ages exist throughout the Arctic region associated with past and current oil and gas development activities. Regular monitoring and maintenance of existing pipelines is needed to prevent future spills from occurring. Further reductions in risk of spills can occur through the use of new technology and equipment instead of relying on old pipelines with unknown maintenance records or corrosion potential.

Traffic Separation Schemes, two-way routes, deep water routes, precautionary areas (where ships must navigate with particular caution), and areas to be avoided (ATBAs) are all techniques that can reduce the risk of spills (IMO 2019). Traffic Separation Schemes identify two "lanes" on applicable maps and charts (e.g., one identified for eastbound traffic and one for westbound, separated by some distance) to reduce collisions along congested routes. Deep-water routes and precautionary areas may help reduce collisions, and they also limit groundings. The establishment of ATBAs is a mechanism by which international vessel traffic routes can be influenced. Three ATBAs have been adopted by the IMO for the Bering Strait region, and additional areas have been proposed for consideration as ATBAs (Huntington et al. 2019).

Additional safety precautions to reduce the risk of vessel collisions and resulting oil spills include the use of pilots with local expertise or escort vessels. Currently implemented in many busy ports (particularly those with tricky approaches), this idea has been proposed and discussed for Arctic areas as well (Beveridge 2017).

4 Risk of Future Spills

There are an estimated 4 million people living permanently in the Arctic today, with approximately 12.5% of those being of indigenous groups. While global population growth is projected to continue increasing from the current total of 7.4 billion to 10 billion in 2055, the population of the broader Arctic region is projected to have just a 1% population increase, with significant regional variability (Emelyanova 2017; Heleniak 2020). Trends expected in nearly all Arctic regions include aging populations, more balanced gender ratios, increased population concentration into larger urban settlements, and depopulation of smaller settlements. Thus, while overall numbers of inhabitants of the Arctic region are not expected to substantially alter, the impacts of the individuals that remain or emigrate into the area may impart significant change if focused on tourism, resource exploration, or other pursuits related to changes noted below.

The Arctic is experiencing rapid climate change, which will continue to accelerate for at least decades to come. The effects of climate change are occurring more quickly in the Arctic than at lower latitudes, in part due to feedbacks between the physical mechanisms, known as "Arctic amplification" (NRC 2014). Changes occurring as the Arctic warms are leading to increasing risk of oil spills in that region, which will also influence the fate and effects of Arctic oil spills (Nordam et al. 2017) and the recovery of the environment after a spill. Retreating and thinning sea ice is increasing access for shipping, oil exploration and extraction, and other resource development; all of which bring increasing risk of oil spills. Rising temperatures are also causing permafrost thaw, which is destabilizing infrastructure in the Arctic, including oil extraction, transport, and storage facilities, and has already been a contributing factor in oil spills. Similarly, climate change is causing increased coastal erosion and exacerbating the frequency and intensity of storms, both of which threaten infrastructure and increase the potential for spills. At the same time that climate change is leading to an increased risk of oil spills, it is causing considerable strain on the Arctic environment. Arctic oil spills may threaten an already stressed system. Warmer water temperatures and reduced sea ice may also make Arctic fish and wildlife, especially ice-associated species, more vulnerable to oil exposure and susceptible to toxicological injuries and latent effects (Laurel et al. 2019).

Vessel traffic in the Arctic is increasing from multiple sources: commercial transport (shipping) due to the ability to use the Northern Sea Route (along coastal northern Russia), and to a lesser extent the Northwest Passage (IPCC 2018); bulk carriers and tug-barge traffic to transport minerals and other commodities as well as bring supplies to remote villages; fishing fleets that will likely follow the northward migration of fish stocks; and cruise ships plying new routes (NRC 2014).

Oil and gas development is increasing throughout circumpolar lands and waters with production and exploration predicted to increase in most arctic countries and a 5-year moratorium on offshore oil and gas development expiring in Canada in 2021. Continued loss of sea ice is expected to result in increases in passage rates and length of the season for transporting oil and natural gas through arctic waters. In January 2021, the first-ever winter voyages of liquid natural gas tankers occurred along the Northern Sea Route from the Barents to the Bering Sea (IUCN 2021), a precursor of future increased operations and spill risk.

5 Conclusion

Prevention of oil spills is the best way to control this stressor in the Arctic. Reducing the chance of a spill occurring should be top priority, particularly for oil and gas development, which has been the source of the largest spills in the Arctic and around the world. However, with increasing oil and gas development and other human uses in this region, accidents and incidental spills of all sizes are bound to happen. With this in mind, preparedness for oil spill response is the next priority. Establishing formal plans that cross geopolitical boundaries and account for the harsh environmental conditions in the Arctic is needed for an agile response to an emergency that could occur along any shipping route, port location, or land-based infrastructure. Cached equipment, trained personnel, mutual aid agreements, and a better understanding of arctic conditions are needed to prepare for the most effective response, recognizing that spill responses in the Arctic will remain extraordinarily challenging under the best circumstances. Community engagement and inclusion of local and traditional knowledge in planning, response, and restoration will greatly increase the applicability of those efforts in this unique region. For large spills, responses may be ineffectual in some cases, and policy-makers should account for the associated risks.

The characteristics of the Arctic, including extreme climate, remoteness, limited infrastructure, dynamic sea ice, unique ecosystems, changing environment, and slower recovery times, amplify the risks and consequences of oil spills and complicate or impede response efforts. The risks and consequences of spills extend to wildlife populations as well as the human communities in the Arctic that rely on a healthy environment and subsistence harvest for food security, culture, and well-being. Incomplete scientific information about many resources in the Arctic and their sensitivity to oil spills, as well as a dynamic and shifting baseline, limit the ability to predict impacts, determine injuries, and evaluate recovery. The increasing risk of oil spills in the Arctic is outpacing efforts to address scientific and oil spill response needs. It is critical to sustain and expand ongoing efforts, from local to international scales, to reduce the risk of spills and identify and address gaps in oil spill preparedness, response, and recovery.

Acknowledgments We thank NOAA and OWCN for supporting our time and efforts dedicated to developing the text and figures for this chapter. We also thank Kate Sweeney for her beautiful illustrations depicting potential impacts of oil spills in the Arctic. We thank Barb Mahoney, Mandy Keogh, and Teri Rowles for their review and comments on illustrations. And we thank Jon Kurland, Aleria Jensen, and Mandy Lindeberg for their review of this chapter.

References

ADEC (2020) SPAR annual report, fiscal year 2019. Alaska Department of Environmental Conservation Spill Prevention and Response. Oil and Hazardous Substance Release Prevention & Response Fund Annual Report, p 42

Aeppli C, Carmichael CA, Nelson RK, Lemkau KL, Graham WM, Redmond MC, Valentine DL, Reddy CM (2012) Oil weathering after the *Deepwater Horizon* disaster led to the formation of oxygenated residues. Environ Sci Technol 46:8799–8807

Agersted MD, Møller EF, Gustavson K (2018) Bioaccumulation of oil compounds in the high-Arctic copepod *Calanus hyperboreus*. Aquat Toxicol 195:8–14

Albers P (1995) Petroleum and individual polycyclic aromatic hydrocarbons. In: Hoffman D, Rattner B, Burton G, Cairns J (eds) Handbook of ecotoxicology. Lewis, Boca Raton, FL

Albers PH, Loughlin TR (2003) Effects of PAHs on birds, mammals and reptiles. In: Douben P (ed) PAHs: an ecotoxicological perspective. Wiley, pp 225–241

Arctic Council (2013) AGREEMENT on cooperation on marine oil pollution preparedness and response in the Arctic

ARRT (2020) Wildlife protection guidelines for oil spill response in Alaska. Alaska Regional Response Team Wildlife Protection Committee. Version 2020.1. August 31, 2020, p 220

ATSDR (1995) Toxicological profile for polycyclic aromatic hydrocarbons. U.S. Department of Health and Human Services, Agency for Toxic Substances and Disease Registry. Atlanta, p 487

ATSDR (2005) Toxicological profile for benzene. U.S. Department of Health and Human Services, Agency for Toxic Substances and Disease Registry

ATSDR (2007a) Toxicological profile for benzene. U.S. Department of Health and Human Services, Agency for Toxic Substances and Disease Registry. Atlanta, p 19

ATSDR (2007b) Toxicological profile for xylene. U.S. Department of Health and Human Services, Agency for Toxic Substances and Disease Registry. Atlanta, p 385

ATSDR (2010) Toxicological profile for ethylbenzene. U.S. Department of Health and Human Services, Agency for Toxic Substances and Disease Registry. Atlanta. p 341

Balseiro A, Espi A, Marquez I, Perez V, Ferreras M, Marin JG, Prieto JM (2005) Pathological features in marine birds affected by the Prestige's oil spill in the north of Spain. J Wildl Dis 41: 371–378

Barron MG (2017) Photoenhanced toxicity of petroleum to aquatic invertebrates and fish. Arch Environ Contam Toxicol 73:40–46

BBC News (2020) Arctic circle oil spill: Russian prosecutors order checks at permafrost sites. Date accessed: March 17, 2021

Bejarano AC, Gardiner WW, Barron MG, Word JQ (2017) Relative sensitivity of Arctic species to physically and chemically dispersed oil determined from three hydrocarbon measures of aquatic toxicity. Mar Pollut Bull 122:316–322

Beveridge L (2017) Arctic Pilots For Canadian Corridors. Canadian Transportation Research Forum 52nd Annual Conference

Biggar KW, Haidar S, Nahir M, Jarrett PM (1998) Site investigations of fuel spill migration into permafrost. J Cold Reg Eng 12:84–104

Bluhm BA, Gradinger R (2008) Regional variability in food availability for Arctic marine mammals. Ecol Appl 18:S77–S96

Bonsdorff E (1981) The *Antonio Gramsci* oil spill impact on the littoral and benthic ecosystems. Mar Pollut Bull 12:301–305

Briggs KT, Yoshida SH, Gershwin ME (1996) The influence of petrochemicals and stress on the immune system of seabirds. Regul Toxicol Pharmacol 23:145–155

Bursian SJ, Alexander C, Cacela D, Cunningham FL, Dean KM, Dorr BS, Ellis CK, Godard-Codding C, Guglielmo CG, Hanson-Dorr KC (2017) Reprint of: overview of avian toxicity studies for the Deepwater horizon natural resource damage assessment. Ecotoxicol Environ Saf 146:4–10

Camus L, Brooks S, Geraudie P, Hjorth M, Nahrgang J, Olsen G, Smit M (2015) Comparison of produced water toxicity to Arctic and temperate species. Ecotoxicol Environ Saf 113:248–258

Chelombitko S (2019) Influence of oil spills on tundra degradation. IOP Publishing

Choy ES, Rosenberg B, Roth JD, Loseto LL (2017) Inter-annual variation in environmental factors affect the prey and body condition of beluga whales in the eastern Beaufort Sea. Mar Ecol Prog Ser 579:213–225

Conlan K, Lenihan H, Kvitek R, Oliver J (1998) Ice scour disturbance to benthic communities in the Canadian high Arctic. Mar Ecol Prog Ser 166:1–16

Daling PS, Brandvik PJ, Mackay D, Johansen O (1990) Characterization of crude oils for environmental purposes. Oil Chem Pollut 7:199–224

Daly KL, Passow U, Chanton J, Hollander D (2016) Assessing the impacts of oil-associated marine snow formation and sedimentation during and after the Deepwater horizon oil spill. Anthropocene 13:18–33

Dickins D, Buist I (1999) Countermeasures for ice covered waters. Pure Appl Chem 71:173–191

Dunton KH, Goodall JL, Schonberg SV, Grebmeier JM, Maidment DR (2005) Multi-decadal synthesis of benthic–pelagic coupling in the western arctic: role of cross-shelf advective processes. Deep-Sea Res II Top Stud Oceanogr 52:3462–3477

Dunton KH, Schonberg SV, Cooper LW (2012) Food web structure of the Alaskan nearshore shelf and estuarine lagoons of the Beaufort Sea. Estuar Coasts 35:416–435

Emelyanova A (2017) Population projections of the Arctic by levels of education. IIASA working paper. International Institute for Applied Systems Analysis. Laxenburg, WP-17-022, p 46

EPA, USCG (2020) Supporting information for the killer whale section of the Northwest Wildlife Response Plan, updated 2014. In: Northwest Area Contingency Plan (section 9310.10.2.4). p 1391

Essaid HI, Bekins BA, Herkelrath WN, Delin GN (2011) Crude oil at the Bemidji site: 25 years of monitoring, modeling, and understanding. Groundwater 49:706–726

Everett K (1978) Some effects of oil on the physical and chemical characteristics of wet tundra soils. Arctic 260–276

Evers K-U, Jensen H, Resby J, Ramstad S, Singsaas I, Dieckmann G, Gerdes B (2004) State of the art report on oil weathering and on the effectiveness of response alternatives. Report of ARCOP Work package 4

EVOSTC (1994) Exxon Valdez oil spill restoration plan. Exxon Valdez oil spill trustee council, Anchorage, Alaska 99501. November 1994, pp 98

Fagoe TE (1991) Coast guard response to the Volgoneft 263 oil spill. American Petroleum Institute

Farrington JW (2020) Need to update human health risk assessment protocols for polycyclic aromatic hydrocarbons in seafood after oil spills. Mar Pollut Bull 150:110744

Fashchuk DY (2011) Wreck of the Tanker *Volgoneft-139* in the Kerch Strait on November 11, 2007. In: Marine ecological geography. Springer, pp 391–427

Ferguson DK, Li C, Jiang C, Chakraborty A, Grasby SE, Hubert CR (2020) Natural attenuation of spilled crude oil by cold-adapted soil bacterial communities at a decommissioned high Arctic oil well site. Sci Total Environ 722:137258

French-McCay DP, Tajalli-Bakhsh T, Jayko K, Spaulding ML, Li Z (2017) Validation of oil spill transport and fate modeling in Arctic ice. Arctic Sci 4:71–97

Fry DM, Lowenstine LJ (1985) Pathology of common murres and Cassin's auklets exposed to oil. Arch Environ Contam Toxicol 14:725–737

Gam KB, Kwok RK, Engel LS, Curry MD, Stewart PA, Stenzel MR, McGrath JA, Jackson WB (2018) Lung function in oil spill response workers 1–3 years after the *Deepwater Horizon* disaster. Epidemiology 29:315–322

Gardiner WW, Word JQ, Word JD, Perkins RA, McFarlin KM, Hester BW, Word LS, Ray CM (2013) The acute toxicity of chemically and physically dispersed crude oil to key Arctic species under Arctic conditions during the open water season. Environ Toxicol Chem 32:2284–2300

Gartrell B, Collen R, Dowding J, Gummer H, Hunter S, King E, Laurenson L, Lilley C, Morgan K, McConnell H (2014) Captive husbandry and veterinary care of northern New Zealand dotterels (*Charadrius obscurus aquilonius*) during the *CV Rena* oil-spill response. Wildl Res 40:624–632

Geraci JR (1990) Physiological and toxic effects on cetaceans. In: Geraci JR, St. Aubin DJ (eds) Sea mammals and oil: confronting the risks. Academic, San Diego, CA, pp 167–197

Gill DA, Ritchie LA, Picou JS (2016) Sociocultural and psychosocial impacts of the Exxon Valdez oil spill: twenty-four years of research in Cordova, Alaska. Extractive Ind Soc 3:1105–1116

Goodchild CG, Grisham K, Belden JB, DuRant SE (2020) Effects of sublethal application of *Deepwater Horizon* oil to bird eggs on embryonic heart and metabolic rate. Conserv Biol 34: 1262–1270

Gorenzel P, Salmon T (2008) Bird hazing manual: techniques and strategies for dispersing birds from spill sites, vol 21638. UCANR

Gundlach ER, Hayes MO (1978) Vulnerability of coastal environments to oil spill impacts. Mar Technol Soc J 12:18–27

Hannam ML, Bamber SD, Moody AJ, Galloway TS, Jones MB (2010) Immunotoxicity and oxidative stress in the Arctic scallop *Chlamys islandica*: effects of acute oil exposure. Ecotoxicol Environ Saf 73:1440–1448

Harr KE, Reavill DR, Bursian SJ, Cacela D, Cunningham FL, Dean KM, Dorr BS, Hanson-Dorr KC, Healy K, Horak K (2017) Organ weights and histopathology of double-crested cormorants (*Phalacrocorax auritus*) dosed orally or dermally with artificially weathered Mississippi Canyon 252 crude oil. Ecotoxicol Environ Saf 146:52–61

Hartung R (1967) Energy metabolism in oil-covered ducks. J Wildlife Manage 798–804

Harvey RG (1991) Polycyclic aromatic hydrocarbons: chemistry and carcinogenicity. CUP Archive

Haynes T, Tibbles M, Robards M, Jones T, Whiting A, Wipfli M (2017) Coastal lagoon community and ecological monitoring in the southern Chukchi Sea National Park units: 2015 field sampling report

Heleniak T (2020) The future of the Arctic populations. Polar Geogr 1–17

Hobson KA, Ambrose WG Jr, Renaud PE (1995) Sources of primary production, benthic-pelagic coupling, and trophic relationships within the northeast water polynya: insights from δ13C and δ15N analysis. Mar Ecol Prog Ser 128:1–10

Hodson PV (2017) The toxicity to fish embryos of PAH in crude and refined oils. Arch Environ Contam Toxicol 73:12–18

Horowitz A, Atlas RM (1977) Continuous open flow-through system as a model for oil degradation in the Arctic Ocean. Appl Environ Microbiol 33:647–653

Huntington HP, Bobbe S, Hartsig A, Knight EJ, Knizhnikov A, Moiseev A, Romanenko O, Smith MA, Sullender BK (2019) The role of areas to be avoided in the governance of shipping in the greater Bering Strait region. Mar Policy 110:103564

Huserbråten MBO, Eriksen E, Gjøsæter H, Vikebø F (2019) Polar cod in jeopardy under the retreating Arctic Sea ice. Commun Biol 2:1–8

IMO (2019) Ships' routeing 2019 Edition. International Maritime Organization

IMO/UNEP (2009) International maritime organization, United Nations environment programme guidance manual on the assessment and restoration of environmental damage following marine oil spills. IMO Publishing

Incardona JP, Carls MG, Holland L, Linbo TL, Baldwin DH, Myers MS, Peck KA, Tagal M, Rice SD, Scholz NL (2015) Very low embryonic crude oil exposures cause lasting cardiac defects in salmon and herring. Sci Rep 5:1–13

IPCC (2018) Global warming of 1.5°C. An IPCC special report on the impacts of global warming of 1.5°C above pre-industrial levels and related global greenhouse gas emission pathways in the context of strengthening the global response to the threat of climate change, sustainable development, and efforts to eradicate poverty

ITOPF (2012) The international tanker owners pollution Federal Limited Country Profiles; a summary of oil spill response arrangements and resources worldwide. London, p 457

ITOPF (2018) Norway. The international tanker owners pollution Federal Limited Country Profiles; a summary of oil spill response arrangements and resources worldwide. London, p 4

IUCN (2021) Baleen whales in the cross hairs: potential for increased ship strike in and near Bering Strait. In: I-SCS Group (ed)

Jenssen BM (1994) Effects of oil pollution, chemically treated oil, and cleaning on thermal balance of birds. Environ Pollut 86:207–215

Jenssen BM, Ekker M (1988) A method for evaluating the cleaning of oiled seabirds. Wildl Soc Bull 16:213–215

Jenssen BM, Ekker M (1991) Dose dependent effects of plumage-oiling on thermoregulation of common eiders *Somateria mollissima* residing in water. Polar Res 10:579–584

Jessup D, Leighton F (1996) Oil pollution and petroleum toxicity to wildlife. In: Fairbrother A, Locke L, Hoff G (eds) Non-infectious diseases of wildlife. Iowa State University, Ames, Iowa, pp 141–157

Kershaw GP, Kershaw LJ (1986) Ecological characteristics of 35-year-old crude-oil spills in tundra plant communities of the Mackenzie Mountains, NWT. Can J Bot 64:2935–2947

Konar B (2013) Lack of recovery from disturbance in high-arctic boulder communities. Polar Biol 36:1205–1214

Laidre KL, Stirling I, Lowry LF, Wiig Ø, Heide-Jørgensen MP, Ferguson SH (2008) Quantifying the sensitivity of Arctic marine mammals to climate-induced habitat change. Ecol Appl 18:S97–S125

Laurel BJ, Copeman LA, Iseri P, Spencer ML, Hutchinson G, Nordtug T, Donald CE, Meier S, Allan SE, Boyd DT (2019) Embryonic crude oil exposure impairs growth and lipid allocation in a keystone Arctic forage fish. Iscience 19:1101–1113

Lawrence KG, Keil AP, Garantziotis S, Umbach DM, Stewart PA, Stenzel MR, McGrath JA, Jackson WB, Kwok RK, Curry MD (2020) Lung function in oil spill responders 4-6 years after the *Deepwater Horizon* disaster. J Toxic Environ Health A 83:233–248

Leighton FA (1993) The toxicity of petroleum oils to birds. Environ Rev 1:92–103

Leighton F, Lee Y, Rahimtula A, O'brien P, Peakall D (1985) Biochemical and functional disturbances in red blood cells of herring gulls ingesting Prudhoe Bay crude oil. Toxicol Appl Pharmacol 81:25–31

Lindeberg MR, Maselko J, Heintz RA, Fugate CJ, Holland L (2018) Conditions of persistent oil on beaches in Prince William sound 26 years after the *Exxon Valdez* spill. Deep-Sea Res II Top Stud Oceanogr 147:9–19

Lowe SR, McGrath JA, Young MN, Kwok RK, Engel LS, Galea S, Sandler DP (2019) Cumulative disaster exposure and mental and physical health symptoms among a large sample of Gulf Coast residents. J Trauma Stress 32:196–205

Mac Innis K (2005) The Canadian SSOP fund and environmental damage assessment (EDA) in Canada. In: Maes F (ed) Marine resource damage assessment; liability and compensation for environmental damage. Springer, pp 67–84

Madeen EP, Williams DE (2017) Environmental PAH exposure and male idiopathic infertility: a review on early life exposures and adult diagnosis. Rev Environ Health 32:73–81

Matkin CO, Saulitis EL, Ellis GM, Olesiuk P, Rice SD (2008) Ongoing population-level impacts on killer whales *Orcinus orca* following the 'Exxon Valdez' oil spill in Prince William Sound, Alaska. Mar Ecol Prog Ser 356:269–281

McEwan E, Koelink A (1973) The heat production of oiled mallards and scaup. Can J Zool 51:27–31

McFarlin KM, Perkins MJ, Field JA, Leigh MB (2018) Biodegradation of crude oil and Corexit 9500 in Arctic seawater. Front Microbiol 9:1788
McKendrick JD, Mitchell WW (1978) Fertilizing and seeding oil-damaged Arctic tundra to effect vegetation recovery Prudhoe Bay, Alaska. Arctic 296–304
Michel J, Rutherford N (2013) Oil spills in marshes: planning & response considerations. Department of Commerce, NOAA National Ocean Service, Office of Response and Restoration, and American Petroleum Institute. September 2013, p 126
Miller MC, Alexander V, Barsdate RJ (1978) The effects of oil spills on phytoplankton in an arctic lake and ponds. Arctic:192–218
Moore SE, Hauser DD (2019) Marine mammal ecology and health: finding common ground between conventional science and indigenous knowledge to track Arctic ecosystem variability. Environ Res Lett 14:075001
Nechepurenko I (2020) Russia declares emergency after Arctic oil spill. New York Times
Newman S, Anderson D, Ziccardi MH, Trupkiewicz J, Tseng F, Christopher MM, Zinkl J (2000) An experimental soft-release of oil-spill rehabilitated American coots (*Fulica americana*): II. Effects on health and blood parameters. Environ Pollut 107:295–304
NMFS (2015) Biological opinion for the Alaska federal/state preparedness plan for response to oil & hazardous substance discharges/releases (unified plan). NMFS# AKR-2014-9361, 173 pp
NMFS (2017) NMFS Arctic marine mammal disaster response guidelines. U.S. Department of Commerce, NOAA Tech. Memo. NMFS-F/AKR-16. 81 pp. https://doi.org/10.7289/V5/TM-F/AKR-16
NOAA (1977) Incident news: Ekofisk Bravo oil field
NOAA (2014) Assessment of marine oil spill risk and environmental vulnerability for the state of Alaska. NOAA contract number: WC133F-11-CQ-0002. NOAA Restoration Center NW, Seattle p 133
Nordam T, Dunnebier DA, Beegle-Krause CJ, Reed M, Slagstad D (2017) Impact of climate change and seasonal trends on the fate of Arctic oil spills. Ambio 46:442–452
Nørregaard RD, Gustavson K, Møller EF, Strand J, Tairova Z, Mosbech A (2015) Ecotoxicological investigation of the effect of accumulation of PAH and possible impact of dispersant in resting high Arctic copepod *Calanus hyperboreus*. Aquat Toxicol 167:1–11
Norwegian Petroleum (2020) Norway's petroleum history. Updated July 10, 2020. https://www.norskpetroleum.no/en/framework/norways-petroleum-history/
NRC (2003) National research council, oil in the sea III. Inputs, fates and effects. National Academies Press, Washington, DC
NRC (2005) Oil spill dispersants: efficacy and effects. The National Academies Press, Washington, DC
NRC (2013) An ecosystem services approach to assessing the impacts of the *Deepwater Horizon* oil spill in the Gulf of Mexico. Committee on the effects of *Deepwater Horizon* Mississippi Canyon-252 oil spill on ecosystem services in the Gulf of Mexico, Ocean Studies Board, National Research Council of the National Academies Press, Washington, DC, National Academies Press
NRC (2014) Responding to oil spills in the U.S. Arctic marine environment. The National Academies Press
O'Hara PD, Morandin LA (2010) Effects of sheens associated with offshore oil and gas development on the feather microstructure of pelagic seabirds. Mar Pollut Bull 60:672–678
Oggier M, Eicken H, Wilkinson J, Petrich C, O'Sadnick M (2020) Crude oil migration in sea-ice: laboratory studies of constraints on oil mobilization and seasonal evolution. Cold Reg Sci Technol 174:102924
Peakall D, Norstrom R, Jeffrey D, Leighton F (1989) Induction of hepatic mixed function oxidases in the herring gull (*Larus argentatus*) by Prudhoe Bay crude oil and its fractions. Comp Biochem Physiol C Toxicol Pharmacol 94:461–463
Percy J (1978) Effects of chronic exposure to petroleum upon the growth and molting of juveniles of the Arctic marine isopod crustacean Mesidotea entomon. J Fish Board Can 35:650–656

Peterson CH, Rice SD, Short JW, Esler D, Bodkin JL, Ballachey BE, Irons DB (2003) Long-term ecosystem response to the *Exxon Valdez* oil spill. Science 302:2082–2086
Petrich C, Karlsson J, Eicken H (2013) Porosity of growing sea ice and potential for oil entrainment. Cold Reg Sci Technol 87:27–32
Piatt JF, Lensink CJ, Butler W, Kendziorek M, Nysewander DR (1990) Immediate impact of the 'Exxon Valdez' oil spill on marine birds. Auk 107:387–397
Prince RC, Owens EH, Sergy GA (2002) Weathering of an Arctic oil spill over 20 years: the BIOS experiment revisited. Mar Pollut Bull 44:1236–1242
Ritchie LA, Long MA (2021) Psychosocial impacts of post-disaster compensation processes: community-wide avoidance behaviors. Soc Sci Med 270:113640
Ross J, Hollander D, Saupe S, Burd AB, Gilbert S, Quigg A (2021) Integrating marine oil snow and MOSSFA into oil spill response and damage assessment. Mar Pollut Bull 165:112025
Sagers MJ (1994) Oil spill in Russian arctic. Polar Geogr 18:95–102
Schwacke LH, Smith CR, Townsend FI, Wells RS, Hart LB, Balmer BC, Collier TK, De Guise S, Fry MM, Guillette LJ Jr (2014) Health of common bottlenose dolphins (*Tursiops truncatus*) in Barataria Bay, Louisiana, following the Deepwater Horizon oil spill. Environ Sci Technol 48: 93–103
Sheppard E, Wells R, Georghiou P (1983) The mutagenicity of a Prudhoe Bay crude oil and its residues from an experimental *in situ* burn. Environ Res 30:427–441
Singsaas I, Leirvik F, Daling PS, Guénette C, Sørheim KR (2020) Fate and behaviour of weathered oil drifting into sea ice, using a novel wave and current flume. Mar Pollut Bull 159:111485
Sørstrøm SE, Brandvik PJ, Buist I, Daling P, Dickins D, Faksness L-G, Potter S, Fritt-Rasmussen J, Singsaas I (2010) Joint industry program on oil spill contingency for Arctic and ice-covered waters: summary report
Spraker TR, Lowry LF, Frost KJ (1994) Gross necropsy and histopathological lesions found in harbor seals. In: Loughlin TR (ed) Marine mammals and the *Exxon Valdez*. Academic, San Diego, pp 281–311
St. Aubin D, Geraci J, Smith T, Friesen T (1985) How do bottlenose dolphins, *Tursiops truncatus*, react to oil films under different light conditions? Can J Fish Aquat Sci 42:430–436
Stephenson R, Andrews CA (1997) The effect of water surface tension on feather wettability in aquatic birds. Can J Zool 75:288–294
Strelitz J, Sandler DP, Keil AP, Richardson DB, Heiss G, Gammon MD, Kwok RK, Stewart PA, Stenzel MR, Engel LS (2019) Exposure to total hydrocarbons during cleanup of the *Deepwater Horizon* oil spill and risk of heart attack across 5 years of follow-up. Am J Epidemiol 188:917–927
Szczybelski AS, Diepens NJ, van den Heuvel-Greve MJ, van den Brink NW, Koelmans AA (2019) Bioaccumulation of polycyclic aromatic hydrocarbons by Arctic and temperate benthic species. Environ Toxicol Chem 38:883–895
Takeshita R, Sullivan L, Smith C, Collier T, Hall A, Brosnan T, Rowles T, Schwacke L (2017) The *Deepwater Horizon* oil spill marine mammal injury assessment. Endanger Species Res 33:95–106
Tarasoff F, Bisaillon A, Piérard J, Whitt AP (1972) Locomotory patterns and external morphology of the river otter, sea otter, and harp seal (Mammalia). Can J Zool 50:915–929
Tibbles M (2018) The seasonal dynamics of coastal Arctic lagoons in Northwest Alaska. University of Alaska Fairbanks
Tseng FS, Ziccardi M (2019) Care of oiled wildlife. In: Hernandez S, Barron H, Miller E, Aguilar R, Yabsley M (eds) Medical management of wildlife species: a guide for practitioners. Wiley, pp 75–84
USCG, EPA (2014) Biological assessment of the Alaska federal/state preparedness plan for response to oil & hazardous substance discharges/releases (unified plan). Prepared by windward environmental, LLC and ERM. Anchorage, AK

USFWS (2015) Oil spill response plan for polar bears in Alaska. U.S. Fish and Wildlife Service, marine mammals management, 1011 East Tudor Road, Anchorage, Alaska 99503. May 2015, p 65

Vermeer K, Vermeer R (1975) Oil threat to birds on the Canadian west coast. Can Field-Nat 89: 278–298

Waldron H (1983) A brief history of scrotal cancer. Br J Ind Med 40:390–401

Werth AJ, Blakeney SM, Cothren AI (2019) Oil adsorption does not structurally or functionally alter whale baleen. R Soc Open Sci 6:182194

Wilkinson J, Beegle-Krause CJ, Evers K-U, Hughes N, Lewis A, Reed M, Wadhams P (2017) Oil spill response capabilities and technologies for ice-covered Arctic marine waters: a review of recent developments and established practices. Ambio 46:423–441

Williams TD, Allen DD, Groff JM, Glass RL (1992) An analysis of California Sea otter (*Enhydra lutris*) pelage and integument. Mar Mamm Sci 8:1–18

Wolfaardt AC, Underhill LG, Altwegg R, Visagie J, Williams AJ (2008) Impact of the *Treasure* oil spill on African penguins *Spheniscus demersus* at Dassen Island: case study of a rescue operation. Afr J Mar Sci 30:405–419

Yamato O, Goto I, Maede Y (1996) Hemolytic anemia in wild seaducks caused by marine oil pollution. J Wildl Dis 32:381–384

Ylitalo GM, Krahn MM, Dickhoff WW, Stein JE, Walker CC, Lassitter CL, Garrett ES, Desfosse LL, Mitchell KM, Noble BT (2012) Federal seafood safety response to the *Deepwater horizon* oil spill. Proc Natl Acad Sci 109:20274–20279

Ziccardi MH, Wilkin SM (2018) Oil spill response and effects. In: Gulland FMD, Dierauf LA, Whitman KL (eds) CRC handbook of marine mammal medicine. CRC Press, Boca Raton, FL, pp 19–36

Ziccardi MH, Wilkin SM, Rowles TK, Johnson S (2015) Pinniped and cetacean oil spill response guidelines. U.S. Department of Commerce, NOAA. NOAA technical memorandum NMFS-OPR-52, pp 138

Zock J-P, Rodríguez-Trigo G, Rodríguez-Rodríguez E, Espinosa A, Pozo-Rodríguez F, Gómez F, Fuster C, Castaño-Vinyals G, Antó JM, Barberà JA (2012) Persistent respiratory symptoms in clean-up workers 5 years after the *Prestige* oil spill. Occup Environ Med 69:508–513

Nuclear Radiation

Birgitta Åhman

1 Introduction

Radioactive substances may originate from both natural and man-made sources. Radioactive material in the environment is of potential concern since it may enter the food chain for animals and humans and persist in the environment for a long time. There is thus a risk that both humans and other organisms may be exposed to radiation.

For most people, cosmic radiation and naturally occurring radionuclides in the environment are the primary sources of radiation exposure. For example, the potassium isotope ^{40}K in our bodies (55 Bq/kg fresh weight in the body of an average adult human) gives an annual radiation dose of about 0.2 mSv (effective dose, millisievert) (AMAP 1998), and cosmic radiation leads to average doses that are about twice as high as that due to ^{40}K (although it is enhanced at high altitudes). Inhalation of decay products of radon (^{222}Rn) and thoron (^{220}Rn), often termed "radon daughters," may cause enhanced doses at locations where the geological profile includes these nuclides. Radon daughters emit alpha radiation, so inhalation of these natural radionuclides may potentially cause lung cancer. Exposure to radiation from inhalation of such decay products is generally low in areas with permafrost (as in the Arctic), and where houses are normally built without a basement.

This chapter focuses on radionuclides that are of anthropogenic (or man-made) origin and also briefly considers naturally present radionuclides which are made available for plant and animal uptake as an effect of human activities. The emphasis of the text is on those radionuclides that have the potential to constitute a health risk for animal populations (with a focus on mammals) or enter the human food chain and create a potential health risk for Arctic and Subarctic people. Most of the focus is on

B. Åhman (✉)
Swedish University of Agricultural Sciences, Uppsala, Sweden
e-mail: birgitta.ahman@slu.se

M. Tryland (ed.), *Arctic One Health*, https://doi.org/10.1007/978-3-030-87853-5_8

terrestrial and freshwater systems since radionuclides accumulate in organisms more easily in these systems compared with those in marine environment. Of special concern are those radionuclides that are either absorbed by inhalation or that contaminate vegetation that is eaten by animals including humans.

2 Radionuclides of Concern

Iodine (^{131}I) will be of particular initial concern after there has been radioactive fallout from sources such as nuclear accidents and atmospheric nuclear weapons testing. Iodine-131 (and other less important shorter-lived radioiodine isotopes) can be inhaled with air, thereby entering the lungs, and may be absorbed into the circulatory blood system and then absorbed by the thyroid where it is accumulated as an essential component of several hormones. At high doses, this may cause thyroid cancer as occurred after the Chernobyl accident (IAEA 2006). For grazing animals, ingested ^{131}I is completely absorbed in the gut and readily transferred to milk. There is a potential transfer of ^{131}I to suckling offspring and to humans by consumption of contaminated milk and milk products. Fortunately, ^{131}I is short lived with a physical half-life of 8 days so it disappears relatively fast from the environment. Therefore, after a single contamination event, the potential exposure from this radioisotope largely exists for the first month, and to a much lesser extent the second month after releases and deposition have ceased.

Potential long-term environmental transfer to humans and other animals may occur for radioactive cesium (largely ^{137}Cs, with a physical half-life of 30 years, and ^{134}Cs, with a 2-year half-life), and strontium (^{90}Sr and ^{89}Sr, with half-lives of 28 years and 50 days, respectively). Cesium belongs to the alkali metals (like sodium and potassium), and when present in the diet, it is readily absorbed from the gut and then transferred to body tissues. The highest activity concentrations are found in soft tissue such as muscle, liver, and kidneys (Åhman 1994; Skuterud et al. 2004). Strontium belongs to the same group as calcium (alkaline earth metals) and is mainly accumulated in bone and milk (Stara et al. 1971). Both of these environmentally mobile radionuclides can be transferred from plants to grazing animals and through the food chain to humans and carnivores. While strontium has a long turnover rate in animals and may remain in the body for long, cesium has a relatively short retention time (a short biological half-life). The reduction of radiocesium is thus rather rapid when there is no further intake of contaminated food. The biological half-life may however differ depending on the organism liveweight and organ considered, and there are generally longer biological half-lives in larger organisms (Beresford et al. 2004; Stara et al. 1971).

Other radionuclides that are of potential concern are the naturally occurring isotopes of polonium (^{210}Po) and lead (^{210}Pb) that mainly accumulate in internal organs like kidney and liver. These radionuclides have a high radiotoxicity and are a potential health risk even at low activity concentrations. In addition, radionuclides such as technetium (^{99}Tc), iodine (^{129}I), americium (^{241}Am), and several plutonium

(Pu) isotopes are emitted into waters around reprocessing plants, thereby entering mainly marine environments.

3 Sources of Radioactivity

Man-made sources of radioactivity are related to either military activities or the production of nuclear power (including local use of radioisotope thermoelectric generators for production of electricity in remote places or energy for use in machinery such as that used in submarines) (AMAP 2010). Most of the man-made radioactive material presently found in Arctic and Subarctic regions derives from atmospheric nuclear weapon testing after the Second World War, the Chernobyl accident in 1986, and the releases from reprocessing plants in the Northern Hemisphere. In addition, there are local sources linked to industrial or military facilities in the Arctic, which may potentially release radionuclides to the environment (AMAP 2010).

Aboveground nuclear bomb testing mainly took place from 1945 to 1960, with additional fewer tests up to about 1980. Many of these tests were made at the USSR test site on Novaya Zemlya in the Arctic Sea. The nuclear weapon testing (NWT) released radioactive isotopes of cesium (^{137}Cs) and strontium (^{90}Sr) high into the atmosphere and stratosphere. These emissions were gradually deposited over subsequent decades after the releases, mainly in a circumpolar belt between the 30° and 60° N latitudes (Langham 1961; UNSCEAR 2000). Most of the radioactive material was deposited with rain or snow, and the highest depositions are, therefore, located in areas with high precipitation (Pálsson et al. 2013).

The accident in the Soviet nuclear power plant in Chernobyl, Ukraine, in 1986, caused substantial releases of several radionuclides ejected upward to 1–10 km in the atmosphere. Subsequent radioactive fallout from the accident was recorded in most European countries (Wilson et al. 1998). Radioactive cesium (^{137}Cs and ^{134}Cs), strontium (^{90}Sr and ^{89}Sr), and iodine (^{131}I) were transported long distances and deposited mainly with precipitation (rain and snow), creating a highly uneven pattern of ground deposition over large areas (Wright et al. 1999). Cesium isotopes dominated the fallout, and ground deposition of ^{137}Cs reached 200 kBq/m^2 in some locations in Scandinavia, just south of the Arctic Circle. As an effect of the prevailing winds after the accident, ground depositions in Arctic Russia were considerably lower (up to 1 kBq/m^2 on the Kola peninsula and declining further east).

The more recent accident at the Fukushima Daiichi nuclear power plant in Japan, 2011, was caused by an earthquake and subsequent tsunami. Although radioisotopes of iodine, cesium, and plutonium were detectable in large areas of the Northern Hemisphere, the global release of radionuclides from Fukushima was too low to cause concern for human or animal health outside Japan, with less than 1% of the total release from Fukushima estimated to be deposited in the Arctic region (above latitude 67°N) (Thakur et al. 2013). Nevertheless, radiocesium from the accident was detected, for instance, in Finland (Koivurova et al. 2015) at very low levels (c.1 Bq of ^{137}Cs per m^2).

Releases of radionuclides from reprocessing plants (e.g., Sellafield in the UK and La Hauge in France) are entering marine environments and include nuclides like ^{99}Tc, ^{129}I, ^{241}Am, several plutonium (Pu) isotopes, and also some ^{137}Cs and ^{90}Sr (AMAP 2010). In the Russian Arctic, there is also operational release of various radionuclides from nuclear power plants (e.g., Kola and Bilibino) and other industry.

Enhanced release of naturally occurring radionuclides may be produced by anthropogenic activities such as mining, oil extraction, and the use of geothermal energy. The radionuclides of concern for these sources belong to either the natural uranium (^{238}U) or thorium (^{232}Th) decay series. Radionuclides belonging to these series have been detected in many living organisms in the Arctic (AMAP 2010). The radionuclides of most concern are ^{210}Po and ^{210}Pb because of their relatively high radiotoxicity, thereby constituting a potential health risk even at low activity concentrations.

4 Contamination of Animals of Concern for the Human Food Chain

4.1 Radionuclides in Terrestrial Environments

Radioactive fallout affects both agricultural food products and food derived from natural ecosystems. Research on the fallout from NWT and from the Chernobyl accident demonstrated that natural and seminatural food chains, together with freshwater systems, are associated with relatively high risk of significant transfer of some radionuclides to humans over a longer timescale (Calmon et al. 2009; Howard et al. 1996). The transfer of radioactive cesium (^{137}Cs) and strontium (^{90}Sr) from lichen to reindeer and caribou (subspecies of *Rangifer tarandus*), and hence to humans, was identified during the 1960s, generating research in North America (Hanson 1967; Holleman et al. 1971), Fennoscandia (Hvinden and Lillegraven 1961; Salo and Miettinen 1964; Svensson and Lidén 1965), and USSR (Nevstrueva et al. 1967; Ramzaev et al. 1970). After the Chernobyl accident in 1986, this food chain gained extensive attention, primarily in Fennoscandia, where reindeer herding is a traditional occupation among the indigenous Sámi population and where parts of the reindeer ranges were contaminated by substantial fallout that was many times higher than that from the previous NWT fallout (De Cort et al. 1998).

Lichens have no roots, but absorb nutrients, including contaminants, directly from air and precipitation (Tuominen and Jaakkola 1973). Reindeer and caribou are particularly vulnerable to radioactive contamination which is because lichens often form a large part of their diet. Close agreement between the measured activity of ^{137}C per m^2 in the lichen carpet and the deposition per m^2 showed that most of deposited radiocesium is captured by ground lichens in places where they cover the ground surface (Svensson and Lidén 1965). Lichens are long-lived and grow slowly; therefore, contaminants remain in the lichen carpet for many years. Consequently, the intake of radionuclides will be exceptionally high for reindeer and caribou, especially in winter when lichens often dominate their diet (Åhman and White 2018). Since meat from reindeer and caribou is a staple food for many indigenous populations in the Arctic and Subarctic, the radiation doses to these people may be much higher than those of people residing in other temperate areas.

Radioactive substances are transferred from fallout to vegetation to reindeer/caribou and other terrestrial species that are consumed by humans. High, long-lasting, contamination with radiocesium has been reported in wild animals such as moose (*Alces alces*), roe deer (*Capreolus capreolus*), Arctic hare (*Lepus arcticus*), and game birds from areas contaminated by the Chernobyl fallout (Howard et al. 1996; Johanson 1994). Much of the vegetation in natural environments, where these herbivores feed, effectively absorb radiocesium, which also seem to persist longer in most natural environments than in agricultural systems (Howard et al. 1996; Pröhl et al. 2006). The specific role of fruit bodies from fungi (mushrooms) was early recognized as responsible for transfer of radiocesium to white-tailed deer (Johnson and Nayfield 1970), and the transfer of radionuclides from soil via consumed fungi to herbivores was further emphasized after the Chernobyl accident (e.g., Avila et al. 1999; Hove et al. 1990).

The NWT fallout was deposited over many years and distributed over the northern hemisphere (relative to the amount of precipitation). In contrast, the fallout from the Chernobyl accident was a 10-day emission into the atmosphere, and the subsequent deposition on terrestrial and aquatic sources was highly heterogeneous even at a local level. Pooled samples from herds of caribou and reindeer at several sites in North America during 1962–1969 showed ^{137}Cs activity concentrations up to 2000–3000 Bq/kg fw (fresh weight) in the meat, with maximum values in caribou of 6000 Bq/kg (Macdonald et al. 2007). There was a trend of increasing ^{137}Cs activity concentration in caribou and reindeer from west to east, correlating with the precipitation density and in accordance with models by Wright et al. (1999). Activity concentrations of ^{137}Cs similar to those reported by Macdonald et al. (2007) were recorded in Fennoscandian reindeer (Rissanen and Rahola 1990; Westerlund et al. 1987), while some herds in Alaska (O'Hara et al. 1999) and Greenland (Aarkrog et al. 2000) were less affected.

In contrast, the Chernobyl fallout resulted in huge variations between regions and sites (several orders of magnitude within a plant or animal species). For example, only moderate elevations (around a doubling) of activity concentrations of ^{137}Cs were recorded in reindeer from the northernmost parts of Norway and Sweden, while reindeer in the southern and central parts of the reindeer herding area were heavily affected (Skuterud and Thørring 2012; Åhman et al. 2001). Activity concentrations of ^{137}Cs in reindeer meat reached around 80,000 Bq/kg fw in the most affected parts in Sweden (Åhman and Åhman 1994), and highest values at about 150,000 Bq/kg fw were recorded for individual reindeer in Norway (Strand et al. 1992). The Finnish reindeer herding area was less affected, with activity concentrations of ^{137}Cs up to around 2000 Bq/kg fw, except for the Halla area in the southeastern corner of the Finnish reindeer herding area, with up to 16,000 Bq/kg fw in reindeer meat (Rissanen and Rahola 1990).

A smaller increase in the activity concentrations of radiocesium was observed in North American caribou after the Chernobyl accident, adding around 100 Bq ^{137}Cs/kg to the remaining contamination from NWT (Macdonald et al. 2007). The "old" ^{137}Cs from NWT, however, still dominated in these animals (on average 80% of total ^{137}Cs).

In 2020, 34 years after the accident, significant levels of radiocesium from Chernobyl still persisted in parts of the reindeer ecosystem in both Norway (Komperød et al. 2017) and Sweden (Åhman and Wiklund 2019). The radiocesium activity concentration in animal tissues reflects that in the diet within a few weeks due to a fairly short biological half-life, and changes in activity concentrations in reindeer are soon to follow that of the forage (Åhman and Åhman 1994). Few reindeer carcasses, however, presently exceed the standard maximum values for sale (3000 Bq/kg in Norway and 1500 Bq/kg in Sweden). In areas where there is possibility that values reindeer will exceed the limit, they are either fed uncontaminated feed to decontaminate the meat for some weeks prior to slaughter or slaughtered in a different season when they have grazed on less contaminated pastures which will then be reflected in the radiocesium activity concentration in the meat (Åhman 1999; Mehli et al. 2000).

Strontium-90 radionuclide has also received attention with respect to human health. The behavior of strontium is very similar to that of its close analogue, calcium. Therefore, the uptake and metabolism of calcium are sometimes expressed in relation to Ca in living organisms. Persson (1971) reported activity concentrations of ^{90}Sr in relation to Ca at 37 Bq/g Ca in bone (recalculated to Bq from original values in pCi) and 6 Bq/g Ca in muscle from Swedish reindeer in 1965. Because muscle, in contrast to bone (and also antlers), contains little Ca, the levels of Sr will be several times lower in muscle compared with bone. Consequently, Hanson et al. (1967) reported activity concentrations of ^{90}Sr between 1173 and 2582 Bq/kg dw (dry weight) in bone, and only about 0.1–1.6 Bq/kg fw in muscle, from Alaskan caribou after the NWT fallout.

The relative proportion of radiostrontium compared with other radionuclides was lower in the Chernobyl fallout than in the fallout from NWT, and therefore it was regarded as a minor problem for human health compared with radiocesium. Strontium-90 was, however, detectable in bone from reindeer in one of the most contaminated reindeer herding districts in Norway (Vågå) in 1988–1989, with average activity concentrations of 1810 Bq/kg dw (Staaland et al. 1991), thus similar to that measured after the NWT fallout. Most of the strontium detected in reindeer bone and antlers from Vågå seemed to originate from the Chernobyl fallout. The ^{90}Sr activity concentrations in bone and antlers collected from both the same site and a more northerly reindeer herding district in 2000–2002 had ^{90}Sr activity concentrations ranging from about 2 to 4 Bq/g Ca, which corresponds to around 300–900 Bq/kg dw in bone or antler tissue (Skuterud et al. 2005b).

Radionuclides in samples from muskox and caribou, collected from 1998 to 2000 from the Aleutian Islands, Alaska, were reported by Hong et al. (2011). Muskox (four animals) had ^{137}Cs and ^{90}Sr activity concentrations in muscle at 5.33–20.4 and 0.28–0.36 Bq/kg fw, respectively. Bone from the same animals contained 36–64 Bq/

kg dw of ^{90}Sr. Cs-137 activity concentrations in caribou, sampled from July to early September, were lower (1–8 Bq/kg fw) than those in muskox, while those of ^{90}Sr in bone were higher (76–154 Bq/kg dw). Lower amounts of ^{137}Cs in caribou than in muskox would be expected at this time of year, because caribou mainly feed on vascular plants. In contrast, due to long biological half-lives, strontium in bone reflects intake over a longer time period and is probably affected by intake during the previous seasons.

The content of several radionuclides from the uranium series in caribou from areas around uranium mines was studied in northern Canada (Thomas and Gates 1999). The highest activity concentrations of ^{226}Ra, ^{210}Pb, and ^{210}Po were in bone with mean values of 72, 669, and 367 Bq/kg fw, respectively. Activity concentrations of ^{226}Ra were much lower in all soft tissues (less than 3% of that in bone), whereas ^{210}Pb and ^{210}Po activity concentrations were relatively high in liver (154 and 286 Bq/kg, respectively) and kidney (169 and 159 Bq/kg, respectively) and low in muscle (1.1 and 12.4 Bq/kg, respectively).

The radionuclide activity concentrations in living organisms in relation to deposition density can differ considerably between species. The relation between deposition and contamination of living organisms is often expressed as aggregated transfer factor "T_{ag}" (m^2/kg), denoting the activity concentration (Bq/kg) in an organism or tissue relative to fallout per area unit (Bq/m^2). Typical T_{ag} values for ^{137}Cs in reindeer in Sweden in the first year after the Chernobyl fallout were around 1 m^2/kg fw in winter and 0.025 m^2/kg in summer (Åhman and Åhman 1990), which can be compared to values for moose around 0.01 m^2/kg fw (Calmon et al. 2009; Howard et al. 1996). Higher T_{ag} values have been reported for Arctic hare (0.03 m^2/kg fw) than for brown hare (*Lepus europaeus*, up to 0.008 m^2/kg), probably reflecting their habitat, as the brown hare is commonly found in agricultural landscapes, while Arctic hare is found in forest and on tundra. T_{ag} values for game birds have been reported to be i 0.01–0.02 m^2/kg (Calmon et al. 2009; Howard et al. 1996).

For reindeer and caribou, which mainly eat lichens in winter and vascular plants in summer, the change in diet, and thus radiocesium intake, creates a pronounced seasonal variation, enhanced by a shorter biological half-life in summer compared to winter (Holleman et al. 1971). Radiocesium activity concentrations in the body of reindeer and caribou are thus typically high in winter and low in summer (Eikelmann et al. 1990; Hanson and Palmer 1965; Åhman and Åhman 1994). However, since the Chernobyl accident, the magnitude of the variation has declined with time (Skuterud et al. 2005a) as radiocesium is lost from the lichens by weathering (or the lichens are eaten by reindeer) and transported into the soil, eventually becoming available for root uptake by vascular plants and, thereby, more evenly distributed in the environment. Seasonal patterns in ^{137}Cs activity concentrations in reindeer may, therefore, gradually change with occasional peaks in autumn if there is a high abundance of fungi (Hove et al. 1990), which has been observed, for instance, for roe deer (Avila et al. 1999).

Long-term decline of radionuclides in the environment, and thus in vegetation and animals, occurs through a number of processes. The most obvious is radioactive decay, which differs for each radioisotope (resulting in physical half-lives discussed earlier in this chapter). There is also an often quoted "ecological half-life" that describes how fast a radionuclide is made unavailable for uptake by animals though a number of processes in the ecosystem and "effective half-life" that describes the combined decline of a radionuclide from an organism or a system that takes into account the physical decay rate of the isotope (Howard et al. 1996). A more rapid reduction of radiocesium activity concentrations in reindeer was observed during the first 10 years after the Chernobyl accident (with effective half-lives at 2–5 years) compared with later years. The early decline to a large extent followed that of lichens (Skuterud et al. 2005a, 2009; Åhman 2007). The decline during latter years approached that of the physical half-life of ^{137}Cs of 30 years.

4.2 Freshwater Systems

The contamination of fish in freshwater systems after radioactive fallout will depend on several factors. For example, lakes that get most of their water from surrounding (contaminated) wetland catchments will receive more contaminants than those that mainly receive water from direct precipitation. In fish, there is often an accumulation with trophic level, so that piscivore fish (fish that feed on other fish) have higher ^{137}Cs activity concentrations than herbivore and planktivore fish (Ishii et al. 2020; Sundbom et al. 2003). After a single deposition event (such as the Chernobyl accident), there will be a delay before maximum contamination of fish is reached. Measurements in three Swedish lakes after the Chernobyl accident showed that, depending on lake and fish species, it took from 56 to 806 days for ^{137}Cs in fish to reach maximum ^{137}Cs activity concentrations (Sundbom et al. 2003). The decline of ^{137}Cs also differed considerably between lakes, but after 10 years, the lakes seemed to have reached a steady state, and the decline of ^{137}Cs activity concentrations in fish was then dominated by physical decay (30 y half-life). Aggregated transfer factors (^{137}Cs activity concentration in fish divided by deposition per m^2) near to steady-state conditions for roach (*Rutilus rutilus*, an omnivore fish) in these lakes varied between 0.0025 and 0.04 m^2/kg. Cs-137 activity concentrations for large perch (*Perca fluviatilis*, a piscivore) were three to six times higher than in roach.

4.3 Radionuclides in the Marine Environment

Radionuclides that have been released to the ocean from nuclear fuel reprocessing plants are partially transported with currents to Arctic waters. Radionuclides that have been deposited with precipitation (e.g., ^{137}Cs and ^{90}Sr) are also dispersed with currents and eventually reach stable, low, mean concentration in seawater (AMAP 2010). Radionuclides are accumulated by marine organisms such as marine fish, seabirds, seals, and wales in Arctic waters. However, the activity concentrations have generally been very low due to large dilution effects.

Several species of marine fish collected around the Arctic and Subarctic had consistently low ^{137}Cs activity concentrations. The highest ^{137}Cs activity concentrations of up to 3 Bq/kg fw were recorded during the 1960s. Marine fish collected in the Barents Sea and from waters along the Norwegian coast in 2005 (NRPA 2007) had activity concentrations that were well below 1 Bq/kg fw. Fish from waters close to the Teriberka settlement in Russia (Kola Penninsula) were below 0.5 Bq/kg fw for both ^{137}Cs and ^{90}Sr, and up to 1.6 Bq/kg fw for $^{239,240}Pu$ (AMAP 2010).

Nevertheless, consumption of seafood may contribute to the radiation dose of humans, depending on the diet. Considering both natural radionuclides and those of anthropogenic origin, and based on recent national dietary surveys, Komperød and Skuterud (2018) concluded that fish and shellfish were currently the most important dietary components contributing to the effective radiation dose received by the general population in Norway. The total effective dose was still low (on average 0.42 mSv) and predominantly caused by the naturally occurring radionuclides.

Samples of muscle, liver, and kidney in several species of seabirds (*Rissa tridactyla*, *Uria lomvia*, *Larus hyperboreus*, *Fulmarus glacialis*, and *Alle alle*) from Svalbard have been analyzed for ^{137}Cs, ^{210}Po, and ^{210}Pb (NRPA 2007). Only low ^{137}Cs activity concentrations were found in these birds (up to about 0.2 Bq/kg fw, but mostly below the detection limit). The highest activity concentrations of ^{210}Po in muscle (on average 14 Bq/kg fw) were found in little auks, while that of ^{210}Pb in muscle were negligible for all species. Both ^{210}Po and ^{210}Pb had the highest activity concentrations in kidney, with an average 6 Bq ^{210}Pb/kg fw in Northern fulmar (*Fulmarus glacialis*), and Bq ^{210}Po /kg fw in the kidney of Brunnich's guillemot (*Uria lomvia*).

Seals (*Pinnipedia*) and wales (*Cetacea*) are top predators in the Arctic marine food chain and are therefore vulnerable to accumulation of contaminants. Seals and wales from Greenland and wales from the Faroe Islands (organs not specified) have been measured for ^{137}Cs from 1960 and onward (AMAP 2010). Activity concentrations declined from generally only between 1 and 10 Bq/kg fw in the late 1960s and stabilized below 1 Bq/kg fw from about 1980 onward. Polonium-210 and ^{210}Pb have been measured in organs from grey seal (*Halichoerus grypus*) from the Baltic Sea and ringed seal (*Pusa hispida*) from Svalbard in 2003 and 2004 (AMAP 2010). Activity concentrations of ^{210}Po in muscle ranged from 3 to 28 Bq/kg fw for ^{210}Po, and 0.14 to 0.45 for ^{210}Pb. Considerably higher amounts were found in kidney and liver from the same animals (up to 215 Bq/kg fw for ^{210}Po and 2 Bq/kg fw for ^{210}Pb).

Activity concentrations of several radionuclides in muscle and other tissue from seals collected in the Canadian Arctic in 2014 have been reported by Chen et al. (2017). Activity concentrations were low, for ^{137}Cs (0.1–0.4 Bq/kg fw) and for ^{210}Pb and ^{226}Ra (mostly below detection limit of 0.3 Bq/kg fw). Higher activity concentrations were found for ^{210}Po (12–39 Bq/kg fw in muscle and 102–396 Bq/kg fw in liver) and ^{40}K (range 50–80 Bq/kg fw). The data for ^{137}Cs and ^{210}Po were similar to those reported for seal and walrus from the Bering Sea in 1996.

4.4 Exposure of Animals and Humans and the Related Health Risks

Intake of food with elevated levels of radioactive substances may involve a health risk for animals, including humans. A general recommendation for humans from international authorities (ICRP 2007) is that the added radiation dose caused by intake of contaminated food products should not exceed 1 mSv per year. This dose limit is reached at different intake rates depending on radioisotope. For ^{137}Cs, the dose is obtained at an intake of about 75,000 Bq (ICRP 1993), while the same dose is reached at an intake of about 36,000 Bq ^{90}Sr, or an intake of only 800 Bq ^{210}Po.

Most dose estimates for populations in general (not for doses achieved from occupation or, e.g., medical treatment) have been carried out for ^{137}Cs, since this is the radionuclide that has entered the human food chain worldwide. For example, the total dose commitment from ^{137}Cs from NWT fallout has been estimated to be 0.19 mSv for an average person living within the northern hemisphere (UNSCEAR 1982). Doses to persons with special food habits, like reindeer herders or caribou hunters, may be considerably higher. The annual dose for a male reindeer herder in Finland during the 1960s was estimated to be 1.2 mSv (Miettinen et al. 1963), and whole-body measurements of reindeer herders in Sweden during the same period gave estimated annual doses around 1 mSv (Falk et al. 1991). An integrated dose at 18 mSv was assessed (based on whole-body monitoring) for an average reindeer herder in Northern Norway during 1950–2010 (Skuterud and Thorring 2015). This means that herders have received higher dose than recommended during several years. Nevertheless, such doses are too low to be expected to significantly affect the risk for cancer (the most likely effect). Similarly, no measurable differences in the frequency of cancer between reindeer herders and the average population have been recorded (Haldorsen and Tynes 2005; Hassler et al. 2008; Kurttio et al. 2010; Tynes and Haldorsen 2007), although Kurttio et al. (2010) suggest that there are some indications, however ambiguous, of raised cancer risk due to elevated radiation doses during childhood.

Despite the high local fallout after the Chernobyl accident, radiation doses to humans have not generally been higher than that after the NWT, even in the most exposed groups. The effective dose during 50 years to a reindeer herder in the most contaminated areas of Norway has been estimated to be 27 mSv (Skuterud and Thørring 2012), which is on average 0.5 mSv/year. Lifetime doses between 4 and 8 mSv were estimated for members of hunter families in the most contaminated areas of Sweden (Tondel et al. 2017). Countermeasures and dietary recommendations limited the intake of highly contaminated food products, and thus the radiation dose to humans (Skuterud and Thørring 2012; Åhman 1999). Meat from game and reindeer, freshwater fish, and wild berries and mushrooms were identified as food products of concern. Countermeasures, such as changing the timing of hunting and reindeer slaughter, clean feeding of reindeer before slaughter, and the use of cesium binders in salt licks or factory-made feeds, were commonly used to reduce radiocesium activity concentrations in animals (Beresford et al. 2016). These measures considerably reduced the radiation dose to humans in Sweden and other countries after the Chernobyl accident (Rääf et al. 2006).

4.5 Radiation Exposure of Wildlife

The importance of protecting both humans and other organism in the environment has been increasingly recognized during the last decades (see, e.g., Howard et al. 2010; ICRP 2014; Pentreath 1999). Transfer factors and dose conversion factors have been presented for a range of radionuclides and living organisms (UNSCEAR 2008). These can be used to estimate expected radiation doses and risk for harm to defined groups of wildlife. Specific reference organisms have been suggested for Arctic environments (Hosseini et al. 2005), based on ecological niche, radiosensitivity, and exposure to radioactive contamination. Proposed representative mammals have been identified for different types of environment. These were seals for the marine system, mink for freshwater systems, and lemmings and voles, reindeer/caribou and foxes for terrestrial environments.

Different dose limits have been suggested by various authors and organizations. Howard et al. (2010) suggested using 10 μGy/h (absorbed dose, micro gray per hour) as a limit below which there is little risk of harm to wild organisms, while above this limit, further assessment is needed to find out if there is a potential risk.

Possible effects of radiation for reindeer were investigated in Norway after the Chernobyl accident by Roed and Jacobsen (1995). The authors did not find any differences between reindeer herds in the pattern of chromosome aberrations (an indicator of damage) that could be attributed to the exposure to radiation. However, in a herd from the area with the highest ground radiocesium deposition, calves born during the spring of 1986, when the accident happened, had significantly more aberrations than expected compared with reindeer born in other years, suggesting a negative effect of radiation exposure to the newborn calves.

5 Summary and Conclusions

There are a number of sources that have released radioactive material into Arctic and Subarctic ecosystems. The resulting radionuclide activity concentrations in animals are, however, generally low, with the exception of reindeer, certain wildlife species and freshwater fish in some areas. The highest contamination of food products for humans was caused by the Chernobyl accident in 1986. However, countermeasures and dietary restrictions and recommendations limited the intake of contaminated food by Arctic and Subarctic populations. No adverse effects of radiation on animal or human health, such as increased frequency of cancer, have been demonstrated in Arctic regions.

Acknowledgments The author would like to thank Lavrans Skuterud and Brenda Howard for valuable comments on this chapter.

References

Aarkrog A, Dahlgaard H, Nielsen SP (2000) Environmental radioactive contamination in Greenland: a 35 years retrospect. Sci Total Environ 245:233–248. https://doi.org/10.1016/s0048-9697(99)00448-9

Åhman B (1994) Body burden and distribution of ^{137}Cs in reindeer. Rangifer 14:23–28. https://doi.org/10.7557/2.14.1.1129

Åhman B (1999) Transfer of radiocesium via reindeer meat to man – effects of countermeasures applied in Sweden following the Chernobyl accident. J Environ Rad 46:113–120. https://doi.org/10.1016/S0265-931X(98)00107-6

Åhman B (2007) Modelling radiocesium transfer and long-term changes in reindeer. J Environ Rad 98:153–165. https://doi.org/10.1016/j.jenvrad.2006.11.011

Åhman B, Åhman G (1990) Levels of ^{137}Cs in reindeer bulls in July/August and September and the effect of early slaughter. Rangifer 20 (Spec Iss 5):34–36. https://doi.org/10.7557/2.10.5.956

Åhman B, Åhman G (1994) Radiocesium in Swedish reindeer after the Chernobyl fallout: seasonal variations and long-term decline. Health Phys 66:503–512. https://doi.org/10.1097/00004032-199405000-00002

Åhman B, White RG (2018) Rangifer diet and nutritional needs. In: Tryland M, Kutz SJ (eds) Reindeer and caribou: health and disease. Taylor & Francis, Boka Raton, pp 107–134

Åhman B, Wiklund E (2019) Radionuclides in reindeer meat; a case report. In: Smulders FJM, Rietjens IMCM, Rose MD (eds) Chemical hazards in foods of animal origin. Wageningen Academic, Wageningen, pp 611–621. https://doi.org/10.3920/978-90-8686-877-3_25

Åhman B, Wright SM, Howard BJ (2001) Effect of origin of radiocesium on the transfer from fallout to reindeer meat. Sci Total Environ 278:171–181. https://doi.org/10.1016/S0048-9697(01)00646-5

AMAP (1998) AMAP assessment report: Arctic pollution issues Arctic monitoring and assessment programme (AMAP), Oslo, Norway

AMAP (2010) AMAP assessment 2009: radioactivity in the Arctic monitoring and assessment programme (AMAP), Oslo, Norway

Avila R, Johanson KJ, Bergström R (1999) Model of the seasonal variations of fungi ingestion and ^{137}Cs activity concentrations in roe deer. J Environ Rad 46:99–112. https://doi.org/10.1016/S0265-931X(98)00108-8

Beresford NA, Broadley MR, Howard BJ et al (2004) Estimating radionuclide transfer to wild species—data requirements and availability for terrestrial ecosystems. J Radiol Prot 24:A89–A103. https://doi.org/10.1088/0952-4746/24/4A/006

Beresford NA, Fesenko S, Konoplev A et al (2016) Thirty years after the Chernobyl accident: what lessons have we learnt? J Environ Rad 157:77–89. https://doi.org/10.1016/j.jenvrad.2016.02.003

Calmon P, Thiry Y, Zibold G et al (2009) Transfer parameter values in temperate forest ecosystems: a review. J Environ Rad 100:757–766. https://doi.org/10.1016/j.jenvrad.2008.11.005

Chen J, Zhang WH, Sadi B et al (2017) Activity concentration measurements of selected radionuclides in seals from Canadian Arctic. J Environ Rad 169:48–55. https://doi.org/10.1016/j.jenvrad.2016.12.015

De Cort M, Dubois G, Fridman SD et al (1998) Atlas of Caesium 137 deposition on Europe after the Chernobyl accident Office for Official Publications of the European Communities, Luxembourg

Eikelmann IMH, Bye K, Sletten HD (1990) Seasonal variation of cesium 134 and cesium 137 in semidomestic reindeer in Norway after the Chernobyl accident. Rangifer 10 (Spec Iss 3):35–38. https://doi.org/10.7557/2.10.3.818

Falk R, Eklund G, Giertz H, Östergren I (1991) Cesium in the Swedish population after Chernobyl: internal radiation, whole-body counting. In: Moberg L (ed) The Chernobyl fallout in Sweden. Results from a research programme on environmental radiology. Swedish Radiation Protection Institute, Stockholm, pp 547–577

Haldorsen T, Tynes T (2005) Cancer in the Sami population of North Norway, 1970-1997. Eur J Cancer Prev 14:63–68. https://doi.org/10.1097/00008469-200502000-00009
Hanson WC (1967) Cesium-137 in Alaskan lichens, caribou and eskimos. Health Phys 13:383–389. https://doi.org/10.1097/00004032-196704000-00007
Hanson WC, Palmer HE (1965) Seasonal cycle of ^{137}Cs in some Arctic natives and animals. Health Phys 11:1401–1406. https://doi.org/10.1097/00004032-196512000-00018
Hanson WC, Watson DG, Perkins RW (1967) Concentration and retention of fallout radionuclides in Alaskan arctic ecosystems. In: Åberg B, Hungate FP (eds) Radioecological concentration processes. Proceedings of an international symposium held in Stockholm 25–29 April, 1966. Pergamon Press, Oxford, pp 233–245. https://doi.org/10.1016/B978-0-08-012122-2.50029-3
Hassler S, Soininen L, Sjölander P, Pukkala E (2008) Cancer among the Sami – a review on the Norwegian, Swedish and Finnish Sami populations. Int J Circumpolar Health 67:421–432. https://doi.org/10.3402/ijch.v67i5.18351
Holleman DF, Luick JR, Whicker FW (1971) Transfer of radiocesium from lichen to reindeer. Health Phys 21:657–666. https://doi.org/10.1097/00004032-197111000-00006
Hong GH, Baskaran M, Molaroni SM et al (2011) Anthropogenic and natural radionuclides in caribou and muskoxen in the Western Alaskan Arctic and marine fish in the Aleutian Islands in the first half of 2000s. Sci Total Environ 409:3638–3648. https://doi.org/10.1016/j.scitotenv.2011.06.044
Hosseini A, Brown J, Thørring H, Børretzen P (2005) Environmental impact assessment in Arctic environments, Strålevern rapport. Norwegian Radiation Protection Authority, Østerås
Hove K, Pedersen Ø, Garmo TH et al (1990) Fungi: a major source of radiocesium contamination of grazing ruminants in Norway. Health Phys 59:189–192. https://doi.org/10.1097/00004032-199008000-00004
Howard BJ, Johanson K, Linsley GS et al (1996) Transfer of radionuclides by terrestrial food products from semi-natural ecosystems to humans. Modelling of radionuclide interception and loss processes in vegetation and of transfer in semi-natural ecosystems. Second report of the VAMP Terrestrial Working Group. International Atomic Energy Agency, Vienna, pp 49–79
Howard BJ, Beresford NA, Andersson P et al (2010) Protection of the environment from ionising radiation in a regulatory context-an overview of the PROTECT coordinated action project. J Radiol Prot 30:195–214. https://doi.org/10.1088/0952-4746/30/2/s01
Hvinden T, Lillegraven A (1961) Caesium-137 and strontium-90 in precipitation, soil and animals in Norway. Nature 192:1144–1146. https://doi.org/10.1038/1921144a0
IAEA (2006) Environmental consequences of the Chernobyl accident and their remediation: twenty years of experience. Report of the Chernobyl forum expert group 'environment', International Atomic Energy Agency, Vienna
ICRP (1993) Age-dependent doses to members of the public from intake of radionuclides: part 2 ingestion dose coefficients, ICRP publication 67. Ann ICRP 23:167
ICRP (2007) The 2007 recommendations of the international commission on radiological protection. ICRP publication 103. Ann ICRP 37:1–332
ICRP (2014) Protection of the environment under different exposure situations, ICRP publication 124. Ann ICRP 43:1–58. https://doi.org/10.1177/0146645313497456
Ishii Y, Matsuzaki S-iS, Hayashi S (2020) Different factors determine Cs-137 concentration factors of freshwater fish and aquatic organisms in lake and river ecosystems. J Environ Rad 213. https://doi.org/10.1016/j.jenvrad.2019.106102
Johanson KJ (1994) Radiocaesium in game animals in the Nordic countries. In: Dahlgaard H (ed) Nordic radioecology. The transfer of radionuclides through nordic ecosystems to man. Elsevier, Amsterdam, pp 287–302
Johnson W, Nayfield CL (1970) Elevated levels of Cesium-137 in common muchrooms (Agaricaceae) with possible relationship to high levels of Cesium-137 in whitetail deer, 1968-1969. Radiol Health Data Rep 11:527–531

Koivurova M, Leppanen AP, Kallio A (2015) Transfer factors and effective half-lives of Cs-134 and Cs-137 in different environmental sample types obtained from Northern Finland: case Fukushima accident. J Environ Rad 146:73–79. https://doi.org/10.1016/j.jenvrad.2015.04.005

Komperød M, Skuterud L (2018) Radiation doses from the Norwegian diet. Health Phys 115:195–202. https://doi.org/10.1097/hp.0000000000000842

Komperød M, Østmo T, Skuterud L (2017) Radioaktivitet i norsk mat – Resultater fra overvåkningen av dyr og næringsmidler 2016 Statens strålevern, Østerås

Kurttio P, Pukkala E, Ilus T et al (2010) Radiation doses from global fallout and cancer incidence among reindeer herders and Sami in northern Finland. Occup Environ Med 67:737–743. https://doi.org/10.1136/oem.2009.048652

Langham WH (1961) Some considerations of present biospheric contamination by radioactive fallout. J Agric Food Chem 9:91–95. https://doi.org/10.1021/jf60114a002

Macdonald CR, Elkin BT, Tracy BL (2007) Radiocesium in caribou and reindeer in northern Canada, Alaska and Greenland from 1958 to 2000. J Environ Rad 93:1–25. https://doi.org/10.1016/j.jenvrad.2006.11.003

Mehli H, Skuterud L, Mosdøl A, Tønnessen A (2000) The impact of Chernobyl fallout in the southern Saami reindeer herders of Norway in 1996. Health Phys 79:682–690. https://doi.org/10.1097/00004032-200012000-00014

Miettinen JK, Jokelainen A, Roine P et al (1963) ^{137}Cs and potassium in people and diet – a study of Finnish Lapps. Ann Acad Sci Fennicae A II Chemia 120:1–46

Nevstrueva MA, Ramzaev PV, Moiseer AA et al (1967) The nature of ^{137}Cs and ^{90}Sr transport over the lichen-reindeer-man food chain. In: Åberg B, Hungate FP (eds) Radioecological concentration processes. Proceedings of an international symposium held in Stockholm 25–29 April, 1966. Pergamon Press, Oxford, pp 209–215

NRPA (2007) Radioactivity in the Marine Environment 2005. Results from the Norwegian marine monitoring programme (RAME), Strålevern Rapport. Norwegian Radiation Protection Authority, Østerås

O'Hara TM, Dasher D, George JC, Woshner V (1999) Radionuclide levels in caribou of northern Alaska in 1995-96. Arctic 52:279–288

Pálsson SE, Howard BJ, Bergan TD et al (2013) A simple model to estimate deposition based on a statistical reassessment of global fallout data. J Environ Rad 121:75–86. https://doi.org/10.1016/j.jenvrad.2012.03.006

Pentreath RJ (1999) A system for radiological protection of the environment: some initial thoughts and ideas. J Radiol Prot 19:117–128. https://doi.org/10.1088/0952-4746/19/2/302

Persson BR (1971) ^{90}Sr in northern Sweden: relationships and annual variation from 1961 to 1969 in lichen, reindeer and man. Health Phys 20:393–402. https://doi.org/10.1097/00004032-197104000-00004

Pröhl G, Ehlken S, Fiedler I et al (2006) Ecological half-lives of ^{90}Sr and ^{137}Cs in terrestrial and aquatic ecosystems. J Environ Rad 91:41–72. https://doi.org/10.1016/j.jenvrad.2006.08.004

Rääf CL, Hubbard L, Falk R et al (2006) Transfer of (CS)-C-137 from Chernobyl debris and nuclear weapons fallout to different Swedish population groups. Sci Total Environ 367:324–340. https://doi.org/10.1016/j.scitotenv.2005.12.006

Ramzaev PV, Troitskaia MN, Ibatullin MS et al (1970) Radioecological parameters of the lichen-northern deer—man chain. Gig Sanit 35:38–42

Rissanen K, Rahola T (1990) Radiocesium in lichens and reindeer after the Chernobyl accident. Rangifer 10 (Spec Iss 3):55–61. https://doi.org/10.7557/2.10.3.826

Roed KH, Jacobsen M (1995) Chromosome aberrations in Norwegian reindeer following the Chernobyl accident. Mutat Res Lett 346:159–165. https://doi.org/10.1016/0165-7992(95)90048-9

Salo A, Miettinen JK (1964) Strontium-90 and caesium-137 in Arctic vegetation during 1961. Nature 201:1177–1179. https://doi.org/10.1038/2011177a0

Skuterud L, Thørring H (2012) Averted doses to Norwegian Sami reindeer herders after the Chernobyl accident. Health Phys 102:208–216. https://doi.org/10.1097/HP.0b013e3182348e12

Skuterud L, Thorring H (2015) Fallout Cs-137 in reindeer herders in Arctic Norway. Environ Sci Technol 49:3145–3149. https://doi.org/10.1021/es506244n
Skuterud L, Pedersen Ø, Staaland H et al (2004) Absorption, retention and tissue distribution of radiocaesium in reindeer: effects of diet and radiocaesium source. Radiat Environ Biophys 43:293–301. https://doi.org/10.1007/s00411-004-0257-4
Skuterud L, Gaare E, Eikelmann IM et al (2005a) Chernobyl radioactivity persists in reindeer. J Environ Rad 83:231–252. https://doi.org/10.1016/j.jenvrad.2005.04.008
Skuterud L, Gwynn JP, Gaare E et al (2005b) ^{90}Sr, ^{210}Po and ^{210}Pb in lichen and reindeer in Norway. J Environ Rad 84:441–456. https://doi.org/10.1016/j.jenvrad.2005.04.016
Skuterud L, Åhman B, Solatie D, Gaare E (2009) Long-term decline of radiocaesium in Fennoscandian reindeer NKS, Roskilde
Staaland H, Bjørnstad H, Pedersen Ø et al (1991) Radiostrontium, radiocesium and stable mineral composition of bones of domestic reindeer from Vågå, Norway. Rangifer 11:17–22. https://doi.org/10.7557/2.11.1.969
Stara JF, Nelson NS, Della Rosa RJ, Bustad LK (1971) Comparative metabolism of radionuclides in mammals: a review. Health Phys 20:113–137. https://doi.org/10.1097/00004032-197102000-00001
Strand P, Selnæs TD, Bøe E et al (1992) Chernobyl fallout: internal doses to the Norwegian population and the effect of dietary advice. Health Phys 63:385–392. https://doi.org/10.1097/00004032-199210000-00001
Sundbom M, Meili M, Andersson E et al (2003) Long-term dynamics of Chernobyl Cs-137 in freshwater fish: quantifying the effect of body size and trophic level. J Appl Ecol 40:228–240. https://doi.org/10.1046/j.1365-2664.2003.00795.x
Svensson GK, Lidén K (1965) The transport of ^{137}Cs from lichen to animal and man. Health Phys 11:1393–1400. https://doi.org/10.1097/00004032-196512000-00017
Thakur P, Ballard S, Nelson R (2013) An overview of Fukushima radionuclides measured in the northern hemisphere. Sci Total Environ 458:577–613. https://doi.org/10.1016/j.scitotenv.2013.03.105
Thomas PA, Gates TE (1999) Radionuclides in the lichen-caribou-human food chain near uranium mining operations in northern Saskatchewan, Canada. Environ Health Perspect 107:527–537. https://doi.org/10.2307/3434394
Tondel M, Raaf C, Walinder R et al (2017) Estimated lifetime effective dose to hunters and their families in the three most contaminated counties in Sweden after the Chernobyl nuclear power plant accident in 1986-a pilot study. J Environ Rad 177:241–249. https://doi.org/10.1016/j.jenvrad.2017.06.017
Tuominen Y, Jaakkola T (1973) Absorption and accumulation of mineral elements and radioactive nuclides. In: Ahmadjian V, Hale ME (eds) The Lichens. Academic, New York, pp 185–223
Tynes T, Haldorsen T (2007) Mortality in the Sami population of North Norway, 1970-98. Scand J Public Health 35:306–312. https://doi.org/10.1080/14034940701226159
UNSCEAR (1982) Ionizing radiation: sources and biological effects. United Nations, New York
UNSCEAR (2000) Sources and effects of ionizing radiation. United Nations, New York
UNSCEAR (2008) Sources and effects of ionizing radiation. United Nations, New York
Westerlund EA, Berthelsen T, Berteig L (1987) Cesium-137 body burdens in Norwegian Lapps, 1965-1983. Health Phys 52:171–177. https://doi.org/10.1097/00004032-198702000-00005
Wilson SJ, Murray JL, Huntington HP (1998) AMAP assessment report: Arctic pollution issues. Arctic Monitoring and Assessment Programme, Oslo
Wright SM, Howard BJ, Strand P et al (1999) Prediction of ^{137}Cs deposition from atmospheric nuclear weapons tests within the Arctic. Environ Pollut 104:131–143. https://doi.org/10.1016/S0269-7491(98)00140-7

Part III

Arctic Zoonoses: Diseases Transmitted from Animals to Man

Rabies in the Arctic

Karsten Hueffer, Morten Tryland, and Svetlana Dresvyanikova

1 Introduction

Rabies is a disease caused by rabies virus in the family *Rhabdoviridae*, genus *Lyssavirus*. The enveloped virus infects warm-blooded animals, with mammals being most often affected. Rabies has been detected in every continent apart from Antarctica and causes severe neurological disease that is nearly uniformly fatal once clinical signs are manifest (Fooks et al. 2014).

Most often the virus is transmitted by bite from an infected animal that secretes the virus with its saliva. After local replication in muscle cells, the virus spreads to nerve cells and travels along neurons to the spinal cord and further to the brain. In the brain, the virus replicates and leads to behavioral changes in infected animals with two clinical manifestations. The paralytic (or dumb) form of rabies is characterized by depression and increasing paralysis prior to death. In the furious form of rabies, the infected animal becomes restless, loses normal fear behavior, and becomes aggressive. Especially the behavioral changes associated with furious rabies aid in transmission of the virus through bites (Jackson 2016). Inflammation in the brain of rabies-infected animals is often limited, and the detailed pathogenesis leading to death is still not very well understood.

K. Hueffer (✉)
Department of Veterinary Medicine, University of Alaska Fairbanks, Fairbanks, AK, USA
e-mail: khueffer@alaska.edu

M. Tryland
Department of Arctic and Marine Biology, UiT The Arctic University of Tromsø, Tromsø, Norway

Department of Forestry and Wildlife Management, Inland Norway University of Applied Sciences, Koppang, Norway
e-mail: morten.tryland@inn.no

S. Dresvyanikova
Federal Research Center for Virology and Microbiology, CVM Pokrov, Vladimir, Russia

M. Tryland (ed.), *Arctic One Health*, https://doi.org/10.1007/978-3-030-87853-5_9

Fig. 1 A red fox (*Vulpes vulpes*) sitting on a pipeline in oilfields in Northern Alaska. Red foxes have expanded their range especially close to human infrastructure and pose a significant threat to human health as a vector for rabies infection (Photo: Garret Savory)

Variants of the rabies virus are adapted to specific reservoir host species. Bats, carnivores, and mustelids are the most important reservoirs for the rabies virus. In the circumpolar North, the arctic rabies virus variant is the most prevalent with bat variants found in more southern regions of some Arctic jurisdictions (Hueffer and Murphy 2018; Mørk and Prestrud 2004). Four different subgroupings have been described for the arctic rabies variant, some of which have been further subdivided into clades using sequence analysis. Rabies virus variant Arctic-1 is associated with an incursion of rabies into Ontario from Northern Canada. Variant Arctic-2 has been isolated in Russia and Alaska's Seward Peninsula. Variant Arctic-3 has circumpolar distribution, while rabies virus variant Arctic-4 has only been isolated from Southwest Alaska. The distribution is quite stable over a decadal timeframe (Goldsmith et al. 2016).

Different mammals have been diagnosed with rabies in the North, ranging from arctic foxes (*Vulpes lagopus*) to red foxes (*Vulpes vulpes*), reindeer and caribou (*Rangifer tarandus*), polar bears (*Ursus maritimus*), and wolverines (*Gulo gulo*), among others. The arctic fox is considered the most important reservoir in the Arctic (Fig. 1), although in some regions, more red foxes (Fig. 2) are diagnosed more frequently with rabies (Hueffer and Murphy 2018; Kim et al. 2014; Mørk and Prestrud 2004). This discrepancy is likely due to closer association of red foxes with human settlements and infrastructure which provide food subsidies. As most surveillance and diagnostic efforts are associated with human exposures, closeness

Fig. 2 The arctic fox (*Vulpes lagopus*) is regarded as the main rabies virus reservoir in arctic regions and may migrate with the virus over long distances. Rabies causes CNS infection and change of behavior. This arctic fox charged and bit the sled dog that was tied up outside the meteorological station at Hopen, Svalbard, in 2011 (Photo: Ragnar Sønsteby, The Norwegian Meteorological Institute)

to human infrastructure leads to increased detection of the disease. The ability of the arctic fox to travel long distances, especially in winter, when sea ice enables access to arctic islands, and the aggregation of large numbers of arctic foxes around available food, such as garbage dumps and carcasses, is likely important for rabies ecology and maintenance in the North.

2 History

Similar to other diseases in the circumpolar North, early reports of rabies are rare, but a rabies—like disease in foxes and sled dogs—has been long known by Indigenous peoples in most of the Arctic. Early written reports from the Arctic about a rabies—like disease in sled dogs and foxes—date from the latter half of the nineteenth century (Colan 1881; Nelson 1887). In addition to lack of written records prior to European colonization, the similarity of rabies to canine distemper in its clinical presentation makes definite diagnosis difficult prior to development of specific antibody-based assays (Secord et al. 1980), and canine distemper should be regarded as a differential diagnoses to rabies (Morten Tryland et al. 2018). The diagnostic challenges together with limited infrastructure in the sparsely populated regions of the circumpolar North introduce challenges in our understanding of historical development of rabies in the far North.

2.1 Alaska

The oldest known description of rabies in Alaska dates from 1887 (Nelson 1887) which describes a disease in dogs in Alaska that is similar to "madness of dogs in lower latitudes…". Other early reports in red foxes from Western Alaska were published in the early twentieth century.

Although rabies is frequently reported in foxes and sled dogs, historical reports of rabies in humans in Alaska are rare with the last reported rabies death dating to the early 1940s in Northern Alaska (Hueffer and Murphy 2018).

The first report of laboratory confirmed rabies dates from the 1940s (Williams 1949), in which the author stated that foxes (both red and arctic foxes) are the main reservoir hosts of the virus for the infection of domestic dogs. However, the detection of Negri bodies in histological samples practiced at the time is not a very sensitive method to detecting infections with the rabies virus, likely leading to underreporting of cases. This outbreak which lasted from1945 to 1947 involved the northern and western coasts of Alaska, as well as Interior Alaska and resulted in the introduction of vaccination of sled dogs.

An additional outbreak in Interior Alaska, the area between the Brooks and Alaska Range bounded by the Yukon Kuskokwim Delta to the West and the Yukon to the East, where rabies is currently not diagnosed, occurred in 1952 and 1953 (Rausch 1958). The northern and western coastal areas of Alaska have been considered endemic for rabies since that time with periodic outbreaks occurring every 3–5 years (Hueffer and Murphy 2018).

2.2 Northern Canada

In Canada's North, a rabies—like disease in dogs—has been known at least since the nineteenth century but was present likely for long periods before that time (Plummer 1932; Secord et al. 1980; Walker and Elkin 2005). Early reports of the disease were focused on sled dogs, due to their importance for transport in northern Canada but foxes where quickly identified as a wildlife reservoir as well (Secord et al. 1980).

In the Northwest Territories, the first official documentation in the medical literature dates from the 1940s. It is considered enzootic throughout the territory (Mair and Epp 2005; Walker and Elkin 2005). Between 1978 and 2002, rabies was reported in every year with numbers of cases varying between 1 and 14 per year (Mitchell and Kandola 2005). As in other arctic regions, the arctic fox is considered the main reservoir host; however, rabies has also been documented in red foxes, dogs, caribou, lynx (*Lynx canadensis*), grizzly bear (*Ursus arctos horribilis*), wolfs (*Canis lupus*), and even in a polar bear (Simon et al. 2020; Taylor et al. 1991; Walker and Elkin 2005).

Arctic fox-associated rabies outbreaks in Northern Canada have repeatedly swept south and have posed significant threats to more populated areas in Southern Canada starting in the 1950s (Tabel et al. 1974). These threats of incursion and establishment of rabies viruses from the far North to lower latitudes in Canada are ongoing as Arctic variants of the rabies virus are still detected in southern Ontario (Nadin-Davis and Fehlner-Gardiner 2019). From 1947 to 2017, a total of 950 rabies cases were reported from Yukon, Northwest Territories, Nunavut and Nunavik with the majority of these cases reported in foxes (Gregory David and Tinline 2020).

2.3 Greenland

The first description of a rabies-like disease in Greenland, called the "Eskimo dog disease," was provided by Colan in 1881 who described an outbreak in sled dogs in Northern Greenland in 1859 that spread throughout the region and also infected arctic foxes (Colan 1881). In 1908, the veterinarian Hjortlund wrote about a contagious and deadly disease among sled dogs, with the onset of disease outbreaks associated with fights between sled dogs and "mad" arctic foxes, the dogs usually suffering from a paralytic form of the disease (Lassen 1962). During the period January to April 1960, more than 1000 dogs died in the region of Egedesminde on the Greenland west coast, threatening local communities with famine due to their dependence of the dogs for hunting (Lassen 1962). The first laboratory confirmation of rabies in this region was reported in 1959, describing rabies in two dogs and two arctic foxes (Jenkins and Wamberg 1960). It was later shown that rabies outbreaks in dogs often coincided with population peaks and migrations patterns of the reservoir species, the arctic fox (Follmann et al. 1992). The serious rabies outbreaks among sled dogs initiated a rabies vaccination program for dogs, which reduced the occurrence of rabies substantially. However, rabies is considered enzootic in arctic foxes in Greenland, causing disease in this species, as well as in dogs, horses, sheep, and caribou, most cases occurring along the west coast (Mansfield et al. 2006). Considering this enzootic status and the close contact with wildlife and numerous dogs, only a single human case of rabies has been reported from Greenland. Associated with the major outbreak in the Egedesminde region in 1960, a 4-year-old non-vaccinated Inuit girl was bitten badly by a pack of dogs. The girl developed early signs of disease, such as nausea and slight headache, about 12 weeks later. She was hospitalized but did not receive a postexposure vaccination treatment. She developed a paralytic form of the disease and succumbed (Lassen 1962).

2.4 Iceland

Rabies has never been diagnosed with certainty on Iceland, but descriptions of a rabies-like outbreak in 1765 (Pálsson Pall 1980) resemble rabies to a large degree. A dog that arrived to the island on a foreign ship became sick and attacked other animals that also became "mad." From Norðfjörður, eastern Iceland, the disease spread to other fjords and in the region Fljótsdalshérað, and the outbreak lasted for about a year. The disease affected dogs, cats, arctic foxes, cattle, sheep, and horses, and possibly also a woman, that was bitten by one of the animal and died. The disease was never observed in other parts of Iceland (Pálsson Pall 1980). Rabies may possibly also enter Iceland with migrating foxes, in years when the pack ice reach the northern region.

2.5 Norway

Mainland Norway has been rabies free since the eradication of canine and fox rabies for over a century (King et al. 2004). According to the Norwegian Institute of Public Health, domestically transmitted rabies in humans and animals was last recorded on the Norwegian mainland in 1815 and 1826, respectively (NPHI 2019). The last imported human rabies case in Norway was recorded in 2019, when a 24-year-old woman returned from the Philippines after having been bitten by a dog. She was hospitalized and died a few weeks later (Euronews 2019). Norway shares a northern border with Russia and Finland, and there might have been imported cases in wildlife that have not been recorded. The first case of bat rabies (European bat lyssavirus type 2; EBLV2) in Norway was detected in a Daubenton's bat (*Myotis daubentonii*) in the southern part of the country in 2015 (Moldal et al. 2017). There have been no thorough investigations of bat species and populations in Norway for the presence of rabies virus, and the epidemiological situation is unknown.

In Svalbard, a Norwegian Archipelago between mainland Norway and the North pole, rabies was first detected 1980 and confirmed in 13 arctic foxes, 3 Svalbard reindeer (*Rangifer tarandus platyrhynchus*), and in a ringed seal (*Pusa hispida*) (Odegaard and Krogsrud 1981). However, as in other regions of the Arctic, rabies may have been present but unnoticed through historical times. Rabies has occurred sporadically on Svalbard since then in arctic foxes and reindeer, although seals have also tested positive (Mørk et al. 2011; Prestrud et al. 1992). A larger screening of 620 trapped and of 22 arctic foxes found dead on Svalbard detected rabies virus antigen in brain tissues in two animals (Mørk et al. 2011). These findings, which are in line with a previous screening of brain tissues from 817 arctic foxes during the period 1980–1989 (Prestrud et al. 1992), indicate that the prevalence of rabies has remained low or that the virus is not enzootic in the arctic fox population. In fact, rabies virus gene sequences have indicated that the virus is brought to the archipelago with migrating arctic foxes from Russia, Greenland, or other Arctic regions (Johnson et al. 2007; Mørk et al. 2011). Outbreaks, also confirmed having involved reindeer, have been documented in 2011 and 2018 (Macdonald et al. 2011; NVI 2018). It is worth noting that serological screenings of arctic foxes ($n = 515$) and polar bears ($n = 266$) from this region did not reveal indications of exposure (Mørk et al. 2011; Tryland et al. 2005), suggesting that surveillance must be conducted on animals with clinical signs and testing of brain tissues of these animals as the utility of rabies serology is questionable to detect exposure.

No human rabies case has ever been recorded on Svalbard, in spite of the fact that arctic foxes have been actively trapped and skinned for fur over centuries. Whereas rabies in the arctic fox has been observed in the furious/rabid form, Svalbard reindeer with the disease have displayed paralysis and other central nervous symptoms and have not actively transmitted the virus. The human population on Svalbard consists of approximately 2500 people, of which about 200–300, including children, were involved in reindeer hunting, harvesting about 200 animals. The hunting season is from August 15th to September 20th, and the first two reindeer rabies cases were diagnosed on September 21st. Thus, many people had already consumed reindeer meat at the time of the outbreak discovery, and many still had reindeer meat in their freezers. The public was advised to cook the reindeer meat well to inactivate the virus and not to consume brain or spinal cord. As a measure against possible exposure, 280 people in the community received postexposure treatment after the outbreak. No human cases have been recorded after the outbreak (Macdonald et al. 2011).

2.6 Sweden and Finland

In Sweden, rabies was historically present in dogs and wild carnivores since at least the middle ages, but the last case was reported in 1886 (King et al. 2004). As in Norway, exposure of Daubenton's bats to bat lyssavirus (EBLV) was recently detected in south and southeast Sweden, in Skåne and Småland (Hammarin et al. 2016). Historically, rabies was less common in Finland but has been reintroduced and maintained longer in Finland than in Sweden. It was eliminated in the 1930s but afterward reintroduced in the 1940s from the USSR but was again eliminated by the early 1960s. After nearly 30 years, however rabies was detected again with the raccoon dog (*Nyctereutes procyonoides*) as the main reservoir. A large oral vaccination field campaign led to the elimination of the disease from Finland with the last case reported in 1989 (King et al. 2004). In 2011, following a rabies outbreak in Karelia, Russia, and Finland extended the vaccination zone along the Russian border, and the annual number of baits distributed has varied from 160,000 to 360,000 per year (Nokireki et al. 2016).

2.7 Russia

Diseases similar to rabies have been known for a long time in circumpolar regions of the Russian territory including the northern regions. Veterinarians and medical workers have described rabies infection since the second part of the nineteenth century, but similar to other regions, it likely occurred prior to that in sparsely populated regions without written records. A widespread nervous disorder ("madness") in the North of Yakutia was first described in 1855 and 1857 (Romanova 2000). The first report of human and animal rabies after wolf bites in the Akshirskii part of the Transbaikalian District dates from 1895. In the first part of the twentieth century, medical specialists in the Baikal and Transbaikalian Region described "natural remedies" for people that were bitten by dogs and wolves suggesting existence of human rabies infections (Sidorova et al. 2007). Human rabies was included in Russia Public Health records from 1887 onward (Pokrovskiy et al. 2003), but literature sources usually provide only generalized statistics of animal and human morbidity and mortality from rabies in Russia without providing detailed geographic information. In 1886–1896, rabies was registered in 52 Russian provinces (Gruzdev and Metlin 2019). In 1895, 2096 animals were reported with rabies in Russia, increasing to 5871 animals in 1905 and 13,911 animals in 1925. The human mortality from "hydrophobia," a common clinical manifestation of rabies in humans, varied from 786 to 1917 per year in the period of 1902–1914 (Savvateyev 1927). Although the proportion of cases in the northern regions of Russia were not reported, rabies treatment stations were established in big cities in the northern regions in the beginning of the twentieth century, suggesting existence of human rabies in the North at that time (Butyagin 1923; Georgiyevskiy 1922; Zhukova-Florensova 1914).

In the circumpolar and arctic territories of the Russian Federation, both the cosmopolitan rabies virus and arctic rabies virus have been isolated. Specialists suggested a hypothesis that the circulation of cosmopolitan rabies virus in Southern Russia was the consequence of outbreaks in the North in the beginning and in the middle of twentieth century. Two outbreaks of rabies in the Northwestern and Northeastern regions of the European part of USSR have been reported. The first outbreak occurred in 1925–1926 and began in the Vologda region and spread to the Archangelsk region. A possible cause of this outbreak was the large dog rabies epizootic in central and southern regions of the USSR in 1921–1924 that spread northward. The peak of another outbreak was reported in 1931. After this outbreak ended in 1932, rabies was not detected in this region until 1955. The Vologda Region, Archangelsk Region, and Komi Autonomous Soviet Socialist Republic were free from animal rabies for 23 years from 1932 to 1954 but were again reported in animals in these regions in 1955–1956 (Romanova 2000).

Throughout the territory of the former USSR, the cases of arctic rabies were reported in Yakutia, Taimyr, Kamchatka, Chukotka, Yamal, Novaya Zemlya, as well as on several islands of Arctic Ocean and covered both tundra, forest tundra, and taiga zones. Similar to other Northern regions, the arctic fox is the main wildlife reservoir of arctic rabies variants in the Russian territory (Romanova 2000).

3 Epidemiology

While the arctic fox is considered the main host species for the rabies virus in the circumpolar Arctic, a variety of other species have also been diagnosed with this disease (Mørk and Prestrud 2004; Simon et al. 2020). Most of these species, however, do not play an important role in the epidemiology of rabies in the North. The arctic fox is the predominant mesocarnivore at high Northern latitudes, and its southern range limit is likely determined by the presence of red foxes. Arctic foxes travel long distances and aggregate at food sources during the cold arctic winter months, a time when rabies is reported more frequently (Pamperin et al. 2008). This behavior and the cyclical nature of its population size play an important role in rabies dynamics. Especially the long-distance travel over sea ice combined with the long and variable incubation period of the virus can likely lead to viral transport over long distances and poses a public health threat to currently rabies-free areas that are connected to endemic areas by sea ice. Among other wildlife species, the red fox is of special importance. Climate change will likely expand the geographical range of this mesocarnivore into northern latitudes (Shirley et al. 2009), which could lead to an even greater importance for this species in maintaining sylvatic replication cycles of the virus, also in northern regions. In some northern regions, the red fox is already diagnosed more often with rabies than the arctic fox (Hueffer and Murphy 2018; Kim et al. 2014). Since the red fox is often found in habitats close to human infrastructure where food and shelter are more readily available at the northern limit of its range, this species can pose an increased risk of exposing humans and their domesticated animals to this important zoonotic disease agent (Hueffer and Murphy 2018; Pamperin 2008).

The interaction of wildlife (especially red and arctic foxes) with domestic animals puts dogs at risk of rabies exposure. In many remote areas of the far North, veterinary services are limited, and rabies vaccination rates are low. This leads to frequent infections of dogs. While dogs do not play a role in the long-term maintenance of the rabies in the Arctic, rabid dogs pose a significant threat to humans (Lassen 1962) due to their close relationships and husbandry practices that often allow for contact with rabid foxes as dogs are often kept outside.

Reindeer in Svalbard (Fig. 3) have repeatedly tested positive for rabies and pose a potential risk to hunters on this most northern inhabited region of the planet (Macdonald et al. 2011; Ytrehus et al. 2011). The relatively high number of rabies positive reindeer during rabies outbreaks among arctic foxes at Svalbard reflects the fact that the reindeer are more easily found and identified with rabies as compared to arctic foxes. Due to rabies associated lameness and paralysis, they are often not able to rise and run when approached by a helicopter. However, they likely do not play any role in maintenance of the virus.

Following the range of its main reservoir host, the arctic fox, rabies is distributed throughout the circumpolar North, with the exception of the Nordic countries of Norway, Sweden, and Finland (WHO 1990), where the arctic fox populations have been critically reduced toward extinction (Dalén et al. 2006). In most other areas, the virus is considered endemic in arctic foxes and possibly red foxes where their ranges overlap, possibly in a multi-host system with the red fox being the main reservoir of the virus (Goldsmith et al. 2016). The status of rabies in Svalbard between outbreaks is not known (Fig. 4).

Even in endemic areas, rabies epidemiology is characterized by a cyclical outbreak dynamic with increased reported cases every 3–5 years (Elton 1931; Hueffer and Murphy 2018). This outbreak pattern of rabies likely follows the cyclical nature of the population density of the main maintenance host, the arctic fox, which is due to changing rodent prey availability. The number of reported red and arctic foxes

Fig. 3 The Svalbard reindeer (*Rangifer tarandus platyrhynchus*) may be regarded as an aberrant (dead end) host for the rabies virus. During a rabies outbreak on Spitsbergen (Svalbard) in September 2011, a female reindeer was found with paralysis in the hind legs, guarded by a male (with wire entangled in its antlers). The animal was euthanized, and diagnostic tests verified the suspicion of rabies (Photo: Arild Lyssand, The Governor of Svalbard/Sysselmannen)

overlap in regions where both species are present, suggesting that they are involved in the same outbreaks and do not support independent epizootics. Especially when both red and arctic foxes are involved in rabies outbreaks, rabies in the far north poses a significant risk to more populated southern regions. Eastern Canada is one example where arctic rabies variants spread south causing multiyear outbreaks that required significant public health response to control (Nadin-Davis and Fehlner-Gardiner 2019).

Despite the long-distance movement of arctic foxes and the resulting potential for rabies virus transport, the geographic distribution of subtypes of the arctic variant is stable over multiple decades (Goldsmith et al. 2016), suggesting that in endemic areas, local maintenance is more important than introduction of new variants. However, Svalbard provides a counterexample where outbreaks of rabies were linked to long-distance transport of virus based on sequence analysis (Johnson et al. 2007). These different dynamics indicate that rabies epidemiology in the arctic is complex and likely differs between areas with endemic rabies and areas characterized by more occasional outbreaks of the disease after introduction (Mørk et al. 2011; Odegaard and Krogsrud 1981).

By the early 1990, nine endemic areas of rabies were identified in the territory of Russia and neighboring countries as a result of long-term comprehensive work:

Fig. 4 During the 2011 rabies outbreak, a Svalbard reindeer with rabies was found with seizures against the rocks, causing damage and bleeding. The animal was euthanized, and rabies was diagnosed (Photo: Per Andreassen)

Baltic, Central Russian, North Caucasian, West Siberian-Kazakhstan, Lower Volga, Middle Volga, Transbaikalian, Manchurian, and Arctic.

During the twenty-first century, endemic areas have expanded northeast as rabies cases are detected in the territories that were previously considered rabies free. In natural foci of the Russian Federation, two main lines of rabies virus are circulating: arctic/arctic-like rabies and cosmopolitan rabies (Kuzmin et al. 2004). Areas of phylogenetic discrete clusters of rabies do not overlap with each other, and in some ecological regions, the genetic diversity of the virus is insignificant. The arctic rabies subgroups (A2 and A3) are common in the arctic territories.

Rabies morbidity in different animal species in the period from 1960 to 2002 shows a difference in disease ecology in the Northern regions of Russia compared to other Russian territories. Up to half of all reported cases in the North were associated with dogs compared to only 10% in other parts of the country.

During the second half of the twentieth century and the beginning of the twenty-first century, the Republic of Yakutia and the Khabarovsk Territory remain affected by rabies. In addition, in the territories of southern Siberia and the Far East, active epizootics among foxes, raccoon dogs, and corsac foxes (*Vulpes corsac*) have been detected for the first time.

In 2018, 3162 cases of animal rabies were reported in a total of 75 regions of Russia, which is similar to the average annual cases for 1960–2018. A total of 28 cases of human rabies were reported in the period of 2012–2018, compared to 67 in 2007–2011. Persistent epizootic outbreaks continue to be registered in Nenets Autonomous District in the North. Thus, there is a risk of human rabies in the Arctic region, especially in the Nenets region.

4 Challenges in Management

Remoteness: The extreme remoteness of many Northern regions and low population density makes rabies surveillance difficult and costly (Hueffer and Murphy 2018). These challenges lead to a lack of knowledge of true distribution of rabies in the circumpolar North, especially in wildlife reservoirs. These challenges together with a lack of urgency due to few human cases and competition with other public health and animal health issues pose significant challenges to rabies management in the North. Logistical and transportation challenges also limit sample quality and timely testing due to transportation times that can exceed 7 days from sample collection to delivery to specialized public health facilities, which are usually located in more southern regions (Gregory David and Tinline 2020).

Lack of veterinary care: Veterinary care is limited in many Northern communities. For example, in the Northwest Territories in Northern Canada, private veterinary praxis is available to provide animal health services only in two communities (Yellowknife and Inuvik) (Brook et al. 2010). Four other communities in the Northwest Territories are serviced regularly by mobile veterinary services. This leaves 27 communities in this remote territory without regular veterinary services. Especially small communities do not have access to regular veterinary services (Brook et al. 2010). A lay vaccinator program was established in the Northwest Territories in 1950 and is operated under the federal *Health of Animal Act*, which allows for rabies vaccination programs by non-veterinarians in "remote areas where veterinary services are not readily available." Currently, this program is overseen by the territory veterinarian (Brook et al. 2010). In a recent survey in communities without regular veterinary services, Brooks and coworkers found that 63% of owned dogs were reported by their owners as not previously vaccinated against rabies (Brook et al. 2010). In addition, over 40% of school students who participated in that study reported been bitten by a dog, which is likely related to challenges in dog population control in remote communities. Similar challenges exist in Alaska and other Northern regions (Hueffer and Murphy 2018).

5 Current Research Topics and Needs

More and better data on rabies prevalence and epidemiology is needed to predict and assess the effects climate change and increased human development might impose on rabies dynamics at high latitudes (Hueffer et al. 2011; Hueffer and Murphy 2018;

Huettmann et al. 2017). Understanding the dynamics of rabies as an example of an infectious disease in a changing Arctic can serve as a model for other diseases in the North and beyond as our planet changes at an ever-increasing rate. Especially, surveillance data that is independent of human exposure will be crucial to better assess the disease ecology in the North beyond the likelihood of human exposure (Huettmann et al. 2017). In this context, the role of the red fox in disease ecology at high latitudes also deserves closer investigations. As a northward expansion of the red fox distribution is predicted based on current climate models, the ability of this host to maintain artic rabies variants at high latitudes as a sole maintenance host is not clear. For example, areas in Interior Alaska that are inhabited by the red fox but not arctic foxes are free of endemic rabies but have experienced sporadic outbreaks that could indicate low levels of endemicity of incursions of rabies virus into non-endemic areas (Hueffer and Murphy 2018; Rausch 1958).

6 Future Outlook

The future of rabies in the North is uncertain due to a rapidly changing environment. Climate change and anticipated increased human encroachment for resource extraction, shipping, tourism, and recreation will drastically change the ecology of arctic and red foxes, the main reservoir hosts for the rabies virus. How these changes will affect rabies that currently is endemic in large areas of the circumpolar North is not clear. Rabies has been identified as a disease of concern in the context of a changing climate (Hueffer et al. 2013; Omazic et al. 2019). For Alaska, three independent approaches of epidemiological data, ecological niche modeling, and conceptual modeling approach point toward a reduced future risk of rabies in that State (Hueffer and Murphy 2018; Huettmann et al. 2017; Kim et al. 2014). These predictions, while all pointing to a similar outcome of reduced rabies, are based on current regimes and do not account for incursion or evolution of rabies virus variants or the expansion of alternate hosts with their associated rabies variants. It is also not clear how transferable these findings are to other regions. Similarly, if the Svalbard archipelago in the future is left ice free most of the year, reduced or no arctic fox immigration to these islands may reduce the introduction of rabies and also other pathogens. In summary, the future outlook of rabies infections in the North is uncertain given the complex mixture of anticipated changes that will likely influence this important zoonotic disease in the far North.

References

Brook RK, Kutz SJ, Millins C et al (2010) Evaluation and delivery of domestic animal health services in remote communities in the Northwest Territories: a case study of status and needs. Can Vet J 51(10):1115–1122

Butyagin PO (1923) About current activities of anti-rabies stations [in Russian]. Siberian Med J 1: 127–131

Colan T (1881) The dog disease, or canine madness of the arctic regions, viewed in connection with hydrophobia; together with the measures used and suggested for its extinction, from information collected and observations made in the country. Vet J Ann Comp Path 8:324–325

Dalén L, Kvaløy K, Linnell JDC et al (2006) Population structure in a critically endangered arctic fox population: does genetics matter? Mol Ecol 15(10):2809–2819. https://doi.org/10.1111/j.1365-294X.2006.02983.x

Elton C (1931) Epidemics among sledge dogs in the Canadian Arctic and their relation to disease in the arctic fox. Can J Res 5(6):673–692

Euronews (2019) Norwegian woman dies from rabies after puppy bite in the Philippines. Retrieved from https://www.euronews.com/2019/05/11/norwegian-woman-dies-from-rabies-after-puppy-bite-in-the-philippines

Follmann EH, Ritter DG, Baer GM (1992) Oral rabies vaccination of arctic foxes (Alopex lagopus) with an attenuated vaccine. Vaccine 10(5):305–308. Retrieved from http://www.ncbi.nlm.nih.gov/entrez/query.fcgi?cmd=Retrieve&db=PubMed&dopt=Citation&list_uids=1574916

Fooks AR, Banyard AC, Horton DL et al (2014) Current status of rabies and prospects for elimination. Lancet 6736(13):1–11. https://doi.org/10.1016/S0140-6736(13)62707-5

Georgiyevskiy A (1922) Anti-rabies station of Chita city for 1918, 1919, 1920, and 1921. [in Russian] Med J Transbaikalian Doctor Soc 1:59–74

Goldsmith EW, Renshaw B, Clement CJ et al (2016) Population structure of two rabies hosts relative to the known distribution of rabies virus variants in Alaska. Mol Ecol 25(3):675–688. https://doi.org/10.1111/mec.13509

Gregory David J, Tinline RR (eds) (2020) Taking the bite out of Rabies; the evolution of rabies Management in Canada, 1st edn. University of Toronto Press, Toronto

Gruzdev KN, Metlin AY (2019) Animal rabies, 364 p. Vladimir City

Hammarin AL, Berndtsson LT, Falk K et al (2016) Lyssavirus-reactive antibodies in Swedish bats. Infect Ecol Epidemiol 6:31262. https://doi.org/10.3402/iee.v6.31262

Hueffer K, Murphy MD (2018) Rabies in Alaska, from the past to an uncertain future. Int J Circumpolar Health 77(1):1475185. https://doi.org/10.1080/22423982.2018.1475185

Hueffer K, O'Hara TM, Follmann EH (2011) Adaptation of mammalian host-pathogen interactions in a changing arctic environment. Acta Vet Scand 53

Hueffer K, Parkinson AJ, Gerlach R et al (2013) Zoonotic infections in Alaska: disease prevalence, potential impact of climate change and recommended actions for earlier disease detection, research, prevention and control. Int J Circumpolar Health 72(1):1–11. https://doi.org/10.3402/ijch.v72i0.19562

Huettmann F, Magnuson EE, Hueffer K (2017) Ecological niche modeling of rabies in the changing Arctic of Alaska. Acta Vet Scand 59(1):1–11. https://doi.org/10.1186/s13028-017-0285-0

Jackson AC (2016) Diabolical effects of rabies encephalitis, 8–13. https://doi.org/10.1128/JVI.01677-07

Jenkins M, Wamberg K (1960) Rabies discovered in Greenland. J Am Vet Med Assoc 137:183–185

Johnson N, Dicker A, Mork T et al (2007) Phylogenetic comparison of rabies viruses from disease outbreaks on the Svalbard Islands. Vector Borne Zoonotic Diseases (Larchmont, N.Y.) 7(3):457–460. https://doi.org/10.1089/vbz.2006.0555

Kim BI, Blanton JD, Gilbert A et al (2014) A conceptual model for the impact of climate change on fox rabies in Alaska, 1980-2010. Zoonoses Public Health 61(1):72–80. https://doi.org/10.1111/zph.12044

King BAA, Fooks AR, Aubert M et al (2004) Historical perspective of rabies in Europe and the Mediterranean Basin a testament to rabies by Dr Arthur A. King. OIE, Paris

Kuzmin IV, Botvinkin AD, McElhinney LM et al (2004) Molecular epidemiology of terrestrial rabies in the former Soviet Union. J Wildl Dis 40(4):617–631. Retrieved from. http://www.ncbi.nlm.nih.gov/pubmed/15650080

Lassen HC (1962) Paralytic human rabies in Greenland. Lancet (London) 1(7223):247–249. https://doi.org/10.1016/s0140-6736(62)91194-7

Macdonald E, Handeland K, Blystad H et al (2011) Euro surveillance: bulletin Européen Sur les maladies Transmissibles = European communicable disease bulletin 16(40):3–6. Retrieved from http://www.ncbi.nlm.nih.gov/pubmed/21996377

Mair N, Epp H (2005) Rabies in the Northwest Territories Part 3: Geographic Information Systems (GIS) and Rabies in the NWT. Epinorth 17(1):7–8

Mansfield KL, Racloz V, Mcelhinney LM et al (2006) Molecular epidemiological study of Arctic rabies virus isolates from Greenland and comparison with isolates from throughout the Arctic and Baltic regions. Virus Res 116:1–10. https://doi.org/10.1016/j.virusres.2005.08.007

Mitchell R, Kandola K (2005) Rabies in the Northwest Territories Part 2: Rabies Surveillance in Northwest Territories. Epinorth 17(1):4–6

Moldal T, Vikøren T, Cliquet F et al (2017) First detection of European bat lyssavirus type 2 (EBLV-2) in Norway. BMC Vet Res 13(216):1–8. https://doi.org/10.1186/s12917-017-1135-z

Mørk T, Prestrud P (2004) Arctic rabies – a review. Acta Vet Scand 45(1–2):1–9. Retrieved from http://www.pubmedcentral.nih.gov/articlerender.fcgi?artid=1820997&tool=pmcentrez&rendertype=abstract

Mørk T, Bohlin J, Fuglei E et al (2011) Rabies in the Arctic fox population, Svalbard, Norway. J Wildl Dis 47(4):945–957. https://doi.org/10.7589/0090-3558-47.4.945

Nadin-Davis S, Fehlner-Gardiner C (2019) Origins of the arctic fox variant rabies viruses responsible for recent cases of the disease in southern Ontario. PLoS Negl Trop Dis 1–17

Nelson EW (1887) Report upon natural history collections made in Alaska: between the years 1877 and 1881. Government printing office, Washington, DC

Nokireki T, Nevalainen M, Sihvonen L, Gadd T (2016) Adverse reactions from consumption of oral rabies vaccine baits in dogs in Finland. Acta Vet Scand 58(1):53

NPHI (2019) Rabies. Retrieved from https://www.fhi.no/nettpub/smittevernveilederen/sykdommer-a-a/rabies%2D%2D-veileder-for-helsepersonel/

NVI (2018) Rabies detected in a Svalbard reindeer. Retrieved November 2, 2020, from https://www.vetinst.no/en/news/rabies-detected-in-a-svalbard-reindeer

Odegaard OA, Krogsrud J (1981) Rabies in Svalbard: infection diagnosed in arctic fox, reindeer and seal. Vet Rec 109(7):141–142. https://doi.org/10.1136/vr.109.7.141

Omazic A, Bylund H, Boqvist S et al (2019) Identifying climate – sensitive infectious diseases in animals and humans in northern regions. Acta Vet Scand 1–12. https://doi.org/10.1186/s13028-019-0490-0

Pálsson Pall A (1980) Hundaæði eða bitæði (rabies). Serprentun Ur FREY 11:1–8

Pamperin NJ (2008) Winter movements of arctic foxes in northern Alaska measured by satellite telemetry. University of Alaska Fairbanks, Master's Thesis

Pamperin NJ, Follmann EH, Person BT (2008) Sea-ice use by arctic foxes in northern Alaska. Polar Biol 31(11):1421–1426

Plummer BY (1932) Preliminary note on Arctic dog disease and its relationship to rabies

Pokrovskiy VI, Onishchenko GG, Cherkasskiy BL (2003) The evolution of infectious diseases in Russia in XX century, 664 p, Meditsina Publishing House, Moscow

Prestrud P, Krogsrud J, Gjertz I (1992) The occurrence of rabies in the Svalbard islands of Norway. J Wildl Dis 28(1):57–63

Rausch R (1958) Some. Observations on Rabies in Alaska, with Special Reference to Wild Canidae 22(3):246–260

Romanova UN (2000) The epizootic peculiarities and diagnostics of rabies in Yakutia Republic (Sakha) Thesis. State Research institution Pokrov

Savvateyev AI (1927) Rabies. 216 p. Leningrad

Secord DC, Bradley JA, Eaton RD et al (1980) Prevalence of rabies virus in foxes trapped in the Canadian. Arctic Can Vet J 3:297–300

Shirley MDF, Elmhagen B, Lurz PWW et al (2009) Modelling the spatial population dynamics of arctic foxes: the effects of red foxes and microtine cycles. Can J Zool 87(12):1170–1183. https://doi.org/10.1139/Z09-104

Sidorova DG, Sidorov GI, Poletschuk EM et al (2007) Rabies in eastern Siberia in XX in the beginning of XXI centuries. Bull VSNTS SO RAMN 3(55):168–172

Simon A, Belanger D, Berteaux D et al (2020) Fox rabies, ecology of rabies ecology in the Arctic fox (Vulpes lagopus). In: Gregory DJ, Tinline R (eds) Taking the bite out of rabies: the evolution of rabies management in Canada, 1st edn. University of Toronto Press, Toronto, pp 453–465

Tabel H, Corner H, Webster W et al (1974) History and epizootiology of rabies in Canada. Can Vet J 15(10):271–281. http://www.pubmedcentral.nih.gov/articlerender.fcgi?artid=1696688&tool=pmcentrez&rendertype=abstract

Taylor M, Elkins B, Maier N et al (1991) Observation of a polar bear with rabies. J Wildl Dis 27: 337–339

Tryland M, Neuvonen E, Huovilainen A et al (2005) Serologic survey for selected virus infections in polar bears at Svalbard. J Wildl Dis 41(2):310–316

Tryland M, Balboni A, Killengreen ST et al (2018) A screening for canine distemper virus, canine adenovirus and carnivore protoparvoviruses in Arctic foxes (*Vulpes lagopus*) and red foxes (*Vulpes vulpes*) from Arctic and sub-Arctic regions of Norway. Polar Res 37

Walker J, Elkin B (2005) Rabies in the Northwest Territories part 1: a historical overview of rabies in NWT. Epinorth 17(1):1–3

WHO (1990) Report of a WHO/NVI workshop on Arctic rabies. Uppsala

Williams RB (1949) Epizootic of rabies in interior Alaska 1945-47. Can J Comp Med 13(6):136–143

Ytrehus B, Handeland K, Nissen S et al (2011) Rabies in an Arctic fox on the Svalbard archipelago, Norway, January 2011. Arctic 4–5

Zhukova-Florensova M (1914) The report of the first year of Krasnoyarsk anti-rabies station. [in Russian] *medical-sanitary chronicle*. Krasnoyarsk 1:1–4.

Brucellosis in the Arctic and Northern Regions

Xavier Fernandez Aguilar, Ingebjørg H. Nymo, Kimberlee Beckmen, Svetlana Dresvyanikova, Irina Egorova, and Susan Kutz

1 General Overview of Brucellosis

1.1 Disease Overview

Brucellosis is a disease of people and domestic and wild animals that is caused by species of bacteria from the genus *Brucella*. It was first described as a disease syndrome in people, "Malta fever," "Mediterranean fever," or "undulant fever," in the 1850s. It was subsequently associated with a micrococcus in 1887 by David Bruce ("brucellosis") and named *Micrococcus melitensis* (Bruce 1887). In 1897, the Danish veterinarian Bernhard L.F. Bang separately described spontaneous abortions in cows related to a *Bacillus* that he named *Bacillus abortus* (Bang's disease) (Bang 1897). It took a couple of decades more until Alice Evans established the etiological link between animal and human brucellosis, being among the first recognized zoonotic diseases (Madkour 2001).

Brucellosis is a major zoonotic disease worldwide. Approximately 500,000 new human cases are documented annually (Pappas et al. 2006), but it is commonly underreported, and the true incidence is unknown (Dean et al. 2012b). Brucellosis causes livestock production losses that have an important impact on livelihoods in low-income countries (McDermott et al. 2013), and huge direct and indirect

X. F. Aguilar (✉) · S. Kutz
Faculty of Veterinary Medicine, Department of Ecosystem and Public Health, University of Calgary, Calgary, AB, Canada

I. H. Nymo
Food Safety and Animal Health Research, The Norwegian Veterinary Institute, Tromsø, Norway

K. Beckmen
Division of Wildlife Conservation, Alaska Department of Fish and Game, Fairbanks, AK, USA

S. Dresvyanikova · I. Egorova
Federal Research Center for Virology and Microbiology, CVM Pokrov, Vladimir, Russia

M. Tryland (ed.), *Arctic One Health*, https://doi.org/10.1007/978-3-030-87853-5_10

economic costs wherever it occurs (Roth et al. 2003). Although brucellosis is an "old" disease, it continues reemerging in different parts of the world and remains one of the most important zoonotic diseases worldwide because of its impacts on the economy, human, and animal health (Seleem et al. 2010). Brucellosis is a reportable disease in most countries, and its management and control is a goal of national and international strategies.

Brucella spp. primarily infect mammals, sometimes at livestock-wildlife interfaces, with humans as spillover hosts. These bacteria have also been isolated from fish (El-Tras et al. 2010), amphibians (Mühldorfer et al. 2017), reptiles (Eisenberg et al. 2020), and nematode and trematode parasites (Dawson et al. 2008), as well as from soil (Scholz et al. 2008a). While *Brucella* spp. differ in their primary hosts, virulence, and zoonotic potential, disease in animals is typically associated with reproductive failure, including abortions, stillbirth, or weak off-spring, as well as lesions in genital organs, joints, or other parts of the body. Negative effects on productivity and population dynamics have been described in enzootically affected livestock and wildlife populations (Fuller et al. 2007; Cotterill et al. 2018).

In Arctic and subarctic regions, wildlife and semidomesticated reindeer (*Rangifer tarandus tarandus*) are the main reservoirs and sources of human infection (Forbes 1991). Indigenous people from northern latitudes have an historical dependence on wild animals and semidomesticated reindeer for subsistence (Nuttall et al. 2004), making brucellosis an important public health concern where it is prevalent. Animal and human brucellosis in Arctic and subarctic regions is relatively common, but its prevalence has changed over time associated with introduction events, abundance of reservoir hosts, and, in semidomesticated reindeer, disease control campaigns.

1.2 Etiology, Taxonomy, and Hosts

Brucella spp. are α2-proteobacteria belonging to the family *Brucellaceae*. There are 12 recognized species in the genus *Brucella* that are further subclassified into biovars, with the first six species described commonly referred to as the "classical" species (Table 1). Full-genome analysis groups *Brucella ceti*, *Brucella pinnipedialis*, and *Brucella microti* with the classical species. In a related but distinct and distant clade, *Brucella vulpis* and *Brucella inopinata* cluster with atypical isolates that also contain genetic material derived from soil-associated *Alphaproteobacteria* (a feature also shared with *B. vulpis*) (Scholz et al. 2016b). Several *Brucella* strains isolated from native rodents from Australia (Cook et al. 1966; Tiller et al. 2010a), frogs worldwide (Eisenberg et al. 2012; Scholz et al. 2016a), a bluespotted ribbontail ray (*Taeniura lymma*) (Eisenberg et al. 2017), or from a complicated pneumonia case in humans (Tiller et al. 2010b), are yet to be fully classified.

Different *Brucella* spp. and their biovars have primary hosts in which they are best adapted and can be maintained enzootically (Vizcaíno et al. 2000; Foster et al.

Table 1 List of formally described species of the *Brucella* genus, biovars, and known hosts with which they are mainly associated

Species	Type	Biovars	Main hosts	Human cases/ zoonotic potential[a]	References
Brucella (B) melitensis	Smooth	1–3	Sheep, goats, camels, wild *Caprinae*	Yes/high	Bruce (1887)
B. abortus	Smooth	1–9	Cattle, yak, camels, elk, bison, African buffalo, and other wild bovids	Yes/ moderate low	Bang (1897)
B. suis	Smooth	1	Pig, wild boar	Yes/ moderate	Cotton (1922)
		2	Pig, wild boar, hares	Yes/very low	
		3	Pig, wild boar	Yes/ moderate	
		4	Reindeer and caribou	Yes/ moderate	
		5	Rodents	None	
B. ovis	Rough		Sheep	None	Simmons and Hall (1953)
B. neotomae	Smooth		Desert wood rats, possibly other rodents	Yes/ unknown[b]	Stoenner and Lackman (1957), Suárez-esquivel et al. (2017)
B. canis	Rough		Dogs	Yes/low	Carmichael and Bruner (1968)
B. ceti	Smooth		Whales, dolphins, and porpoises	Yes only ST27/ unknown[b]	Foster et al. (2007), Whatmore et al. (2008)
B. pinnipedialis	Smooth		Pinnipeds	None	Foster et al. (2007)
B. microti	Smooth		Common vole, red fox[c]	None	Scholz et al. (2008b)
B. inopinata	Smooth		Unknown	Yes/ unknown	Scholz et al. (2010)
B. papionis	Smooth		Baboon	None	Whatmore et al. (2014)
B. vulpis	Smooth		Red fox[c]	None	Scholz et al. (2016b)

[a]Informed by Moreno (2014)
[b]Few but severe human cases described
[c]Uncertain and defined by few isolation cases
Scientific names: African buffalo (*Syncerus caffer*), baboon (*Papio* spp.), bison (*Bison bison*), common vole (*Microtus arvalis*), desert wood rats (*Neotoma lepida*), elk (*Cervus canadensis*), hares (*Lepus* spp.), red fox (*Vulpes vulpes*), reindeer and caribou (*Rangifer tarandus*), yak (*Bos grunniens*), wild boar (*Sus scrofa*)

2009). While the mechanism of this host preference is unclear, the genome of *Brucella* is remarkably constant with only minor nucleotide differences among species, indicating that relatively small changes are responsible for host adaptations and altered virulence (Paulsen et al. 2002; Scholz et al. 2016b). Many of the classical *Brucella* spp. are associated with livestock or human brucellosis, but their host range may also involve wild hosts. The ecology, host range, and zoonotic potential of *Brucella* are best known from these species, whereas the remaining species, which have been described more recently, are less well characterized (Table 1). It is noteworthy that atypical *Brucella* strains share genetics with soil-living bacteria, suggesting a different ecology than the classical species (Scholz et al. 2016b). Furthermore, *B. microti*, which has atypical phenotypic characteristics, has been isolated form the soil and recovered 6 months later from the same soil samples (stored at 4 °C), indicating a long-term soil survival (Scholz et al. 2008a).

1.3 Pathobiology and Pathogenesis

The current understanding of infection, replication, and disease outcomes for *Brucella* is derived primarily from studies of the classical *Brucella* species in domestic animals or humans. One of the most important aspects of *Brucella* is its ability to replicate and persist within cells (Gorvel and Moreno 2002). This stealthy intracellular lifestyle drives the specific features of the pathology in hosts, including the chronic course of the disease, as well as dictating specific therapeutic needs. Indeed, *Brucella* spp. do not have classical virulence factors, rather their virulence is associated with mechanisms that allow them to invade, survive, and replicate in host cells (Seleem et al. 2008).

Brucella is internalized in *Brucella*-containing vacuoles (BCV) and modifies intracellular trafficking to survive, ultimately fusing with the endoplasmic reticulum (ER) to seize membranes and components (Pizarro-Cerdá et al. 2000). These ER-associated compartments are the *Brucella*-induced niche for intracellular replication and have been demonstrated in macrophages, trophoblasts, dendritic cells, and epithelial cells, although other intracellular replication niches may remain to be identified (von Bargen et al. 2012). The infection process can be divided into different stages, the incubation period, followed by an acute phase when the bacteria invades and disseminates in host tissues, and a chronic stage when severe pathological changes and organ damage may occur.

The infective dose of *Brucella* spp. is relatively low (10–100 organisms for *Brucella melitensis* or *Brucella suis*) (Miller and Neiland 1980; Mantur et al. 2007). Entry sites include the oral mucosae, nasopharynx, conjunctiva, digestive and genitourinary tracts, or directly through damaged skin. In the digestive tract, the epithelial transmigration occurs through M cells, and the jejunal-ileal Peyer's patches are among the most important entry sites (Salcedo et al. 2008; Rossetti et al. 2013). The transepithelial migration of *Brucella* spp. may also occur through phagocytes (Ackermann et al. 1988) and is facilitated with the functional subversion of the mucosal barrier (Rossetti et al. 2013). *Brucella* spp. are transported in the infected cells to the regional lymph nodes where they can remain for weeks before further dissemination to different organs. During infection, *Brucella* can naturally differentiate into "smooth" and "rough" mutants according to their cell surface, which may have an important biological role for *Brucella* dissemination (Turse et al. 2011). Rough mutants induce higher macrophage cytotoxicity that helps smooth variants to egress from the infected host cells and disseminate (Pei et al. 2014). The major target cells of *Brucella* spp. are macrophages, dendritic cells, and trophoblasts, but replication can also occur in other cell types like microglia, fibroblasts, epithelial cells, endothelial cells, or erythrocytes (Vitry et al. 2014). *Brucella abortus* bacteremia in calves was reported as fast as 30 minutes after intraluminal inoculation of the ileum (Rossetti et al. 2013), suggesting that invasion and dissemination of *Brucella* spp. may differ between species, route of infection, or the bacterial load (Demars et al. 2019).

After the bacteremia phase and dissemination, *Brucella* spp. can establish chronic infections that may last for several months or years (Castaño and Solera 2009). Persistence may occur in the joints, liver, or spleen and may remain sequestered in white blood cells from the lymph nodes, mammary gland (Harmon et al. 1988), and bone marrow (Gutiérrez-Jiménez et al. 2018).

1.4 Gross and Microscopic Pathology

Brucella may be associated with gross pathological lesions in the lymph nodes, liver, kidneys, spleen, testes and the accessory glands, uterus and mammary glands, synovial structures, and bone marrow (Saegerman et al. 2010). In females, brucellosis is primarily manifested with abortions, stillbirths, weak offspring, placental retentions, or endometritis resulting in transient or permanent infertility. The endometritis may have necrotic and ulcerative lesions that affect the luminal epithelium and caruncular surfaces. Mucopurulent exudate with focal granulomas and lymphoid nodules in the endometrium have also been described in swine (Foster 2017). In more chronic stages, brucellosis can cause interstitial mastitis (Meador et al. 1988; Xavier et al. 2009). In males, epididymitis and orchitis are common manifestations along with seminal vasculitis, prostatitis, or inflammation of the bulbourethral glands, which can also cause infertility (Fig. 1). A lymphoid hyperplasia with enlarged lymph nodes and splenomegaly may occur. Arthritis and bursitis may also appear, which can lead to conspicuous hygromas in chronic stages (Forbes 1991, Fig. 1). More rare pathological outcomes include spondylitis or discospondylitis and meningoencephalitis (Megid et al. 2010; Dean et al. 2012a).

Brucellosis lesions can be suppurative or nonsuppurative, with multifocal or diffuse lymphocytic or lymphohistiocytic inflammatory infiltrates as well as neutrophilic infiltrates (Forbes 1991). Granulomatous and necrotic lesions may be found in affected organs and tissues, as well as abscesses of different sizes, more frequently occurring with *B. suis* (Braude 1951).

Fetal lesions include lymphoid hyperplasia, pleuropneumonia, and pericarditis (Lopez et al. 1984) but may also have focal necrotizing granulomas in the liver,

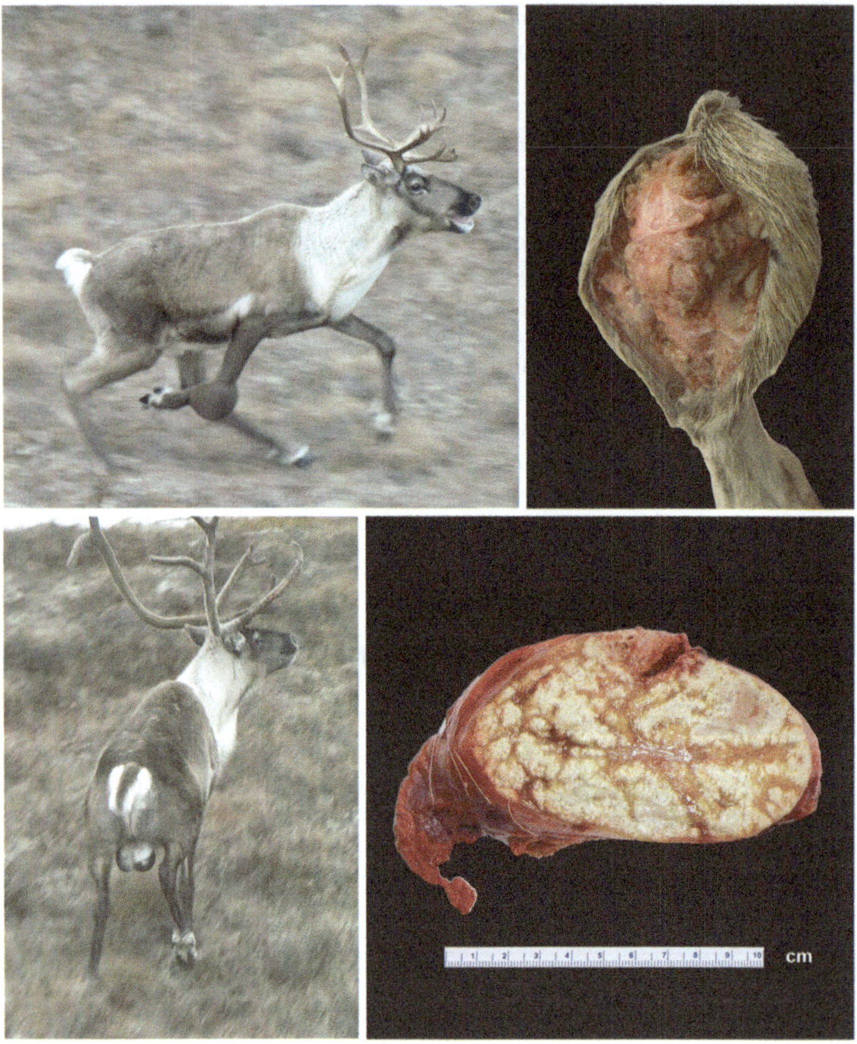

Fig. 1 Brucellosis gross lesions in migratory tundra caribou. Top left: caribou from the Mulchatna herd (Alaska) with a hygroma affecting the right carpal joint, a typical sign observed in *Rangifer* herds with brucellosis. Top right: opened hygroma in a swollen carpal joint from a Dolphin and Union caribou caused by *Brucella suis* biovar 4, filled with a translucent, watery, pink liquid with smooth, nodular suspended aggregates of fibrin. Bottom left: abnormal big testes indicating orchitis in a male caribou from the Mulchatna herd, Alaska. Bottom right: longitudinal section of a testicle showing a severe granulomatous orchitis caused by *B. suis* biovar 4 in a Dolphin and Union caribou. Photo credits: alive caribou photos from Dominic Demma, Alaska Department of Fish and Game. Brucellosis gross lesions from Jamie Rothenburger and Dayna Goldsmith, Diagnostic Service Unit, Faculty of Veterinary Medicine, University of Calgary

kidneys, and spleen (Gorham et al. 1986; Meador et al. 1988; Xavier et al. 2009). Fibrinous exudate may be present in pleural surfaces, pericardium, or abdominal organs (Xavier et al. 2009; Poester et al. 2013).

1.5 Diagnosis

None of the signs of brucellosis are pathognomonic, and diagnosis is hence based on the epidemiological history and laboratory analyses. Diagnostic tests can directly target *Brucella* spp. or may provide indirect evidence of the infection (i.e., antibodies). The isolation of *Brucella* spp. is the gold standard for diagnostic cases (OIE 2019). Microbiological cultures also enable comprehensive subtyping and further research on the isolates; however, culture is time-consuming and requires appropriate facilities with a high degree of biosafety. Laboratory-acquired brucellosis in people is not uncommon (Traxler et al. 2013). DNA-based approaches are highly specific and sensitive and in combination with culture methods offer an appropriate diagnostic approach (Leyla et al. 2003). However, serological methods can be more practical depending on the objectives (diagnostic case vs. population surveys), case-specific constraints, or epidemiological contexts.

1.5.1 Bacteriology

Brucella spp. are Gram-negative, nonmotile, and facultative intracellular coccobacillus 0.6–1.5 μm long and 0.5–0.7 μm wide (Corbel and Brinley-Moran 2005). Smears from pathological material such as abortions or biological fluids can be fixed with heat or ethanol and stained to identify *Brucella*-like bacteria. *Brucella* appear red with the Stamp's modification of the Ziehl–Neelsen's stain and typically arrange single or less frequently in pairs, chains, or small groups (Alton et al. 1988). Other pathogens like *Coxiella burnetii* or *Chlamydia abortus* may appear similar in the Stamp stain, and this method only provides presumptive evidence of brucellosis (Alton et al. 1988). A fluorochrome or peroxidase-labeled antibody conjugate can also be used to detect *Brucella* in smears (OIE 2019).

The isolation of *Brucella* can be performed in live animals from aborted material and vaginal secretions, milk, sperm, and fluids from arthritis and hygromas. In *postmortem* examinations, the preferred tissues are those with lesions, and spleen, reproductive organs, bone marrow, or lymph nodes. In cetaceans, *B. ceti* has also been isolated numerous times from the central nervous system (Guzman-Verri et al. 2012). There are many culture media suited for *Brucella*; however, field samples are often contaminated, and thus selective media should be used (Poester et al. 2010). The modified Farrell's medium (FM) is highly selective but inhibits the growth of *Brucella ovis* and some strains of *B. melitensis, B. abortus, B. suis,* and *B. pinnipedialis.* The use of FM together with a less selective media such as the Thayer–Martin's modified culture media (mTM) (Foster et al. 2002) or conventional blood agar is a classical approach for field samples. However, the more recently described CITA medium offers a better success in field samples and can also be used simultaneously with FM (De Miguel et al. 2011). After 48–72 h at 37 °C, the colonies are 0.5–1.0 mm in diameter with a convex, circular outline. Typically, smooth strains are transparent yellowish, while rough colonies are opaque and granular. Some primary smooth cultures may rapidly dissociate and appear as rough phenotype (Foster et al. 2011).

The typing of *Brucella* in biovars has classically been defined based on phenotypic characteristics, pathogenicity, or primary hosts. Classical microbiological tests assess around 25 phenotypic characteristics for typing, predominantly the requirement for CO_2, H_2S production, urease activity, sensitivity to dyes and phages, and agglutination with specific antisera (Alton et al. 1988).

1.5.2 DNA-Based Methods

There are numerous molecular methods to detect *Brucella* genus, species, biovars, or sequence types (STs). In field samples, these methods may have higher sensitivity than conventional bacteriology and the specificity of PCR-based methods for *Brucella* detection approaches 100% (Yu and Nielsen 2010).

Brucella genus can be confirmed with the 16S rRNA (Gee et al. 2004), or by conventional and real-time PCRs that target *Brucella*-specific markers (Yu and Nielsen 2010). For the identification of *Brucella* species or subtypes, the AMOS-PCR uses the polymorphism of the insertion sequence IS*711* to differentiate by size the amplicons of *B. abortus*, *B. melitensis*, *B. ovis,* and *B. suis*, including few of the biovars, and some of the *B. abortus* vaccine strains (Bricker and Halling 1995). A more recent multiplex PCR (Bruce-ladder), and its subsequent advancements, discriminates in the same test most of the *Brucella* species, including classical *Brucella* with their biovars, *B. microti, B. inopinata, B. ceti,* and *B. pinnipedialis*, and the vaccine strains *B. abortus* S19, *B. abortus* RB51, and *B. melitensis* Rev 1 (López-Goñi et al. 2011). Additionally, several real-time PCRs based on similar principles or single-nucleotide polymorphism (SNP) have also been developed for the detection of species and biovars (Yu and Nielsen 2010), including for the marine-zoonotic ST27 (Wu et al. 2014). The Bruce-ladder multiplex PCR and PCR-restriction fragment length polymorphism (RFLP) are suitable for *Brucella* spp. subtyping into biovars and can be applied together with conventional bacteriology (Al Dahouk et al. 2005).

Genotyping methods provide better insights into the genetic structure of *Brucella* populations and a higher resolution for epidemiological inference than the classical biovar subtyping. Multi-locus sequence analyses (MLSA) on housekeeping genes characterize sequence-type profiles according to allelic variation, which gives a better understanding on *Brucella* populations and phylogenetic relationships (Whatmore et al. 2016). Multi-locus sequencing of rapidly evolving markers, such as the variable-number tandem repeats (VNTR or MLVA), provides fingerprint profiles that can be applied for outbreak investigations or trace epidemiological roles (Valdezate et al. 2007; Higgins et al. 2012).

1.5.3 Serological Tests

Serological diagnosis is a presumptive evidence of infection. The first serological test for brucellosis was described in 1897 (Wright et al. 1897), and since then a large number of tests have been developed. There are two main groups of serological tests, the conventional tests relying on the antibodies performing a secondary function (e.g., agglutination tests, complement fixation tests, and precipitation tests), and the primary binding assays where the antibodies solely react with an antigen (e.g., various enzyme-linked-immunosorbent assays [ELISA] and fluorescence immunoassays) (Nielsen 2002; Poester et al. 2010). Antibodies against *Brucella* can be detected in serum, milk, whole blood, or eluates from blood collected on filter paper (Curry et al. 2011; OIE 2019). The immunogenic characteristics of the outer membrane lipopolysaccharide (LPS), which may (smooth; S-LPS) or may not (rough) contain the surface-exposed O-polysaccharide chain (OPS), is an important feature for *Brucella* serology. Most of *Brucella* species are naturally occurring as smooth strains (Table 1). The OPS epitopes are similar across species and cross-react among S-LPS; therefore, primary binding assays for smooth *Brucella* can be developed with non-species-specific antigens, typically from *B. abortus* or *B. melitensis* reference strains. For rough brucellae (*B. ovis* and *B. canis*), some secondary function tests can be used (Kimura et al. 2008), but agglutination tests may not perform well due to autoagglutination of whole-cell antigens.

Because of differences in performance, it is common to use a panel of serological tests as indication of exposure. Among conventional tests, the slow agglutination, rose bengal and buffered antigen plate agglutination tests are simple and cost-effective methods for *Brucella* antibody screening, but can result in more false positives and negatives than other techniques (Forbes and Tessaro 2003). The complement fixation test (CFT) is a laborious technique that has higher specificity (Sp) and it is considered a confirmatory test, classically used in combination with agglutination tests (OIE 2019). Indirect ELISAs, and to a less extent competitive ELISAs, offer an excellent balance between sensitivity and specificity, can be designed to detect specific immunoglobulins, and can be used for both population surveys and confirmatory tests. Fluorescence polarization assays (FPA) also have similar or higher sensitivity and specificity than iELISAs and have extensively been validated for the detection of *Brucella* antibodies (Minas et al. 2005). A review of the methods according to their characteristics for different purposes and epidemiological situations is provided by the World Organization for Animal Health (OIE 2019).

The attenuated vaccines *B. abortus* strain 19 or *B. melitensis* Rev.1 (Schurig et al. 2002) and some members of the families *Enterobacteriaceae* and *Vibrionaceae* can generate cross-reactivity in OPS-LPS-based tests (Gee et al. 2004), but only *Yersinia enterocolitica* O:9 may represent a significant source of false positives in ruminants (Kittelberger et al. 1995). The quality of the serum, like hemolysis or high lipid content, affects the performance of the classical agglutination tests and CFT (Poester et al. 2010). A 30-min chloroform/centrifugation cleanup (Castro et al. 2000) can remove false positives caused by lipemic serum in certain serological tests (Blanchet et al. 2014; Godfroid et al. 2016; Nymo et al. 2018).

In wildlife, many of the test procedures have been directly transferred from domestic animals, however, and ideally, a proper validation of the sensitivity and specificity should be performed in each species with bacteriological gold standards (Godfroid 2002). Competitive ELISAs or the use of protein A/G may solve the lack of species-specific secondary antibodies. An iELISA has been validated for the detection of anti-*Brucella* antibodies in reindeer and caribou and has shown good coherence with agglutination tests in other Arctic species (Nymo et al. 2013a, 2018; Godfroid et al. 2016). Marine mammal conjugated antibodies and ELISAs based on *B. ceti* and *B. pinnipedialis* antigens have also been developed (Hernandez-Mora et al. 2009; Meegan et al. 2010, 2012).

2 Brucellosis in Terrestrial Ecosystems

Brucellosis has been long recognized in terrestrial ecosystems from Alaska, the northern territories of Canada and Russia. It was first described in the Arctic in 1939, infecting Indigenous people from Alaska (Huntley et al. 1963), and in 1930–1935 in livestock and people from the northern territories of Krasnoyarsk Krai and Yakutia (current Republic of Sakha) in Russia (Pinigin 1959). The main etiologic agent of brucellosis in Arctic terrestrial ecosystems, in both humans and animals, is *B. suis* biovar 4. *Brucella abortus* and *B. melitensis* are much less common in most of the Arctic. This is because these species are mostly present in livestock, which is uncommon in the Arctic, and are also target of eradication programs. Other less studied *Brucella* occur depending on the presence of the reservoir hosts. Here we focus mainly on *B. suis* biovar 4 due to its primary role in brucellosis in the terrestrial Arctic and subarctic.

2.1 *Brucella* Species and Hosts

Brucella suis biovar 4 is found throughout the Canadian, Alaskan, and Russian Arctic mainly in migratory tundra caribou (*Rangifer tarandus granti; R. t. groenlandicus*; *R. t. groenlandicus* x *pearyi*) and semidomesticated and wild reindeer (*Rangifer t. tarandus*). It is apparently absent from Greenland, Iceland, Norway (including Svalbard), Sweden, Finland, and some eastern regions of Russia. In Russia, brucellosis in reindeer was first described in 1948 in the Taimyr national district (Davydov 1967), and since then it has been recorded in the Magadan Oblast, Murmansk Oblast, Tyumen Oblast, Yamalo-Nenets Autonomous Okrug, Chukotka Autonomous Okrug, Kamchatka Krai, Khabarovsk Krai, Krasnoyarsk Krai, and the Republic of Sakha (former Yakutia) and Tuva (Sleptsov et al. 2017).

Brucella suis biovar 4 was first characterized in isolates from Alaskan Indigenous people (Meyer 1964) and subsequently found to be the same as reindeer and caribou isolates, formerly referred to as *Brucella rangiferi* (Parnas 1964, 1966; Meyer 1966). Because it is mainly isolated from caribou and reindeer, it is commonly referred to as rangiferine brucellosis. However, *B. suis* biovar 4 can also be found in muskoxen (Tomaselli et al. 2019), and less commonly, in moose and carnivores from the Arctic and subarctic (Morton 1986; Honour and Hickling 1993) (Table 2). Experimental infections demonstrated that cattle (Forbes and Tessaro 2003), bison (*Bison bison*) (Bevins et al. 1996), Dall's sheep (*Ovis dalli dalli*) (Rausch and Huntley 1978), white-tailed deer (*Odocoileus virginianus*) (Qureshi et al. 1999), a variety of northern rodents, snowshoe hares (*Lepus americanus*) (Miller and Neiland 1980), red foxes (*Vulpes vulpes*) (Morton 1986), wolves (*Canis lupus*), black bears (*Ursus americanus*), and grizzly bears (*Ursus arctos horribilis*) can be infected with *B. suis* biovar 4 (Neiland and Miller 1981).

It is hypothesized that *B. suis* biovar 4 was introduced along with imported semidomesticated reindeer to North America; however, this remains to be proven. Reindeer were introduced to Alaska and Canada from Norway and Siberia between 1892 and 1935, for food security, and to promote the settlement of nomadic Inuit and Inupiaq people (Lantis 1950; Scotter 1972). Brucellosis was recognized by traditional knowledge as a new syndrome in caribou in some areas of the central Canadian Arctic in the late 1980s (Gunn et al. 1991). However, parasite-host assemblages from the Eurasian and American continent are similar in *Rangifer* spp. (Hoberg et al. 2012), and rangiferine brucellosis may have been historically present in North America. Regardless, reindeer movements within North America, such as the land transportation of Siberian reindeer from Alaska to Canada (1929–1935) (Scotter 1972), may have favored the spread of *B. suis* biovar 4 strains across regions.

Brucella abortus is the causative agent of bovine brucellosis. In northern regions, *B. abortus* is present in cattle from Russia and in wood bison (*Bison b. athabascae*) from Canada. It is believed to have African origins from where it was introduced into Europe (Whatmore et al. 2016), and most probably into North America with European cattle during colonization in the 1500s (Aguirre and Starkey 1994; Moreno 2014). Costly national eradication programs for bovine brucellosis in livestock have succeeded, leading to declaration of bovine-brucellosis free status in Norway (1952), Sweden (1957), Finland (1960), Alaska (1982), and Canada (1985) (Elberg et al. 1977; State of Alaska Section of Epidemiology 2011; Canadian Food Inspection Agency 2018), and no cases have ever been reported in Iceland or Greenland (OIE 2013). In Russia, in 1957, 55% of the brucellosis cases recorded in livestock were from subarctic regions (Verteletsky 1959). However, the disease decreased in the subsequent decades with the implementation of vaccination programs (Denisov et al. 2013).

The most northern extent of bovine brucellosis in North America is the boreal population of wood bison, in the greater Wood Buffalo National Park area (WBNP, Alberta and Northwest Territories, Canada), where it has been detected since 1955 (Choquette et al. 1978; Tessaro et al. 1990). Brucellosis was most probably introduced to this population with the translocation of plains bison (*Bison b. bison*) from the now-closed Buffalo National Park in the 1920s (Alberta, Canada) (Tessaro et al. 1990). Both *B. abortus* biovar 1 and 2 have been isolated in wildlife from the WBNP region (Tessaro 1986) (Table 2). Bovine brucellosis has also been reported in foxes and rodents from Russia (Neiland 1970). At more southern latitudes in North America, *B. abortus* is maintained by plains bison and elk in the Greater Yellowstone Ecosystem (Cheville et al. 1998) and has also been isolated from moose, feral swine (*Sus scrofa*), and coyotes (*Canis latrans*) (Corner and Connell 1958; Davis et al. 1979; Higgins et al. 2012). Wild rodents and lagomorphs are also susceptible to *B. abortus* infection (Thorpe et al. 1967).

Table 2 *Brucella* spp. isolation and detection of antibodies in mammals from terrestrial ecosystems from the Arctic and Subarctic. Additional references may exist for several of the species

Species *scientific name*	Detection methods		*Brucella* species	References
	Antibodies[a]	Agent[b]		
Arctic ground squirrel *Urocitellus parryii*	BPAT, SPAT		*Brucella* spp.	Morton (1986)
Arctic fox *Vulpes lagopus*	BPAT, CFT, ME, Riv, SPAT, STAT	BA	*B. suis* biovar 4	Pinigin et al. (1970), Morton (1986), Zheludkov and Tsirelson (2010)
Caribou *Rangifer t. groenlandicus/ granti/ groenlandicus pearyi*	BPAT, CAR, cELISA, iELISA, RBT, SPAT, STAT	BA	*B. suis* biovar 4	Broughton et al. (1970), Forbes (1991), Zarnke et al. (2006), Nymo et al. (2013a), Carlsson et al. (2019)
Dall's sheep *Ovis dalli dalli*	SPAT		*Brucella* spp.	Foreyt et al. (1983)
Domestic dog *Canis lupus familiaris*	CFT, STAT	BA	*B. suis* biovar 4	Neiland (1970, 1975)
Ermine *Mustela erminea*		BA	*B. suis* biovar 4	Zheludkov and Tsirelson (2010)
Grizzly bear *Ursus arctos horribilis*	BPAT, CAR, CFT, iELISA, RBT, SPAT, STAT		*Brucella* spp.	Morton (1986), Zarnke et al. (2006), Godfroid et al. (2016)
Moose *Alces alces*	BPAT, iELISA, RBT, WIA	BA	*B. suis* biovar 4	Honour and Hickling (1993), Edmonds et al. (1999), Nymo et al. (2016b)
Muskox *Ovibos moschatus*	BPAT, CAR, CFT, iELISA, cELISA, RBT, STAT	BA	*B. suis* biovar 4	Gates et al. (1984), Nymo et al. (2016b), Tomaselli et al. (2019)
Red fox *Vulpes vulpes*	BPAT, CFT, ME, Riv STAT, SPAT	BA	*B. suis* biovar 4 *B. abortus* biovar 1	Neiland (1975), Morton (1986), Tessaro (1986)
Reindeer *Rangifer t. tarandus*	STAT	BA	*B. suis* biovar 4 *B. abortus* biovar 1	Broughton et al. (1970), Forbes (1991), Higgins et al. (2012)
Wolf *Canis lupus*	BPAT, STAT, SPAT, CAR, CFT	BA	*B. suis* biovar 4 *B. abortus* biovar 1	Neiland (1970, 1975), Pinigin and Zabrodin (1970), Tessaro (1986), Zarnke et al. (2006), Zheludkov and Tsirelson (2010)
Wolverine *Gulo gulo*		BA	*B. suis* biovar 4	Pinigin and Zabrodin (1970), Zheludkov and Tsirelson (2010)

(continued)

Table 2 (continued)

Species *scientific name*	Detection methods		*Brucella* species	References
	Antibodies[a]	Agent[b]		
Wood bison *Bison b. athabascae*	STAT, CFT, BPAT	BA	*B. abortus* biovar 1 *B. abortus* biovar 2	Choquette et al. (1978), Tessaro et al. (1990)

[a]Abbreviations used: *BPAT* buffered *Brucella* antigen test, *CAR* card test, *CFT* complement fixation test, *ME, 2*-mercaptoethanol, *RBT* rose Bengal test, *Riv* rivanol, *SPAT* standard plate agglutination test, *STAT* standard tube agglutination test, *WIA* western immunoblot analyses
[b]Abbreviations used: *BA* bacteriology isolation

Brucella canis is sporadically detected in domestic dogs from Alaska (State of Alaska Section of Epidemiology 2011) and has caused cases and outbreaks in kennels from southern Canada and the Nordic countries of Europe (Forbes and Pantekoek 1988; Brennan et al. 2008). No cases of *B. canis* have been reported in wildlife from the Arctic and subarctic. Serosurveys suggest that it is uncommon in dogs in most northern countries (Buhmann et al. 2019). However, it has been detected in imported dogs to Canada, Sweden, and Finland (Forbes and Pantekoek 1988; Holst et al. 2012), and increasing international movement of dogs may lead to a higher risk of emergence (Buhmann et al. 2019).

Other typical brucellosis agents in livestock, such as *B. suis* biovar 1, 2, and 3 or *B. melitensis,* do not typically occur in arctic or subarctic regions, but cases may be imported (Holst et al. 2012; Garofolo et al. 2016). The less pathogenic *Brucella suis* biovar 2 is, however, widely present in Eurasian wild boar (*Sus scrofa*) and European brown hares (*Lepus europaeus*) that range in adjacent subarctic regions (Fretin et al. 2008; Gyuranecz et al. 2011).

2.2 Pathology of *B. suis* biovar 4

Brucellosis in caribou or reindeer is similar to the reproductive disease described in livestock with a higher frequency of abscesses in tissues and may also progress to osteoarthritis and granulomatous lesions in different organs (Dieterich 1981; Forbes 1991). In Siberian reindeer herds, 1–5% of the animals show evident brucellosis signs at some point, and morbidity may be up to 15% in animals older than 6 months (Neiland et al. 1968). However, animals with visible clinical signs may represent only a small portion of the infected animals in caribou or reindeer herds (Dieterich 1981).

The reproductive effects of *B. suis* biovar 4 in *Rangifer* females mainly occur during the first pregnancy following infection (Rausch and Huntley 1978; Dieterich 1981; Cotterill et al. 2018). Brucellosis can cause metritis and mastitis that may lead to reproductive failure about 2 months following infection (Neiland et al. 1968; Rausch and Huntley 1978). The outcome of abortion in experimentally infected reindeer is dependent on the stage of pregnancy and the bacterial dose (Rausch and

Huntley 1978). In an experimental setting, among seven reindeer cows naturally infected with *B. suis* biovar 4, two aborted (Forbes and Tessaro 1993). The infection can also cause retention of placenta and excessive bleeding during parturition that may be associated with stillborn and weak calves with short survival (Rausch and Huntley 1978; Forbes and Tessaro 1993). In the Western Arctic caribou herd (Alaska), excessive bleeding and retention of placenta were observed in 3–5% of the cows when *Brucella* seroprevalence in this herd was estimated to be 8–12%, and 40% of the cows analyzed with these signs had antibodies against *Brucella* (Neiland et al. 1968). Neiland (1972) also reported that about 50% of the cows experiencing these signs would lose their calves within a few days after birth.

In males, epididymitis or epididymo-orchitis appears as enlarged testes with purulent content that may be partly calcified (Neiland et al. 1968) (Fig. 1). Although testicles may be enlarged to 12–20 cm in diameter, a swollen epididymis is only detectable through palpation or careful examination (Dieterich 1981). Lymphadenopathy is common, and abscesses (described up to 70 mm) can occur in different organs, including the spleen, liver, lung, or omentum from 45 days to several months following infection (Rausch and Huntley 1978), or less commonly in muscle, subcutaneously and in other organs and tissues (Dieterich 1981, Forbes 1991). Granulomatous lesions can also occur in the kidneys, liver, or spleen (Tessaro and Forbes 1986; Forbes 1991). Bursitis-synovitis are commonly reported in carpal and tarsal joints and may lead to conspicuous hygromas and lameness (Neiland et al. 1968; Forbes 1991). Bursitis in carpal joints is the most commonly reported visible sign of rangiferine brucellosis in wild populations (Gunn et al. 1991; Tomaselli et al. 2018) (Fig. 1).

Brucella suis biovar 4 occurs less commonly in other arctic ungulates. In muskoxen, severe disease, including bursitis, orchitis, lymphadenitis, splenitis, nephritis, endometritis, granulomatous mastitis, and intramedullary and vaginal abscesses, has been reported (Gates et al. 1984; Forbes 1991; Chopra et al. 2016; Tomaselli et al. 2019). In moose, clinical and pathological presentations vary among the few cases examined. A wasting and chronic disease with bursitis and osteomyelitis in carpal joints and phalanges was observed in a naturally infected moose (Honour and Hickling 1993). Severe disease in moose was also reproduced in an experimental infection with *B. suis* biovar 4 at a high dose (1.7×10^7-colony forming units, cfu, conjunctival sac inoculation), involving long-lasting bacteremia from day 28 to 103 postinfection (p.i.) and a 2-week febrile syndrome starting at 42 days p.i. Lymphadenitis, epididymis, and splenitis with multiple white-necrotic foci (0.1–0.2 cm) were detected at necropsy (Dieterich et al. 1991). In contrast, a moose challenged with a different strain and at a lower dose (2.2×10^3 cfu, conjunctival sac inoculation), seroconverted 1 month p.i., antibodies peaked at 2 months and declined thereafter, but no lesions attributable to brucellosis were observed 1 year later (Rausch and Huntley 1978).

Bovines may be more resistant to *B. suis* biovar 4 infection than *Rangifer*, muskoxen, or moose. In experimental infections, cattle, and bison had no clinical signs, lesions, or abortion attributable to *Brucella*. However, the infection was confirmed by seroconversion of cattle and *B. suis* biovar 4 was recovered from regional lymph nodes in both species, cattle and bison, several months after the challenge (Forbes and Tessaro 1993, 2003; Bevins et al. 1996).

In carnivores, *B. suis* biovar 4 can cause systemic infections but usually without major lesions. Lymphadenitis and splenomegaly are the only signs occasionally found in natural infections of arctic and red foxes (Pinigin et al. 1970; Morton 1989), and in experimental infections in wolves and red foxes (Neiland and Miller 1981; Morton 1989). Reproductive failure in carnivores is probable but has not been yet demonstrated (Morton 1989).

Some endemic rodents from the Arctic and subarctic, including the northern flying squirrel (*Glaucomys sabrinus*), Arctic ground squirrel (*Urocitellus parryii*; former *Citellus* and *Spermophilus* genus), northern red-backed vole (*Clethrionomys rutilus*), taiga vole (*Microtus xanthognathus*), Scandinavian lemming (*Lemmus lemmus*), and Siberian lemming (*L. sibiricus*), and snowshoe hare, have been experimentally infected with intraperitoneal inoculations of *B. suis* biovar 4 (Miller and Neiland 1980). No marked pathology was found for most of the species, but the bacteria was recovered from the spleens, livers, and other organs to the end of the studies (2–5 months). Collared lemmings from North America (*Dicrostonyx* spp.), however, were especially sensitive to *B. suis* biovar 4. Disease was established with intraperitoneal inoculation with as few as 2 cfu, and all challenged animals died from 14 to 126 days p.i. with enlarged spleen, epididymitis, suppurative metritis, and abscesses in the livers, kidneys, and abdominal cavity (Miller and Neiland 1980). Russian reports also described the susceptibility of field voles (*Microtus arvalis*) to *Brucella* with as few as 50 cfu, reviewed in Miller and Neiland (1980).

2.3 Ecology and Transmission of *B. suis* biovar 4

Terrestrial brucellae is typically maintained enzootically in one or a few primary hosts, but spillover and cases of brucellosis may happen in other susceptible species. The functional role of the different hosts in *Brucella* ecology does not only depend on their susceptibility or capacity to excrete *Brucella* (intrinsic traits) but also on patterns of aggregation and effective contacts for transmission within and between populations (Cross et al. 2010). For example, *B. suis* biovar 4 is maintained in northern regions by reindeer and migratory tundra caribou, but exposure also happens in mountain ecotype herds from Alaska that are in contact with tundra caribou (Zarnke 2000).

Transmission routes of brucellae in domestic ruminants include horizontal (around abortion, stillbirth, and parturition events), congenital (transplacental), pseudo-vertical (milk or feces in neonates), and less likely venereal transmission (Aparicio 2013). *Brucella suis* biovar 4 in *Rangifer* is probably transmitted similarly to other classical *Brucella*, mainly through horizontal transmission (Cheville et al. 1998). Aborted fetuses and placental membranes/fluids in parturition contain high numbers of bacteria that contaminate the environment or can directly infect other animals that sniff or lick the remains (Maichak et al. 2009). *Brucella* spp. is also excreted in lower amounts through urine and feces (Bicknell and Bell 1979; Cheville et al. 1998). Horizontal transmission of *B. suis* biovar 4 has been demonstrated under controlled conditions among reindeer and from reindeer to cattle (Rausch and Huntley 1978; Forbes and Tessaro 1993). Classical *Brucella* do not multiply outside the host and are susceptible to UV radiation but may persist for years in frozen aborted material or for months in moist conditions at 10–15 °C (Crawford et al. 1990; Aune et al. 2012). The transplacental transmission is well-known for classical brucellae and occurs for *B. suis* biovar 4, from which some calves may be seronegative (Wilesmith 1978; Dieterich et al. 1987; Bercovich et al. 1990). If *Brucella* colonize the mammary gland and the supramammary lymph nodes, it can be excreted in milk intermittently and infect neonates (Forbes and Tessaro 1993). In Alaska, *B. suis* biovar 4 was detected in a mediastinal lymph node of a calf of approximately 4 days of age (Beckmen, unpublished results). The venereal route is also likely, as brucellae can be found in the urogenital tract and semen of bulls, and reindeer females are susceptible to *B. suis* biovar 4 infection via the genital tract (Rausch and Huntley 1978). This venereal route, even at low frequencies, can have important consequences in free-ranging wildlife (Lambert et al. 2020).

Brucellae can be vectored by arthropod insects until approximately 72 h after feeding on contaminated blood or tissues (Cheville et al. 1989), and experimental evidence suggests that it may also be transmitted by ticks (Pinigin and Zabrodin 1970; Rementsova 1987). *Brucella suis* biovar 4 has been isolated from larvae, pupae, and adults of the *Rangifer* warble fly (*Hypoderma tarandi*) from *Brucella*-infected reindeer, and its persistence in pupae has been documented for up to 6–8 weeks (Nilssen 2006), yet the significance of these findings remain unknown.

It is not known how *B. suis* biovar 4 is maintained in migratory tundra caribou, but it is commonly detected at low seroprevalence (2–5%) in some herds (Broughton et al. 1970; Zarnke 2000; Carlsson et al. 2019). The prevalence of brucellosis may fluctuate within herds, and higher seroprevalences have been reported in caribou on Baffin Island (15–43%, 1983–1986) (Ferguson 1997), Southampton Island (59%, 2011) (Campbell 2013), and the Dolphin and Union caribou herd (14%, 2015–2019) from Canada (Carlsson et al. 2019; Aguilar and Kutz 2020) and in the Western Arctic herd (12–30%, 1962–1965; 8%, 1975–2000) from Alaska (Neiland et al. 1968; Zarnke 2000). However, sampling biases (e.g., captured vs. harvested) and differences in diagnostic tests mean that seroprevalence across herds and time are not always comparable. The age of animals is an important factor to consider in serosurveys as younger animals (calves or yearlings) are generally seronegative (Neiland et al. 1968; Ferguson 1997; Carlsson et al. 2019; Aguilar and Kutz 2020). Factors influencing the dynamics of *B. suis* biovar 4 in caribou are unknown, but existing data demonstrating high and maintained prevalence during population declines suggest that that transmission is not entirely density-dependent (Campbell 2013).

In muskoxen or moose, the exposure to *Brucella* is generally absent or low (Bourque and Higgins 1984; Tessaro et al. 1984; Gunn et al. 1991). However, sporadic cases of brucellosis have been reported in these species in areas where the disease is prevalent in tundra caribou or reindeer, suggesting that are spillover hosts for *B. suis* biovar 4 (Tomaselli et al. 2016; Afema et al. 2017). Higher exposure was also reported in muskoxen and moose from the eastern North Slope (Alaska) (Edmonds et al. 1999; Nymo et al. 2016b; Afema et al. 2017), and muskoxen from Victoria island (Canada) (Tomaselli et al. 2019). The continued and increasing detection of *B. suis* biovar 4 in muskoxen from Victoria island suggests that it may be contributing to its maintenance in that region.

Northern rodents or hares can also be infected and excrete *Brucella* in urine and feces and may occasionally play a role in the spread (Miller and Neiland 1980). *Brucella* can persist for months in rodent species without marked pathology (Miller and Neiland 1980; Noi et al. 2009) and has survived for nearly 2 years in Eurasian ground squirrels (*Spermophilus* spp.) (Rementsova 1987). Northern rodents or hares may get infected on calving grounds of caribou and reindeer (Morton 1986), but their role in the ecology of this disease in the Arctic is unknown.

Carnivores are exposed to *Brucella* through infected prey species, and in an experimental setting, feeding of food contaminated with *B. suis* biovar 4 resulted in infection of bears, red foxes, wolves, and dogs (Neiland and Miller 1981; Morton 1986), and in farmed arctic foxes (Pinigin et al. 1970). Indeed, the exposure of arctic and red foxes, wolves, and grizzly bears spatially overlaps with the presence of brucellosis in caribou and reindeer (Pinigin et al. 1970; Morton 1986; Zarnke et al. 2006), as does seroprevalence in dogs fed with barren-ground caribou remains (Neiland 1970, 1975). In a study from Alaska, *B. suis* biovar 4 was isolated from 3 out of 34 red foxes and 1 out of 4 arctic foxes that were captured in a calving ground of a brucellosis-infected reindeer herd (Morton 1989). Serosurveys also indicate that grizzly bears have higher frequency of exposure to *Brucella* than wolves in Alaska (Neiland 1975; Zarnke et al. 2006; Godfroid et al. 2016), possibly linked to the bears' higher density and opportunistic feeding behavior on caribou calving grounds (Adams et al. 1995).

Brucella suis biovar 4 has been intermittently isolated from the urine, saliva, and milk from experimentally infected dogs and wolves, from the urine of experimentally infected black bears and a red fox (Neiland and Miller 1981; Morton 1989), and the feces from experimentally infected red foxes (Morton 1986). Horizontal transmission was confirmed experimentally in red foxes (Morton 1989) and may also occur among carnivores naturally. Experimental infections in red foxes and wolves failed to conclusively demonstrate vertical transmission or reproductive failure, but *B. suis* biovar 4 was recovered experimentally from stillborn, neonates, and pups (Neiland and Miller 1981, Morton 1989). Spill-back of *Brucella* from carnivores to primary hosts can happen (Davis et al. 1988), and was confirmed, under controlled conditions, from *B. suis* biovar 4 infected red foxes to reindeer (Morton 1986, 1989), although the significance of this for the epidemiology in *Rangifer* is unknown.

In Russia, natural foci of brucellosis among wild reindeer or other wildlife are believed to be of great importance to maintain brucellosis in the northern regions. Brucellosis is more often registered in semidomesticated reindeer on territories of tundra and taiga zones where the five major wild reindeer herds migrate (northern Krasnoyarsk Krai, Republic of Sakha or Yakutia and Chukotka Autonomous Okrug). Factors that help to maintain the circulation of *B. suis* biovar 4 in semidomesticated reindeer include late detection and elimination, movement of animals between herds, and interactions with wild reindeer (Vinokurov et al. 2019). Wild and domestic carnivores are also believed to contribute to the spread of brucellosis (Vinokurov et al. 2019).

2.4 Impact on Animal Populations

Brucellosis is considered primarily a reproductive disease that can affect the productivity of *Rangifer* populations by reproductive failure. *Brucella suis* biovar 4 affects the reproduction of *Rangifer* through infertility, fetal loss, abortion, stillbirths, and neonate-reduced survival, yet the overall impact to recruitment for a given exposure rate remains unknown. Such effects are better known for other *Brucella* spp. in livestock or other wildlife populations where brucellosis is also maintained enzootically (Cotterill et al. 2018). *Brucella suis* biovar 4 can also cause severe pathology in some animals, which has been more or less frequently reported in free-ranging populations of caribou or muskoxen (Forbes 1991; Ferguson 1997; Tomaselli et al. 2019). These findings raise the question of whether rangiferine brucellosis may also affect survival, such as in chronic infections when body condition is compromised and articular or systemic pathology may be involved (Neiland et al. 1968; Gates et al. 1984). Considering these severe effects, caribou performance during migration or predator escape, and resilience under unfavorable environmental conditions or other stressors, may be affected. Post-rut mortality associated with a chronic infection (antibodies detected for more than 7 months) was reported in a reindeer bull in captivity (Rausch and Huntley 1978).

High exposure to *B. suis* biovar 4 has been documented during some caribou herd declines (Ferguson 1997; Aguilar and Kutz 2020). A remarkable case is caribou from Southampton Island, in which *B. suis* biovar 4 was first detected in 2000. Its spread within the population (from 2% to 59% seroprevalence, 2000–2011) was associated with a reduction of pregnancy rates (from 93% to 37%, 2001–2011) and herd size (from 30,381 to 7286, 1997–2013) (Campbell 2013). Increased seroprevalence of *Brucella* in muskoxen, together with other pathogens and trace mineral imbalances, has also been reported in declining muskox populations (Afema et al. 2017; Tomaselli et al. 2019). Further research is needed to understand better the full array of population consequences of rangiferine brucellosis and its potential role as a contributing factor in population declines.

2.5 Management and Control

The control of infectious diseases in wildlife is challenging, requires complex logistical efforts, and is, in general, unfeasible. However, brucellosis control programs have been implemented in semidomesticated reindeer from Russia and the introduced herds in North America. The management in reindeer generally consists of early detection of disease, vaccination, and test and cull of positive animals (Dieterich 1981).

The commercially available vaccines *B. melitensis* strain H-38 and *B. abortus* strain 19 are not protective against the reproductive disease caused by *B. suis* biovar 4 (Dieterich et al. 1980; Dieterich and Morton 1987), and *B. abortus* strain 19 is potentially pathogenic for reindeer (Dieterich and Morton 1987). In controlled conditions, the killed *B. abortus* strain 45/20 yielded increased resistance to *B. suis* biovar 4 in reindeer and prevented reproductive losses (Dieterich et al. 1981). A heat-killed *B. suis* biovar 4 vaccine developed by Dieterich et al. in 1985–1986 was successful in controlled and field conditions. Bacteremia and the spread to different tissues, including detection in vaginal swabs, were significantly reduced, and antibodies were confirmed for up to 43 months (Dieterich et al. 1987). In reindeer herds, this vaccination reduced brucellosis lesions, and the exposure in sentinel animals from one monitored herd decreased from 30% to less than 5% (Dieterich pers. comm., Bevins 1993). Another heat-killed *Brucella suis* biovar 3 vaccine also has protective effects against the infection and reproductive failure, and it is distinguishable by serology from natural infections (Bevins 1993).

Since the 1980s, brucellosis control and monitoring studies have been implemented by the Reindeer Research Program of the University of Alaska Fairbanks for semidomesticated reindeer herds from the Seward Peninsula in northwestern Alaska. These reindeer move freely on unfenced ranges up to 4000 km^2 and are managed in June by foot and helicopter for ear tagging of fawns, vaccination, castration, and selection of animals for slaughter. Vaccination against brucellosis with the heat-killed *B. suis* biovar 4 vaccine was more intensively done in 1990s (Dieterich et al. 1987), but its use declined in the following decades and was changed to a commercial vaccine developed and provided by a private company around 2012–2013 (Dieterich and Finstad pers. comm.). However, vaccination rates are declining due to logistical constrains and reduced economic viability of reindeer herding in Alaska (Finstad pers. comm.).

In Russia, a set of organizational, economic, and preventive measures are being implemented to control brucellosis, including mass vaccination of healthy animals and culling of diseased animals. Semidomesticated reindeer have been vaccinated with the *B. abortus* strain 82, and in the Republic of Sakha (Yakutia), with the *non-abortogenic B. abortus* s75/79-AB vaccine (Vinokurov et al. 2019). According to Russian specialists, the low efficiency of control programs is due to the lack of compliance of reindeer herders with sanitary rules for the movement of animals (Gordienko et al. 2015; Vinokurov et al. 2019). Regardless, control strategies in semidomesticated reindeer succeeded at a regional level, and morbidity rates were directly dependent on the implementation of measures. Morbidity in reindeer herds from Yamalo-Nenets, Chukotka Autonomous Okrug, Taymyr Peninsula, and Republic of Sakha (Yakutia) was reduced from up to 50% in the 1950s to 0.1–10% in the 2000s. Since 2009, these measures have most effectively been implemented in herding farms from the Republic of Sakha (Yakutia), where rangiferine brucellosis was declared absent in 2019.

3 Brucellosis in Marine Ecosystems

Brucella were initially isolated from marine ecosystems in 1994 from a harbor seal (*Phoca vitulina*), a harbor porpoise (*Phocoena phocoena*), and a common dolphin (*Delphinus delphis*) in Scotland (Ross et al. 1994), and from a captive bottlenose dolphin (*Tursiops truncatus*) in the USA (Ewalt et al. 1994). These findings changed the historical paradigm of brucellosis as a disease exclusively affecting terrestrial hosts and expanded the knowledge on the complexity of *Brucella* spp. ecology. In 2007, Foster and colleagues formally described two novel species for *Brucella*; *B. ceti*, and *B. pinnipedialis*, with cetaceans and seals as their natural hosts, respectively (Foster et al. 2007). Marine mammal brucellae have since been serologically indicated and isolated from multiple species of marine mammals from various locations around the world, including in the Arctic and subarctic regions (Table 3).

3.1 *Brucella* Species and Hosts

Brucella ceti has been isolated from dolphins, porpoises, whales (odontocetes and mysticetes), and sea lions (Table 3). Genotyping of *B. ceti* isolates with MLSA identify two major distinct clades, one group more commonly associated with dolphins, and less extensively with beaked whales, and another one associated with harbor porpoises (Whatmore et al. 2016, 2017). Isolates from the *B. ceti* genotype ST27 form a distinct and separate group and are of particular interest because all naturally acquired human brucellosis associated with marine environments belongs to this genotype (Whatmore et al. 2008). ST27 has also been detected in bottlenose dolphins (Mackie et al. 2020; Ueno et al. 2020), a dwarf sperm whale (*Kogia sima*) (Suárez-Esquivel et al. 2019), minke whales (*Balaenoptera acutorostrata*) (Ohishi et al. 2004), California sea lions (*Zalophus californianus*) (Whatmore et al. 2017), and Steller sea lions (*Eumetopias jubatus*) (Esquible et al. 2019).

Brucella pinnipedialis has been isolated from true seals, eared seals, sea otters (*Enhydra lutris*), and beluga whales (*Delphinapterus leucas*) (Table 3). In MLSA, most *B. pinnipedialis* strains fell into the same cluster of STs, including strains from true seals, eared seals, and sea otters, but *B. pinnipedialis* from hooded and harp seals fell into a separate cluster in multiple studies (Nymo et al. 2011; Whatmore et al. 2017). *Brucella pinnipedialis* isolated from beluga whales represented a number of distinct STs (Whatmore et al. 2017).

Other *Brucella*: The host range of aquatic *Brucella* has also been extended to include ectotherms with the isolation of novel brucellae from frogs (Eisenberg et al. 2012; Fischer et al. 2012). The first report of a natural *Brucella* infection in a saltwater fish was described when a novel and atypical *Brucella* strain was isolated from a wild-caught bluespotted ribbontail ray (*Taeniura lymma*) (Eisenberg et al. 2017). Smooth *Brucella* antibodies have been detected in polar bears (*Ursus maritimus*) (Tryland et al. 2001), although it is not clear if polar bears are exposed

Table 3 Non-exhaustive list of *Brucella* spp. isolation and detection of antibodies in marine mammals that are present in Arctic and subarctic waters. Additional references exist for several of the mentioned species

Species *scientific name*	Detection methods		*Brucella* species	References
	Antibodies[a]	Agent[b]		
Atlantic walrus *Odobenus r. rosmarus*	iELISA			Scotter et al. (2019)
Baltic ringed seal *Pusa hispida botanica*	cELISA, RBT			Sonne et al. (2018)
Bearded seal *Erignathus barbatus*	iELISA	BA	*B. pinnipedialis*	Foster et al. (2018)
Grey seal *Halichoerus grypus*	cELISA	BA, PCR	*B. pinnipedialis*	Hirvela-Koski et al. (2017), Sauvé et al. (2020)
Harbor seal *Phoca vitulina*	iELISA	BA	*B. pinnipedialis*	Foster et al. (2002), Nymo et al. (2018)
Harp seal *Pagophilus groenlandicus*	Anticomplement ELISA, CFT, iELISA, RBT, SAT, SAT-EDTA	BA	*Brucella* spp.	Tryland et al. (1999), Forbes et al. (2000)
Hooded seal *Cystophora cristata*	Anticomplement ELISA, CFT, iELISA, RBT, SAT, SAT-EDTA	BA	*B. pinnipedialis*	Tryland et al. (1999), Nymo et al. (2013b)
Northern fur seals *Callorhinus ursinus*	BMAT	PCR	*B. pinnipedialis*	Duncan et al. (2014)
Ribbon seals *Histriophoca fasciata*	iELISA, RBT			Nymo et al. (2018)
Ringed seal *Phoca hispida*	iELISA, RBT	BA	*Brucella* spp.	Forbes et al. (2000), Nymo et al. (2018)
Steller Sea lion *Eumetopias jubatus*	iELISA, RBT	PCR	*Brucella* spp. *B. ceti* (ST27)	Nymo et al. (2018), Esquible et al. (2019)
Spotted seal *Phoca largha*	iELISA, RBT			Nymo et al. (2018)
Western Pacific harbor seals *Phoca vitulina stejnegeri*	iELISA			Abe et al. (2017)
Atlantic white-sided dolphin *Lagenorhynchus acutus*		BA	*B. ceti*	Foster et al. (2002)

(continued)

Table 3 (continued)

Species *scientific name*	Detection methods		*Brucella* species	References
	Antibodies[a]	Agent[b]		
Beluga *Delphinapterus leucas*	iELISA	BA, PCR	*B. pinnipedialis*	Whatmore et al. (2017), Ohishi et al. (2018)
Fin whale *Balaenoptera physalus*	Anticomplement ELISA, CFT, iELISA, RBT, SAT, SAT-EDTA			Tryland et al. (1999)
Harbor porpoise *Phocoena phocoena*	cELISA, FPA, iELISA	BA	*B. ceti*	Foster et al. (2002), Neimanis et al. (2008)
Killer whale *Orcinus orca*		PCR	*Brucella* spp.	Raverty et al. (2002)
Long-finned pilot whale *Globicephala melas*		BA	*B. ceti*	Foster et al. (2015)
Minke whale *Balaenoptera acutorostrata*	Anticomplement ELISA, CFT, iELISA, RBT, SAT, SAT-EDTA	BA	*Brucella* spp.	Tryland et al. (1999)
Narwhale *Monodon monoceros*	cELISA			Nielsen et al. (2001)
Sowerby's beaked whale *Mesoplodon bidens*		BA	*B. ceti*	Foster et al. (2015)
Striped dolphins *Stenella coeruleoalba*	cELISA	BA, PCR	*Brucella* spp.	Gonzalez et al. (2002)
White-beaked dolphin *Lagenorhynchus albirostris*		BA	*B. ceti*	Foster et al. (2002)
European otter *Lutra lutra*		BA	*Brucella* spp.	Foster et al. (1996)
Asian sea otter *Enhydra lutris lutris*		PCR	*B. abortus*, *B. melitensis*, *B. pinnipedialis*	Burgess et al. (2017)
Polar bear *Ursus maritimus*	iELISA, RBT, SAT, SAT-EDTA			Tryland et al. (2001)

[a]Abbreviations used: *BMAT Brucella* microagglutination test, *cELISA* competitive enzyme-linked immunosorbent assay, *CFT* complement fixation test, *FPA* fluorescence polarization assay, *iELISA* indirect enzyme-linked immunosorbent assay, *RBT* rose Bengal test, *SAT* standard agglutination test, *SAT-EDTA* Wright's slow agglutination test with ethylenediaminetetraacetic acid disodium salt dehydrate

[b]Abbreviations used: *BA* bacteriology isolation, *PCR* polymerase chain reaction

from *Brucella*-infected seals, other sources, or harbor different *Brucella* species (O'Hara et al. 2010).

3.2 Pathology of *B. ceti* and *B. pinnipedialis*

In many cases, *Brucella* spp. have been isolated from stranded marine mammals in poor health or already dead (Davison et al. 2011). It is thus sometimes difficult to determine what clinical signs and pathological lesions are primarily related to *Brucella* infection (González-Barrientos et al. 2010). Although some findings may be related to the primary disease, others may be due to underlying or secondary factors, such as persistent organic pollutants or trauma associated with the stranding (Davison et al. 2009, 2011; Oliveira et al. 2011). Often other bacteria, fungi, or viruses are detected in the lesions in coinfection with *Brucella,* and detailed investigation is required to understand primary and opportunistic roles (González-Barrientos et al. 2010; Guzman-Verri et al. 2012; West et al. 2015).

The isolation of *Brucella ceti* from the central nervous system, associated with neurological symptoms, is common, often found in stranded striped dolphins (*Stenella coeruleoalba*). Pathological changes include spinal discospondylitis, meningeal hyperemia, cerebral edema, meningoencephalitis, meningitis, choroiditis, altered cerebrospinal fluid, and remodeling of the occipital condyles (Nymo et al. 2011; Guzman-Verri et al. 2012). Among the most common clinical signs observed in relation to this pathology are opisthotonus, tremors, seizures, disorientation, and a general inability to maintain buoyancy (Guzman-Verri et al. 2012). Neurological pathology may increase the risk of stranding (Pintore et al. 2018) and thus possibly be overrepresented in known brucellosis cases from marine mammals. Neurobrucellosis is not a common feature in bovine, caprine, ovine, swine, or canine hosts (Radostits et al. 2000; Wanke 2004), but it is more common in untreated human brucellosis (Maji et al. 2020). Neurobrucellosis was also observed in the three human cases from which the ST27 marine mammal brucellae was isolated and raises the question whether these strains also have special tropisms to neurological tissues (Sohn et al. 2003; McDonald et al. 2006). Osteolytic lesions were associated with *Brucella* infection in a wild southern sea otter (*Enhydra lutris nereis*) (Miller et al. 2017).

Brucella ceti has been isolated from aborted fetuses and reproductive organs in bottlenose dolphins with placentitis (Miller et al. 1999). Isolation has also been successful from the reproductive organs, milk, and fetus of stranded striped dolphins (Hernandez-Mora et al. 2008), as well as from the uterus and a dead fetus of a stranded striped dolphin with placentitis (González-Barrientos et al. 2010). In utero transmission with associated pathology in mother and calf was reported in sperm whale (*Physeter macrocephalus*) (West et al. 2015) and endangered Hector's dolphins (*Cephalorhynchus hectori*) (Buckle et al. 2017). Immunohistochemical investigations revealed *B. ceti* in a genital ulcer, uterus, mammary gland, and milk from a stranded harbor porpoise with endometritis and signs of recent pregnancy (Jauniaux et al. 2010). Further, *B. ceti* was isolated in association with mastitis and endometritis in cetaceans (Foster et al. 2002), and from a testicular abscess (Dagleish et al. 2008). Suppurative granulomatous lesions were found in both female and male reproductive organs in seropositive baleen whales (*Mysticeti*) (Ohishi et al. 2003), and *B. ceti* was isolated from a large abscess in a common minke whale (Davison et al. 2017).

Brucella pinnipedialis has been isolated from a variety of organs in numerous apparently healthy true seals (family *Phocidae*), and the infection does not seem to cause pathology in these species (Nymo et al. 2011). In eared seals (family *Otariidae*), however, signs of disease are reported. Inflammation and multifocal acute necrosis in placentas of California sea lion were identified in two out of four positive *B. pinnipedialis* cases (Sidor et al. 2008; Goldstein et al. 2009). *Brucella pinnipedialis* was detected by PCR in six Northern fur seal (*Callorhinus ursinus*) placentas, of which one had severe placentitis (Duncan et al. 2014). Although the low number of isolates or PCR-positive cases obtained from eared seals limits our knowledge on *Brucella* pathology in this group of hosts, it is worth noticing that the few cases reported were often associated with pathological findings in the reproductive organs (Goldstein et al. 2009; Duncan et al. 2014).

3.3 Ecology and Transmission

In some cetaceans, there is considerable support for vertical transmission of *Brucella* (Guzman-Verri et al. 2012). Transplacental transmission was demonstrated in California sea lions when brucellae strains belonging to the zoonotic ST27 were detected by PCR in three placentas and multiple fetal tissues in parallel (Sidor et al. 2008). Similarly, *Brucella* spp. was detected by PCR in lung samples from three aborted and premature-born Steller sea lions (Esquible et al. 2019). Furthermore, *B. pinnipedialis* was detected by PCR in Northern fur seal placentas, in some cases associated with inflammation (Duncan et al. 2014). Therefore, it seems that vertical transmission may be taking place in certain eared seal species. An extension of the vertical transmission may be horizontal transmission through infected aborted tissues (Ewalt et al. 1994), since several cetacean species are known to assist births and be in close contact with the newborn calf and the placental tissues of other individuals (Béland et al. 1990). Close contact between individuals may be another possibility of *Brucella* transmission, for example, during mating or maternal feeding (Jauniaux et al. 2010). Terrestrial brucellae of unknown origin have been detected by PCR in blood and milk from two apparently healthy wild California sea lions, and marine mammal brucellae were detected in blood and milk from one California sea lion (Avalos-Tellez et al. 2014).

Reproductive pathology or vertical transmission of *B. pinnipedialis* has not yet been described in true seals (Nymo et al. 2011). The limited number of serologically and bacteriologically positive true seals, specifically at reproductive age (Lambourn et al. 2013; Nymo et al. 2013b, 2018), further suggests that maternal transmission is unlikely. The mean probability of being seropositive increased from pups to juveniles, suggesting that exposure to *B. pinnipedialis* is primarily during the postweaning period and during the first few years of life, thus not being transmitted in utero or to neonates. Furthermore, when females reach sexual maturity, they are generally seronegative. Although the underlying reason for this pattern is unknown, it might be related to changes in the diet. Stable isotopes and mercury biomarkers have indicated that in general, adult harbor, ringed, ribbon, and spotted seals feed at a higher trophic level than pups (Dehn et al. 2006; Bertolino et al. 2009). Hence, there may be a general shift in diet composition toward higher trophic levels with increasing age, which coincides with the age of seroconversion of juveniles (Lambourn et al. 2013; Nymo et al. 2013b, 2018). This hypothesis implies that possible reservoirs of *B. pinnipedialis* may be more represented in one or more lower trophic level of their prey species.

Brucella ceti and *B. pinnipedialis* have also been isolated in nematode lungworms (*Pseudalius inflexus* and *Parafilaroides*), from cetacean (Dawson et al. 2008) and pinniped lungs (Rhyan et al. 2018) and in seal liver flukes (*Pseudamphistomum truncatum*) (Hirvela-Koski et al. 2017). These findings suggest that parasites could be an important host of *B. pinnipedialis* and that possible transmission pathways in marine mammals may occur through parasitic cycles (Rhyan et al. 2018).

Experimental infections in Atlantic cod suggest that this species could act as asymptomatic carriers of *B. pinnipedialis* (Nymo et al. 2016c; Larsen et al. 2018). In other systems, naturally occurring *Brucella* spp. infections in wild fish were reported. These include the isolation of *B. melitensis* biovar 3 from skin swabs of the freshwater Nile catfish (*Clarias gariepinus*) (El-Tras et al. 2010), and a novel *Brucella* sp. in a wild-caught saltwater bluespotted ribbontail ray (Eisenberg et al. 2017). *Brucella* was also retrieved from internal organs after a challenge of Nile catfish with *B. melitensis biovar* 3 (Salem and Mohsen 1997). Despite these experimental and field findings, the role of fish in the ecology of *Brucella* in aquatic environments is still not clear.

3.4 Impact on Marine Mammal Populations

It is highly likely that *B. ceti* has some effect on cetacean populations as numerous cases of reproductive pathology have been reported in association with infection, including abortions (Guzman-Verri et al. 2012). The effects of reproductive losses may be magnified in cetacean hosts that already have low reproductive rates; however, the impact of *Brucella* on Arctic cetacean population dynamics is difficult to assess, and it is currently unknown.

Considering the reports of *B. pinnipedialis* pathology in eared seals (Goldstein et al. 2009; Duncan et al. 2014), morbidity and/or mortality in these species in the Arctic due to infection is possible. Taking into account the lack of impact on harbor seal populations (Nymo et al. 2018), the age-dependent serological patterns, the lack of *Brucella*-associated pathology in true seals (Nymo et al. 2013b), and the lack of multiplication in established in vitro (Larsen et al. 2013a, b, 2016) and in vivo models (Nymo et al. 2014, 2016a), *B. pinnipedialis* may not be causing any significant morbidity or mortality in true seals.

4 Brucellosis in People

4.1 Importance and Burdens in the Arctic

Terrestrial and marine mammals have historically played, and continue to play, an important role for Indigenous peoples around the Arctic with respect to food security, income, and continuance of cultural traditions. Brucellosis is an important zoonosis that both pose a real threat to peoples' health as well as threatening peoples' confidence in country foods. The main etiological agents for brucellosis in people from the most northern regions are *B. suis* biovar 4, associated primarily with caribou and reindeer and, to a lesser extent, *B. abortus* in bison in North America and *B. melitensis* associated with livestock in Russian regions. However, people from the Arctic and subarctic may also be exposed to brucellae through muskoxen, moose, other spillover hosts or marine mammals (Tables 2 and 3) (Tessaro et al. 1990; Forbes 1991; Tomaselli et al. 2019).

Brucella suis biovar 4 has been identified in people of the Arctic since the 60s (Pinigin 1959; Huntley et al. 1963); however, brucellosis was recognized decades before. In Alaska, between 1939 and 1961, 53 cases of brucellosis in people were reported (Huntley et al. 1963), with a subsequent average detection of 1 case every 2 years between 1958 and 2010 (State of Alaska Section of Epidemiology 2011). In the 50s and 60s in the Central Arctic of Canada, antibodies against *Brucella* spp. were detected in people (Greenberg et al. 1958; Corrigan et al. 1962; Davies and Hanson 1965), and brucellosis was also diagnosed in seven Inuit people from 1953 to 1961 (Toshach 1963). An average of one case of *B. suis* biovar 4 per year was reported between 1982 and 1990 in Canada. However, these numbers came from sporadic sampling and are not necessarily representative of the disease incidence (Forbes 1991). In Russia, human brucellosis is mainly associated with reindeer herding in the Arctic and with livestock in the more southern regions. In the period 2009–2018, 3832 new cases of brucellosis were reported in people from the Russian federation, with a decreasing trend during that period. The annual incidence per 100,000 inhabitants was 0.27, and among children under 17, it was 0.13.

All human brucellosis with a marine origin has been related so far to the ST27 and the consumption or contact with raw fish and shellfish at temperate latitudes (Sohn et al. 2003; McDonald et al. 2006). To date, no marine mammal-associated cases have been reported in the Arctic peer-reviewed literature. Although ST27 has only been isolated in southern distributions, its detection in Steller sea lions and minke whales suggests that northern people can potentially be exposed to this zoonotic genotype (Ohishi et al. 2004; Esquible et al. 2019).

4.2 Risks and Route of Transmission

Brucellosis in people is closely linked to contact with infected animals or their food products; horizontal transmission between humans is rare (Meltzer et al. 2010). People can get brucellosis by consuming unpasteurized dairy products, eating uncooked infected tissues, bone marrow or blood, by contacting with the organism through their mucous membranes or skin abrasions (e.g., from an abscess or amniotic fluids), or perhaps through aerosols from abortions or opened lesions. In the Arctic, there is an increased risk of infection among Indigenous people that harvest caribou, and among reindeer herders, veterinary specialists and food processing workers. Exposure to infected reindeer and caribou is the most common risk (Forbes 1991; State of Alaska Section of Epidemiology 2011). The consumption of raw bone marrow or other raw and dry tissues from harvested animals is a common practice amongst Indigenous people and has directly been associated with brucellosis (Chan et al. 1989; Kanji and Saxinger 2018). Studies from the 1960s in Alaska showed a higher exposure to *Brucella* spp. in communities from the North Slope where barren-ground caribou was an important component of the local diet, with a detection of antibodies in approximately 5–20%, and as high as 30%, of the people (Huntley et al. 1963; Brody et al. 1966). Historical surveys from Russia also reported that about 23% of people from areas where reindeer are abundant had antibodies against *Brucella*. However, the source and route of transmission of brucellosis in Russia in general remain mostly unrecognized (65% of the cases), with 25% of the cases associated with contacts with infected animals, 7% associated with the consumption of *Brucella*-contaminated products, and 3% with a mixed route of transmission. While dogs may become infected with brucellae if they are fed with contaminated meat or carcass remains (Neiland 1970, 1975), there is no evidence of dog-to-human transmission in Arctic regions.

4.3 Signs, Diagnosis, and Treatment

In people, the clinical signs of infection depend on the stage of the disease, the organs involved, and the species of *Brucella* (Franco et al. 2007). The disease is often systemic with a tendency to become a chronic granulomatous disease that can affect almost any organ system (Matas and Corrigan 1953; Huntley et al. 1963). For *Brucella* species in general, the onset of the disease can be sudden or more insidious, with an incubation period that is typically about 2–4 weeks but that can range from 5 days to several months. Intermittent fever is the most common feature of brucellosis, followed by chills, sweats, lack of energy, joint and back pain, headache, loss of appetite, and epididymo-orchitis (Corbel et al. 2006). The disease can progress to severe complications, such as endocarditis or neurological brucellosis (Dean et al. 2012a). On physical examination, the most common findings are hepatomegaly and splenomegaly. Signs reported in people infected with *Brucella* can include fevers and chills, anorexia, malaise, abdominal pain, back pain, splenomegaly, hepatomegaly, arthralgia, myalgia, endocarditis, orchitis, and skin rashes, among others (Dean et al. 2012a; Kanji and Saxinger 2018). Brucellae can be isolated from the blood, bone marrow, or affected tissues (Franco et al. 2007); however, culture sensitivity from the blood may particularly be low in chronic cases (Franco et al. 2007; Kanji and Saxinger 2018). Diagnosis of brucellosis is informed by clinical signs and history (such as exposure to infected wildlife) as well as by serology and *Brucella* detection or isolation (Kanji and Saxinger 2018).

Treatment requires long-term antibiotic therapy including doxycycline and aminoglycosides. It can be costly, and initial treatment failure and relapse may occur, although complete resolution of infection is possible (Pappas et al. 2005; Kanji and Saxinger 2018). There is no licensed vaccine for humans (Lalsiamthara and Lee 2017).

4.4 Prevention

Guidelines to harvesters to prevent infection include discarding diseased parts of the animal without contaminating the meat, and thoroughly cooking any suspect meat as *Brucella* can survive freezing, smoking, or drying (Corbel et al. 2006). Extreme care is needed when handling any suspected brucellosis case (including fetal membranes and abortions) by thoroughly washing hands, knives, or any other tool after butchering. Through traditional knowledge passed down over generations, Indigenous peoples have learned a variety of practices that optimize food safety. This starts with identification of "healthy" animals to harvest and includes external and internal examinations of the harvested animal to determine condition and presence of abnormalities, and then specific butchering, meat preparation, and storage practices to ensure food safety. In regions where there is a long history of brucellosis in wildlife, this knowledge and practices, supplemented by public health messaging, include how to recognize animals with *Brucella* and precautions necessary in butchering and preparation of meat and cleaning of tools and selves to prevent infection. However, brucellosis is not present uniformly across species or geographic regions of the Arctic, where it can emerge or reemerge (Tomaselli et al. 2019). Therefore, knowledge of how the disease may present in wildlife, and knowledge of how to prevent zoonotic transmission, may be lacking, and human health may be more at risk.

Acknowledgments The authors would like to thank Anne Gunn for her comments on historical insights of brucellosis in caribou, Jacques Godfroid for his comments on earlier versions of the chapter, and Gregory Finstad, Darrell Blodgett, Robert Dieterich, and Julia Bevins for the information they provided on vaccines and the Reindeer Research Program in Alaska.

References

Abe E, Ohishi K, Ishinazaka T, Fujii K, Maruyama T (2017) Serologic evidence of *Brucella* infection in pinnipeds along the coast of Hokkaido, the northernmost main island of Japan. Microbiol Immunol 61:114–122

Ackermann MR, Cheville NF, Deyoe BL (1988) Bovine Ileal dome Lymphoepithelial cells: endocytosis and transport of *Brucella abortus* strain 19. Vet Pathol 25:28–35

Adams LG, Singer FJ, Dale BW (1995) Caribou calf mortality in Denali National Park, Alaska. J Wildl Manag 59:584

Afema JA, Beckmen KB, Arthur SM, Huntington KB, Mazet JAK (2017) Disease complexity in a declining alaskan muskox (*Ovibos moschatus*) population. J Wildl Dis 53:311–329
Aguilar XF, Kutz S (2020) Dolphin and union Caribou health assessment. In: Interim results. University of Calgary. July 2020
Aguirre AA, Starkey EE (1994) Wildlife disease in U.S. National Parks: historical and coevolutionary perspectives. Conserv Biol 8:654–661
Al Dahouk S, Tomaso H, Prenger-Berninghoff E, Splettstoesser WD, Scholz HC, Neubauer H (2005) Identification of *Brucella* species and biotypes using polymerase chain reaction-restriction fragment length polymorphism (PCR-RFLP). Crit Rev Microbiol 31:191–196
Alton GG, Jones LM, Angus R, Verger JM (1988) Techniques for the brucellosis laboratory. Institut National de la Recher- che Agronomique, Paris
Aparicio ED (2013) Epidemiology of brucellosis in domestic animals caused by *Brucella melitensis*, *Brucella suis* and *Brucella abortus*. Rev Sci Techniq de l'OIE 32:53–60
Aune K, Rhyan JC, Russell R, Roffe TJ, Corso B (2012) Environmental persistence of *Brucella abortus* in the greater Yellowstone area. J Wildl Manag 76:253–261
Avalos-Tellez R, Ramirez-Pfeiffer C, Hernandez-Castro R, Diaz-Aparicio E, Sanchez-Dominguez-C, Zavala-Norzagaray A, Arellano-Reynoso B, Suarez-Guemes F, Aguirre AA, Aurioles-Gamboa D (2014) Infection of California Sea lions (*Zalophus californianus*) with terrestrial *Brucella* spp. Vet J 202:198–200
Bang B (1897) The etiology of epizootic abortion. J Comp Pathol Therap 10:125–149
Béland P, Faucher A, Corbeil P (1990) Observations on the birth of a beluga whale (*Delphinapterus leucas*) in the St. Lawrence Estuary, Quebec, Canada. Can J Zool 68:1327–1329
Bercovich Z, Haagsma J, ter Laak EA (1990) Use of delayed-type hypersensitivity test to diagnose brucellosis in calves born to infected dams. Vet Q 12:231–237
Bertolino S, Di Montezemolo NC, Bassano B (2009) Food-niche relationships within a guild of alpine ungulates including an introduced species. J Zool 277:63–69
Bevins JS (1993) Thesis (Ph.D.): detection an control of brucellosis in reindeer vaccinated with *Brucella suis* biovar 3. University of Alaska Fairbanks
Bevins JS, Blake JE, Adams LG, Templeton JW, Morton JK, Davis DS (1996) The pathogenicity of *Brucella suis* biovar 4 for bison. J Wildl Dis 32:581–585
Bicknell SR, Bell RA (1979) *Brucella abortus* in the bitch: subclinical infection associated with urinary excretion. J Hyg 82:249–254
Blanchet MA, Godfroid J, Breines EM, Heide-Jørgensen MP, Nielsen NH, Hasselmeier I, Iversen M, Jensen SK, Asbakk K (2014) West Greenland harbour porpoises assayed for antibodies against *toxoplasma gondii*: false positives with the direct agglutination method. Dis Aquat Org 108:181–186
Bourque M, Higgins R (1984) Serologic studies on brucellosis, leptospirosis and tularemia in moose (*Alces alces*) in Quebec. J Wildl Dis 20:95–99
Braude AI (1951) Studies in the pathology and pathogenesis of experimental brucellosis: I. A comparison of the pathogenicity of *Brucella abortus*, *Brucella melitensis*, and *Brucella suis* for Guinea pigs. J Infect Dis 89:76–86
Brennan SJ, Ngeleka M, Philibert HM, Forbes LB, Allen AL (2008) Canine brucellosis in a Saskatchewan kennel. Can Vet J 49:703–708
Bricker BJ, Halling SM (1995) Enhancement of the *Brucella* AMOS PCR assay for differentiation of *Brucella abortus* vaccine strains S19 and RB51. J Clin Microbiol 33:1640–1642
Brody JA, Huntley B, Overfield TM, Maynard J (1966) Studies of human brucellosis in Alaska. J Infect Dis 116:263–269
Broughton E, Choquette LP, Cousineau JG, Miller FL (1970) Brucellosis in reindeer, *Rangifer tarandus* L., and the migratory barren-ground caribou, *Rangifer tarandus groenlandicus* (L.), in Canada. Can J Zool 48:1023–1027
Bruce D (1887) Note on the recovery of a microorganism in Malta fever. Practitioner 29:161
Buckle K, Roe WD, Howe L, Michael S, Duignan PJ, Burrows E, Ha HJ, Humphrey S, McDonald WL (2017) Brucellosis in endangered Hector's dolphins (*Cephalorhynchus hectori*). Vet Pathol 54:838–845

Buhmann G, Paul F, Herbst W, Melzer F, Wolf G, Hartmann K, Fischer A (2019) Canine brucellosis: insights into the epidemiologic situation in Europe. Front Vet Sci 6:1–9

Burgess TL, Johnson CK, Burdin A, Gill VA, Doroff AM, Tuomi P, Smith WA, Goldstein T (2017) *Brucella* infection in Asian Sea Otters (*Enhydra lutris lutris*) on Bering Island, Russia. J Wildl Dis 53:864–868

Campbell M (2013) Population estimate of a declining population of island bound barren-ground Caribou (*Rangifer tarandus groenlandicus*), Southampton Island NU. Arviat, Nunavut Department of Environment

Canadian Food Inspection Agency (2018) [Internet]. https://www.inspection.gc.ca/animal-health/terrestrial-animals/diseases/accredited-veterinarian-s-manual/chapter-8/eng/1345476743300/1345476855757?chap=2

Carlsson AM, Curry P, Elkin B, Russell D, Veitch A, Branigan M, Campbell M, Croft B, Cuyler C, Côté SD, Leclerc L-M, Tryland M, Nymo IH, Kutz SJ (2019) Multi-pathogen serological survey of migratory caribou herds: a snapshot in time. PLoS One 14:e0219838

Carmichael L, Bruner D (1968) Characteristics of a newly-recognized species of *Brucella* responsible for infectious canine abortions. Cornell Vet 48:579–592

Castaño MJ, Solera J (2009) Chronic brucellosis and persistence of *Brucella melitensis* DNA. J Clin Microbiol 47:2084–2089

Castro AR, Morrill WE, Pope V (2000) Lipid removal from human serum samples. Clin Diagn Lab Immunol 7:197–199

Chan J, Baxter C, Wenman WM (1989) Brucellosis in an Inuit child, probably related to caribou meat consumption. Scand J Infect Dis 21:337–338

Cheville NF, Rogers DG, Deyoe WL, Krafsur ES, Cheville JC (1989) Uptake and excretion of *Brucella abortus* in tissues of the face fly (*Musca autumnalis*). Am J Vet Res 50:1302–1306

Cheville NF, McCullough DR, Paulson LR (1998) Brucellosis in the greater yellowstone area. The National Academies Press, Washington, DC. https://doi.org/10.17226/5957

Chopra S, Mavrot F, Sharpe S, Gunn A, Gamberg M, Tomaselli M, Behrens S, Kutz SJ (2016) Investigating the dental health of muskox and caribou populations in the Canadian Arctic, pp 28–29

Choquette LPE, Broughton E, Cousineau JG (1978) Parasites and diseases of bison in Canada IV. Serologic survey for brucellosis in Bison in Northern Canada. J Wildl Dis 14:329–332

Cook I, Campbell RW, Barrow G (1966) Brucellosis in North Queensland rodents. Aust Vet J 42: 5–8

Corbel MJ, Brinley-Moran WJ (2005) Genus *Brucella*. In: Brenner DJ, Krieg NR, Stanley JT (eds) Bergey's manual of systematic bacteriology, vol II. The proteobacteria. Part C The alpha-, beta-, delta-, and epsilonproteobacteria. Springer, London, pp 371–386

Corbel MJ, Food and Agriculture Organization of the United Nations, World Health Organization & World Organisation for Animal Health (2006) Brucellosis in humans and animals. World Health Organization. https://apps.who.int/iris/handle/10665/43597

Corner AH, Connell R (1958) Brucellosis in bison, elk, and moose in Elk Island National Park, Alberta, Canada. Can J Comp Med Vet Sci 22:9–21

Corrigan C, Penikett EJ, Williams ME (1962) Antibodies to viral and other antigens in blood serum of children resident in the northwest territories. Can J Public Health Rev 53:284–289

Cotterill GG, Cross PC, Middleton AD, Rogerson JD, Scurlock BM, du Toit JT (2018) Hidden cost of disease in a free-ranging ungulate: brucellosis reduces mid-winter pregnancy in elk. Ecol Evol 8:10733–10742

Cotton WE (1922) The character and possible significance of the Bang abortion bacillus that attacks swine. J Am Vet Med Assoc 62:179

Crawford RP, Huber JD, Adams BS (1990) Epidemiology and surveillance. In: Nielsen K, Duncan JR (eds) Animal Brucellosis. CRC Press, Boca Raton, pp 131–151

Cross PC, Cole EK, Dobson AP, Edwards WH, Hamlin KL, Luikart G, Middleton AD, Scurlock BM, White PJ (2010) Probable causes of increasing brucellosis in free-ranging elk of the greater yellowstone ecosystem. Ecol Appl 20:278–288

Curry PS, Elkin BT, Campbell M, Nielsen K, Hutchins W, Ribble C, Kutz SJ (2011) Filter-paper blood samples for Elisa detection of Brucella antibodies in Caribou. J Wildl Dis 47:12–20
Dagleish MP, Barley J, Finlayson J, Reid RJ, Foster G (2008) *Brucella ceti* associated pathology in the testicle of a harbour porpoise (*Phocoena phocoena*). J Comp Pathol 139:54–59
Davies LE, Hanson S (1965) The Eskimos of the Northwest passage: a survey of dietary composition and various blood and metabolic measurements. Can Med Assoc J 92:205–216
Davis DS, Boeer WJ, Mims JP, Heck FC, Adams LG (1979) *Brucella abortus* in coyotes. I A serologic and bacteriologic survey in Eastern Texas. J Wildl Dis 15:367–372
Davis DS, Heck FC, Williams JD, Simpson TR, Adams LG (1988) Interspecific transmission of *Brucella abortus* from experimentally infected coyotes (*Canis latrans*) to parturient cattle. J Wildl Dis 24:533–537
Davison NJ, Cranwell MP, Perrett LL, Dawson CE, Deaville R, Stubberfield EJ, Jarvis DS, Jepson PD (2009) Meningoencephalitis associated with *Brucella* species in a live-stranded striped dolphin (*Stenella coeruleoalba*) in south-west England. Vet Rec 165:86–89
Davison NJ, Perrett LL, Law RJ, Dawson CE, Stubberfield EJ, Monies RJ, Deaville R, Jepson PD (2011) Infection with *Brucella ceti* and high levels of polychlorinated biphenyls in bottlenose dolphins (*Tursiops truncatus*) stranded in south-west England. Vet Rec 169:14
Davison NJ, Perrett LL, Dawson C, Dagleish MP, Haskins G, Muchowski J, Whatmore AM (2017) *Brucella ceti* infection in a common Minke whale (*Balaenoptera acutorostrata*) with associated pathology. J Wildl Dis 53:572–576
Davydov NN (1967) Brucellosis of reindeer [in Russian], pp 79–85
Dawson CE, Perrett LL, Stubberfield EJ, Stack JA, Farrelly SSJ, Cooley WA, Davison NJ, Quinney S (2008) Isolation and characterization of Brucella from the lungworms of a harbor porpoise (Phocoena phocoena). J Wildl Dis 44:237–246
De Miguel MJ, Marín CM, Muñoz PM, Dieste L, Grilló MJ, Blasco JM (2011) Development of a selective culture medium for primary isolation of the main *Brucella* species. J Clin Microbiol 49:1458–1463
Dean AS, Crump L, Greter H, Hattendorf J, Schelling E, Zinsstag J (2012a) Clinical manifestations of human brucellosis: a systematic review and meta analysis. PLoS Negl Trop Dis 6(12):e1929
Dean AS, Crump L, Greter H, Schelling E, Zinsstag J (2012b) Global burden of human brucellosis: a systematic review of disease frequency. PLoS Negl Trop Dis 6(10):e1865
Dehn L-A, Sheffield GG, Follmann EH, Duffy LK, Thomas DL, O'Hara TM (2006) Feeding ecology of phocid seals and some walrus in the Alaskan and Canadian Arctic as determined by stomach contents and stable isotope analysis. Polar Biol 30:167–181
Demars A, Lison A, Machelart A, Van Vyve M, Potemberg G, Vanderwinden JM, De Bolle X, Letesson JJ, Muraille E (2019) Route of infection strongly impacts the host-pathogen relationship. Front Immunol 10:1–23
Denisov AA, Sclyarov OD, Salmakov KM, Shumilov KV (2013) The Russian experience in brucellosis veterinary public health. OIE Rev Sci Techniq 32:229–237
Dieterich RA (1981) Brucellosis. In: Dieterich RA (ed) Alaskan wildlife diseases Fairbanks, Alaska: Institute of Arctic biology, University of Alaska, pp 53–58
Dieterich RA, Morton JK (1987) Effects of live *Brucella abortus* strain 19 vaccine on reindeer later challenged exposed with *Brucella suis* type 4. Rangifer 7:33–36
Dieterich RA, Morton-Dieterich JK, Deyoe BL (1980) Observations on reindeer vaccinated with *Brucella melitensis* strain H-38 and challenged with *Brucella suis* type 4. pp 442–448
Dieterich RA, Devoe BL, Morton JK (1981) Effects of killed *Brucella abortus* strain 45/20 vaccine on reindeer later challenge exposed with *Brucella suis* type 4. Am J Vet Res 42:131–140
Dieterich RA, Morton JK, Deyoe BL (1987) Effects of killed *Brucella suis* biovar 4 vaccine on reindeer later challenge exposed with *Brucella suis* biovar 4. Internal report. Institute of Arctic Biology, University of Alaska, Fairbanks
Dieterich RA, Morton JK, Zanke RL (1991) Experimental *Brucella suis* Biovar 4 infection in a moose. J Wildl Dis 27:470–472

Duncan CG, Tiller R, Mathis D, Stoddard R, Kersh GJ, Dickerson B, Gelatt T (2014) *Brucella placentitis* and seroprevalence in northern fur seals (*Callorhinus ursinus*) of the Pribilof Islands, Alaska. J Vet Diagn Investig 26:507–512

Edmonds MD, Ward FM, O'Hara TM, Elzer PH (1999) Use of western immunoblot analysis for testing moose serum for *Brucella suis* biovar 4 specific antibodies. J Wildl Dis 35:591–595

Eisenberg T, Hamann HP, Kaim U, Schlez K, Seeger H, Schauerte N, Melzer F, Tomaso H, Scholz HC, Koylass MS, Whatmore AM, Zschock M, Zschöck M (2012) Isolation of potentially novel *Brucella* spp. from frogs. Appl Environ Microbiol 78:3753–3755

Eisenberg T, Riße K, Schauerte N, Geiger C, Blom J, Scholz HC (2017) Isolation of a novel 'atypical' *Brucella* strain from a bluespotted ribbontail ray (*Taeniura lymma*). Anton Leeuw Int J Gen Mol Microbiol 110:221–234

Eisenberg T, Schlez K, Fawzy A, Völker I, Hechinger S, Curić M, Schauerte N, Geiger C, Blom J, Scholz HC (2020) Expanding the host range: infection of a reptilian host (*Furcifer pardalis*) by an atypical *Brucella* strain. Antonie Van Leeuwenhoek 113:1531–1537

Elberg SS, Fitzhugh HA, King NB, Live I, Robertson DC, Schnurrenberger PR, Shook JC (1977) Brucellosis research: an evaluation. Whashington, National Academy of Sciences, Rpt.:140

El-Tras WF, Tayel AA, Eltholth MM, Guitian J (2010) Brucella infection in fresh water fish: evidence for natural infection of Nile catfish, *Clarias gariepinus*, with Brucella melitensis. Vet Microbiol 141:321–325

Esquible JA, Burek-Huntington K, Atkinson S, Klink AC, Bortz E, Goldstein TA, Beckmen K, Pabilonia K, Tiller R (2019) Pathological findings and survey for pathogens associated with reproductive failure in perinatal Steller Sea lions *Eumetopias jubatus*. Dis Aquat Org 137:131–144

Ewalt DR, Payeur JB, Martin BM, Cummins DR, Miller WG (1994) Characteristics of a *Brucella* species from a bottle-nosed-dolphin (*Tursiops-Truncatus*). J Vet Diagn Investig 6:448–452

Ferguson MAD (1997) Rangiferine brucellosis on Baffin Island. J Wildl Dis 33:536–543

Fischer D, Lorenz N, Heuser W, Kampfer P, Scholz HC, Lierz M (2012) Abscesses associated with a *Brucella inopinata*-like bacterium in a big-eyed tree frog (*Leptopelis vermiculatus*). J Zoo Wildl Med 43:625–628

Forbes LB (1991) Isolates of *Brucella suis* biovar 4 from animals and humans in Canada, 1982-1990. Can Vet J 32:686–688

Forbes LB, Pantekoek JF (1988) *Brucella canis* isolates from Canadian dogs. Can Vet J 29:149–14952

Forbes LB, Tessaro SV (1993) Transmission of brucellosis from reindeer to cattle. J Am Vet Med Assoc 203:289–294

Forbes LB, Tessaro SV (2003) Evaluation of cattle for experimental infection with and transmission of *Brucella suis* biovar 4. J Am Vet Med Assoc 222:1252–1256

Forbes LB, Nielsen O, Measures L, Ewalt DR (2000) Brucellosis in ringed seals and harp seals from Canada. J Wildl Dis 36:595–598

Foreyt WJ, Smith TC, Evermann JF, Heimer WE (1983) Hematologic, serum chemistry and serologic values of Dall's sheep (*Ovis dalli dalli*) in Alaska. J Wildl Dis 19:136–139

Foster RA (2017) Female reproductive system and mammae. In: Zachary JF (ed) Pathologic basis of veterinary disease, 6th edn. Elsevier, St. Louis, Missouri, pp 1147–1193

Foster G, Jahans KL, Reid RJ, Ross HM (1996) Isolation of *Brucella* species from cetaceans, seals and an otter. Vet Rec 138:583–586

Foster G, MacMillan AP, Godfroid J, Howie F, Ross HM, Cloeckaert A, Reid RJ, Brew S, Patterson IAP (2002) A review of *Brucella* sp. infection of sea mammals with particular emphasis on isolates from Scotland. Vet Microbiol 90:563–580

Foster G, Osterman BS, Godfroid J, Jacques I, Cloeckert A (2007) *Brucella ceti* sp. nov. and *Brucella pinnipedialis* sp. nov. for *Brucella* strains with cetaceans and seals as their preferred hosts. Int J Syst Evol Microbiol 57:2688–2693

Foster JT, Beckstrom-Sternberg SM, Pearson T, Beckstrom-Sternberg JS, Chain PSG, Roberto FF, Hnath J, Brettin T, Keim P (2009) Whole-genome-based phylogeny and divergence of the genus *Brucella*. J Bacteriol 191:2864–2870

Foster G, Whatmore AM, Koylass MS, Perrett L, Dagleish MP, Reid RJ (2011) Smooth and rough strains of *Brucella ceti* recovered from different tissues in a stranded harbour porpoise (*Phocoena phocoena*). Brucellosis 2011 International Research Conference, Buenos Aires

Foster G, Whatmore AM, Dagleish MP, Baily JL, Deaville R, Davison NJ, Koylass MS, Perrett LL, Stubberfield EJ, Reid RJ, Brownlow AC (2015) Isolation of *Brucella ceti* from a long-finned pilot whale (*Globicephala melas*) and a Sowerby's beaked whale (*Mesoploden bidens*). J Wildl Dis 51:868–871

Foster G, Nymo IH, Kovacs KM, Beckmen KB, Brownlow AC, Baily JL, Dagleish MP, Muchowski J, Perrett LL, Tryland M, Lydersen C, Godfroid J, McGovern B, Whatmore AM (2018) First isolation of *Brucella pinnipedialis* and detection of *Brucella* antibodies from bearded seals *Erignathus barbatus*. Dis Aquat Org 128:13–20

Franco MP, Mulder M, Gilman RH, Smits HL (2007) Human brucellosis. Lancet Infect Dis 7:775–786

Fretin D, Whatmore AM, Al Dahouk S, Neubauer H, Garin-Bastuji B, Albert D, Van Hessche M, Ménart M, Godfroid J, Walravens K, Wattiau P (2008) *Brucella suis* identification and biovar typing by real-time PCR. Vet Microbiol 131:376–385

Fuller JA, Garrot RA, White PJ, Aune KE, Roffe TJ, RHYAN JC (2007) Reproduction and survival of Yellowstone bison. J Wildl Manag 71:2365

Garofolo G, Fasanella A, Di Giannatale E, Platone I, Sacchini L, Persiani T, Boskani T, Rizzardi K, Wahab T (2016) Cases of human brucellosis in Sweden linked to Middle East and Africa. BMC Res Notes 9:1–5

Gates CC, Wobeser G, Forbes LB (1984) Rangiferine brucellosis in a muskox, *Ovibos moschatus moschatus* (Zimmermann). J Wildl Dis 20:233–234

Gee JE, De BK, Levett PN, Whitney AM, Novak RT, Popovic T (2004) Use of 16S rRNA gene sequencing for rapid confirmatory identification of *Brucella* isolates. Gene 42:3649–3654

Godfroid J (2002) Brucellosis in wildlife. Rev Sci Techniq l'Office Int Épizooties 21:277 286

Godfroid J, Beckmen K, Nymo IH (2016) Removal of lipid from serum increases coherence between brucellosis rapid agglutination test and enzyme-linked immunosorbent assay in bears in Alaska, USA. J Wildl Dis 52:912–915

Goldstein T, Zabka TS, Delong RL, Wheeler EA, Ylitalo G, Bargu S, Silver M, Leighfield T, Van Dolah F, Langlois G, Sidor I, Dunn JL, Gulland FM (2009) The role of domoic acid in abortion and premature parturition of California Sea lions (*Zalophus californianus*) on San Miguel Island, California. J Wildl Dis 45:91–108

Gonzalez L, Patterson IA, Reid RJ, Foster G, Barberan M, Blasco JM, Kennedy S, Howie FE, Godfroid J, MacMillan AP, Schock A, Buxton D (2002) Chronic meningoencephalitis associated with *Brucella* sp. infection in live-stranded striped dolphins (*Stenella coeruleoalba*). J Comp Pathol 126:147–152

González-Barrientos R, Morales JA, Hernández-Mora G, Barquero-Calvo E, Guzmán-Verri C, Chaves-Olarte E, Moreno E (2010) Pathology of striped dolphins (*Stenella coeruleoalba*) infected with *Brucella ceti*. J Comp Pathol 142:347–352

Gordienko LN, Novikov AN, Kulikova EV, Gaidutskaya GM, Elantseva NB, Guskova TV, Volkova TI (2015) Manifestation of infection in a novel focus of reindeer brucellosis [in Russian]. Achieve Sci Technol AIC 4:77–79

Gorham SL, Enright FM, Snider TG, Roberts ED (1986) Morphologic lesions in *Brucella abortus* infected ovine fetuses. Vet Pathol 23:331–332

Gorvel JP, Moreno E (2002) Brucella intracellular life: from invasion to intracellular replication. Vet Microbiol 90:281–297

Greenberg L, Blacke JD, Gorman FM (1958) An immunological study of the Canadian Eskimo. Can Med Assoc J 78:27–31

Gunn A, Leighton T, Wobeser G (1991) Wildlife diseases and parasites in the Kitikmeot region, 1984-90. File Report N°104. 60

Gutiérrez-Jiménez C, Hysenaj L, Alfaro-Alarcón A, Mora-Cartín R, Arce-Gorvel V, Moreno E, Gorvel JP, Barquero-Calvo E (2018) Persistence of *Brucella abortus* in the bone marrow of infected mice. J Immunol Res 2018:1–8

Guzman-Verri C, Gonzalez-Barrientos R, Hernandez-Mora G, Morales JA, Baquero-Calvo E, Chaves-Olarte E, Moreno E (2012) *Brucella ceti* and brucellosis in cetaceans. Front Cell Infect Microbiol 2:3

Gyuranecz M, Erdélyi K, Makrai L, Fodor L, Szépe B, Mészáros ÁR, Dán A, Dencso L, Fassang E, Szeredi L (2011) Brucellosis of the European Brown Hare (*Lepus europaeus*). J Comp Pathol 145:1–5

Harmon B, Adams L, Frey M (1988) Survival of rough and smooth strains of *Brucella abortus* in bovine mammary gland macrophages. Am J Vet Res 49:1092–1097

Hernandez-Mora G, Gonzalez-Barrientos R, Morales JA, Chaves-Olarte E, Guzman-Verri C, Barquero-Calvo E, De-Miguel MJ, Marin CM, Blasco JM, Moreno E (2008) Neurobrucellosis in stranded dolphins, Costa Rica. Emerg Infect Dis 14:1430–1433

Hernandez-Mora G, Manire CA, Gonzalez-Barrientos R, Barquero-Calvo E, Guzman-Verri C, Staggs L, Thompson R, Chaves-Olarte E, Moreno E (2009) Serological diagnosis of *Brucella* infections in odontocetes. Clin Vaccine Immunol 16:906–915

Higgins J, Stuber T, Quance C, Edwards WH, Tiller RV, Linfield T, Rhyan J, Berte A, Harris B (2012) Molecular epidemiology of *Brucella abortus* isolates from cattle, elk, and bison in the United States, 1998 to 2011. Appl Environ Microbiol 78:3674–3684

Hirvela-Koski V, Nylund M, Skrzypczak T, Heikkinen P, Kauhala K, Jay M, Isomursu M (2017) Isolation of *Brucella pinnipedialis* from Grey seals (*Halichoerus grypus*) in the Baltic Sea. J Wildl Dis 53:850–853

Hoberg EP, Galbreath KE, Cook JA, Kutz SJ, Polley L (2012) Northern host-parasite assemblages. History and biogeography on the borderlands of episodic climate and environmental transition. Adv Parasitol 79:1–97

Holst BS, Löfqvist K, Ernholm L, Eld K, Cedersmyg M, Hallgren G (2012) The first case of *Brucella canis* in Sweden: background, case report and recommendations from a northern European perspective. Acta Vet Scand 54:1–9

Honour S, Hickling KM (1993) Naturally occurring *Brucella suis* biovar 4 infection in a moose (*Alces alces*). J Wildl Dis 29:596–598

Huntley BE, Philip RN, Maynard JE (1963) Survey of brucellosis in Alaska. J Infect Dis 112: 100–104

Jauniaux TP, Brenez C, Fretin D, Godfroid J, Haelters J, Jacques T, Kerckhof F, Mast J, Sarlet M, Coignoul FL (2010) *Brucella ceti* infection in harbor porpoise (*Phocoena phocoena*). Emerg Infect Dis 16:1966–1968

Kanji JN, Saxinger L (2018) Brucella infection at cardiac pacemaker site in a patient who had consumed raw caribou meat in northern Canada. Can Med Assoc J 190:E1108–E1110

Kimura M, Imaoka K, Suzuki M, Kamiyama T, Yamada A (2008) Evaluation of a microplate agglutination test (MAT) for serological diagnosis of canine brucellosis. J Vet Med Sci 70: 707–709

Kittelberger R, Hilbink F, Hasen MF, Ross GP, Joyce MA, Fenwick S, Heesemann J, Wolf-Watz H, Nielsen K (1995) Serological crossreactivity between *Brucella abortus* and *Yersinia enterocolitica* 0:9 II the use of *Yersinia* outer proteins for the specific detection of *Yersinia enterocolitica* infections in ruminants. Vet Microbiol 47:271–280

Lalsiamthara J, Lee JH (2017) Development and trial of vaccines against *Brucella*. J Vet Sci 18: 281–290

Lambert S, Gilot-Fromont E, Toïgo C, Marchand P, Petit E, Garin-Bastuji B, Gauthier D, Gaillard JM, Rossi S, Thébault A (2020) An individual-based model to assess the spatial and individual heterogeneity of *Brucella melitensis* transmission in Alpine ibex. Ecol Model 425:109009

Lambourn DM, Garner M, Ewalt D, Raverty S, Sidor I, Jeffries SJ, Rhyan J, Gaydos JK (2013) *Brucella pinnipedialis* infections in Pacific harbor seals (*Phoca vitulina richardsi*) from Washington State, USA. J Wildl Dis 49:802–815

Lantis M (1950) The reindeer industry in Alaska. Arctic 3:27–44

Larsen AK, Nymo IH, Boysen P, Tryland M, Godfroid J (2013a) Entry and elimination of marine mammal *Brucella* spp. by hooded seal (*Cystophora cristata*) alveolar macrophages in vitro. PLoS One 8:e70186

Larsen AK, Nymo IH, Briquemont B, Sorensen KK, Godfroid J (2013b) Entrance and survival of *Brucella pinnipedialis* hooded seal strain in human macrophages and epithelial cells. PLoS One 8:e84861

Larsen AK, Godfroid J, Nymo IH (2016) *Brucella pinnipedialis* in hooded seal (*Cystophora cristata*) primary epithelial cells. Acta Vet Scand 58:9

Larsen AK, Nymo IH, Sorensen KK, Seppola M, Rodven R, Jimenez de Bagues MP, Al Dahouk S, Godfroid J (2018) Concomitant temperature stress and immune activation may increase mortality despite efficient clearance of an intracellular bacterial infection in Atlantic Cod. Front Microbiol 9:2963

Leyla G, Kadri G, Ümran O (2003) Comparison of polymerase chain reaction and bacteriological culture for the diagnosis of sheep brucellosis using aborted fetus samples. Vet Microbiol 93: 53–61

Lopez A, Hitos F, Perez A, Navarro-Fierro RR (1984) Lung lesions in bovine fetuses aborted by *Brucella abortus*. Can J Comp Med 48:275–277

López-Goñi I, García-Yoldi D, Marín CM, de Miguel MJ, Barquero-Calvo E, Guzmán-Verri C, Albert D, Garin-Bastuji B (2011) New Bruce-ladder multiplex PCR assay for the biovar typing of *Brucella suis* and the discrimination of *Brucella suis* and *Brucella canis*. Vet Microbiol 154: 152–155

Mackie JT, Blyde D, Harris L, Roe WD, Keyburn AL (2020) Brucellosis associated with stillbirth in a bottlenose dolphin in Australia. Aust Vet J 98:92–95

Madkour MM (2001) Historical aspects of brucellosis. Madkour's Brucellosis Springer, Berlin, pp 15–20

Maichak EJ, Scurlock BM, Rogerson JD, Meadows LL, Barbknecht AE, Edwards WH, Cross PC (2009) Effects of management, behavior, and scavenging on risk of brucellosis transmission in elk of western Wyoming. J Wildl Dis 45:398–410

Maji S, Manjunath N, Bahubali VH, Shome R, Palaniappan M, Lahiri S, Ravikumar R, Parthasarathi S, Chandrashekar N (2020) Neurobrucellosis: A neglected entity? An update from tertiary care Neurocentre of South East Asia. J Neurol Sci 411:116683

Mantur B, Amarnath S, Shinde R (2007) Review of clinical and laboratory features of human brucellosis. Indian J Med Microbiol 25:188–202

Matas M, Corrigan C (1953) Brucellosis in an Eskimo boy. Can Med Assoc J 69:531

McDermott J, Grace D, Zinsstag J (2013) Economics of brucellosis impact and control in low-income countries. OIE Rev Sci Techniq 32:249–261

McDonald WL, Jamaludin R, Mackereth G, Hansen M, Humphrey S, Short P, Taylor T, Swingler J, Dawson CE, Whatmore AM, Stubberfield E, Perrett LL, Simmons G (2006) Characterization of a Brucella sp. strain as a marine-mammal type despite isolation from a patient with spinal osteomyelitis in New Zealand. J Clin Microbiol 44:4363–4370

Meador VP, Hagemoser WA, Deyoe BL (1988) Histopathologic findings in *Brucella abortus*-infected, pregnant goats. Am J Vet Res 49:274–280

Meegan J, Field C, Sidor I, Romano T, Casinghino S, Smith CR, Kashinsky L, Fair PA, Bossart G, Wells R, Dunn JL (2010) Development, validation, and utilization of a competitive enzyme-linked immunosorbent assay for the detection of antibodies against *Brucella* species in marine mammals. J Vet Diagn Investig 22:856–862

Meegan J, Dunn JL, Venn-Watson SK, Smith CR, Sidor I, Jensen ED, Van Bonn WG, Pugh R, Ficht T, Adams LG, Nielsen K, Romano TA (2012) Serologic response in bottlenose dolphins *Tursiops truncatus* infected with *Brucella* sp using a dolphin-specific indirect ELISA. Dis Aquat Org 102:73–85

Megid J, Antonio Mathias L, A. Robles C. (2010) Clinical manifestations of brucellosis in domestic animals and humans. Open Vet Sci J 4:119–126

Meltzer E, Sidi Y, Smolen G, Banai M, Bardenstein S, Schwartz E (2010) Sexually transmitted brucellosis in humans. Clin Infect Dis 51:e12–e15

Meyer ME (1964) Species identity and epidemiology of *Brucella* strains isolated from Alaskan Eskimos. J Infect Dis 114:169–173

Meyer ME (1966) Identification and virulence studies of *Brucella* strains isolated from Eskimos and reindeer in Alaska, Canada, and Russia. Am J Vet Res 27:353–358

Miller LG, Neiland KA (1980) Experimental infections by *Brucella suis* type 4 in Alaskan rodents. J Wildl Dis 16:457–464

Miller WG, Adams LG, Ficht TA, Cheville NF, Payeur JP, Harley DR, House C, Ridgway SH (1999) *Brucella*-induced abortions and infection in bottlenose dolphins (*Tursiops truncatus*). J Zoo Wildl Med 30:100–110

Miller MA, Burgess TL, Dodd EM, Rhyan JC, Jang SS, Byrne BA, Gulland FM, Murray MJ, Toy-Choutka S, Conrad PA, Field CL, Sidor IF, Smith WA (2017) Isolation and characterization of a novel marine *Brucella* from a Southern Sea otter (*Enhydra Lutris Nereis*), California, USA. J Wildl Dis 53:215–227

Minas A, Stournara A, Minas M, Papaioannou A, Krikelis V, Tselepidis S (2005) Validation of fluorescence polarization assay (FPA) and comparison with other tests used for diagnosis of *B. melitensis* infection in sheep. Vet Microbiol 111:211–221

Moreno E (2014) Retrospective and prospective perspectives on zoonotic brucellosis. Front Microbiol 5:1–18

Morton JK (1986) Role of predators in reindeer brucellosis in Alaska. Rangifer 6:368

Morton JK (1989) Thesis (Ph.D.): *Brucella suis* type 4 in foxes and their role as reservoirs/vectors among reindeer. University of Alaska, Faribanks

Mühldorfer K, Wibbelt G, Szentiks CA, Fischer D, Scholz HC, Zschöck M, Eisenberg T (2017) The role of 'atypical' *Brucella* in amphibians: are we facing novel emerging pathogens? J Appl Microbiol 122:40–53

Neiland KA (1970) Rangiferine Alaskan brucellosis canids. J Wildl Dis 6:136–139

Neiland KA (1975) Further observations on Rangiferine brucellosis in Alaskan carnivores. J Wildl Dis 11:45–53

Neiland KA, Miller LG (1981) Experimental *Brucella suis* type 4 infections in domestic and wild Alaskan carnivores. J Wildl Dis 17:183–189

Neiland KA, King JA, Huntley BE, Skoog RO (1968) The diseases and parasites of Alaskan wildlife populations, Part I Some observations on brucellosis in caribou. Bull Wildl Dis Assoc 4: 27–36

Neiland KA (1972) Caribou disease studies. Final Report Federal Aid in Wildlife Restoration Projects W-17-2 and W-17-3, Job 18.lR. Alska Department of Fish and Game

Neimanis AS, Koopman HN, Westgate AJ, Nielsen K, Leighton FA (2008) Evidence of exposure to *Brucella* sp. in harbor porpoises (*Phocoena phocoena*) from the Bay of Fundy, Canada. J Wildl Dis 44:480–485

Nielsen K (2002) Diagnosis of brucellosis by serology. Vet Microbiol 90:447–459

Nielsen O, Stewart REA, Nielsen K, Measures L, Duignan P (2001) Serologic survey of *Brucella* spp. antibodies in some marine mammals of North America. J Wildl Dis 37:89–100

Nilssen AC (2006) Pupal biology and metamorphosis behaviour. In: Colwell DD, Hall MJR, Scholl PJ (eds) The Oestrid flies. Biology, host–parasite relationships, impact and management. CABI, Cambridge, pp 124–139

Noi P, Olsen SC, Rhyan JC (2009) Experimental infection of Richardson's ground squirrels (*Spermophilus richardsonii*) with attenuated and virulent strains of *Brucella abortus*. J Wildl Dis 45:189–195

Nuttall M, Berkes F, Forbes B, Kofinas G, Vlassova T, Wenzel G (2004) Hunting, herding, fishing and gathering: indigenous peoples and renewable resource use in the Arctic. ACIA, impacts of a warming Arctic: Arctic climate impact assessment. Cambridge University Press, Cambridge, pp 649–690

Nymo IH, Tryland M, Godfroid J (2011) A review of *Brucella* infection in marine mammals, with special emphasis on *Brucella pinnipedialis* in the hooded seal (*Cystophora cristata*). Vet Res 42:93

Nymo IH, Godfroid J, Åsbakk K, Larsen AK, das Neves CG, Rødven R, Tryland M. (2013a) A protein A/G indirect enzyme-linked immunosorbent assay for the detection of anti-*Brucella* antibodies in Arctic wildlife. J Vet Diagn Investig 25:369–375

Nymo IH, Tryland M, Frie AK, Haug T, Foster G, Rodven R, Godfroid J (2013b) Age-dependent prevalence of anti-*Brucella* antibodies in hooded seals *Cystophora cristata*. Dis Aquat Organ 106:187–196

Nymo IH, das Neves CG, Tryland M, Bardsen BJ, Santos RL, Turchetti AP, Janczak AM, Djonne B, Lie E, Berg V, Godfroid J (2014) *Brucella pinnipedialis* hooded seal (*Cystophora cristata*) strain in the mouse model with concurrent exposure to PCB 153. Comp Immunol Microbiol Infect Dis 37:195–204

Nymo IH, Arias MA, Pardo J, Alvarez MP, Alcaraz A, Godfroid J, Jimenez de Bagues MP (2016a) Marine mammal *Brucella* reference strains are attenuated in a BALB/c mouse model. PLoS One 11:e0150432

Nymo IH, Beckmen K, Godfroid J (2016b) Anti-*Brucella* antibodies in moose (*Alces alces gigas*), muskoxen (*Ovibos moschatus*), and plains bison (*Bison bison bison*) in Alaska, USA. J Wildl Dis 52:96–99

Nymo IH, Seppola M, Al Dahouk S, Bakkemo KR, Jimenez de Bagues MP, Godfroid J, Larsen AK (2016c) Experimental challenge of Atlantic cod (*Gadus morhua*) with a *Brucella pinnipedialis* strain from hooded seal (*Cystophora cristata*). PLoS One 11:e0159272

Nymo IH, Rodven R, Beckmen K, Larsen AK, Tryland M, Quakenbush L, Godfroid J (2018) *Brucella* antibodies in Alaskan true seals and eared seals-two different stories. Front Vet Sci 5:8

O'Hara TM, Holcomb D, Elzer P, Estepp J, Perry Q, Hagius S, Kirk C (2010) *Brucella* species survey in polar bears (*Ursus maritimus*) of northern Alaska. J Wildl Dis 46:687–694

Ohishi K, Zenitani R, Bando T, Goto Y, Uchida K, Maruyama T, Yamamoto S, Miyazaki N, Fujise Y (2003) Pathological and serological evidence of *Brucella*-infection in baleen whales (*Mysticeti*) in the western North Pacific. Comp Immunol Microbiol Infect Dis 26:125–136

Ohishi K, Takishita K, Kawato M, Zenitani R, Bando T, Fujise Y, Goto Y, Yamamoto S, Maruyama T (2004) Molecular evidence of new variant *Brucella* in North Pacific common minke whales. Microbes Infect 6:1199–1204

Ohishi K, Abe E, Amano M, Miyazaki N, Boltunov A, Katsumata E, Maruyama T (2018) Detection of serum antibodies to *Brucella* in Russian aquatic mammals. J Vet Med Sci 80:1696–1701

OIE (2013) [Internet]. Available from: https://www.oie.int/wahis_2/public/wahid.php/Diseaseinformation/Diseasedistributionmap

OIE (2019) *Brucellosis* (*Brucella abortus*, *B. melitensis* and *B. suis*) (infection with *B. abortus*, *B. melitensis* and *B. suis*). Manual of diagnostic tests and vaccines for terrestrial animals 2019, pp 355–395

Oliveira JB, Morales JA, Gonzalez-Barrientos RC, Hernandez-Gamboa J, Hernandez-Mora G (2011) Parasites of cetaceans stranded on the Pacific coast of Costa Rica. Vet Parasitol 182: 319–328

Pappas G, Akritidis N, Bosilkoki M, Tsianos E (2005) Brucellosis. N Engl J Med 352:2325–2336

Pappas G, Papadimitriou P, Akritidis N, Christou L, Tsianos EV (2006) The new global map of human brucellosis. Lancet Infect Dis 6:91–99

Parnas J (1964) A new species of Brucella: *B. rangiferi*. Nature 202:1242

Parnas J (1966) The taxonomic properties of Brucellae isolated from reindeer (*Brucella rangiferi* Pinigin-Meyer). Zentralbl Veterinarmed B 13:690–697

Paulsen IT, Seshadri R, Nelson KE, Eisen JA, Heidelberg JF, Read TD, Dodson RJ, Umayam L, Brinkac LM, Beanan MJ, Daugherty SC, Deboy RT, Durkin AS, Kolonay JF, Madupu R, Nelson WC, Ayodeji B, Kraul M, Shetty J, Malek J, Van Aken SE, Riedmuller S, Tettelin H, Gill SR, White O, Salzberg SL, Hoover DL, Lindler LE, Halling SM, Boyle SM, Fraser CM

(2002) The *Brucella suis* genome reveals fundamental similarities between animal and plant pathogens and symbionts. Proc Natl Acad Sci U S A 99:13148–13153

Pei J, Kahl-McDonagh M, Ficht TA (2014) *Brucella* dissociation is essential for macrophage egress and bacterial dissemination. Front Cell Infect Microbiol 5:1–9

Pinigin AF (1959) Brucellosis infection in eastern Siberia and Far East. Brucellosis of livestock and swine fever [in Russian], pp 65–68

Pinigin A, Zabrodin VA (1970) Natural foci of brucellosis. Vest Selskok Nauki Mosk 7:96–99

Pinigin AF, Zabrodin VA, Nikulina VI (1970) Brucellosis in the arctic fox. Krol Zver 5:39–40

Pintore MD, Mignone W, Di Guardo G, Mazzariol S, Ballardini M, Florio CL, Goria M, Romano A, Caracappa S, Giorda F, Serracca L, Pautasso A, Tittarelli C, Petrella A, Lucifora G, Di Nocera F, Uberti BD, Corona C, Casalone C, Iulini B (2018) Neuropathologic findings in cetaceans stranded in Italy (2002-14). J Wildl Dis 54:295–303

Pizarro-Cerdá J, Moreno E, Gorvel JP (2000) Invasion and intracellular trafficking of *Brucella abortus* in nonphagocytic cells. Microbes Infect 2:829–835

Poester P, Nielsen K, Samartino LE, Yu WL (2010) Diagnosis of brucellosis. Open Vet Sci J 4: 46–60

Poester FP, Samartino LE, Santos RI (2013) Pathogenesis and pathobiology of brucellosis in livestock. OIE Rev Sci Techniq 32:105–115

Qureshi T, Stittmatter J, Turner K, Davis DS (1999) Experimental infection of white-tailed deer with rangiferine brucellosis. J Wildl Dis 35:388–391

Radostits OM, Gay CC, Blood DC, Hinchcliff KW (2000) Diseases caused by *Brucella* spp. In: Radostits OM, Gay CC, Blood DC, Hinchcliff KW (eds) Veterinary medicine. A textbook of the diseases of cattle, sheep, pigs, goats and horses. Elsevier, Edinburgh, pp 867–882

Rausch RL, Huntley BE (1978) Brucellosis in reindeer, *Rangifer tarandus* L., inoculated experimentally with *Brucella suis*, type 4. Can J Microbiol 24:129–135

Raverty SA, Gaydos JK, Nielsen K, Nielsen O, Ross PS, Lambourn DM, Jeffries SJ (2002) An overview of marine mammal diagnoses in the Pacific northwest from 1999 to 2004

Rementsova MM (1987) Brucellosis in wild animals. Oxonian Press, New Delhi

Rhyan J, Garner M, Spraker T, Lambourn D, Cheville N (2018) *Brucella pinnipedialis* in lungworms *Parafilaroides* sp. and Pacific harbor seals *Phoca vitulina richardsi*: proposed pathogenesis. Dis Aquat Org 131: 87–94

Ross HM, Foster G, Reid RJ, Jahans KL, MacMillan AP (1994) *Brucella* species infection in sea-mammals. Vet Rec 134:359

Rossetti CA, Drake KL, Siddavatam P, Lawhon SD, Nunes JES, Gull T, Khare S, Everts RE, Lewin HA, Adams LG (2013) Systems biology analysis of *Brucella* infected Peyer's patch reveals rapid invasion with modest transient perturbations of the host transcriptome. PLoS One 8

Roth F, Zinsstag J, Orkhon D, Chimed-Ochir G, Hutton G, Cosivi O, Carrin G, Otte J (2003) Human health benefits from livestock vaccination for brucellosis: case study. Bull World Health Organ 81:867–876

Saegerman C, Berkvens D, Godfroid J, Walravens K (2010) Bovine brucellosis. In: Lefèvre P-C, Blancou J, Chermette R, Uilenberg G (eds) Infectious and parasitic diseases of livestock. Lavoisier, Paris, p 2080

Salcedo SP, Marchesini MI, Lelouard H, Fugier E, Jolly G, Balor S, Muller A, Lapaque N, Demaria O, Alexopoulou L, Comerci DJ, Ugalde RA, Pierre P, Gorvel JP (2008) Brucella control of dendritic cell maturation is dependent on the TIR-containing protein Btp1. PLoS Pathog 4:e21

Salem SF, Mohsen A (1997) Brucellosis in fish. Vet Med (Praha) 42:5–7

Sauvé CC, Hernández-Ortiz A, Jenkins E, Mavrot F, Schneider A, Kutz S, Saliki JT, Daoust P-Y (2020) Exposure of the Gulf of St. Lawrence grey seal *Halichoerus grypus* population to potentially zoonotic infectious agents. Dis Aquat Organ 142:105–118

Scholz HC, Hubalek Z, Nesvadbova J, Tomaso H, Vergnaud G, Le Flèche P, Whatmore AM, Dahouk S Al, Krüger M, Lodri C, Pfeffer M. (2008a) Isolation of *Brucella* microti from soil. Emerg Infect Dis 14:1316–1317

Scholz HC, Hubalek Z, Sedláček I, Vergnaud G, Tomaso H, Al Dahouk S, Melzer F, Kämpfer P, Heubauer H, Cloeckaert A, Maquart M, Zygmunt MS, Whatmore AM, Falsen E, Bahn P, Göllner C, Pfeffer M, Huber B, Busse HJ, Nöckler K (2008b) *Brucella microti* sp. nov., isolated from the common vole *Microtus arvalis*. Int J Syst Evol Microbiol 58:375–382

Scholz HC, Nöckler K, Llner CG, Bahn P, Vergnaud G, Tomaso H, Al Dahouk S, Kämpfer P, Cloeckaert A, Maquart M, Zygmunt MS, Whatmore AM, Pfeffer M, Huber B, Busse HJ, De BK (2010) *Brucella inopinata* sp. nov., isolated from a breast implant infection. Int J Syst Evol Microbiol 60:801–808

Scholz HC, Mühldorfer K, Shilton C, Benedict S, Whatmore AM, Blom J, Eisenberg T (2016a) The change of a medically important genus: worldwide occurrence of genetically diverse novel brucella species in exotic frogs. PLoS One 11:1–11

Scholz HC, Revilla-Fernández S, Dahouk SA, Hammerl JA, Zygmunt MS, Cloeckaert A, Koylass M, Whatmore AM, Blom J, Vergnaud G, Witte A, Aistleitner K, Hofer E (2016b) *Brucella vulpis* sp. Nov., isolated from mandibular lymph nodes of red foxes (*Vulpes vulpes*). Int J Syst Evol Microbiol 66:2090–2098

Schurig GG, Sriranganathan N, Corbel MJ (2002) Brucellosis vaccines: past, present and future. Vet Microbiol 90:479–496

Scotter GW (1972) Reindeer Ranching in Canada. J Range Manag 25:167

Scotter SE, Tryland M, Nymo IH, Hanssen L, Harju M, Lydersen C, Kovacs K, Klein J, Fisk AT, Routti H (2019) Contaminants in Atlantic walruses in Svalbard part 1: relationships between exposure, diet and pathogen prevalence. Environ Pollut 244:9–18

Seleem MN, Boyle SM, Sriranganathan N (2008) *Brucella*: a pathogen without classic virulence genes. Vet Microbiol 129:1–14

Seleem MN, Boyle SM, Sriranganathan N (2010) Brucellosis: a re-emerging zoonosis. Vet Microbiol 140:392–398

Sidor I, Goldstein T, Whatmore AM, Zabka TS, Frasca S, Gulland FMD, Dunn JL (2008) Transplacental infection with *Brucella* in California Sea lion (*Zalophus californianus*) fetus. Brucellosis 2008 International Research Conference, Chicago

Simmons GC, Hall WTK (1953) Epididymitis of rams. Preliminary studies on the occurrence and pathogenicity of a brucella-like organism. Aust Vet J 29:33–40

Sleptsov ES, Evgrafov GG, Vinokurov NV, Laishev KA, Fedorov VI, Iskandarov MI, Zakharova OI (2017) Brucellosis of reindeer and measures to control it in the extreme north of the Russian Federation [in Russian], Novosibirsk

Sohn AH, Probert WS, Glaser CA, Gupta N, Bollen AW, Wong JD, Grace EM, McDonald WC (2003) Human neurobrucellosis with intracerebral granuloma caused by a marine mammal *Brucella* spp. Emerg Infect Dis 9:485–488

Sonne C, Andersen-Ranberg E, Rajala EL, Agerholm JS, Bonefeld-Jorgensen E, Desforges JP, Eulaers I, Jenssen BM, Koch A, Rosing-Asvid A, Siebert U, Tryland M, Mulvad G, Harkonen T, Acquarone M, Nordoy ES, Dietz R, Magnusson U (2018) Seroprevalence for *Brucella* spp. in Baltic ringed seals (*Phoca hispida*) and East Greenland harp (*Pagophilus groenlandicus*) and hooded (*Cystophora cristata*) seals. Vet Immunol Immunopathol 198: 14–18

State of Alaska Section of Epidemiology (2011) Human and animal brucellosis in Alaska. Bulletin No 31

Stoenner HG, Lackman DB (1957) A new species of *Brucella* isolated from the desert wood rat, *Neotoma lepida* Thomas. Am J Vet Res 18:947–951

Suárez-esquivel M, Ruiz-villalobos N, Jiménez-rojas C, Barquero-calvo E, Chacón-díaz C, Víquez-ruiz E, Rojas-campos N, Baker KS, Oviedo-sánchez G, Amuy E, Chaves-olarte E, Thomson NR, Moreno E, Guzmán-verri C (2017) *Brucella neotomae* infection in humans, Costa Rica. Emerg Infect Dis 23:997–1000

Suárez-Esquivel M, Ruiz-Villalobos N, Hernández-Mora G, González-Barrientos R, David Palacios-Alfaro J, Barquero-Calvo E, Chaves-Olarte E, Thomson N, Moreno E, Guzman-

Verri C (2019) *Brucella* sequence type 27 isolated from dwarf sperm whale (*Kogia sima*) stranded in the costa Rican Pacific coast. Access Microbiol 1:1A

Tessaro SV (1986) The existing and potential importance of brucellosis and tuberculosis in Canadian wildlife: a review. Can Vet J 27:119–124

Tessaro SV, Forbes LB (1986) Brucella suis biotype 4: a case of granulomatous nephritis in a barren ground caribou (*Rangifer tarandus groenlandicus* L.) with a review of the distribution of rangiferine brucellosis in Canada. J Wildl Dis 22:479–483

Tessaro SV, Rowell EJ, Cawthorn RPL (1984) Banks Island muskox harvest, proceedings of the first international muskox symposium. Biol Pap Rep No 4. University of Alaska, pp 177–180

Tessaro SV, Forbes LB, Turcotte C (1990) A survey of brucellosis and tuberculosis in bison in and around Wood Buffalo National Park, Canada. Can Vet J 31:174–180

Thorpe BD, Sidwell RW, Lundgren DL (1967) Experimental studies with four species of Brucella in selected wildlife, laboratory, and domestic animals. Am J Trop Med Hyg 16:665–674

Tiller RV, Gee JE, Frace MA, Taylor TK, Setubal JC, Hoffmaster AR, De BK (2010a) Characterization of novel *Brucella* strains originating from wild native rodent species in North Queensland, Australia. Appl Environ Microbiol 76:5837–5845

Tiller RV, Gee JE, Lonsway DR, Gribble S, Bell SC, Jennison AV, Bates J, Coulter C, Hoffmaster AR, De BK (2010b) Identification of an unusual *Brucella* strain (BO2) from a lung biopsy in a 52 year-old patient with chronic destructive pneumonia. BMC Microbiol 10:23. https://doi.org/10.1186/1471-2180-10-23

Tomaselli M, Dalton C, Duignan PJ, Kutz S, van der Meer F, Kafle P, Surujballi O, Turcotte C, Checkley S (2016) Contagious ecthyma, rangiferine brucellosis, and lungworm infection in a muskox (*Ovibos moschatus*) from the Canadian Arctic, 2014. J Wildl Dis 52:719–724

Tomaselli M, Kutz S, Gerlach C, Checkley S (2018) Local knowledge to enhance wildlife population health surveillance: conserving muskoxen and caribou in the Canadian Arctic. Biol Conserv 217:337–348

Tomaselli M, Elkin B, Kutz S, Harms NJ, Nymo IH, Davison T, Leclerc LM, Branigan M, Dumond M, Tryland M, Checkley S (2019) A transdisciplinary approach to *Brucella* in muskoxen of the Western Canadian Arctic 1989–2016. EcoHealth 16:488–501

Toshach S (1963) Brucellosis in the Canadian Arctic. Can J Public Health 54:271–275

Traxler RM, Lehman MW, Bosserman EA, Guerra MA, Smith TL (2013) A literature review of laboratory-acquired brucellosis. J Clin Microbiol 51:3055–3062

Tryland M, Kleivane L, Alfredsson A, Kjeld M, Arnason A, Stuen S, Godfroid J (1999) Evidence of *Brucella* infection in marine mammals in the North Atlantic Ocean. Vet Rec 144:588–592

Tryland M, Derocher AE, Wiig Y, Godfroid J (2001) *Brucella* sp. antibodies in polar bears from Svalbard and the Barents Sea. J Wildl Dis 37:523–531

Turse JE, Pei J, Ficht TA (2011) Lipopolysaccharide-deficient *Brucella* variants arise spontaneously during infection. Front Microbiol 2:1–12

Ueno Y, Yanagisawa M, Kino S, Shigeno S, Osaki M, Takamatsu D, Katsuda K, Maruyama T, Ohishi K (2020) Molecular characterization of *Brucella ceti* from a bottlenose dolphin (*Tursiops truncatus*) with osteomyelitis in the western Pacific. J Vet Med Sci 82:754–758

Valdezate S, Cervera I, Hernandez P, Navarro A, Saéz Nieto JA (2007) Characterisation of human outbreaks of brucellosis and sporadic cases by the use of hyper-variable octameric oligonucleotide fingerprint (HOOF) variable number tandem repeats. Clin Microbiol Infect 13:887–892

Verteletsky LL (1959) Epizootic situation on brucellosis of cattle and small ruminants in areas of Ural, Siberia and Far East and organization of measures for prevention and liquidation of this disease. Brucellosis of livestock and swine fever [in Russian], Moscow, pp 9–17

Vinokurov N, Iskandarov MI, Laishev KA, Fedorov AI, Iskandarova SS (2019) The main reasons for the long-term brucellosis of reindeer in Yakutia [In Russian]. Hippol Vet 3:96–100

Vitry MA, Mambres DH, Deghelt M, Hack K, Machelart A, Lhomme F, Vanderwinden JM, Vermeersch M, De Trez C, Pérez-Morga D, Letesson JJ, Muraille E (2014) *Brucella melitensis* invades murine erythrocytes during infection. Infect Immun 82:3927–3938

Vizcaíno N, Cloeckaert A, Verger JM, Grayon M, Fernández-Lago L (2000) DNA polymorphism in the genus *Brucella*. Microbes Infect 2:1089–1100
von Bargen K, Gorvel JP, Salcedo SP (2012) Internal affairs: investigating the Brucella intracellular lifestyle. FEMS Microbiol Rev 36:533–562
Wanke MM (2004) Canine brucellosis. Anim Reprod Sci 82–83:195–207
West KL, Levine G, Jacob J, Jensen B, Sanchez S, Colegrove K, Rotstein D (2015) Coinfection and vertical transmission of *Brucella* and morbillivirus in a neonatal sperm whale (*Physeter macrocephalus*) in Hawaii, USA. J Wildl Dis 51:227–232
Whatmore AM, Dawson CE, Groussaud P, Koylass MS, King AC, Shankster SJ, Sohn AH, Probert WS, McDonald WL (2008) Marine mammal Brucella genotype associated with zoonotic infection. Emerg Infect Dis 14:517–518
Whatmore AM, Davison N, Cloeckaert A, Al Dahouk S, Zygmunt MS, Brew SD, Perrett LL, Koylass MS, Vergnaud G, Quance C, Scholz HC, Dick EJ, Hubbard G, Schlabritz-Loutsevitch NE (2014) *Brucella papionis* sp. nov., isolated from baboons (*Papio* spp.). Int J Syst Evol Microbiol 64:4120–4128
Whatmore AM, Koylass MS, Muchowski J, Edwards-Smallbone J, Gopaul KK, Perrett LL (2016) Extended multilocus sequence analysis to describe the global population structure of the genus *Brucella*: phylogeography and relationship to biovars. Front Microbiol 7:1–14
Whatmore AM, Dawson C, Muchowski J, Perrett LL, Stubberfield E, Koylass M, Foster G, Davison NJ, Quance C, Sidor IF, Field CL, St Leger J (2017) Characterisation of North American *Brucella isolates* from marine mammals. PLoS One 12:e0184758
Wilesmith JW (1978) The persistence of *Brucella abortus* infection in calves: a retrospective study of heavily infected herds. Vet Rec 103:149–153
Wright AE, Dub MD, Smith F (1897) On the application of the serum test to the differential diagnosis of typhoid fever and Malta fever. Lancet 149:656–659
Wu Q, McFee WE, Goldstein T, Tiller RV, Schwacke L (2014) Real-time PCR assays for detection of *Brucella* spp. and the identification of genotype ST27 in bottlenose dolphins (*Tursiops truncatus*). J Microbiol Methods 100:99–104
Xavier MN, Paixão TA, Poester FP, Lage AP, Santos RL (2009) Pathological, Immunohistochemi cal and bacteriological study of tissues and Milk of cows and fetuses experimentally infected with *Brucella abortus*. J Comp Pathol 140:149–157
Yu WL, Nielsen K (2010) Review of detection of *Brucella* spp. by polymerase chain reaction. Croat Med J 51:306–313
Zarnke RL (2000) Alaska wildlife serologic survey, 1975-2000. Alaska Department of Fish. Federal aid in wildlife restoration. Research final report. Grants W-24-5 and W-27-1 through W-27-4. Study 18.71
Zarnke RL, Ver Hoef JM, DeLong RA (2006) Geographic pattern of serum antibody prevalence for *Brucella* spp. in caribou, grizzly bears, and wolves from Alaska, 1975-1998. J Wildl Dis 42: 570–577
Zheludkov MM, Tsirelson LE (2010) Reservoirs of *Brucella* infection in nature. Biol Bull 37:709–715

Anthrax in the North

Karsten Hueffer, Svetlana Dresvyanikova, and Irina Egorova

1 Introduction

Anthrax is caused by the Gram-positive aerobic bacterial pathogen *Bacillus anthracis*, which causes a systemic infection in animals, including people. It is one of the oldest diseases of livestock that poses a significant threat to people. The two fathers of modern microbiology, Louis Pasteur (Pasteur 1881) and Robert Koch (Koch 1876), focused much of their work on this organism. Ruminants such as reindeer and cattle are extremely susceptible, while horses, pigs, and humans are moderately susceptible to the bacterium. Carnivores are relatively resistant (Quinn et al. 2011). It has been one of the pathogens of interest for bioweapon development in the twentieth century and in 2001 gained infamy through the anthrax-laced letters that were sent to politicians and media outlets in the USA after the terror attacks of 9/11. This act of bioterrorism caused a renewed interest in both the scientific community and public health and biodefense agencies in anthrax research, prevention, and the distribution of the bacterium in the environment.

Like other members of the genus *Bacillus*, the causative agent of anthrax is an aerobic bacterium that requires oxygen in its environment to thrive and divide (Quinn et al. 2011; WHO 2008). Such an oxygen-rich, or aerobic, environment is provided by an infected host through breathing and oxygen transport via blood vessels throughout the body. After an animal is infected, the bacterium spreads through the body and produces several toxins. These toxins lead to the disruption of endothelial cells in the blood vessels' linings and result in leakage of fluid, including blood from small vessels. The toxins that cause the major clinical signs are also the

K. Hueffer (✉)
Department of Veterinary Medicine, University of Alaska Fairbanks, Fairbanks, AK, USA
e-mail: khueffer@alaska.edu

S. Dresvyanikova · I. Egorova
Federal Research Center for Virology and Microbiology, CVM Pokrov, Vladimir, Russia

M. Tryland (ed.), *Arctic One Health*, https://doi.org/10.1007/978-3-030-87853-5_11

target of effective vaccines that have been used successfully to protect animals at risk. The toxin-induced fluid loss throughout the body ultimately leads to hemorrhaging, shock, and ultimately death of infected animal in many cases (Quinn et al. 2011; WHO 2008).

The widespread leakage of fluid, including blood, leads to widespread bleeding and frothing from body openings. This bleeding immediately prior to death can lead to the spread of the bacterium in the environment around a sick, dying, or dead animal infected with anthrax.

When an animal dies and oxygen concentrations in the body drop, many biological processes at the cell level continue after death and utilize the remaining oxygen in the carcass. *Bacillus anthracis* does not thrive in this oxygen-deprived, or anaerobic, environment. In order to survive, it forms endospores that are metabolically inactive, and therefore, they do not need oxygen to survive. These spores can survive in the soil for decades or longer, depending on the conditions. Especially permafrost has been suggested as a potential substrate for long-term survival of spores. In addition, these spores are generally not susceptible to antibiotics and are hard to kill using many common disinfectants. When these spores are ingested, inhaled, or otherwise introduced into a new host animal, they germinate and start dividing to start a new infection cycle in the newly infected animal. In this infection and transmission cycle, the spore phase is crucial in the long-term survival of the bacterium in a local associated with anthrax (Quinn et al. 2011; WHO 2008). However, insect transmission has also been described both experimentally and through epidemiological investigations (WHO 2008).

Disposal methods of carcasses are important in the long-term maintenance of the infectious agent and therefore are key in the control of anthrax in livestock. Burials of carcasses prevent immediate infection of new hosts, but especially shallow burial leads to long-term contamination of the ground and can serve as a focal point for new infections decades after the carcass was buried. Current standards of carcass disposal include burning of unopened carcasses to destroy spores and prevent the spread of the infectious agent. Other prophylactic measures include vaccination of animals with purified parts of the anthrax toxin or killed bacteria and avoidance of possibly contaminated areas (Quinn et al. 2011).

2 History of Anthrax in the North

Anthrax has been described in the Yamal regions since at least the nineteenth century (Nansen 2011). It has been widely described throughout Siberia since the 1920s and was likely present for long times before that time but was not described in the western scientific literature. In North America, it has had a more limited distribution, especially in the North. Reports of anthrax in Northern Canada (latitude >60°N) focus on wood bison. A search in the infectious disease reporting service Promed (http://www.promedmail.org) using "anthrax" and "Canada" as keywords results in 125 hits, the majority focusing on Northwest Territories wood bison outbreaks with some cases in Western provinces of Alberta, Saskatchewan, Manitoba and British

Columbia. The cases in the Northwest Territory occurred at its Southern border. Prior to 1960, however, the majority of cases in Canada occurred in Eastern Canada (Ontario and Quebec) where cases used to be associated with tanneries before public health measures stopped the spread through infected hides (Moynihan 1963).

Anthrax has not been reported in Alaska. Searches on Pubmed and Promed with the search terms "anthrax" and "Alaska" result in zero relevant hits. No records exist in the Office of the State Veterinarian on anthrax in Alaska (R. Gerlach, personal communication).

Anthrax activity in Scandinavia is also low with only one case reported in Norway in Promed, which was a human case associated with intravenous drug use. A search in Pubmed using "anthrax" and "Norway" or "Finland" as search terms does not result in any relevant articles describing anthrax outbreaks in Norway or Finland. Two small outbreaks on the same farm are reported in Promed from Finland. In Sweden, 13 Promed reports discussing 4 small outbreaks are listed. Sweden's National Veterinary Institute reported 15 animal deaths due to anthrax (3 moose, 9 cows, sheep, horse) in 2016 in Omberg, a region with a history of animal anthrax outbreaks (Bröjer et al. 2016). Historically anthrax was widespread in Sweden but after 1957 has been rare (Elvander et al. 2017; Lewerin et al. 2010). This recent event in Omberg demonstrates the capacity for anthrax spores to persist in the environment for years and to cause outbreaks widely spaced in time. A recent modeling study also identified anthrax cases as reported to the World Organization for Animal Health (OIE) and Promed mainly in southern areas of the circumpolar North with the exception of the outbreak on the Yamal Peninsula discussed in detail below (Walsh et al. 2018).

The first mention of anthrax in Russia in the literature dates back to the eighteenth century, and the disease was described in detail in Western Siberia. Since 1760 there is information about large anthrax outbreaks in tundra regions of the current Yamalsky district of the Yamalo-Nenets Autonomous District (Popova and Kulichenko 2017). The ecology and epidemiological situation of anthrax in the Russian Federation is characterized by sporadic cases among animals and people, and the Arctic region is one of the regions with sporadic outbreaks. A literal translation of the Russian name of the disease as "Siberian ulcer" indicates the historical territorial connection between the disease and Siberia. Not only in Siberia but also in the subarctic and arctic regions of Russia in general, anthrax is considered endemic.

This endemic status of anthrax in the Russian Arctic depends on the large number of soil foci (livestock burials, biothermal pits (beccari pits), and other burial sites for animals that previously died of anthrax) and "dead fields" that do not have clear focal boundaries of burial sites or places where the disease was previously reported. They occurred during outbreaks when reindeer herders, with a lack of means and methods to control it, left the dead and infected animals and when the remaining reindeer migrated away until the outbreak stopped. Therefore, a large number of dead unburied animals remained along the way of reindeer herders' migration paths, forming these so-called "dead fields." These regional areas present a high risk of infecting animals and humans to this day. In the territory of the Arkhangelsk region and the Komi Republic in Northwestern Russia, more than 100 of these dead fields have been described, more than 60 dead fields are known in Yamal, about 4 in Taymyr, and more than 200 in Yakutia.

During the late nineteenth to early twentieth centuries, mass anthrax outbreaks occurred in most northern areas of the Russian European North and Siberia. During the period from 1896 to 1917 in the Bolshozemelskaya and Malozemelskaya tundra, more than one million reindeer died during the anthrax outbreaks. Sometimes, reindeer herders and their families got infected and died at the same time as the reindeer (Popova and Kulichenko 2017; Egorov et al. 2002; Layshev and Zelensky 2003; Cherkasskiy 2002; Cherkasskiy 2003). The last major outbreak of the twentieth century was reported in 1941, involving 6700 reindeer in Antipoyutinskaya tundra and on the Tazovsky Peninsula of Maly (Cherkasskiy 2003).

As a result of vaccination in the twentieth century, the epizootic situation in this region has significantly improved. In 2016, according to the data of the Regional Veterinary Service in the Yamalo-Nenets Autonomous District, 39 affected foci ("dead fields") were registered on an area of over 2 million hectares (Ministry of Agriculture 1976; Department of Natural Resource Regulation 2015).

The last cases of anthrax in the Republic of Sakha (Yakutia) in livestock and wildlife were reported from 1986 to 1993. Horses, pigs, moose, reindeer, and bears were involved in the epizootic process. The most massive outbreak was observed in 1986, when 432 animals died of anthrax.

New outbreaks of anthrax after several decades of few infections are due to discontinued vaccination of reindeer herds. For example, in 2016, in the Yamalo-Nenets Autonomous District, after being free from anthrax for 75 years, an anthrax outbreak involving 2349 reindeer was reported (see also discussion below).

On the territory of the Taymyr Peninsula, the first outbreak of anthrax was officially reported in 1899, and the last case of anthrax in Taymyr occurred in 1977 at the Pelyatkinskoye field. The outbreak (86 animals died) occurred among domesticated reindeer that had not been vaccinated against anthrax. The route of summer grazing areas was through pastures, where the remains of reindeer carcasses that died of anthrax in 1931 were partially preserved. At present, according to official data, there are 39 registered places with anthrax outbreaks in the Taymyr Autonomous District. Active work on natural gas production and associated soil disturbances on the Taymyr Peninsula increases the risks of anthrax infection in both people and animals (Layshev and Zabrodin 2012). The largest population of wild reindeer in Eurasia in Taymyr, reaching 10,000 animals, is in a risk zone of anthrax. This population migrates northward in the spring and southward over 1500 km in autumn posing a significant threat of dispersal of the pathogen.

A common feature of anthrax outbreaks in livestock and wildlife in 1986 in the Republic of Sakha (Yakutia), on the population of reindeer in 1977 on the Taymyr Peninsula, and in 2016 in the Yamalo-Nenets Autonomous District was abnormally high summer temperatures, which caused the thawing of perennial cryolitozone to a depth of more than 40 cm and the involvement of dogs in the epizootic process that was confirmed by the isolation of *Bacillus anthracis*. Likely dogs got infected while eating meat from infected animals.

In the Kamchatka region and the Komi Republic, anthrax was reported once in 1974 and 1993, respectively (Dugarzhapova et al. 2017). On the territory of Khanty-Mansi Autonomous Okrug, the infection was detected twice—in 1925 and 1944. In the territory of the Arkhangelsk region, anthrax was reported before the Second World War and also in 1981 and 1984.

Over the past 20 years, the epidemiological situation for anthrax in the Russian Federation has been characterized as concerning; however, the morbidity rate decreased in recent years. In 2001–2010, 326 cases of animal anthrax and 104 human cases were reported. Compared to the previous decade, the morbidity rate of animals decreased by more than threefold (Federal Service 2008).

Between 2009 and 2018, anthrax infection of people was reported in six federal districts. Twenty-three human anthrax outbreaks were detected, resulting in 90 human cases, three of which were lethal. During an outbreak of anthrax in the Yamalo-Nenets Autonomous District in the year 2016, contact with infected and dead animals resulted in 36 human infections with one lethal case (Popova and Kulichenko 2017). Compared to the previous 10-year period (1999–2008), the number of cases decreased by 1.6-fold. Since the 2016 anthrax outbreak on the Yamal Peninsula, this disease has not been reported in the arctic regions, but the risk remains high (Figs. 1 and 2).

In summary, it is clear that distribution of anthrax in the North is largely limited to the Eurasian continent, and the association with northern permafrost-rich regions seems to be limited largely to Russia where anthrax is widespread, both historically and in contemporary times. In Fenno-Scandinavia and North America, anthrax has

Fig. 1 Reindeer that died during anthrax outbreak on the Yamal Peninsula. The bleeding of moribund animals contaminates the soil around carcasses that can initiate outbreaks many years later. Photo Yuri Selyaninov

Fig. 2 A group of reindeer and a herding dog that died of anthrax during the 2016 outbreak on the Yamal Peninsula. These images show some of the scale of an anthrax outbreak in herding communities. Photo Yuri Selyaninov

been described in more southern regions that are not associated with permafrost. This lack of reports of anthrax in many areas in the far north outside of Russia could be due to either lack of occurrence or lack of detection and description in the scientific literature. The generally acidic soil overlaying permafrost could limit the survival of spores (WHO 2008; Quinn et al. 2011; Walker et al. 2009). In many northern regions, systematic surveillance for the disease, including anthrax, is still limited due to high cost and logistical constrains such as low population density and a harsh environment (Hueffer et al. 2013).

3 Explaining the 2016 Yamal Outbreak

The anthrax outbreak on the Yamal Peninsula in 2016 after decades without reported anthrax cases in this northern region created significant interest in the scientific community and public media. In the popular press, the theory of anthrax released from thawing permafrost and infecting animals and people was most often discussed as a reason for the reemergence of this zoonotic disease. The climate change related to thawing permafrost releasing old and dangerous pathogens aligns well with a storyline that creates readership and fits with the concerns of climate change–

induced negative changes; however, alternative explanations have been put forward as well (Hueffer et al. 2020).

Permafrost is ground with a temperature below 0°C for at least 2 consecutive years (French 2013b). Above the permanently frozen permafrost, an active layer freezes and thaws annually. The upper layer of the permafrost is sometimes referred to as the transient layer, which thaws and freezes on a decadal or century timescale, while the "true" permafrost is frozen for millennia or even hundreds of thousands of years. Seasonal warming can reach to different depths, depending on local conditions, but most often occurs in the top 20 m of the soil. Regular freezing and thawing generally occurs only in the upper 1–2 m (French 2013b). Temperatures must be very high and heat conductivity must also be high in the ground for a single warm year to affect the true permafrost below the transient layer. The seasonal heat generally does not reach significantly below the normal level of the active layer. Even projected long-term changes in temperature regime are not predicted to alter the characterization of the soil underlaying the Yamal Peninsula as continuous permafrost until about 2100 (French 2013a).

In addition, anthrax should be a relatively new phenomenon if it was solely the result of thawing permafrost as a result of anthropogenic climate change because anthropogenic climate change is a relatively new environmental change that is largely confined to the second half of the twentieth century and beyond. However, descriptions regarding an anthrax-like disease exist from the nineteenth century and possibly even earlier. One example of such early mention of anthrax in the Yamal region comes from Nansen's report on his first voyage on the Fram. Nansen writes about a disease that was known to the indigenous reindeer herders to affect reindeer that travel in an area previously affected by an outbreak such as dead fields discussed above. Nansen suggests this disease to likely be anthrax (Nansen 2011). Anthrax is also called Yamal disease (Gainer 2016) or Siberian plague or ulcer (Arkhangelskaya 2016) in Russia, suggesting a long-standing connection between anthrax and Siberia and the Yamal region in particular that predates recent anthropogenic climate change.

As mentioned above, discontinuation of vaccination has been put forth as an alternative explanation (Arkhangelskaya 2016) together with an increasing number of reindeer (Bogdanov and Golovatin 2017). In 2007 routine vaccination of reindeer was discontinued on the Yamal Peninsula, which was considered free of anthrax at that time. This discontinuation of vaccination created a naïve population through natural population turnover and can aid the spread of an infectious agent.

Large increases in reindeer numbers on the Yamal Peninsula have been proposed as another alternative explanation for the Yamal outbreak. In many diseases, certain density of host animals is necessary to sustain long-term maintenance of an infectious agent over time to ensure sufficient numbers of susceptible animals are available as potential hosts (Anderson and May 1978). In addition to the density-dependent disease spread, overpopulation could contribute to increased disease through stress, resulting in increased susceptibility to infection (Gainer 2016). On the Yamal Peninsula, the number of reindeer has increased from about 75,000 in 1950 to over 250,000 in 2010 (Bogdanov and Golovatin 2017). High population densities of reindeer can also exacerbate outbreaks through increased grazing that causes soil disruption and reduces the thermal protection of permafrost and increases erosion, which in turn can lead to the release of spores from the soil (Bogdanov and Golovatin 2017).

4 Summary

The ecology of anthrax in the circumpolar North is complex, differs by specific region, and likely is not primarily driven by the presence of anthrax spores in permafrost but is limited to processes above the true permafrost layers. A recent study predicted an increase in anthrax suitability of Northern regions based on a warming climate (Walsh et al. 2018); however, for anthrax as well as for many infectious diseases, the consequences of a warming climate are complex and not well understood (Hueffer et al. 2011). Predictions of future dynamics of specific diseases in the far North in the face of a warming planet should be treated with caution, given the lack of data and solid understanding of disease dynamics in this rapidly changing environment. While extended periods of thawing in the active layer are more likely to make contaminated soil and carcasses available for longer and different times of the year, the role of true permafrost in lower layers is less clear. These changes in the active layer, together with increased reindeer numbers and discontinued vaccination, more likely contributed to the 2016 outbreak on the Yamal and the potential for other resurgence of this important zoonotic disease (Hueffer et al. 2020). Taken together the information presented in this chapter suggests that anthrax is of concern mainly in the Eurasian and southern part of the circumpolar North with the North American High North generally not affected by this disease. However, information might be limited based on sparse population and limited livestock industry outside of traditional reindeer herding areas in Fenno-Scandinavia and Northern Russia.

References

Anderson RM, May R (1978) Regulation and stability of host-parasite population interactions: I. Regulatory processes. J Anim Ecol 47:219–247. https://doi.org/10.2307/3933

Arkhangelskaya S (2016) The Siberian plague. Russia Beyond The Headlines 2016. https://www.rbth.com/longreads/siberian_plague/

Bogdanov VD, Golovatin MG (2017) Anthrax in Yamal: an ecological view on traditional reindeer husbandry. Russ J Ecol 48(2):95–100. https://doi.org/10.1134/S1067413617020059

Bröjer C, Hestvik G, Neimanis A, Uhlhorn H, Ågren E (2016) Wildlife disease surveillance in Sweden 2016. Uppsala

Cherkasskiy BL (2002) Epidemiology and prevention of anthrax. [in Russian]. Moscow

Cherkasskiy BL (2003) Epidemiologist journey in time and space. [in Russian] Voronezh

Department of Natural Resource Regulation, forestry relations, oil and gas development in Yamal-Nenets Autonomous District (2015) Report on the environmental situation in the Yamal-Nenets Autonomous District in 2015. [in Russian]

Dugarzhapova ZF, Chesnokova MV, Goldapel EG (2017) Anthrax in the Asian part of the Russian Federation. [in Russian]. Probl Highly Infect Dis 1:54–58

Egorov IY, Maramovich AS, Botvinkin AD (2002) Epidemiological monitoring of highly infectious and natural-focal diseases in the Far North. [in Russian] Yakutsk

Elvander M, Persson B, Sternberg Lewerin S (2017) Historical cases of anthrax in Sweden 1916–1961. Transbound Emerg Dis 64(3):892–898. https://doi.org/10.1111/tbed.12456

Federal Service for Supervision of Consumer Rights Protection and Human Welfare (2008) On organization of measures for anthrax prevention. [in Russian] Letter of Federal Service for Supervision of Consumer Rights Protection and Human Welfare № 01/2139-8-32

French HM (2013a) Climate change and periglacial environments this. In: The periglacial environment. Wiley, pp 373–387. https://doi.org/10.2307/1796697

French HM (2013b) Permafrost. In: The periglacial environment. Wiley, pp 143:108. https://doi.org/10.2307/1796697

Gainer R (2016) Practitioners' Corner Le Coin Des Praticiens Yamal and Anthrax. Can Vet J 57 (1):985–987

Hueffer K, O'Hara TM, Follmann EH (2011) Adaptation of mammalian host-pathogen interactions in a changing Arctic environment. Acta Vet Scand 53

Hueffer K, Parkinson AJ, Gerlach R, Berner J (2013) Zoonotic infections in Alaska: disease prevalence, potential impact of climate change and recommended actions for earlier disease detection, research, prevention and control. Int J Circumpolar Health 72(1):1–11. https://doi.org/10.3402/ijch.v72i0.19562

Hueffer K, Drown D, Romanovsky V, Hennessy T (2020) Factors contributing to anthrax outbreaks in the circumpolar north. EcoHealth 17(1):174–180. https://doi.org/10.1007/s10393-020-01474-z

Koch R (1876) Die Aetiologie Der Milzbrand-Krankheit, Begründet Auf Die Entwicklungsgeschichte Des Bacillus Anthracis. Beitraege Biol Pflanzen 2(2):277–310

Layshev KA, Zabrodin VA (2012) Problems of veterinary status on infectious diseases in reindeer herding. [in Russian] Vetpharma Farm Anim 1:36–40

Layshev KA, Zelensky VM (2003) Current status and ways of traditional activities renewal of nature management in Taymyr Autonomous District. [in Russian] Siberian Bull Agric Sci 3:65–69

Lewerin SS, Elvander M, Westermark T, Hartzell LN, Norström AK, Ehrs S, Knutsson R et al (2010) Anthrax outbreak in a Swedish beef cattle herd – 1st case in 27 years: case report. Acta Vet Scand 52(1):1–8. https://doi.org/10.1074/jbc.M111.315762

Ministry of Agriculture of the RSFSR (1976) Guideline of the settlements of the RSFSR affected by anthrax. [in Russian] Moscow

Moynihan WA (1963) Anthrax in Canada. Can Vet J 4(11):283–287. http://www.pubmedcentral.nih.gov/articlerender.fcgi?artid=1695464&tool=pmcentrez&rendertype=abstract

Nansen F (2011) Farthest north. Being the record of a voyage of exploration of the ship Fram, 1893–96, and of a fifteen Months' sleigh journey. Cambridge University Press, Cambridge. https://doi.org/10.1017/CBO9781139014076

Pasteur L (1881) Sur Les Virus-Vaccins Du Cholera Des Poules et Dy Charbon. Comptes Rendus Des Traveaux Du Congres International Des Directeurs Des Stations Agronomiques, Session Versailles, pp 151–162

Popova AY, Kulichenko AN (eds) (2017) Experience of anthrax outbreak elimination in Yamal in 2016. [in Russian] Stavropol Plague Research Institute. https://doi.org/10.23648/PRNT.2184

Quinn PJ, Markey BK, Leonard FC, FitzPatrick ES, Fanning S, Hartigan PJ (2011) Bacillus species. In: Veterianry microbiology and microbial disease, pp 227–32

Walker DA, Epstein HE, Leibman ME, Moskalenko NG, Orekhov P, Kuss JP, Matyshak GV et al (2009) Data report of the 2007 and 2008 Yamal expeditions. Fairbankls

Walsh MG, de Smalen AW, Mor SM (2018) Climatic influence on anthrax suitability in warming northern latitudes. Sci Rep 8:9269

WHO (2008) Anthrax in humans and animals. In: Turnbull P (ed) Anthrax guidelines, 4th edn. WHO Press. https://doi.org/10.2105/AJPH.30.3.299

Cystic and Alveolar Echinococcosis Caused by *Echinococcus canadensis* and *E. multilocularis* in the Arctic

Temitope U. Kolapo, Antti Oksanen, Rebecca Davidson, and Emily J. Jenkins

1 Introduction

Echinococcus is the genus name given to a group of cestode parasites belonging to the Taeniidae family. Before the nineteenth century, over 85 Latin names had been published for this genus based on the host origin and morphological appearance of its larval stage, or metacestode (Abuladze 1964). The name *Echinococcus* was eventually established in 1801 by Karl Asmund Rudolphi (1771–1832). Important species which are generally well recognised include *E. multilocularis*, *E. granulosus* species complex (which includes *E. granulosus* sensu stricto, *E. equinus*, *E. ortleppi* and *E. canadensis*), *E. oligarthra*, *E. vogeli* and *E. shiquicus*, which is a newly described species from China (Boufana et al. 2013).

In this chapter, the Arctic includes arctic regions of Alaska in the United States, Canada, the Kingdom of Denmark (Greenland and the Faroe Islands), Finland, Iceland, Norway (including the Svalbard archipelago), Russia and Sweden (Arctic Council 2021). In the Arctic, *E. multilocularis* and cervid genotypes of the *E. granulosus* complex (now known as *E. canadensis*) have established sylvatic cycles in wildlife, leading to public and animal health consequences as well as economic implications. Globally, these two species of the parasite were ranked the second and third most important food-borne parasites according to FAO/WHO (FAO/WHO 2014). Considerations for this ranking included public health importance, global

T. U. Kolapo (✉) · E. J. Jenkins
Department of Veterinary Microbiology, WCVM, University of Saskatchewan, Saskatoon, SK, Canada
e-mail: tuk781@mail.usask.ca

A. Oksanen
University of Helsinki, FINPAR, Finnish Food Authority, Animal Parasitology and Wildlife Diseases, Oulu, Finland

R. Davidson
Norwegian Veterinary Institute, Tromsø, Norway

M. Tryland (ed.), *Arctic One Health*, https://doi.org/10.1007/978-3-030-87853-5_12

distribution, potential for increases in disease rates, trade impacts and socioeconomic effects of infection.

Definitive hosts of *E. multilocularis* are wild and domestic canids which harbour the adult worms in their intestines and are almost invariably asymptomatic. These adult worms shed eggs along with feces. These eggs are environmentally resistant and immediately infective for intermediate hosts, who inadvertently consume eggs when grazing or drinking. In these intermediate hosts, the larval stage, called metacestode, of *E. multilocularis* initially establishes in the liver and is referred to as an alveolar hydatid cyst, or alveolar echinococcosis (AE). AE is often debilitating in its intermediate hosts and case lethality can reach up to 100% when left untreated (Ammann and Eckert 1996; Eckert 1998). The natural intermediate hosts are rodents, while accidental intermediate hosts like humans, non-human primates, pigs and dogs have been recorded (Deplazes and Eckert 2001). Affected humans and dogs often require radical surgical resection and drug treatment which might even be lifelong for a good prognosis. This often leads to a financial burden and reduced quality of life (Torgerson et al. 2010; Torgerson and MacPherson 2011).

Echinococcus canadensis is the primary member of the *E. granulosus* complex established in the Arctic (Rausch 2003). While *E. granulosus* sensu stricto may have been present in sheep in some regions of the Palearctic, it has subsequently been eradicated, for example, in Iceland. Historically a reindeer/dog cycle of what would now be called *E. canadensis* was found across the Nordkalotten region (Cap of the North) encompassing Nordland and Troms and Finnmark counties (Norway), Norrbotten (Sweden), Lappi (Finland) and Murmansk Oblast (Russia). Wild and domestic canids serve as definitive hosts for *E. canadensis*, for which they are almost invariably subclinically affected while contaminating the environment with eggs immediately infective to the intermediate hosts. The larval stage, or metacestode, of the parasite is referred to as cystic hydatid, or cystic echinococcosis (CE), and affects a wide range of intermediate hosts, primarily cervids for *E. canadensis*. In the Arctic, caribou/reindeer (*Rangifer tarandus*) and elk/moose (*Alces alces*) are the major intermediate hosts (Rausch 2003). Humans can also serve as accidental or aberrant intermediate hosts, as is the case with AE. Infected intermediate hosts develop fluid-filled, discrete cysts primarily in the lungs, as well as in the liver and other organs. These space-occupying lesions can affect the health of the animal and economic value in terms of carcass rejection; however, more commonly, only the affected organs (lungs, liver) are discarded, which if consumed by domestic dogs or wild canids may perpetuate the life cycle. Humans with CE caused by *E. canadensis* often remain asymptomatic unless complications such as cystic rupture occur (Ammann and Eckert 1996), making it a milder infection compared to CE caused by *E. granulosus* sensu stricto and AE caused by *E. multilocularis*.

In the Arctic, prevalence of *Echinococcus* infection remains high in wildlife, while prevalence of AE and CE in humans may be trending downward, at least in some regions (Jenkins et al. 2013). Despite opportunities from improved diagnosis through modern medical imaging, emerging and re-emerging human cases of AE and CE may be missed in arctic regions due to the lack of access to medical imaging, the subclinical nature of the disease (CE), prolonged latency and the lack of index of suspicion in health care personnel trained in developed countries. In some regions, introduction of strains of *E. multilocularis* with more zoonotic potential may be driving emergence of human AE (Houston et al. 2021). While dogs are no longer a primary means of transportation in the North American Arctic, dogs remain very popular companion animals, are often free-roaming and fed harvested wildlife. They are subclinically affected as definitive hosts, and there is limited veterinary infrastructure in many rural and remote communities to detect and prophylactically treat intestinal worm infections in them (Hotez 2010). The implication of this is that dogs may continue to serve as a source of human exposure to CE and AE in the Arctic. As well, as is increasingly recognised, dogs also serve as aberrant intermediate hosts for AE, especially for European-type strains in North America (Deplazes et al. 2017). As these parasites are well established in wildlife reservoirs, the interface between people and wildlife continues to shift as wildlife venture into human living areas and humans encroach into the wild for various activities such as deforestation and the Arctic continues to warm at an unprecedented rate. Management of *Echinococcus* spp. requires a One Health approach to minimise the impact on public and animal health in the Arctic.

1.1 Biology and Basic Life Cycle of *Echinococcus* spp.

These endoparasitic flatworms belong to the class Cestoda. They are considered true tapeworms of unusually small size, with the entire adult cestode only 2–10 mm long and composed of five or less segments (proglottids) (Thompson 2017). Like other tapeworms, they are dorsoventrally flattened, with adults having a characteristic elongated body known as strobila, demarcated by externally visible segmentation into proglottids, which are in various stages of development from immature to mature to gravid as one moves from anterior to posterior. They have a specialised anterior organ of attachment referred to as the scolex. Like most taenids, the scolex has four muscular suckers and a rostellum armed with hooks. A narrow 'neck' separates the scolex from the strobila. The sexually mature adult is a hermaphrodite and the larval metacestode stages proliferate asexually. This asexual reproduction in the intermediate host results in numerous protoscoleces within a cyst, and each protoscolex has the potential of becoming an adult cestode (Thompson 2017). The cycle becomes complete when intermediate hosts are preyed upon by definitive hosts or have access to infected organs of the definitive host, thus ingesting the cyst containing the protoscoleces which evert, attach to the intestinal epithelium and begin to bud off proglottids. The pre-patent period (time between ingestion of metacestodes and shedding of eggs in feces) is thought to be approximately 4 weeks for *E. multilocularis* and 6 weeks for *E. canadensis* (Thompson and Eckert 1983; Thompson 2017).

In CE, thousands of protoscoleces (each an immature, inverted scolex) bud off the germinal membrane and float freely (sometimes clumping together to form hydatid sand) in a fluid-filled, discrete, thick-walled cyst which may grow to over 10 cm in diameter. In AE, the germinal membrane forms multi-chambered cysts (hence multilocular) which resemble a cluster of grapes, containing thousands of protoscoleces floating freely in fluid. In aberrant (dead-end) intermediate hosts, the germinal membrane may proliferate, without forming discrete cysts or protoscoleces.

2 Life Cycle and Hosts in the Arctic

In the Arctic, *Echinococcus* spp. circulate between carnivore definitive hosts (usually canids, potentially felids for *E. multilocularis*) which harbour the adult tapeworms in the small intestine and the intermediate hosts, usually cervids (for *E. canadensis*) or rodents (for *E. multilocularis*) who serve as prey of the definitive hosts, with the asexually reproducing metacestode stage in their viscera (Eckert and Deplazes 2004). Humans are dead-end intermediate hosts who do not play a role in the life cycle propagation but can be significantly clinically affected by the zoonotic disease (CE and AE). Furthermore, in recent times, dogs, which are usually definitive hosts, have been diagnosed with the metacestode stage of the parasite, serving as intermediate hosts—sometimes with what appears to be viable protoscoleces (Frey et al. 2017; Skelding et al. 2014; Jenkins et al. 2012).

2.1 *Echinococcus multilocularis*

In North America, the northern tundra zone (NTZ) of Alaska and Canada and the north central region (NCR), including four Canadian provinces and 13 US states (Massolo et al. 2014; Deplazes et al. 2017), arctic and red foxes (*Vulpes lagopus* and *Vulpes vulpes*), coyotes (*Canis latrans*) and wolves (*C. lupus*) are primary definitive hosts of *E. multilocularis* in the sylvatic cycle (Liccioli et al. 2014; Schurer et al. 2014) (Fig. 1). In northern Europe, arctic and red fox are the primary hosts, with wolves and raccoon dogs (*Nyctereutes procyonoides*) serving as secondary definitive hosts in regions where fox serve as primary definitive hosts (EFSA and Zancanaro 2021). Domestic dogs also serve as definitive hosts, and historically, sled dogs served as bridging hosts between sylvatic cycles and people; prevalence was 5–13% in sled dogs in a hyperendemic region on St. Lawrence Island, Alaska (reviewed in Jenkins et al. 2013). Although prevalence is generally thought to be much lower in pet dogs in contemporary times in North America, studies using highly sensitive molecular methods are now able to detect DNA of *E. multilocularis* in dogs, even in urban regions of North America. The increasing proximity of wildlife to urban settlements might be a factor in the transmission to domestic dogs (Massolo et al. 2014). Domestic and wild felids (such as Canadian lynx, *Lynx canadensis*) are potential definitive hosts, but the adult cestodes may not fully develop or reproduce to their maximum potential as compared to canids.

Major intermediate hosts in arctic regions of North America are northern red-backed vole (*Myodes*, previously *Clethrionomys*, *rutilus*), tundra vole (*Microtus oeconomus*) and, to a lesser degree, brown lemming (*Lemmus trimucronatus*)

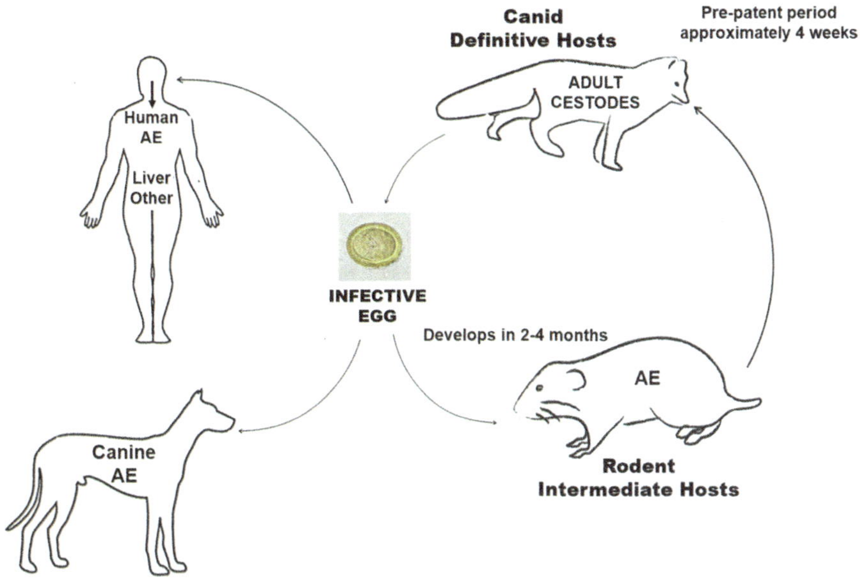

Fig. 1 Life cycle of *Echinococcus multilocularis*. Canid definitive hosts include foxes (*Vulpes vulpes* or *Vulpes lagopus*), coyotes (*Canis latrans*), wolves (*Canis lupus*) and dogs (*Canis lupus familiaris*), while rodent intermediate hosts include voles, lemmings, shrews and muskrats. AE, alveolar echinococcosis, or alveolar hydatid. Photo Credit (for infective egg): Brent Wagner

(Jenkins et al. 2013), whereas in central North America, the primary intermediate hosts are meadow vole (*Microtus pennsylvanicus*), southern red-backed vole (*Myodes gapperi*) and deer mouse (*Peromyscus maniculatus*). Multiple species of shrews, ground squirrels and muskrat (*Ondatra zibethicus*) are also reported as occasional intermediate hosts in North America. Prevailing intermediate host and transmission potential depends on the overall prey–predator association within the community and ecology of the habitat (Liccioli et al. 2014).

In Russia, red fox and wolves are the definitive hosts and various species of arvicoline rodents (*Microtus gregalis*, *M. oeconomus* and *Myodes rufocanus*) are the major intermediate hosts (Konyaev et al. 2013). In Sweden and the Norwegian archipelago of Svalbard, red and arctic foxes are considered definitive hosts, while rodents such as voles (European water vole *Arvicola amphibius*, field vole *Microtus agrestis* and sibling vole *Microtus levis*) are the intermediate hosts (Wahlström et al. 2015; Knapp et al. 2012; Stien et al. 2010). In Sweden, a national screening in 2012–2014 found 3 positive foxes out of 2779, as well as the detection of the metacestode in intermediate hosts, field and water voles (Axén et al. 2019). Introduction of sibling vole (*Microtus epiroticus* and *M. levis*) into the archipelago of Svalbard appears to have recently enabled the life cycle in this region (Henttonen et al. 2001). Finland and mainland Norway are considered free of *Echinococcus multilocularis* infection; in Norway to date, over 6500 red foxes and 18 wolves tested since 2020 have all been negative for *E. multilocularis* (Wahlström et al. 2015;

Hamnes et al. 2020; EFSA and Zancanaro 2021). Aberrant intermediate hosts include dogs, pigs, horses, non-human primates and people.

2.2 *Echinococcus canadensis*

In North America, wolves, coyotes and domestic dogs are considered the major definitive hosts for *E. canadensis*, although there are occasional reports of similar findings in arctic and red foxes (*Vulpes lagopus* and *Vulpes vulpes*) (Fig. 2). Elk (wapiti, *Cervus canadensis*), moose, caribou and deer (*Odocoileus* spp.) are the known intermediate hosts (Rausch 2003); however, the parasite has also been reported in muskoxen (*Ovibos moschatus*) (Schurer et al. 2013). In Russia, wolves are the major definitive hosts, while moose and reindeer have been identified as sylvatic intermediate hosts; reports in sheep likely reflect *E. granulosus* sensu stricto (G1–3) since *E. canadensis* does not establish well in domestic livestock (Sweatman and Williams 1963; Konyaev et al. 2013). In Fennoscandia countries, wolves were considered the only definitive hosts until recently when DNA of *E. granulosus* sensu lato was detected in feces of domestic dogs (Hämäläinen et al. 2015); moose and reindeer are known intermediate hosts (Lavikainen et al. 2006). *Echinococcus canadensis* was last detected in moose and reindeer in northern Sweden in the late 1990s and in reindeer in Norway in 1990 and 2003 and is present at a low prevalence

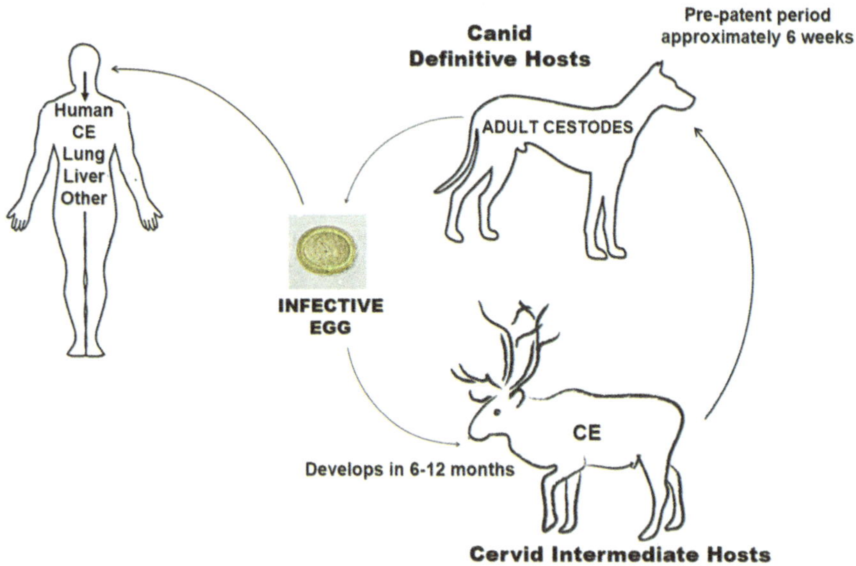

Fig. 2 Life cycle of *Echinococcus canadensis* (cervid genotypes G8 and G10 of *E. granulosus* species complex). Canid definitive hosts include wolves (*Canis lupus*), coyotes (*Canis latrans*) and dogs, while cervid intermediate hosts include moose (*Alces alces*), caribou/reindeer (*Rangifer tarandus*) and elk (wapiti, *Cervus canadensis*). CE, cystic echinococcosis, or cystic hydatid. Photo Credit (for infective egg): Brent Wagner

in wildlife (wolves, moose and reindeer) in Finland (Jørgensen et al. 2016; Axén et al. 2019). Iceland and Greenland are considered free of the infection.

2.3 Genetic Diversity Within *Echinococcus* spp.

This genus has several species, naming of which remains controversial and is based on morphological differences of the adult worms, geographical distribution, host preferences and, increasingly, molecular differences (Romig et al. 2015). Four undisputed species were recognised by the beginning of 1980—*E. granulosus*, *E. multilocularis*, *E. oligarthra* and *E. vogeli* (Kumaratilake et al. 1986). Recognising diversity within *E. granulosus*, attempts were made at creating various subspecies, hence the use of 'strain' to describe variants which differ in terms of epidemiological significance (Thompson and McManus 2001) and based initially on non-genetic characters like host spectrum, geography and morphology. However, by the early 1990s, gene sequence data became important in species definition. This brought up the 'genotype' nomenclature, which is sometimes used interchangeably with 'strain'. The outcome of molecular studies on nuclear and mitochondrial loci within the *E. granulosus* complex has led to the taxonomic revisions of the *E. granulosus* sensu lato to *E. granulosus* sensu stricto and at least four other species (Romig et al. 2015). Currently, ten species and strains/genotypes are recognised (not without controversy) and they include *E. granulosus* sensu stricto (G1–G3), *E. equinus* (G4), *E. ortleppi* (G5), *E. canadensis* (G8, G10), *E. intermedius* (G6, G7), *E. felidis* (Lion strain), *E. multilocularis*, *E. shiquicus*, *E. vogeli* and *E. oligarthra* (Thompson 2017, 2020; Romig et al. 2017). In the Arctic, only genotypes G6, G8 and G10 have been identified so far: G8 and G10 in North America (Schurer et al. 2018), G6 in Russia and G10 in Sweden and Finland (Deplazes et al. 2017).

2.4 Genetic Diversity in *E. multilocularis*

There is a much less genetic diversity within *E. multilocularis*, generally recognised as haplotypes based on mitochondrial loci (Nakao et al. 2009) or microsatellites (Knapp et al. 2012). Initially only 2 haplotypes (M1 and M2) were described based on single mitochondrial loci (Okamoto et al. 1995; Rinder et al. 1997; Haag et al. 1997). Using three mtDNA loci, 18 haplotypes from 76 isolates from different parts of the world were described (Nakao et al. 2009), including European haplotypes (E1–E5), Asian haplotypes (A1–A9), North American haplotypes (N1 and N2) and O1 from an inner Mongolian isolate.

In addition to the N1 (Alaska isolates) and N2 (Indiana and South Dakota isolates) haplotypes which are localised in North America, European haplotypes of the parasite, especially E3 and E4, have been consistently isolated in wild canids (Gesy et al. 2013), in dogs as intermediate hosts (Zajac et al. 2020; Skelding et al. 2014) and in humans (Massolo et al. 2019; Schurer et al. 2021) in North America. In

Scandinavian countries of Denmark and Sweden, isolates from wild canids were consistent with previously identified European haplotypes (Knapp et al. 2019). Further diversity linked to biogeographic differences will likely be recognised as we move to full genome approaches, but for now only a single species of *E. multilocularis* is recognised across its circumpolar distribution, albeit with potential differences among haplotypes in zoonotic potential, host specificity and pathogenicity which require further investigation. With the knowledge of genetic diversity, diagnostic tools can be better tailored towards identification of the correct species of interest, by the development of relevant markers. Thus, clinical interventions and management can be more direct and public health measures can be specific. From the taxonomic point of view, future evolutionary changes, genetic differentiation and speciation can also be predicted through the knowledge of genetic diversity (Thompson and Lymbery 1995).

3 Epidemiology

3.1 Veterinary and Public Health Significance of Alveolar Echinococcosis (AE)

Alveolar echinococcosis (AE) occurs in both normal and aberrant intermediate hosts of the parasite, the latter including humans and dogs. AE in humans is a life-threatening debilitating condition whose importance is reflected by WHO's ranking as the third most important food-borne parasite in terms of global impact (FAO/WHO 2014). Clinical disease in humans takes between 5 and 15 years to develop, often mimicking a neoplasia on initial clinical presentation and medical imaging. Similar to neoplasia, the germinal membrane has the ability to metastasize to other organs of the body, often becoming debilitating, especially in immunocompromised patients, if not detected and treated aggressively (Kern et al. 2004; Houston et al. 2021). Infection can occur in phases with an asymptomatic initial phase leading to a spontaneous recovery or abortive phase, or a progressive infection where liver function becomes greatly impaired (Eckert and Deplazes 2004). The progressive phase is characterised by hepatomegaly, abdominal pain, jaundice, fever anaemia and weight loss, leading to portal hypertension in the advanced phase of the infection. The initiation of parasitostatic chemotherapy can bring about the stable phase of infection where parasite proliferation is inhibited (Brunetti et al. 2010).

Dogs are considered natural definitive hosts for *Echinococcus multilocularis*; however, over the last few decades, dogs in Europe and North America have been increasingly recognised as intermediate hosts capable of developing AE (Frey et al. 2017; Skelding et al. 2014). Considering that the development of AE infection in dogs is rapid compared to humans, with symptoms expressed as early as a year (Frey et al. 2017), it can easily be fatal if detected in the late stages of disease (Fig. 3). Symptoms are usually nonspecific; thus, infection can be misdiagnosed or diagnosed late, leading to a poor prognosis and expensive treatment (Corsini et al. 2015). Dogs become infected through coprophagy of infected wild canid feces or their own feces

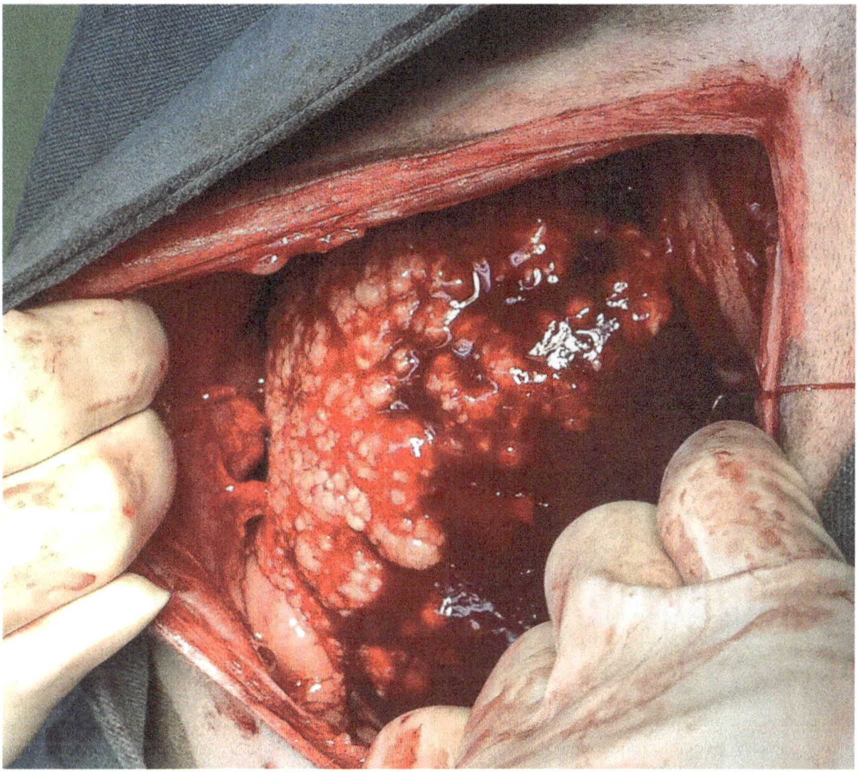

Fig. 3 Alveolar echinococcosis in the liver of a dog, which may appear as multiple grape-like clusters of fluid-filled cysts or mimic an invasive neoplasm without discrete cyst formation. Photo Credit: Audrey Tataryn

when infected by the adult cestodes, or possibly through auto infection internally through reflux of immediately infective eggs from the intestine into the stomach or bile ducts (Haller et al. 1998; Staebler et al. 2006). AE in animals and humans is nationally notifiable in Sweden but not in Finland and Norway. In North America, it is not nationally notifiable but required by the public health authorities of Alaska USA, Northwest Territories and recently Ontario in Canada.

3.2 Veterinary and Public Health Significance of Cystic Echinococcosis (CE)

In the Arctic, cystic echinococcosis (CE) is caused by the larval stage of *E. canadensis* in cervid intermediate hosts (Fig. 4) as well as accidental hosts like humans. In cervids, infection is largely asymptomatic and often goes unnoticed; however, high intensities of cysts in the lungs may reduce respiratory fitness and render cervids more likely to be preyed on by chase predators such as wolves (Joly

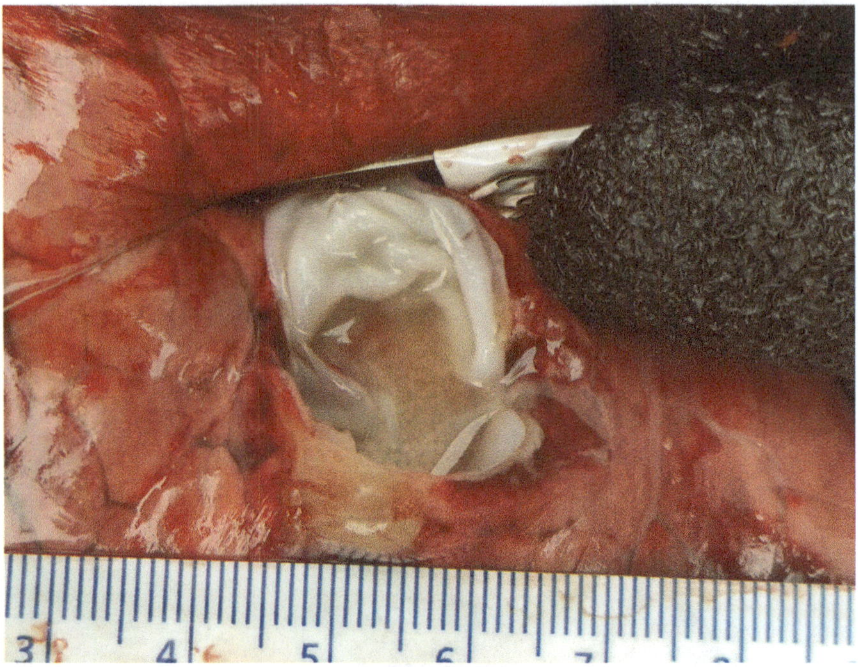

Fig. 4 Cystic echinococcosis in the lungs of a moose (*Alces alces*), cut open and showing the discrete cyst filled with fluid and hydatid sand (multiple protoscoleces) typical of the metacestode of *E. canadensis*. Ruler in cm. Photo Credit: David McRuer

and Messier 2004). In wild or captive cervids harvested as food animals, cysts may raise aesthetic concerns leading to carcass condemnation with the exception of Finland where only affected organs are condemned. Furthermore, affected organs should not be fed to dogs or left for scavenging by wild canids so as not to continue perpetuating the life cycle (Rausch 2003; Eckert and Deplazes 2004). Cysts of *E. canadensis* are more often found in the lungs (which is unusual for other members of the *E. granulosus* species complex), followed by the liver and other organs including the heart, spleen, kidneys, skin, muscles, brain, bones and abdominal cavity. Infection which occurs directly from ingestion of eggs is referred to as primary CE, while secondary CE results from spontaneous cyst rupture or trauma-induced rupture with the release of protoscoleces within the abdominal cavity (Ammann and Eckert 1996). Infected people are initially asymptomatic and may remain so for years, with progression to symptomatic disease dependent on cyst size, number and location as well as immune status of the affected individual (Eckert and Deplazes 2004). Fatality is relatively rare but can occur quickly due to cyst rupture leading to anaphylaxis and/or seeding of protoscoleces throughout the peritoneal cavity. In most of northern Europe and North America, human CE is sporadic; endemically acquired CE is most likely caused by *E. canadensis*, which is relatively benign compared to foreign acquired cases of *E. granulosus* s.s. Dogs do not serve as intermediate hosts for CE, and as definitive hosts, they are almost invariably asymptomatic even with heavy infection intensities; however, treatment and control in dogs is of importance because of the zoonotic risk. CE in human is notifiable to the public health authorities in Alaska and Fennoscandia countries (where cases in animals are notifiable as well) at a national level but not in Russia and Canada (except the Northwest Territories).

3.3 Diagnosis

Diagnosis of echinococcosis in humans is based on imaging (ultrasonography, computed tomography, x-ray examinations) and immunodiagnostic tests such as ELISA which detect serum antibodies to *Echinococcus* spp., ideally followed by confirmatory immunoblotting, which is more specific (Müller et al. 2007). Due to unresolved issues of specificity and sensitivity with serological assays, polymerase chain reaction (PCR) can be used to check for the presence of protoscoleces, either on material sampled by fine needle aspiration of hydatid cysts guided by ultrasonography or on cysts removed at surgery, as a confirmatory diagnosis (Schurer et al. 2021).

4 Source and Modes of Transmission in the North

4.1 Food- and Waterborne Transmission

Echinococcus infection occurs following oral ingestion of infective stages, for both definitive and intermediate hosts. Definitive hosts become infected when they ingest metacestodes in organs from infected intermediate hosts. Various cycles of host–prey interaction enable the transmission and perpetuation of *E. canadensis* in the North: the synanthropic cycle involving domesticated definitive and intermediate hosts (dogs and semi-domesticated reindeer), semi-synanthropic cycle involving domestic definitive hosts and wild intermediate hosts (dogs and wild cervids) and a purely sylvatic cycle involving wild definitive and intermediate hosts (wild canid and wild cervid). Furthermore, for *E. canadensis*, dogs may scavenge or be fed organs (including the lungs and liver) of wild cervids harvested by hunters in the north, a situation which in turn perpetuates human exposure (Rausch 2003). Intermediate hosts, including humans, become infected when they ingest eggs which are passed along with feces of infected definitive hosts; eggs are immediately infective and can survive for long periods of time (months to years) even under arctic conditions. Eggs in the environment can contaminate vegetables and fruits leading to infection when consumed unwashed (Kern et al. 2004). In the environment, eggs can easily be dispersed by rain or run-off water, which creates a possibility of contaminating potable water sources resulting in waterborne transmission of infection when such water is consumed without boiling or filtration (Yamamoto et al. 2001; Davidson et al. 2016). Eggs of arctic species of *Echinococcus* are not reliably inactivated by routine disinfectants, fixatives (ethanol or formalin) or freezing at −20 °C or warmer; inactivation requires heating to temperatures greater than 60 °C for 5 min, desiccation, freezing at −70 °C for 96 h or −80 °C for 48 h or application of strong bleach or glutaraldehyde solutions after cleaning to remove excess organic material (WHO/OIE 2001). Infection in small children has also been linked to ingestion of contaminated sand or soil in playgrounds frequented by wild and domestic canids (Eckert and Deplazes 2004). Eggs of *Echinococcus* can adhere to hair of dogs either directly from their own feces (if they have a patent infection with the adult cestode) or by rolling in grass or sand containing eggs from dogs or wild canids, thereby serving as a potential source of infection for people who pet these dogs without washing their hands before eating.

4.2 Prevention and Control

Surveillance programs for *E. multilocularis* in northern Europe monitor the prevalence of infection in wild canids (historically through recovery of adult cestodes from intestines of harvested carcasses, more recently through coproPCR methods) to enhance decision and policy making towards control and prevention in humans and dogs and to demonstrate country freedom status (EFSA and Zancanaro 2021). Such surveillance has detected an emergence of *E. multilocularis* in new regions of Europe and an increased prevalence in wild canids in endemic regions, possibly in response to increased wild canid abundance due to rabies control and decreased hunting pressure. There is often little formal surveillance in countries considered endemic for *E. multilocularis* (such as Canada). Surveillance for *E. canadensis* could capitalise on inspection of organs of semi-domesticated reindeer harvested for commercial purposes in Fennoscandia (Kautto et al. 2017) and captive cervids in game farms in North America, although the North American cervid farming industry has been severely affected by chronic wasting disease (CWD). Harvest of wild cervids in Fennoscandia and North America does not occur in regulated abattoirs, and thus formal surveillance mechanisms have not been established (Axén et al. 2019). In addition, echinococcosis in animals (especially wildlife) is not always a reportable animal disease across the global Arctic, although human cases are often notifiable to public health. Recent advances in non-invasive fecal testing using sensitive molecular techniques could make routine surveillance for both species of *Echinococcus* in wild canids and domestic dogs much more feasible (compared to adult cestode recovery, which is fatal for the host; arecoline purging, which has been used in domestic dogs but can have variable results and sensitivities <80%; and detection of eggs in feces, which has low sensitivity, 20–30%) (WHO/OIE 2001; Kolapo et al. 2021). Investment in surveillance could allow countries to demonstrate freedom status, prevent introduction of foreign strains and species of *Echinococcus* and identify focal hotspots of increased risk of human and canine transmission, even in endemic countries.

Control of *Echinococcus* spp. transmission in wildlife hosts is usually an expensive adventure which mostly cannot be sustained. Efforts to treat wild canids with praziquantel baits have short-term efficacy, high risk of consumption by non-target wildlife and significant expense (Hegglin et al. 2003). Culls of wild canids and rodents may be unacceptable, ineffective and even counterproductive, creating empty niches which will be invaded by younger, more marginal hosts that may incur higher infection rates (Comte et al. 2017). The best target for minimising the public health risks associated with *Echinococcus* relies on veterinary interventions—hence the need for a One Health approach. Owners of pet dogs should avoid walking them off leash to control indiscriminate defecation and prevent coprophagy and ensure regular deworming with an effective cestocide (monthly for dogs at high risk of consuming wild rodents and within 6 weeks of access to cervid carcasses) (Gottstein et al. 2015). As many dogs in northern communities are free-roaming, control programs should include measures to reduce stray dogs, promote routine deworming of community and owned dogs and prevent dogs from eating offal of harvested cervids (for example, by burying or burning carcasses and/or fencing landfill sites), similar to the holistic approach used in the 1950s with great success to reduce *E. canadensis* in reindeer in northern Norway. Control of *E. multilocularis* is more difficult, as the parasite transmits primarily among wildlife, and it is difficult to prevent dogs from hunting or scavenging wild rodents. Animal control and veterinary infrastructure remain logistical, economic and cultural challenges in many rural and remote communities in the Arctic. However, investment is important to benefit both animal and human health through education, regulation of dog importation and wildlife translocation (including mandatory testing and/or treatment) and enhanced capacity for surveillance in the Arctic.

5 Future Trends and Challenges

Prompt recognition and diagnosis of both AE and CE in people and dogs remain a challenge in the Arctic. Signs and symptoms are usually not specific, and imaging (if available) results are often mistaken for neoplasia, especially by medical and veterinary clinicians unfamiliar with these relatively rare infections. As well, it is difficult to correctly diagnose walled off or aborted human infections. Delayed diagnosis is strongly linked to poor prognosis, especially for AE, resulting in additional financial and health burdens for the public health system and affected individuals. Given the recent establishment of European strains of *E. multilocularis* with a greater zoonotic potential in North America, establishment of *E. multilocularis* in regions previously considered free (Svalbard) and challenges with poor sensitivity and specificity of serological tests, optimisation of non-invasive diagnostic tools along with increased awareness campaigns within the public and among clinicians is critically needed. Problems of sensitivity and specificity, such as the inability to distinguish between early stages of CE and AE, or eggs of *E. canadensis*, *E. multilocularis* and *Taenia* spp., are especially important in arctic regions where co-infection is a possibility (Schurer et al. 2014) and can be addressed by increasing utility of molecular diagnostic tools (e.g. PCR). The development of vaccines against AE in dogs and people and against CE in semi-domesticated reindeer could be considered, similar to those developed for CE caused by *E. granulosus* s.s. in sheep in some countries (Gauci et al. 2005; Wang et al. 2016); however, vaccination of wild rodents and cervids is not highly feasible.

The understanding of the ecology and interplay of wildlife, domestic and human hosts for transmission of *Echinococcus* spp. remains a challenge due to the complexity of habitat, climate, vegetation and anthropogenic issues, especially in the Arctic, which is experiencing climate change at much higher rates than the rest of the globe. Rapid climate change may enhance survival and distribution of eggs in the environment (for example, through increased run-off), alter the ecology and distribution of hosts and alter the interface of humans, wildlife and domestic animals, especially if livestock production becomes more feasible in northern regions. With lifting of climate barriers and increased pressure for resource extraction and shipping, monitoring and proactive management are needed to prevent introduction and movement of introduced hosts and species of *Echinococcus* within the Arctic. Detection in wild hosts is often research driven and cited as evidence of emergence (Bessonov 1998; Henttonen et al. 2001; Gesy et al. 2013, 2014; Gesy and Jenkins 2015); however, strategic surveillance programs incorporating a centralised data reporting system across different stakeholders, including indigenous peoples, are still lacking. This will be highly beneficial for ensuring informed policy making and establishing culturally appropriate and sustainable control and prevention measures.

References

Abuladze KI (1964) Taeniata of animals and man and diseases cause by them. In: Skrjabin AKI (ed) Essentials of cestodology, vol IV. Israel program for scientific translations, 1970, Jerusalem

Ammann RW, Eckert J (1996) Cestodes. *Echinococcus*. Gastroenterol Clin N Am 25:655–689. https://doi.org/10.1016/s0889-8553(05)70268-5

Arctic Council. https://arctic-council.org/en/about/states/. Accessed 07 May 2021

Axén C, Brytting M, Bujila I et al (2019) Surveillance of infectious diseases in animals and humans in Sweden 2019, National Veterinary Institute (SVA), Uppsala, Sweden. SVA:s rapportserie 64 1654–7098. https://www.sva.se/media/sgafld4h/rapport_surveillance-2019_kapitel_echinococcosis.pdf

Bessonov AS (1998) *Echinococcus multilocularis* infection in Russia and neighbouring countries. Helminthologia 35(2):73–78

Boufana B, Umhang G, Qiu J et al (2013) Development of three PCR assays for the differentiation between *Echinococcus shiquicus*, *E. granulosus* (G1 genotype), and *E. multilocularis* DNA in the co-endemic region of Qinghai-Tibet plateau, China. Am J Trop Med Hyg 88:795–802

Brunetti E, Kern P, Vuitton DA (2010) Writing panel for the WHO-IWGE. Expert consensus for the diagnosis and treatment of cystic and alveolar echinococcosis in humans. Acta Trop 114:1–16

Comte S, Umhang G, Raton V, Raoul F, Giraudoux P, Combes B, Boué F (2017) *Echinococcus multilocularis* management by fox culling: an inappropriate paradigm. Prev Vet Med 147:178–185

Corsini M, Geissbühler U, Howard J et al (2015) Clinical presentation, diagnosis, therapy and outcome of alveolar echinococcosis in dogs. Vet Rec 177:569

Davidson RK, Lavikainen A, Konyaev S et al (2016) *Echinococcus* across the north: current knowledge, future challenges, food and waterborne. Parasitology 4:39–53. https://doi.org/10.1016/j.fawpar.2016.08.001

Deplazes P, Eckert J (2001) Veterinary aspects of alveolar echinococcosis – a zoonosis of public health significance. Vet Parasitol 98:65–87

Deplazes P, Rinaldi L, Alvarez Rojas CA et al (2017) Global distribution of alveolar and cystic echinococcosis. Adv Parasitol 95:315–493

Eckert J (1998) Alveolar echinococcosis (*Echinococcus multilocularis*) and other forms of echinococcosis (*Echinococcus oligarthrus and Echinococcus vogeli*). In: Palmer SR, Soulsby EJL, Simpson DIH (eds) Zoonosis. Oxford University Press, Oxford, pp 689–716

Eckert J, Deplazes P (2004) Biological, epidemiological, and clinical aspects of echinococcosis, a zoonosis of increasing concern. Clin Microbiol Rev 17:107–135

EFSA; European Food Safety Authority, Zancanaro G (2021) Annual assessment of *Echinococcus multilocularis* surveillance reports submitted in 2020 in the context of commission delegated regulation (EU) 2018/772. EFSA J 19:e06382. https://doi.org/10.2903/j.efsa.2021.6382

FAO/WHO (2014) Multicriteria-based ranking for risk management of food-borne parasites. Microbiological Risk Assessment Series. Food and Agriculture Organization of the United Nations/World Health Organization, Rome, p 302

Frey CF, Marreros N, Renneker S et al (2017) Dogs as victims of their own worms: Serodiagnosis of canine alveolar echinococcosis. Parasit Vectors 10:422. https://doi.org/10.1186/s13071-017-2369-0

Gauci C, Heath D, Chow C et al (2005) Hydatid disease: vaccinology and development of the EG95 recombinant vaccine. Expert Rev Vaccines 4(1):103–112

Gesy KM, Jenkins EJ (2015) Introduced and native haplotypes of *Echinococcus multilocularis* in wildlife in Saskatchewan, Canada. J Wildl Dis 51:743–748. https://doi.org/10.7589/2014-08-214

Gesy K, Hill JE, Schwantje H et al (2013) Establishment of a European-type strain of *Echinococcus multilocularis* in Canadian wildlife. Parasitology 140:1133–1137. https://doi.org/10.1017/S0031182013000607

Gesy KM, Schurer JM, Massolo A et al (2014) Unexpected diversity of the cestode *Echinococcus multilocularis* in wildlife in Canada. Int J Parasitol Parasites Wildl 3:81–87. https://doi.org/10.1016/j.ijppaw.2014.03.002

Gottstein B, Stojkovic M, Vuitton DA et al (2015) Threat of alveolar echinococcosis to public health - a challenge for Europe. Trends Parasitol 31:407–412

Haag KL, Zaha A, Araújo AM et al (1997) Reduced genetic variability within coding and non-coding regions of the *Echinococcus multilocularis* genome. Parasitology 115:521–529

Haller M, Deplazes P, Guscetti F et al (1998) Surgical and chemotherapeutic treatment of alveolar echinococcosis in a dog. J Am Anim Hosp Assoc 34:309–314. https://doi.org/10.5326/15473317-34-4-309

Hämäläinen S, Kantele A, Arvonen M et al (2015) An autochthonous case of cystic echinococcosis in Finland. Eur Secur 20(42):2–5

Hamnes IS, Enemark HL, Henriksen K, Madslien K, Øines Ø, Er C (2020) The surveillance program for *Echinococcus multilocularis* in red foxes (*Vulpes vulpes*) in Norway 2019. Annual Report. Norwegian Veterinary Institute, Oslo, 7 p

Hegglin D, Ward PI, Deplazes P (2003) Anthelmintic baiting of foxes against urban contamination with *Echinococcus multilocularis*. Emerg Infect Dis 9(10):1266–1272. https://doi.org/10.3201/eid0910.030138

Henttonen H, Fuglei E, Gower CN et al (2001) *Echinococcus multilocularis* on Svalbard: introduction of an intermediate host has enabled the local life-cycle. Parasitology 123:547–552

Hotez PJ (2010) Neglected infections of poverty among the indigenous peoples of the Arctic. PLoS Negl Trop Dis 4(1):e606. https://doi.org/10.1371/journal.pntd.0000606

Houston S, Belga S, Buttenschoen K et al (2021) Epidemiological and clinical characteristics of alveolar Echinococcosis: an emerging infectious disease in Alberta, Canada. Am J Trop Med Hyg. https://doi.org/10.4269/ajtmh.20-1577

Jenkins EJ, Peregrine AS, Hill JE et al (2012) Detection of a European strain of *Echinococcus multilocularis* in North America. Emerg Infect Dis 18:1010–1012. https://doi.org/10.3201/eid1806.111420

Jenkins EJ, Castrodale LJ, de Rosemond SJ et al (2013) Tradition and transition: parasitic zoonoses of people and animals in Alaska, northern Canada, and Greenland. Adv Parasitol 82:33–204

Joly DO, Messier F (2004) The distribution of *Echinococcus granulosus* in moose: evidence for parasite-induced vulnerability to predation by wolves? Oecologia 140:586–590

Jørgensen HJ, Heier B, Hauge K, Guzmán Herrador B, Hofshagen M (2016) The Norwegian zoonoses report 2015. Norwegian Veterinary Institute, p 15

Kautto AH, Vagsholm I, Niskanen R (2017) Meat inspection of reindeer – a rich source of data for monitoring food safety and animal and environmental health in Sweden. Infect Ecol Epidemiol 7(1):1340695

Kern P, Ammon A, Kron M et al (2004) Risk factors for alveolar echinococcosis in humans. Emerg Infect Dis 10(12):2088–2093

Knapp J, Staebler S, Bart JM et al (2012) *Echinococcus multilocularis* in Svalbard, Norway: microsatellite genotyping to investigate the origin of a highly focal contamination. Infect Genet Evol 12(6):1270–1274

Knapp J, Umhang G, Wahlström H et al (2019) Genetic diversity of *Echinococcus multilocularis* in red foxes from two Scandinavian countries: Denmark and Sweden FAWPAR 14:e00045

Kolapo TU, Bouchard É, Wu J et al (2021) Copro-polymerase chain reaction has higher sensitivity compared to centrifugal fecal flotation in the diagnosis of taeniid cestodes, especially *Echinococcus* spp, in canids. Vet Parasitol 292:109400. https://doi.org/10.1016/j.vetpar.2021.109400

Konyaev S, Yanagida T, Nakao M et al (2013) Genetic diversity of *Echinococcus* spp. in Russia. Parasitology 140:1637–1647

Kumaratilake LM, Thompson RC, Eckert J (1986) *Echinococcus granulosus* of equine origin from different countries possess uniform morphological characteristics. Int J Parasitol 16(5):529–540. https://doi.org/10.1016/0020-7519(86)90089-5

Lavikainen A, Lehtinen MJ, Laaksonen S et al (2006) Molecular characterization of *Echinococcus* isolates of cervid origin from Finland and Sweden. Parasitology 133:565–570

Liccioli S, Kutz SJ, Ruckstuhl KE et al (2014) Spatial heterogeneity and temporal variations in *Echinococcus multilocularis* infections in wild hosts in a north American urban setting. Int J Parasitol 44:457–465

Massolo A, Liccioli S, Budke C et al (2014) *Echinococcus multilocularis* in North America: the great unknown. Parasite 21:73. https://doi.org/10.1051/parasite/2014069

Massolo A, Klein C, Kowalewska-Grochowska K et al (2019) European *Echinococcus multilocularis* identified in patients in Canada. N Engl J Med 381:384–385

Müller N, Frei E, Nuñez S et al (2007) Improved serodiagnosis of alveolar echinococcosis of humans using an in vitro-produced *Echinococcus multilocularis* antigen. Parasitology 134:879. https://doi.org/10.1017/S0031182006002083

Nakao M, Xiao N, Okamoto M et al (2009) Geographic pattern of genetic variation in the fox tapeworm *Echinococcus multilocularis*. Parasitol Int 58:384–389

Okamoto M, Bessho Y, Kamiya M et al (1995) Phylogenetic relationships within *Taenia taeniaeformis* variants and other taeniid cestodes inferred from the nucleotide sequence of the cytochrome c oxidase subunit I gene. Parasitol Res 81(6):451–458. https://doi.org/10.1007/BF00931785

Rausch RL (2003) Cystic echinococcosis in the Arctic and sub-Arctic. Parasitology 127:S73–S85. https://doi.org/10.1017/s0031182003003664

Rinder H, Rausch RL, Takahashi K et al (1997) Limited range of genetic variation in *Echinococcus multilocularis*. J Parasitol 83:1045–1050

Romig T, Ebi D, Wassermann M (2015) Taxonomy and molecular epidemiology of *Echinococcus sensu lato*. Vet Parasitol 213:76–84

Romig T, Deplazes P, Jenkins D et al (2017) Ecology and life cycle patterns of *Echinococcus* species. Adv Parasitol 95:213–314. https://doi.org/10.1016/bs.apar.2016.11.002

Schurer J, Shury T, Leighton F et al (2013) Surveillance for *Echinococcus canadensis* genotypes in Canadian ungulates. Int J Parasitol Parasites Wildl 2:97–101
Schurer JM, Gesy KM, Elkin BT et al (2014) *Echinococcus multilocularis* and *Echinococcus canadensis* in wolves from western Canada. Parasitology 141:159–163. https://doi.org/10.1017/S0031182013001716
Schurer JM, Émilie B, Bryant A et al (2018) *Echinococcus* in wild canids in Québec (Canada) and Maine (USA). PLoS Negl Trop Dis 12:e0006712. https://doi.org/10.1371/journal.pntd.0006712
Schurer JM, Tsybina P, Gesy KM et al (2021) Molecular evidence for local acquisition of human alveolar echinococcosis in Saskatchewan, Canada. J Infect Dis 223(6):1015–1018. https://doi.org/10.1093/infdis/jiaa473
Skelding A, Brooks A, Stalker M et al (2014) Hepatic alveolar hydatid disease (*Echinococcus multilocularis*) in a boxer dog from southern Ontario. Can Vet J 55:551–553
Staebler S, Grimm F, Glaus T et al (2006) Serological diagnosis of canine alveolar echinococcosis. Vet Parasitol 141:243–250. https://doi.org/10.1016/j.vetpar.2006.05.026
Stien A, Voutilainen L, Haukisalmi V et al (2010) Intestinal parasites of the Arctic fox in relation to the abundance and distribution of intermediate hosts. Parasitology 137:149–157
Sweatman G, Williams J (1963) Comparative studies on the biology and morphology of *Echinococcus granulosus* from domestic livestock, moose and reindeer. Parasitology 53:493–514
Thompson RCA (2017) Biology and systematics of *Echinococcus*. Adv Parasitol 95:65–109. https://doi.org/10.1016/bs.apar.2016.07.001
Thompson RCA (2020) The molecular epidemiology of *Echinococcus* infections. Pathogens 9(6):453. https://doi.org/10.3390/pathogens9060453
Thompson RCA, Eckert J (1983) Observations on *Echinococcus multilocularis* in the definitive host. Z Parasitenk 69(3):335–345. https://doi.org/10.1007/BF00927875
Thompson RCA, Lymbery AJ (1995) *Echinococcus* and hydatid disease. CAB International, Wallingford
Thompson RC, McManus D (2001) Aetiology: parasites and life cycles. In: Eckert J, Gemmell MA, Meslin FX, Pawowski ZS (eds) WHO/OIE manual on echinococcosis in humans and animals: a public health problem of global concern. World Health Organisation/World Organisation of Animal Health, Paris, pp 1–16
Torgerson PR, MacPherson CNL (2011) The socioeconomic burden of parasitic zoonoses: global trends. Vet Parasitol 182:79–95. https://doi.org/10.1016/j.vetpar.2011.07.017
Torgerson PR, Keller K, Magnotta M et al (2010) The global burden of alveolar echinococcosis. PLoS Neg Trop Dis 4(6):e722. https://doi.org/10.1371/journal.pntd.0000722
Wahlström H, Enemark HL, Davidson RK et al (2015) Present status, actions taken and future considerations due to the findings of *Echinococcus multilocularis* in two Scandinavian countries. Vet Parasitol 213:172–181
Wang H, Li Z, Gao F et al (2016) Immunoprotection of recombinant Eg. P29 against *Echinococcus granulosus* in sheep. Vet Res Commun 40(2):73–79
WHO/OIE (2001) WHO/OIE manual on Echinococcosis in humans and animals. In: Eckert J, Gemmell MA, Meslin F-X, Pawlowski Z (eds) Office International des Épizooties, Paris (in press)
Yamamoto N, Katakura Y, Miyake H (2001) Risk factors for human alveolar echinococcosis: a case-control study in Hokkaido, Japan. Ann Trop Med Parasitol 95(7):689–696
Zajac A, Fairman D, McGee E et al (2020) Alveolar echinococcosis in a dog in the eastern United States. J Vet Diagn Investig 32:742–746

Toxoplasmosis in Northern Regions

Émilie Bouchard, Pikka Jokelainen, Rajnish Sharma, Heather Fenton, and Emily J. Jenkins

1 Introduction

Toxoplasma gondii, one of the most successful parasites in the world, is capable of infecting nearly all warm-blooded animals, including over 350 mammalian and avian species (Tenter et al. 2000; Lindsay and Dubey 2007; Dubey 2010). It is estimated that a third of the world human population has been exposed to and may be chronically infected with *T. gondii* (Tenter et al. 2000). *Toxoplasma gondii* is ranked in the second highest category of biological agents that could cause serious epidemics in both human and animal populations (Gajadhar and Allen 2004) and the fourth most important food-borne parasite globally after *Taenia solium*, *Echinococcus granulosus*, and *E. multilocularis* (FAO and WHO 2014). *Toxoplasma gondii* can cause disease, toxoplasmosis, in both humans and animals. Economic

É. Bouchard (✉) · E. J. Jenkins
Department of Veterinary Microbiology, Western College of Veterinary Medicine, University of Saskatchewan, Saskatoon, SK, Canada
e-mail: emb232@mail.usask.ca; emily.jenkins@usask.ca

P. Jokelainen
Laboratory of Parasitology, Infectious Disease Preparedness, Statens Serum Institut, Copenhagen, Denmark
e-mail: pijo@ssi.dk

R. Sharma
Department of Veterinary Microbiology, Western College of Veterinary Medicine, University of Saskatchewan, Saskatoon, SK, Canada

Centre for One Health, College of Veterinary Science, Guru Angad Dev Veterinary and Animal Sciences University, Ludhiana, Punjab, India
e-mail: ras863@mail.usask.ca

H. Fenton
Ross University School of Veterinary Medicine, Basseterre, St. Kitts, West Indies
e-mail: hfenton@rossvet.edu.kn

M. Tryland (ed.), *Arctic One Health*, https://doi.org/10.1007/978-3-030-87853-5_13

losses due to abortions that the parasite causes in domestic animals can be substantial (Dubey 2009a). While the infection may cause only mild symptoms and clinical signs, it can also cause severe, even life-threatening manifestations, especially in immunocompromised people and developing fetuses (Desmonts and Couvreur 1974; Montoya and Liesenfeld 2004; Dubey 2010). Therefore, *T. gondii* has both public health and veterinary importance, and a One Health approach is needed in risk mitigation.

Toxoplasma gondii successfully transmits worldwide, including cold climates and northern latitudes (defined here as the Arctic and sub-Arctic regions), where it survives inside its hosts as tissue cysts and in the environment as sporulated oocysts (Dubey 2010). Serological studies in animals and humans in the northern hemisphere have revealed an increasing exposure to the parasite along a north-to-south gradient (Messier et al. 2009; Jokelainen et al. 2010; Malmsten et al. 2011; Suvisaari et al. 2017). As the world changes, the ecology of many diseases is expected to shift (Patz et al. 2000; Deksne et al. 2020). With changing ecological factors and expansion of the geographical range of parasitic diseases due to natural and man-made causes, exposure rates to *T. gondii* in the circumpolar North are likely to increase (Pilfold et al. 2021).

2 Life Cycle and Transmission

The life cycle of *T. gondii* is divided into two phases: a sexual phase that occurs naturally only within felids (Martorelli Di Genova et al. 2019) and an asexual phase that can occur within all warm-blooded animals acting as intermediate hosts (Cleary et al. 2002). To the best of our current knowledge, only felids act as definitive hosts (Dubey et al. 2004). The three infective stages of *T. gondii* are as follows: (1) a rapidly dividing, invasive tachyzoite systemically distributed in hosts, including across the placenta, in the acute phase of the infection; (2) a slowly dividing bradyzoite forming tissue cysts, especially within muscle and nervous tissue (Fig. 1); and (3) an environmental stage, the sporozoite, which develops in the environment inside an oocyst produced by sexual reproduction in the intestinal epithelium of the definitive host (Robert-Gangneux and Darde 2012). Tachyzoites and bradyzoites result from asexual reproduction in all hosts, while oocysts are the result of sexual reproduction and are naturally produced in the intestinal epithelium of the definitive hosts only (Dubey 2010). If infected, over the course of few weeks, a cat may shed millions of unsporulated oocysts in feces into the environment, and the shedding of oocysts can occur more than once in the lifetime of the cat following re-infection (Dubey 1995; Zulpo et al. 2018). The unsporulated oocysts sporulate and become infective within 1–5 days, faster in humid conditions and warmer temperature (Dubey et al. 1970). Three major routes of transmission are fecal–oral transmission (environmental route, through soil, fresh produce, or water contaminated with oocysts), transmission by carnivorism (through raw/undercooked meat or other tissues containing bradyzoites), and transplacental transmission (through tachyzoites via vertical route) (Dubey 2009b). The consequences of

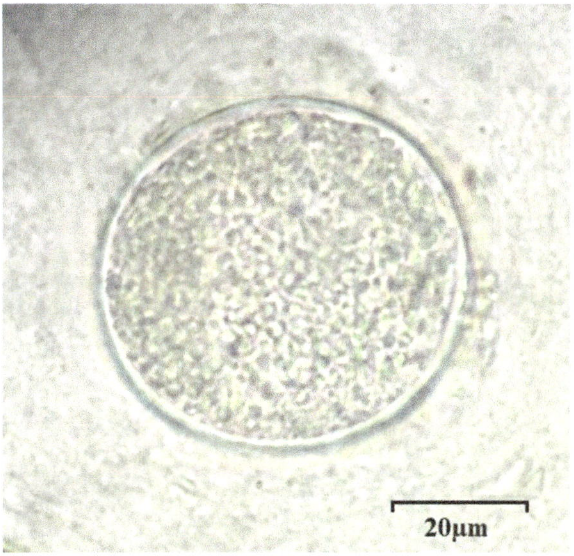

Fig. 1 *Toxoplasma gondii* tissue cyst in the brain of an experimentally infected reindeer (*Rangifer tarandus*), visualized on compound microscopy (60×). (Reproduced from Bouchard et al. 2017)

transplacental transmission for the fetus are usually more severe if infection occurs in early pregnancy and may lead to stillbirth or severe neurological complications. Most human infections are acquired later in life, via the environmental or the meat-borne routes (Tenter et al. 2000). The relative importance of these two transmission routes is largely unknown.

Herbivores acquire *T. gondii* infection typically by ingesting oocysts from the environment, while carnivores and omnivores can become infected either by ingesting oocysts from the environment or by ingesting tissue cysts containing bradyzoites in the tissues of infected animals. Once introduced into a food web, *T. gondii* can be maintained by hosts ingesting tissue cysts from infected hosts, via carnivorism, cannibalism, and scavenging (Dubey 2010; Wilson et al. 2020). In addition to humans, vertical transmission has also been reported in several domestic and free-ranging animal hosts (Calero-Bernal et al. 2013; Vargas-Villavicencio et al. 2016; Verma et al. 2016) and could play a role in transmission in a population (Hide et al. 2009). The importance of the vertical transmission route in the Arctic and sub-Arctic ecosystems remains to be better elucidated.

In the North, environmental contamination with *T. gondii* oocysts could occur in boreal and sub-Arctic regions where domestic cats and free-ranging wild felids are present (Simon et al. 2013a). However, free-ranging wild and domestic felids are largely absent above the treeline (Baker et al. 2018), which makes the transmission of *T. gondii* enigmatic in this environment. Migratory hosts, such as Arctic nesting birds, ungulates, and marine mammals can become infected through consumption of oocysts in sub-Arctic environments (Elmore et al. 2014). In Arctic regions, resident carnivores and humans could become infected through consumption of tissue cysts in migratory animals (Fig. 2) (Bachand et al. 2019).

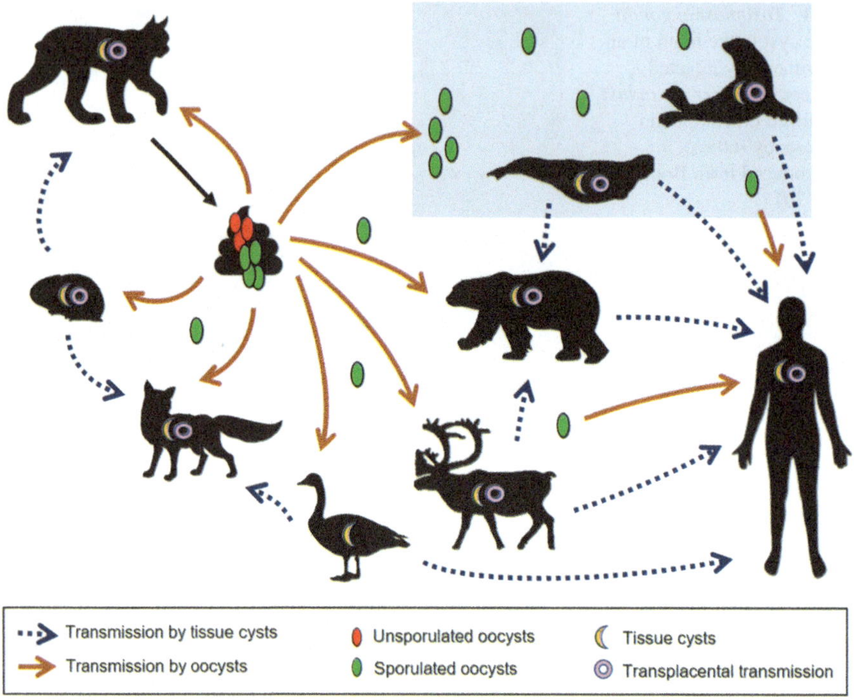

Fig. 2 Potential routes of transmission of *Toxoplasma gondii* in the North, with focus on free-ranging hosts and the shared environment

Food-borne transmission is likely a significant route of *T. gondii* infection among northern peoples, given their close contact with the land and wildlife and particularly due to dietary preferences for raw, fermented, or dried meat (Messier et al. 2009; Jenkins et al. 2013). Around the world, cultural habits affect the acquisition of *T. gondii* via ingestion of tissue cysts in undercooked meat (Hill and Dubey 2002). In the North, seal, ptarmigan, and caribou are often consumed without cooking (Mcdonald et al. 1990). Consumption of marine mammals and seafood has been found to be a risk factor for *T. gondii* seropositivity in epidemiological studies in the Canadian North (Goyette et al. 2014). Filter-feeding invertebrates and fish may filter and concentrate the oocysts from the marine environment (Massie et al. 2010). Moreover, skinning of animals for fur has been identified as a risk factor for *T. gondii* seropositivity (Mcdonald et al. 1990). An epidemiological investigation of an outbreak of toxoplasmosis in Nunavik, northern Québec (Canada), identified skinning of animals and consumption of raw caribou as potential risk factors (Mcdonald et al. 1990). In response to this outbreak, guidelines and a screening program to prevent *T. gondii* infection in pregnant women were developed in the region (Lavoie et al. 2008). Outbreaks of acute toxoplasmosis are rare in North America, but both food- and waterborne outbreaks have been reported, and outbreaks might occur more often in the future (Aramini et al. 1999; Gaulin et al.

2020). According to the Nunavik Inuit Health Survey in 2004, frequent cleaning of water reservoirs and consuming untreated surface water are risk factors for *T. gondii* seropositivity in northern Canada (Martin et al. 2007; Messier et al. 2007). Overall, the relative contributions of the different transmission routes, including food-borne and waterborne routes, are still unknown, and source attribution is challenging for this ubiquitous parasite (Koutsoumanis et al. 2018).

3 Genetic Diversity

Toxoplasma gondii is a single species of the genus *Toxoplasma*, with multiple genotypes (Galal et al. 2019). Studies in wildlife have demonstrated a wide host and geographic range as well as wider than previously known genetic diversity of *T. gondii* (Grigg and Sundar 2009; Wendte et al. 2011; Shwab et al. 2014). Knowledge of genotypes and strains circulating in a particular geographical region is important to better understand the epidemiology, population structure, and phylogeny. In laboratory mice, virulence varies greatly depending on the strain of the parasite (Saeij et al. 2005; Dubremetz and Lebrun 2012). Disease-producing strains have also been found in sylvatic cycles, with parasite genotype associated with disease severity (Boothroyd and Grigg 2002). Highly virulent strains can potentially pose a threat to wildlife health and a higher risk of severe toxoplasmosis if transmitted to humans (Gerhold et al. 2017). Several methods are used to identify *T. gondii* genotypes, including multilocus polymerase chain reaction-restriction fragment length polymorphism (PCR-RFLP), microsatellite analysis, and multilocus sequence typing (Su et al. 2006).

A limited number of studies have identified genotypes of *T. gondii* in the circumpolar regions. In North American wildlife, type 12 strains dominate, followed by strains of types II and III. In Alaska (USA), type II and atypical genotypes were described in red fox (*Vulpes vulpes*) and black bear (*Ursus americanus*) (Dubey et al. 2010, 2011). In northern Québec (Canada), type II was described in Canada geese (*Branta canadensis*), lesser snow geese (*Chen caerulescens*), and black bear (Dubey et al. 2008; Bachand et al. 2019). Type II was also described in beluga (*Delphinapterus leucas*) from the St. Lawrence Estuary in Québec (Canada) (Iqbal et al. 2018). Two new atypical genotypes of *T. gondii* have been detected (but not fully described) in beluga in the Eastern Beaufort Sea (Haman et al. 2013).

In Finland, type II strains have caused mortality in wildlife (Jokelainen et al. 2011; Jokelainen and Nylund 2012), as well as in domestic hosts (Jokelainen et al. 2012). In Denmark, type II was detected in a captive harbor porpoise (*Phocoena phocoena*) (Herder et al. 2015). In clinical samples from humans in Denmark, type II was predominant, while a substantial proportion of other genotypes were also described (Jokelainen et al. 2018). In Norway, type II and III and atypical genotypes have been described in wildlife from the high Arctic archipelago of Svalbard (Prestrud et al. 2008a, b).

Type II *T. gondii* strains, which are common in Europe, appear widespread in wildlife in the circumpolar region, which may reflect spillover from domestic cycles to wildlife as well as transport from sub-Arctic regions via migratory birds or in freshwater runoff. However, other genotypes are also circulating in northern latitudes and some may reflect a more localized circulation in wild felids and their prey. The genotyping data from some Arctic regions are sparse and scattered—in particular, no genotype information is available from Russia. Further studies are required to understand the molecular epidemiology and biological significance of genetic diversity of this zoonotic parasite in the Arctic and connected ecosystems.

4 Hosts

4.1 *Toxoplasma gondii* and Toxoplasmosis in Humans

Human toxoplasmosis is generally thought to be underreported due to lacking or limited formal surveillance, and underdiagnosed due to often mild or nonspecific symptoms (Barry et al. 2013). Toxoplasmosis can be life-threatening, in particular in immunocompromised individuals, and is a leading food-borne cause of hospitalization and death even in developed countries (Montoya and Liesenfeld 2004; Scallan et al. 2011). Congenital infection can cause devastating manifestations, ranging from ocular disease to severe hydrocephalus (Desmonts and Couvreur 1974; Hill and Dubey 2002), and there may be sequelae later in life (Wilson et al. 1980). More recently, *T. gondii* infection has been found to be associated with schizophrenia and other psychiatric and mental health issues (Yolken and Torrey 2008; Weiss and Dubey 2009).

Toxoplasma gondii infection has been named as the most important parasitic infection in the North American Arctic in terms of public health impact (Hotez 2010). Cultural practices such as food preparation and dietary traditions further influence the risk of acquiring the infection in northern regions (Jenkins et al. 2013). For instance, *T. gondii* seroprevalence in Inuit communities in Nunavik in 2004 was approximately 60%, which is significantly higher than the North American average of 10–20% and twice the global average of approximately 30% (Tenter et al. 2000; Robert-Gangneux and Darde 2012; Jones et al. 2018; Molan et al. 2019). Higher seroprevalences in Inuit than in Cree in shared communities in northern Québec have been reported, possibly due to Inuit preferences for uncooked meat since other risk factors such as water sources were considered shared (Sampasa-Kanyinga et al. 2012). A lower seroprevalence of 8% was found in other Inuit regions, such as the Inuvialuit Settlement Region (NT), 11% in Nunatsiavut (NL), and 32% in Nunavut (NU) (Goyette et al. 2014), which could reflect variation in dietary preferences (harvested wildlife vs. store bought, raw vs. cooked, terrestrial vs. marine wildlife) and differences in environmental transmission across the Canadian Arctic. In Alaska, a serosurvey in the 1960s and 1970s showed that 16% of Alaska Natives had antibodies to *T. gondii*, very similar to the North American average (Peterson et al. 1974); a more recent study showed substantially lower seroprevalence in Alaskans (3% of 887) (Miernyk et al. 2019), similar to declining trends observed in the contiguous USA. In Finland, the seroprevalence was lower in the north (11%) than in other parts of the country (18–23%) (Suvisaari et al. 2017). A similar gradient was reported in Nunavik, Canada (Messier et al. 2009) and also in pregnant women from Sweden (Ljungström et al. 1995). *Toxoplasma gondii* seroprevalence may be higher in more southern communities due to higher levels of environmental contamination with oocysts, increased survival and development rate of oocysts at warmer temperatures, and/or higher prevalence in animal hosts locally raised or hunted for consumption.

4.2 *Toxoplasma gondii* and Toxoplasmosis in Wildlife

In the circumpolar Arctic, a substantial proportion of terrestrial and marine mammals, as well as some avian species, have antibodies to *T. gondii*, with seroprevalences ranging from 10 to 60% in carnivores and 1 to 40% in herbivores (Jenkins et al. 2013). In Alaska, antibodies against *T. gondii* have been detected in a wide variety of species, including lynx (*Felis lynx*), black bears, grizzly bears (*Ursus arctos horribilis*), wolves (*Canis lupus*), and herbivores (Zarnke et al. 1997, 2000, 2001). Seropositivity in polar bears (*Ursus maritimus*) has been documented in Eastern Greenland, Russia, Alaska, Canada, and Norway (Rah et al. 2005; Oksanen et al. 2009), with a seroprevalence as high as 70% in western Hudson Bay, Canada (Pilfold et al. 2021). Antibodies have also been detected in Antarctic pinnipeds, demonstrating how well-established *T. gondii* is in marine ecosystems (Rengifo-Herrera et al. 2012). In Alaska, among marine mammals, antibodies have been detected in walrus (*Odobenus rosmarus*), Steller sea lions (*Eumetopias jubatus*), harbor seal (*Phoca vitulina*), ringed seal (*Pusa hispida*), spotted seal (*Phoca largha*), and bearded seal (*Erignathus barbatus*) (Dubey et al. 2003). In the high Arctic of Svalbard, ringed seals and bearded seals showed high seroprevalence of 19% and 67%, respectively (Jensen et al. 2010); however, another study did not detect any seropositive pinnipeds and whales from the North Atlantic (Oksanen et al. 1998). It should be emphasized that serology results of marine mammals may need to be interpreted with some caution.

Transmission of oocysts via freshwater runoff could contribute to the contamination of marine ecosystems and has been viewed as the most likely route of infection in sea otters off the coast of California (Conrad et al. 2005). As well, oocysts may travel via ocean currents from southern latitudes and infect northern marine wildlife.

It has also been suggested that *T. gondii* may enter the terrestrial Arctic ecosystem of northern Norway via migratory birds (Sandström et al. 2013), since no wild or domestic felids are present and an estimated 7% of barnacle geese (*Branta leucopsis*) on Svalbard are exposed to the parasite (Prestrud et al. 2007). Similarly, seroprevalence was 11% in migratory geese (primarily *Branta canadensis*) in Nunavik, eastern Canadian Arctic (Bachand et al. 2019), and over 30% in Ross's geese (*Anser rossii*) and lesser snow geese in Nunavut, central Canadian Arctic (Elmore et al. 2014). As sentinel host species (i.e., indicators of the presence of a pathogen within an ecosystem), high seroprevalence for antibodies to *T. gondii* was observed in wolverines (*Gulo gulo*) from the Northwest Territories (62%) in the western

Canadian Arctic (Sharma et al. 2019b) and in Arctic foxes (*Vulpes lagopus*) from Nunavut (39–64%), Nunavik (41%), and Norway (43%) (Prestrud et al. 2007; Elmore et al. 2016; Bachand et al. 2018, 2019).

While prevalence is generally higher in carnivores than herbivores, muskoxen (*Ovibos moschatus*) and mainland caribou (*Rangifer tarandus groenlandicus*) in the Northwest Territories and Nunavut, Canada, had an overall seroprevalence of 6.4% and 37%, respectively (Kutz et al. 2000, 2001). There are several reported seroprevalence estimates for moose (*Alces alces*): 15% in Canada/Nova Scotia (Siepierski et al. 1990), 1–23% in USA/Alaska (Kocan et al. 1986; Zarnke et al. 2000), 10% in USA/Minnesota (Verma et al. 2016), 20% in Sweden (Malmsten et al. 2011), 13% in Norway (Vikoren et al. 2004), and 10% in Finland (Jokelainen et al. 2010). Antibodies for *T. gondii* have also been reported in caribou in Alaska (USA) and northern Québec (Canada) (Mcdonald et al. 1990; Zarnke et al. 2000) and reindeer in Fennoscandia (Oksanen et al. 1997).

Experimental infection with oocysts delivered by intraruminal inoculation in reindeer led to fatal enteritis (Oksanen et al. 1996), while exposure to oocysts delivered by stomach tube led to largely subclinical infection with minor behavioral and respiratory changes (Bouchard et al. 2017). The dose of oocysts given, strain virulence, and host age could also explain the differences in pathology observed. No report of clinical toxoplasmosis has been described in naturally infected reindeer or caribou in the wild. However, encephalitis and placentitis associated with *T. gondii* was diagnosed in a full-term, stillborn reindeer fetus from the Houston Zoo, Texas (Dubey et al. 2002).

Although clinical disease associated with *T. gondii* is rarely reported in free-ranging wildlife, acute disseminated toxoplasmosis in Arctic foxes has been reported in Svalbard (Sorensen et al. 2005). Susceptibility to severe toxoplasmosis has been reported in Australian marsupials, Hawaiian birds, European brown hare (*Lepus europaeus*), and Eurasian red squirrels (*Sciurus vulgaris*) (Sedlak et al. 2000; Parameswaran et al. 2009; Jokelainen et al. 2011; Jokelainen and Nylund 2012; Work et al. 2016). Reasons for this remain largely unknown but may involve genetics, immune response, and a lack of coevolution with domestic cats (Ketz-Riley et al. 2003; Maubon et al. 2008; Shwab et al. 2018; Shapiro et al. 2019b).

Little is known about the population-level impact of *T. gondii* on free-ranging wildlife in the circumpolar Arctic; however, *T. gondii* is known to have an impact on free-ranging wildlife populations in Australia and the Pacific Northwest and can cause abortion and congenital disease in domestic livestock. Since *T. gondii* is capable of negatively affecting the reproductive success, it is considered to be of greater concern in declining populations, such as caribou and reindeer herds worldwide (Vors and Boyce 2009; Festa-Bianchet et al. 2011). Parasitic diseases have been identified as a threat within recovery strategies and management plans for different herds in Canada (Environment Canada 2012). For conservation as well as food safety, it is crucial to monitor the distribution and prevalence of *T. gondii* in Arctic wildlife.

Monitoring requires effective diagnostics. *Toxoplasma gondii* has developmental stages that live in tissues, and definitive diagnosis requires invasive biopsy or postmortem techniques. Therefore, antemortem diagnosis frequently relies on collection of blood and serological techniques to determine host response to exposure. Serology is suitable for screening to obtain population-level estimates of exposure to *T. gondii*, but definitive diagnosis of clinical toxoplasmosis should not rely on a single serological test but ideally look at rising titers, IgM and IgG levels, and the clinical picture (Li et al. 2016). Postmortem diagnosis can involve detection of antibodies in the heart's blood, chest fluid, or filter paper (i.e., fluid dried on filter paper strips) or direct detection methods: polymerase chain reaction (PCR) on tissues, histological demonstration of the parasite and/or antigens via immunohistochemistry, or isolation of the organism itself using mouse bioassay or cell cultures (Dubey 2010). Serological studies, especially in wildlife for which these tests are rarely optimized or validated (by comparing with reference tests, such as in Sharma et al. 2019a), are inherently limited; even a positive result indicates a life-time exposure, not necessarily a recent infection. False-positive results in sera from marine mammals, which often are rich in lipids, have been reported (Blanchet et al. 2014). Cross-reaction with other coccidian species might also occur with serological methods (Dubey and Lindsay 1996). Ideally, using a combination of serological and molecular methods can provide complementary information; serology overestimates the levels of active infection, while molecular methods frequently underestimate true infection status. This information is vital to develop a balanced approach to risks to human and animal health posed by this zoonotic parasite in the Arctic ecosystem.

4.3 *Toxoplasma gondii* in the Environment

Toxoplasma gondii is widely prevalent around the world, in part, due to the survival of sporulated oocysts in the environment (Dubey et al. 1998), broad intermediate host range, and multiple methods of transmission. This environmental reservoir of oocysts constitutes a considerable source of *T. gondii* infection, especially for herbivores (Dubey 2010). Herbivores can therefore be used as sentinels or indicators of environmental contamination with oocysts of *T. gondii*, and seroepidemiological studies can be helpful in detecting geographical patterns (Elmore et al. 2015). For example, similar to humans, there is an increasing north-to-south gradient in *T. gondii* seroprevalence in moose in Finland and Sweden, indicating a higher environmental oocyst contamination in the southern parts of these countries (Jokelainen et al. 2010; Malmsten et al. 2011).

Sporulated oocysts of *T. gondii* can survive for months in the soil in moist conditions (Frenkel et al. 1975; Lindsay and Dubey 2009). Environmental contamination with infective sporulated oocysts from felid feces has been determined to be the source of large waterborne outbreaks of toxoplasmosis in the Panama Canal and in southwestern Canada (Benenson et al. 1982; Bowie et al. 1997). Fewer oocysts survive in the soil under dry conditions (Lelu et al. 2012). In northern climates, snow cover likely increases oocyst survival as it reduces exposure to UV sunlight, insulates against extreme temperature, and maintains high humidity levels (Simon et al. 2013b). Oocysts can be transported from terrestrial to aquatic environments via snowmelt and runoff and thereby be disseminated in fresh- or seawater (Simon et al. 2013b). The circumpolar North is undergoing rapid warming, with a projected temperature increase of 4 to 7 °C over the next 100 years (Acia 2004). As a result, warmer and wetter conditions could affect the transmission of *T. gondii* in animals and people through enhanced survival and transport of oocysts from water sources flowing from south to north, as well as northward movement of felid definitive hosts. Although currently not common, domestic cats are present in some northern regions (Baker et al. 2018) and could play a role in environmental contamination, especially if immigration of people and their pet cats to the North increases along with resource extraction (such as oil, gas, and minerals) and employment opportunities due to shipping in the increasingly open waters of the Arctic.

5 Management

To minimize the risk of human exposure to *T. gondii,* recommendations include cooking meat, washing fruits and vegetables prior to consumption, washing kitchen utensils thoroughly with soap and water, and washing hands after contact with soil, carcasses, raw meat, and cat litter. Freezing meat for at least 3 days at −12 °C (or colder) may kill the parasite and could be considered as an additional step to other processing, such as drying or fermenting (Djurkovik-Djakovik and Milenkovic 2000). However, *T. gondii* survived in reindeer meat for up to 28 days at −11 to −56 °C (Kolychev 1969); more work is needed to determine if genotypes of *T. gondii* present in the Arctic have a higher freeze tolerance than genotypes present in more temperate regions, as what has been observed in other food-borne parasites in the genus *Trichinella* (Pozio 2016). Wearing gloves and thorough cleaning and disinfection of traps, knives, and other tools utilized in dressing, skinning, and processing hides and furs from animals that may harbor *T. gondii* are also recommended (Hueffer et al. 2013). To decrease risk of transmission, carcasses should be disposed of in a way that scavengers have no access to them, by burning or scavenger-proof fencing at landfills, which can be challenging in remote communities and when many animal carcasses are processed. Lynx may be a potential source of oocysts, although there are as yet no reports of oocyst shedding by Eurasian or North American lynx (Ryser-Degiorgis et al. 2006; Jokelainen et al. 2013; Simon et al. 2013a). A potential source of infection for people is drinking contaminated surface water or eating unwashed produce in areas that could be contaminated with sporulated oocysts from older feces or carcasses of felids. Litter boxes of domestic cats should be cleaned daily, because oocysts require 1–5 days to become infective, and litter should be disposed of in a way that would not contaminate watersheds. Mature cats not fed with raw meat in their diet and with no access to the outdoors are considered low risk for shedding oocysts of *T. gondii* (Jokelainen et al. 2012; Must et al. 2017; Shapiro et al. 2019a).

Some of these recommendations may not be achievable or culturally appropriate in remote areas of the circumpolar North, and some cultural traditions—such as

Fig. 3 Caribou meat in a community freezer in Nunavik, Canada

consumption of dry or fermented meat and drinking untreated water—may be potential risks. Other challenges include the microscopic size of the parasite, often the subclinical nature of the infection, lack of formal screening of harvested wildlife for *T. gondii*, and, potentially, lack of awareness of risks posed by undercooked meat and organs from harvested wildlife (Fig. 3). These can be addressed through developing northern capacity for veterinary public health measures. There is a need for more public education, particularly among high-risk populations such as pregnant women and immunocompromised individuals in Arctic regions. Targeting outreach efforts first to at-risk groups (e.g., pregnant women, immunocompromised) may more effectively mitigate risk while respecting traditional practices (DHSS 2005). Conversely, risk communication should not deter the general population from consuming harvested wildlife, which are highly nutritious sources of food and contribute to the cultural well-being of northern communities (Pufall et al. 2011; FAO 2013). Any specific public health recommendations for hunters, trappers, gatherers, and food preparers should be done in consultation and coordination with local and traditional knowledge holders and should respect the traditional cultural practices. Developing a better understanding of the impact of zoonotic diseases on Indigenous health would assist in determining the need for community-based monitoring programs and public health recommendations.

6 Future Challenges

Despite the absence of definitive hosts, *T. gondii* can thrive even in remote Arctic regions (Reiling and Dixon 2019). While felines play a crucial role in the ecology of *T. gondii*, the parasite can persist in their absence and be transported with wildlife and water even into the farthest reaches of the circumpolar North. Climate change is leading to more frequent and extreme rainfall events, which, in turn, can lead to increased risk for freshwater contamination with *T. gondii* via runoff (Charron et al. 2004). The North is undergoing rapid climate change and anthropogenic disturbance (Acia 2004), making it vulnerable to emergence of infectious diseases at the interface between people and wildlife. Wildlife are an inherent part of the Arctic and a cultural and socioeconomic cornerstone for northern peoples. Ongoing monitoring, particularly if there is a community interest, would assist in better understanding the ecology of the pathogen and methods to mitigate risks to both animals and people. While *T. gondii* is rarely mentioned specifically in species at risk recovery strategies and management plans, parasites and disease have been listed as threats to a number of species. As access to medical and veterinary care can be limited in remote regions, monitoring for pathogens of public health significance is challenging. More information is needed to better understand the ecology, epidemiology, and impact of *T. gondii* in Arctic regions, in order to put into practice culturally appropriate and effective measures to protect human, animal, and ecosystem health—One Health.

References

Acia (2004) Impacts of a warming Arctic: Arctic climate impact assessment. Cambridge University Press, Cambridge

Aramini JJ, Stephen C, Dubey JP et al (1999) Potential contamination of drinking water with *Toxoplasma gondii* oocysts. Epidemiol Infect 122(2):305–315

Bachand N, Ravel A, Leighton P et al (2018) Foxes (*Vulpes vulpes*) as sentinels for parasitic zoonoses, *Toxoplasma gondii* and *Trichinella nativa*, in the northeastern Canadian Arctic. Int J Parasitol Parasites Wildl 7(3):391–397

Bachand N, Ravel A, Leighton P et al (2019) Serological and molecular detection of *Toxoplasma gondii* in terrestrial and marine wildlife harvested for food in Nunavik, Canada. Parasit Vectors 12(1):155

Baker T, Flaig J, Shillingford M et al (2018) Ice road vets: perspectives on the role of veterinarians in northern community health. Can Vet J 59(6):668–672

Barry MA, Weatherhead JE, Hotez PJ et al (2013) Childhood parasitic infections endemic to the United States. Pediatr Clin N Am 60(2):471–485

Benenson MW, Takafuji ET, Lemon SM et al (1982) Oocyst-transmitted toxoplasmosis associated with ingestion of contaminated water. N Engl J Med 307(11):666–669

Blanchet MA, Godfroid J, Breines EM et al (2014) West Greenland harbour porpoises assayed for antibodies against *Toxoplasma gondii*: false positives with the direct agglutination method. Dis Aquat Org 108(3):181–186

Boothroyd JC, Grigg ME (2002) Population biology of *Toxoplasma gondii* and its relevance to human infection: do different strains cause different disease? Curr Opin Microbiol 5(4):438–442

Bouchard E, Sharma R, Bachand N et al (2017) Pathology, clinical signs, and tissue distribution of *Toxoplasma gondii* in experimentally infected reindeer (*Rangifer tarandus*). Int J Parasitol Parasites Wildl 6(3):234–240

Bouchard E, Elmore SA, Alisauskas RT et al (2019) Transmission dynamics of *Toxoplasma gondii* in Arctic foxes (*Vulpes lagopus*): a long-term mark-recapture serologic study at Karrak Lake, Nunavut, Canada. J Wildl Dis 55(3):619–626

Bowie WR, King AS, Werker DH et al (1997) Outbreak of toxoplasmosis associated with municipal drinking water. The BC Toxoplasma investigation team. Lancet 350(9072):173–177

Calero-Bernal R, Gomez-Gordo L, Saugar JM et al (2013) Congenital toxoplasmosis in wild boar (*Sus scrofa*) and identification of the *Toxoplasma gondii* types involved. J Wildl Dis 49(4):1019–1023

Canada E (2012) Recovery strategy for the Woodland Caribou, boreal population (*Rangifer tarandus* caribou) in Canada. Species at Risk Act Recovery Strategy Series. Environment Canada, Ottawa, p 138

Charron D, Thomas M, Waltner-Toews D et al (2004) Vulnerability of waterborne diseases to climate change in Canada: a review. J Toxicol Environ Health A 67(20–22):1667–1677

Cleary MD, Singh U, Blader IJ et al (2002) *Toxoplasma gondii* asexual development: identification of developmentally regulated genes and distinct patterns of gene expression. Eukaryot Cell 1(3):329–340

Conrad PA, Miller MA, Kreuder C et al (2005) Transmission of Toxoplasma: clues from the study of sea otters as sentinels of *Toxoplasma gondii* flow into the marine environment. Int J Parasitol 35(11–12):1155–1168

Deksne G, Davidson RK, Buchmann K et al (2020) Parasites in the changing world – ten timely examples from the Nordic-Baltic region. Parasite Epidemiol Control 10:e00150

Desmonts G, Couvreur J (1974) Congenital toxoplasmosis – prospective study of 378 pregnancies. N Engl J Med 290(20):1110–1116

DHSS (2005) Preventing infection with *Toxoplasma gondii*. Section of Epidemiology. State of Alaska Epidemiology, Bulletin 7

Djurkovik-Djakovik O, Milenkovic V (2000) Effect of refrigeration and freezing on survival of *Toxoplasma gondii* tissue cysts. Acta Vet (Beograd) 50(5/6):375–380

Dubey JP (1995) Duration of immunity to shedding of *Toxoplasma gondii* oocysts by cats. J Parasitol 81(3):410–415

Dubey JP (2009a) Toxoplasmosis in sheep—the last 20 years. Vet Parasitol 163(1–2):1–14

Dubey JP (2009b) History of the discovery of the life cycle of *Toxoplasma gondii*. Int J Parasitol 39(8):877–882

Dubey JP (2010) Toxoplasmosis of animals and humans. CRC Press, Boca Raton

Dubey JP, Lindsay DS (1996) A review of *Neospora caninum* and neosporosis. Vet Parasitol 67(1–2):1–59

Dubey JP, Miller NL, Frenkel JK et al (1970) The *Toxoplasma gondii* oocyst from cat feces. J Exp Med 132(4):636–662

Dubey JP, Lindsay DS, Speer CA (1998) Structures of *Toxoplasma gondii* tachyzoites, bradyzoites, and sporozoites and biology and development of tissue cysts. Clin Microbiol Rev 11(2):267–299

Dubey JP, Lewis B, Beam K et al (2002) Transplacental toxoplasmosis in a reindeer (*Rangifer tarandus*) fetus. Vet Parasitol 110(1–2):131–135

Dubey JP, Zarnke R, Thomas NJ et al (2003) *Toxoplasma gondii*, *Neospora caninum*, *Sarcocystis neurona*, and *Sarcocystis canis*-like infections in marine mammals. Vet Parasitol 116(4):275–296

Dubey JP, Graham DH, De Young RW et al (2004) Molecular and biologic characteristics of *Toxoplasma gondii* isolates from wildlife in the United States. J Parasitol 90(1):67–71

Dubey JP, Quirk T, Pittt JA et al (2008) Isolation and genetic characterization of *Toxoplasma gondii* from raccoons (*Procyon lotor*), cats (*Felis domesticus*), striped skunk (*Mephitis mephitis*), black bear (*Ursus americanus*), and cougar (*Puma concolor*) from Canada. J Parasitol 94(1):42–45

Dubey JP, Rajendran C, Ferreira LR et al (2010) A new atypical highly mouse virulent *Toxoplasma gondii* genotype isolated from a wild black bear in Alaska. J Parasitol 96(4):713–716

Dubey JP, Velmurugan GV, Rajendran C et al (2011) Genetic characterisation of *Toxoplasma gondii* in wildlife from North America revealed widespread and high prevalence of the fourth clonal type. Int J Parasitol 41(11):1139–1147

Dubremetz JF, Lebrun M (2012) Virulence factors of *Toxoplasma gondii*. Microbes Infect 14(15):1403–1410

Elmore SA, Huyvaert KP, Bailey LL et al (2014) *Toxoplasma gondii* exposure in arctic-nesting geese: a multi-state occupancy framework and comparison of serological assays. Int J Parasitol Parasites Wildl 3(2):147–153

Elmore SA, Samelius G, Fernando C et al (2015) Evidence for *Toxoplasma gondii* in migratory vs. nonmigratory herbivores in a terrestrial arctic ecosystem. Can J Zool 93(8):671–675

Elmore SA, Samelius G, Al-Adhami B et al (2016) Estimating *Toxoplasma gondii* exposure in Arctic foxes (*Vulpes lagopus*) while navigating the imperfect world of wildlife serology. J Wildl Dis 52(1):47–56

FAO (2013) Indigenous Peoples' food systems & well-being: interventions & policies for healthy communities. Food and Agriculture Organization of the United Nations, p 398

FAO, WHO (2014) Multicriteria-based ranking for risk management of foodborne parasites. Report of a joint FAO/WHO expert meeting, p 287

Festa-Bianchet M, Ray JC, Boutin S et al (2011) Conservation of caribou (*Rangifer tarandus*) in Canada: an uncertain future. Can J Zool 89(5):419–434

Frenkel JK, Ruiz A, Chinchilla M (1975) Soil survival of Toxoplasma oocysts in Kansas and Costa Rica. Am J Trop Med Hyg 24(3):439–443

Gajadhar AA, Allen JR (2004) Factors contributing to the public health and economic importance of waterborne zoonotic parasites. Vet Parasitol 126(1–2):3–14

Galal L, Hamidovic A, Darde ML et al (2019) Diversity of *Toxoplasma gondii* strains at the global level and its determinants. Food Waterborne Parasitol 15:e00052

Gaulin C, Ramsay D, Thivierge K et al (2020) Acute toxoplasmosis among Canadian deer hunters associated with consumption of undercooked deer meat hunted in the United States. Emerg Infect Dis 26(2):199–205

Gerhold RW, Saraf P, Chapman A et al (2017) *Toxoplasma gondii* seroprevalence and genotype diversity in select wildlife species from the southeastern United States. Parasit Vectors 10(1):508

Goyette S, Cao ZR, Libman M et al (2014) Seroprevalence of parasitic zoonoses and their relationship with social factors among the Canadian Inuit in Arctic regions. Diagn Microbiol Infect Dis 78(4):404–410

Grigg ME, Sundar N (2009) Sexual recombination punctuated by outbreaks and clonal expansions predicts *Toxoplasma gondii* population genetics. Int J Parasitol 39(8):925–933

Haman KH, Raverty S, Wendte JM et al (2013) Infected tissues from hunter harvested beluga (*Delphinapterus leucas*) in the Western Canadian Arctic. In: 44th Annual IAAAM conference, the marine mammal Center Sausalito, California, 2013

Herder V, Van De Velde N, Hojer Kristensen J et al (2015) Fatal disseminated *Toxoplasma gondii* infection in a captive harbour porpoise (*Phocoena phocoena*). J Comp Pathol 153(4):357–362

Hide G, Morley EK, Hughes JM et al (2009) Evidence for high levels of vertical transmission in *Toxoplasma gondii*. Parasitology 136(14):1877–1885

Hill D, Dubey JP (2002) *Toxoplasma gondii*: transmission, diagnosis and prevention. Clin Microbiol Infect 8(10):634–640

Hotez PJ (2010) Neglected infections of poverty among the indigenous peoples of the Arctic. PLoS Negl Trop Dis 4(1)

Hueffer K, Parkinson AJ, Gerlach R et al (2013) Zoonotic infections in Alaska: disease prevalence, potential impact of climate change and recommended actions for earlier disease detection, research, prevention and control. Int J Circumpolar Health 72(1)

Iqbal A, Measures L, Lair S et al (2018) *Toxoplasma gondii* infection in stranded St. Lawrence Estuary beluga *Delphinapterus leucas* in Quebec. Can Dis Aquat Organ 130(3):165–175

Jenkins EJ, Castrodale LJ, De Rosemond SJC et al (2013) Tradition and transition: parasitic zoonoses of people and animals in Alaska, Northern Canada, and Greenland. In: Rollinson D (ed) Advances in parasitology. Academic, pp 33–204

Jensen SK, Aars J, Lydersen C et al (2010) The prevalence of *Toxoplasma gondii* in polar bears and their marine mammal prey: evidence for a marine transmission pathway? Polar Biol 33(5):599–606

Jokelainen P, Nylund M (2012) Acute fatal toxoplasmosis in three Eurasian red squirrels (*Sciurus vulgaris*) caused by genotype II of *Toxoplasma gondii*. J Wildl Dis 48(2):454–457

Jokelainen P, Nareaho A, Knaapi S et al (2010) *Toxoplasma gondii* in wild cervids and sheep in Finland: north-south gradient in seroprevalence. Vet Parasitol 171(3–4):331–336

Jokelainen P, Isomursu M, Nareaho A et al (2011) Natural *Toxoplasma gondii* infections in European brown hares and mountain hares in Finland: proportional mortality rate, antibody prevalence, and genetic characterization. J Wildl Dis 47(1):154–163

Jokelainen P, Simola O, Rantanen E et al (2012) Feline toxoplasmosis in Finland: cross-sectional epidemiological study and case series study. J Vet Diagn Investig 24(6):1115–1124

Jokelainen P, Deksne G, Holmala K et al (2013) Free-ranging Eurasian lynx (*Lynx lynx*) as host of *Toxoplasma gondii* in Finland. J Wildl Dis 49(3):527–534

Jokelainen P, Murat JB, Nielsen HV (2018) Direct genetic characterization of *Toxoplasma gondii* from clinical samples from Denmark: not only genotypes II and III. Eur J Clin Microbiol Infect Dis 37(3):579–586

Jones JL, Kruszon-Moran D, Elder S et al (2018) *Toxoplasma gondii* infection in the United States, 2011-2014. Am J Trop Med Hyg 98(2):551–557

Ketz-Riley CJ, Ritchey JW, Hoover JP et al (2003) Immunodeficiency associated with multiple concurrent infections in captive Pallas' cats (*Otocolobus manul*). J Zoo Wildl Med 34(3):239–245

Kocan AA, Barron SJ, Fox JC et al (1986) Antibodies to Toxoplasma-*gondii* in Moose (*Alces-alces* L) from Alaska. J Wildl Dis 22(3):432–432

Kolychev VV (1969) Viability of Toxoplasma in reindeer meat. Veterinariia 46(8):99–100

Koutsoumanis K, Allende A, Alvarez-Ordóñez A et al (2018) Public health risks associated with food-borne parasites. EFSA J 16(12):113

Kutz SJ, Elkin B, Gunn A et al (2000) Prevalence of *Toxoplasma gondii* antibodies in muskox (*Ovibos moschatus*) sera from northern Canada. J Parasitol 86(4):879–882

Kutz SJ, Elkin BT, Panayi D et al (2001) Prevalence of *Toxoplasma gondii* antibodies in barren-ground caribou (*Rangifer tarandus* groenlandicus) from the Canadian Arctic. J Parasitol 87(2):439–442

Lavoie E, Lévesque B, Proulx J-F et al (2008) Evaluation of the efficacy of the *Toxoplasma gondii* screening program among pregnant women in Nunavik, 1994-2003. Can J Public Health 99(5):397–400

Lelu M, Villena I, Darde ML et al (2012) Quantitative estimation of the viability of *Toxoplasma gondii* oocysts in soil. Appl Environ Microbiol 78(15):5127–5132

Li X, Pomares C, Gonfrier G et al (2016) Multiplexed anti-Toxoplasma IgG, IgM, and IgA assay on plasmonic gold chips: towards making mass screening possible with dye test precision. J Clin Microbiol 54(7):1726–1733

Lindsay DS, Dubey JP (2007) Toxoplasmosis in wild and domestic animals. In: Weiss LM, Kim K (eds) *Toxoplasma gondii*: the model apicomplexan. Perspectives and methods. Academic, London, pp 133–152

Lindsay DS, Dubey JP (2009) Long-term survival of *Toxoplasma gondii* sporulated oocysts in seawater. J Parasitol 95(4):1019–1020

Ljungström I, Gille E, Nokes J et al (1995) Seroepidemiology of *Toxoplasma gondii* among pregnant women in different parts of Sweden. Eur J Epidemiol 11(2):149–156

Malmsten J, Jakubek EB, Bjorkman C (2011) Prevalence of antibodies against *Toxoplasma gondii* and *Neospora caninum* in moose (*Alces alces*) and roe deer (*Capreolus capreolus*) in Sweden. Vet Parasitol 177(3-4):275–280

Martin D, Belanger D, Gosselin P et al (2007) Drinking water and potential threats to human health in Nunavik: adaptation strategies under climate change conditions. Arctic 60:195–202

Martorelli Di Genova B, Wilson SK, Dubey JP et al (2019) Intestinal delta-6-desaturase activity determines host range for Toxoplasma sexual reproduction. PLoS Biol 17(8):e3000364

Massie GN, Ware MW, Villegas EN et al (2010) Uptake and transmission of *Toxoplasma gondii* oocysts by migratory, filter-feeding fish. Vet Parasitol 169(3–4):296–303

Maubon D, Ajzenberg D, Brenier-Pinchart MP et al (2008) What are the respective host and parasite contributions to toxoplasmosis? Trends Parasitol 24(7):299–303

Mcdonald JC, Gyorkos TW, Alberton B et al (1990) An outbreak of toxoplasmosis in pregnant women in northern Quebec. J Infect Dis 161(4):769–774

Messier V, Lévesque B, Proulx JF et al (2007) Zoonotic diseases, drinking water and gastroenteritis in Nunavik: a brief portrait. Nunavik Inuit Health Survey 2004, Institut national de santé publique du Québec

Messier V, Levesque B, Proulx JF et al (2009) Seroprevalence of *Toxoplasma gondii* among Nunavik Inuit (Canada). Zoonoses Public Health 56(4):188–197

Miernyk KM, Bruden D, Parkinson AJ et al (2019) Human Seroprevalence to 11 zoonotic pathogens in the U.S. Arctic, Alaska. Vector Borne Zoonotic Dis 19(8):563–575

Molan A, Nosaka K, Hunter M et al (2019) Global status of *Toxoplasma gondii* infection: systematic review and prevalence snapshots. Trop Biomed 36(4):898–925

Montoya JG, Liesenfeld O (2004) Toxoplasmosis. Lancet 363(9425):1965–1976

Must K, Hytönen MK, Orro T et al (2017) *Toxoplasma gondii* seroprevalence varies by cat breed. PLoS One 12(9):e0184659

Oksanen A, Gustafsson K, Lunden A et al (1996) Experimental *Toxoplasma gondii* infection leading to fatal enteritis in reindeer (*Rangifer tarandus*). J Parasitol 82(5):843–845

Oksanen A, Åsbakk K, Nieminen M et al (1997) Antibodies against *Toxoplasma gondii* in Fennoscandian reindeer – association with the degree of domestication. Parasitol Int 46:255–261

Oksanen A, Tryland M, Johnsen K et al (1998) Serosurvey of *Toxoplasma gondii* in North Atlantic marine mammals by the use of agglutination test employing whole tachyzoites and dithiothreitol. Comp Immunol Microbiol Infect Dis 21(2):107–114

Oksanen A, Asbakk K, Prestrud KW et al (2009) Prevalence of antibodies against *Toxoplasma gondii* in polar bears (*Ursus maritimus*) from Svalbard and East Greenland. J Parasitol 95(1):89–94

Parameswaran N, O'handley RM, Grigg ME et al (2009) Vertical transmission of *Toxoplasma gondii* in Australian marsupials. Parasitology 136(9):939–944

Patz JA, Graczyk TK, Geller N et al (2000) Effects of environmental change on emerging parasitic diseases. Int J Parasitol 30(12–13):1395–1405

Peterson DR, Cooney MK, Beasley RP (1974) Prevalence of antibody to Toxoplasma among Alaskan natives – relation to exposure to Felidae. J Infect Dis 130(6):557–563

Pilfold NW, Richardson ES, Ellis J et al (2021) Long-term increases in pathogen seroprevalence in polar bears (*Ursus maritimus*) influenced by climate change. Glob Change Biol. https://doi.org/10.1111/gcb.15537

Pozio E (2016) Adaptation of Trichinella spp. for survival in cold climates. Food Waterborne Parasitol 4:4–12

Prestrud KW, Asbakk K, Fuglei E et al (2007) Serosurvey for *Toxoplasma gondii* in Arctic foxes and possible sources of infection in the high Arctic of Svalbard. Vet Parasitol 150(1–2):6–12

Prestrud KW, Asbakk K, Mork T et al (2008a) Direct high-resolution genotyping of *Toxoplasma gondii* in arctic foxes (*Vulpes lagopus*) in the remote arctic Svalbard archipelago reveals widespread clonal type II lineage. Vet Parasitol 158(1–2):121–128

Prestrud KW, Dubey JP, Asbakk K et al (2008b) First isolate of *Toxoplasma gondii* from arctic fox (*Vulpes lagopus*) from Svalbard. Vet Parasitol 151(2–4):110–114

Pufall EL, Jones AQ, Mcewen SA et al (2011) Perception of the importance of traditional country foods to the physical, mental, and spiritual health of Labrador Inuit. Arctic 64(2):242–250

Rah H, Chomel BB, Follmann EH et al (2005) Serosurvey of selected zoonotic agents in polar bears (*Ursus maritimus*). Vet Rec 156(1):7–13

Reiling SJ, Dixon BR (2019) *Toxoplasma gondii*: how an Amazonian parasite became an Inuit health issue. Can Commun Dis Rep 45(7–8):183–190

Rengifo-Herrera C, Ortega-Mora LM, Alvarez-Garcia G et al (2012) Detection of *Toxoplasma gondii* antibodies in Antarctic pinnipeds. Vet Parasitol 190(1–2):259–262

Robert-Gangneux F, Darde ML (2012) Epidemiology of and diagnostic strategies for toxoplasmosis. Clin Microbiol Rev 25(2):264–296

Ryser-Degiorgis MP, Jakubek EB, Af Segerstad CH et al (2006) Serological survey of *Toxoplasma gondii* infection in free-ranging Eurasian lynx (*Lynx lynx*) from Sweden. J Wildl Dis 42(1):182–187

Saeij JP, Boyle JP, Boothroyd JC (2005) Differences among the three major strains of *Toxoplasma gondii* and their specific interactions with the infected host. Trends Parasitol 21(10):476–481

Sampasa-Kanyinga H, Levesque B, Anassour-Laouan-Sidi E et al (2012) Zoonotic infections in native communities of James Bay, Canada. Vector Borne Zoonotic Dis 12(6):473–481

Sandström CA, Buma AG, Hoye BJ et al (2013) Latitudinal variability in the seroprevalence of antibodies against *Toxoplasma gondii* in non-migrant and Arctic migratory geese. Vet Parasitol 194(1):9–15

Scallan E, Hoekstra RM, Angulo FJ et al (2011) Foodborne illness acquired in the United States—major pathogens. Emerg Infect Dis 17(1):7–15

Sedlak K, Literak I, Faldyna M et al (2000) Fatal toxoplasmosis in brown hares (*Lepus europaeus*): possible reasons of their high susceptibility to the infection. Vet Parasitol 93(1):13–28

Shapiro K, Bahia-Oliveira L, Dixon B et al (2019a) Environmental transmission of *Toxoplasma gondii*: oocysts in water, soil and food. Food Waterborne Parasitol 15:e00049

Shapiro K, Vanwormer E, Packham A et al (2019b) Type X strains of *Toxoplasma gondii* are virulent for southern sea otters (*Enhydra lutris* nereis) and present in felids from nearby watersheds. Proc Biol Sci 286(1909):20191334

Sharma R, Parker S, Al-Adhami B et al (2019a) Comparison of tissues (heart vs. brain) and serological tests (MAT, ELISA and IFAT) for detection of *Toxoplasma gondii* in naturally infected wolverines (*Gulo gulo*) from the Yukon, Canada. Food Waterborne Parasitol 15: e00046

Sharma R, Parker S, Elkin B et al (2019b) Risk factors and prevalence of antibodies for *Toxoplasma gondii* in diaphragmatic fluid in wolverines (*Gulo gulo*) from the Northwest Territories, Canada. Food Waterborne Parasitol 15:e00056

Shwab EK, Zhu XQ, Majumdar D et al (2014) Geographical patterns of *Toxoplasma gondii* genetic diversity revealed by multilocus PCR-RFLP genotyping. Parasitology 141(4):453–461

Shwab EK, Saraf P, Zhu XQ et al (2018) Human impact on the diversity and virulence of the ubiquitous zoonotic parasite *Toxoplasma gondii*. Proc Natl Acad Sci USA 115(29):E6956–E6963

Siepierski SJ, Tanner CE, Embil JA (1990) Prevalence of antibody to *Toxoplasma gondii* in the moose (*Alces alces* americana Clinton) of Nova Scotia, Canada. J Parasitol 76(1):136–138

Simon A, Bigras Poulin M, Rousseau AN et al (2013a) Spatiotemporal dynamics of *Toxoplasma gondii* infection in Canadian lynx (*Lynx canadensis*) in western Quebec, Canada. J Wildl Dis 49(1):39–48

Simon A, Rousseau AN, Savary S et al (2013b) Hydrological modelling of *Toxoplasma gondii* oocysts transport to investigate contaminated snowmelt runoff as a potential source of infection for marine mammals in the Canadian Arctic. J Environ Manag 127:150–161

Sorensen KK, Mork T, Sigurdardottir OG et al (2005) Acute toxoplasmosis in three wild arctic foxes (*Alopex lagopus*) from Svalbard; one with co-infections of *Salmonella enteritidis* PT1 and Yersinia pseudotuberculosis serotype 2b. Res Vet Sci 78(2):161–167

Su C, Zhang X, Dubey JP (2006) Genotyping of *Toxoplasma gondii* by multilocus PCR-RFLP markers: a high resolution and simple method for identification of parasites. Int J Parasitol 36(7):841–848

Suvisaari J, Torniainen-Holm M, Lindgren M et al (2017) *Toxoplasma gondii* infection and common mental disorders in the Finnish general population. J Affect Disord 223:20–25

Tenter AM, Heckeroth AR, Weiss LM (2000) *Toxoplasma gondii*: from animals to humans. Int J Parasitol 30(12–13):1217–1258

Vargas-Villavicencio JA, Besne-Merida A, Correa D (2016) Vertical transmission and fetal damage in animal models of congenital toxoplasmosis: a systematic review. Vet Parasitol 223:195–204

Verma SK, Carstensen M, Calero-Bernal R et al (2016) Seroprevalence, isolation, first genetic characterization of *Toxoplasma gondii*, and possible congenital transmission in wild moose from Minnesota, USA. Parasitol Res 115(2):687–690

Vikoren T, Tharaldsen J, Fredriksen B et al (2004) Prevalence of *Toxoplasma gondii* antibodies in wild red deer, roe deer, moose, and reindeer from Norway. Vet Parasitol 120(3):159–169

Vors LS, Boyce MS (2009) Global declines of caribou and reindeer. Glob Chang Biol 15(11):2626–2633

Weiss LM, Dubey JP (2009) Toxoplasmosis: a history of clinical observations. Int J Parasitol 39(8):895–901

Wendte JM, Gibson AK, Grigg ME (2011) Population genetics of *Toxoplasma gondii*: new perspectives from parasite genotypes in wildlife. Vet Parasitol 182(1):96–111

Wilson CB, Remington JS, Stagno S et al (1980) Development of adverse sequelae in children born with subclinical congenital Toxoplasma infection. Pediatrics 66(5):767–774

Wilson AG, Lapen DR, Mitchell GW et al (2020) Interaction of diet and habitat predicts *Toxoplasma gondii* infection rates in wild birds at a global scale. Glob Ecol Biogeogr 00:1–10

Work TM, Verma SK, Su C et al (2016) *Toxoplasma gondii* antibody prevalence and two new genotypes of the parasite in endangered Hawaiian geese (Nene: *Branta sandvicensis*). J Wildl Dis 52(2):253–257

Yolken RH, Torrey EF (2008) Are some cases of psychosis caused by microbial agents? A review of the evidence. Mol Psychiatry 13(5):470–479

Zarnke RL, Dubey JP, Kwok OC et al (1997) Serologic survey for *Toxoplasma gondii* in grizzly bears from Alaska. J Wildl Dis 33(2):267–270

Zarnke RL, Dubey JP, Kwok OCH et al (2000) Serologic survey for *Toxoplasma gondii* in selected wildlife species from Alaska. J Wildl Dis 36(2):219–224

Zarnke RL, Dubey JP, Ver Hoef JM et al (2001) Serologic survey for *Toxoplasma gondii* in lynx from interior Alaska. J Wildl Dis 37(1):36–38

Zulpo DL, Sammi AS, Dos Santos JR et al (2018) *Toxoplasma gondii*: a study of oocyst re-shedding in domestic cats. Vet Parasitol 249:17–20

Trichinella spp. in the North

Rajnish Sharma, Edoardo Pozio, Émilie Bouchard, and Emily J. Jenkins

1 Introduction

Trichinellosis (formerly known as trichiniasis or trichinosis), an important foodborne human disease worldwide, is caused by the zoonotic nematodes of the *Trichinella* genus (Gottstein et al. 2009). Species of *Trichinella* do not only have an impact on public health but also on the global meat trade (Pozio 2015). Among the 24 most significant global foodborne parasites listed by the World Health Organization/United Nations Food and Agriculture Organization, *Trichinella spiralis* ranked first and seventh based on global trade and public health significance, respectively (FAO/WHO 2014). Every year, 1 billion USD are spent on the prevention of swine and human infections caused by these nematodes, especially on health regulatory activities in the USA and European Union (Murrell and Pozio 2000; Kapel 2005). More than 65,000 human cases of trichinellosis and 42 deaths from 41 countries were reported during 1986–2009 (Murrell and Pozio 2011). The global disability-adjusted life years due to trichinellosis were 76 per billion persons per year (Devleesschauwer et al. 2015). *Trichinella* spp. are also an important foodborne

R. Sharma
Department of Veterinary Microbiology, Western College of Veterinary Medicine, University of Saskatchewan, Saskatoon, SK, Canada

Centre for One Health, College of Veterinary Science, Guru Angad Dev Veterinary and Animal Sciences University, Ludhiana, Punjab, India
e-mail: ras863@mail.usask.ca

E. Pozio
Department of Infectious Diseases, Istituto Superiore di Sanità, Rome, Italy
e-mail: edoardo.pozio@iss.it

É. Bouchard (✉) · E. J. Jenkins
Department of Veterinary Microbiology, Western College of Veterinary Medicine, University of Saskatchewan, Saskatoon, SK, Canada
e-mail: emb232@mail.usask.ca; emily.jenkins@usask.ca

M. Tryland (ed.), *Arctic One Health*, https://doi.org/10.1007/978-3-030-87853-5_14

parasite in the Arctic, as several outbreaks of human trichinellosis have been reported and almost all were linked to consumption of game meat.

2 History and Taxonomy

The genus *Trichinella* belongs to phylum Nematoda, class Enoplea, order Trichocephalida, and family Trichinellidae. *Trichinella* spp. were first discovered by James Paget in an Italian man who died of pulmonary tuberculosis in the UK; however, this nematode was first described and named *Trichina spiralis* by Richard Owen in 1835 (Owen 1835). The life cycle of the parasite was described by Rudolf Virchow in 1859 (Virchow 1859). *Trichinella* was considered monospecific until 1972. In the last 50 years, 12 more taxa have been discovered, recently reviewed in Pozio (2020) (Fig. 1). Currently, there are 13 taxa of *Trichinella*, classified as encapsulated or non-encapsulated based on the presence or absence of a collagen layer surrounding the first-stage larva (L1) in the striated muscle cell. The encapsulated group infectious only for mammals includes seven named species: *T. spiralis* (genotype T1), *T. nativa* (T2), *T. britovi* (T3), *T. murrelli* (T5), *T. nelsoni* (T7), *T. patagoniensis* (T12), and *T. chanchalensis* (T13). It also includes three genotypes: *Trichinella* T6, *Trichinella* T8, and *Trichinella* T9. The non-encapsulated group includes three species: *T. pseudospiralis* (T4), *T. papuae* (T10), and *T. zimbabwensis* (T11) (Pozio and Zarlenga 2013; Sharma et al. 2020). In this group, *T. pseudospiralis* infects mammals and birds, whereas *T. papuae* and *T. zimbabwensis* can infect mammals and reptiles (Fig. 1).

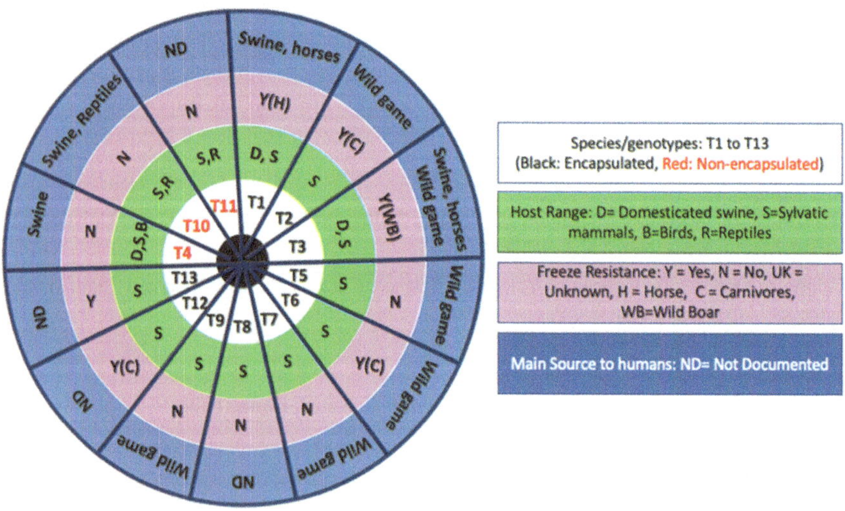

Fig. 1 Features of species/genotypes of *Trichinella* (based on data from Pozio and Murrell 2006; Pozio and Zarlenga 2013; Sharma et al. 2020; Tada et al. 2018)

3 Host Range

Trichinella persists in sylvatic life cycles involving omnivorous and carnivorous vertebrates (mammals, birds, and reptiles). When humans fail in the management and biosecurity of domestic animals, *Trichinella* spp. (especially *T. spiralis*) infection may transmit from the sylvatic environment to the domestic one (spillover), sometimes through synanthropic animals (i.e., rats). The domestic cycle involves mainly pigs and secondarily other domestic animals such as dogs, cats, and horses. In addition, some species can spill back from domestic animals to wildlife (Pozio and Murrell 2006; Oksanen et al. 2018). Domestic life cycles persist when pigs are fed uncooked pork or game scraps, and free-ranging and backyard pigs can feed on carcasses or meat scraps (Pozio 2014). *Trichinella spiralis* is mainly adapted to domestic and wild swine, where it must be distinguished from *T. britovi* in wild boar. As well, *T. spiralis* has been reported in other domestic animals (dogs, cats, and horses) and wild carnivores. *Trichinella nativa* and *Trichinella* T6 (the most common taxa in the Arctic and subarctic regions) have been detected in the range of wild carnivores, including American mink (*Neovison vison*), Arctic fox (*Vulpes lagopus*), Asian badger (*Meles leucurus*), badger (*Meles meles*), beaver (*Castor fiber*), black bears (*Ursus americanus*, *Ursus thibetanus*), brown bear (*Ursus arctos*), beluga (*Delphinapterus leucas*), brown rat (*Rattus norvegicus*), coyote (*Canis latrans*), dog (*Canis familiaris*), golden jackal (*Canis aureus*), grizzly bear (*Ursus arctos*), least weasel (*Mustela nivalis*), lynx (*Lynx canadensis* and *Lynx lynx*), marten (*Martes americana*), minks (*Mustela lutreola*), mountain lion (*Puma concolor*), otters (*Lutra lutra*), pine martens (*Martes martes*), polar bear (*Ursus maritimus*), raccoon dogs (*Nyctereutes procyonoides*), red fox (*Vulpes vulpes*), sea lion (*Eumetopias jubatus*), Siberian weasels (*Mustela sibirica*), skunk (*Mephitis mephitis*), seal (different species), snowshoe hare (*Lepus americanus*), stone martens (*Martes foina*), tiger (*Panthera tigris*), white-breasted hedgehog (*Erinaceus roumanicus*), wolf (*Canis lupus*), wolverine (*Gulo gulo*), and walrus (*Odobenus rosmarus*).

4 Biology and Life Cycle

Trichinella spp. have a unique life cycle, as larvae establish in the muscle cells and all stages (larval and adult) of the parasite occur in the same host (Pozio 2007a). In Fig. 2, (1) animals become infected after consuming flesh containing L1 of *Trichinella*. (2) In the stomach, larvae are released by the action of gastric acid and enzymes (3) and pass into the small intestine. (4) Larvae enter the epithelial wall and undergo four moults before developing into dioecious adult nematodes in 2 days. (5) Adults mate in the intestine, and females release newborn larvae around 6–7 days after infection. (6) Newborn larvae penetrate the sub-mucosa and reach the striated muscle via lymphatic and then blood vessels. (7) Larvae enter the muscle cells and (8) induce the development of a collagen layer to create the nurse cell–larva complex in taxa of the encapsulated clade. (9) In the nurse cell, a larva increases up to 600 times in volume without moulting, becomes infective within 21–30 days, and

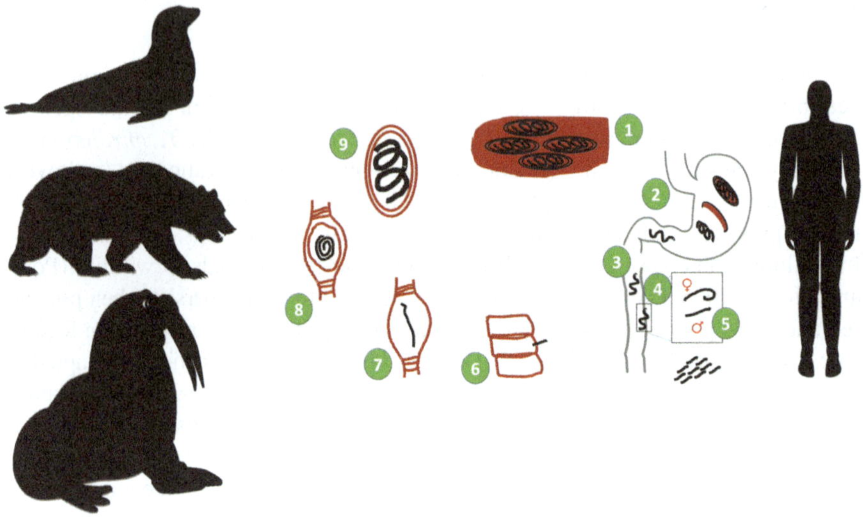

Fig. 2 Life cycle of *Trichinella* spp. and sources of infection to humans in the Arctic

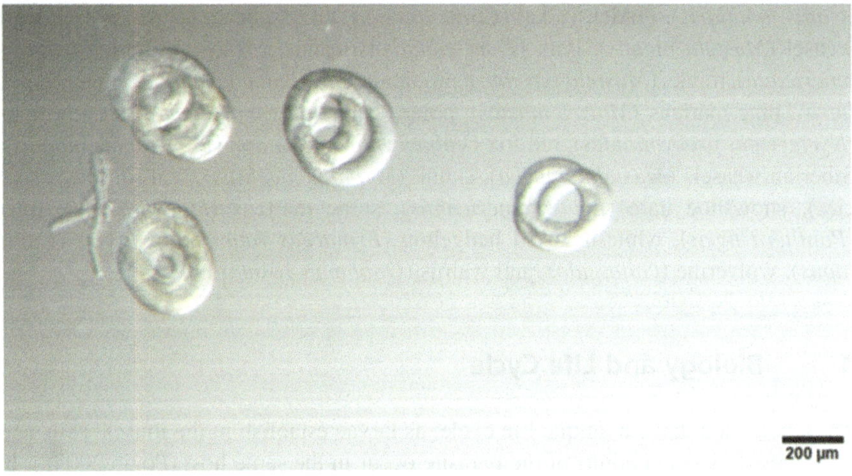

Fig. 3 Tightly coiled larvae of *Trichinella* spp. recovered from digestion of the tongue of a wolverine (*Gulo gulo*)

can remain viable for years (Bruschi and Dupouy-Camet 2014; Bruschi and Murrell 2002): up to 20 years in polar bears and 40 years in humans (Fröscher et al. 1988; Kumar et al. 1990). The infectious dose for people is not known; however, Dupouy-Camet and Bruschi (2007) estimated approximately 100 to 300 larvae of *T. spiralis*. First-stage larvae of all encapsulated taxa of *Trichinella* are coiled in a characteristic spiral fashion (Fig. 3). Females of *T. spiralis* are from 1/3 to two times longer than males; males measure between 1.4 and 1.6 mm long (Roberts and John 2005). *Trichinella* taxa are indistinguishable morphologically and can only be differentiated by biochemical or molecular analyses (Pozio 2007a). Freeze resistance is an important Arctic adaptation, that is, *Trichinella nativa*, *Trichinella* T6, and *T. chanchalensis* in sylvatic carnivorous mammals, *T. spiralis* in horses only and *T. britovi* in wild boar and carnivores (Fig. 1). Freeze tolerance can place consumers at risk if they rely on freezing of meat for inactivation of larvae of *Trichinella* spp.

5 Epidemiology

5.1 Transmission

Animals acquire the infection by the consumption of flesh of other infected animals, through scavenging, predation, and/or cannibalism. Sylvatic life cycle plays a pivotal role in northern environment, where domestic animal hosts (pigs) are rare. People become infected after consuming raw or undercooked meat of infected animals. Most of the human cases occurring globally are due to pork consumption (Murrell and Pozio 2011). In Arctic regions, outbreaks of human trichinellosis are usually linked to consumption of game meat, often walrus, black bear, and polar bear (Fig. 2). Walrus meat is considered as the most common cause of trichinellosis in the Arctic.

5.2 Prevalence and Distribution

Trichinella sp. infection has been reported in animals in all continents except Antarctica (Pozio 2007b). The prevalence and distribution of *Trichinella* in the Arctic and subarctic regions is reported below.

5.2.1 Nearctic Region

The Arctic and subarctic areas of the Nearctic region include northern Canada, Alaska (USA), and Greenland. The five taxa, namely, *T. spiralis*, *T. nativa*, *T. pseudospiralis*, *T. chanchalensis*, and *Trichinella* T6, have been documented in animals (Gajadhar and Forbes 2010; Jenkins et al. 2013; Sharma et al. 2020). Of these, *T. nativa* and *Trichinella* T6 are the most common, with T6 more commonly detected in the western Canadian Arctic and T2 in the eastern Canadian Arctic.

Alaska

In Alaska, *Trichinella* sp. infection was detected in Arctic fox, black bear, coyote, dog, ermine, grizzly bear, least weasel, lynx, polar bear, red fox, wolf, wolverine, beluga whale, snowshoe hare, seals, and walrus (Jenkins et al. 2013; Seymour et al. 2014). The prevalence varied between <1 and 93%. Less than 1% prevalence was observed in seals ($n = 310$), ground squirrel (*Citellus undulatus*) ($n = 129$), and muskrat (*Ondatra zibethicus*) ($n = 113$) (Rausch et al. 1956). The highest prevalence was reported in dogs (93%, $n = 41$) (Rausch et al. 1956). Until 2000, no reports identified the taxon of *Trichinella*. Of 57 wolves, 16 (28%) were positive for *Trichinella* spp., and the predominant species was *T. nativa* (69%) followed by *Trichinella* T6 (31%) (La Rosa et al. 2003). *Trichinella nativa* was also documented in Arctic fox (originated from Barrow and Prudhoe Bay, Alaska), polar bear (Barrow, Alaska), and ringed seal (Barrow, Alaska) (Seymour et al. 2014). In a recent survey, antibodies to *Trichinella* were not detected in any of the 155 brown bears from Alaska (Ramey et al. 2019).

Wild game is the major, if not only, source of infection for people in Alaska. Since 1975, 94% (227/241) of human trichinellosis infections were linked to consumption of wild game (Springer et al. 2017). Of these 227 cases, 100 and 99 infected people had consumed meat of walrus and bear, respectively. Twenty-four patients ate both walrus and seal, whereas four consumed both bear and seal. However, a decline in walrus-associated trichinellosis has been reported; average annual cases decreased from 6.3 (1975–1992) to 0.5 (1993–2017) (Springer et al. 2017).

Northern Canada

Northern Canada comprises the three northern territories (Yukon, Northwest Territories, and Nunavut) and regions north of the southern limit of discontinuous permafrost in the following provinces: Alberta, British Columbia, Labrador, Manitoba, Ontario, Québec (Nunavik and the James Bay region), and Saskatchewan (Jenkins et al. 2013). In northern Canada, the prevalence of *Trichinella* sp. infection varied from 1 to 78% in several species of carnivore mammals (Gajadhar and Forbes 2010; Jenkins et al. 2013; Sharma et al. 2021). All six species of *Trichinella* reported from the Nearctic region have been detected in wildlife in northern Canada (Gajadhar and Forbes 2010; Jenkins et al. 2013; Sharma et al. 2020). *Trichinella nativa* and *Trichinella* T6 are the most common species in the Arctic and subarctic regions of Canada.

In the Yukon (a territory bordered with Alaska), the prevalence of *Trichinella* spp. (taxa not identified) based on larval recovery ranged from 11 to 78% in Arctic fox, grizzly bear, wolf, and wolverine (Choquette et al. 1969; Gajadhar and Forbes 2010; Sharma et al. 2018). The predominant species of *Trichinella* identified in wolverine in the Yukon was *Trichinella* T6, followed by *T. nativa* and *T. chanchalensis* (Sharma et al. 2020, 2021). Nine polar bears tested seropositive by ELISA (Rah et al. 2005).

In the Northwest Territories, the prevalence based on larval recovery varied from 3 to 77% in Arctic fox, black bear, grizzly bear, polar bear, red fox, wolf, and wolverine (Larter et al. 2011; reviewed in Jenkins et al. 2013; Gajadhar and Forbes 2010). Both *T. nativa* and *Trichinella* T6 were documented in black bear, grizzly bear, and wolverine (Gajadhar and Forbes 2010; Larter et al. 2011; Sharma et al. 2018, 2021). *Trichinella chanchalensis*, a recently discovered species, has been detected in wolverine, but none in other carnivores, from the Yukon and Northwest Territories (Sharma et al. 2020, 2021).

In Nunavut, the prevalence of *Trichinella* spp. ranged from 3 to 43% in Arctic foxes, grizzly bear, lynx, marten, and polar bear (reviewed in Jenkins et al. 2013;

Gajadhar and Forbes 2010). *Trichinella nativa* was common in polar bear, wolf, and walrus, whereas *Trichinella* T6 was detected only in lynx (Gajadhar and Forbes 2010).

In Québec, *Trichinella* spp. were documented in black bear, dog, polar bear, red fox, wolf, and walrus (Gajadhar and Forbes 2010; reviewed in Jenkins et al. 2013; Bachand et al. 2018), and the prevalence varied from 7 to 66%. *Trichinella nativa* was identified in black bear, red fox, and polar bear (Gajadhar and Forbes 2010; Bachand et al. 2018).

In Newfoundland and Labrador, *Trichinella* spp. were documented in black bear, polar bear, wolf, and walrus (Butler and Khan 1992; Smith and Snowdon 1988; Thorshaug and Rosted 1956); there is a single report from black bear (Butler and Khan 1992), whereas the prevalence in other host species varied from 4% to 59% (Smith and Snowdon 1988; Thorshaug and Rosted 1956). Species of *Trichinella* was not identified in wild animals from this region, as no investigations were carried out in the last 30 years.

In Alberta, British Columbia, Manitoba, and Saskatchewan, the prevalence of 1–33% was reported in black bear, grizzly bear, lynx, marten, polar bear, red fox, skunk, wolf, and wolverine (Gajadhar and Forbes 2010; Jenkins et al. 2013). *Trichinella nativa* and *Trichinella* T6 were detected in a North American cougar (*Puma concolor couguar*) and a grizzly bear of British Columbia, respectively, and *T. nativa* in black bears of British Columbia and Saskatchewan (Gajadhar and Forbes 2010; Jenkins et al. 2013).

In parts of Alaska and northern Canada, *T. nativa* and *Trichinella* T6 may occur in sympatry in wolves and wolverines (La Rosa et al. 2003; Reichard et al. 2008). Recently, sympatry among *T. chanchalensis*, *T. nativa*, and *Trichinella* T6 was reported in wolverines from the Yukon and Northwest Territories (Sharma et al. 2020).

Trichinella spiralis and *T. pseudospiralis* are rarely reported in the Canadian North. *Trichinella spiralis* has been eradicated from commercial confinement-raised pigs in Canada and is rarely reported in backyard pigs (Appleyard and Gajadhar 2000; Appleyard et al. 2002; Gajadhar et al. 1997; Gajadhar and Forbes 2010; Newman 2014). Wild animals are also considered free of *T. spiralis*; however, infection has been seen in red foxes and coyotes from Prince Edward Island (Appleyard et al. 1998). In Canadian Arctic and subarctic regions, *T. spiralis* had never been reported until recent detection in a wolverine harvested from the Klondike region of the Yukon, near the border of Alaska (Sharma et al. 2019). Only two cases of *T. pseudospiralis* have been reported in Canadian wildlife. The first case was reported from a mountain lion (*Puma concolor*) from Vancouver Island in the southwestern region of the country (Gajadhar and Forbes 2010). The second case, and the only case of *T. pseudospiralis* reported in subarctic Canada, was from a wolverine harvested in the Northwest Territories (Sharma et al. 2019). This was the northernmost observation of *T. pseudospiralis* from North America.

Human cases and outbreaks of trichinellosis were historically common in Northern Canada, almost all associated with the consumption of wild game (walrus or black bear) (MacLean et al. 1989). Traditional methods of walrus preparation (fermentation, freezing, and drying) are not sufficient to inactivate *Trichinella* larvae (Proulx et al. 2002), whereas Inuit usually thoroughly cook bear meat (Forbes et al. 2003). In Canada, one of the largest outbreaks of trichinellosis occurred in Salluit, Québec, Canada; after consuming walrus meat, 42 of 68 individuals became infected (MacLean et al. 1992). Cases and outbreaks associated with walrus consumption continue to occur in Nunavut and Nunavik and elsewhere in Canada. In northern Saskatchewan, the attack rate was 39.7% (31/78) in people consuming bear meat infected by *T. nativa* (Schellenberg et al. 2003). A recent outbreak involving 10 persons occurred after consuming jerky (dried meat) prepared from black bear meat in northern Ontario (Dalcin et al. 2017).

Serosurveys conducted in communities from Nunatsiavut, Nunavik (Québec), and Inuvialuit regions (Northwest Territories) reported low seroprevalence of antibodies to *Trichinella* spp. (<1%) (Egeland and Committee 2010; Egeland and Inuvialuit Settlement Region Steering Committee 2010; Messier et al. 2012). In James Bay (Québec), none of the 267 individuals tested were seropositive (Sampasa-Kanyinga et al. 2012). However, in an extensive survey conducted in 36 Inuit communities (involving 2212 individuals) across the Inuvialuit Settlement Region (ISR), Nunavut, and Nunatsiavut, the overall seroprevalence was 18.6%, with the highest seroprevalence in Nunavut (24%) followed by Nunatsiavut (11%) and Inuvialuit regions (7.5%) (Goyette et al. 2014). Age (over the age of 50 years), education level, and consumption of marine mammals and seafood (consisting primarily of Arctic char, as well as other fish and clams) were associated with *Trichinella* seropositivity (Goyette et al. 2014). A higher seroprevalence (16%) was also observed in a Dene population in northern Saskatchewan (Schurer et al. 2013).

Greenland

In Greenland, *Trichinella* infection was detected in Arctic fox, dog, polar bear, wolf, seals, and walrus, with prevalence ranging from 2 to 76% (Jenkins et al. 2013). The prevalence was 10–15 times higher in polar bear than walrus, and *Trichinella nativa* is the only species reported in the region (La Rosa and Pozio 1990; Kapel 1997). A survey in an eastern Greenlandic hunting community showed a prevalence of 3% (Møller et al. 2010). Prevalence was higher in persons aged >60 years than in <40 years (12% vs. 1.4%), which could be attributed to a shift from traditional to western diets. Age, hunting, and consumption of polar bear meat were associated risk factors (Møller et al. 2010).

5.2.2 Arctic and Subarctic in the Palearctic Region

The Arctic (largely defined by tundra) and subarctic (largely defined by taiga) Palearctic ranges from the North Atlantic Ocean to the Pacific Ocean over 11 countries, namely, Iceland, Norway, Sweden, Finland, Estonia, Latvia, Lithuania, Belarus, the Russian Federation, Kazakhstan, and Mongolia. During the Pliocene (from 5332 to 2588 million years ago), these areas were affected by a progressive cooling that culminated with the glaciations in the Quaternary era.

Cooling during this period is selected for the development of physiological mechanisms to survive freezing, that is, the longer the survival of larvae of *Trichinella* spp. in host carcasses, the higher the probability to be ingested by a new host, especially in regions with low density of carnivores. This selective pressure favored the speciation of freeze-tolerant *T. nativa* and its separation from the more temperate species, such as *T. britovi* (Korhonen et al. 2016; Pozio 2016a).

In the Northern Palearctic region, *T. nativa* is thought to be the predominant species circulating among wild animals, even if in-depth molecular investigations on *Trichinella* isolates from the great North of the Palearctic region are lacking. Therefore, it cannot be excluded that there is a circulation of other taxa as observed in the Arctic areas of the Nearctic region. The southern distribution boundary has been tentatively identified between the isotherms −5 °C and −4 °C in January (Shaikenov and Boev 1983; Shaikenov 1992; Pozio et al. 1998). The second species circulating in subarctic regions of the Palearctic is *T. britovi*, which is sympatric with *T. nativa* between the isotherms −4 °C and −6 °C, and there are many reports of mixed infections in the same host from Estonia, Finland, Lithuania, Latvia, and Sweden.

Less commonly, two other species, *T. spiralis* and *T. pseudospiralis*, have been documented in the northern areas of the Palearctic region (Pozio 2016b; Oksanen et al. 2018).

Iceland

In Iceland, *Trichinella* sp. has never been documented in humans or in endemic domestic or wild animals, despite the presence of suitable wildlife hosts such as Arctic fox and American mink (Skírnisson et al. 2003, 2010); all horses and pigs intended for human consumption are tested, but wildlife species are not routinely monitored. Periodically, polar bears arrive when pack ice is stranded ashore or comes into close proximity of the northern or northwestern coasts; however, according to Icelandic law, soon after their arrival, the bears must be shot and burned to prevent scavenging by terrestrial carnivores. Over the centuries, more than 500 polar bears have been recorded, 50 in the past 100 years. Many were killed, although in some cases they returned to the ice and disappeared (Haraldsson and Hersteinsson 2004). In 2008, a polar bear that had swum to Iceland was found to have been infected with *T. nativa* (Skírnisson et al. 2010). Due to global warming and the consequent reduction of the Arctic pack ice, a decrease in polar bears swimming to Iceland may result in reduced risk of *T. nativa* introduction.

Norway

Trichinella nativa is the prevalent species among red foxes (5% prevalence) and minks (11% prevalence) of mainland Norway. *T. britovi* has been also documented in a red fox in the southern region of mainland Norway (Davidson et al. 2006). Anti-*Trichinella* sp. antibodies were detected in 7.3% (24/328) of red fox in mainland Norway (Davidson et al. 2009). Infections have also been documented in dogs and cats. An important *T. nativa* focus is present in the Svalbard archipelago (Kapel et al. 1999), where anti-*Trichinella* antibodies were detected in 78% of polar bears

(Asbakk et al. 2010). No human cases have been documented in Norway since 1940 (Pozio 2007b).

Sweden

In Sweden, *T. nativa* is the predominant species (97/185, 52%), detected in 10 brown bears, 2 raccoon dogs, 55 lynxes, 8 red foxes, 16 wolves, and 5 wolverines across the whole country. Less frequently, *T. britovi* (61/185, 33%) has been reported in 20 lynxes, 10 red foxes, 16 wild boars, 13 wolves, and 2 wolverines of central-southern counties (Pozio et al. 2004; International Trichinella Reference Center 2019). Finally, *T. pseudospiralis* (21/185, 11%) has been reported in 1 lynx and 20 wild boars of central-southern counties (Pozio et al. 2004; International Trichinella Reference Center 2019), and *T. spiralis* (9/185, 5%) has been documented in 5 red foxes (one co-infected with *T. nativa*), 2 domestic pigs, and 2 wild boars in the south of the country. In the past 40 years, among the approximately 3.5–4 million pigs slaughtered per year, the number of infected pigs progressively reduced from 67 in 1980–1989, eight in 1990–1999, and no cases since 2000 (Pozio et al. 2004). Ten outbreaks of trichinellosis, involving a total of 504 persons, were reported between 1917 and 1969, and a few sporadic cases were also reported until 1978. The source of infection was pork from backyard pigs and, in one outbreak, a hunted wild boar (Pozio et al. 2004). In 2013, a probable case of trichinellosis caused by wild boar meat consumption was documented (https://www.folkhalsomyndigheten.se/folkhalsorapportering-statistik/statistik-a-o/sjukdomsstatistik/trikinos-/?p=16732).

Finland

Trichinella spp. are common in suitable wildlife hosts in Finland. The predominant species is *T. nativa* (89.2%), followed by *T. britovi* (9.5%), *T. pseudospiralis* (1.8%), and *T. spiralis* (1.7%) detected as single or double infections (Oivanen et al. 2002; Airas et al. 2010; Oksanen et al. 2018). *T. nativa* is the most prevalent species in the whole country, and the other three species circulate only in the southern part of the country. The highest prevalence has been detected in lynx (45.5%–47.8%), followed by raccoon dogs (28.1%–33.2%), red foxes (18.7%–25.3%), wolves (34.1%–39.2%), pine martens (9.3%–11.6%), badgers (7.5%), brown bears (5.6%–6.1%), and otters 2.0%–3.2%) (Airas et al. 2010; Oksanen et al. 2018). The prevalence of *T. spiralis* in Finnish pigs has been decreasing, with the last cases detected in 2004 (Oivanen and Oksanen 2009). *Trichinella* sp. has been reported in hunted wild boar sporadically in Finland (Oivanen et al. 2002; European Food Safety Authority 2019). In spite of the presence of domestic foci, human trichinellosis is rare in Finland, with a total of eight cases reported prior to 1977 (Oivanen and Oksanen 2009).

Estonia

In Estonia, *T. britovi* is the most common species (159/226, 70%) among susceptible wildlife (brown bear, raccoon dog, lynx, red fox, wild boar, and wolf), followed by *T. nativa* (83/226, 37%) detected in the previous reported host species as well as a pine marten and a badger (International Trichinella Reference Center 2019; Järvis et al. 2001). Mixed infections with *T. britovi* and *T. nativa* have been detected in about 8% of infected wildlife. *T. spiralis* (14/226, 6%) has been detected in lynx, wild boar, a domestic pig, brown rats, and farmed fur animals (International Trichinella Reference Center 2019; Järvis et al. 2001), as well as *T. pseudospiralis* (2/226, 0.9%) in two wild boars (International Trichinella Reference Center 2019). The prevalence of *Trichinella* spp. was 42%–57.5% in raccoon dogs, 40%–69.0% in red foxes, 50% in lynxes, and 30.4% in pine martens (Kärssin et al. 2017; Malakauskas et al. 2007). Seroprevalence was 17% in hunted wild boars (Kärssin et al. 2016), while earlier studies based on larval detection reported a prevalence of 0.3% (Malakauskas et al. 2007). In the twentieth century, human trichinellosis associated with consumption of game meat has been reported in Estonia almost every year (Järvis et al. 2001). In the last decade, human infections were documented sporadically, with the last outbreak involving two persons in 2015 (EFSA 2016).

Latvia

In Latvia, *T. britovi* is the most prevalent species (421/439, 96%) among susceptible wildlife, that is, lynx, raccoon dog, wolf, red fox, wild boar, pine marten, beaver, and golden jackal. *Trichinella nativa* (17/439, 3.9%) has been detected in red foxes, wild boar, wolves, and a lynx. *Trichinella spiralis* has been detected in four domestic pigs and a wild boar and in five red foxes as single or mixed infections with *T. britovi* (Malakauskas et al. 2007; International Trichinella Reference Center 2019). According to Malakauskas et al. (2007), the prevalence of *Trichinella* sp. infection among 1955 tested wild animals was 28.9% in red foxes, 17% in raccoon dogs, 28.6% in pine martens, 88.9% in lynx, and 1.3% in wild boar. In a more recent study, which screened 1286 wild animals, the prevalence of *Trichinella* spp. was 56.2% in pine martens, 45.8% in stone martens, 100% in wolves, 37.3% in raccoon dogs, 50.6% in red foxes, and 100% in lynxes (Deksne et al. 2016). In the period 2002–2018, 283 cases of trichinellosis in humans were documented in Latvia, with a sharply decreasing trend in the last years (European Food Safety Authority 2007, 2014, 2019). The main source of infection was meat from hunted wild boar.

Lithuania

In Lithuania, *T. britovi* is the most prevalent species (279/388, 71.9%) among susceptible wildlife (red fox, raccoon dog, pine marten, and wild boar); *T. spiralis* was detected mainly in wild boar and domestic pigs but also in foxes and raccoon dogs (99/388, 25.5%); *T. nativa* was detected in red foxes, raccoon dogs, and wild boar (9/388, 2.3%); and *T. pseudospiralis* has been reported in a red fox (1/388, 0.2%) (Malakauskas et al. 2007; ITRC 2019). The prevalence of *Trichinella* sp. infection was 40% in red foxes, 40% in pine martens, 32.5% in raccoon dogs, and 0.51% in wild boars (Malakauskas et al. 2007). In the period 2002–2018, 331 human cases of trichinellosis were documented in Lithuania, with a sharply decreasing trend in the last years (European Food Safety Authority 2007, 2014, 2019). The main source of infection was meat from hunted wild boar.

Belarus

There is very little information on *Trichinella* spp. in animals and humans of Belarus, most dating back to the twentieth century. Both *T. spiralis* and *T. britovi* have been reported in domestic pigs (Shaikenov and Boev 1983). Between 1980 and 1989, the prevalence of human trichinellosis was 0.55 per 100,000 inhabitants (Skripova and Kovchur 1994).

Russian Federation

The Russian Federation has the largest surface area in the Palearctic (and, indeed, the world); however, the number of contemporary investigations into the epidemiology of *Trichinella* spp. in animals and humans is limited. The most common *Trichinella* species circulating among wildlife is *T. nativa*. Out of 46 regions (oblasts) of European Russia, epidemiological information on *Trichinella* spp. is available for 9 (19.5%) oblasts. In the Voronezh and Lipetsk oblasts, *T. nativa* has been detected in 28% of wolves and 28% of red foxes, with prevalence as high as 54% in foxes from protected ecosystems, for example, the Voronezh Reserve (Romashov et al. 2018a, b). In the Kirov oblast, *T. nativa* has been detected in 3% (4/134) of American minks (Maslennikova et al. 2017). In the Kursk oblast, *T. nativa* has been detected in American minks (36.6%), red foxes (34.9%), domestic dogs (4%), and a northern white-breasted hedgehog (Vagin and Malisheva 2010; Vagin et al. 2016). *Trichinella nativa* and *T. spiralis* were found in wildlife of the Ryazanskaya oblast. *Trichinella spiralis* was isolated from a badger, raccoon dog, and foxes of the Nizhegorodskaya and Ryazanskaya oblasts (Konyaev et al. 2012). In the Tver and Smolensk oblasts, the highest prevalence (97.5%) was found among wolves, followed by the red foxes (48.3%), the raccoon dogs (20.0%), and mustelids (12.5%). *Trichinella nativa* was the most common species detected (98%), whereas *T. britovi* was only detected in one wolf and one domestic dog (Pozio et al. 2001). In the Vologda oblast, *T. nativa* has been detected in badgers (20%), brown bears (34% %), pine martens (14.3%), raccoon dogs (33%), red foxes (43.2%), tigers (100%), wolves (58.8%), sables (2.1%), and Siberian weasels (5.2%) (Glushnev et al. 2012; Gorodovich and Gorodovich 2010). Mixed infections of *T. nativa* and *T. pseudospiralis* have been reported in a wild boar, a red fox, and a raccoon dog (Gorodovich and Gorodovich 2010). In central Siberia and the Russian Far East, *Trichinella* spp. have been investigated only in Altai Krai, the Altai Republic, Yakutia, the Amur region, and the Chukotka peninsula. *T. nativa* has been detected in foxes, wolves, brown and black bears, and Asian badger of Yakutia, Altai Krai, and the Altai Republic (Konyaev et al. 2012). For the first time, *T. pseudospiralis* has been detected in a polar bear of Yakutia (Kokolova 2017). In the Amur region from 2011 to 2016, *T. nativa* has been detected in badgers, brown bears, lynx, red foxes, and Siberian weasels (Solovyeva et al. 2017).

In the Amur oblast, *T. nativa* has been detected in lynx, red foxes, Siberian weasels, and raccoon dogs (Guba 2010a, b, c). In Chukotka, *T. nativa* has been detected in a polar bear, wolverine, Arctic fox, and a sled dog, *T. spiralis* in a domestic cat, *T. pseudospiralis* in a ringed seal (*Pusa hispida*) and in a domestic pig, and mixed infections of *T. spiralis* and *T. nativa* in an arctic fox and a northern sea lion (Goździk et al. 2017). In the Russian Federation, the incidence of human trichinellosis was 0.12 per 100,000 inhabitants in 2010 and 0.06 per 100,000 inhabitants in 2011 (Anonymous 2012).

6 Clinical Manifestations of Infection with *Trichinella*

6.1 In Animals

Clinical manifestations due to *Trichinella* spp. in naturally infected, free-ranging Arctic and subarctic wildlife are unknown; however, clinical signs were seen in grey seals (*Halichoerus grypus*) and raccoon dogs experimentally infected with *T. nativa* (Kapel et al. 2003; Nareaho et al. 2000). Experimentally infected seals showed anorexia 17 days post infection and then resumed eating (Kapel et al. 2003). Experimentally infected raccoon dogs showed anorexia, diarrhea, weight loss, and mild anemia (Nareaho et al. 2000).

6.2 In Humans

The incubation period can last between 7 and 30 days depending on the species of *Trichinella*, number of larvae, and host factors (age, sex, and immune status) (Pozio and Bruschi 2018). Seroconversion occurs between 12 (in severe or moderate cases) and 60 (in mild or asymptomatic cases) days post infection (Pozio and Bruschi 2018). Clinical signs and symptoms of trichinellosis are non-specific and can be confused with many other diseases (Kocięcka 2000). Acute infection has two phases: intestinal (enteral) and muscular (parenteral). In the initial intestinal stage, symptoms may include abdominal pain, diarrhea (commonly), malaise, anorexia, nausea, and vomiting, which is difficult to distinguish from other causes of foodborne illness or viral infection. In the parenteral phase associated with larval establishment, fever, facial edema (classically, periorbital), and myalgia may occur. In addition, anorexia, headache, conjunctivitis, and urticaria may also be present (Kocięcka 2000). Rarely, infection may lead to complications like myocarditis, pneumonitis, and ultimately death if not diagnosed rapidly and treated appropriately (Bruschi and Murrell 2002), often quite difficult in remote northern regions.

7 Diagnosis and Detection

7.1 In Animals

In animals, detection of *Trichinella* can be direct or indirect. Direct detection methods are important for safe consumption of domestic pigs or wild game and to determine wild animal reservoirs and indicators of a two-way transmission between wild and domestic animals (Gottstein et al. 2009). The double separatory digestion method is a highly sensitive and widely accepted gold standard method to detect and recover larvae of *Trichinella* spp. (ISO 2015; European Commission 2015; Rossi et al. 2017), and this has been deployed in laboratories in northern Canada (Proulx et al. 2002). Sensitivity is higher with large sample volumes (at least 10 g), especially at low larval burdens (e.g., < 1 larva per gram) (Gamble et al. 2000). Predilection sites for *Trichinella* sp. larvae depend on host and parasite species (Kapel et al. 2005; Serrano et al. 1999). Usually, tongue, diaphragm and masseter muscles act as predilection sites for *Trichinella spiralis* in pigs (*Sus scrofa*) and horses (*Equus caballus*) (Kapel et al. 2005). Sampling sites in wild animals include the forearm, diaphragm, and tongue in wild boar; the diaphragm, masseter, and tongue in bear; the tongue, diaphragm, flippers, and masseter in walrus and seal; the diaphragm, forearm, and tongue in fox; the diaphragm, forearm, and tongue in raccoon dog; and the tongue and diaphragm in wolverine (Gamble et al. 2000; Kapel 2000; Kapel et al. 2005; Sharma et al. 2018, reviewed in Gottstein et al. 2009).

As larvae are morphologically similar, molecular methods are needed to determine species/genotypes of *Trichinella* circulating in domestic and wild animals (Gottstein et al. 2009). Multiplex PCR is an inexpensive, rapid, and sensitive method of identification of species/genotype of *Trichinella* (Zarlenga et al. 1999). Recently, *T. chanchalensis*, a newly recognized species, had been misdiagnosed as *T. nativa* using multiplex PCR, which showed similar band patterns for both *T. chanchalensis* and *T. nativa*. Therefore, it is recommended to sequence PCR amplicons or conduct an inexpensive and rapid PCR-RFLP to differentiate between *T. nativa* and *T. chanchalensis* (Sharma et al. 2020). This finding also calls into question previous identifications of *T. nativa* across the circumpolar region.

Serology, an indirect method that detects antibodies, is used for epidemiological investigations in animals and is not recommended by the International Commission on Trichinellosis (ICT) for meat inspection as it does not necessarily indicate active infection (Gottstein et al. 2009; Gamble et al. 2000; Bruschi et al. 2019). On the other hand, antibody detection may have a higher sensitivity in early infection before larvae are sufficiently developed to be detected by digestion and microscopy (OIE 2004). ELISA on serum, plasma, or meat juice/tissue fluid is the most commonly used serological method, often using the excretory/secretory (E/S) antigens.

7.2 In Humans

Sporadic human infections can be difficult to diagnose (Bruschi and Murrell 2002) and rely on both clinical (symptoms and dietary history) and laboratory diagnosis. Clinical manifestations (diarrhea, fever, muscle pain, facial edema, periorbital edema), exposure to raw/undercooked meat, and eosinophilia suggest further investigation (serology or biopsy) for trichinellosis (Kocięcka 2000). Seroconversion generally occurs between the third and fifth week after infection, but in mild/asymptomatic infections, seroconversion can occur up to 60 days post infection, and antibodies persist for a year or more (Bruschi et al. 2019). Several serological methods were used in the past (e.g., indirect immunofluorescence, indirect hemagglutination, latex agglutination). Today, the most widely used screening test is an enzyme-linked immunosorbent assay (ELISA) followed by Western blot as a confirmatory test (Gómez-Morales et al. 2012; Bruschi et al. 2019). Recently, a serological method to distinguish infections caused by encapsulated and non-encapsulated taxa has been developed (Gómez-Morales et al. 2018). For muscle biopsy, samples are obtained from the deltoid muscle and then processed for trichinoscopy, histopathology, or muscle digestion. Although invasive, muscle biopsy is useful to stage infection, determine pathology, and identify species/genotype (Bandi et al. 1995).

8 Prevention and Control

Although numbers of human cases have been continuously declining, *Trichinella* spp. are widely established in northern wildlife, and human trichinellosis is still a food safety issue in the Arctic and subarctic regions, often with large outbreaks associated with communal food sharing. A "One Health" approach is required to prevent and control *Trichinella* in the Arctic and subarctic regions (Fig. 4).

8.1 Harvest Practices

In wild animals, eradication is not feasible, but the larval burden of *Trichinella* can be reduced by preventing scavenging of carcasses of hunted and trapped wild carnivores, not using carcasses of potential wildlife hosts as bait (Pozio et al. 2001), and cooking meat thoroughly before feeding to dogs.

8.2 Wildlife Surveillance

Monitoring of reservoir/sentinel wildlife hosts (generally high-trophic-level carnivores and scavengers such as foxes, raccoon dogs, wolves, jackals, and wolverines) provides information on the circulation of *Trichinella* in harvested wildlife and domestic animal hosts. Regular monitoring of wildlife indicator hosts would be especially helpful in areas currently considered free of *Trichinella* spp. (e.g., Iceland) and in regions where domestic pigs are supposed to be free of infection (e.g., Canada) to determine the potential of spillover from wildlife to backyard/outdoor swine production systems and spillback from livestock to wildlife. Hunting dogs can also act as sentinels for monitoring infections with *Trichinella* spp. in wildlife (Gómez-Morales et al. 2016). The selection of indicator host species depends on factors such as host distribution, life span, availability, and position in food chain.

8.3 Meat Inspection

Consumption of harvested wildlife is integral to many Arctic communities, often shared among members of the communities and consumed raw or frozen; therefore, inspection of *Trichinella* susceptible animals before their consumption or distribution to the community members should be considered. In 1992, a program known as the Nunavik Trichinellosis Prevention Program (NTPP) was started in Québec, Canada, involving multiple stakeholders (Larrat et al. 2012). Samples (walrus tongue) are transported to a northern laboratory, tested using the digestion method, and results are available within 24 h, ideally before distribution or consumption. This program has resulted in the decline in human outbreaks and has now been deployed elsewhere in the Canadian North. In Greenland, meat inspection of walrus and polar bear is mandatory before consumption. Testing of harvested wildlife for *Trichinella* sp. infections should be considered to protect health of Arctic indigenous populations, but also in other countries, due to the illegal introduction of meat of bears hunted in the Arctic and consumed by foreign hunters on their return (Rostami et al. 2017).

8.4 Culinary Practices

Proper cooking of game meat can prevent human infections but is not always culturally acceptable. Smoking, curing, drying, fermenting, and microwave cooking are not considered as safe for killing *Trichinella* sp. larvae (Gamble et al. 2000, 2007; Noeckler et al. 2019). Freezing of meat can inactivate/kill larvae of some species/genotypes of *Trichinella* (Pozio et al. 2006) but not the most common taxa in wildlife in the Arctic and subarctic regions (*T. nativa*, *Trichinella* T6, *T. chanchalensis*, and *T. britovi*). Therefore, an increased northern capacity to test wildlife before consumption and cooking untested or known positive meat to an internal temperature of 71 °C for 1 min are recommended.

8.5 Awareness

Engaging community members, hunters, trappers, and consumers is key to preventing trichinellosis. In addition to good hygiene (washing hands after handling/processing carcasses and before consumption), more work is needed to determine and share traditional knowledge about targeted wildlife (species, age, and season) and harvest practices (not consuming raw meat of some wildlife species) that can decrease transmission of *Trichinella* spp. in the circumpolar North.

8.6 Medical Facilities

Health facilities (medical and veterinary) are not available or limited in remote areas in the Arctic. A decreased access to such facilities delays outbreak investigations and rapid diagnosis and treatment of human cases. Financial assistance and infrastructure development is only one component of building northern capacity, as well as technological advances that enable carcass-side testing and training of northerners to conduct surveillance in wildlife, domestic animals, and people.

8.7 Reporting System

Trichinella sp. is a reportable infection in susceptible livestock in Canada and the USA, but not in wildlife. This seriously disadvantages the understanding of trichinellosis in northern regions. In the European Union, domestic (pigs and horses) and wild (wild boar, bears, etc.) *Trichinella*-susceptible animals intended for human consumption must be tested by digestion; furthermore, the prevalence of *Trichinella* sp. infection in wild susceptible animals must be monitored in EU countries (European Commission 2015). From a One Health perspective, an exchange of information between animal and human health agencies regarding disease reporting and outbreak investigations is needed, as well as close communication among circumpolar countries, many of whom share connected wildlife populations (such as polar bears, fox, migratory birds, and marine mammals) that reservoir species of *Trichinella*.

References

Airas N, Saari S, Mikkonen T, Virtala AM, Pellikka J, Oksanen A, Isomursu M, Kilpelä SS, Lim CW, Sukura A (2010) Sylvatic *Trichinella* spp. infection in Finland. J Parasitol 96:67–76

Anonymous (2012) On the incidence of trichinellosis in the Russian Federation in 2011. Official letter of Federal Service for Surveillance on Consumer Rights Protection and Human Wellbeing No. 01/10535-12-32 of 17.09.2012. http://89.rospotrebnadzor.ru/documents/ros/letters/89696/ [in Russian]

Appleyard GD, Gajadhar AA (2000) A review of trichinellosis in people and wildlife in Canada. Can J Public Health 91(4):293–297

Appleyard GD, Conboy G, Gajadhar AA (1998) *Trichinella spiralis* in sylvatic hosts from Prince Edward Island. J Wildl Dis 34(1):158–160

Appleyard GD, Forbes LB, Gajadhar AA (2002) National serologic survey for trichinellosis in sows in Canada 1996-1997. Can Vet J 43(4):271–273

Asbakk K, Aars J, Derocher AE, Wiig O, Oksanen A, Born EW, Dietz R, Sonne C, Godfroid J, Kapel CM (2010) Serosurvey for *Trichinella* in polar bears (*Ursus maritimus*) from Svalbard and the Barents Sea. Vet Parasitol 172:256–263

Authority EFS (2014) The European Union summary report on trends and sources of zoonoses, zoonotic agents and food-borne outbreaks in 2012. EFSA J 12:3547

Authority EFS (2016) The European Union summary report on trends and sources of zoonoses, zoonotic agents and food-borne outbreaks in 2015. EFSA J 14:4634

Authority EFS (2019) The European Union one health 2018 zoonoses report. EFSA J 17:5926

Bachand N, Ravel A, Leighton P, Stephen C, Iqbal A, Ndao M, Konecsni K, Fernando C, Jenkins E (2018) Foxes (*Vulpes vulpes*) as sentinels for parasitic zoonoses, *toxoplasma gondii* and *Trichinella nativa*, in the northeastern Canadian Arctic. Int J Parasitol Parasites Wildl 7:391–397

Bandi C, La Rosa G, Bardin MG, Damiani G, Comincini S, Tasciotti L, Pozio E (1995) Random amplified polymorphic DNA fingerprints of the eight taxa of *Trichinella* and their comparison with allozyme analysis. Parasitology 110:401–407

Bruschi F, Dupouy-Camet J (2014) Trichinellosis. In: Bruschi F (ed) Helminth infections and their impact on global public health, p 229. Springer, Wien. https://doi.org/10.1007/978-3-7091-1782-8_8

Bruschi F, Murrell KD (2002) New aspects of human trichinellosis: the impact of new *Trichinella* species. Postgrad Med J 78:15–22

Bruschi F, Gómez-Morales MA, Hill DE (2019) International commission on Trichinellosis: recommendations on the use of serological tests for the detection of *Trichinella* infection in animals and humans. Food Waterborne Parasitol 12:e00032

Butler CE, Khan RA (1992) Prevalence of *Trichinella spiralis* in black bears (*Ursus americanus*) from Newfoundland and Labrador, Canada. J Wildl Dis 28:474–475

Choquette LP, Gibson GG, Pearson AM (1969) Helminths of the grizzly bear, *Ursus arctos* L., in northern Canada. Can J Zool 47:167–170

Dalcin D, Zarlenga DS, Larter NC, Hoberg E, Boucher DA, Merrifield S, Lau R, Ralevski F, Cheema K, Schwartz KL, Boggild AK (2017) *Trichinella nativa* outbreak with rare thrombotic complications associated with meat from a black bear hunted in northern Ontario. Clin Infect Dis 64(10):1367–1373

Davidson RK, Gjerde B, Vikøren T, Lillehaug A, Handeland K (2006) Prevalence of *Trichinella* larvae and extra-intestinal nematodes in Norwegian red foxes (*Vulpes vulpes*). Vet Parasitol 136: 307–316

Davidson RK, Ørpetveit I, Møller L, Kapel CM (2009) Serological detection of anti-*Trichinella* antibodies in wild foxes and experimentally infected farmed foxes in Norway. Vet Parasitol 163: 93–100

Deksne G, Segliņa Z, Jahundoviča I, Esīte Z, Bakasejevs E, Bagrade G, Keidāne D, Interisano M, Marucci G, Tonanzi D, Pozio E, Kirjušina M (2016) High prevalence of *Trichinella* spp. in sylvatic carnivore mammals of Latvia. Vet Parasitol 231:118–123

Devleesschauwer B, Praet N, Speybroeck N, Torgerson PR, Haagsma JA, De Smet K, Murrell KD, Pozio E, Dorny P (2015) The low global burden of trichinellosis: evidence and implications. Int J Parasitol 45(2–3):95–99. https://doi.org/10.1016/j.ijpara.2014.05.006

Dupouy-Camet J, Bruschi F (2007) Management and diagnosis of human trichinellosis. In: Dupouy-Camet J, Murrell KD (eds) FAO/WHO/OIE guidelines for the surveillance, management, prevention and control of trichinellosis. World Organisation for Animal Health Press, Paris, pp 37–68

Egeland GM, Committee NS (2010) Inuit health survey 2007–2008 Nunatsiavut. McGill University, Ste-Anne-de-Bellevue, QC

Egeland GM, Inuvialuit Settlement Region Steering Committee (2010) Inuit health survey 2007–2008. Inuvialuit settlement region. McGill University, Ste-Anne-de-Bel-levue, QC. http://www.irc.inuvialuit.com/publications/pdf/ihs-report-final.pdf

European Commission (2015) Commission implementing Regulation (EU) 2015/1375 of 10 August 2015 laying down specific rules on official controls for *Trichinella* in meat. Off J Eur Union L 212:7–34

European Food Safety Authority (2007) The community summary report on trends and sources of zoonoses, zoonotic agents, antimicrobial resistance and foodborne outbreaks in the European Union in 2006. EFSA J 130:2–352

FAO/WHO (2014) Multicriteria-Based Ranking for Risk Management of Food-borne Parasites. Report of a Joint FAO/WHO Expert Meeting, 3–7 September 2012. FAO Headquarters, Rome. http://www.fao.org/3/a-i3649e.pdf. Accessed 17 Aug 2018

Forbes LB, Measures L, Gajadhar A, Kapel C (2003) Infectivity of *Trichinella nativa* in traditional northern (country) foods prepared with meat from experimentally infected seals. J Food Prot 66(10):1857–1863

Fröscher W, Gullotta F, Saathoff M, Tackmann M (1988) Chronic trichinosis. Clinical, bioptic, serological and electromyographic observations. Eur Neurol 28:221–226

Gajadhar AA, Forbes LB (2010) A 10-year wildlife survey of 15 species of Canadian carnivores identifies new hosts or geographic locations for *Trichinella* genotypes T2, T4, T5, and T6. Vet Parasitol 168:78–83

Gajadhar AA, Bisaillon JR, Appleyard GD (1997) Status of *Trichinella spiralis* in domestic swine and wild boar in Canada. Can J Vet Res 61(4):256–259

Gamble HR, Bessonov AS, Cuperlovic K, Gajadhar AA, van Knapen F, Noeckler K, Schenone H, Zhu X (2000) International commission on Trichinellosis: re- commendations on methods for the control of *Trichinella* in domestic and wild animals intended for human consumption. Vet Parasitol 93:393–408

Gamble HR, Boireau P, Nöckler K, Kapel CMO (2007) Prevention of *Trichinella* infection in the domestic pig. In: Dupouy-Camet J, Murrell KD (eds) FAO/WHO/OIE guidelines for the surveillance, management, prevention and control of trichinellosis. World Organisation for Animal Health Press, Paris, pp 99–108

Glushnev AG, Voevodina YA, Simanova IA (2012) Prevalence of trichinosis among synanthropic and hunting animals in the Vologda Region. In: Collection of Scientific Articles adapted from the International Scientific Conference, vol 10, pp 126–129. (in Russian)

Gómez-Morales MA, Ludovisi A, Amati M, Blaga R, Zivojinovic M, Ribicich M, Pozio E (2012) A distinctive Western blot pattern to recognize *Trichinella* infections in humans and pigs. Int J Parasitol 42:1017–1023

Gómez-Morales MA, Selmi M, Ludovisi A, Amati M, Fiorentino E, Breviglieri L, Poglayen G, Pozio E (2016) Hunting dogs as sentinel animals for monitoring infections with *Trichinella* spp in wildlife. Parasite Vec 9:154. https://doi.org/10.1186/s13071-016-1437-1

Gómez-Morales MA, Ludovisi A, Amati M, Cherchi S, Tonanzi D, Pozio E (2018) Differentiation of *Trichinella* species (*Trichinella spiralis/Trichinella britovi* versus *Trichinella pseudospiralis*) using western blot. Parasite Vec 11:631

Gorodovich NM, Gorodovich SN (2010) Monitoring of *Trichinella* infection at the Far-East of Russia. In: Theory and practice of parasitic disease control. Collection of Scientific Articles adapted from the International Scientific Conference, vol 10, pp 129–131 [in Russian]

Gottstein B, Pozio E, Nockler K (2009) Epidemiology, diagnosis, treatment, and control of trichinellosis. Clin Microbiol Rev 22:127–145

Goyette S, Cao Z, Libman M, Ndao M, Ward BJ (2014) Seroprevalence of parasitic zoonoses and their relationship with social factors among the Canadian Inuit in Arctic regions. Diagn Microbiol Infect Dis 78(4):404–410

Goździk K, Odoevskaya IM, Movsesyan SO, Cabaj W (2017) Molecular identification of *Trichinella* isolates from wildlife animals of the Russian Arctic territories. Helminthologia 54: 11–16

Guba LA (2010a) Prevalence of *Trichinella* infection in the Seryshevsky Area of the Amur Region. In: Theory and practice of parasitic disease control. Collection of scientific articles adapted from the international scientific conference 11, 135. [in Russian]

Guba LA (2010b) Recovery of *Trichinella* in lynx (*Felis linx*) in the Amur Region of the Far-East. In: Theory and practice of parasitic disease control. Collection of Scientific Articles adapted from the International Scientific Conference, vol 11, pp 136–137 [in Russian]

Guba LA (2010c) Recovery of *Trichinella* larvae in Siberian weasels (*Mustela sibirica*) in the Amur Region of the Far-East. In: Theory and practice of parasitic disease control. Collection of Scientific Articles adapted from the International Scientific Conference, vol 10, pp 141–142 [in Russian]

Haraldsson P, Hersteinsson P (2004) Hvítabjörn. In: Hersteinsson P (ed) Íslenskspendýr. Reykjavík, Vaka-Helgafell, pp 102–107 [in Icelandic]

International Organization for Standardization (2015) ISO 18743: microbiology of the food chain—detection of *Trichinella* larvae in meat by artificial digestion method. Geneva

International Trichinella Reference Center (2019) Downloaded on December 20, 2019. www.iss.it/site/Trichinella/index.asp

Järvis T, Miller I, Pozio E (2001) Epidemiological studies on animal and human trichinellosis in Estonia. Parasite 8(2S):S86–S87

Jenkins EJ, Castrodale LJ, de Rosemond SJC, Dixon BR, Elmore SA, Gesy KM, Hoberg EP, Polley L, Schurer JM, Simard M, Thompson RCA (2013) Tradition and transition: parasitic zoonoses of people and animals in Alaska, Northern Canada, and Greenland. Adv Parasitol 33–204

Kapel CMO (1997) *Trichinella* in arctic, subarctic and temperate regions: Greenland, the Scandinavian countries and the Baltic States. Southeast Asian J Trop Med Public Health 28(Suppl 1): 14–19

Kapel CMO (2000) Host diversity and biological characteristics of the Trichinella genotypes and their effect on transmission. Vet Parasitol 93:263–278

Kapel CMO (2005) Changes in the EU legislation on *Trichinella* inspection—new challenges in the epidemiology. Vet Parasitol 132:189–194

Kapel CM, Pozio E, Sacchi L, Prestrud P (1999) Freeze tolerance, morphology, and RAPD-PCR identification of *Trichinella nativa* in naturally infected arctic foxes. J Parasitol 85(1):144–147

Kapel CM, Measures L, Moller LN, Forbes L, Gajadhar A (2003) Experimental *Trichinella* infection in seals. Int J Parasitol 33(13):1463–1470

Kapel CM, Webster P, Gamble HR (2005) Muscle distribution of sylvatic and domestic *Trichinella* larvae in production animals and wildlife. Vet Parasitol 132:101–105

Kärssin A, Velström K, Gómez-Morales MA, Saar T, Jokelainen P, Lassen B (2016) Cross-sectional study of anti-*Trichinella* antibody prevalence in domestic pigs and hunted wild boars in Estonia. Vector Borne Zoonotic Dis 16:604–610

Kärssin A, Häkkinen L, Niin E, Peik K, Vilem A, Jokelainen P, Lassen B (2017) *Trichinella* spp. biomass has increased in raccoon dogs (*Nyctereutes procyonoides*) and red foxes (*Vulpes vulpes*) in Estonia. Parasite Vec 10:609

Kocięcka W (2000) Trichinellosis: human disease, diagnosis and treatment. Vet Parasitol 93:365–383

Kokolova LM (2017) *Trichinella* spp. infection among wild animals in Yakutia. In: Theory and practice of parasitic disease control. Collection of Scientific Articles adapted from the International Scientific Conference, vol 18, pp 204–206 [in Russian]

Konyaev S V, Krivopalov AV, Yanagida T, Nakao M, SakoYa, Ito A, Malkina AV, Andreyanov ON, Odnokurtsev VA, Esaulova NV, Seredkin IV, Bondarev AY, Tkachenko LV (2012) Molecular genetic studies of *Trichinella* spp. in Russia: first results. In: Proceedings of the International scientific conference: "Modern problems of General Parasitology", October 30–November 1, 2012, Moscow, pp 171–174. [in Russian]

Korhonen PK, Pozio E, La Rosa G, Chang BC, Koehler AV, Hoberg EP, Boag PR, Tan P, Jex AR, Hofmann A, Sternberg PW, Young ND, Gasser RB (2016) Phylogenomic and biogeographic reconstruction of the *Trichinella* complex. Nat Commun 7:10513

Kumar V, Pozio E, De Borchgrave J, Mortelmans J, Meurichy WD (1990) Characterization of a *Trichinella* isolate from polar bear. Ann Soc Belge M ed Trop 70:131–135

La Rosa G, Pozio E (1990) Biochemical characterization of *Trichinella* in Greenland. Acta Vet Scand 31:381–383

La Rosa G, Marucci G, Zarlenga DS, Casulli A, Zarnke RL, Pozio E (2003) Molecular identification of natural hybrids between *Trichinella nativa* and *Trichinella* T6 provides evidence of gene flow and ongoing genetic divergence. Int J Parasitol 33:209–216

Larrat S, Simard M, Lair S, Bélanger D, Proulx JF (2012) From science to action and from action to science: the Nunavik Trichinellosis prevention program. Int J Circumpolar Health 71:18595. https://doi.org/10.3402/ijch.v71i0.18595

Larter NC, Forbes LB, Elkin BT, Allaire DG (2011) Prevalence of *Trichinella* spp. in black bears, grizzly bears, and wolves in the Dehcho region, Northwest Territories, Canada, including the first report of *T. nativa* in a grizzly bear from Canada. J Wildl Dis 47(3):745–749

MacLean JD, Viallet J, Law C, Staudt M (1989) Trichinosis in the Canadian Arctic: report of five outbreaks and a new clinical syndrome. J Infect Dis 160(3):513–520

MacLean JD, Poirier L, Gyorkos TW, Proulx JF, Bourgeault J, Corriveau A, Illisituk S, Staudt M (1992) Epidemiologic and serologic definition of primary and secondary trichinosis in the Arctic. J Infect Dis 165(5):908–912. https://doi.org/10.1093/infdis/160.3.513

Malakauskas A, Paulauskas V, Järvis T, Keidans P, Eddi C, Kapel CM (2007) Molecular epidemiology of *Trichinella* spp. in three Baltic countries: Lithuania, Latvia, and Estonia. Parasitol Res 100:687–693

Maslennikova OU, Strelnikov DP, Zhitkov BM (2017) *Trichinella* infection in American mink at the north of Nechernozemje. In: Theory and practice of parasitic disease control. Collection of Scientific Articles adapted from the International Scientific Conference, vol 18, pp 259–261 [in Russian]

Messier V, Lévesque B, Proulx JF, Rochette L, Srehir B, Couillard M, Ward BJ, Libman MD, Dewailly E, Déry S (2012) Seroprevalence of seven zoonotic infections in Nunavik, Quebec (Canada). Zoonoses Public Health 59(2):107–117

Møller LN, Koch A, Petersen E, Hjuler T, Kapel CM, Andersen A, Melbye M (2010) *Trichinella* infection in a hunting community in East Greenland. Epidemiol Infect 138(9):1252–1256. https://doi.org/10.1017/S0950268810000282

Murrell KD, Pozio E (2000) Trichinellosis: the zoonosis that won't go quietly. Inter J Parasitol 30: 1339–1349

Murrell KD, Pozio E (2011) Worldwide occurrence and impact of human trichinellosis, 1986-2009. Emerg Infect Dis 17(12):2194–2202

Nareaho A, Sankari S, Mikkonen T, Oivanen L, Sukura A (2000) Clinical features of experimental trichinellosis in the raccoon dog (*Nyctereutes procyonoides*). Vet Parasitol 91(1–2):79–91

Newman A (2014) Investigation of a human case of trichinellosis on a farm in Southwest Ontario. Environ Health Rev 57(4):93–96

Noeckler K, Pozio E, van der Giessen J, Hill DE, Gamble HR (2019) International Commission on Trichinellosis: recommendations on post-harvest control of *Trichinella* in food animals. Food Waterborne Parasitol 14:e00041

Office International des Epizooties (2004) Trichinellosis, chapter 2.2.9. In: Manual of standards for diagnostic tests and vaccines, 5th ed. Office International des Epizooties, Paris

Oivanen L, Oksanen A (2009) Synanthropic *Trichinella* infection in Finland. Vet Parasitol 159: 281–284

Oivanen L, Kapel CM, Pozio E, La Rosa G, Mikkonen T, Sukura A (2002) Associations between *Trichinella* species and host species in Finland. J Parasitol 88:84–88

Oksanen A, Interisano M, Isomursu M, Heikkinen P, Tonanzi D, Oivanen L, Pozio E (2018) *Trichinella spiralis* prevalence among wildlife of a boreal region rapidly reduced in the absence of spillover from the domestic cycle. Vet Parasitol 262:1–5

Owen R (1835) Description of a microscopic entozoon infesting the muscles of the human body. Trans Zool Sov 1:315–324

Pozio E (2007a) Taxonomy, biology and epidemiology of *Trichinella* parasites. In: Dupouy-Camet J, Murrell KD (eds) FAO/WHO/OIE guidelines for the surveillance, management, prevention and control of trichinellosis. World Organisation for Animal Health Press, Paris, pp 1–35

Pozio E (2007b) World distribution of *Trichinella* spp. infections in animals and humans. Vet Parasitol 149:3–21

Pozio E (2014) Searching for *Trichinella*: not all pigs are created equal. Trends Parasitol 30:4–11

Pozio E (2015) *Trichinella* spp. imported with live animals and meat. Vet Parasitol 213:46–55
Pozio E (2016a) Adaptation of *Trichinella* spp. for survival in cold climates. Food Waterborne Parasitol 4:4–12
Pozio E (2016b) *Trichinella pseudospiralis* an elusive nematode. Vet Parasitol 231:97–101
Pozio E (2020) Scientific achievements of the last 60 years: from a single to a multispecies concept of the genus Trichinella. Vet Parasitol 30:109042. https://doi.org/10.1016/j.vetpar.2020.109042
Pozio E, Bruschi F (2018) Trichinella. In: Liu D (ed) Handbook of foodborne diseases. CRC Press, pp 885–896
Pozio E, Murrell KD (2006) Systematics and epidemiology of *Trichinella*. Adv Parasitol 63:367–439
Pozio E, Zarlenga DS (2013) New pieces of the *Trichinella* puzzle. Int J Parasitol 43:983–997
Pozio E, Miller I, Järvis T, Kapel CMO, La Rosa G (1998) Distribution of sylvatic species of *Trichinella* in Estonia according to climate zones. J Parasitol 84:193–195
Pozio E, Casulli A, Bologov VV, Marucci G, La Rosa G (2001) Hunting practices increase the prevalence of *Trichinella* infection in wolves from European Russia 87:1498–1501
Pozio E, Christensson D, Steen M, Marucci G, La Rosa G, Brojer C, Morner T, Uhlhorn H, Agren E, Hall M (2004) *Trichinella pseudospiralis* foci in Sweden. Vet Parasitol 125:335–342
Pozio E, Kapel CM, Gajadhar AA, Boireau P, Dupouy-Camet J, Gamble HR (2006) *Trichinella* in pork: current knowledge on the suitability of freezing as a public health measure. Euro Surveill 11:E061116.1
Proulx JF, MacLean JD, Gyorkos TW, Leclair D, Richter AK, Serhir B, Forbes L, Gajadhar AA (2002) Novel prevention program for trichinellosis in Inuit communities. Clin Infect Dis 34(11): 1508–1514
Rah H, Chomel BB, Follmann EH, Kasten RW, Hew CH, Farver TB, Garner GW, Amstrup SC (2005) Serosurvey of selected zoonotic agents in polar bears (*Ursus maritimus*). Vet Rec 156:7–13
Ramey AM, Cleveland CA, Hilderbrand GV, Joly K, Gustine DD, Mangipane B, Leacock WB, Crupi AP, Hill DE, Dubey JP, Yabsley MJ (2019) Exposure of Alaska brown bears (*Ursus arctos*) to bacterial, viral, and parasitic agents varies spatiotemporally and may be influenced by age. J Wildl Dis 55(3):576–588
Rausch R, Babero B, Rausch RV, Schiller E (1956) Studies on the helminth fauna of Alaska. XXVII. The occurrence of larvae of Trichinella spiralis in Alaskan mammals. J Parasitol 42:259–271. https://doi.org/10.2307/3274850
Reichard MV, Torretti L, Snider TA, Garvon JM, Marucci G, Pozio E (2008) *Trichinella* T6 and *Trichinella nativa* in wolverines (*Gulo gulo*) from Nunavut, Canada. Parasitol Res 103:657–661
Roberts LS, John J (2005) Foundations of parasitology, 7th edn. McGraw-Hill, New York, pp 405–407
Romashov EB, Odoevskaya IM, Romashova BV (2018a) Features of trichinellosis circulation in natural conditions of the Voronezh region. In: Proceedings of the all-Russian Research Institute of experimental veterinary medicine, vol 80, pp 312–316. http://viev.ru/wordpress/wp-content/uploads/2019/01/Trudyi-VIEV-tom-80-chast-2.pdf [in Russian]
Romashov BV, Breslavtsev SA, Manzhurina OA, Skokoreva AM, Romashova EN, Popova OV, Golubova NA, Nikulin PA (2018b) The role of carnivorous mammals in circulation of natural focal helminthoses in conditions of the Central Chernozem region. In: Theory and practice of parasitic disease control. Collection of Scientific Articles adapted from the International Scientific Conference, vol 19, pp 413–416 [in Russian]
Rossi P, de Smet K, Pozio E (2017) Detection of *Trichinella* larvae in meat: comparison of ISO 18743:2015 with regulation (EU) 2015/1375. Food Anal Methods 10:634–639
Rostami A, Gamble HR, Dupouy-Camet J, Khazan H, Bruschi F (2017) Meat sources of infection for outbreaks of human trichinellosis. Food Microbiol 64:65–71
Sampasa-Kanyinga H, Lévesque B, Anassour-Laouan-Sidi E, Côté S, Serhir B, Ward BJ, Libman MD, Drebot MA, Ndao M, Dewailly É (2012) Zoonotic infections in native communities of

James Bay, Canada. Vector-borne Zoonotic Dis 12(6):473–481. https://doi.org/10.1089/vbz.2011.0739

Schellenberg RS, Tan BJ, Irvine JD, Stockdale DR, Gajadhar AA, Serhir B, Botha J, Armstrong CA, Woods SA, Blondeau JM, McNab TL (2003) An outbreak of trichinellosis due to consumption of bear meat infected with *Trichinella nativa*, in 2 northern Saskatchewan communities. J Infect Dis 188:835–843

Schurer JM, Ndao M, Skinner S, Irvine J, Elmore SA, Epp T, Jenkins EJ (2013) Parasitic zoonoses: one health surveillance in northern Saskatchewan. PLoS Negl Trop Dis 7(3):e2141. https://doi.org/10.1371/journal.pntd.0002141

Serrano FJ, Perez-Martin JE, Reina D, Navarrete I, Kapel CM (1999) Influence of infection intensity on predilection sites in swine trichinellosis. J Helminthol 73:251–254

Seymour J, Horstmann-Dehn L, Rosa C, Lopez JA (2014) Occurrence and genotypic analysis of *Trichinella* species in Alaska marine-associated mammals of the Bering and Chukchi seas. Vet Parasitol 200(1–2):153–164

Shaikenov BS (1992) Ecological border of distribution of *Trichinella nativa* Britov and Boev 1972 and *T. nelsoni* Britov and Boev 1972. Wiadomosci Parazytologiczne 38:85–91

Shaikenov BS, Boev SN (1983) Distribution of *Trichinella* species in the old world. Wiadomosci Parazytologiczne 29:595–608

Sharma R, Harms NJ, Kukka PM, Parker SE, Gajadhar AA, Jung TS, Jenkins EJ (2018) Tongue has higher larval burden of Trichinella spp. than diaphragm in wolverines (Gulo gulo). Vet Parasitol 253:94–97. https://doi.org/10.1016/j.vetpar.2018.02.032. Epub 2018 Feb 21. PMID: 29605011

Sharma R, Thompson P, Elkin B, Mulders R, Branigan M, Pongracz J, Wagner B, Scandrett B, Hoberg E, Rosenthal B, Jenkins E (2019) *Trichinella pseudospiralis* in a wolverine (*Gulo gulo*) from the Canadian North. Inter J Parasitol Parasite Vec 9:274–280

Sharma R, Thompson P, Hoberg EP, Scandrett B, Konecsni K, Harms NJ, Kukka PM, Jung TS, Elkin B, Mulders R, Larter NC, Branigan M, Pongracz J, Wagner B, Kafle P, Lobanov VA, Rosenthal B, Jenkins EJ (2020) Hiding in plain sight: discovery and phylogeography of a cryptic species of *Trichinella* (Nematoda: Trichinellidae) in wolverine (*Gulo gulo*). Inter J Parasitol. https://doi.org/10.1016/j.ijpara.2020.01.003

Sharma R, Harms NJ, Kukka PM, Jung TS, Parker SE, Ross S, Thompson P, Rosenthal B, Hoberg EP, Jenkins EJ (2021) High prevalence, intensity, and genetic diversity of *Trichinella* spp. in wolverine (*Gulo gulo*) from Yukon, Canada

Skírnisson K, Richter SH, Eydal M (2003) Prevalence of human parasites in Iceland: past and present status. In: Akuffo H, Linder E, Ljungström I, Wahlgren M (eds) Parasites of the colder climates. Taylor & Francis, London, pp 34–44

Skírnisson K, Marucci G, Pozio E (2010) *Trichinella nativa* in Iceland: an example of *Trichinella* dispersion in a frigid zone. J Helminthol 84:182–185

Skripova LV, Kovchur VN (1994) Trichinellosis in Byelorussia. Wiad Parazytol 40:389–391

Smith HJ, Snowdon KE (1988) Sylvatic trichinosis in Canada. Can J Vet Res 52:488–489

Solovyeva IA, Bondarenko GA, Truchina TI, Ivanov DA (2017) Spreading of *Trichinella* infection causative agent among wild animals in the Amur region. In: Theory and practice of parasitic disease control. Collection of Scientific Articles adapted from the International Scientific Conference, vol 18, pp 467–469 [in Russian]

Springer YP, Casillas S, Helfrich K, Mocan D, Smith M, Arriaga G, Mixson L, Castrodale L, McLaughlin J (2017) Two outbreaks of Trichinellosis linked to consumption of Walrus Meat — Alaska, 2016–2017. MMWR Morb Mortal Wkly Rep 66:692. https://doi.org/10.15585/MMWR.MM6626A3

Tada K, Suzuki H, Sato Y, Morishima Y, Nagano I, Ishioka H, Gomi H (2018) Outbreak of *Trichinella* T9 infections associated with consumption of bear meat, Japan. Emerg Infect Dis 24:1532–1535

Thorshaug KN, Rosted KF (1956) Researches on the prevalences of trichinosis in animals in Arctic and Antarctic waters. Nord Vet Med 8:115–129

Vagin NA, Malisheva NS (2010) Prevalence of *Trichinella* infection in mammals at the territory of the Kursk region. In: Theory and practice of parasitic disease control. Collection of Scientific Articles adapted from the International Scientific Conference, vol 11, pp 91–92 [in Russian]

Vagin NA, Malisheva NS, Vlasov EA, Samofalova NA, Uspensky AV, Odoyevskaya IM, Vagin Yu A (2016) Detection of *Trichinella* in northern white-breasted hedgehog (*Erinaceus roumanicus*) in the Kursk region. In: Theory and practice of parasitic disease control collection of scientific articles adapted from the international scientific conference, vol 17, pp 100–102 [in Russian]

Virchow R (1859) Recherches sur le developpement du Trichina spiralis. CR Acad Sci 49:660–662

Zarlenga DS, Chute MB, Martin A, Kapel CMO (1999) A multiplex PCR for unequivocal differentiation of all encapsulated and non-encapsulated genotypes of Trichinella. Inter J Parasitol 29:1859–1867

Cryptosporidiosis and Giardiosis in the Arctic: Increasing Threats in a Warmer World?

Lucy J. Robertson and John J. Debenham

1 Introduction

Cryptosporidiosis and giardiosis are protozoan diseases, usually gastrointestinal, with diarrhoea as a common clinical sign. The parasites responsible, *Cryptosporidium* spp. and *Giardia duodenalis*, have direct faecal-oral lifecycles, often via contaminated water or food. The transmission stages, which are infectious upon excretion, can be shed in high quantities and are very robust. These are both characteristics that facilitate transmission via environmental contamination.

These similarities have meant that these diseases are often considered together. However, there are many aspects of these infections and of the parasites that cause them that are quite unalike, as indicated by the huge phylogenetic distance between these genera. Therefore, it is important that, despite their similarities and some aspects of probable convergent evolution, they are considered individually. The differences between these genera should be recognised as they may affect many aspects of both the parasites and their diseases, including their biology, epidemiology, evolution, and pathology. These aspects may impact on the suitability of these parasites for establishing and disseminating in Arctic conditions. Thus, a brief overview of these aspects is provided in the following sections.

1.1 Taxonomy and Phylogeny of *Cryptosporidium* and *Giardia*

Cryptosporidium and *Giardia* are both unicellular parasitic organisms (protozoa) belonging, respectively, in the phyla Apicomplexa and Metamonada. Although *Cryptosporidium* is traditionally considered a member of coccidia, evidence

L. J. Robertson (✉) · J. J. Debenham
Department of Paraclinical Sciences, Faculty of Veterinary Medicine, Norwegian University of Life Sciences, Ås, Norway
e-mail: Lucy.robertson@nmbu.no

M. Tryland (ed.), *Arctic One Health*, https://doi.org/10.1007/978-3-030-87853-5_15

indicates that it has a closer affinity with gregarines, which are a large group of Apicomplexa that are considered particularly primitive (Ryan et al. 2016). This classification has implications for the survival and spread of this parasite. To date, over 30 species of *Cryptosporidium* have been identified, some of which are host specific, whereas others are more promiscuous regarding host infectivity. Whereas some species of *Cryptosporidium* are associated with only mild symptoms or signs, or none at all, other species are particularly pathogenic and are associated with severe symptoms. These may even result in mortality, particularly if the host has other health challenges.

In contrast, for *Giardia*, which is considered an early-diverging eukaryote in the Diplomonadida group, around 40 species have been described. However, only 6 are currently recognised as distinct species, with the others considered likely to be synonyms. Of these, *Giardia duodenalis* (also known as *G. lamblia* or *G. intestinalis*) is the species generally associated with the disease giardiosis. However, *G. duodenalis* is itself considered a species complex, divided into a several morphologically identical, but genetically distinct Assemblages (also known as subtypes of genotypes). Some of these Assemblages are host-specific, but others are more promiscuous, infecting a range of host groups. Again, there seems to be variation in disease severity, but it has been difficult to associate the different Assemblages with a particular symptom spectrum or level of virulence.

1.2 Lifecycles and Transmission of *Cryptosporidium* and *Giardia*

Although both *Cryptosporidium* and *Giardia* have a direct, predominantly faecal-oral lifecycle (for some *Cryptosporidium* species and hosts, respiratory cryptosporidiosis is also relevant), other facets of their transmission and development are very different from each other. For *Cryptosporidium*, when an infective oocyst is ingested, it excysts in the small intestine, and the released sporozoites invade the cells of the epithelium. Here, in an epicellular location (intracellular but extracytoplasmic), asexual multiplication occurs. The resulting merozoites invade neighbouring cells, and sexual multiplication occurs with the production of microgamonts and macrogamonts. Following fertilization of the macrogamonts, oocysts are produced that sporulate within the host and are excreted in the host faeces.

In contrast, for *Giardia*, only two distinct forms occur in its lifecycle; the cyst and the trophozoite. An infective cyst hatches in the duodenum releasing two trophozoites that remain in the host's intestinal lumen. The trophozoites tend to adhere to the intestinal epithelium, but are not intracellular and may move around within the intestinal lumen. Replication is asexual, by binary fission, which limits the opportunity for exchange of genetic data. Distally, in the ileum, the trophozoites form the non-replicating cyst stages that are subsequently passed in the faeces.

Thus, both *Cryptosporidium* and *Giardia* infection involves multiplication by asexual reproduction, but *Cryptosporidium* also has a sexual stage within the lifecycle. The robust transmission stages of both parasites (oocyst for *Cryptosporidium*, cyst for *Cryptosporidiosis and sporidiosis in the Arctic*) are shed in huge quantities in the faeces. These stages are already infective to a susceptible host upon excretion and without any requirement for further maturation in the environment. Both *Cryptosporidium* oocysts and *Giardia* cysts can survive for prolonged periods (weeks or months) in the environments, especially in cool, damp conditions. However, in general, *Cryptosporidium* oocysts are more robust than *Giardia* cysts, particularly regarding tolerance towards freezing and drying (Fayer and Nerad 1996; Robertson and Gjerde 2004, 2006; Utaaker et al. 2017).

1.3 Pathogenesis and Signs of Infection

A symptom is a manifestation of disease apparent to the patient himself, while a sign is a manifestation of disease that the physician perceives Suggest changing symptoms to signs. The pathogenesis and symptoms for both parasites are predominantly gastrointestinal, as evinced by their location in the infected host. The pathogenesis of both tends to be associated with damage and changes to the cells of the intestinal wall, although there are differences between the mechanisms for each of the parasites (Certad et al. 2017). The interaction of *Cryptosporidium* sporozoites and merozoites with the host cells is assumed to result in signalling cascades with different resultant molecules (such as proteases and haemolysins) damaging the cells, increasing fluid secretion, and causing malabsorption. For *Giardia*, the attachment to the walls of the epithelial cells with the trophozoite ventral disc is considered also to activate signalling molecules, with caspases considered to activate apoptosis and break down junctions between the epithelial cells resulting in barrier dysfunction. This, in turn, leads to enhanced secretion of electrolytes and malabsorption. Thus, the predominant feature of both infections is diarrhoea (along with other gastrointestinal symptoms such as abdominal pain and nausea); however, the mechanisms by which the diarrhoea occurs seem to differ, and the diarrhoea in itself is different, tending to be watery for *Cryptosporidium* and fatty for *Giardia*. Post-infection sequelae have also been reported for both infections (in humans) and may reflect host immune responses rather than infection per se. In addition, for both parasites, interactions with the microbiota in the host intestine may also be of relevance. One interesting observation in human and some animal hosts is that occasionally *Giardia* infection may be associated with reduced probability or reduced severity of some other infections, presumably associated with different immunological pathways (Manko et al. 2017). Thus, in some instances, *Giardia* may be considered to have a protective effect against other pathogens and associated diseases.

1.4 Host Range and Zoonotic Potential

Both parasites have a wide range of potential hosts, although not all species or assemblages are infectious to all hosts. Among the 30 or so *Cryptosporidium* species identified, *C. parvum* is considered of most relevance to young livestock and is also very important as a zoonotic species. Other zoonotic species include *C. meleagridis* (commonly found in poultry), *C. cuniculus* (commonly found in rabbits), and

C. ubiquitum (commonly found in sheep); *C. hominis* largely infects humans, and many other species also are host specific. It is assumed that all vertebrate species can be infected by at least one species of *Cryptosporidium.*

Among the seven recognised Assemblages of the *G. duodenalis* species complex, Assemblages A (in particular) and B (to a lesser extent) are considered to have limited zoonotic potential, infecting humans and a range of other animals (such as livestock, some deer, and canids). Despite this host flexibility, reports of zoonotic transmission are not common. Furthermore, there are sub-types within Assemblage A with different host specificities such that some strains of Assemblage A found in humans may not necessarily infect animals, and vice versa (Sprong et al. 2009). The other Assemblages have been associated with particular mammal groups, such as dogs (Assemblages C and D), sheep and cattle (Assemblage E), cats (F), rodents (G), and pinnipeds (H); other vertebrates tend to be hosts for other species of *Giardia.*

2 Occurrence of *Cryptosporidium* and *Giardia* in Arctic Regions

As many of the countries of the Arctic (around 66.5 °N and northwards from the equator) are contiguous with non-Arctic regions, determining from publications when these parasites have been found in Arctic regions and in which host species is not always straightforward. However, it is clear that although both parasites may be better adapted to more temperate climates (as their transmission stages tend not to survive prolonged freezing or repeated freeze-thawing; Robertson et al. 1992), both parasites do occur in this region.

2.1 Occurrence in Wild and Domestic Animals in Arctic and Subarctic Regions

Various surveys of animals for infection with *Cryptosporidium* and/or *Giardia* have been conducted in both Arctic (66.5 °N and north) and subarctic (50 °N and up to the Arctic) regions. In general, immunofluorescent antibody tests (IFAT) have been used for the detection of cysts or oocysts in faeces or intestinal material, although older studies have used other microscopy techniques (e.g. lugol for *Giardia*). More recent studies have often used molecular techniques (e.g. PCR), both for detection and for genetic characterisation to the species, assemblage, or genotype level. Results of wildlife studies using IFAT and PCR from Arctic regions alone are summarised in Tables 1 (*Cryptosporidium*) and 2 (*Giardia*).

Results obtained by Roach et al. (1993) are not included in these tables as they were based on low sample numbers (from 1 to 14), and IFAT was only used for *Cryptosporidium* (all samples negative) and not *Giardia*. Furthermore, it was not entirely clear from the publication whether the samples were from Arctic or subarctic regions, although the indication is that they were subarctic (around Dawson and Whitehorse, at, 64.1 °N and 60.7 °N, respectively). There is a strong geographical

Table 1 Reports of *Cryptosporidium* spp. amongst Arctic wildlife

Species	Country	Detection method	Prevalence % (no. of samples)	Excretion rate median opg (range)	Species/genotype	References
Order Carnivora						
Ringed seal *Phoca hispida*	Canada	IFAT	0 (15)	–		Olson et al. (1997)
	USA—Alaska	IFAT	23 (31)	–		Hughes-Hanks et al. (2005)
	Canada	IFAT	9 (55)	–	*C. muris*, 2 novel genotypes	Dixon et al. (2008) Santín et al. (2005)
Bearded seal *Erignathus barbatus*	USA—Alaska	IFAT	0 (22)	–		Hughes-Hanks et al. (2005)
	Canada	IFAT	0 (4)	–		Dixon et al. (2008) Santín et al. (2005)
Arctic fox *Vulpes lagopus*	Canada	IFAT	9 (95)	77 (18–146)		Elmore et al. (2013)
	Norway	PCR	0 (62)	–		Myšková et al. (2019)
Order Rodentia						
Sibling vole *Microtus levis*	Norway	PCR	1.6 (63)	–	*C. alticolis*	Myšková et al. (2019)
Order Cetacea						
Beluga whale *Delphinapterus leucas*	Canada	IFAT	0 (16)	–		Olson et al. (1997)
	USA—Alaska	IFAT	0 (18)	–		Hughes-Hanks et al. (2005)

(continued)

Table 1 (continued)

Species	Country	Detection method	Prevalence % (no. of samples)	Excretion rate median opg (range)	Species/genotype	References
Bowhead whale *Balaena mysticetus*	USA—Alaska	IFAT	5 (39)	–		Hughes-Hanks et al. (2005)
North Atlantic right whale *Eubalaena glacialis*	USA—Alaska	IFAT	24 (49)	–		Hughes-Hanks et al. (2005)
Order Artiodactyla						
Muskoxen *Ovibos moschatus*	Canada	IFAT	0 (72)	–		Kutz et al. (2008)
Caribou *Rangifer tarandus*	USA—Alaska	IFAT, PCR	6 (49)	–	*Cryptosporidium* sp., caribou genotype	Siefker et al. (2002)
	Canada	IFAT	1.3 (149)	–		Johnson et al. (2010)
Moose *Alces alces*	USA—Alaska	IFAT	0 (42)	–		Siefker et al. (2002)

Opg oocysts per gram faeces; *IFAT* Immunofluorescence antibody test

Table 2 Reports of *Giardia duodenalis* amongst Arctic wildlife

Species	Country	Detection method	Prevalence % (no. of samples)	Excretion rate median cpg (range)	Assemblage	References
Order Carnivora						
Ringed seal *Phoca hispida*	Canada	IFAT	20 (15)	–		Olson et al. (1997)
	USA—Alaska	IFAT	65 (31)	–		Hughes-Hanks et al. (2005)
	Canada	IFAT PCR	80 (55)	–	B	Dixon et al. (2008)
Bearded seal *Erignathus barbatus*	USA—Alaska	IFAT	0 (22)	–		Hughes-Hanks et al. (2005)
	Canada	IFAT	75 (4)	–		Dixon et al. (2008)
Arctic fox *Vulpes lagopus*	Canada	IFAT	16 (95)	162 (2–12,800)		Elmore et al. (2013)
	Norway	PCR	0 (62)			Myšková et al. (2019)
Order Rodentia						
Sibling vole *Microtus levis*	Norway	PCR	0 (63)	–		Myšková et al. (2019)
Order Cetacea						
Beluga whale *Delphinapterus leucas*	Canada	IFAT	0 (16)	–		Olson et al. (1997)
	USA—Alaska	IFAT	0 (18)	–		Hughes-Hanks et al. (2005)
Bowhead whale *Balaena mysticetus*	USA—Alaska	IFAT	33 (39)	–		Hughes-Hanks et al. (2005)
North Atlantic right whale *Eubalaena glacialis*	USA—Alaska	IFAT	71 (49)	–		Hughes-Hanks et al. (2005)

(continued)

Table 2 (continued)

Species	Country	Detection method	Prevalence % (no. of samples)	Excretion rate median cpg (range)	Assemblage	References
Order Artiodactyla						
Muskoxen *Ovibos moschatus*	Canada	IFAT	21 (72)	6933 (133–348,533)	A	Kutz et al. (2008)
Caribou *Rangifer tarandus*	USA—Alaska	IFAT	0 (49)	–		Siefker et al. (2002)
	Canada	Sucrose gradient	2 (149)	–		Johnson et al. (2010)
Moose *Alces alces*	USA—Alaska	IFAT	0 (42)	–		Siefker et al. (2002)

Cpg cysts per gram of faeces; *IFAT* Immunofluorescence antibody test bias to the reports included in these tables, with most studies on these protozoa in Arctic wildlife coming from the USA (Alaska) or Canada, a single study from Norway, and no published studies in English investigating Arctic wildlife from Denmark (Greenland), Finland, Iceland, Russia, or Sweden. It is worth noting, however, that studies from the Russian Arctic may well be published in Russian. Nevertheless, it is apparent that the occurrence of these parasites in wildlife remains largely unknown and unexplored in the majority of Arctic regions.

In general, reports suggest that the occurrence of both *Cryptosporidium* and *Giardia* is low in wildlife in those Arctic regions where such surveys have been performed compared with data from elsewhere. This appears particularly true for *Cryptosporidium*, which, to date, does not appear to be well established within Arctic wildlife. Interestingly, the two wildlife species reporting the highest prevalence of *Cryptosporidium* are the ringed seal (*Phoca hispida*) and the North Atlantic Right Whale (*Eubalaena glacialis*). Both are migratory species that can venture out of the Arctic, and infection may therefore represent parasite-host life cycles based in the subarctic.

In contrast, *Giardia* appears to be more widespread in the Far North, with infections detected in 13 of the 16 Arctic wildlife species that have been examined. Given that *Cryptosporidium* oocysts seem better able to survive in the environment than *Giardia* cysts, particularly under freezing conditions, this distribution is perhaps surprising and may reflect other more important epidemiological features in this environment.

Although the occurrence of *Cryptosporidium* and *Giardia* appears to be lower in wildlife in the Arctic than the subarctic, inter-regional comparisons are difficult since many of the animals studied are relatively Arctic-specific and therefore have not been investigated elsewhere. Muskoxen, however, have been investigated for both *Cryptosporidium* and *Giardia* in both Arctic (Banks Island, Northwest Territories, Canada—73 °N; Kutz et al. 2008) and subarctic regions (Dovrefjell, Norway—62 °N; Davidson et al. 2014). In the Banks Island survey (Kutz et al. 2008), in which IFAT was used for detection, none of the samples were positive for *Cryptosporidium*, but 15/72 (21%) of samples were positive for *Giardia*. In contrast, in the Norwegian study (Davidson et al. 2014), which also used IFAT for detection, *Cryptosporidium* oocysts were detected in 17/167 (10%) of muskox samples. Intriguingly, however, *Giardia* was detected in a considerably lower proportion of animals in the Norwegian Dovrefjell study than in the Banks Island study, being detected in only 7/167 (4%) of samples. In the Norwegian study, it was speculated that the muskox infections may have resulted from sympatric grazing with sheep. On Banks Island where mammalian biodiversity is low (Peary caribou, Arctic hares, Arctic fox, lemmings, polar bears, and grizzly bears), such sympatric grazing would not occur. Although the absence or presence of sympatric grazing may be one reason for the lack of similarity in results, the lack of analysis of samples from other potential hosts means that supportive evidence is missing, and the difference in results could reflect other aspects of the study design.

Similarly, the occurrence of these parasites in moose has also been investigated in both Arctic and subarctic regions. Whereas neither *Cryptosporidium* nor *Giardia* were detected (using PCR methods) in 42 moose samples from parts of Northern Alaska within the Arctic (Siefker et al. 2002), a study from subArctic Norway (Hamnes et al. 2006) reported *Cryptosporidium* from 15 (3%) and *Giardia* from 56 of 455 (12%) of moose samples using IFAT. However, there are many differences between the studies, including sampling regime, detection method, and age of animal sampled. Thus, again, there is no reason to assume that differences in results are due solely to sampling location, although this may also be of relevance.

Molecular data characterising *Cryptosporidium* found in wildlife in the Arctic is sparse and therefore it is difficult to draw firm conclusions on epidemiological patterns of transmission. Barren-ground Caribou (*Rangifer tarandus groenlandicus*) from Northern Alaska were found to be infected with a previously uncharacterised genotype, named *Cryptosporidium* sp. caribou isolate, based on PCR at the 18S rRNA and heat shock protein (HSP-70; Siefker et al. 2002). *Cryptosporidium* isolates from Canadian ringed seals were characterised based on PCR at the 18S rRNA, HSP-70, and actin gene loci, with two novel seal genotypes and *Cryptosporidium muris* identified (Santín et al. 2005; Dixon et al. 2008). In Svalbard (Norway), *Cryptosporidium* found in sibling vole (Myšková et al. 2019; a rodent species considered to have been introduced from Russia, and of relevance in maintaining the *Echinococcus multilocularis* cycle there (Davidson et al. 2012)) was identified as *C. alticolis*. This species has previously been identified in voles in the Czech Republic (Horčičková et al. 2019), and it seems likely that this *Cryptosporidium* species was introduced to Svalbard simultaneously with the sibling voles that have established there. The data suggest that this *Cryptosporidium* infection is maintained at a low density (only 1 of 63 vole samples was found positive) in that population, although this was not discussed by the authors. This lack of *Cryptosporidium* species associated with human infections in Arctic wildlife suggests that zoonotic transmission may be unlikely.

As with *Cryptosporidium*, most of the studies investigating *Giardia* in wildlife in the Arctic have not attempted to characterise the isolates identified using molecular methods. However, those results that are available are more indicative of the potential for transmission between wildlife and humans. In the Banks Island survey on muskoxen (Kutz et al. 2008), 4 *Giardia* isolates were identified as belonging to Assemblage A by PCR targeting the 18S rRNA gene. Similar results were gained from subarctic muskoxen in Dovrefjell, Norway (Davidson et al. 2014), where one sample was classified as Assemblage AI based on PCR targeting both the glutamate dehydrogenase and beta giardin genes. It is interesting that these geographically distinct populations of muskoxen were both found to be infected by the non-host specific, and potentially zoonotic, Assemblage A, and not Assemblage E, which occurs most frequently with their taxonomically close relatives, sheep and goats.

Ringed seals (*Phoca hispida*) in Arctic Canada were found to be infected with *Giardia* Assemblage B, based on PCR at the small subunit rRNA (SSU-rRNA) and triosephosphate isomerase (TPI) genes. This differed from results of molecular characterisation of pinnipeds from the subarctic Gulf of St Lawrence, where *Giardia* Assemblage A was identified in harp seals (*Pagophilus groenlandicus*) and hooded seals (*Cystophora cristata*) based on PCR at the SSU-rRNA gene (Appelbee et al.

2010). Whilst this difference may be due to a variety of factors, including host species, rather than geographical location, it is interesting that, as with the muskoxen, neither population of pinnipeds were infected by their host-adapted *Giardia* Assemblage (Assemblage H).

Unlike with *Cryptosporidium*, these results raise questions over the source of these infections, and possible links to human activity. Further research is needed to elucidate the background for this: have humans, venturing further and further north in larger numbers, brought with them their *Giardia* parasites, that now spill over into previously naïve wildlife populations? It also raises the question of whether these populations could serve as a wildlife reserve for zoonotic strains of *Giardia*.

For some wildlife species whose population ranges across the subarctic and Arctic, data are available from subarctic populations, although it is unknown whether the same species or genotypes of parasites would be found in the same host species living further north. For example, molecular analysis at the SSU rRNA gene of two *Cryptosporidium* isolates from muskoxen in the subarctic (Dovrefjell, Norway, Davidson et al. 2014) demonstrated the presence of *C. xiaoi* (possibly reflecting sympatric grazing with Norwegian sheep; Robertson et al. 2010) and *Giardia* isolates from subarctic moose in Norway were genotyped as Assemblages AI and AII, depending on target gene (Robertson et al. 2007).

Intensity of infection is also of relevance for considering these parasites, although data on cyst/oocyst quantity is not always recorded. This is particularly evident when the detection techniques used are not appropriate for quantification or semi-quantification of excretion rate. It should also be borne in mind that *Giardia* excretion is recognised as varying over time, and thus, intensity of excretion may differ on a daily basis. Nevertheless, when high numbers of cysts or oocysts are detected in a faecal sample, then it is more likely to reflect productive infection than carriage without infection. In addition, high excretion rates indicate the likelihood of a more contaminated environment, thus increasing the risk of onward transmission. This could be either to the same or different species of host, depending on the host-species flexibility of the parasite.

Several of the studies reported in Tables 1 and 2 indicate low levels of excretion.

Domesticated/semi-domesticated animals in the Arctic are predominantly dogs (hunting dogs and sled dogs, in particular) and reindeer. Although data on the occurrence of *Cryptosporidium* and *Giardia* in domestic animals in the Arctic is sparse, several studies have been conducted in the northern subarctic. In Arctic Finland (Kaamanen, 69.1 °N), reindeer calves from an experimental herd were found to have a very high occurrence of both *Cryptosporidium* and *Giardia* with a cumulative prevalence (based on IFAT detection) of 100% for *Giardia* and 23% for *Cryptosporidium* among 54 reindeer calves from aged 0 to 33 months (Niine et al. 2017). An earlier study from Arctic Norway and Finland had not detected *Cryptosporidium* in 2243 samples from semi-domesticated reindeer (using immunomagnetic separation followed by IFAT) and did not investigate *Giardia* (Kemper et al. 2006). However, these were adult slaughter animals, and therefore probably less likely to be infected with *Cryptosporidium* than reindeer calves. Similarly, a later study (Idland et al. 2021) also investigating slaughter reindeer (in Arctic Norway), did not find any *Cryptosporidium*, but reported a 5% occurrence of *Giardia*, with genotyping indicating the presence of zoonotic Assemblage AI (Idland et al. 2021).

From the subarctic, the presence of both *Cryptosporidium* and *Giardia* has been investigated in various populations of domestic dogs in Canada. Reports of 3% occurrence of *Cryptosporidium* shedding in dogs was reported in 3 different studies; 5/155 (Himsworth et al. 2010), 2/75 (Bryan et al. 2011), 8/231 (Schurer et al. 2012). Unfortunately, molecular data were not available on these isolates. Each of these studies also found *Giardia* in domestic dogs to be relatively common, and occurring considerably more frequently than *Cryptosporidium* oocysts. In these studies, *Giardia* shedding was reported from 95/155 (61%; Himsworth et al. 2010), 10/75 (13%; Bryan et al. 2011) and 40/231 (17%; Schurer et al. 2012) of the dog samples. Molecular analysis of these isolates revealed Assemblage A in all three studies, and this was also supported by Salb et al. (2008). It is of note that the dog-specific *Giardia* Assemblages, C and D, were not reported, perhaps suggesting that the *Giardia* occurring in these dogs has not originated from canine hosts.

2.2 Occurrence in Human Hosts in Arctic Regions and Lack of Evidence of Zoonotic Transmission

Although cases of cryptosporidiosis are considered rare in Alaska, the annual rates of giardiosis cases are usually higher than elsewhere in the USA (Hueffer et al. 2013). Given that the transmission routes to humans are similar, these data suggest that Arctic communities in North America are affected by particular variables that do not occur further south. However, this could also reflect an absence of appropriate methods for diagnosis of cryptosporidiosis in diagnostic laboratories, as more sensitive methods have indicated that perhaps *Cryptosporidium* is more common than previously assumed (Yansouni et al. 2016). Indeed, diverse methods have been used to investigate the prevalence of *Cryptosporidium* and *Giardia* among people living in Arctic regions, including microscopy with staining, such as IFAT and acid-fast stains, molecular methods, and serology. As with human population surveys in other world regions, study cohorts have varied according to a range of variables including age group, presence or absence of diarrhoea, and occupation. Results are summarised in Tables 3 (*Cryptosporidium*) and 4 (*Giardia*).

Perhaps of particular interest are the associations made between activities that involve associations with wildlife. For example, in a survey of 887 Alaskan residents, Mosites et al. (2018) found no association with *Cryptosporidium* infection and different variables, but did note a higher seroprevalence against *Giardia* in subsistence hunters and their families, that they associate with lack of running water in the home, with water commonly collected from rain, ice or snow. Given that *Cryptosporidium* is also considered to be a waterborne parasite, it is possible that some other associated factor may be relevant, or that *Cryptosporidium* is more widely distributed in the environment, including the municipal water supply, such that exposure is not greater among subsistence hunters and their families than any

Table 3 Reports of *Cryptosporidium* spp. amongst humans living in the Arctic

Country	Detection method	Prevalence % (no. of samples)	Species/genotype	References
Canada	qPCR	20 (86)[a]		Goldfarb et al. (2013)
Canada	IFAT, PCR	16 (108)[a]	*C. parvum* genotype IIa	Iqbal et al. (2015)
Canada	Acid-fast, PCR	18 (283)	*C. hominis* genotype Id	Thivierge et al. (2016)
Alaska	Serology	29 (887)		Mosites et al. (2018)

IFAT immunofluorescence antibody test; *qPCR* quantitative PCR
[a]Samples from patients with diarrhoea; overlap in samples between Goldfarb et al. (2013) and Iqbal et al. (2015)

Table 4 Reports of *Giardia duodenalis* amongst humans living in the Arctic

Country	Detection method	Prevalence % (no. of samples)	Assemblage	References
Russia	Not provided	1.6 (378,000)		Dudarev et al. (2013)
Canada	IFAT, PCR	5 (108)[a]	B	Iqbal et al. (2015)
Canada	qPCR	1 (86)[a]		Goldfarb et al. (2013)
Alaska	Serology	19 (887)		Mosites et al. (2018)

IFAT immunofluorescence antibody test; *qPCR* quantitative PCR
[a]Samples from patients with diarrhoea

other group. Whether zoonotic transmission has occurred in any of the cases reported is not possible to ascertain with the data available.

The majority of human cases of *Cryptosporidium* infection in the Arctic where species identification has been conducted indicate *C. hominis* subtypes, arguing against zoonotic transmission, although *C. parvum* genotype IIA has also been reported (Iqbal et al. 2015). Furthermore, as detailed earlier, none of the few *Cryptosporidium* species identified in Arctic wildlife are those associated with human infection. It is perhaps worthy of note that a recent systematic review and meta-analysis (King et al. 2019) suggested that human-to-human transmission of *Cryptosporidium* may predominate in countries with poorer sanitation; however, as that review focuses specifically on *C. parvum* (gp60 subtype IIc) there is no mention of Arctic communities. Indeed, it has been shown that *C. parvum* divides into two clades, of which one, designated *C. parvum parvum* is a zoonotic generalist, whereas the other, *C. parvum anthroponosum,* transmits only between humans (Nader et al. 2019). A further review of the evidence for zoonotic *Cryptosporidium* transmission in Africa has also argued against this transmission route occurring widely (Robertson et al. 2020). Although it would seem, intuitively, that there is little in common between Arctic and African communities, there are several commonalities that are of relevance to the transmission of *Cryptosporidium* infections. These include suboptimal water treatment and sewage disposal, along with the potential for environmental extremes.

Although the presence of *Giardia* assemblages with zoonotic potential in Arctic wildlife would suggest that wildlife-to-human transmission is possible, where efforts have been made to characterise the *Giardia* Assemblages found in humans infections, Assemblage B has been identified (Iqbal et al. 2015). This Assemblage tends to be less associated with infections in animals (Sprong et al. 2009), indicating that, as with *Cryptosporidium*, zoonotic transmission may not be occurring extensively. However, given there are few data available on the Assemblages of *Giardia* found in human infections in the Arctic and Assemblage A *Giardia* is reported from wild and domestic animals in this region, the potential for zoonotic transmission should not be overlooked. It should also be noted that animals may themselves be infected by contamination of the environment by faecal material from humans.

2.3 Occurrence in the Environment in Arctic Regions

Although both *Cryptosporidium* and *Giardia* can be transmitted directly by contact with the faeces of an infected host as both oocysts and cysts are immediately infectious upon excretion, foodborne and waterborne transmission following contamination of the environment can also occur. Fresh produce, or other food that is consumed raw, has frequently been implicated in outbreaks of cryptosporidiosis (EFSA 2018). Although no cases have been identified, shellfish that concentrate the parasites via filter feeding have also been suggested as potential vehicles of infection (Robertson 2007). Contamination of water, fresh produce, and molluscan shellfish has been extensively studied in some parts of the world, but relatively few studies are reported from Arctic regions (see Tables 5 and 6).

Among the studies of protozoan parasites contaminating environmental matrices in Arctic regions, one study was conducted on water (Masina et al. 2019) and reported a relatively high level of contamination with both parasites (Tables 5 and 6) using standard techniques and detection by IFAT, with *Giardia* contamination

Table 5 Reports of *Cryptosporidium* spp. from environmental samples in the Arctic

Country	Source	Detection method	Prevalence % (no. of samples)	References
	Water			
Canada	Untreated surface water	IFAT	2 (55)	Masina et al. (2019)
	Shellfish			
Canada	Blue mussels *Mytilus edulis*	IFAT, PCR	72 (11)	Lévesque et al. (2010)
Canada	Clams *Mya truncata*	PCR	0 (404)	Manore et al. (2020)

IFAT immunofluorescence antibody test

Table 6 Reports of *Giardia duodenalis* from environmental samples in the Arctic

Country	Source	Detection method	Prevalence % (no. of samples)	References
	Water			
Canada	Untreated surface water	IFAT	20 (55)	Masina et al. (2019)
	Shellfish			
Canada	Blue mussels *Mytilus edulis*	IFAT, PCR	18 (11)	Lévesque et al. (2010)
Canada	Clams *Mya truncata*	PCR	0.5% (404)	Manore et al. (2020)

predominating. Similar results have been obtained in water samples from the subarctic (Roach et al. 1993). About one-third of Finland is in the Arctic, and in a study from there on water contamination (with specific locations of sampling sites not provided), of 54 surface water samples, 4 were found positive for *Cryptosporidium* and 1 for *Giardia* using a molecular methodology for detection (Rimhanen-Finne et al. 2002).

In addition, two studies have investigated shellfish. One of these investigated contamination of blue mussels (*Mytilus edulis*) from Nunavik, Canada (Lévesque et al. 2010), and found both *Cryptosporidium* and *Giardia* relatively frequently, with *Cryptosporidium* being detected in 8/11 samples (73%) and *Giardia* in 2/11 samples (18%). The other study (Manore et al. 2020) investigated 404 samples of clams (*Mya truncata*) from Nunavut, Canada, using a molecular method; *Cryptosporidium* contamination was not detected, but *Giardia* DNA (Assemblage B) was identified in the haemolymph of two clams.

3 Effects of Infection on Animal Hosts in the Arctic

The majority of studies that have investigated infections with *Cryptosporidium* and *Giardia* in animals in the Arctic do not report clinical signs or faecal morphology from the host. This may be because the nature of sampling makes it difficult to determine whether the animals have symptomatic infections or may reflect the age group of animals sampled. In general, cryptosporidiosis is a disease associated with younger animals (often within a week of birth for domestic ruminants; Robertson et al. 2014), and these may be less accessible for sampling in the Arctic. Although severe gastrointestinal cryptosporidiosis has been reported from captive muskox calves, the species of *Cryptosporidium* was not identified (Kutz et al. 2012). *Cryptosporidium xiaoi* was identified in the muskox in the Norwegian study (Davidson et al. 2012). This species tends not to be associated with symptoms in other related ruminants such as sheep, but it is possible that the symptomatic muskox calves were infected with another species. Lambs and calves are often infected with *C. parvum* and develop symptomatic cryptosporidiosis, whereas older sheep and cattle are infected with other species of *Cryptosporidium* and are largely asymptomatic (e.g., Santín et al. 2004). It should be noted that research on ruminants has indicated that acute episodes of cryptosporidiosis or giardiosis early in life may result in long-term health impacts regarding growth and weight gain (Sweeny et al. 2011; Jacobson et al. 2016; Shaw et al. 2020). Under Arctic conditions, where the availability of nutrition may already be critical, such a disadvantage may make a significant difference to survival.

Many Arctic wildlife species also face a multitude of other threats, such as human use of biological resources limiting habitat availability and affecting food supply, mining and other human intrusions resulting in habitat fragmentation, and pollution and climate change affecting health parameters, as well as altering the ecological web. All of these, which may themselves interact and affect each other, may result in greater susceptibility to diseases, including those due to parasitic infection (Bradley et al. 2005).

4 Potential Effects of Climate Change

Climate change in the Arctic region is already involving (and predicted to involve) the following events: rising air and water temperatures, loss of sea ice and melting of the glacial and snow cover, ocean circulation changes, with increased fresh water inputs and oceanic acidification, and a greater frequency of extreme weather events. Currently, warming of the Arctic is occurring approximately twice as fast as elsewhere in the world. These changes are of particular relevance to environmentally transmitted pathogens, especially those with waterborne transmission, due to the changes in hydrology associated with climate change. Indeed, both *Cryptosporidium* and *Giardia* have been labelled as potentially climate-sensitive zoonotic pathogens of circumpolar concern (Parkinson et al. 2014). How these various climate change phenomena may affect *Cryptosporidium* and *Giardia* are illustrated in Fig. 1 (Davidson et al. 2011). In brief, survival of oocysts and cysts in the environment may be affected by rising temperatures; both parasite transmission stages are negatively affected by freezing and desiccation, and a warmer, wetter environment may increase persistence in the environment. This may be of particular relevance for *Giardia*, as *Giardia* cysts are less capable of surviving cold or dry conditions than *Cryptosporidium* oocysts.

Melting of ice and extreme weather events may result in run-off into fresh water, potentially resulting in a greater frequency of water contamination events, and thus increasing the likelihood for outbreaks in the human population. This could then result in greater contamination of the marine environment, increasing the exposure of different marine animals, from bivalve molluscs to whales, to these parasites. Marine bivalves may concentrate the parasites, providing a transmission route to those animals, including humans, that consume them, and larger marine animals may transport the parasites to other areas.

All the environmental transformations that are associated with climate change are likely to result in alterations in habitats and vegetation shifts. This may result in some

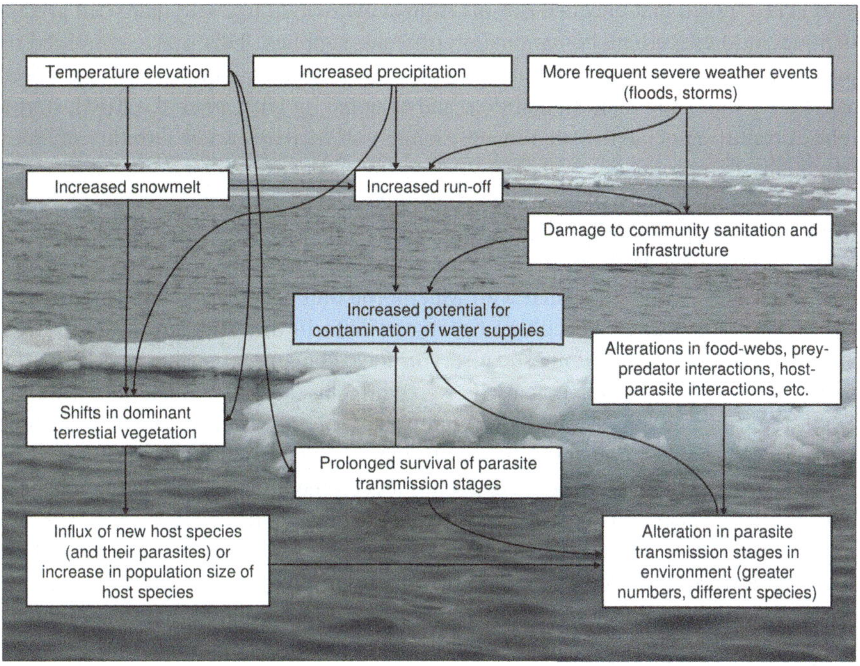

Fig. 1 Potential climate change effects on water contamination by parasites. Predicted and current climate changes are likely to affect contamination of water supplies with parasite transmission stages in the Arctic. Only changes that could increase contamination potential are included in this figure; some changes that may reduce contamination potential (e.g. temperature changes can decrease survival of parasite transmission stages) are not included (after Davidson et al. 2011).

animal populations being lost, but may also be associated with incursions of new populations. These may bring their own parasites with them or amplify parasites already there by becoming infected and contaminating the environment. However, the outcomes associated with the myriad interactions and adaptations resulting from climate change are complex, heterogeneous, and potentially unexpected; without integrated investigations, it is impossible to predict outcomes (Bjorkman et al. 2020).

All these potential phenomena have previously been discussed in the literature. A systematic review to identify the most climate-sensitive diseases and vulnerable populations in the Arctic and subarctic regions found strong evidence for an association between climatic factors and waterborne diseases (Hedlund et al. 2014). However, with a lack of studies from Greenland and Iceland, and only a few from Siberia, there is a clear need for more high-quality studies to investigate the adverse health impacts of weather and climatic factors in the Arctic and subarctic region (Hedlund et al. 2014). Parkinson and Butler (2005) noted the potential for heavy rain, snowfall, melting permafrost, and flooding, which could result in widespread contamination of traditional water sources and damage to community sanitation infrastructures (Warren et al. 2005). Similarly, the article by Davidson et al.

(2011) also noted that elevated run-off from snowmelt, along with increased precipitation, could exacerbate contamination of water supplies. This was exemplified by documented outbreaks of waterborne cryptosporidiosis and giardiosis that have been associated with snowmelt. In addition, and as noted by Hueffer et al. (2013), storm-related runoff associated with climate change will increase water turbidity, thereby potentially compromising water purification filter systems.

The movement north of the treeline that is associated with climate change has also been suggested to have resulted in the northward movement of rodents, such as beavers and muskrats. Both of these are known to harbour *Giardia* infections and have been previously associated with waterborne outbreaks of infections (Brubaker et al. 2010). Indeed, as agriculture becomes viable further and further north, livestock will form a suitable domestic reservoir for pathogens, contributing to environmental contamination and potentially spreading disease to humans and wildlife. Semi-domesticated reindeer in Arctic regions have already been shown to be suitable hosts for *Cryptosporidium* and *Giardia* (Kemper et al. 2006; Idland et al. 2021), and the reduction of their grazing lands due to development and infrastructure changes may result in increased infection pressures. Climate change may have a negative impact on reindeer husbandry, as, although plant productivity may improve and the growing season be prolonged, ice-crust formation may prevent grazing during the colder times of years, competitive interactions associated with higher forest productivity may reduce lichen availability, and insect harassment may increase (Moen 2008). How these potential changes may interact with and affect the distribution of parasites such as *Cryptosporidium* and *Giardia* is difficult to predict.

Whilst the Arctic may seem a geographically isolated environment, many wildlife species migrate great distances to spend variable amounts of time there, some coming from as far as the tropics. Naturally, parasites take these journeys together with their hosts and thus have the opportunity to attempt to establish themselves in these northern regions. Birds in particular have been suggested to be important in the dissemination of waterborne enteropathogens like *Giardia* and *Cryyptosporidium* (Graczyk et al. 2008). These parasites also have the potential to spread themselves into the Arctic through the movement of humans and domestic animals. These processes would suggest that geographical isolation is not a significant impediment to the establishment of *Giardia* or *Cryptosporidium* amongst Arctic wildlife species. Rather, low animal population densities and harsh environmental conditions could be considered more likely to be barriers to them gaining a strong foothold in the Far North. However, as climate change modifies the Arctic environment, likely to better suit the survival and transmission of both *Giardia* and *Cryptosporidium*, these obstacles will become less important, potentially paving the way for disease on an individual or even population level. This is concerning, given some species such as the harp seal (*Pagophilus groenlandicus*) have been shown experimentally to have the ability to be infected by *C. parvum* and transmit it horizontally (Appelbee 2006), whilst wild populations have been shown to be naïve to this pathogen (Appelbee et al. 2010). In general, parasites that co-evolve with their hosts do so in a manner that does not result in severe disease to that host, as this is often not a successful survival strategy. However, given the longevity of the transmission stages of both

Cryptosporidium and *Giardia* in the environment, coupled with their relative host-flexibility, acute cryptosporidiosis or giardiosis in animal populations that have previously had minimal exposure may be expected. Indeed, among eukaryotic pathogens, *Cryptosporidium* and *Giardia* could be considered to be likely candidates for the "sit-and-wait" hypothesis (Walther and Ewald 2004), in which it is postulated that the virulence of a pathogen should have a positive correlation with its survival in the environment due to the reduced requirement on host mobility for transmission. Although this has not, to our knowledge, been closely discussed for these parasites, it is clear that infections with parasites that result in more severe and prolonged symptoms will result in greater contamination of the environment with the parasites' robust transmission stages. This will, thereby, feed into the cycle of these infections that currently seem to occur relatively rarely among animal populations in the Arctic, becoming established, and potentially spreading more amongst the human populations also. Given that there is already concern regarding how cryptosporidiosis may synergise with challenges, such as food insecurity and overcrowding faced by children in remote Arctic communities (Yansouni et al. 2016), similar to children in countries such as in sub-Saharan Africa where cryptosporidiosis has an alarming toll on child health (Khalil et al. 2018), this is a relevant issue.

5 Conclusions

Cryptosporidiosis and giardiosis have an impact on both animal health and human health globally. Although both parasites seem to be relatively well established in human populations in the Arctic, in animal populations they seem to be relatively rare compared with other areas of the world, particularly for *Cryptosporidium*. The very few genotyping studies that have been conducted on *Giardia* in Arctic wildlife suggest that Assemblage A and B may be more common in wildlife here than elsewhere. However, it is not clear if the Assemblage A isolates are AI, AII, or AIII; AI and AII occur commonly in humans, but also in wildlife; in contrast, AIII occurs commonly in wildlife, but very rarely in humans (Sprong et al. 2009). Nevertheless, although these data may raise concerns that wildlife may act as an infection source of *Giardia* for human populations, it may also well reflect that humans brought the parasites to the Arctic initially, and spillover has been from humans to wildlife rather than in the opposite direction. This possibility has been raised regarding *Giardia* on numerous other occasions already (e.g., Kutz et al. 2009). However, studies are relatively scattered so it is difficult to reach a clear overview.

Although extrapolation from Arctic animals in subarctic locations may provide an indication of what may be found in the same animals in the Arctic, this cannot be extrapolated to the likelihood of transmission in Arctic conditions. However, climate change is bringing a new set of environmental determinants. Although it is impossible to predict how the "new" Arctic will appear, some elements that are expected or are already occurring, such as warmer, wetter weather, flooding, and incursion of new species, are all likely to facilitate survival of *Cryptosporidium* oocysts and

Giardia cysts in the environment and increase the likelihood of transmission to new hosts. In combination with other challenges, the establishment and expansion of these parasites among new hosts will provide an additional burden for Arctic communities, both human and animal. In order to be able to instigate appropriate mitigation measures, it is important that the occurrence of these parasites among different Arctic populations and in the environment is monitored. Use of sensitive and appropriate techniques will not only be able to indicate the presence of the parasites, but could also provide some clues regarding their epidemiology in the altered Arctic environment.

References

Appelbee AJ (2006) Giardia and Cryptosporidium in Pinnipeds from the Gulf of St. Lawrence, Canada. Ph.D. Thesis. Murdoch University, Murdoch, Western Australia, p 218

Appelbee AJ, Thompson RCA, Measures LM, Olson ME (2010) *Giardia* and *Cryptosporidium* in harp and hooded seals from the Gulf of St. Lawrence, Canada. Vet Parasitol 173(1–2):19–23. https://doi.org/10.1016/j.vetpar.2010.06.001

Bjorkman AD, García Criado M, Myers-Smith IH, Ravolainen V, Jónsdóttir IS, Westergaard KB, Lawler JP, Aronsson M, Bennett B, Gardfjell H, Heiðmarsson S, Stewart L, Normand S (2020) Status and trends in Arctic vegetation: evidence from experimental warming and long-term monitoring. Ambio 49(3):678–692

Bradley M, Kutz SJ, Jenkins E, O'Hara TM (2005) The potential impact of climate change on infectious diseases of Arctic fauna. Int J Circumpolar Health 64(5):468–477. https://doi.org/10.3402/ijch.v64i5.18028

Brubaker M, Berner J, Bell J, Warren J (2010) Climate change in Kivalina, Alaska – strategies for community health. ANTHC. Available from http://www.anthc.org/chs/ces/climate/climateandhealthreports.cfm

Bryan HM, Darimont CT, Paquet PC, Ellis JA, Goji N, Gouix M, Smits JE (2011) Exposure to infectious agents in dogs in remote coastal British Columbia: possible sentinels of diseases in wildlife and humans. Can JVet Res 75(1):11–17

Certad G, Viscogliosi E, Chabé M, Cacciò SM (2017) Pathogenic mechanisms of *Cryptosporidium* and *Giardia*. Trends Parasitol 33(7):561–576. https://doi.org/10.1016/j.pt.2017.02.006

Davidson R, Simard M, Kutz SJ, Kapel CM, Hamnes IS, Robertson LJ (2011) Arctic parasitology: why should we care? Trends Parasitol 27(6):239–245

Davidson RK, Romig T, Jenkins E, Tryland M, Robertson LJ (2012) The impact of globalisation on the distribution of *Echinococcus multilocularis*. Trends Parasitol 28(6):239–247

Davidson RK, Amundsen H, Lie NO, Luyckx K, Robertson LJ, Verocai GG, Kutz SJ, Ytrehus B (2014) Sentinels in a climatic outpost: endoparasites in the introduced muskox (*Ovibos moschatus wardi*) population of Dovrefjell, Norway. Int J Parasitol Parasites Wildl 3(2): 154–160. https://doi.org/10.1016/j.ijppaw.2014.06.002

Dixon BR, Parrington LJ, Parenteau M, Leclair D, Santín M, Fayer R (2008) *Giardia duodenalis* and *Cryptosporidium* spp. in the intestinal contents of ringed seals (*Phoca hispida*) and bearded seals (*Erignathus barbatus*) in Nunavik, Quebec, Canada. J Parasitol 94(5):1161–1163. https://doi.org/10.1645/ge-1485.1

Dudarev AA, Dushkina EV, Sladkova YN, Alloyarov PR, Chupakhin VS, Dorofeyev VM, Kolesnikova TA, Fridman KB, Evengard B, Nilsson LM (2013) Food and water security issues in Russia II: water security in general population of Russian Arctic, Siberia and Far East, 2000–2011. Int J Circumpolar Health 72:22646. https://doi.org/10.3402/ijch.v72i0.22646

EFSA BIOHAZ Panel (EFSA Panel on Biological Hazards), Koutsoumanis K, Allende A, Alvarez-Ordonez A, Bolton D, Bover-Cid S, Chemaly M, Davies R, De Cesare A, Herman L, Hilbert F,

Lindqvist R, Nauta M, Peixe L, Ru G, Simmons M, Skandamis P, Suffredini E, Caccio S, Chalmers R, Deplazes P, Devleesschauwer B, Innes E, Romig T, van der Giessen J, Hempen M, Van der Stede Y, Robertson L (2018) Scientific opinion on the public health risks associated with food-borne parasites. EFSA J 16(12):5495

Elmore SA, Lalonde LF, Samelius G, Alisauskas RT, Gajadhar AA, Jenkins EJ (2013) Endoparasites in the feces of arctic foxes in a terrestrial ecosystem in Canada. Int J Parasitol Parasites Wildl 2:90–96. https://doi.org/10.1016/j.ijppaw.2013.02.005

Fayer R, Nerad T (1996) Effects of low temperatures on viability of *Cryptosporidium* parvum oocysts. Appl Environ Microbiol 62(4):1431–1433

Goldfarb DM, Dixon B, Moldovan I, Barrowman N, Mattison K, Zentner C et al (2013) Nanolitre real-time PCR detection of bacterial, parasitic, and viral agents from patients with diarrhoea in Nunavut, Canada. Int J Circumpolar Health 72(1):19903. https://doi.org/10.3402/ijch.v72i0.19903

Graczyk TK, Majewska AC, Schwab KJ (2008) The role of birds in dissemination of human waterborne enteropathogens. Trends Parasitol 24(2):55–59. https://doi.org/10.1016/j.pt.2007.10.007

Hamnes IS, Gjerde B, Robertson L, Vikøren T, Handeland K (2006) Prevalence of *Cryptosporidium* and *Giardia* in free-ranging wild cervids in Norway. Vet Parasitol 141(1–2):30–41. https://doi.org/10.1016/j.vetpar.2006.05.004

Hedlund C, Blomstedt Y, Schumann B (2014) Association of climatic factors with infectious diseases in the Arctic and subarctic region: a systematic review. Glob Health Action 7:24161

Himsworth CG, Thompson RCA, Chaban B, Wagner BA, Jenkins E, Skinner S et al (2010) Multiple zoonotic pathogens identified in canine feces collected from a remote Canadian indigenous community. Am J Trop Med Hygiene 83(2):338–341. https://doi.org/10.4269/ajtmh.2010.10-0137

Horčičková M, Čondlová Š, Holubová N, Sak B, Květoňová D, Hlásková L, Konečný R, Sedláček F, Clark M, Giddings C, McEvoy J, Kváč M (2019) Diversity of *Cryptosporidium* in common voles and description of *Cryptosporidium alticolis* sp. n. and *Cryptosporidium microti* sp. n. (Apicomplexa: Cryptosporidiidae). Parasitology 146(2):220–233. https://doi.org/10.1017/S0031182018001142

Hueffer K, Parkinson AJ, Gerlach R, Berner J (2013) Zoonotic infections in Alaska: disease prevalence, potential impact of climate change and recommended actions for earlier disease detection, research, prevention and control. Int J Circumpolar Health 72. https://doi.org/10.3402/ijch.v72i0.19562

Hughes-Hanks JM, Rickard LG, Panuska C, Saucier JR, O'Hara TM, Dehn L, Rolland RM (2005) Prevalence of *Cryptosporidium* spp. and *Giardia* spp. in five marine mammal species. J Parasitol 91(5):1225–1228. https://doi.org/10.1645/ge-545r.1

Idland L, Juul AM, Solevåg EK, Tysnes KR, Robertson LJ, Utaaker KS (2021) Occurrence of endoparasites identified in faecal samples from reindeer in selected grazing areas in northern Norway. Acta Vet Scand 63:13. https://doi.org/10.1186/s13028-021-00578-y

Iqbal A, Goldfarb DM, Slinger R, Dixon BR (2015) Prevalence and molecular characterization of *Cryptosporidium* spp. and *Giardia duodenalis* in diarrhoeic patients in the Qikiqtani region, Nunavut, Canada. Int J Circumpolar Health 74(1):27713. https://doi.org/10.3402/ijch.v74.27713

Jacobson C, Williams A, Yang R, Ryan U, Carmichael I, Campbell AJ, Gardner GE (2016) Greater intensity and frequency of *Cryptosporidium* and *Giardia* oocyst shedding beyond the neonatal period is associated with reductions in growth, carcase weight and dressing efficiency in sheep. Vet Parasitol 228:42–51

Johnson D, Harms NJ, Larter NC, Elkin BT, Tabel H, Wei G (2010) Serum biochemistry, serology, and parasitology of boreal caribou (*Rangifer tarandus caribou*) in the Northwest Territories, Canada. J Wildl Dis 46(4):1096–1107. https://doi.org/10.7589/0090-3558-46.4.1096

Kemper N, Aschfalk A, Höller C (2006) *Campylobacter* spp., *Enterococcus* spp., *Escherichia coli*, *Salmonella* spp., *Yersinia* spp., and *Cryptosporidium* oocysts in semi-domesticated reindeer (*Rangifer tarandus tarandus*) in Northern Finland and Norway. Acta Vet Scand 48:7

Khalil IA, Troeger C, Rao PC, Blacker BF, Brown A, Brewer TG, Colombara DV, De Hostos EL, Engmann C, Guerrant RL, Haque R (2018) Morbidity, mortality, and long-term consequences associated with diarrhoea from *Cryptosporidium* infection in children younger than 5 years: a meta-analyses study. Lancet Glob Health 6(7):e758–e768

King P, Tyler KM, Hunter PR (2019) Anthroponotic transmission of *Cryptosporidium parvum* predominates in countries with poorer sanitation: a systematic review and meta-analysis. Parasit Vectors 12(1):16. https://doi.org/10.1186/s13071-018-3263-0

Kutz SJ, Thompson RA, Polley L, Kandola K, Nagy J, Wielinga CM, Elkin BT (2008) *Giardia* assemblage A: human genotype in muskoxen in the Canadian Arctic. Parasit Vectors 1(1):32. https://doi.org/10.1186/1756-3305-1-32

Kutz SJ, Thompson RCA, Polley L (2009) Wildlife with *Giardia:* villain, or victim and vector? In: Ortega-Pierres G, Cacciò S, Fayer R, Mank TG, Smith HV, Thompson RCA (eds) *Giardia* and *Cryptosporidium:* from molecules to disease. CABI, New York

Kutz SJ, Ducrocq J, Verocai GG, Hoar BM, Colwell DD, Beckmen KB et al (2012) Parasites in ungulates of Arctic North America and Greenland. Adv Parasitol 2012:99–252. https://doi.org/10.1016/b978-0-12-398457-9.00002-0

Lévesque B, Barthe C, Dixon BR, Parrington LJ, Martin D, Doidge B et al (2010) Microbiological quality of blue mussels (*Mytilus edulis*) in Nunavik, Quebec: a pilot study. Can J Microbiol 56(11):968–977. https://doi.org/10.1139/w10-078

Manko A, Motta JP, Cotton JA, Feener T, Oyeyemi A, Vallance BA, Wallace JL, Buret AG (2017) *Giardia* co-infection promotes the secretion of antimicrobial peptides beta-defensin 2 and trefoil factor 3 and attenuates attaching and effacing bacteriainduced intestinal disease. PLoS One 12 (6):e0178647. https://doi.org/10.1371/journal.pone.0178647

Manore AJW, Harper SL, Sargeant JM, Weese JS, Cunsolo A, Bunce A, Shirley J, Sudlovenick E, Shapiro K (2020 Feb 17) Cryptosporidium and Giardia in locally harvested clams in Iqaluit, Nunavut. Zoonoses Public Health. https://doi.org/10.1111/zph.12693

Masina S, Shirley J, Allen J, Sargeant JM, Guy RA, Wallis PM et al (2019) Weather, environmental conditions, and waterborne *Giardia* and *Cryptosporidium* in Iqaluit. Nunavut J Water Health 17 (1):84–97. https://doi.org/10.2166/wh.2018.323

Moen J (2008) Climate change: effects on the ecological basis for reindeer husbandry in Sweden. Ambio 37(4):304–311

Mosites E, Miernyk K, Priest JW, Bruden D, Hurlburt D, Parkinson A, Klejka J, Hennessy T, Bruce MG (2018) *Giardia* and *Cryptosporidium* antibody prevalence and correlates of exposure among Alaska residents, 2007–2008. Epidemiol Infect 146:888–894. https://doi.org/10.1017/S095026881800078X

Myšková E, Brož M, Fuglei E, Kvičerová J, Mácová A, Sak B, Kváč M, Ditrich O (2019) Gastrointestinal parasites of Arctic foxes (*Vulpes lagopus*) and sibling voles (*Microtus levis*) in Spitsbergen. Svalbard Parasitol Res 118(12):3409–3418. https://doi.org/10.1007/s00436-019-06502-8

Nader JL, Mathers TC, Ward BJ, Pachebat JA, Swain MT, Robinson G, Chalmers RM, Hunter PR, van Oosterhout C, Tyler KM (2019) Evolutionary genomics of anthroponosis in *Cryptosporidium*. Nat Microbiol 4(5):826–836. https://doi.org/10.1038/s41564-019-0377-x

Niine T, Peetsalu K, Nieminen M, Oksanen A, Soveri T, Orro T (2017) *Giardia* and *Cryptosporidium* infections in neonatal reindeer calves: relation to the acute phase response. Comp Immunol Microbiol Infect Dis 54:45–50. https://doi.org/10.1016/j.cimid.2017.08.001

Olson ME, Roach PD, Stabler M, Chan W (1997) Giardiasis in ringed seals from the western Arctic. J Wildl Dis 33(3):646–648

Parkinson AJ, Butler JC (2005) Potential impacts of climate change on infectious diseases in the Arctic. Int J Circumpolar Health 64(5):478–478

Parkinson AJ, Evengard B, Semenza JC, Ogden N, Børresen ML, Berner J, Brubaker M, Sjöstedt A, Evander M, Hondula DM, Menne B, Pshenichnaya N, Gounder P, Larose T, Revich B, Hueffer K, Albihn A (2014) Climate change and infectious diseases in the Arctic: establishment of a circumpolar working group. Int J Circumpolar Health 73(1):25163. https://doi.org/10.3402/ijch.v73.25163

Rimhanen-Finne R, Hörman A, Ronkainen P, Hänninen ML (2002) An IC-PCR method for detection of *Cryptosporidium* and *Giardia* in natural surface waters in Finland. J Microbiol Methods 50(3):299–303

Roach PD, Olson ME, Whitley G, Wallis PM (1993) Waterborne *Giardia* cysts and *Cryptosporidium* oocysts in the Yukon, Canada. Appl Environ Microbiol 59(1):67–73

Robertson LJ (2007) The potential for marine bivalve shellfish to act as transmission vehicles for outbreaks of protozoan infections in humans: a review. Int J Food Microbiol 120(3):201–216

Robertson LJ, Gjerde BK (2004) Effects of the Norwegian winter environment on *Giardia* cysts and *Cryptosporidium* oocysts. Microb Ecol 47(4):359–365

Robertson LJ, Gjerde BK (2006) Fate of *Cryptosporidium* oocysts and *Giardia* cysts in the Norwegian aquatic environment over winter. Microb Ecol 52(4):597–602

Robertson LJ, Campbell AT, Smith HV (1992) Survival of *Cryptosporidium parvum* oocysts under various environmental pressures. Appl Environ Microbiol 58(11):3494–3500

Robertson LJ, Forberg T, Hermansen L, Hamnes IS, Gjerde B (2007) *Giardia duodenalis* cysts isolated from wild moose and reindeer in Norway: genetic characterization by PCR-rflp and sequence analysis at two genes. J Wildl Dis 43(4):576–585

Robertson LJ, Gjerde BK, Furuseth HE (2010) The zoonotic potential of *Giardia* and *Cryptosporidium* in Norwegian sheep: a longitudinal investigation of 6 flocks of lambs. Vet Parasitol 171 (1–2):140–145

Robertson LJ, Björkman C, Silverlås C, Fayer R (2014) Chapter 4: cryptosporidiosis in farmed animals. In: Caccio S, Widmer G (eds) *Cryptosporidium:* parasite and disease. Springer, New York

Robertson LJ, Johansen ØH, Kifleyohannes T, Efunshile AM, Terefe G (2020) *Cryptosporidium* infections in Africa—how important is zoonotic transmission? A review of the evidence. Front Vet Sci 7:575881. https://doi.org/10.3389/fvets.2020.575881

Ryan U, Paparini A, Monis P, Hijjawi N (2016) It's official – *Cryptosporidium* is a gregarine: what are the implications for the water industry? Water Res 15(105):305–313. https://doi.org/10.1016/j.watres.2016.09.013

Salb AL, Barkema HW, Elkin BT, Thompson RCA, Whiteside DP, Black SR et al (2008) Dogs as sources and sentinels of parasites in humans and wildlife, Northern Canada. Emerg Infect Dis 14(1):60–63. https://doi.org/10.3201/eid1401.071113

Santín M, Trout JM, Xiao L, Zhou L, Greiner E, Fayer R (2004) Prevalence and age-related variation of *Cryptosporidium* species and genotypes in dairy calves. Vet Parasitol 122 (2):103–117

Santín M, Dixon BR, Fayer R (2005) Genetic characterization of *Cryptosporidium* isolates from ringed seals (*Phoca hispida*) in northern Quebec. Canada J Parasitol 91(3):712–716. https://doi.org/10.1645/ge-3438rn

Schurer JM, Hill JE, Fernando C, Jenkins EJ (2012) Sentinel surveillance for zoonotic parasites in companion animals in indigenous communities of Saskatchewan. Am J Trop Med Hygiene 87(3):495–498. https://doi.org/10.4269/ajtmh.2012.12-0273

Shaw HJ, Innes EA, Morrison LJ, Katzer F, Wells B (2020) Long-term production effects of clinical cryptosporidiosis in neonatal calves. Int J Parasitol. https://doi.org/10.1016/j.ijpara.2020.03.002

Siefker C, Rickard LG, Pharr GT, Simmons JS, O'Hara TM (2002) Molecular characterization of *Cryptosporidium* sp. Isolated from northern Alaskan caribou (*Rangifer tarandus)*. J Parasitol 88 (1):213. https://doi.org/10.2307/3285424

Sprong H, Cacciò SM, van der Giessen JW, ZOOPNET network and partners (2009) Identification of zoonotic genotypes of *Giardia duodenalis*. PLoS Negl Trop Dis 3(12):e558

Sweeny JP, Ryan UM, Robertson ID, Jacobson C (2011) *Cryptosporidium* and *Giardia* associated with reduced lamb carcase productivity. Vet Parasitol 182(2–4):127–139

Thivierge K, Iqbal A, Dixon B, Dion R, Levesque B, Cantin P et al (2016) *Cryptosporidium hominis* is a newly recognized pathogen in the Arctic region of Nunavik, Canada: molecular characterization of an outbreak. PLoS Negl Trop Dis 10(4):e0004534. https://doi.org/10.1371/journal.pntd.0004534

Utaaker KS, Skjerve E, Robertson LJ (2017) Keeping it cool: survival of *Giardia* cysts and *Cryptosporidium* oocysts on lettuce leaves. Int J Food Microbiol 255:51–57

Walther BA, Ewald PW (2004) Pathogen survival in the external environment and the evolution of virulence. Biol Rev Camb Philos Soc 79:849–869. https://doi.org/10.1017/S1464793104006475

Warren JA, Berner JE, Curtis T (2005) Climate change and human health: infrastructure impacts to small remote communities in the north. Int J Circumpolar Health 64(5):487–497

Yansouni CP, Pernica JM, Goldfarb D (2016) Enteric parasites in Arctic communities: tip of the iceberg? Trends Parasitol 32(11):834–838

Erysipelas in Arctic and Northern Regions

Fabien Mavrot, O. Alejandro Aleuy, Taya Forde, and Susan J. Kutz

1 Introduction

Erysipelothrix rhusiopathiae [from the Greek 'erysipelas'—red skin, 'thrix'—a hair or thread, 'rhusius'—reddish and 'pathus'—disease (Woodbine 1950)], is a non-sporulating, facultative anaerobic, Gram-positive bacillus that infects people, domestic animals, and wildlife species worldwide (Brooke and Riley 1999). This organism was first documented as a human pathogen in the nineteenth century and since then, more than 15 different serovars have been described (Wang et al. 2010). Erysipelas, caused by *E. rhusiopathiae*, is one of the most prevalent and economically important diseases in the swine industry and is commonly described in farmed turkeys, chickens, ducks, and sheep. In recent years, it has been detected as an important disease-causing agent in arctic wildlife (Kutz et al. 2015; Spraker and White 2017; Mavrot et al. 2020) and people (Groeschel et al. 2019). *Erysipelothrix rhusiopathiae* is ubiquitous in nature, has a worldwide distribution, and can persist for long periods of time in the environment, including in marine environments (Wood 1994; Wang et al. 2010).

Fabrien Mavrot and O. Alejandro Aleuy contributed euqally to this chapter.

F. Mavrot (✉) · O. A. Aleuy · S. J. Kutz
Department of Ecosystem and Public Health, University of Calgary, Calgary, AB, Canada
e-mail: fabien.mavrot@ucalgary.ca; oaleuy@ucalgary.ca

T. Forde
Institute of Biodiversity, Animal Health & Comparative Medicine, University of Glasgow, Glasgow, UK

M. Tryland (ed.), *Arctic One Health*, https://doi.org/10.1007/978-3-030-87853-5_16

2 Host Range

Although domestic pigs are considered the most important reservoir, *E. rhusiopathiae* is a generalist and is reported from over 30 species of wild birds, at least 50 species of wild mammals, at least four species of fish, and from people (Suer et al. 1988; Wood 1994; Wolcott 2007; Opriessnig et al. 2013; Chong et al. 2015). In the arctic and boreal regions, *E. rhusiopathiae* has been isolated from muskoxen (*Ovibos moschatus*) (Mavrot et al. 2020) (Fig. 1), Pribilof arctic foxes (*Alopex lagopus pribilofensis*) (Spraker and White 2017), barren-ground caribou (*Rangifer tarandus groenlandicus*) (CWHC 2016), boreal woodland caribou (*Rangifer tarandus caribou*) (Bondo et al. 2019), moose (*Alces alces*) (Forde et al. 2016b), wolves (*Canis lupus*), and bison (*Bison bison*) (Langford and Dorward 1977). It was also reported as a cause of morbidity in a captive beluga (*Delphinapterus leucas*) (Calle et al. 1993). Its widespread and historic distribution in the North American Arctic is supported by seropositivity in all muskox and barren-ground caribou populations tested to date (Mavrot et al. 2020; Aleuy et al. unpublished data), as well as in ringed seals (*Pusa hispida*) in the Eastern Canadian Arctic (Sudlovenick 2019). There is broad literature on this bacterium in domestic and wild species from lower latitudes that will not be covered here (see Reboli and Farrar 1989; Wang et al. 2010 for further information).

Fig. 1 Carcass of a muskox infected with *Erysipelothrix rhusiopathiae*. (Photo: David Pritchard)

3 Taxonomy

Erysipelothrix rhusiopathiae can be divided into three main clades—1, 2, and 3—based on phylogenetic relatedness, with several isolates also falling intermediate to Clades 2 and 3 (Forde et al. 2016a). Serotype, which has been traditionally used to classify isolates, is not correlated with this population structure (Forde et al. 2016a, 2020). Isolates from Clade 1 are rarely reported and have been obtained mostly from marine mammals but with occasional reports in different wildlife species including boreal caribou (Forde et al. 2016b). Clade 1 isolates have not yet been observed in production species (i.e., pigs or poultry). Clade 2 is the predominant clade of *E. rhusiopathiae* in Europe (Forde et al. 2020), whereas the majority of isolates from domestic and wild animals in North America have been from Clade 3 and the intermediate clade. The level of *E. rhusiopathiae* diversity has been found to vary geographically: high levels of diversity have been observed among pigs in Great Britain (Forde et al. 2020) and in poultry in Germany (Janßen et al. 2015) but with limited diversity observed in pigs in Japan (Ogawa et al. 2017). Diverse strains of *E. rhusiopathiae* have been isolated from wild birds and ungulates in North America, including several observations of multiple strains co-infecting individual hosts (Forde et al. 2016b). One important exception is the almost exclusive detection of a single genotype in muskoxen across two large islands in the western Canadian Arctic Archipelago (Forde et al. 2016b). There appears to be limited host-specificity among *E. rhusiopathiae* strains.

4 Epidemiology

Erysipelothrix rhusiopathiae has been associated with individual cases, clusters, and large-scale morbidity and mortality events in wildlife (Jensen et al. 1976; Campbell et al. 1994; Melero et al. 2011; Kutz et al. 2015; Fiorito et al. 2016; Spraker and White 2017). It is an opportunist, and outbreaks of disease may be associated with other underlying factors or stressors.

In domestic animals, transmission of *E. rhusiopathiae* occurs through direct exposure to feces, urine, saliva, and nasal secretions from infected animals. It can persist for prolonged periods in the environment, and contaminated soil, bedding, food, and water are important sources of infection (Wood 1973). The transmission of *E. rhusiopathiae* among wildlife is not well understood but is believed to occur through ingestion of contaminated material (e.g., plants, meat) and wound infections (Wolcott 2007). Its persistence and circulation in wildlife are likely a result of asymptomatic infections/carrier animals that are capable of maintaining and disseminating viable organisms (Wood 1974). Aquatic environments are an important source of *E. rhusiopathiae*, and it has been isolated both from the surface and from the organs of fish (Suer et al. 1988; Opriessnig et al. 2013; Chong et al. 2015). Mortality events resulting from *E. rhusiopathiae* infection have been reported in brown pelicans (*Pelecanus occidentalis*) (Wolcott 2007) and a captive bald eagle (*Haliaeetus leucocephalus*) (Franson et al. 1994) fed with contaminated fish. Insects, including flies, mites, and ticks, also carry *E. rhusiopathiae* and can act as vectors (Stickdorn 1936; Wood 1975).

The infection of *E. rhusiopathiae* in people and animals seems to have a seasonal pattern with a higher prevalence during warmer months in temperate regions (Proctor and Richardson 1954; Wood 1975). Very little is known about the natural history of *E. rhusiopathiae,* but potential explanations for this seasonal pattern include an interaction between *E. rhusiopathiae* and ambient temperature, with disease occurring following periods of high temperatures promoting pathogen development (Proctor and Richardson 1954); a seasonal increase in the contact rates between hosts and sources of infection during warmer months (Wood 1975); and/or heat-related stress (Goerttler et al. 1960; Opriessnig and Coutinho 2019).

A seasonal pattern of infection also occurs in arctic muskoxen and caribou. Traditional knowledge studies on muskoxen in the Kitikmeot region of Nunavut, Canada, documented a four-year period of unusual mortality events, with peak mortality occurring in July and August (Tomaselli et al. 2018). A single genotype of *E. rhusiopathiae* was isolated from all muskox carcasses sampled during this time (Forde et al. 2016b;). Similar unusual summer mortality was observed in an Alaskan population of muskoxen during this time, with *E. rhusiopathiae* isolated from the bone marrow of several animals found dead during the summer (Forde et al. 2016b). In migratory tundra caribou from Alaska, seroprevalence studies support a seasonal pattern with higher seropositivity in mid-late summer. This seasonality is thought to be associated with predictable caribou life history stressors, with the peak of intensity associated with specific weather conditions in the preceding 12 months (Aleuy et al. unpublished data).

Wildlife mortality events associated with *E. rhusiopathiae* are often linked with a variety of underlying factors or stressors. For instance, cumulative stressors, including winter migration and an unexpected snowstorm, were implicated in the *E. rhusiopathiae*-associated mortality of an estimated 5000 aquatic birds, mostly eared grebes (*Podiceps nigricollis*), in western North America (Jensen et al. 1976). Similarly, *E. rhusiopathiae* mortality events of the critically endangered kakapo (*Strigops habroptilus*) (Gartrell et al. 2005) and malleefowl (*Leipoa ocellata*) (Blyde and Woods 1999) were triggered by environmental stressors. In mammals, mortality due to *E. rhusiopathiae* septicemia was observed in moose from Algonquin Provincial Park, Ontario, Canada, and was associated with poor nutrition and tick infestation (Campbell et al. 1994). Underlying factors such as climate change, excessive heat, and low genetic diversity have been proposed to be associated with the widespread unusual mortality and population declines observed in muskoxen of the western Canadian Arctic archipelago from 2010 to 2014: however, these linkages remain to be established (Kutz et al. 2015; Mavrot et al. 2020).

5 Pathogenesis and Clinical Manifestations

The pathogenesis of *E. rhusiopathiae* is characterized by an initial infection through the tonsils, gastrointestinal mucosa, damaged skin barrier, or arthropod bite, followed by dissemination via blood (Opriessnig and Coutinho 2019). The bacterium has an affinity for endothelial cells (Nakato et al. 1987; Zhu et al. 2017) and typically invades predilection sites such as skin capillaries, endocardium, or synovial membranes (Opriessnig and Coutinho 2019). As a result, the clinical manifestations of *E. rhusiopathiae* infection are direct consequences of these pathomechanisms (e.g., cutaneous vasculitis resulting in the typical diamond-shaped skin lesions in pigs [Reboli and Farrar 1989]).

Several virulence factors contribute to the pathogenicity of *E. rhusiopathiae*. Surface protective antigen A (spaA) increases the adhesion to the endothelium, neuraminidase enhances cellular invasion, and the capsular polysaccharide improves immune evasion (Ogawa et al. 2011; Opriessnig and Coutinho 2019). In general, the genes coding for known virulence factors and immunogenic surface proteins are present among most *E. rhusiopathiae* genomes (Janßen et al. 2015; Forde et al. 2020).

In domestic mammals, *E. rhusiopathiae* is mainly described in pigs and sheep where it can cause subacute dermatitis, acute and often fatal sepsis with multiple organ involvement/failure, chronic arthritis, endocarditis, or abortion (Lamont 1979; Fthenakis et al. 2006; Atyabi et al. 2012; Opriessnig and Coutinho 2019).

In wild mammals, *E. rhusiopathiae*-associated pathologies have also been reported in a variety of species. It has been associated with large-scale mortality events in muskoxen with animals in good body condition showing signs of acute death (Kutz et al. 2015). Various pathological findings, including disseminated petechial hemorrhages, accumulation of serosanguinous fluid in body cavities, peritonitis, typhlitis, and diarrhea, were documented in these muskoxen, but the necropsies were impaired by the advanced autolytic state of most examined carcasses. A case of multiple mortalities due to *E. rhusiopathiae* was also reported in moose (Campbell et al. 1994). In contrast to muskoxen, the animals were in poor body condition, and some animals had severe tick infestations. Lesions included enlarged lymph nodes, bacterial emboli, and myocarditis.

Erysipelothrix rhusiopathiae infections in free-ranging ungulates have been further described in white-tailed deer (*Odocoileus virginianus*) (Bruner et al. 1984), Alpine ibex (*Capra ibex*) (Domenis et al. 2017), and roe deer (*Capreolus capreolus*) (Pewsner et al. 2017). Lesions included, but were not limited to, serosanguinous fluid accumulation and edema, fibrinous peritonitis, diarrhea, enteritis, liver necrosis, amyloidosis, and epicardial hemorrhages.

In Pribilof Island arctic foxes, *E. rhusiopathiae* has been described as the causative agent of the shaggy lame fox syndrome (Spraker and White 2017). The syndrome is characterized by polyarthritis and lameness, poor body condition, and failure to shed the winter coat (conferring a shaggy appearance to affected individuals).

In marine mammals, *E. rhusiopathiae*-associated pathologies are well described since it is a recurring problem for captive cetaceans. In both captive and free-ranging cetaceans, the pathogen has been associated with either acute septicemia and intravascular bacterial emboli or subacute cutaneous vasculitis with diamond-shaped lesions similar to those described in pigs (Melero et al. 2011; Waltzek et al. 2012; Fiorito et al. 2016). An atypical presentation of *E. rhusiopathiae* infection in a killer whale (*Orcinus orca*), resulting in vesicles on the tongue, has also been reported (Bossart and Eimstad 1988).

In birds, *E. rhusiopathiae* is typically associated with peracute septicemia with multiple organ failures and disseminated hemorrhages and thrombi. Chronic forms of the disease with skin lesions and arthritis have also been reported (Jensen et al. 1976; Bobrek et al. 2013; Silva et al. 2020). Other notable clinical manifestations include digital necrosis in a vulture (Ramsay and Baumeister 1986) and respiratory distress and diarrhea in an emu (Jones et al. 1999). Finally, a case of *E. rhusiopathiae* infection in farmed eels constitutes the first report of clinical manifestation in fish, causing acute septicemia and multiple disseminated hemorrhages (Chong et al. 2015).

6 Diagnosis and Detection

Bacterial culture is the most common method for diagnosis of *E. rhusiopathiae* as it can be conducted with basic lab equipment and is often part of the routine processing of diagnostic cases (Clark 2015; Opriessnig and Coutinho 2019; Silva et al. 2020). *Erysipelothrix rhusiopathiae* can usually be isolated using a general enrichment medium. However, in case of suspected contamination (e.g., postmortem bacterial growth in an autolytic sample), culture in several steps using a selective medium increases the likelihood of isolating *E. rhusiopathiae* (Bender et al. 2009) (Fig. 2). Identification from culture plates can be achieved using morphological and biochemical identification or through polymerase chain reaction (PCR) (Wang et al. 2010). Detection can also be achieved through direct PCR testing (Kutz et al. 2015). This approach is suitable to detect DNA in the absence of live bacteria (McLelland et al. 2011). PCR is less labor-intensive, highly sensitive, and less impacted by contaminated samples than culture and can distinguish among *Erysipelothrix* species. However, PCR cannot demonstrate the viability of *Erysipelothrix* detected in a sample (Brooke and Riley 1999; Pal et al. 2010). Serotyping, as a method for characterizing *E. rhusiopathiae* isolates, has been traditionally done through agglutination testing against a panel of serovar-specific antisera (Kucsera 1973), for which culture is a prerequisite. More recently, however, PCR assays have been developed to distinguish among the most common serotypes (1a, 1b, 2, and 5) (Shiraiwa et al. 2018).

Ideally, organs or tissues with lesions indicative of *E. rhusiopathiae* infection are the best candidates for the isolation or detection of the pathogen (Opriessnig and Coutinho 2019). In the absence of detectable lesions, tonsils and lymph nodes have been used to investigate *E. rhusiopathiae* infection in various species including

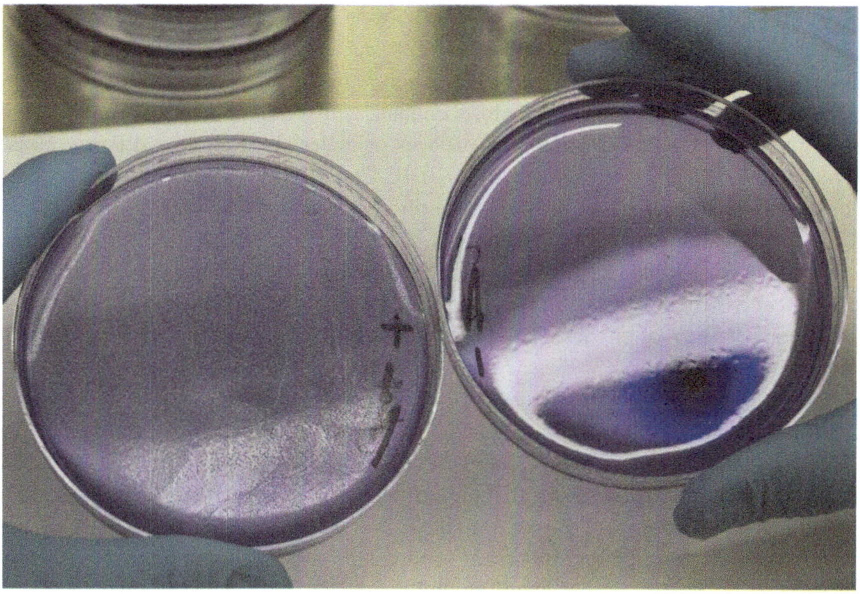

Fig. 2 *Erysipelothrix rhusiopathiae* grown on selective enrichment medium (crystal violet and sodium azide). Left: positive culture result, with pin-prick-sized colonies visible; Right: negative culture result. (Photo: Fabien Mavrot)

muskoxen and wolves (Stephenson and Berman 1978; Hassanein et al. 2001; Forde et al. 2016b). As *E. rhusiopathiae* spreads in infected individuals through blood (Opriessnig and Coutinho 2019), highly vascularized organs such as the spleen, liver, and kidney are also appropriate samples for bacteriological investigation. Bone marrow has also been successfully used for *E. rhusiopathiae* detection in both mammals (Kutz et al. 2015; Forde et al. 2016b) and birds (Alley et al. 2005; McLelland et al. 2011). Bone marrow has the advantage of being preserved longer than other tissues from autolysis and scavenging, making it an appropriate sample for disease surveillance of wildlife carcasses (Rouquet et al. 2005; Butkus et al. 2017).

Finally, serology has also been used to diagnose exposure to *E. rhusiopathiae*. This is particularly relevant for free-ranging species for which monitoring and management occur mostly at the population level. An enzyme-linked immunoassay (ELISA) originally developed for pigs (Giménez-Lirola et al. 2012) has been successfully adapted for monitoring *E. rhusiopathiae* prevalence in caribou, muskoxen, and grey seals (Bondo et al. 2019; Mavrot et al. 2020; Sauvé et al. 2020). Specific serological tools have also been developed to detect antibodies against *E. rhusiopathiae* in birds and cetaceans (Livingston et al. 2013; Nollens et al. 2016).

7 *Erysipelothrix* in Arctic People

Infection of people with *E. rhusiopathiae* is considered an occupational hazard occurring in workers that have close contact with contaminated animals and/or their products or with contaminated surfaces or materials (Reboli and Farrar 1989; Brooke and Riley 1999; Wang et al. 2010; Clark 2015). Infection often occurs through skin abrasions or wounds (Reboli and Farrar 1989; Wang et al. 2010; Clark 2015). The presentation in people can take three different forms: localized cutaneous lesions (i.e., erysipeloid, not to be confused with human erysipelas, a *Streptococcus*-associated skin disorder), a generalized cutaneous form, and a septicaemic form often associated with endocarditis (Brooke and Riley 1999; Wang et al. 2010).

Although the disease is usually self-limiting, cases of generalized or septicaemic *E. rhusiopathiae* infection can be readily treated with penicillin, cephalosporins, or clindamycin (Brooke and Riley 1999; Wang et al. 2010). For populations at risk (e.g., pig and poultry owners, fishermen, slaughterhouse workers, hunters, and veterinarians), reduction of exposure to the pathogen can occur through management of live animals (vaccination, treatment or removal of infected animals), disinfection of surfaces and tools, proper handling and disposal of animal carcasses and waste, and the use of protective equipment such as gloves (Brooke and Riley 1999; Ugochukwu et al. 2019).

Erysipelothrix rhusiopathiae has also been suspected of being the cause of "seal finger," "whale finger," or "fish finger." This condition, reminiscent of erysipeloid, is frequently associated with handling seals and other marine mammals and/or fish (Hillenbrand 1953; Reboli and Farrar 1989). However, the etiology of the condition is debated (Beck and Smith 1976; Mass et al. 1981; Tryland et al. 2014), and in recent years, *Mycoplasma* spp. has been widely accepted as the causative agent of "seal finger" (Baker et al. 1998; White and Jewer 2009; Tryland et al. 2014; Westley et al. 2016).

Food is also considered a possible source for *Erysipelothrix* transmission in humans and wildlife (Wolcott 2007). In the Arctic, where people rely heavily on aquatic and terrestrial wildlife species for their cultural and economic wellbeing, and for food safety, *E. rhusiopathiae* is a growing concern. Recently, an unusual case of *E. rhusiopathiae* infection in a prosthetic joint was described in a patient from the Canadian Arctic (Groeschel et al. 2019). Although this individual had considerable occupational exposure to the meat and tissues of a variety of wildlife species, whole genome sequencing did not identify a known wildlife source. Nevertheless, knowledge of the occurrence of *E. rhusiopathiae* in arctic species and increased awareness by the local population and health personnel of clinical signs associated with infection in people are central to promoting protective behaviors, reporting suspected human cases, and ensuring timely recognition and appropriate treatment if warranted.

8 Conclusion

Erysipelothrix rhusiopathiae is an old pathogen gaining renewed importance. Once considered a serious pathogen of domestic swine, its significance diminished with the advent of effective vaccines. It remains, however, an important pathogen in wildlife globally, with spillover to people being a very real threat. In the Arctic, recent emergence of this opportunistic bacterium as a significant cause of mortality in muskoxen, and morbidity in Pribilof foxes, alerts us to the vulnerability of the arctic ecosystems: ecosystems that are experiencing unprecedented and accelerating changes and stressors due to climate change. Given the broad host range and opportunistic nature of *E. rhusiopathiae*, the impacts on wildlife health, and the associated human health concerns, increased efforts to better understand the epidemiology and significance of this pathogen in the changing Arctic are warranted.

References

Alley MR, Gartrell BD, Mack HJ, McInnes CM (2005) *Erysipelothrix rhusiopathiae* septicaemia in translocated kakapo (*Strigops habroptilus*). N Z Vet J 53(1):94–94. https://doi.org/10.1080/00480169.2005.36481

Atyabi N, Youssefi R, Javdani G, Tavasoli A, Vojgani M, Gharegozloo F (2012) Isolation of *Erysipelothrix rhusiopathiae* from aborted lambs in Iran: a case report. Iran J Vet Med 6(2): 129–132

Baker AS, Ruoff KL, Madoff S (1998) Isolation of *Mycoplasma* species from a patient with seal finger. Clin Infect Dis 27(5):1168–1170. https://doi.org/10.1086/514980

Beck B, Smith TG (1976) Letter: seal finger: an unsolved medical problem in Canada. Can Med Assoc J 115(2):105–106

Bender JS, Kinyon JM, Kariyawasam S, Halbur PG, Opriessnig T (2009) Comparison of conventional direct and enrichment culture methods for *Erysipelothrix* spp. from experimentally and naturally infected swine. J Vet Diagn Investig 21(6):863–868. https://doi.org/10.1177/104063870902100617

Blyde DJ, Woods R (1999) Erysipelas in malleefowl. Aust Vet J 77(7):434–435. https://doi.org/10.1111/j.1751-0813.1999.tb12084.x

Bobrek K, Gaweł A, Mazurkiewicz M (2013) Infections with *Erysipelothrix rhusiopathiae* in poultry flocks. Worlds Poult Sci J 69(4):803–812. https://doi.org/10.1017/S0043933913000822

Bondo KJ, Macbeth B, Schwantje H, Orsel K, Culling D, Culling B, Tryland M, Nymo IH, Kutz S (2019) Health survey of boreal caribou (*Rangifer tarandus caribou*) in northeastern British Columbia, Canada. J Wildl Dis 55(3):544–562. https://doi.org/10.7589/2018-01-018

Bossart GD, Eimstad EA (1988) *Erysipelothrix* vesicular glossitis in a killer whale (*Orcinus orca*). J Zoo Anim Med 19(1/2):42–47. https://doi.org/10.2307/20094851

Brooke CJ, Riley TV (1999) *Erysipelothrix rhusiopathiae*: bacteriology, epidemiology and clinical manifestations of an occupational pathogen. J Med Microbiol 48(9):789–799. https://doi.org/10.1099/00222615-48-9-789

Bruner JA, Griffith RW, Greve JH, Wood RL (1984) *Erysipelothrix rhusiopathiae* serotype 5 isolated from a white-tailed deer in Iowa. J Wildl Dis 20(3):235–236. https://doi.org/10.7589/0090-3558-20.3.235

Butkus CE, Allender MC, Phillips CA, Adamovicz LA (2017) Detection of ranavirus using bone marros harvested from mortality events in eastern box turtles (*Terrapene carolina carolina*). J Zoo Wildl Med 48(4):1210–1214. https://doi.org/10.1638/2017-0098.1

Calle PP, Kenny DE, Cook RA (1993) Successful treatment of suspected erysipelas septicemia in a beluga whale (*Delphinapterus leucas*). Zoo Biol 12(5):483–490. https://doi.org/10.1002/zoo.1430120510

Campbell GD, Addison EM, Barker IK, Rosendal S (1994) *Erysipelothrix rhusiopathiae*, serotype septicemia in moose (*Alces alces*) from Algonquin Park, Ontario. J Wildl Dis 30(3):436–438

Chong RS-M, Shinwari MW, Amigh MJ, Aravena-Roman M, Riley TV (2015) First report of *Erysipelothrix rhusiopathiae*-associated septicaemia and histologic changes in cultured Australian eels, *Anguilla reinhardtii* (Steindachner, 1867) and *A. australis* (Richardson, 1841). J Fish Dis 38(9):839–847. https://doi.org/10.1111/jfd.12293

Clark AE (2015) The occupational opportunist: an update on *Erysipelothrix rhusiopathiae* infection, disease pathogenesis, and microbiology. Clin Microbiol Newsl 37(18):143–151. https://doi.org/10.1016/j.clinmicnews.2015.09.001

CWHC (2016) Report CWHC # 124628. Canadian Wildlife Health Cooperative

Domenis L, Spedicato R, Orusa R, Robetto S (2017) *Erysipelothrix rhusiopathiae* infection in an Alpine ibex (*Capra ibex*). J Comp Pathol 156(1):101

Fiorito CD, Bentancor A, Lombardo D, Bertellotti M (2016) *Erysipelothrix rhusiopathiae* isolated from gull-inflicted wounds in southern right whale calves. Dis Aquat Org 121(1):67–73. https://doi.org/10.3354/dao03041

Forde T, Biek R, Zadoks R, Workentine ML, De Buck J, Kutz S, Opriessnig T, Trewby H, van der Meer F, Orsel K (2016a) Genomic analysis of the multi-host pathogen *Erysipelothrix rhusiopathiae* reveals extensive recombination as well as the existence of three generalist clades with wide geographic distribution. BMC Genomics 17(1):461. https://doi.org/10.1186/s12864-016-2643-0

Forde TL, Orsel K, Zadoks RN, Biek R, Adams LG, Checkley SL, Davison T, De Buck J, Dumond M, Elkin BT, Finnegan L, Macbeth BJ, Nelson C, Niptanatiak A, Sather S, Schwantje HM, van der Meer F, Kutz SJ (2016b) Bacterial genomics reveal the complex epidemiology of an emerging pathogen in Arctic and Boreal ungulates. Front Microbiol 7:1759. https://doi.org/10.3389/fmicb.2016.01759

Forde TL, Kollanandi Ratheesh N, Harvey WT, Thomson JR, Williamson S, Biek R, Opriessnig T (2020) Genomic and immunogenic protein diversity of *Erysipelothrix rhusiopathiae* isolated from pigs in Great Britain: implications for vaccine protection. Front Microbiol 11:418. https://doi.org/10.3389/fmicb.2020.00418

Franson JC, Galbreath EJ, Wiemeyer SN, Abell JM (1994) *Erysipelothrix rhusiopathiae* infection in a captive bald eagle (*Haliaeetus leucocephalus*). J Zoo Wildl Med 25(3):446–448

Fthenakis GC, Christodoulopoulos G, Leontides L, Tzora A (2006) Abortion in ewes associated with *Erysipelothrix rhusiopathiae*. Small Rumin Res 63(1):183–188. https://doi.org/10.1016/j.smallrumres.2005.01.017

Gartrell BD, Alley MR, Mack H, Donald J, McInnes K, Jansen P (2005) Erysipelas in the critically endangered kakapo (*Strigops habroptilus*). Avian Pathol 34(5):383–387. https://doi.org/10.1080/03079450500268583

Giménez-Lirola LG, Xiao CT, Halbur PG, Opriessnig T (2012) Development of a novel fluorescent microbead-based immunoassay and comparison with three enzyme-linked immunoassays for detection of anti-*Erysipelothrix* spp. IgG antibodies in pigs with known and unknown exposure. J Microbiol Methods 91(1):73–79. https://doi.org/10.1016/j.mimet.2012.07.014

Goerttler V, Hubrig T, Others (1960) Pathogenesis of swine erysipelas. Zentralbl Veterinaermed 7:364–391

Groeschel M, Forde T, Turvey S, Joffe AM, Hui C, Naidu P, Mavrot F, Kutz S, Singh AE (2019) An unusual case of *Erysipelothrix rhusiopathiae* prosthetic joint infection from the Canadian Arctic: whole genome sequencing unable to identify a zoonotic source. BMC Infect Dis 19(1):282. https://doi.org/10.1186/s12879-019-3913-7

Hassanein R, Sawada T, Kataoka Y, Itoh K, Suzuki Y (2001) Serovars of *Erysipelothrix* species isolated from the tonsils of healthy cattle in Japan. Vet Microbiol 82(1):97–100. https://doi.org/10.1016/s0378-1135(01)00379-0

Hillenbrand FKM (1953) Whale finger and seal finger; their relation to erysipeloid. Lancet 1(6762): 680–681. https://doi.org/10.1016/s0140-6736(53)91807-8

Janßen T, Voss M, Kühl M, Semmler T, Philipp H-C, Ewers C (2015) A combinational approach of multilocus sequence typing and other molecular typing methods in unravelling the epidemiology of *Erysipelothrix rhusiopathiae* strains from poultry and mammals. Vet Res 46:84. https://doi.org/10.1186/s13567-015-0216-x

Jensen WI, Cotter SE, Fish and Wildlife Service US (1976) An outbreak of erysipelas in eared grebes (*Podiceps nigricollis*). J Wildl Dis 12:583–586

Jones MP, Orosz SE, Finnegan MV, Sleeman JM, Bemis DA (1999) *Erysipelothrix rhusiopathiae* infection in an emu (*Dromaius novaehollandiae*). J Avian Med Surg 13(2):104–107

Kucsera G (1973) Proposal for standardization of the designations used for serotypes of *Erysipelothrix rhusiopathiae* (migula) Buchanan. Int J Syst Bacteriol 23(2):184–188. https://doi.org/10.1099/00207713-23-2-184

Kutz S, Bollinger T, Branigan M, Checkley S, Davison T, Dumond M, Elkin B, Forde T, Hutchins W, Niptanatiak A, Orsel K (2015) *Erysipelothrix rhusiopathiae* associated with recent widespread muskox mortalities in the Canadian Arctic. Can Vet J 56(6):560–563

Lamont MH (1979) *Erysipelothrix rhusiopathiae*: epidemiology and infection in sheep. Vet Bull 49:479–495

Langford EV, Dorward WJ (1977) *Erysipelothrix insidiosa* recovered from sylvatic mammals in northwestern Canada during examinations for rabies and anthrax. Can Vet J 18(4):101–104

Livingston M, Fidler A, Mellor D, de Kloet S, Eason D, Elliott G, Moorhouse R (2013) Prevalence of IgY antibodies against *Erysipelothrix rhusiopathiae* in a critically endangered parrot (kakapo, *Strigops habroptilus*) and associated responses to vaccination. Avian Pathol 42(5):502–507. https://doi.org/10.1080/03079457.2013.832146

Mass DP, Newmeyer WL, Kilgore ES Jr (1981) Seal finger. J Hand Surg Am 6(6):610–612. https://doi.org/10.1016/s0363-5023(81)80144-x

Mavrot F, Orsel K, Hutchins W, Adams LG, Beckmen K, Blake JE, Checkley SL, Davison T, Di Francesco J, Elkin B, Leclerc L-M, Schneider A, Tomaselli M, Kutz SJ (2020) Novel insights into serodiagnosis and epidemiology of *Erysipelothrix rhusiopathiae*, a newly recognized pathogen in muskoxen (*Ovibos moschatus*). PLoS One 15(4):e0231724. https://doi.org/10.1371/journal.pone.0231724

McLelland JM, Gartrell BD, Roe WD (2011) A retrospective study of post-mortem examination findings in takahe (*Porphyrio hochstetteri*). N Z Vet J 59(4):160–165. https://doi.org/10.1080/00480169.2011.579243

Melero M, Rubio-Guerri C, Crespo JL, Arbelo M, Vela AI, García-Párraga D, Sierra E, Domínguez L, Sánchez-Vizcaíno JM (2011) First case of erysipelas in a free-ranging bottlenose dolphin (*Tursiops truncatus*) stranded in the Mediterranean Sea. Dis Aquat Org 97(2):167–170. https://doi.org/10.3354/dao02412

Nakato H, Shinomiya K, Mikawa H (1987) Adhesion of *Erysipelothrix rhusiopathiae* to cultured rat aortic endothelial cells. Role of bacterial neuraminidase in the induction of arteritis. Pathol Res Pract 182(2):255–260. https://doi.org/10.1016/S0344-0338(87)80114-0

Nollens HH, Giménez-Lirola LG, Robeck TR, Schmitt TL, DiRocco S, Opriessnig T (2016) Evaluation of anti-*Erysipelothrix rhusiopathiae* IgG response in bottlenose dolphins *Tursiops truncatus* to a commercial pig vaccine. Dis Aquat Org 121(3):249–256. https://doi.org/10.3354/dao03061

Ogawa Y, Ooka T, Shi F, Ogura Y, Nakayama K, Hayashi T, Shimoji Y (2011) The genome of *Erysipelothrix rhusiopathiae*, the causative agent of swine erysipelas, reveals new insights into the evolution of firmicutes and the organism's intracellular adaptations. J Bacteriol 193(12): 2959–2971. https://doi.org/10.1128/JB.01500-10

Ogawa Y, Shiraiwa K, Ogura Y, Ooka T, Nishikawa S, Eguchi M, Hayashi T, Shimoji Y (2017) Clonal lineages of *Erysipelothrix rhusiopathiae* responsible for acute swine erysipelas in Japan identified by using genome-wide single-nucleotide polymorphism analysis. Appl Environ Microbiol 83(11):e00130–e001317. https://doi.org/10.1128/AEM.00130-17

Opriessnig T, Coutinho TA (2019) Erysipelas. In: Zimmerman J (ed) Diseases of swine. Wiley, Boca Raton, FL, pp 835–843

Opriessnig T, Shen HG, Bender JS, Boehm JR, Halbur PG (2013) *Erysipelothrix rhusiopathiae* isolates recovered from fish, a harbour seal (*Phoca vitulina*) and the marine environment are capable of inducing characteristic cutaneous lesions in pigs. J Comp Pathol 148(4):365–372. https://doi.org/10.1016/j.jcpa.2012.08.004

Pal N, Bender JS, Opriessnig T (2010) Rapid detection and differentiation of *Erysipelothrix* spp. by a novel multiplex real-time PCR assay. J Appl Microbiol 108(3):1083–1093. https://doi.org/10.1111/j.1365-2672.2009.04560.x

Pewsner M, Origgi FC, Frey J, Ryser-Degiorgis M-P (2017) Assessing fifty years of general health surveillance of roe deer in Switzerland: a retrospective analysis of necropsy reports. PLoS One 12(1):e0170338. https://doi.org/10.1371/journal.pone.0170338

Proctor DM, Richardson IM (1954) A report on 235 cases of erysipeloid in Aberdeen. Br J Ind Med 11(3):175–179. https://doi.org/10.1136/oem.11.3.175

Ramsay EC, Baumeister BM (1986) Isolation of *Erysipelothrix rhusiopathiae* from lesions of distal extremity necrosis in a captive king vulture. J Wildl Dis 22(3):430–431. https://doi.org/10.7589/0090-3558-22.3.430

Reboli AC, Farrar WE (1989) *Erysipelothrix rhusiopathiae*: an occupational pathogen. Clin Microbiol Rev 2(4):354–359. https://doi.org/10.1128/cmr.2.4.354

Rouquet P, Froment J-M, Bermejo M, Kilbourn A, Karesh W, Reed P, Kumulungui B, Yaba P, Délicat A, Rollin PE, Leroy EM (2005) Wild animal mortality monitoring and human Ebola outbreaks, Gabon and Republic of Congo, 2001-2003. Emerg Infect Dis 11(2):283–290. https://doi.org/10.3201/eid1102.040533

Sauvé CC, Hernández-Ortiz A, Jenkins E, Mavrot F, Schneider A, Kutz S, Saliki JT, Daoust P-Y (2020) Exposure of the Gulf of St. Lawrence grey seal *Halichoerus grypus* population to potentially zoonotic infectious agents. Dis Aquat Org 142:105–118. https://doi.org/10.3354/dao03536

Shiraiwa K, Ogawa Y, Nishikawa S, Eguchi M, Shimoji Y (2018) Identification of serovar 1a, 1b, 2, and 5 strains of *Erysipelothrix rhusiopathiae* by a conventional gel-based PCR. Vet Microbiol 225:101–104. https://doi.org/10.1016/j.vetmic.2018.09.014

Silva AP, Cooper G, Blakey J, Jerry C, Shivaprasad HL, Stoute S (2020) Retrospective summary of *Erysipelothrix rhusiopathiae* diagnosed in avian species in California (2000–19). Avian Dis 64(4):499 506. https://doi.org/10.1637/aviandiseases-d20-00038

Spraker TR, White PA (2017) Shaggy Lame Fox Syndrome in Pribilof Island Arctic Foxes (*Alopex lagopus pribilofensis*), Alaska. Vet Pathol 54(2):258–268. https://doi.org/10.1177/0300985816660745

Stephenson EH, Berman DT (1978) Isolation of *Erysipelothrix rhusiopathiae* from tonsils of apparently normal swine by two methods. Am J Vet Res 39(1):187–188

Stickdorn H (1936) Versuche zur Übertragung von Rotlaufbakterien durch die Schweinelaus (*Haematopinus suis*). Z Parasitenkd 8(4):492–503. https://doi.org/10.1007/bf02163070

Sudlovenick E (2019) A serological survey and Inuit Qaujimajatuqangit of Ringed seals (Nattiit) in Frobisher Bay, Nunavut. Masters Thesis, University of Prince Edward Island

Suer LD, Vedros NA, Schroeder JP, Dunn JL (1988) *Erysipelothrix rhusiopathiae*. I. Isolation and characterization from pinnipeds and bite/abrasion wounds in humans. Dis Aquat Org 5:1–5

Tomaselli M, Kutz S, Gerlach C, Checkley S (2018) Local knowledge to enhance wildlife population health surveillance: conserving muskoxen and caribou in the Canadian Arctic. Biol Conserv 217:337–348. https://doi.org/10.1016/j.biocon.2017.11.010

Tryland M, Nesbakken T, Robertson L, Grahek-Ogden D, Lunestad BT (2014) Human pathogens in marine mammal meat – a northern perspective. Zoonoses Public Health 61(6):377–394. https://doi.org/10.1111/zph.12080

Ugochukwu ICI, Samuel F, Orakpoghenor O, Nwobi OC, Anyaoha CO, Majesty-Alukagberie LO, Ugochukwu MO, Ugochukwu EI (2019) Erysipelas, the opportunistic zoonotic disease: history, epidemiology, pathology, and diagnosis–a review. Comp Clin Pathol 28(3):853–859. https://doi.org/10.1007/s00580-018-2856-5

Waltzek TB, Cortés-Hinojosa G, Wellehan JFX Jr, Gray GC (2012) Marine mammal zoonoses: a review of disease manifestations. Zoonoses Public Health 59(8):521–535. https://doi.org/10.1111/j.1863-2378.2012.01492.x

Wang Q, Chang BJ, Riley TV (2010) Erysipelothrix rhusiopathiae. Vet Microbiol 140(3–4): 405–417. https://doi.org/10.1016/j.vetmic.2009.08.012

Westley BP, Horazdovsky RD, Michaels DL, Brown DR (2016) Identification of a novel *Mycoplasma* species in a patient with septic arthritis of the hip and seal finger. Clin Infect Dis 62(4): 491–493. https://doi.org/10.1093/cid/civ875

White CP, Jewer DD (2009) Seal finger: a case report and review of the literature. Can J Plast Surg 17(4):133–135. https://doi.org/10.1177/229255030901700415

Wolcott MJ (2007) Erysipelas. In: Hunter NJ, Hunter DB, Atkinson CT (eds) Infectious diseases of wild birds. Blackwell, New York, pp 332–340

Wood RL (1973) Survival of *Erysipelothrix rhusiopathiae* in soil under various environmental conditions. Cornell Vet 63(3):390–410

Wood RL (1974) Isolation of pathogenic *Erysipelothrix rhusiopathiae* from feces of apparently healthy swine. Am J Vet Res 35(1):41–43

Wood RL (1975) Erysipelothrix infection. In: Hubbert WT (ed) Diseases transmitted from animals to man, 6th edn. Thomas, Springfield, IL

Wood RL (1994) *Erysipelothrix* infections. In: Beran GW (ed) Handbook of zoonoses. Section A: bacterial, rickettsial, chlamydial, and mycotic, 2nd edn. CRC Press, Boca Raton, FL

Woodbine M (1950) *Erysipelothrix rhusiopathiae*. Bacteriology and chemotherapy. Bacteriol Rev 14(2):161–178

Zhu W, Cai C, Wang Y, Li J, Wu C, Kang C, Sun X, Jin M (2017) Characterization of roles of SpaA in *Erysipelothrix rhusiopathiae* adhesion to porcine endothelial cells. Microb Pathog 113:176–180. https://doi.org/10.1016/j.micpath.2017.10.020

Tularemia in the Arctic

Cristina M. Hansen and Svetlana Dresvyannikova

1 Introduction

While studying an outbreak of a plague-like disease in ground squirrels (*Citellus beecheyi*) in Tulare County, California (USA), in 1911, George McCoy and Charles Chapin isolated a unique bacterial organism and named it *Bacterium tularense*, after Tulare County (McCoy 1911; McCoy and Chapin 1912). In 1919, while investigating "deer-fly fever," Edward Francis determined that it was the same disease and named it "tularemia" (Hirschmann 2018). The organism was eventually renamed *Francisella tularensis*, after Edward Francis. The first reported laboratory-confirmed isolation of this organism from a human was in 1914 in Ohio (USA) (Wherry and Lamb 1914). However, descriptions of a similar plague-like disease predate this first isolation by almost 100 years (Sjostedt 2007). *Francisella tularensis* is zoonotic and, since its first description, has been isolated from more than 300 species of mammals, birds, amphibians, and invertebrates (Keim et al. 2007).

Francisella tularensis is a highly virulent Gram-negative aerobic coccobacillus. It is reportable in every Arctic nation and is classified as a category A select bioterrorism agent in the United States due to its virulence, zoonotic potential, ability to spread via aerosol, and up to 30% fatality rate if not treated. In fact, *F. tularensis* has been weaponized by several countries, and since the 2001 anthrax attacks in the United States, tularemia research has been rejuvenated (Sjostedt 2007).

Francisella tularensis is ubiquitous in the Northern hemisphere (Fig. 1). There are currently three recognized subspecies: *tularensis*, *holarctica*, and *mediasiatica*.

C. M. Hansen (✉)
Department of Veterinary Medicine, University of Alaska Fairbanks, Fairbanks, AK, USA
e-mail: cmhansen@alaska.edu

S. Dresvyannikova
Federal Research Center of Virology and Microbiology, Pokrov, Russia

M. Tryland (ed.), *Arctic One Health*, https://doi.org/10.1007/978-3-030-87853-5_17

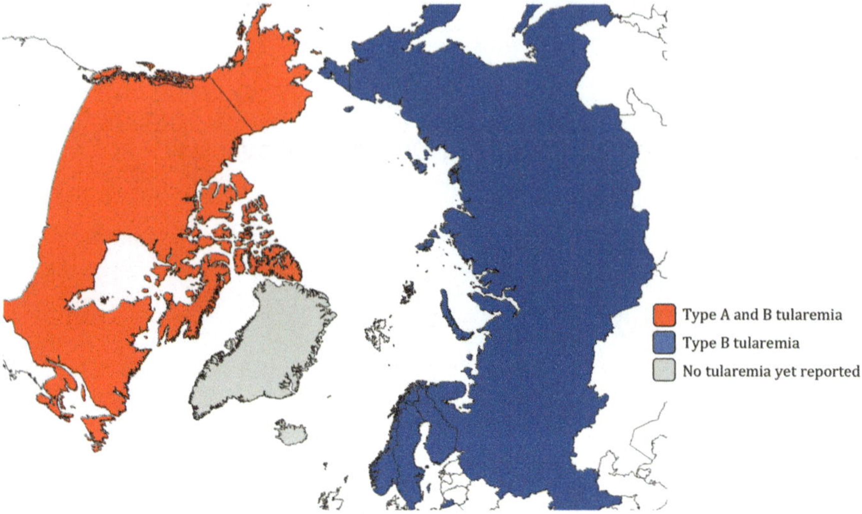

Fig. 1 A map of the Arctic showing the distribution of types A and B tularemia. Tularemia has not yet been reported in Greenland (Denmark) or Iceland

F. tularensis subsp. *tularensis* (also called type A) is found in North America and is the most virulent subspecies. *F. tularensis* subsp. *holarctica* (also called type B) is found in Europe, Japan, and North America. It is less virulent than type A tularemia. *F. tularensis* subsp. *mediasiatica* is found in central Asia and very rarely causes human disease (Challacombe et al. 2017).

There are six clinical forms of tularemia, and the route of entry determines which will manifest (Maurin and Gyuranecz 2016). Pneumonic tularemia is the most severe and results from the inhalation of aerosolized bacteria. If untreated, it can be up to 30% fatal (Sjostedt 2007). The most common form is ulceroglandular, which results from percutaneous exposure via a wound or arthropod bite. Ulceroglandular tularemia results in an ulcer at the site of infection (usually a skin wound) followed by localized lymphadenopathy and is very rarely fatal. The glandular form presents as lymphadenopathy without an ulcer. Other less common forms of the disease include oculoglandular (exposure via the eye), oropharyngeal (exposure via ingestion), and typhoidal (systemic disease without a primary ulcer). In humans, symptoms generally appear 3-5 days after infection and are often influenza-like including fever, chills, headache, myalgia, and arthralgia. In animals, susceptibility varies widely. Tularemia is generally considered to be a disease of lagomorphs and rodents, as they are highly susceptible and when infected are often found moribund or dead (Hopla and Hopla 1994; Maurin and Gyuranecz 2016). Dogs, on the other hand, are considered relatively resistant to tularemia, and cases are relatively rare (Foley and Nieto 2010).

The World Health Organization (WHO) has proposed a case definition as a person with an exposure history, symptoms consistent with tularemia, and a clinical sample that tests positive for antigen or antibody (Maurin and Gyuranecz 2016).

There is currently no gold standard for tularemia diagnosis (Maurin 2020). In humans and animals, during active infection, DNA may be detected by polymerase chain reaction (PCR) of blood, lymph node, or other tissues. Antibodies peak 3-4 weeks after infection and can be detected serologically (Maurin and Gyuranecz 2016). Bacterial isolation is complicated by the fact that culturing *F. tularensis* requires biosafety level 3 containment facilities and that the organism is fastidious (Maurin and Gyuranecz 2016). Intradermal skin testing, in which a small amount of chemically inactivated *F. tularensis* is injected into the skin and monitored for inflammation, is described in older literature (Foshay 1932) and is cited later in this chapter but is no longer a common diagnostic test. Tularemia can be treated with aminoglycoside, fluoroquinolone, or tetracycline antibiotics. Acquired antibiotic resistance in clinical isolates has not yet been reported (Maurin and Gyuranecz 2016), but strains of *F. tularensis* subsp. *holarctica* have been documented to be erythromycin resistant (Keim et al. 2007; Maurin and Gyuranecz 2016).

The high virulence of *Francisella tularensis* is due in part to its ability to evade the immune system. The lipopolysaccharide (LPS) on its outer membrane is not highly inflammatory (i.e., does not incite a large immune response). In addition, it leads an intracellular lifecycle allowing it to escape immune detection (Sjostedt 2007). When a *Francisella* bacterium is engulfed by a phagocytic cell, it quickly escapes the phagosome and resides and replicates in the cytosol. Additionally, it has numerous virulence factors (including a type VI secretion system) that are encoded by a pathogenicity island (a group of genes encoding virulence factors) within its genome (Nano et al. 2004).

The tularemia transmission cycle is depicted in Fig. 2. Tularemia can be transmitted by contact with infected carcasses (usually resulting in ulceroglandular tularemia), inhalation of aerosolized bacteria (resulting in pneumonic tularemia), ingestion of contaminated water (usually resulting in oropharyngeal tularemia), or insect vectors (resulting in ulceroglandular, glandular, or typhoidal tularemia). Mosquito-borne transmission has been hypothesized and is theoretically possible, but has not been proven (Maurin and Gyuranecz 2016; Triebenbach et al. 2010). Similarly, a hydro-telluric reservoir has been suggested but isn't well studied (Hennebique et al. 2019). There is even evidence that this tenacious organism can be disseminated by migratory birds (Padeshi et al. 2010; Lopes de Carvalho et al. 2012). *Francisella* organisms can survive for up to 70 days in water depending on temperature and salinity. It also may survive in amoebae and mosquito larvae present in aquatic environments (Hennebique et al. 2019; Telford and Goethert 2020). Interestingly, human tularemia cases have been reported after near-drowning incidents (Ughetto et al. 2015).

2 United States (Alaska)

Tularemia is a reportable disease in the United States and in Alaska. The first report of *F. tularensis* (then *B. tularense*) in Alaska was in 1937, when the organism was isolated from a rabbit tick (*Haemophysalis leporis-palustris*) that was removed from

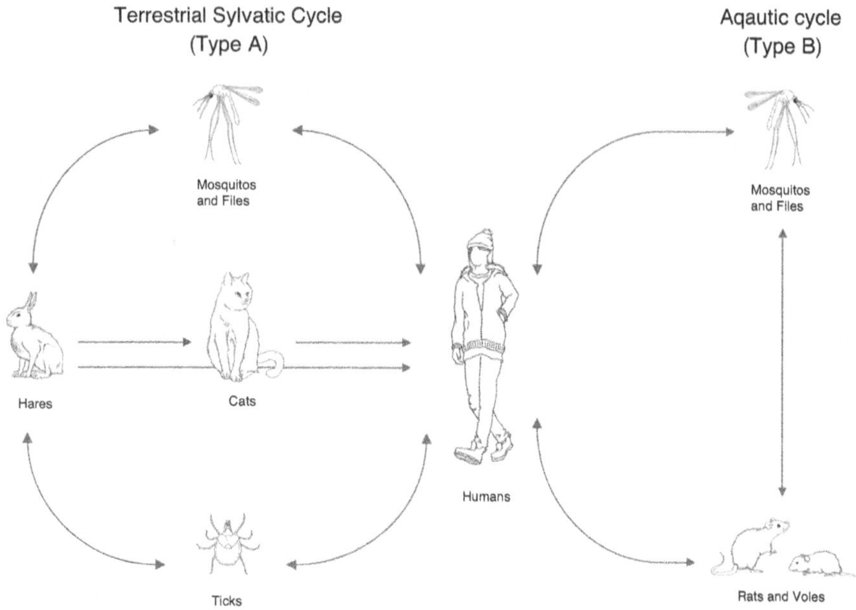

Fig. 2 The terrestrial and aquatic tularemia transmission cycles. *F. tularensis* has been isolated from more than 300 host species; common vectors are pictured here. (© Erika Gangware, used with permission)

a varying hare (*Lepus americanus*) that was trapped near Fairbanks in July of 1937 (Philip and Parker 1938). The isolate in that case was highly virulent in both rabbits and guinea pigs, suggesting that it was type A tularemia. Later, less virulent isolates were identified in ticks removed from willow ptarmigan (*Lagopus lagopus*; Hopla 1965) and in tundra voles (*Microtus oeconomus*; Rausch et al. 1969). In the 1970s, the first side-by-side biochemical and animal challenge studies were conducted to confirm the coexistence of type A and type B tularemia in Alaska (Miller 1974). Hansen et al. (2010) used molecular subtyping to further confirm this coexistence and identified types A.I., A.II, and B from Alaskan *F. tularensis* isolates, showing that not two but at least three genetic groups of this organism are present in the state.

A number of surveys for tularemia have been conducted in wild and domestic animals in Alaska, and more than 25 different wildlife species have been reported as seropositive (Hopla 1965, 1968; Zarnke and Ballard 1987; Chomel et al. 1998; Zarnke et al. 2004). Seroprevalences in these studies ranged from 0.2% (1/376) in Northern red-backed voles (*Myodes rutilus*) from Delta Creek to 100% (1/1) in a Northern shrike (*Lanius borealis*) from Southcentral Alaska. Notable high prevalence in larger sample sizes include 23.8% (16/67) in wolves (*Canis lupus*) from Southcentral Alaska (Zarnke and Ballard 1987), 33% (13/40) in grizzly bears from Interior Alaska, 33% (13/40) in black bears from Interior Alaska, and a surprising 35% (34/96) from grizzly bears from Arctic Northwestern Alaska. Zarnke and Ballard (1987) and Zarnke et al. (2004) noted that tularemia peaks in predators followed peaks in snowshoe hare populations, suggesting the possibility of hares as a reservoir.

It is also possible that mosquitos are vectors for tularemia in Alaska, but this has not been proven. During the summer of 2006, 2,610 Interior Alaskan mosquitos (Diptera: Culicidae) were captured and 30% of pooled samples (pools of 10 mosquitos) tested positive for the *F. tularensis FopA* gene, but the authors were unable to transmit tularemia via a bite by an Alaskan mosquito species (Triebenbach et al. 2010).

The first probable human case of tularemia in Alaska was documented in 1938, in a 62-year-old man from Wiseman, north of the Arctic Circle (Philip 1939). The first laboratory-confirmed case (via serology) was documented in 1946 in a 31-year-old trapper from Northway (Interior) Alaska (Williams 1946). The patient had a history of skinning muskrat, had reported an ulcer on his finger, and presented with headache, orbital pain, general malaise, fever, and swollen lymph nodes. The first culture-positive human infection in Alaska was documented in 1974 when *F. tularensis* was recovered from a laboratory worker's pleural fluid (Miller 1974).

Several human tularemia surveys have been conducted in Alaska, but none recently. Between 1954 and 1957, 816 skin tests were performed on inhabitants of Alaskan villages. Of those, 64 (8%) tested positive, with the highest incidence reported in 50–59-years -old (Hopla 1960). The highest incidence was reported in central Alaska and corresponds with rich trapping areas. In a 1960 seroprevalence survey, Aleut, Indian, and Eskimo men, 18% (139 of 793) had antibody titers ranging from 1:20 to 1:640 (Philip et al. 1962). A subset of those 793 ($n = 115$) were also skin tested, with 51 (44%) being positive.

Between 1946 and 2010, 38 human tularemia cases have been reported to the state of Alaska, of which 23 were laboratory confirmed. Of the confirmed cases, 19 had detailed information available. The median age of patients was 39 years. Most patients were male (22 of 30), were Caucasian (27 of 38), and had signs that presented between June and August (20 of 29). Most (26 of 38) were from Central Alaska. The most common disease form reported was ulceroglandular (19 of 27), followed by typhoidal (3 of 27) and pneumonic (2 of 27). No cases were fatal. Notably, 79% (19 of 24) had reported direct contact with wild or domestic animals (Hansen et al. 2010), including two human cases directly related to housecats (Liles and Burger 1993).

Between 2011 and 2020, a total of 25 animal cases were reported to the Office of the State Veterinarian (Gerlach personal communication). Nine of those cases were reported in wildlife (8 in rabbits or hares, and 1 in a squirrel), and 14 were reported in domestic animals (10 dogs and 4 cats). Two reported cases were in unknown animal species. Figure 3 shows the liver of a housecat that was submitted to the University of Alaska Fairbanks Department of Veterinary Medicine and was determined to have died from tularemia.

Tularemia has clearly been endemic in Alaska for some time, and the impacts of climate change on its ecology and epidemiology are unknown. The reservoirs in Alaska are still unknown, though it's suspected to be hares or muskrats. To gain

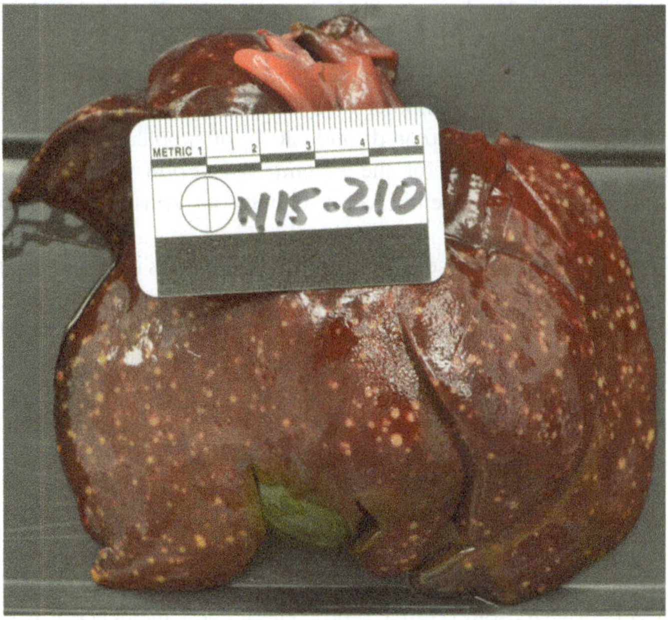

Fig. 3 The liver of a 3-year-old indoor/outdoor housecat that was submitted as a necropsy specimen to the University of Alaska Fairbanks. The liver is mildly enlarged and has scattered pinpoint white foci of necrosis. The presence of *F. tularensis* DNA was confirmed by polymerase chain reaction. (© Molly Murphy, used with permission)

further understanding of this important disease in Alaska, renewed research is needed.

3 Canada (Yukon, Northwest Territory, Nunavut)

The first reported case of human tularemia in Canada was in 1929 in Ontario and was linked to contact with "rabbits" (most likely the snowshoe hare, *Lepus americanus*). Since that time, tularemia has been reported in every province and territory except Yukon (Wobeser et al. 2009), though reports in the Northwest Territories (NT) and Nunavut are rare.

Human tularemia is currently a nationally reportable disease in Canada, animal tularemia is not, except for the provinces Yukon, Northwest Territories, and Nunavut.

A relatively large outbreak of tularemia was reported in Fort Resolution NT in 1976. During that outbreak, 33 individuals (from a population of less than 300) were infected (Lantis 1981). That outbreak was related to a tularemia epizootic in muskrats that put trappers and fur dealers at the highest risk. Two more human cases were described the following year in the Mackenzie district, NT (Lantis 1981). There have been three reported cases of human tularemia in NT that reported direct contact with wild rodents or lagomorphs: one with a beaver (*Castor canadensis*), one with a muskrat (*Ondatra zibethicus*), and one with a red squirrel (*Tamiasciurus hudsonicus*; Wobeser et al. 2009). There has only been one reported case in a human above treeline (by latitude), in a 14-year-old boy from Arviat, Nunavut (formerly NT), and that case was presumed to be due to an insect vector (Silverman et al. 1991). Though it is assumed that wild lagomorphs and rodents are reservoirs for tularemia in Arctic Canada, there has been only one confirmed case of tularemia in a beaver in NT (Wobeser et al. 2009).

Very few isolates from Canada, and none from Arctic Canada, have been molecularly typed. To the South and East, *F. t.* subsp. *holarctica* (type B tularemia) has been isolated from deer mice (*Peromyscus maniculatus*) and house mice (*Mus musculus*) in Saskatchewan (Wobesor et al. 2007) and Manitoba (Wobeser et al. 2009) and was also identified in ticks removed from snowshoe hares in Ontario (Ditchfield et al. 1960). However, a human sample from Kamloops, British Columbia, was identified as type A.1., and another from Muskoka Ontario was typed as A. II (Farlow et al. 2004). To the West, type A and type B tularemia have been confirmed to coexist in Alaska (Hansen et al. 2010), and it is this author's hypothesis that both type A and B tularemia exist also in Arctic Canada and that notable outbreaks like the 1976 NT outbreak were caused by type A tularemia (*F. t.* subsp. *tularensis*).

4 Iceland and Greenland (Denmark)

Tularemia has not yet been reported in wild or domestic animals or humans in Iceland or Greenland (Maurin and Gyuranecz 2016).

5 Norway

In contrast to North America, only type B tularemia is reported in Eurasia. Scandinavia has a long-documented history with tularemia, with Norway documenting the lowest prevalence. Tularemia is a notifiable disease in Norway, and all disease forms are reported in humans, but the oropharyngeal form is most common and is thought to be due to drinking wells contaminated with dead rodents (Afset et al. 2015), though cases of pulmonary tularemia appear to be on the rise, so much so that it is now considered as a differential diagnosis to lung cancer (Kravdal et al. 2020).

There is literature in Norway describing a disease called "lemming fever" dating as far back as 1532 (Pearson 1975), but the first official case of human tularemia was documented in 1929 in Telemark (Thjøtta 1931). That case was in a physician who had most likely contracted the disease while skinning a rabbit he had hunted. Between that first case and 1996, an average of one to four human cases were documented per year, with occasional outbreaks occurring (Pearson 1975; Berdal et al. 1996), though tularemia has only been reportable in Norway since 1979, so many of these early cases are likely not documented. Between 2006 and 2010, an average of 34 human cases was reported annually (range 11–66) (Afset et al. 2015), and in 2011, a very large outbreak of 180 human cases was recorded (suggest using Larssen et al. 2014 as a reference here since it is the first description of the outbreak; see also remark on page 13 when referring to the same outbreak is made) (Afset et al. 2015).

Despite increasing numbers of reported human cases of tularemia in Norway, only one human serosurvey has been conducted; it was of school children in 4 different regions of Norway. That survey documented between 0% and 4.7% prevalence, indicating that tularemia is likely widespread (Berdal et al. 1996). All children were seronegative in Trysil, in central Eastern Norway. There have been no reported human deaths from tularemia in Norway.

Tularemia prevalence in Norway mirrors rodent and rabbit populations, and this was indeed the case during a large outbreak of human tularemia in 2011 that coincided with the highest density of lemmings (*Lemmus lemmus*) recorded in at least 40 years (Larssen et al. 2014). The outbreak started in January with a cluster of oropharyngeal tularemia in central Norway and was associated with contaminated drinking wells. From there, it progressed northward, and it ended in December in Northern Norway as a mixture of oropharyngeal, glandular, typhoidal, and pneumonic tularemia (Afset et al. 2015). *F. tularensis* was cultured from 18 out of 180 human patients during that outbreak, DNA sequences identified all 18 as type B tularemia (clades B.4, B.7, and B.12). This outbreak was notable in that 62 cases were reported in county Finnmark, in northern Norway. Prior to this, and since 1977, only five cases of tularemia had been reported in Finnmark (Afset et al. 2015). Perhaps even more interestingly, migratory birds are thought to have potentially disseminated tularemia to Northern Norway in this instance.

There are few confirmed cases of wild or domestic animal tularemia in Norway. Hare and small rodent populations fluctuate, and peaks occur every 3-4 years. When human tularemia is documented in a region, hare and rodent populations seem to decline and are often suspected to be associated with tularemia (Berdal et al. 1996). Following a lemming peak in 1978/79, several people in Bogen became ill with tularemia, and lemming populations declined. Immunofluorescent assay confirmed tularemia in some carcasses (Berdal et al. 1996). Following this, four species of small mammals were trapped (six red-backed voles, *Clethrionomys rutilus*; one ermine, *Mustela nivalis*; three field voles, *Microtus agrestis*; and three grey red-backed voles *Clethrionomys rufocanus*) for a serosurvey in Northern Norway. Two animals (a red-backed vole and an ermine) had detectable antibody titers.

Dogs are considered relatively resistant to tularemia (Foley and Nieto 2010), but there are some data available from Norway. During the large human outbreak in 2011, numerous dog owners brought their dogs to veterinarians because they showed nonspecific signs of disease. One of the dogs, a hunting dog, had hunted and killed a mountain hare (*Lepus timidus*), and the owner of the dog dressed and hung the hare. Days later, both the dog and the owner became sick. The owner sought treatment for himself and his dog. Both the dog and hunter had seroconverted, and the rabbit's bone marrow was positive for *F. tularensis* by

PCR. Following this, a small cohort of hunting dogs ($n = 11$) that had become ill after exposure to lemmings had blood samples collected for tularemia serology. Ten of 11 dogs had agglutinating antibodies to *F. tularensis* (Nordstoga et al. 2014).

Tularemia is well documented in humans in Norway, and its prevalence has increased since 2011. The reasons for this increase are unclear but may have to do with more awareness of the disease by the public or due to changes in insect and rodent vector populations (Larssen et al. 2014).

6 Sweden

The first confirmed cases of tularemia in Sweden were also described in 1931 and were seen in a mountain hare and in three humans who had been in contact with the hare (Granström 1931). It has been a nationally notifiable disease in Sweden since 1968 (Eliasson et al. 2002). Sweden often sees (the) most human tularemia cases in all of Europe and experiences between 0 and 12 local outbreaks each year. Most cases are seen in July and August, and most cases are reported in adult men (Desvars et al. 2015). Between 1931 and 2000, more than 6000 cases of human tularemia were reported in Sweden (an average of ~87 cases per year; Eliasson et al. 2002). Incidence has risen since that time, and between 2000 and 2018, 4422 cases were reported (an average of ~246 cases per year; Dryselius et al. 2019), and in 2019, the country experienced its largest outbreak in more than 50 years (979 cases; Dryselius et al. 2019). The most common disease form reported is ulceroglandular, but a large outbreak of pneumonic tularemia that occurred in 1967–1968 was attributed to inhalation of hay dust that was contaminated with vole feces (Eliasson et al. 2002). There have been no reported fatalities from tularemia in Sweden, likely because only type B exists there.

Prior to 2000, most cases were reported in Central Sweden, which is considered an endemic zone, but since then, more and more cases have occurred in other regions (Desvars et al. 2015). In 2000, an outbreak of 464 cases was reported, and 187 (40%) of those were outside the endemic area. Research concluded that owning a cat, farming, and being bitten by mosquitos were risk factors during this outbreak (Eliasson et al. 2002). In 2013, a small outbreak (6 cases) occurred in southwestern Sweden and was connected to a contaminated drinking well (Lindé et al. 2018). It has been shown that Swedish outbreaks are highly localized and likely the result of point sources of infection (Svensson et al. 2009). This seems to be the case for the most recent outbreak as well, in which all cases occurred in residents of and visitors to Ljusdal, in central Sweden (Dryselius et al. 2019). In central Sweden, where tularemia is endemic, approximately 20% of the human population had antibodies to *F. tularensis*, whereas seroprevalence in non-endemic areas was 1-2% (Christenson 1984).

Most literature implicates mosquito vectors as the most common mode of transmission in Sweden (Christenson 1984; Dryselius et al. 2019), but there is evidence that supports rodents as a source of human infection also (Pearson 1975; Eliasson et al. 2002; Lindé et al. 2018). Dead hares are routinely collected and analyzed for tularemia by the Swedish government, and an average of approximately nine infected hares are identified each year (Dryselius et al. 2019). Infected hares are increasingly being reported outside of the endemic zone, which fits with human outbreaks occurring more frequently outside this zone also. Antibodies to *F. tularensis* have been identified in Swedish predators as well, with reported seroprevalence ranging from 2.9% (1/34) in Eurasian lynx (*Lynx lynx*) to 10% (2/20) in wolverines (*Gulo gulo*; Hestvik et al. 2019).

Sweden may be the sentinel for the effects of climate change and tularemia. Desvers and coworkers (2015) showed that cases of tularemia increased from 0.26 cases/100,000 people/year from 1984 to 1998 to 2.47 cases/100,000 people/year from 1999 to 2012. Incidence data also showed that tularemia occurred over a much wider geographic area during the second half of the study period and included more southern regions. In fact, the rate of increase in cases in the south was 9.6 times higher than in the north. During the study period, the authors also noted a strong correlation between the incidence of tularemia and the proportion of municipality covered by inland water, and they found that tularemia cases were overrepresented in boreal forest and alpine regions of Sweden (ref).

7 Finland

Human tularemia due to *F. tularensis* type B was first confirmed in Finland in 1932 (Koskela and Herva 1982), but was not listed as a nationally notifiable disease until 1995 (Rossow et al. 2014b). Typically, dozens to several hundreds of human cases are reported annually and with marked geographic variation. The highest number of cases are typically reported in Northern and Central Finland (Rossow et al. 2015). Between 1996 and 2004, an average of 37 cases/100,000 people were reported, which is the highest of all EU member states (Rossow et al. 2014a). Between 1995 and 2013, the average rate dropped to 5.1/100,000 people (Rossow et al. 2015). Most human cases reported in Finland are ulceroglandular, followed by the pneumonic type. The strongest risk factor for ulceroglandular tularemia is insect bites, and for pneumonic tularemia it is exposure to dry hay (Rossow et al. 2014b; Väyrynen et al. 2017). Epidemics in Finland are strongly seasonal, with the highest occurrence of cases occurring in the summer and early fall. Outbreaks in humans are documented in Finland every 3 to 5 years (Rossow et al. 2015).

The first serosurvey of humans was done in the early 1980s, when testing 1072 Finish Red Cross blood donors from 12 rural Finnish communities in Northern Finland (Oulu). Sera from 168 (15.7%) of the volunteers were positive for antibodies to *F. tularensis*, though most had low titers. Only seven donors (0.7%) had titers $\geq$80 (Koskela and Herva 1982). A subset of samples (n = 60, including 46 with antibody titers) were tested for cell-mediated immunity using blast transformation, and lymphocytes from only five donors responded (all five were people with antibody titers $\geq$80). None of the seropositive or cell-immunity positive subjects were aware of ever having had tularemia; however, one person reported having been sick with a tularemia-like illness 15 years prior (previously?). That person reported fever and enlarged lymph nodes (one of which was surgically removed). A second serosurvey of 1045 people was conducted in 2015 and reported 1.5% (16/1045) seroprevalence (Rossow et al. 2015).

Voles (*Microtus agrestis* and *M. glareolus*) have been documented with active tularemia infections in Finland, and experimental studies have been conducted and have shown that they are highly susceptible to infection, develop lethal disease, and shed live bacteria in their urine and feces (Rossow et al. 2014a). Rossow et al. (2015) concluded that tularemia peaks in vole populations predicted human outbreaks, with human outbreaks occurring 1 year after a peak in voles.

8 Russia

In the Russian Federation tularemia occurs sporadically, as a group of cases or as epidemics (Popova et al. 2016). In the USSR, tularemia was diagnosed for the first time in 1926, when S. V. Suvorov, A. A. Volferts, and M. M. Voronkova isolated the causative agent of tularemia from infected people during an outbreak on the Volga River Delta. In the 1940s, up to 100.000–140.000 cases of tularemia among people were detected in the USSR. During 1946–1956, mass immunization against tularemia reduced the incidence to 1000–2000 cases per year (Olsufiev and Dunaeva 1971; Cherkassky 1996).

Natural foci of tularemia in the far east were studied and described in the early 60s, when the first cases of human tularemia were detected in 1963 in Khabarovsk Krai and then in 1965 in the Amur River region. Natural foci of this infection are mainly located in river valleys. The greatest number of foci was found in the Amur River region and also in Khabarovsk territory (Demidov et al. 2019).

The epizootological investigations of natural tularemia foci in Yakutia began in the early 60s after an outbreak with over 800 human cases in 1959–1960. In 1944–1995, 2526 patients were registered, and the predominant form of tularemia was ulceroglandular. Natural foci of tularemia above the Arctic Circle, including Chukotka Autonomous District, were detected and characterized in the late 1970s to early 1980s. The first case of ulceroglandular tularemia was registered on Wrangel Island in 1984 and was likely transmitted from a lemming (Petrischeva and Olsufiev 1969).

In Russia, tularemia is not often registered due to the use of live tularemia vaccine and revaccination of the population in enzootic areas and risk groups (Kudryavtseva et al. 2020). In 2019, 42 cases of human tularemia were registered in the Russian Federation, 85% of which occurred in three federal districts: Northwestern, Central, and Siberian (Kudryavtseva et al. 2018).

During the 10-year period from 2009 to 2018, 1944 cases of human tularemia were reported in Russia despite vaccination, 1049 of which were in the Urals Federal District and 177 in the Siberian Federal District. During this period, one outbreak occurred due to a decrease in the vaccination level, affecting 1005 people in the Khanty-Mansi Autonomous Okrug. There is a trend towards tularemia urbanization in Russia. Recent data show that more than 80% of cases now occur in people living in urban areas (Kudryavtseva et al. 2020). This does not mean, however, that people are infected in urban regions; it is possible that people acquire infections while visiting less inhabited areas while recreating. Additionally, there is a trend towards diagnosis in younger human patients, and cases in children (under 17 years of age) diagnosed have increased to 30% (Kudryavtseva et al. 2020).

Tularemia is not a typical disease for high latitudes. However, the natural foci of tularemia in the Arctic and a lot of rodents in the tundra and forest tundra that are the main vectors threaten people's (or public health) health. In Taimyr, two naturally occurring foci of tularemia are well known. One of them is located in the Norilsk basin in the valley of the Norilka River. Vectors are different types of voles and Arctic shrew. The other focus is located in the valley of the Khatangy River near Novoribnoye settlement. Here lemmings are the vectors of infection (Popova et al. 2016).

In Russia, the main species composition of infected animals are voles (red and Grey red-backed vole) and common shrews. Ticks, regurgitates of predatory birds, excrements of predators, as well as water from open reservoirs were also contaminated (Kudryavtseva et al. 2018, 2020). The spread of infection is associated with climatic conditions such as rivers flooding, water stagnation, as well as reduced level of vaccination.

9 Disease Management, Control, and the Future

Tularemia is endemic to the Northern hemisphere, can infect more than 300 animal species, and can survive in water. It is a disease that has been and always will be a part of the northern landscape. However, relative to other zoonotic diseases, it is also relatively rare and does not take many human lives, though it is likely underdiagnosed (Afset et al. 2015). Experiences with both animal and human outbreak investigations are scarce, and little surveillance is done.

Climate change will likely impact tularemia prevalence in humans and animals by altering the distributions of rodent reservoirs and insect vectors and by increasing liquid water (vs. ice) on the landscape (Hueffer et al. 2013). Thus, control should focus on the monitoring and control of insect vectors and rodent reservoirs and on access to safe drinking water (Hennebique et al. 2019). However, there is not a clear enough picture of the current status of tularemia in the Arctic to be able to detect the impact of climate change. Increased surveillance is needed, as tularemia is thought to be underreported worldwide due in part to its often vague clinical signs and symptoms. It is also important that practitioners (physicians, veterinarians, etc.) work to get proper diagnoses when presented with patients with febrile illnesses. Should prevalence rise and tularemia become a more urgent threat, vaccine development may be necessary. A human vaccine was developed by the Soviet Union in the 1940s, but there is currently no licensed vaccine in the United States or European Union (Sunager et al. 2016). There are no animal vaccines for tularemia.

References

Afset JE, Larssen KW, Bergh K, Lärkeryd A, Sjödin A, Johansson A, Forsman M (2015) Phylogeographical pattern of *Francisella tularensis* in a nationwide outbreak of tularaemia in Norway, 2011. Euro Surveill 20(19):21125

Berdal BP, Mehl R, Meidell NK, Lorentzen-Styr AM, Scheel O (1996) Field investigations of tularemia in Norway. FEMS Immunol Med Microbiol 13:191–195

Challacombe JF, Petersen JM, Gallegos-Graves L, Hodge D, Pillai S, Kuske CR (2017) Whole-genome relationships among *Francisella* bacteria of diverse origins define new species and provide specific regions for detection. Appl Environ Microbiol 83:e02589–e02516. https://doi.org/10.1128/AEM.02589-16

Cherkassky BL (1996) Particularly dangerous infections. Reference edition. M. Medicine 160s

Chomel BB, Karsten RW, Chappuis G, Soulier M, Kikuchi Y (1998) Serological survey of selected canine viral pathogens and zoonoses in grizzly bears (*Ursus horribilis*) and black bears (*Ursus americanus*) from Alaska. Rev Sci Tech Off Int Epiz 17:756–766

Christenson B (1984) An outbreak of tularemia in the northern part of Central Sweden. Scand J Infect Dis 16:285–290

Demidov TH, Popov N, Mikhailova T. In Semihina A with Podobedova YS (2019) Epidemiology of tularemia in the far eastern federal district (2000-2017). Epidemiology and Vaccine Prevention

Desvars A, Furberg M, Hjertqvist M, Vidman L, Sjöstedt A, Rydén P, Johansson A (2015) Epidemiology and ecology of tularemia in Sweden, 1984-2012. Emerg Infect Dis 21(1): 32–39. https://doi.org/10.3201/eid2101.14096

Ditchfield J, Meads EB, Julian RJ (1960) Tularemia of muskrats in eastern Ontario. Can J Public Health 54:474–478

Dryselius R, Hjertqvist M, Mäkitalo S, Lindblom A, Lilja T, Eklöf D, Lindström A (2019) Large outbreak of tularaemia, Central Sweden, July to September 2019. Euro Surveill 24(42): 1900603. https://doi.org/10.2807/1560-7917.ES.2019.24.42.1900603

Eliasson H, Lindbäck J, Nuorti JP, Arneborn M, Giesecke J, Tegnell A (2002) The 2000 tularemia outbreak: a case-control study of risk factors in disease-endemic and emergent areas, Sweden. Emerg Infect Dis 8(9):956–960

Farlow J, Wagner DM, Dukerich M (2004) *Francisella tularensis* in the United States. Emerg Infect Dis 11:1835–1841

Foley JE, Nieto NC (2010) Tularemia. Vet Microbiol 140:332–338

Foshay L (1932) Tularemia: accurate and earlier diagnosis by means of the intradermal reaction. J Infect Dis 51(2):286–291

Granström KO (1931) Tre fall av tularemi. Sven Lakartidn 28:641–646

Hansen CM, Vogler AJ, Keim P, Wagner DM, Hueffer K (2010) Tularemia in Alaska, 1938-2010. Acta Vet Scand 53:61. https://doi.org/10.1186/1751-0147-53-61

Hennebique A, Boisset S, Maurin M (2019) Tularemia as a waterborne disease: a review. Emerg Microbes Infect 8(1):1027–1042. https://doi.org/10.1080/22221751.2019.1638734

Hestvik G, Uhlhorn H, Koene M, Åkerström S, Malmsten A, Dahl F, Åhlén P-A, Dalin A-M, Gavier-Widén D (2019) *Francisella tularensis* in Swedish predators and scavengers. Epidemiol Infect 147:e293. https://doi.org/10.1017/S0950268819001808

Hirschmann JV (2018) From squirrels to biological weapons: the early history of tularemia. Am J Med Sci 356(4):319–328. https://doi.org/10.1016/j.amjms.2018.06.006

Hopla CE (1960) Epidemiology of tularemia in Alaska. USAF Arctic Aeromedical Laboratory Technical Report 59(1):1–42

Hopla CE (1965) Ecology and epidemiology research studies in Alaska: a report of field collections and laboratory diagnostic assays. Project 1471 University of Oklahoma Research Institute, Norman, OK

Hopla CE (1968) Ecology and epidemiology research studies in Alaska: a report of field collections and laboratory diagnostic assays. Univ Okla Res Instk Norman OK

Hopla CE, Hopla AK (1994) Tularemia. In: Beran GW (ed) Handbook of zoonoses, 2nd edn. CRC Press, Boca Raton, FL, pp 113–125

Hueffer K, Parkinson AJ, Gerlach R, Berner J (2013) Zoonotic infections in Alaska: disease prevalence, potential impact of climate change and recommended actions for earlier disease detection, research, prevention, and control. Int J Circum Health 72(1):19652. https://doi.org/10.3402/ijch.v.72i0.19562

Keim P, Johansson A, Wagner DM (2007) Molecular epidemiology, evolution, and ecology of *Francisella*. Ann N Y Acad Sci 1105:30–66. https://doi.org/10.1196/annals.1409.011

Koskela P, Herva E (1982) Immunity against *Francisella tularensis* in northern Finland. Scand J Infect Dis 14:195–199

Kravdal A, Stubhaug ØO, Wågo AG, Sætereng MS, Amundsen D, Piekuviene R, Kristiansen A (2020) Pulmonary tularemia: a differential diagnosis to lung cancer. Resp Infect 6:00093. https://doi.org/10.1183/23120541.00093-2019

Kudryavtseva TY, Popov VP, Mokrievich AN, Pakskina ND, Kholin AV, Mazepa AV, Kulikalova ES, Trankvilevsky DV, Khramov MVV, Dyatlov IA (2018) Tularemia: topical issues and forecast of the epidemic situation in the Russian federation in 2018. Prob Especially Danger Infect 1:22–29. https://doi.org/10.21055/0370-1069-2018-1-22-29

Kudryavtseva TY, Popov VP, Mokrievich AN, Pakskina ND, Kholin AV, Mazepa AV, Kulikalova ES, Kosilko SA, Birkovskaya YAA, Trankvilevsky DV, Khramov MV, Dyatlov IA (2020) Epidemic activity of natural foci of tularemia on the territory of the Russian Federation in 2019 and forecast for 2020 year. Prob Especially Danger Infect. https://doi.org/10.21055/0370-1069-2020-1-21-32

Lantis M (1981) Zoonotic diseases in the Canadian and Alaskan north. Inuit Stud 5:83–108

Larssen KW, Bergh K, Heier BT, Vold L, Afset JE (2014) All-time high tularemia incidence in Norway in 2011: report from the national surveillance. Eur J Clin Microbiol Infect Dis 16(13): 10–12

Liles WC, Burger RJ (1993) Tularemia from domestic cats. Western J Med 159:619–622

Lindé EL, Hjertqvist M, Wahab T (2018) Outbreak of tularaemia connected to a contaminated well in the Västra Götaland region in Sweden. Zoo Pub Health 65:142–146. https://doi.org/10.1111/zph.12382

Lopes de Carvalho I, Zé-Zé L, Alves AS, Pardal S, Lopes RI, Mendes L (2012) *Borrelia garinii* and *Francisella tularensis* subsp. *holarctic*a detected in migratory shorebirds in Portugal. Eur J Wildl Res 58(5):857–861

Maurin M (2020) *Francisella tularensis*, Tularemia and Serological Diagnosis. Front Cell Infect Microbiol 10:512090. https://doi.org/10.3389/fcimb.2020.512090

Maurin M, Gyuranecz M (2016) tularaemia: clinical aspects in Europe. Lancet Infect Dis 16:113–124

McCoy GW (1911) A plague-like disease of rodents. Public Health Bull 43:53–71

McCoy GW, Chapin CW (1912) Further observations on a plague-like disease of rodents with a preliminary note on the causative agent, *Bacterium tularense*. J Infect Dis 10:61–72. https://doi.org/10.1093/infdis/10.1.61

Miller LG (1974) Further studies of tularemia in Alaska: virulence and biochemical characteristics of indigenous strains. Can J Microbiol 20:1585–1590

Nano FE, Zhang N, Cowley SC, Klose KE, Cheung KKM, Roberts MJ, Ludu JS, Letendre GW, Meierovics AI, Stephens G, Elkins K (2004) A *Francesella tularensis* pathogenicity island required for intramacrophage growth. J Bacteriol 186(19):6430–6436. https://doi.org/10.1128/JB.186.19.6430-6436.2004

Nordstoga A, Handeland K, Johansen TB, Iversen L, Gavier-Widen D, Mattsson R, Larssen K, Afset JE, Næverdal LA (2014) Tularaemia in Norwegian dogs. Vet Microbiol 173:318–321. https://doi.org/10.1016/j.vetmic.2014.06.031

Olsufiev NG, Dunaeva TN (1971) Natural focality, epidemiology and prevention of tularemia. Medicina, Moscow, p 261

Padeshi PI, Ivanov IN, Popov B, Kantardijev RV (2010) The role of birds in dissemination of *Francisella tularensis*: first direct molecular evidence for bird-to-human transmission. Epidemiol Infect 138(3):376–379

Pearson AD (1975) Epidemiology of rodent tularemia in Norway and Sweden. Ecol Bull 19:99–111

Petrischeva PA, Olsufiev NG (1969) Geography of natural focal human diseases in connection with the tasks of their prevention. Medicina, Moscow, p 312

Philip CB (1939) Tularemia in Alaska. In: Proceedings of the sixth Pacific science congress of the Pacific science association. 24 July–12 Aug, Berkely, CA

Philip CB, Parker RR (1938) Occurrence of tularemia in the rabbit tick (*Haemaphysalis leporis-palustris*) in Alaska. Pub Health Rep 53:574–575

Philip RN, Huntley B, Lackman DB, Comstock GW (1962) Serologic and skin test evidence of tularemia infection among Alaskan Eskimos, Indians, and Aleuts. J Infect Dis 110:220–230

Popova A, Methodiev VV, Stepanova TF, Ezhlova EB, Demina YV, Marchenko AN (2016) Epidemiology and prevention of tularemia in the endemic territories of Russia. Izhevsk OOO Print-2

Rausch RL, Huntley BE, Bridgens JG (1969) Notes on *Pasteurella tularensis* isolated from a vole, *Microtus oeconomous* Pallas, in Alaska. Can J Microbiol 15:47–55

Rossow H, Forbes KM, Tarkka E, Kinnunen PM, Hemmila H, Huitu O, Nikkari S, Henttonen H, Kipar A, Vapalahti O (2014a) Experimental infection of voles with *Francisella tularensis* indicates their amplification role in tularemia outbreaks. PLoS One 9(10):e108864. https://doi.org/10.1371/journal.pone.0108864

Rossow H, Ollgren J, Klemets P, Pietarinen I, Saikku J, Pekkanen E, Nikkari S, Syrjälä H, Kuusi M, Nuorti JP (2014b) Risk factors for pneumonic and ulceroglandular tularemia in Finland: a population-based case-control study. Epidemiol Infect 142:2207–2216. https://doi.org/10.1017/S0950268813002999

Rossow H, Ollgren J, Hytönen J, Rissanen H, Huitu O, Henttonen H, Kuusi M, Vapalahti O (2015) Incidence and seroprevalence of tularemia in Finland, 1995 to 2013: regional epidemics with cyclic patterns. Euro Surveill 20(33):21209

Silverman M, Law B, Carson J (1991) A case of insect borne tularemia above the tree line. Arctic Med Res Suppl:377–379

Sjostedt A (2007) Tularemia: history, epidemiology, pathogen physiology, and clinical manifestations. Ann N Y Acad Sci 1105:1–29

Sunager R, Kumar S, Franz BJ, Gosselin EJ (2016) Tularemia vaccine development: paralysis or progress? Vaccine 6:9–23. https://doi.org/10.2147/VDT.S85545

Svensson K, Bäck E, Eliasson H, Berglund L, Granberg M, Karlsson L, Larsson P, Forsman M, Johansson A (2009) Landscape epidemiology of tularemia outbreaks in Sweden. Emerg Infect Dis 15(12):1937–1947. https://doi.org/10.3201/eid1512.090487

Telford SR, Goethert HK (2020) Ecology of *Francisella tularensis*. Annu Rev Entomol 65:351–372. https://doi.org/10.1146/annurev-ento-011019-025134

Thjøtta T (1931) Tularemia in Norway. J Infect Dis 49(2):99–103

Triebenbach AN, Vogl SJ, Lotspeich-Cole L, Sikes DS, Happ GM, Hueffer K (2010) Detection of *Francisella tularensis* in Alaskan mosquitoes (Diptera: Culicidae) and assessment of a laboratory model for transmission. J Med Entomol 47(4):639–648

Ughetto E, Héry-Arnaud G, Cariou ME, Pelloux I, Maurin M, Caillon J, Moreau JFY, Corvec S (2015) An original case of *Francisella tularensis* subsp. *holarctica* bacteremia after a near-drowning accident. Infect Dis 47(8):588–590. https://doi.org/10.3109/23744235.2015.1028099

Väyrynen SA, Saarela E, Henry J, Lahti S, Harju T, Kauma H (2017) Pneumonic tularemia: experience of 58 cases from 200 to 2012 in northern Finland. Infect Dis 49(10):758–764. https://doi.org/10.1080/23744235.2017.1341054

Wherry WB, Lamb BH (1914) Infection of man with *Bacterium tularense*. J Infect Dis 15:331–340. https://doi.org/10.1093/infdis/15.2.331

Williams R (1946) Tularemia: first case to be reported in Alaska. Pub Health Rep 61:875–876

Wobeser G, Campbell GD, Dallaire A, McBurney S (2009) Tularemia, plague, and Tyzzer's disease in wild rodents and lagomorphs in Canada: a review. Can Vet J 50:1251–1256

Wobesor G, Ngeleka M, Appleyard G, Bryden L, Mulvey MR (2007) Tularemia in deer mice (*Peromyscus maniculatus*) during a population irruption in Saskatchewan, Canada. J Wildl Dis 43:23–31

Zarnke RL, Ballard WB (1987) Serologic survey for selected microbial pathogens of wolves in Alaska, 1975-1982. J Wildl Dis 23:77–85

Zarnke RL, Ver Hoef J, DeLong RA (2004) Serologic survey for selected disease agents in wolves (*Canis lupus*) from Alaska and the Yukon Territory, 1984-2000. J Wildl Dis 40:632–638

Orthohantaviruses in the Arctic: Present and Future

Frauke Ecke, Magnus Magnusson, Barbara A. Han, and Magnus Evander

This chapter focuses on zoonotic orthohantaviruses, their distribution and reservoir hosts (nomenclature following www.itis.gov), epidemiology, clinical manifestations of infection, diagnosis and detection, as well as the predicted future of orthohantaviruses in reservoirs in light of climate change and exploitation of natural resources in the Arctic. We put special emphasis on the ecology and transmission of the widespread and zoonotic Puumala orthohantavirus (PUUV) and its reservoir host the bank vole (*Myodes glareolus*); a model system in disease ecology.

1 Orthohantaviruses

Orthohantaviruses are RNA viruses in the genus *Orthohantavirus* in the order *Bunyavirales*, family *Hantaviridae*. At present, >30 species are known (Adams et al. 2017), and new species are constantly being detected including epizootic and zoonotic ones. Orthohantaviruses are globally distributed except for Antarctica where they are absent. The viral reservoirs are mainly rodents but also include bats

F. Ecke (✉)
Department of Wildlife, Fish, and Environmental Studies, Swedish University of Agricultural Sciences, Umeå, Sweden
e-mail: frauke.ecke@slu.se

M. Magnusson
Department of Wildlife, Fish, and Environmental Studies, Swedish University of Agricultural Sciences, Umeå, Sweden

Swedish Forest Agency, Umeå, Sweden

B. A. Han
Cary Institute of Ecosystem Studies, Millbrook, NY, USA

M. Evander
Department of Clinical Microbiology, Umeå University, Umeå, Sweden

M. Tryland (ed.), *Arctic One Health*, https://doi.org/10.1007/978-3-030-87853-5_18

and insectivores amongst others (Vapalahti et al. 2003; Zhang 2014; Arai and Yanagihara 2020).

2 Epidemiology

2.1 Transmission

In animals, orthohantaviruses are transmitted horizontally, either directly through aggressive interactions and grooming or by inhaling infectious particles shed from urine, feces, or saliva in the environment (Yanagihara et al. 1985; Hardestam et al. 2008). Humans become infected by inhaling aerosols of the virus-contaminated rodent excretions (Tsai 1987). In addition, infection through rodent bites or tissue handling in laboratory settings has also been reported (Tsai 1987; Hart and Bennett 1999; Schultze et al. 2002). Person-to-person transmission has been shown only for Andes orthohantavirus (Martínez et al. 2020), although PUUV has been detected in human saliva (Pettersson et al. 2008b).

2.2 Reservoir Species in the Arctic

Within the current Arctic and extended Arctic region (300 km buffer), 16 rodent species (order Rodentia), five shrew species (order Eulipotyphla), two even-toed ungulate species (order Artiodactyla), one hare species (order Lagomorpha), and one carnivore species (order Carnivora) have been identified as reservoirs for zoonotic orthohantaviruses according to one global human disease database (Global Infectious Diseases and Epidemiology Network, GIDEON) (Berger 2005) (Table 1). However, the reservoir status of reported even-toed ungulate, hare, and carnivore hosts remains inconclusive as exposure of these species to orthohantaviruses so far only has been identified by the detection of antibodies (i.e., serology) (Table 1). The majority of confirmed reservoir species carry Old World orthohantaviruses causing hemorrhagic fever with renal syndrome (HFRS) and are distributed in the Eurasian part of the Arctic. Only three species (*Rattus rattus, Rattus norvegicus*, and *Peromyscus maniculatus*) are known to carry New World orthohantaviruses that may cause the more severe hantavirus pulmonary syndrome (HPS). The highest reservoir richness is found within the Eurasian part of the Arctic (Fig. 1), while human morbidity caused by orthohantaviral infections is likely higher in the North American Artic compared to the Eurasian Arctic due to the severity of HPS in humans caused by Sin Nombre orthohantavirus (SNV), which is transmitted by *P. maniculatus* (Table 1).

Table 1 Mammal species occurring in the Arctic that have been identified in GIDEON (Berger 2005) as reservoirs for zoonotic orthohantaviruses via RNA detection (Reservoir = RNA) or antibody detection (Reservoir = AB). Species given in bold have been found carrying orthohantavirus or antibodies against orthohantavirus in the Arctic (Reference = *) or human infections (HOW hantaviruses – Old World, HPS hantavirus pulmonary syndrome) within the Arctic have been found identifying the species as reservoir (Reference = **). Virus species with unknown zoonotic status are given in italic

Host species	Order	Distribution[a]	Human infection[b]	Virus species[c]	Reservoir	References[d]
Alces alces	Artiodactyla	A	HOW	PUUV	AB	Ahlm et al. (2000)*
Apodemus agrarius	Rodentia	E	HOW	DOBV HTNV PUUV SAAV	RNA RNA RNA RNA	Jakab et al. (2007) Song et al. (2000) Michalski et al. (2014) Korva et al. (2009)
Apodemus flavicollis	Rodentia	E	HOW	DOBV	RNA	Avsic-Zupanc et al. (1992)
Apodemus peninsulae	Rodentia	A	HOW	HTNV SOO[e]	RNA RNA	Jiang et al. (2007) Baek et al. (2006)
Apodemus sylvaticus	Rodentia	A	HOW	DOBV	RNA	Weidmann et al. (2005)
Arvicola amphibius	Rodentia	A	HOW	TULV	RNA	Schlegel et al. (2012)
Blarina brevicauda	Eulipotyphla	E		Camp Ripley virus[e]	RNA	Arai et al. (2007)
Capreolus capreolus	Artiodactyla	A	HOW	PUUV	AB	Danes et al. (1992)
Lemmus sibiricus	Rodentia	A	HOW	KBRV	RNA	Vapalahti et al. (1999)*
Lepus europaeus	Lagomorpha	A	HOW	PUUV	AB	Danes et al. (1992)
Micromys minutus	Rodentia	A	HOW	HTNV	AB	Klein et al. (2012)
Microtus agrestis	Rodentia	A	HOW	*Tatenale virus*[e] TULV	RNA RNA	Thomason et al. (2017) Schmidt-Chanasit et al. (2010)

(continued)

Table 1 (continued)

Host species	Order	Distribution[a]	Human infection[b]	Virus species[c]	Reservoir	References[d]
Microtus arvalis	Rodentia	E	HOW	DOBV TULV	RNA RNA	Garanina et al. (2009) Sibold et al. (1995)
Mus musculus	Rodentia	A	HOW	DOBV SEOV HTNV	RNA RNA RNA	Weidmann et al. (2005); Garanina et al. (2009) Zuo et al. (2008); Kang et al. (2012) Wichmann et al. (2002)
Myodes glareolus	Rodentia	A	HOW	PUUV DOBV	RNA RNA	Bowen et al. (1997); Johansson et al. (2008)*; Khalil et al. (2016b)* Garanina et al. (2009); Kariwa et al. (2009)
Myodes rufocanus	Rodentia	A	HOW	PUUV (HOKV)	RNA	Kariwa et al. (1995)
Neomys fodiens	Eulipotyphla	A	HOW	*BOGV*	RNA	Gu et al. (2013)
Ondatra zibethicus	Rodentia	A	HOW	PUUV	RNA	Vahlenkamp et al. (1998)
Peromyscus maniculatus	Rodentia	A	HPS	SNV	RNA	Nichol et al. (1993); Warner et al. (2020)**
Rattus norvegicus	Rodentia	A	HOW, HPS	HTNV SEOV	RNA RNA	Lu et al. (2017) Kariwa et al. (1996)
Rattus rattus	Rodentia	A	HOW, HPS	THAIV SEOV	RNA RNA	Raharinosy et al. (2018) Reynes et al. (2003)
Sorex araneus	Eulipotyphla	A	HOW	DOBV *SWSV*	RNA RNA	Garanina et al. (2009) Song et al. (2007); Lwande et al. (2020)*

Sorex cinereus	Eulipotyphla	A		JMSV (ARRV)	RNA	Arai et al. (2008)
Sorex monticola	Eulipotyphla	A		JMSV	RNA	Arai et al. (2008); Liphardt et al. (2019)*
Vulpes vulpes	Carnivora	A	HOW	PUUV	AB	Escutenaire et al. (2000)

[a]A Arctic, E extended Arctic (300 km buffer)
[b]Human infection caused by reservoir according to GIDEON (Berger 2005)
[c]Nomenclature of viruses follows (ICTV 2020). Virus species: BOGV (Boginia virus)[e], DOBV (Dobrava-Belgrade orthohantavirus), HOKV (Hokkaido virus)[e], HTNV (Hantaan orthohantavirus), JMSV (Jemez Springs virus)[e], KBRV (Khabarovsk orthohantavirus), PUUV (Puumala orthohantavirus), SAAV (Saaremaa virus)[e], SNV (Sin Nombre orthohantabvirus), SEOV (Seoul orthohantavirus), SOO (Soochong virus)[e], SWSV (Seewis orthohantavirus), THAIV (Thailand orthohantavirus), TULV (Tula orthohantavirus)
[d]Search in Web of Science combining the topic search terms for host species name with "hantavirus" and "PCR". Only first published study and studies detecting the virus in the Arctic are given
[e]Not official orthohantavirus according to (ICTV 2020); also referred to as ARRV (Ash River virus) (Arai et al. 2008)

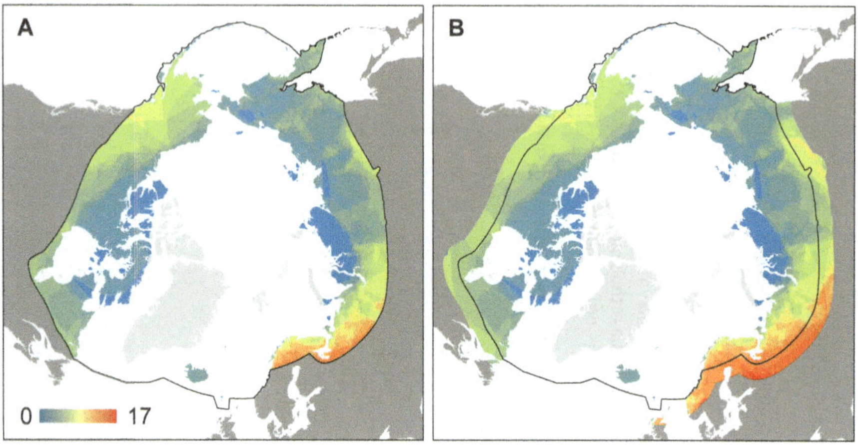

Fig. 1 Distribution of the number of host species for orthohantavirus (orders Rodentia, Artiodactyla, Carnivora, Eulipotyphla, and Lagomorpha) within the Arctic (black line) (**a**) and the Arctic including a 300 km buffer (**b**). Light grey areas represent land areas in the Arctic where reservoir species currently are absent. Distribution areas are obtained from International Union for Conservation of Nature (IUCN) Red List of Threatened Species and Global Biodiversity Information Facility (GBIF). Dark grey areas represent land areas outside the Arctic. The Arctic was delineated as a union between the areas defined by the Arctic Biodiversity Assessment and the Arctic Assessment and Monitoring Programme (AMAP). Species richness was mapped within hexagons of 784 km^2

2.3 Puumala Orthohantavirus (PUUV) and Bank Vole Ecology

In northern Europe, Siberia, and eastern Russia, the bank vole is the only known reservoir host for the zoonotic PUUV causing HFRS in humans (reviewed in Vapalahti et al. 2003). Within this vast area, Fennoscandia stands out as a high endemic region for HFRS. The reservoir host animal, the bank vole, experiences lifelong infection and excretes viral particles throughout its lifespan (Meyer and Schmaljohn 2000). Bank voles are distributed throughout large parts of Europe, and in Sweden and Finland, the species is apparently missing only in the northernmost regions (Mitchell-Jones et al. 1999). Overall, bank voles occur approximately up to the Arctic Circle (ca 66.3 °N; IUCN 2021). There is a spatial mismatch between bank vole and PUUV distribution. For example, PUUV occurrence in Sweden appears restricted to northern Sweden despite the fact that the bank vole is also occurring in southern Sweden (Olsson et al. 2010); see Lõhmus et al. (2016) and Borg et al. (2017) for a potential southwards expansion of the distribution range of PUUV. These spatial mismatches are likely related to the presence of distinct genetic lineages of the host and the virus due to different recolonization routes of bank voles after the last glaciation (Razzauti et al. 2009). Host genetics have been suggested to be important for the diversity of immune-related genes, which could affect the infection tolerance of orthohantavirus reservoirs (Rohfritsch et al. 2018; Vulin et al. 2021).

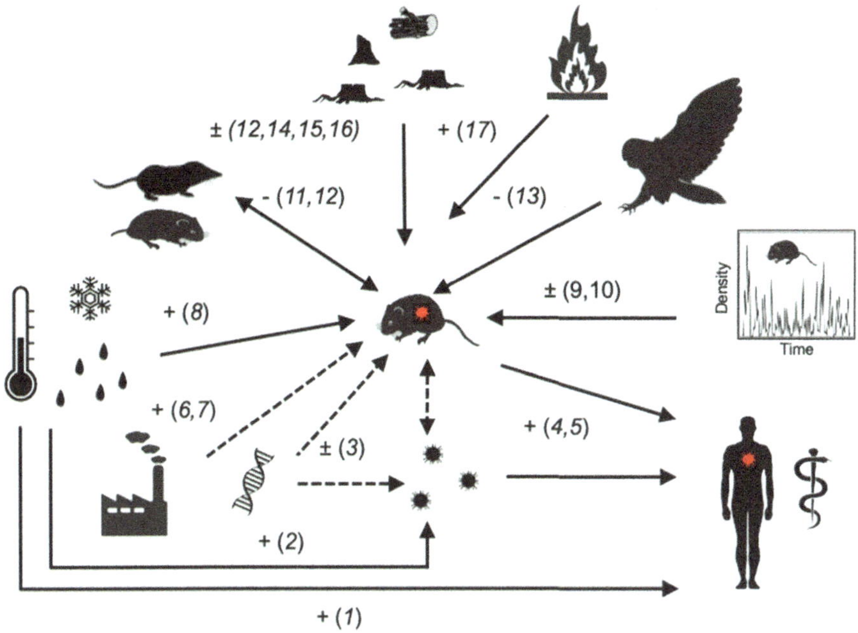

Fig. 2 Schematic illustration of the effect of abiotic and biotic factors on Puumala orthohantavirus (PUUV) transmission in the competent host species, bank vole (*Myodes glareolus*), and transmission of the virus to humans. Drivers of virus transmission include land-use change (e.g., intensive forestry and wildfires), global warming, environmental pollution, the "dilution effect" (high species diversity reduces pathogen transmission; Keesing et al. 2006), predator control and timing, and dynamics of the vole population cycle. Human infection generally occurs via aerosol inhalation of viruses from rodent secretions. For several of the identified relationships, we largely lack a mechanistic understanding (e.g., role of competition, habitat disturbance, and contaminants). So far, no direct link between environmental pollution and PUUV infection in bank voles has been established. However, environmental contaminants impair the health status of the host (Ecke et al. 2018) and hence likely increase the susceptibility of the host to infection. The dashed arrows identify generally understudied factors. The signs ("+" and "−") indicate a positive and negative relationship, respectively. Numbers refer to references: 1: Khalil et al. (2014b), 2: Kallio et al. (2006), 3: Razzauti et al. (2009), 4: Vapalahti et al. (2003), 5: Ermonval et al. (2016), 6: Ecke et al. (2018), 7: Easterbrook and Klein 2008), 8: Sipari et al. (2021), 9: Niklasson et al. (1995), 10: Khalil et al. (2019), 11: Khalil et al. (2016b), 12: Ecke et al. (2017), 13: Khalil et al. (2016a), 14: Voutilainen et al. (2012), 15: Magnusson et al. (2015a), 16: Ecke et al. (2013), 17: Ecke et al. (2019)

PUUV prevalence in bank voles shows seasonal and between-years dynamics with higher overall prevalence in spring (commonly >40%) compared to fall and a positive correlation with bank vole population density (Voutilainen et al. 2012; Khalil et al. 2016b, 2019). Climate and land-use change affect transmission within bank vole populations and animal-human transmission of PUUV (summarized in Fig. 2). This includes increased animal-human transmission following rain-on-snow in early winter, a phenomenon that is expected to increase with climate warming (Khalil et al. 2014b) and during high peaks of the vole population cycle (Niklasson et al. 1995). Land-use changes such as clear-cutting and overall habitat disturbance affect virus transmission in the bank vole population via multiple pathways (Voutilainen et al. 2012; Magnusson et al. 2015a; Ecke et al. 2017, 2019; Khalil et al. 2019) (Fig. 2). In addition, predators such as owls selectively prey on infected animals (Khalil et al. 2016a) and hence potentially alter transmission in bank vole populations (Fig. 2). Stress can increase susceptibility to infection (Padgett and Glaser 2003), and environmental pollution (metals and organic contaminants) can impair host animal health (Ecke et al. 2018). It is therefore likely that environmental pollution can promote susceptibility to orthohantavirus infection in its hosts.

The "dilution effect" (Keesing et al. 2006), a concept where high biodiversity reduces pathogen transmission, has been demonstrated for PUUV in both Swedish and Finnish bank vole populations (Voutilainen et al. 2012; Khalil et al. 2016b; Ecke et al. 2017). In northern Sweden, the composition and structure of the small mammal community have changed since the 1970s including local extinction of the previous common grey-sided vole (*Myodes rufocanus*) following clear-cutting at a regional scale (Ecke et al. 2013; Magnusson et al. 2015b). Consequently, and as a result of decreased biodiversity, the proportion of bank voles within the small mammal community increased and, with it, PUUV transmission (Ecke et al. 2017).

There is uncertainty concerning transmission of PUUV in the extended Arctic outside of the distribution range of the bank vole. Here, two other *Myodes* species might be of relevance: the northern red-backed vole (*M. rutilus*) and the grey-sided vole with the grey-sided vole occurring throughout the Eurasian Arctic and the northern red-backed vole occurring in both the North American and Eurasian Arctic (IUCN 2021).

The northern red-backed vole is closely related to the bank vole. The species can hybridize in an overlapping "hybrid zone" and have done so during different events in overlapping distribution areas since the last glaciation (Tegelström 1987; Melnikova et al. 2012). Close to Alta, northern Norway, one study describes human cases of HFRS outside of the known distribution range of the bank vole (Traavik et al. 1984). These human cases are likely directly linked to the presence of red-backed voles or/and grey-sided voles. The involvement of northern red-backed voles in HFRS is further supported by human cases recently diagnosed in regions of northern Norway outside of the bank vole distribution range (Milhano et al. 2017) even though there is uncertainty of where the patients have contracted the disease (MM personal contact with Tromsö University hospital). Both the northern red-backed vole and the grey-sided vole carry their individual orthohantaviruses (Kariwa et al. 1995; Vapalahti et al. 1999; Zhang et al. 2007; Yashina et al. 2015; Lu et al. 2017; Arai and Yanagihara 2020), but the Puumala strain in grey-sided voles (PUUV (HOKV), Table 1) appears to be non-zoonotic, at least in the northern part of Japan (Miyamoto et al. 2003). The northern red-backed vole is currently not classified as a zoonotic reservoir in GIDEON (Table. 1) but is listed as a reservoir in other classifications (Shaw et al. 2020).

2.4 Other Orthohantaviruses and Ecology of Their Reservoirs

Apart from PUUV, the main orthohantavirus posing major zoonotic risk in the Arctic is SNV with North American deermouse (*Peromyscus maniculatus*) as reservoir host (Nichol et al. 1993). North American deermouse, even though generally reported as a forest-dwelling species, is a habitat generalist like the bank vole and occurs in multiple natural and artificial habitats (Cassola 2016). Like PUUV, SNV transmission occurs horizontally among North American deermice, and high biodiversity can not only reduce but also amplify transmission with a net dilution effect (Luis et al. 2018). In contrast to PUUV in bank voles, prevalence of SNV in North American deermice is generally lower (up to 31.1%; Luis et al. 2018), which is also reflected in lower human incidence of HPS (Warner et al. 2020) compared to HFRS (Khalil et al. 2019).

Much less is known about the disease ecology of the Khabarovsk orthohantavirus (KBRV) in the brown lemming *Lemmus sibiricus* that occurs in the Eurasian Arctic. The brown lemming in contrast to both the bank vole and North American deermouse does not occur in human-modified habitat (Tsytsulina et al. 2007). Among shrews, Seewis orthohantavirus (SWSV) and Jemez Springs virus (JMSV) have been found in the Arctic (Table 1). While the common shrew (*Sorex araneus*), the reservoir of SWSV, can be found in artificial habitats (Hutterer and Kryštufek 2016), the dusky shrew (*Sorex monticola*) only occurs in natural habitats (Matson et al. 2016). For both KBRV, SWSV, and JMSV, the transmission pathways from the reservoir host to humans have still to be identified. Considering that especially the brown lemming and the dusky shrew only have been documented in natural habitats, humans might get infected when visiting, for example, shrubby tundra and wetlands.

2.5 Prevalence and Distribution in Humans

2.5.1 Nearctic Region

The Nearctic region includes Alaska (USA), northern Canada, and Greenland.

Alaska

In Alaska, antibodies against orthohantaviruses have been detected in 2.5% of Forestry Service and Geological Service personnel (Forthal et al. 1987), but without accompanying signs of disease (e.g., renal disease, hemorrhagic phenomenon, or unidentified febrile illnesses). This study did not rule out that the people could have been infected outside Alaska; thus, the detected antibodies cannot confirm that orthohantavirus is present in Alaska.

Canada

Annually, there is an average of four to five human cases in Canada. Until January 1, 2020, a total of 143 cases of HPS caused by SNV had been laboratory-confirmed in Canada, including three cases retrospectively identified since 1993 that were diagnosed in 1989, 1990, and 1992 (Warner et al. 2020). Most cases have occurred in Alberta, but cases have also been reported in British Columbia, Saskatchewan, Manitoba, and Québec (Drebot et al. 2015). The northernmost case of HPS in Canada was identified north of 59°N in British Columbia, <20 km from the border of Yukon Territory (Warner et al. 2020). Most cases occurred in rural settings and approximately 70% of the cases have been associated with domestic and farming activities. The transmission pathways are most probably from wild rodent secretions to humans, but other pathways have also been described such as Seoul orthohantavirus via contact with excreta from imported pet rats (Himsworth et al. 2015; Kerins et al. 2018). SNV-infected mice were detected in every Canadian province except Prince Edward Island and Nova Scotia and the Northwest territories (Drebot et al. 2000, 2001). However, only limited numbers of North American deermice have been tested from the Northwest Territories; therefore, SNV may be circulating in this and other regions of Canada.

Greenland

No orthohantavirus has been detected.

2.5.2 Arctic in the Palearctic Region

The Arctic (largely dominated by tundra) in the Palearctic region ranges from the North Atlantic Ocean to the Pacific Ocean including Iceland, Norway, Sweden, Finland, and the Russian Federation.

Iceland

No orthohantavirus has been detected.

Norway

In Norway, HFRS cases are diagnosed all over the country. The causative virus is PUUV and most likely spreads from bank voles to humans via rodent excreta. There are cases as far north as in Troms and Finnmark County (ca. 70°N), as well as in the region just south of these areas, Nordland fylke, which has the fifth highest incidence of HFRS in Norway (Milhano et al. 2017).

Sweden

In Sweden, the reported incidence rate of human PUUV infection varies from 20 to 313 per 100,000 persons each year, but the real figure is considered to be seven to eight times higher (Ahlm et al. 1994; Pettersson et al. 2008a). The endemic area in northern Sweden largely overlaps with the extended Arctic (Fig. 1b) with overall seroprevalence in 2009 estimated to be 13.4% (Bergstedt Oscarsson et al. 2016) in the two northernmost counties (between 63 °N and 67 °N), a number that has increased from 5% to 9% since 1994 (Ahlm et al. 1994).

Finland

Finland has one of the highest prevalences of HFRS in the world. Cases are widespread across the country, which has an estimated 5% seroprevalence in the overall population, with seroprevalence being highest (11%) in eastern Finland (Vapalahti et al. 2003). Like in Sweden, the causative virus is PUUV and likely spreads from bank voles via rodent excreta. There are cases as far north as Lapland, the northernmost healthcare district located around 67°N where there was an average incidence of 56.2 cases/100,000 between 1995 and 2008 (Makary et al. 2010).

The Russian Federation

HFRS is found mainly in Western Russia, and in this region, it is caused by PUUV and DOBV. However, in northwest Russia, only PUUV is present and mainly in the northernmost regions of Murmansk oblast (68°N), republic of Karelia, and Arkhangelsk oblast (Tkachenko et al. 2019).

3 Clinical Manifestations of Infection with *Orthohantavirus*

3.1 In Animals

Orthohantaviruses cause persistent infections in animal hosts that may remain for months (Voutilainen et al. 2015). The viruses have been found in the saliva, urine, and feces of rodents (Hardestam et al. 2008) and can persist for several weeks in the environment (Kallio et al. 2006), which increases the risk of transmission. The virus spreads by inhalation of secretions or bites (reviewed by Khalil et al. 2014a).

Orthohantaviruses replicate in their animal hosts despite an adaptive immune response with production of antibodies. Thus, orthohantaviruses appear to escape the immune response, establishing a long-term productive infection with no apparent signs of disease (Easterbrook and Klein 2008).

3.2 In Humans

In humans, infection occurs via the inhalation of aerosolized viruses from rodent secretions (Ermonval et al. 2016). Known risk factors for contracting the infection include occupations that may enhance contact with reservoir excreta, especially agricultural and forest work (Bergstedt Oscarsson et al. 2016), construction working (Abu Sin et al. 2007), and wood handling (Ahlm et al. 1998; Olsson et al. 2003; Richardson et al. 2013; Gherasim et al. 2015), and activities in peridomestic areas such as cleaning of sheds or summer cottages. Other risk factors are related to increased susceptibility to infection following contact with reservoir excreta, for instance, smoking (Vapalahti et al. 2010; Bergstedt Oscarsson et al. 2016). At present, over 28 disease-causing orthohantaviruses have been identified around the world, all with rodents as reservoirs, and causing disease ranging from acute renal failure (mainly HFRS) to pulmonary edema and severe hemorrhagic illness (mainly HPS). In many countries, orthohantaviral infections are undetected and not reported due to lack of surveillance, so it is likely that additional orthohantaviruses remain undiscovered (Avšič-Županc et al. 2019). There is geographic variation in orthohantaviral disease observed in humans, with HFRS occurring primarily in Europe and Asia and HPS in the Americas. The clinical presentation of HFRS (ranging from subclinical to severe) varies depending in part on the causative agent of the disease. In general, HFRS caused by Hantaan virus, Amur virus, and Dobrava virus cause more severe disease, with mortality rates ranging from 5% to 15%, whereas Seoul virus causes moderate disease, and PUUV causes mild forms of disease with mortality rates <1% (Avšič-Županc et al. 2019). The orthohantaviruses causing HPS in the Americas have a mortality rate up to 40% but are less common (MacNeil et al. 2011). In both HFRS and HPS patients, the virus infects endothelial cells throughout the body, although injury is most prominent in the lungs and kidneys. The infection is non-lytic, but the altered function of the infected endothelium accounts for increased vascular permeability, hemorrhage, acute thrombocytopenia and pulmonary edemas or kidney failure (Rasmuson et al. 2011; Mackow et al. 2014). There is currently no effective treatment or vaccine approved, and public awareness and preventive measures remain the only ways to minimize the risk of hantavirus disease.

4 Diagnosis and Detection

An infection can be diagnosed by detecting anti-orthohantavirus antibodies or orthohantavirus RNA. Antibodies against one orthohantavirus can cross-react against other viruses depending on how closely related they are. Detection of viral RNA is based on reverse transcription polymerase chain reaction (RT-PCR). The design of primers for the RT-PCR determines the specificity of detection, targeting either a specific orthohantavirus (Evander et al. 2007) or many orthohantaviruses (Klempa et al. 2006; Mohamed et al. 2013). Viral detection by RT-PCR can be followed by a sequencing step, where the sequence of the amplified viral fragment is compared to existing virus sequences. Although not in common use, it is also possible to directly sequence the sample and apply bioinformatics approaches for orthohantavirus detection.

4.1 In Humans

Diagnosis of acute orthohantavirus infections is based on serology, as virtually all patients have immunoglobulin (Ig) M and IgG antibodies present in serum at the onset of symptoms. The presence of IgM antibodies is a sign of acute infection and disappears relatively quickly. IgG antibodies appear slightly later than IgM but will remain for a long time, usually life-long. Thus, the detection of IgG in humans is most commonly used for seroprevalence studies, where the interest lies in assessing the prevalence of disease at the population level rather than identifying acute infections. The most common tests are indirect, using enzyme linked immunosorbent assays (ELISAs). Indirect immunofluorescence assays are also used for diagnostics but have lower specificity (Jonsson et al. 2010). In addition, rapid

5-min user-friendly immunochromatographic IgM-antibody tests have been developed for the detection of acute disease and are available commercially (Hujakka et al. 2001). The infection can also be confirmed by the detection of orthohantavirus genome in blood or serum samples by RT-PCR. Both traditional and quantitative RT-PCR are used to detect viremia (Evander et al. 2007; Jonsson et al. 2010). Although the presence of viremia varies, viral RNA can usually be detected if an acute sample is available. In humans, viral RNA is usually not detectable in blood after 1–2 weeks of disease symptoms (Evander et al. 2007); orthohantavirus RNA has been confirmed even before the presence of specific antibodies (Ferres et al. 2007; Padula et al. 2007; Pettersson et al. 2011).

4.2 In Animals

Orthohantavirus infection in animals can be detected by serology or RT-PCR, in the same way as for humans. In the reservoir animals, the orthohantavirus infection is persistent, which means that virus RNA can be detected together with IgG antibodies, although the virus RNA levels can vary (Madai et al. 2021). In non-reservoir animals, the situation is the same as for humans, where virus RNA disappears after the acute infection while IgG remains (Jonsson et al. 2010). Sampling is an important issue to consider for detection for orthohantavirus in animals. It can take place directly in the field or, after euthanasia, in the laboratory. For antibody analysis, the most common source of antibodies are blood or tissue samples such as lung, heart, or liver, which contain blood vessels, and thus antibodies. For virus RNA detection, many different types of samples can be utilized, such as blood, saliva, lung, and kidney tissue (Hardestam et al. 2008; Madai et al. 2021). It is important to take into account the high sensitivity of the RT-PCR assay and to avoid contamination of samples and false positives.

5 Effects of Climate Change and Exploitation of Natural Resources in the Arctic on Zoonotic Orthohantaviruses and Reservoirs

Global warming has frequently been projected to increase pathogen prevalence and zoonotic risk in general (Lindgren et al. 2012; Pecl et al. 2017), especially in the Arctic (Revich et al. 2012; Bebber et al. 2013). The driving mechanisms include expanding ranges of pathogens, vectors, and/or reservoirs, and altered disease dynamics as well as land use change due to exploitation of natural resources.

In Fennoscandia, field mice (*Apodemus flavicollis* and *A. sylvaticus*) have increased their distribution range into the Arctic during the last 30 years (GBIF.org 2021a, b). Both species are reservoirs for DOBV, and this virus might therefore emerge in the Arctic in the future. Onshore exploitation of natural resources (ore, oil, and gas) is projected to cover vast areas in the Arctic (Turunen 2019). Such exploration implies not only the exposure of land areas that can be populated by reservoirs of orthohantaviruses but also the need for bigger and more harbors. Historically, shipping has been the major cause of introduction and long-distance dispersal of reservoir hosts and their pathogens, identifying harbors as point sources of further dispersal (Rothenburger et al. 2017). Most famous examples for such dispersal are the house rat (*R. rattus*), brown rat (ship rat, *R. norvegicus*), and house mouse (*Mus musculus*) and their pathogens (Rothenburger et al. 2017). All three species are already widely distributed in the coastal Arctic. Any further exploitation of the Arctic is therefore likely to increase their distribution range and their associated orthohantaviruses (Table 1) with the potential risk of introducing New World orthohanatviruses to areas where so far only Old World orthohantaviruses occur and vice versa.

Warmer climate not only increases the distribution range of reservoirs and their orthohantavirus. Rain-on-snow (ROS) events are expected to increase in frequency and lead to cascading ecological effects (Berger et al. 2018). Wetter winters have already increased PUUV transmission in bank voles within its endemic area (Sipari et al. 2021). Suggested mechanisms driving such an increase include aggregation and greater social contact of the reservoir in less abundant suitable microhabitats, impaired reservoir health, and increased environmental persistence of orthohantavirus (Sipari et al. 2021). These factors, either isolated or in combination, support that ROS events will increase the risk of outbreaks of human orthohantavirus diseases in the future.

The expected changes in both the expanding distribution and prevalence of orthohantaviruses in their hosts and associated zoonotic risk for humans together call for increased surveillance and preparedness to mitigate potential outbreaks of orthohantavirus diseases in humans. Recently, an international circumpolar working group of subject matter experts from public health and academic institutions was established to assess the potential emergence and health impact of climate-sensitive infectious diseases in northern human and animal populations and to identify activities that may minimize the risks of disease emergence (see Parkinson et al. 2014). This working group is an important first step towards surveillance of and preparedness for emerging and re-emerging diseases in the Arctic.

References

Abu Sin M, Stark K, van Treeck U, Dieckmann H, Uphoff H, Hautmann W et al (2007) Risk factors for hantavirus infection in Germany, 2005. Emerg Infect Dis 13(9):1364–1366. https://doi.org/10.3201/eid1309.070552

Adams MJ, Lefkowitz EJ, King AMQ, Harrach B, Harrison RL, Knowles NJ et al (2017) Changes to taxonomy and the international code of virus classification and nomenclature ratified by the International Committee on Taxonomy of Viruses (2017). Arch Virol 162(8):2505–2538. https://doi.org/10.1007/s00705-017-3358-5

Ahlm C, Linderholm M, Juto P, Stegmayr B, Settergren B (1994) Prevalence of serum IgG antibodies to Puumala virus (haemorrhagic fever with renal syndrome) in Northern Sweden. Epidemiol Infect 113. https://doi.org/10.1017/s0950268800051542

Ahlm C, Thelin A, Elgh F, Juto P, Stiernström EL, Holmberg S et al (1998) Prevalence of antibodies specific to Puumala virus among farmers in Sweden. Scand J Work Environ Health 2:104–108. https://doi.org/10.5271/sjweh.286

Ahlm C, Wallin K, Lundkvist A, Elgh F, Juto P, Merza M et al (2000) Serologic evidence of Puumala virus infection in wild moose in northern Sweden. Am J Trop Med Hyg 62(1):106–111

Arai S, Yanagihara R (2020) Genetic diversity and geographic distribution of bat-borne Hantaviruses. Curr Issues Mol Biol 39:1–28. https://doi.org/10.21775/cimb.039.001

Arai S, Song J-W, Sumibcay L, Bennett SN, Nerurkar VR, Parmenter C et al (2007) Hantavirus in northern short-tailed shrew, United States. Emerg Infect Dis 13(9):1420. https://doi.org/10.3201/eid1309.070484

Arai S, Bennett SN, Sumibcay L, Cook JA, Song J-W, Hope A et al (2008) Phylogenetically distinct hantaviruses in the masked shrew (*Sorex cinereus*) and dusky shrew (*Sorex monticolus*) in the United States. Am J Trop Med Hyg 78(2):348–351

Avsic-Zupanc T, Xiao S-Y, Stojanovic R, Gligic A, van der Groen G, Leduc JW (1992) Characterization of Dobrava virus: a hantavirus from Slovenia, Yugoslavia. J Med Virol 38(2):132–137. https://doi.org/10.1002/jmv.1890380211

Avšič-Županc T, Saksida A, Korva M (2019) Hantavirus infections. Clin Microbiol Infect 21:e6–e16. https://doi.org/10.1111/1469-0691.12291

Baek LJ, Kariwa H, Lokugamage K, Yoshimatsu K, Arikawa J, Takashima I et al (2006) Soochong virus: an antigenically and genetically distinct hantavirus isolated from *Apodemus peninsulae* in Korea. J Med Virol 78(2):290–297. https://doi.org/10.1002/jmv.20538

Bebber DP, Ramotowski MAT, Gurr SJ (2013) Crop pests and pathogens move polewards in a warming world. Nat Clim Chang 3(11):985–988. https://doi.org/10.1038/nclimate1990

Berger SA (2005) GIDEON: a comprehensive web-based resource for geographic medicine. Int J Health Geogr 4(1):10. https://doi.org/10.1186/1476-072X-4-10

Berger J, Hartway C, Gruzdev A, Johnson M (2018) Climate degradation and extreme icing events constrain life in cold-adapted mammals. Sci Rep 8(1):1156. https://doi.org/10.1038/s41598-018-19416-9

Bergstedt Oscarsson K, Brorstad A, Baudin M, Lindberg A, Forssén A, Evander M et al (2016) Human Puumala hantavirus infection in northern Sweden; increased seroprevalence and association to risk and health factors. BMC Infect Dis 16(1):566. https://doi.org/10.1186/s12879-016-1879-2

Borg O, Wille M, Kjellander P, Bergvall U, Lindgren P, Chirico J et al (2017) Expansion of spatial and host range of Puumala virus in Sweden: an increasing threat for humans? Epidemiol Infect 145:1642–1648. https://doi.org/10.1017/S0950268817000346

Bowen MD, Gelbmann W, Ksiazek TG, Nichol ST, Nowotny N (1997) Puumala virus and two genetic variants of tula virus are present in Austrian rodents. J Med Virol 53(2):174–181. https://doi.org/10.1002/(SICI)1096-9071(199710)53:2<174::AID-JMV11>3.0.CO;2-J

Cassola, F. (2016). *Peromyscus maniculatus*. The IUCN red list of threatened species e. T16672A22360898. https://doi.org/10.2305/IUCN.UK.2016-2.RLTS.T16672A22360898.en. Accessed 6 May 2021

Danes L, Pejcoch M, Bukovjan K, Veleba J, Halacková M (1992) Antibodies against hantaviruses in game and domestic oxen in the Czech Republic. Ceskoslovenska epidemiologie, mikrobiologie, imunologie 41(1):15–18. http://europepmc.org/abstract/MED/1534516

Drebot MA, Artsob H, Werker D (2000) Hantavirus pulmonary syndrome in Canada, 1989-1999. Canada Communicable Dis Rep 26(8):65–69

Drebot MA, Gavrilovskaya I, Mackow ER, Chen ZX, Lindsay R, Sanchez AJ et al (2001) Genetic and serotypic characterization of Sin Nombre-like viruses in Canadian *Peromyscus maniculatus* mice. Virus Res 75(1):75–86. https://doi.org/10.1016/s0168-1702(01)00227-1

Drebot MA, Jones S, Grolla A, Safronetz D, Strong JE, Kobinger G et al (2015) Hantavirus pulmonary syndrome in Canada: an overview of clinical features, diagnostics, epidemiology and prevention. Canada Commun Dis Rep 41(6):124–131

Easterbrook JD, Klein SL (2008) Immunological mechanisms mediating hantavirus persistence in rodent reservoirs. PLoS Pathog 4(11). https://doi.org/10.1371/journal.ppat.1000172

Ecke F, Magnusson M, Hörnfeldt B (2013) Spatiotemporal changes in the landscape structure of forests in northern Sweden. Scand J Forest Res 28(7):651–667. https://doi.org/10.1080/02827581.2013.822090

Ecke F, Angeler DG, Magnusson M, Khalil H, Hörnfeldt B (2017) Dampening of population cycles in voles affects small mammal community structure, decreases diversity, and increases prevalence of a zoonotic disease. Ecol Evol 7(14):5331–5342. https://doi.org/10.1002/ece3.3074

Ecke F, Berglund ÅMM, Rodushkin I, Engström E, Pallavicini N, Sörlin D et al (2018) Seasonal shift of diet in bank voles explains trophic fate of anthropogenic osmium? Sci Total Environ 624:1634–1639. https://doi.org/10.1016/j.scitotenv.2017.10.056

Ecke F, Nematollahi Mahani SA, Evander M, Hörnfeldt B, Khalil H (2019) Wildfire-induced short-term changes in a small mammal community increase prevalence of a zoonotic pathogen? Ecol Evol 9(22):12459–12470. https://doi.org/10.1002/ece3.5688

Ermonval M, Baychelier F, Tordo N (2016) What do we know about how hantaviruses interact with their different hosts? Viruses 8(8):223

Escutenaire S, Pastoret PP, Sjolander KB, Heyman P, Brochier B, Lundkvist A (2000) Evidence of Puumala Hantavirus infection in red foxes (*Vulpes vulpes*) in Belgium. Vet Rec 147 (13):365–366. https://doi.org/10.1136/vr.147.13.365

Evander M, Eriksson I, Pettersson L, Juto P, Ahlm C, Olsson GE et al (2007) Puumala hantavirus viremia diagnosed by real-time reverse transcriptase PCR using samples from patients with hemorrhagic fever and renal syndrome. J Clin Microbiol 45(8):2491–2497. https://doi.org/10.1128/jcm.01902-06

Ferres M, Vial P, Marco C, Yanez L, Godoy P, Castillo C et al (2007) Prospective evaluation of household contacts of persons with hantavirus cardiopulmonary syndrome in Chile. J Infect Dis 195(11):1563–1571. https://doi.org/10.1086/516786

Forthal DN, Bauer SP, McCormick JB (1987) Antibody to hemorrhagic-fever with renal syndrome viruses (hantaviruses) in the United-States. Am J Epidemiol 126(6):1210–1213. https://doi.org/10.1093/oxfordjournals.aje.a114760

Garanina SB, Platonov AE, Zhuravlev VI, Murashkina AN, Yakimenko VV, Korneev AG et al (2009) Genetic diversity and geographic distribution of hantaviruses in Russia. Zoonoses Public Health 56(6–7):297–309. https://doi.org/10.1111/j.1863-2378.2008.01210.x

GBIF.org (2021a) GBIF Occurrence Download. https://doi.org/10.15468/dl.ckwfk4 (*Apodemus sylvaticus*). Accessed 07 May 2021

GBIF.org (2021b) GBIF Occurrence Download https://doi.org/10.15468/dl.d6s96d (*Apodemus flavicollis*). Accessed 07 May 2021

Gherasim A, Hjertqvist M, Lundkvist Å, Kühlmann-Berenzon S, Carlson JV, Stenmark S et al (2015) Risk factors and potential preventive measures for nephropatia epidemica in Sweden 2011–2012: a case–control study. Infect Ecol Epidemiol 5(1):27698

Gu SH, Markowski J, Kang HJ, Hejduk J, Sikorska B, Liberski PP et al (2013) Boginia virus, a newfound hantavirus harbored by the Eurasian water shrew (*Neomys fodiens*) in Poland. Virol J 10:160. https://doi.org/10.1186/1743-422x-10-160

Hardestam J, Karlsson M, Falk KI, Olsson G, Klingström J, Lundkvist A (2008) Puumala hantavirus excretion kinetics in bank voles (*Myodes glareolus*). Emerg Infect Dis 14(8):1209–1215. https://doi.org/10.3201/eid1408.080221

Hart CA, Bennett M (1999) Hantavirus infections: epidemiology and pathogenesis. Microbes Infect 1(14):1229–1237. https://doi.org/10.1016/s1286-4579(99)00238-5

Himsworth CG, Bai Y, Kosoy MY, Wood H, DiBernardo A, Lindsay R et al (2015) An investigation of Bartonella spp., Rickettsia typhi, and Seoul Hantavirus in iats (*Rattus* spp.) from an inner-city neighborhood of Vancouver, Canada: is pathogen presence a reflection of global and local rat population structure? Vector Borne Zoonotic Dis 15(1):21–26. https://doi.org/10.1089/vbz.2014.1657

Hujakka H, Koistinen V, Eerikainen P, Kuronen I, Mononen I, Parviainen M et al (2001) New immunochromatographic rapid test for diagnosis of acute Puumala virus infection. J Clin Microbiol 39(6):2146–2150. https://doi.org/10.1128/jcm.39.6.2146-2150.2001

Hutterer R, Kryštufek B (2016) *Sorex araneus* (errata version published in 2017). The IUCN red list of threatened species e.T29661A115170489. https://doi.org/10.2305/IUCN.UK.2016-3.RLTS.T29661A22315145.en. Accessed 7 May 2021

ICTV (2020) International committee on taxonomy of viruses (ICTV) 2019 master species list (MSL35)

IUCN (2021) International Union for Conservation of Nature's Red List of Threatened Species. https://www.iucnredlist.org/. Accessed 11 May 2021

Jakab F, Horváth G, Ferenczi E, Sebők J, Varecza Z, Szűcs G (2007) Detection of Dobrava hantaviruses in *Apodemus agrarius* mice in the Transdanubian region of Hungary. Virus Res 128(1):149–152. https://doi.org/10.1016/j.virusres.2007.04.015

Jiang J-F, Zhang W-Y, Yao K, Wu X-M, Zuo S-Q, Zhan L et al (2007) A new Hantaan-like virus in rodents (*Apodemus peninsulae*) from northeastern China. Virus Res 130(1):292–295. https://doi.org/10.1016/j.virusres.2007.05.021

Johansson P, Olsson GE, Low H-T, Bucht G, Ahlm C, Juto P et al (2008) Puumala hantavirus genetic variability in an endemic region (northern Sweden). Infect Genet Evol 8:286–296

Jonsson CB, Figueiredo LTM, Vapalahti O (2010) A global perspective on hantavirus ecology, epidemiology, and disease. Clin Microbiol Rev 23(2):412. https://doi.org/10.1128/cmr.00062-09

Kallio E, Klingstrom J, Gustafsson E, Manni T, Vaheri A, Henttonen H et al (2006) Prolonged survival of Puumala hantavirus outside the host: evidence for indirect transmission via the environment. J Gen Virol:87. https://doi.org/10.1099/vir.0.81643-0

Kang Y-J, Zhou D-J, Tian J-H, Yu B, Guo W-P, Wang W et al (2012) Dynamics of hantavirus infections in humans and animals in Wuhan city, Hubei, China. Infect Genet Evol 12(8):1614–1621. https://doi.org/10.1016/j.meegid.2012.07.017

Kariwa H, Yoshizumi S, Arikawa J, Yoshimatsu K, Takahashi K, Takashima I et al (1995) Evidence for the existence of Puumula-related virus among *Clethrionomys rufocanus* in Hokkaido, Japan. Am J Trop Med Hyg 53(3):222–227

Kariwa H, Kimura M, Yoshizumi S, Arikawa J, Yoshimatsu K, Takashima I et al (1996) Modes of Seoul virus infections: persistency in newborn rats and transiency in adult rats. Arch Virol 141(12):2327–2338. https://doi.org/10.1007/bf01718634

Kariwa H, Tkachenko EA, Morozov VG, Seto T, Tanikawa Y, Kolominov SI et al (2009) Epidemiological study of hantavirus infection in the Samara region of European Russia. J Vet Med Sci 71(12):1569–1578. https://doi.org/10.1292/jvms.001569

Keesing F, Holt RD, Ostfeld RS (2006) Effects of species diversity on disease risk. Ecol Lett 9(4):485–498. https://doi.org/10.1111/j.1461-0248.2006.00885.x

Kerins JL, Koske SE, Kazmierczak J, Austin C, Gowdy K, Dibernardo A (2018) Outbreak of Seoul virus among rats and rat owners - United States and Canada, 2017. MMWR Morb Mortal Wkly Rep 67(4):131–134

Khalil H, Hörnfeldt B, Evander M, Magnusson M, Olsson G, Ecke F (2014a) Dynamics and drivers of hantavirus prevalence in rodent populations. Vector Borne Zoonotic Dis 14(8):537–551. https://doi.org/10.1089/vbz.2013.1562

Khalil H, Olsson G, Ecke F, Evander M, Hjertqvist M, Magnusson M et al (2014b) The importance of bank vole density and rainy winters in predicting nephropathia epidemica incidence in northern Sweden. PLoS One 9(11):e111663. https://doi.org/10.1371/journal.pone.0111663

Khalil H, Ecke F, Evander M, Hörnfeldt B (2016a) Selective predation on hantavirus-infected voles by owls and confounding effects from landscape properties. Oecologia 181:597–606

Khalil H, Ecke F, Evander M, Magnusson M, Hörnfeldt B (2016b) Declining ecosystem health and the dilution effect. Sci Rep 6:31314. https://doi.org/10.1038/srep31314

Khalil H, Ecke F, Evander M, Bucht G, Hörnfeldt B (2019) Population dynamics of bank voles predict human Puumala hantavirus risk. EcoHealth 16(3):545–555. https://doi.org/10.1007/s10393-019-01424-4

Klein TA, Kim H-C, Chong S-T, O'Guinn ML, Lee JS, Turell MJ et al (2012) Hantaan virus surveillance in small mammals at Firing Points 10 and 60, Yeoncheon, Gyeonggi Province, Republic of Korea. Vector Borne Zoonotic Dis 12(8):674–682. https://doi.org/10.1089/vbz.2011.0618

Klempa B, Fichet-Calvet E, Lecompte E, Auste B, Aniskin V, Meisel H et al (2006) Hantavirus in African wood mouse, Guinea. Emerg Infect Dis 12(5):838–840. https://doi.org/10.3201/eid1205.051487

Korva M, Duh D, Saksida A, Trilar T, Avšič-Županc T (2009) The hantaviral load in tissues of naturally infected rodents. Microbes Infect 11(3):344–351. https://doi.org/10.1016/j.micinf.2008.12.016

Lindgren E, Andersson Y, Suk JE, Sudre B, Semenza JC (2012) Monitoring EU emerging infectious disease risk due to climate change. Science 336(6080):418–419. https://doi.org/10.1126/science.1215735

Liphardt SW, Kang HJ, Dizney LJ, Ruedas LA, Cook JA, Yanagihara R (2019) Complex history of codiversification and host switching of a newfound soricid-borne orthohantavirus in North America. Viruses Basel 11(7). https://doi.org/10.3390/v11070637

Lõhmus M, Verner-Carlsson J, Borg O, Albihn A, Lundkvist Å (2016) Hantavirus in new geographic regions. Sweden. Infect Ecol Epidemiol 6. https://doi.org/10.3402/iee.v6.31465

Lu T, Fu Y, Hou Y, Yang Y, Liu L, Liang H et al (2017) Hantavirus RNA prevalence in myomorph rodents on Bolshoy Ussuriysky Island at the Sino-Russian border. Vector Borne Zoonotic Di 17 (8):588–595. https://doi.org/10.1089/vbz.2016.1953

Luis AD, Kuenzi AJ, Mills JN (2018) Species diversity concurrently dilutes and amplifies transmission in a zoonotic host–pathogen system through competing mechanisms. Proc Natl Acad Sci 115(31):7979–7984. http://www.pnas.org/content/early/2018/07/10/1807106115.abstract

Lwande OW, Mohamed N, Bucht G, Ahlm C, Olsson G, Evander M (2020) Seewis hantavirus in common shrew (*Sorex araneus*) in Sweden. Virol J 17(1):198. https://doi.org/10.1186/s12985-020-01477-w

Mackow ER, Dalrymple NA, Cimica V, Matthys V, Gorbunova E, Gavrilovskaya I (2014) Hantavirus interferon regulation and virulence determinants. Virus Res 187:65–71. https://doi.org/10.1016/j.virusres.2013.12.041

MacNeil A, Nichol ST, Spiropoulou CF (2011) Hantavirus pulmonary syndrome. Virus Res 162 (1–2):138–147. https://doi.org/10.1016/j.virusres.2011.09.017

Madai M, Horváth G, Herczeg R, Somogyi B, Zana B, Földes F et al (2021) Effectiveness regarding hantavirus detection in rodent tissue samples and urine. Viruses 13(4):570. https://doi.org/10.3390/v13040570

Magnusson M, Ecke F, Khalil H, Olsson G, Evander M, Niklasson B et al (2015a) Spatial and temporal variation of hantavirus bank vole infection in managed forest landscapes. Ecosphere 6 (9):163. https://doi.org/10.1890/es15-00039.1

Magnusson M, Hörnfeldt B, Ecke F (2015b) Evidence for different drivers behind long-term decline and depression of density in cyclic voles. Popul Ecol 57(4):569–580. https://doi.org/10.1007/s10144-015-0512-3

Makary P, Kanerva M, Ollgren J, Virtanen MJ, Vapalahti O, Lyytikainen O (2010) Disease burden of Puumala virus infections, 1995-2008. Epidemiol Infect 138(10):1484–1492. https://doi.org/10.1017/s0950268810000087

Martínez VP, Di Paola N, Alonso DO, Pérez-Sautu U, Bellomo CM, Iglesias AA et al (2020) "Super-spreaders" and person-to-person transmission of Andes virus in Argentina. N Engl J Med 383(23):2230–2241

Matson J, Woodman N, Castro-Arellano I, de Grammont PC (2016) *Sorex monticola* (errata version published in 2021). The IUCN red list of threatened species e.T41405A194054902. https://doi.org/10.2305/IUCN.UK.2016-3.RLTS.T41405A194054902.en. Accessed 7 May 2021

Melnikova E, Kshnyasev I, Bodrov SY, Mukhacheva S, Davydova YA, Abramson N (2012) Sympatric area of *Myodes glareolus* and *M. rutilus* (Rodentia, Cricetidae): historic and recent hybridization. Proc Zool Inst RAS 2012:307–323

Meyer BJ, Schmaljohn CS (2000) Persistent hantavirus infections: characteristics and mechanisms. Trends Microbiol 8(2):61–67. https://doi.org/10.1016/s0966-842x(99)01658-3

Michalski A, Niemcewicz M, Bielawska-Drózd A, Nowakowska A, Gaweł J, Pitucha G et al (2014) Surveillance of hantaviruses in Poland: a study of animal reservoirs and human hantavirus disease in Subcarpathia. Vector Borne Zoonotic Dis 14(7):514–522. https://doi.org/10.1089/vbz.2013.1468

Milhano N, Korslund L, Evander M, Ahlm C, Vainio K, Dudman SG et al (2017) Circulation and diagnostics of Puumala virus in Norway: nephropatia epidemica incidence and rodent population dynamics. APMIS 125(8):732–742. https://doi.org/10.1111/apm.12712

Mitchell-Jones AJ, Amori G, Bogdanowicz W, Kryštufek B, Reijnders PJH, Spitzenberger F et al (1999) The atlas of European mammals. Poyser Natural History, London

Miyamoto H, Kariwa H, Araki K, Lokugamage K, Hayasaka D, Cui B et al (2003) Serological analysis of hemorrhagic fever with renal syndrome (HFRS) patients in far eastern Russia and identification of the causative hantavirus genotype. Arch Virol 148(8):1543–1556

Mohamed N, Nilsson E, Johansson P, Klingstrom J, Evander M, Ahlm C et al (2013) Development and evaluation of a broad reacting SYBR-green based quantitative real-time PCR for the detection of different hantaviruses. J Clin Virol 56(4):280–285. https://doi.org/10.1016/j.jcv.2012.12.001

Nichol S, Spiropoulou C, Morzunov S, Rollin P, Ksiazek T, Feldmann H et al (1993) Genetic identification of a hantavirus associated with an outbreak of acute respiratory illness. Science 262(5135):914–917. https://doi.org/10.1126/science.8235615

Niklasson B, Hörnfeldt B, Lundkvist Å, Björsten S, LeDuc J (1995) Temporal dynamics of Puumala virus antibody prevalence in voles and of nephropathia epidemica incidence in humans. Am J Trop Med Hyg 53:134–140

Olsson G, Dalerum F, Hörnfeldt B, Elgh F, Palo T, Juto P et al (2003) Human hantavirus infections, Sweden. Emerg Infect Dis 9. https://doi.org/10.3201/eid0911.030275

Olsson G, Leirs H, Henttonen H (2010) Hantaviruses and their hosts in Europe: reservoirs here and there, but not everywhere? Vector Borne Zoonotic Dis 10:549–561. https://doi.org/10.1089/vbz.2009.0138

Padgett DA, Glaser R (2003) How stress influences the immune response. Trends Immunol 24 (8):444–448. https://doi.org/10.1016/S1471-4906(03)00173-X

Padula P, Martinez VP, Bellomo C, Maidana S, San Juan J, Tagliaferri P et al (2007) Pathogenic hantaviruses, northeastern Argentina and eastern Paraguay. Emerg Infect Dis 13(8):1211–1214. https://doi.org/10.3201/eid1308.061090

Parkinson AJ, Evengard B, Semenza JC, Ogden N, Børresen ML, Berner J et al (2014) Climate change and infectious diseases in the Arctic: establishment of a circumpolar working group. Int J Circumpolar Health 73:25163. https://doi.org/10.3402/ijch.v73.25163

Pecl GT, Araújo MB, Bell JD, Blanchard J, Bonebrake TC, Chen I-C et al (2017) Biodiversity redistribution under climate change: impacts on ecosystems and human well-being. Science 355 (6332). https://doi.org/10.1126/science.aai9214

Pettersson L, Boman J, Juto P, Evander M, Ahlm C (2008a) Outbreak of Puumala virus infection, Sweden. Emerg Infect Dis 14(5):808–810

Pettersson L, Klingstrom J, Hardestam J, Lundkvist A, Ahlm C, Evander M (2008b) Hantavirus RNA in saliva from patients with hemorrhagic fever with renal syndrome. Emerg Infect Dis 14 (3):406–411. https://doi.org/10.3201/eid1403.071242

Pettersson L, Rasmuson J, Andersson C, Ahlm C, Evander M (2011) Hantavirus-specific IgA in saliva and viral antigen in the parotid gland in patients with hemorrhagic fever with renal syndrome. J Med Virol 83(5):864–870. https://doi.org/10.1002/jmv.22040

Raharinosy V, Olive MM, Andriamiarimanana FM, Andriamandimby SF, Ravalohery JP, Andriamamonjy S et al (2018) Geographical distribution and relative risk of Anjozorobe virus

(Thailand orthohantavirus) infection in black rats (*Rattus rattus*) in Madagascar. Virol J 15. https://doi.org/10.1186/s12985-018-0992-9

Rasmuson J, Andersson C, Norrman E, Haney M, Evander M, Ahlm C (2011) Time to revise the paradigm of hantavirus syndromes? Hantavirus pulmonary syndrome caused by European hantavirus. Eur J Clin Microbiol Infect Dis 30(5):685–690

Razzauti M, Plyusnina A, Sironen T, Henttonen H, Plyusnin A (2009) Analysis of Puumala hantavirus in a bank vole population in northern Finland: evidence for co-circulation of two genetic lineages and frequent reassortment between strains. J Gen Virol 90(8):1923–1931. https://doi.org/10.1099/vir.0.011304-0

Revich B, Tokarevich N, Parkinson AJ (2012) Climate change and zoonotic infections in the Russian Arctic. Int J Circumpolar Health 71(1):18792

Reynes J-M, Soares J-L, Hüe T, Bouloy M, Sun S, Kruy SL et al (2003) Evidence of the presence of Seoul virus in Cambodia. Microbes Infect 5(9):769–773. https://doi.org/10.1016/S1286-4579(03)00149-7

Richardson KS, Kuenzi A, Douglass RJ, Hart J, Carver S (2013) Human exposure to particulate matter potentially contaminated with Sin Nombre virus. EcoHealth 10(2):159–165. https://doi.org/10.1007/s10393-013-0830-x

Rohfritsch A, Galan M, Gautier M, Gharbi K, Olsson G, Gschloessl B et al (2018) Preliminary insights into the genetics of bank vole tolerance to Puumala hantavirus in Sweden. Ecol Evol 8(22):11273–11292. https://doi.org/10.1002/ece3.4603

Rothenburger JL, Himsworth CH, Nemeth NM, Pearl DL, Jardine CM (2017) Environmental factors and zoonotic pathogen ecology in urban exploiter species. EcoHealth 14(3):630–641

Schlegel M, Kindler E, Essbauer SS, Wolf R, Thiel J, Groschup MH et al (2012) Tula virus infections in the Eurasian water vole in Central Europe. Vector Borne Zoonotic Dis 12(6):503–513. https://doi.org/10.1089/vbz.2011.0784

Schmidt-Chanasit J, Essbauer S, Petraityte R, Yoshimatsu K, Tackmann K, Conraths FJ et al (2010) Extensive host sharing of central European Tula virus. J Virol 84(1):459–474. https://doi.org/10.1128/jvi.01226-09

Schultze D, Lundkvist A, Blauenstein U, Heyman P (2002) Tula virus infection associated with fever and exanthema after a wild rodent bite. Eur J Clin Microbiol Infect Dis 21(4):304–306. https://doi.org/10.1007/s10096-002-0705-5

Shaw LP, Wang AD, Dylus D, Meier M, Pogacnik G, Dessimoz C et al (2020) The phylogenetic range of bacterial and viral pathogens of vertebrates. Mol Ecol 29(17):3361–3379. https://doi.org/10.1111/mec.15463

Sibold C, Sparr S, Schulz A, Labuda M, Kozuch O, Lysý J et al (1995) Genetic characterization of a new hantavirus detected in Microtus arvalis from Slovakia. Virus Genes 10(3):277–281. https://doi.org/10.1007/BF01701817

Sipari S, Khalil H, Magnusson M, Evander M, Hörnfeldt B, Ecke F (2021) Climate change accelerates winter transmission of a zoonotic pathogen. Ambio. https://doi.org/10.1007/s13280-021-01594-y

Song JW, Baek LJ, Kim SH, Kho EY, Kim JH, Yanagihara R et al (2000) Genetic diversity of *Apodemus agrarius*-borne Hantaan virus in Korea. Virus Genes 21(3):227–232. https://doi.org/10.1023/a:1008199800011

Song JW, Gu SH, Bennett SN, Arai S, Puorger M, Hilbe M et al (2007) Seewis virus, a genetically distinct hantavirus in the Eurasian common shrew (*Sorex araneus*). Virol J 4:114. https://doi.org/10.1186/1743-422x-4-114

Tegelström H (1987) Transfer of mitochondrial DNA from the northern red-backed vole (*Clethrionomys rutilus*) to the bank vole (*C. glareolus*). J Mol Evol 24(3):218–227

Thomason A, Begon M, Bradley J, Paterson S, Jackson J (2017) Endemic hantavirus in field voles, northern England. Emerg Infect Dis 23(6):1033. https://doi.org/10.3201/eid2306.161607

Tkachenko EA, Ishmukhametov AA, Dzagurova TK, Bernshtein AD, Morozov VG, Siniugina AA et al (2019) Hemorrhagic fever with renal syndrome, Russia. Emerg Infect Dis 25(12):2325–2328. https://doi.org/10.3201/eid2512.181649

Traavik T, Sommer AI, Mehl R, Berdal BP, Stavem K, Hunderi OH et al (1984) Nephropathia epidemica in Norway - antigen and antibodies in rodent reservoirs and antibodies in selected human-populations. J Hyg 93(1):139–146. https://doi.org/10.1017/s0022172400061027

Tsai TF (1987) Hemorrhagic-fever with renal syndrome - clinical aspects. Lab Anim Sci 37 (4):419–427

Tsytsulina K, Formozov N, Sheftel B (2007) *Lemmus sibiricus*. The IUCN red list of threatened species e.T11482A3283383. Accessed 7 May 2021

Turunen E (2019) Resources in the Arctic 2019. https://nordregio.org/maps/resources-in-the-arctic-2019/

Vahlenkamp M, Müller T, Tackmann K, Löschner U, Schmitz H, Schreiber M (1998) The muskrat (*Ondatra zibethicus*) as a new reservoir for puumala-like hantavirus strains in Europe. Virus Res 57(2):139–150. https://doi.org/10.1016/S0168-1702(98)00084-7

Vapalahti O, Lundkvist Å, Fedorov V, Conroy CJ, Hirvonen S, Plyusnina A et al (1999) Isolation and characterization of a hantavirus from Lemmus sibiricus: evidence for host switch during hantavirus evolution. J Virol 73(7):5586–5592

Vapalahti O, Mustonen J, Lundkvist A, Henttonen H, Plyusnin A, Vaheri A (2003) Hantavirus infections in Europe. Lancet Infect Dis 3(12):753–754. https://doi.org/10.1016/s1473-3099(03)00828-4

Vapalahti K, Virtala AM, Vaheri A, Vapalahti O (2010) Case-control study on Puumala virus infection: smoking is a risk factor. Epidemiol Infect 138(4):576–584. https://doi.org/10.1017/s095026880999077x

Voutilainen L, Savola S, Kallio ER, Laakkonen J, Vaheri A, Vapalahti O et al (2012) Environmental change and disease dynamics: effects of intensive forest management on Puumala hantavirus infection in boreal bank vole populations. PLoS One 7(6):e39452. https://doi.org/10.1371/journal.pone.0039452

Voutilainen L, Sironen T, Tonteri E, Tuiskunen Bäck A, Razzauti M, Karlsson M et al (2015) Lifelong shedding of Puumala hantavirus in wild bank voles (*Myodes glareolus*). J Gen Virol 96:1238–1247. https://doi.org/10.1099/vir.0.000076

Vulin J, Murri S, Madrieres S, Galan M, Tatard C, Piry S et al (2021) Isolation and genetic characterization of Puumala orthohantavirus strains from France. Pathogens 10(3). https://doi.org/10.3390/pathogens10030349

Warner BM, Dowhanik S, Audet J, Grolla A, Dick D, Strong JE et al (2020) Hantavirus cardiopulmonary syndrome in Canada. Emerg Infect Dis 26(12):3020–3024. https://doi.org/10.3201/eid2612.202808

Weidmann M, Schmidt P, Vackova A, Krivanec K, Munclinger P, Hufert FT (2005) Identification of genetic evidence for Dobrava virus spillover in rodents by nested reverse transcription (RT)-PCR and TaqMan RT-PCR. J Clin Microbiol 43(2):808–812. https://doi.org/10.1128/jcm.43.2.808-812.2005

Wichmann D, Grone HJ, Frese M, Pavlovic J, Anheier B, Haller O et al (2002) Hantaan virus infection causes an acute neurological disease that is fatal in adult laboratory mice. J Virol 76 (17):8890–8899. https://doi.org/10.1128/jvi.76.17.8890-8899.2002

Yanagihara R, Amyx HL, Gajdusek DC (1985) Experimental-infection with Puumala virus, the etiologic agent of nephropathia epidemica, in bank voles (*Clethrionomys glareolus*). J Virol 55 (1):34–38. https://doi.org/10.1128/jvi.55.1.34-38.1985

Yashina LN, Abramov SA, Dupal TA, Danchinova GA, Malyshev BS, Hay J et al (2015) Hokkaido genotype of Puumala virus in the grey red-backed vole (*Myodes rufocanus*) and northern red-backed vole (*Myodes rutilus*) in Siberia. Infect Genet Evol 33:304–313. https://doi.org/10.1016/j.meegid.2015.05.021

Zhang Y-Z (2014) Discovery of hantaviruses in bats and insectivores and the evolution of the genus hantavirus. Virus Res 187:15–21. https://doi.org/10.1016/j.virusres.2013.12.035

Zhang YZ, Zou Y, Yan YZ, Hu GW, Yao LS, Du ZS et al (2007) Detection of phylogenetically distinct Puumala-like viruses from red-grey vole *Clethrionomys rufocanus* in China. J Med Virol 79(8):1208–1218. https://doi.org/10.1002/jmv.20871

Zuo SQ, Zhang PH, Jiang JF, Zhan L, Wu XM, Zhao WJ et al (2008) Seoul virus in patients and rodents from Beijing, China. Am J Trop Med Hyg 78(5):833–837. https://doi.org/10.4269/ajtmh.2008.78.833

Zoonotic Marine Helminths: Anisakid Nematodes and Diphyllobothriid Cestodes

Heather Fenton

1 Zoonotic Marine Nematodes: *Anisakis* Spp., *Pseudoterranova* Spp., and *Contracaecum* Spp.

1.1 Morphological and Genetic Information

The genus *Anisakis* (Dujardin 1845) refers to a group of nematodes with shared morphological characteristics. Recent molecular technological advances recognize species belonging to this genus: *A. pegreffii*, *A. simplex* s.s., *A. simplex* C, *A. typica*, *A. ziphidarum*, *A. physeteris*, *A. brevispiculata*, and *A. paggiae* with two distinct clades molecularly and based on morphological features (Matiiucci and Nascetti 2006). Clade I includes *A. simplex* sensu stricto, *A. pegreffii, A. simplex C, A. typica*, and *A. ziphidarum*, and Clade II includes *A. physeteris, A. brevispiculata*, and *A. paggiae*. The zoonotic condition "anisakiasis" is herein defined as human disease associated with nematode larvae and has been associated with a number of marine nematodes in addition to *Anisakis* spp., e.g., *Pseudoterranova* spp. and *Contracaecum* spp. (Myers 1976; Mehrdana et al. 2014). In tissue sections, ascarids such as *Anisakis* spp. and *Pseudoterranova* spp. are large nematodes with adults found within the stomach of definitive hosts and larvae present in any tissue section. Lateral alae and a thick cuticle are characteristic features, although not always present in larval ascarids, along with coelomyarian musculature that projects far into the body cavity (pseudocoelom), prominent lateral chords, and an esophagus lined by a uninucleate cuboidal to columnar layer of cells with a brush border (Gardiner and Poynton 1999). Species determination often requires microscopic and/or molecular testing.

H. Fenton (✉)
Ross University School of Veterinary Medicine, Basseterre, St. Kitts, West Indies
e-mail: HFenton@rossvet.edu.kn

M. Tryland (ed.), *Arctic One Health*, https://doi.org/10.1007/978-3-030-87853-5_19

1.2 Life Cycle

The typical location of adult *Anisakis* spp. and *Anisakis*-like parasites is within the stomach of marine mammals (Fig. 1) from ingestion of larval stages within paratenic or intermediate invertebrate (e.g., squid and crustaceans) and vertebrate (e.g., fish) hosts (Smith and Wooten 1978). Definitive hosts of *Anisakis* spp. and *Pseudoterranova* spp. are typically marine mammals (e.g., cetaceans and pinnipeds), whereas birds are more commonly hosts to *Contracaecum* spp., where the adult nematodes develop within the digestive tract and adult female nematodes shed eggs in the feces of the definitive hosts. The eggs embryonate within the aquatic environment and the second-stage larvae are consumed by crustaceans (e.g., krill) that act as

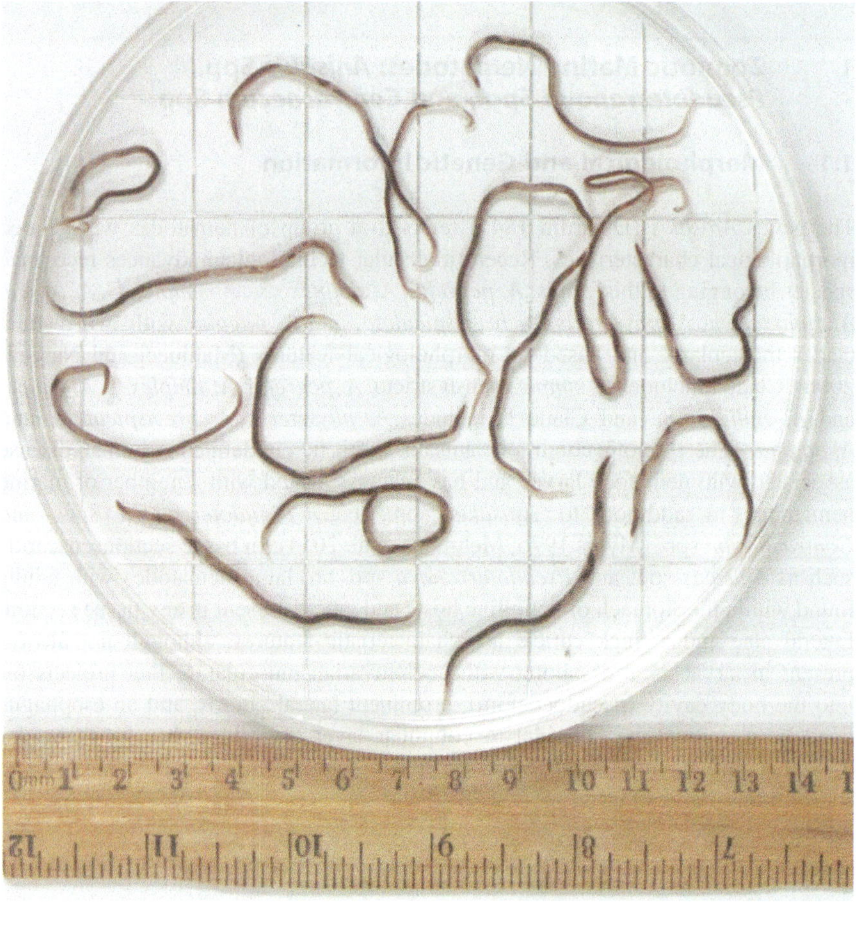

Fig. 1 Examples of adult *Anisakis* sp. nematodes recovered from the stomach of a definitive host, the harbor porpoise (*Phocoena phocoena*), demonstrating the size and gross morphology. The nematodes were not associated with significant pathology in the porpoise. (photo courtesy of Dr. Gary Conboy, Atlantic Veterinary College)

the first intermediate hosts. The crustaceans are then consumed by the second intermediate hosts (i.e., fish or squid) where the third larval stage of the parasite encysts to be consumed by the definitive host to complete the cycle (Sakanari and McKerrow 1989). Humans can act as dead-end hosts for the parasite from ingestion of the infective larvae in the intermediate hosts, but the nematodes do not mature within the intestines of humans (Arriaza et al. 2010).

1.3 Pathology in Wild Animals

Severe ulceration and chronic inflammation of the gastrointestinal tract has been reported in many free-ranging species infected with ascarid nematodes with occasional gastric perforation, peritonitis, and death associated with infection, presumably in an aberrant or nontypical host (Fig. 2) (Nemeth et al. 2012; Wagner et al. 2012). Changes in species distributions related to climate change and other factors could result in an expansion of the host range of these parasites (Shamsi et al. 2017). Severe infection of *A. simplex* larvae in Atlantic salmon can be associated with hemorrhage and inflammation around the vent that has been termed "red vent syndrome," with anglers first noticing and reporting the condition (Larrat et al. 2013; Noguera et al. 2009). The presence of the parasite within the muscles of

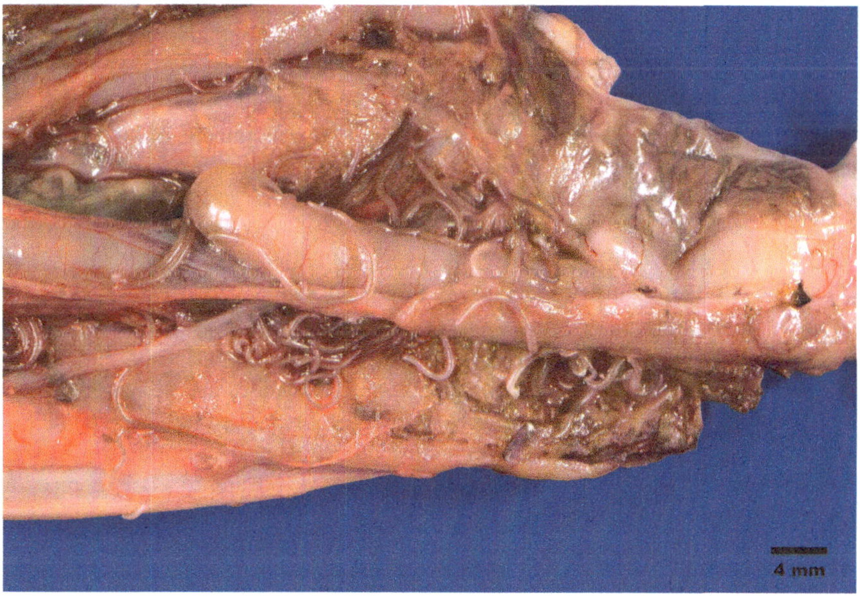

Fig. 2 Gross image of the gastrointestinal tract from a greater shearwater (*Puffinus gravis*) found dead off the coast of Georgia, U.S.A., that suffered fatal peritonitis due to an *Anisakis* sp. infection. Multiple adult nematodes are present in the image associated with proventricular rupture and coelomitis (inflammation within the abdominal cavity). (photo courtesy of Dr. Kevin Keel and the Southeastern Cooperative Wildlife Disease Study)

European smelt (*Osmerus eperlanus*) and eel (*Anguilla anguilla*) has been associated with increased mortality, presumably due to impairment of swimming ability caused by the nematodes (Sprengel and Lüchtenberg 1991).

1.4 Human Health Impacts

Humans develop anisakiasis from the consumption of raw, undercooked, smoked, or dried second intermediate hosts. Symptoms in humans can range from minimal to severe gastrointestinal disease that includes abdominal pain, nausea, vomiting, abdominal distention, diarrhea, and fever with occasional migration of larvae through other body systems such as the respiratory system with coughing, pharyngeal pain, and even the presence of larvae within sputum (Noh et al. 2003). Nasal and anal itching (pruritus) have also been reported. Severe infections can result in obstructions of the alimentary tract that have to be corrected surgically (Yoon et al. 2004). Gastrointestinal disease is also reported with *Pseudoterranova* sp. and *Contracaecum* sp. infections in humans; although these genera have not been as well studied as *Anisakis* sp., the clinical disease in humans is considered to be similar to that observed with *Anisakis* sp. infection, and all nematodes are lumped into the generic syndrome of "anisakiasis" with speciation likely requiring molecular analysis of the parasites (Weitzel et al. 2015). The *Anisakis* sp. nematodes can induce allergic reactions associated with exposure, and infected human patients are known to have elevations in serum IgE and eosinophilia, as well as potentially suffering from skin rashes, urticaria (hives), airway obstruction, and anaphylaxis (Perteguer et al. 2000; Nieuwenhuizen et al. 2006; Ludovisi et al. 2017; Mehrdana and Buchmann 2017). The pathogenesis of clinical disease in humans of both gastrointestinal and allergic reactions is thought to be associated with specific excretory and secretory proteins produced by the parasite (Mehrdana and Buchmann 2017). *Anisakis* spp. can be inactivated with prolonged freezing periods (i.e., below −20 °C for 7 days or below −35 °C for 15 h) and with cooking (56 °C for 5 min) (FDA 1998; Wharton and Aalders 2002). Prolonged salting can kill anisakid larvae, but some marinating procedures may not be sufficient for inactivation (Karl et al. 1994). Specific information on the effect of drying and smoking techniques on the survival of these parasites is not available. Larval nematodes can be removed from fillets, but sometimes, the fillets need to be cut deep or candled to detect the larvae. The identified allergens associated with anisakid nematodes are not destroyed with cooking or freezing and do not require previous exposure to elicit a reaction (Audicana et al. 1997).

1.5 Geographical Information

Members of the *Anisakis* genus, as well as similar nematodes of the *Pseudoterranova* and *Contracaecum* genera, can be found globally in many oceans worldwide, especially in areas with colder water. The species that is most common within circumpolar oceans is *Anisakis simplex* s. s. Human cases of anisakiasis have been reported in Europe, North America, Asia, and South America (Oshima 1972).

1.6 Relevance to Arctic Ecosystems

Climatic changes are warming Arctic Ocean temperatures, which are changing marine ecosystems in multiple ways, including an increase in the amount of freshwater. This can decrease salinity and alter pH, oxygenation, and other parameters thought to influence the environmental survival and development of anisakid eggs and larvae (Rokiki 2009). As the parasites can be found worldwide and infect a wide range of migratory hosts (fish, mammals, and birds), there are likely to be changes in species distributions and potential for emergence and re-emergence of zoonotic diseases associated with the consumption of raw or undercooked fish. The presence of the parasites would also be relevant to any commercial harvest of marine species since a high prevalence of these parasites would likely result in decreased quality of the products. Although these nematodes are not easily detected with the naked eye, processors are trained to detect them, and may utilize techniques such as candling or trimming fillets and removing the larvae manually (Levsen et al. 2005).

2 Zoonotic Marine Diphyllobothriid Cestodes; *Dibothriocephalus* spp. (Formerly *Diphyllobothrium* spp.)

2.1 Morphological and Genetic Information

There are approximately 80 species of *Dibothriocephalus* spp. (formerly *Diphyllobothrium* spp.) cestodes or tapeworms documented worldwide, with 15 species known to infect humans that are often referred to as the "fish tapeworm" or the "broad tapeworm." Infection in humans as a zoonosis is associated with the consumption of larvae (plerocercoids) within the flesh of marine and freshwater fish (Chai et al. 2005). The most common species associated with human infections worldwide are *D. latum* and *D. dendriticum* from freshwater fish and *D. pacificum* from saltwater fish (Sagua et al. 2001). The most common species with a holarctic distribution is *D. dendriticum*, with a number of other species documented in humans, specific documented infections in Alaska have included *D. alascense*, *D. dalliae*, and *D. ursi.* In addition to *D. dendriticum* and *D. latum*, infections with *D. nihonkaiense*, and *D. lanceolatum* and related tapeworms from other genera (*Pyramicocephalus phocarum* and *Schistocephalus solidus*) are known to infect humans living in cold climates (Scholz et al. 2009; Scholz and Kuchta 2016). Parasites are not always recovered and speciated from all infections, particularly in more remote and lower-income areas. Anadromous fish (fish that have freshwater and saltwater components in their life cycles), such as salmonids, can also act as intermediate hosts of the parasite and are commonly consumed by people. In tissue sections, cestodes have characteristic features that include segmentation into proglottids with both male and female components (hermaphrodite), as well as numerous eggs present in the adult worms, the absence of a body cavity or pseudocoelom, and instead the presence of a parenchymatous body with embedded calcareous corpuscles that is surrounded by a thin tegument (Gardiner and Poynton 1999). The larval and adult stages of *Diphyllobothrium* spp. have a scolex with grooves (referred to as bothria) that the parasite uses to attach to host tissue. Variability in the number and position of the scolex and bothria are often used to differentiate different species. The operculate, unembryonated eggs are 55–75 by 40–55 μm and have a small lobe at the end opposite the operculum.

2.2 Life Cycle

Dibothriocephalus spp. (formerly *Diphyllobothrium* spp.) begin as eggs that are shed into water bodies mature under appropriate environmental conditions, a process that takes approximately 18–20 days, and then develop into the free-swimming first larval stage (a coracidium) that, when ingested by a crustacean intermediate host (often a copepod), develops into the second larval stage (a procercoid). The crustacean host is then ingested by a second intermediate host (typically a freshwater or anadromous salmonid fish), and the tapeworm matures into a plerocercoid within the flesh, often muscle tissue of that fish. When the flesh of the second intermediate host is ingested by a definitive host thought to be a fish-eating (piscivorous) carnivore (e.g., cetacean, pinniped, felid, or canid) or bird, the cestode matures into an adult tapeworm within the small intestine (Bazsalovicsová et al. 2020). Other predatory fish (e.g., sharks) can act as parantenic hosts and develop plerocercoids within the flesh that could also be consumed by a piscivorous mammal. Humans can act as a definitive host of the parasite and are infected by consuming raw or undercooked (including smoked or dried) fish meat that is infected with plerocercoid larvae (Meyer 1970). The adult cestodes can grow very long (Fig. 3) (up to 11 m from a human in Chile) (Cabello 2007).

2.3 Pathology in Wild Animals

The intermediate and parantenic hosts (e.g. free-ranging fish) will have subcutaneous and intramuscular tissue cysts containing a plerocercoid larva that can be detected at necropsy or in harvested animals. These cysts may be surrounded by variable amounts of fibrous connective tissue and small amounts of inflammatory cells. In addition to higher trophic level marine mammals, birds, and humans, other potential definitive hosts can include other carnivores such as canids, bears, felines, and otters (Bazsalovicsová et al. 2020; Cabrera et al. 2001). Limited specific information on the impact of cestode infections on free-ranging mammals exists as it is likely that many species have coevolved for millennia with these parasites. However, the clinical syndromes reported in people, including gastrointestinal signs and competition for host nutrients, could presumably also negatively impact other mammalian hosts in

Fig. 3 Preserved adult diphyllobothriid cestode removed from a human patient from the mainland United States with no travel history. (Photo courtesy of Dr. Sarah G. H. Sapp, Parasitic Diseases Branch, Centers for Disease Control and Prevention)

cases of severe infection. This may be relevant in the face of other cumulative impacts associated with food availability, stress, pollutants, other infectious diseases, habitat loss, climate change, and resource development.

2.4 Human Health Impacts

Zoonotic cestodes thought to be *D. pacificum* have been detected in the intestinal tract of mummified ancient humans and coprolites (fossilized feces) from Huaca, Peru, from approximately 4500 BP, from coastal Chile from approximately

6060–3900 BP, and in Germany in 6000 BP (Callen and Cameron 1960; Reinhard and Aufderheide 1990; Reinhard and Urban 2003; Le Bailly and Bouchet 2013). In 1973, it was estimated that 9 million persons were infected globally with known endemic regions of high prevalence in Sweden, Finland, and Russia (von Bonsdorff 1977). Diphyllobothriasis is currently considered the most significant food-borne parasite from fish with current estimates of worldwide human cases exceeding 20 million. Infections by zoonotic nematodes in humans can range from asymptomatic to severe flu-like illness with abdominal cramps, flatulence, abdominal distention, diarrhea, and induced vitamin B_{12} insufficiency due to malabsorption and even megaloblastic anemia (Baer 1969; von Bonsdorff 1977; Ito and Budke 2014). Although diphyllobothriasis can affect any age and sex, middle-aged men are overrepresented. A few studies have specifically examined the number of cases in Arctic peoples in Finland and Alaska and claim declining case numbers in more recent years in specific locations despite the global increase (Von Bonsdorf 1964). Without targeted studies, the infections could be overlooked as the symptoms are nonspecific. *Dibothriocephalus* spp. can be inactivated with prolonged freezing periods (i.e., below −20 °C for 7 days or below −35 °C for 15 h) and with cooking (56 °C for 5 min) (FDA 1998; Wharton and Aalders 2002).

2.5 Geographic Information

Dibothriocephalus spp. (formerly *Diphyllobothrium* spp.) are found worldwide with documented human infections in Europe, Asia, and North and South America (Oshima 1972; Sagua et al. 2001; Dupouy-Camet and Peduzzi 2004). In many cases, the parasite is considered endemic within the aquatic life, but there has been at least one instance thought to have been an anthropogenic introduction into South America from European immigrants (Semenas and Úbeda 1997). In Finland, although the parasite is present throughout the country and environmental conditions are similar, the prevalence of human infections is much higher in eastern Finland where consumption of raw fish is an ancestral practice compared with more western regions of the country where the customs differ (von Bonsdorff 1977).

2.6 Relevance to Arctic Ecosystems

There are concerns about the potential exposure of humans to zoonotic parasites associated with changes in global climate patterns, ocean nutrient levels, and species distributions of intermediate and definitive hosts. Human infections have been associated with imported fish species with molecular analysis necessary for teasing out sources of infection (Greigert et al. 2020). Changes in fish species distributions are currently being observed in the Arctic, especially for salmonids that could have implications for the emergence of human infections (Bilous and Dunmall 2020). Associations between zoonotic marine parasite levels and climatic features, such as the *El Niño*-Southern Oscillation, have been found in coastal regions of South

America (Arriaza et al. 2010). As the climate warms in the Arctic, changes in species distributions are being observed, including the introduction of new potential intermediate hosts and new species of parasites into that region (Chueng et al. 2009). It is likely that human and animal cases of diphyllobothriasis are underdiagnosed, and this zoonosis is expected to continue to be considered a (re)-emerging disease of concern associated with human preferences of consuming raw or undercooked fish, global trade of fish products, and globalized human movement (Scholz and Kuchta 2016).

3 Conclusions

The human incidence of anisakiasis and diphyllobothriasis in relation to the consumption of traditionally uncooked fish (i.e., *civeche* in Spanish) has been more extensively studied in South American human populations (Arriaza et al. 2010), as well as targeted studies from Europe (Dupouy-Camet and Peduzzi 2004), Asia (i.e., sushi in Japan; Nawa et al. 2001), and parts of North America (from travel and consumption of wild and reared fish (Deardorff and Overstreet 1991; Dick et al. 2001; Dick 2008). The significance of these parasites to humans and wildlife in the Arctic is currently poorly understood. In general, research gaps exist to understand their molecular identity, host range, geographic range, prevalence, incidence, and significance in Arctic environments. Effective treatments (e.g., anthelminthics such as praziquantel) are available for diagnosed infections in people and domestic animals. Avoiding consumption of raw or undercooked meat (e.g., dried, smoked, or pickled) would prevent human disease, and avoiding feeding raw fish to domestic animals, such as dogs and cats, would also decrease risks for people. However, many of these foods have important sociocultural benefits, and the development of public health messaging regarding the risks of consumption requires collaboration with traditional knowledge holders, communities, and local health authorities. Hunters and fishers may detect the parasites at harvest, with the potential for scientists and governments to work with communities on zoonotic disease education and scientific investigation. Regular deworming is often recommended to dog handlers in communities that feed raw fish to dogs, but many communities do not have regular access to veterinary care. Although the relevance of these guidelines to all cold-adapted species of these parasites is unknown, further investigation and collaborative community-based food safety projects are needed to better meet the needs of northern communities. Knowledge about these parasites will be relevant to any communities wishing to participate in the commercial harvest of marine species with consideration for risk mitigation measures related to sale of the products and protection of the workers from allergic reactions. Control of fecal material from animals and humans in waterways would also be important to limit the number of eggs of these parasites being shed in the environment, as well as multiple other water-borne zoonoses. Human and animal health providers should be aware of these zoonoses and their preventive measures and treatments, as increased detection will improve our understanding of the importance of these parasites in communities and ecosystems.

References

Arriaza BT et al (2010) Possible influence of the ENSO phenomenon on the pathoecology of diphyllobothriasis and anisakiasis in ancient Chinchoor populations. Mem Ins Oswaldo Cruz 105:66–72

Audicana L et al (1997) Cooking and freezing may not protect against allergic reactions to ingested *Anisakis simplex* antigens in humans. Vet Rec 140:235

Baer JG (1969) *Diphyllobothrium pacificum*, a tapeworm of sealions endemic in along the coastal area of Peru. J Fish Rest Canada 26:717–772

Bazsalovicsová E et al (2020) Development of 14 Microsatellite Markers for Zoonotic Tapeworm *Dibothriocephalus dendriticus* (Cestoda: Diphyllobothriidea). Genes 11:782–790

Bilous M, Dunmall K (2020) Atlantic salmon in the Canadian Arctic: potential dispersal, establishment, and interaction with Arctic char. Rev Fish Biol Fisheries 30:463–483

Cabello FC (2007) Aquaculture and public health. The emergence of diphyllobothriasis in Chile and the world. Rev Med Chile 135:1064–1071

Cabrera R et al.(2001) *Diphyllobothrium pacificum* (Nybelin, 1931) Margolis, 1956 en Canis familiaris de la ciudad de Chincha, Peru. Bol Chil Parasitol 56:26–28

Callen EO, Cameron TW (1960) A prehistoric diet revealed in coprolites. New Sci 8:35–40

Chai JY et al (2005) Fish-borne parasitic zoonoses: status and issues. Int J Parasitol 35:1233–1254

Chueng WWL et al (2009) Projecting global marine biodiversity impacts under climate change scenarios. Fish Fish 10:235–251

Deardorff TL, Overstreet RM (1991) Seafood-transmitted zoonoses in the United States: the fishes, the dishes, and the worms, microbiology of marine food products. Van Nostrand Reinhold, New York

Dick T (2008) Diphyllobothriasis: the *Diphyllobothrium latum* human infection conundrum and reconciliation with a worldwide zoonosis. In: Murrell KD, Fried B (eds) Food-borne parasitic zoonoses: fish and plant-borne parasites (world class parasites). Springer, London, pp 151–184

Dick TA et al (2001) Diphyllobothriasis: update on human cases, foci, patterns and sources of human infections and future considerations. Southeast Asian J Trop Med Public Health 32 (Suppl 2):59–76

Dupouy-Camet J, Peduzzi R (2004) Current situation of human diphyllobothriasis in Europe. Euro Surveill 9:31–34

Dujardin F (1845) Histoire Naturelle des Helminthes ou vers intestinaux. Paris, xvi + 654 pp

Food and Drug Administration (FDA) (1998) Fish and fisheries products hazards and controls guide. FDA, Washington

Gardiner CH, Poynton SL (1999) An atlas of metazoan parasites in animal tissues, vol 19–21. Registry of Veterinary Pathology, Armed Forces Institute of Pathology, American Registry of Pathology, Washington, pp 50–55

Greigert V et al (2020) Locally acquired infection with *Dibothriocephalus nihonkaiense* (*Diphyllobothrium nihonkaiense*) in France: the importance of molecular diagnosis. Parasitol Res 119:513–518

Ito A, Budke CM (2014) Culinary delights and travel? A review of zoonotic cestodiases and metacestodiases. Travel Med Infect Dis 12:582–591

Karl H et al (1994) Survival of *Anisakis* larvae in marinated herring fillets. Int J Food Sci Technol 29:661–670

Larrat S et al (2013) Relationship between red vent syndrome and anisakid larvae burden in wild Atlantic salmon (*Salmo salar*). J Wildl Dis 49:229–234

Le Bailly M, Bouchet F (2013) Diphyllobothrium in the past: review and new records. Int J Paleopathol 3:182–187

Levsen A et al (2005) Low detection efficiency of candling as a commonly recommended inspection method for nematode larvae in the flesh of pelagic fish. J Food Prot 68:828–832

Ludovisi A et al (2017) Allergenic activity of *Pseudoterranova decipiens* (Nematoda: Anisakidae) in BALB/c mice. Parasites Vectors 10:290

Matiiucci M, Nascetti G (2006) Molecular systematics, phylogeny, and ecology of anisakid nematodes of the genus *Anisakis* Dujarin, 1845: an update. Parasite 13:99–113

Mehrdana F, Buchmann K (2017) Excretory/secretory products of anisakid nematodes: biological and pathological roles. Acta Vet Scand 59:42

Mehrdana F, Bahlool Q, Skov J et al (2014) Occurrence of zoonotic nematodes *Pseudoterranova decipiens*, *Contracaecum osculatum* and *Anisakis simplex* in cod (*Gadus morhua*) from the Baltic Sea. Vet Parasitol 205:581–587

Meyer M (1970) Cestode zoonoses of aquatic animals. J Wildl Dis 6:249–254

Myers BJ (1976) Research then and now on the Anisakidae nematodes. Trans Am Microsc Soc 95: 137–142

Nawa Y, Noda S, Uchiyama-Nakamura F et al (2001) Current status of food-borne parasitic zoonoses in Japan. Southeast Asian J Trop Med Public Health 32(Suppl 2):4–7

Nemeth NM, Yabsley M, Keel MK (2012) Anisakiasis with proventricular perforation in a greater shearwater (*Puffinus gravis*) off the coast of Georgia, United States. J Zoo Wildl Med 43:412–415

Nieuwenhuizen N, Lopata AL et al (2006) Exposure to the fish parasite *Anisakis* causes allergic airway hyperreactivity and dermatitis. J Allergy Clin Immunol 117:1098–1105

Noguera P, Collins C et al (2009) Red vent syndrome in wild Atlantic salmon *Salmo salar* in Scotland is associated with *Anisakis simplex sensu stricto* (Nematoda: Anisakidae). Dis Aquat Org 87:199–215

Noh JH, Kim B-J et al (2003) A case of acute gastric anisakiasis provoking severe clinical problems by multiple infections. Korean J Parasitol 41:97–100

Oshima T (1972) Anisakis and anisakiasis in Japan and adjacent areas. Prog Med Parasitol Jpn 4: 305–393

Perteguer MJ, Chivato T et al (2000) Specific and total IgE in patients with recurrent, acute urticarial caused by *Anisakis simplex*. Ann Trop Med Parasitol 94:259–268

Reinhard KJ, Aufderheide AC (1990) Diphyllobothriasis in prehistoric Chile and Peru: adaptive radiation of a helminth species to native American populations. Paleopathol News 72:18–19

Reinhard K, Urban O (2003) Diagnosing ancient diphyllobothriasis from Chinchorro mummies. Mem Inst Oswaldo Cruz 98:191–193

Rokiki J (2009) Effects of climatic changes on anisakid nematodes in polar regions. Pol Sci 3:197–201

Sagua H, Neira I et al (2001) New cases of *Diphyllobothrium pacificum* (Nybelin, 1931) Margolis, 1956 human infection in North of Chile, probably related with El Niño phenomenon, 1975-2000. Bol Chil Parasitol 56:22–25

Sakanari J, McKerrow J (1989) Anisakiasis. Clin Microbiol Rev 2:278–284

Scholz T, Kuchta R (2016) Fish-borne, zoonotic cestodes (*Diphyllobothrium* and relatives) in cold climates: a never-ending story of neglected and (re)-emergent parasites. Food Waterborne Parasitol 4:23–38

Scholz T, Garcia HH et al (2009) Update on the human broad tapeworm (genus *Diphyllobothrium*), including clinical relevance. Clin Microbiol Rev 22:146–160

Semenas L, Úbeda C (1997) Difilobotriasis humana en la Patagonia, Argentina. Rev Saude Publica 31:302–307

Shamsi S, Briand MJ, Justine JL (2017) Occurrence of Anisakis (Nematoda: Anisakidae) larvae in unusual hosts in southern hemisphere. Parasitol Int 66:837–840

Smith JW, Wooten R (1978) Anisakis and Anisakiasis. Adv Parasit 16:93–163

Sprengel G, Lüchtenberg H (1991) Infection by endoparasites reduces swimming speed of European smelt *Osmerus eperlanus* and European eel *Anguilla Anguilla*. Dis Aquat Org 11: 31–35

von Bonsdorf B (1964) The fish tapeworm, *Diphyllobothrium latum*; a major health problem in Finland. World Med J 11:170–172

von Bonsdorff G (1977) Diphyllobothriasis in man. Academic Press, London, 189 pp

Wagner BA, Hoberg EB et al (2012) Gastrointestinal helminth parasites of double-crested cormorants (*Phalacrocorax auritus*) at four sites in Saskatchewan, Canada, 2006-2007. Comp Parasitol 79:275–282

Weitzel T, Sugiyama H et al (2015) Human infections with *Pseudoterranova cattani* Nematodes, Chile. Emerg Infect Dis 21:1874–1875

Wharton DA, Aalders O (2002) The response of *Anisakis* larvae to freezing. J Helminthol 76:363–368

Yoon SW, Yu JS et al (2004) CT findings of surgically verified acute invasive small bowel anisakiasis resulting in small bowel obstruction. Yonsei Med 45:739–742

Parapoxvirus Infections in Northern Species and Populations

Morten Tryland

1 Introduction

Parapoxviruses belong to the genus *Parapoxvirus*, which is one of nine genera in the subfamily *Chordopoxvirinae*, of the *Poxviridae* family (Skinner et al. 2012). Parapoxviruses cause proliferative processes in skin and mucosal membranes, often on the mucocutaneous junctions, in a wide range of host species. Orf virus (ORFV) is the prototype virus species of the genus and is distributed worldwide with sheep and goats, causing the disease contagious ecthyma (syn. contagious pustular dermatitis, contagious stomatitis, "scabby mouth"). However, ORFV has a broad host range and may cause infections and disease in a wide range of animal species, including wildlife. ORFV is also zoonotic, causing skin lesions, usually on the fingers and hands of people that have been in contact with infected animals or animal products. In addition, other parapoxvirus species cause disease. Bovine papular stomatitis virus (BPSV) and pseudocowpox virus (PCPV) are close relatives to ORFV and may cause disease similar to contagious ecthyma, but after exposure to infected cattle, and both are zoonotic. Parapoxviruses also infect a wide range of wild animals, such as ORFV in reindeer and caribou (*Rangifer tarandus*) and PCPV in reindeer, ORFV in muskoxen (*Ovibos moschatus*), parapoxvirus in red deer of New Zealand (PVNZ) in red deer (*Cervus elaphus*), squirrelpox virus (SQPV) in red squirrels (*Sciurus vulgaris*), and several parapoxvirus species and tentative members of the genus that infect seals and other mammals.

Viruses of the *Parapoxvirus* genus have similar morphology and are genetically and antigenically related (Fig. 1). They are large (140–170 nm wide, 220–300 nm

M. Tryland (✉)
Department of Arctic and Marine Biology, UiT The Arctic University of Tromsø, Tromsø, Norway

Department of Forestry and Wildlife Management, Inland Norway University of Applied Sciences, Koppang, Norway
e-mail: morten.tryland@inn.no

M. Tryland (ed.), *Arctic One Health*, https://doi.org/10.1007/978-3-030-87853-5_20

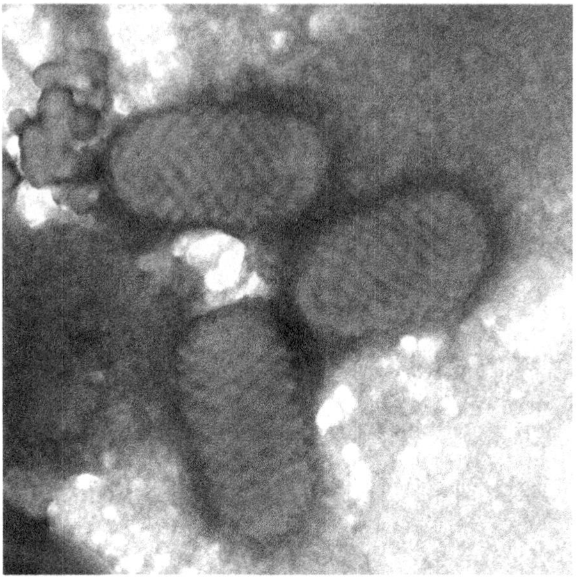

Fig. 1 Parapoxvirus particles as visualized in the electron microscope (negative staining), obtained from a muskox with contagious ecthyma. The virus particles are 140–170 nm wide and 220–300 nm long, with the typical criss-cross pattern of the outer protein tubuli (Photo: Morten Tryland)

long), double-stranded DNA viruses that replicate in the cytoplasm of the host cell. The classification of parapoxviruses was based on natural host range and pathology, but molecular methods, such as restriction enzyme analyses, hybridization, DNA sequencing, and phylogenetic analysis, have provided tools for studying the different virus species and their epidemiological characteristics.

Contagious ecthyma lesions usually start as a papule that develops through vesicular and pustular stages before it is normally covered by crusts. Several lesions may coalesce with the formation of large crusts or scabs, with proliferative tissue formation underneath. It has been documented that ORFV encodes proteins homologous to the endothelial growth factor and a vascular permeability factor, which plays critical roles during the formation of new and highly vascularized tissue, often referred to as having a "cauliflower-like" appearance. Such tissues are easily exposed to trauma and bleeding, and in the skin, such lesions are often covered by thick, black crusts that can be affected by secondary bacterial infections (Ueda et al. 2007). Histological changes of the epithelium are characterized by ballooning degeneration, reticular degeneration, marked epidermal proliferation, and epidermal microabscesses (Kitching 2004).

2 Parapoxvirus Infections in Reindeer and Caribou

In Eurasian tundra reindeer (*R. t. tarandus*), parapoxvirus infections resemble contagious ecthyma in sheep. It was first reported in semidomesticated reindeer in Sweden (Nordkvist 1973) and Finland (Büttner et al. 1995). In Norway, the disease was first observed in reindeer under natural herding conditions in 2000 (Tryland

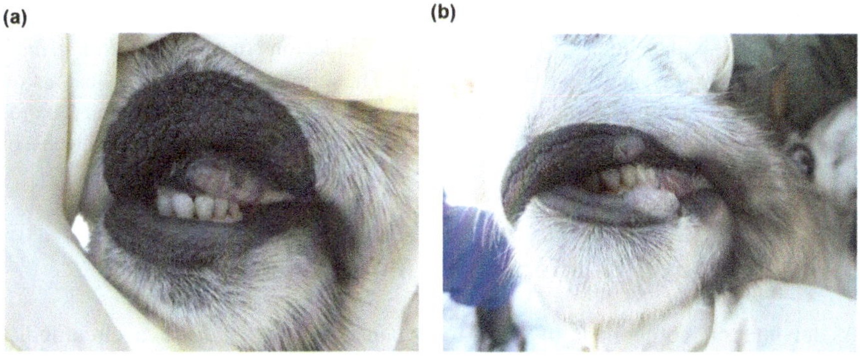

Fig. 2 Semi-domesticated reindeer in Sweden with contagious ecthyma caused by the parapoxvirus Orf (ORFV). (**a**) In some animals, a close inspection is needed to detect contagious ecthyma lesions in the oral mucosa. (**b**) Contagious ecthyma lesions are often visible on the skin or at the junction between skin and the mucosal membrane. (Photo **a**, **b** Ingebjørg H. Nymo)

et al. 2001). In contrast to the situation in Norway and Sweden, where the disease appears only sporadically, contagious ecthyma has appeared regularly in Finland since the first outbreak during winter 1992–1993 when about 400 reindeer died and more than 2800 was affected. Severe outbreaks were also reported in Finland in 1999–2000 and in late winter of 2007 (Hautaniemi et al. 2011). In addition to ORFV, also PCPV has been involved in recent outbreaks in Finland (Tikkanen et al. 2004).

The virus enters the skin and mucosal membranes through abrasions or small lesions on the muzzle or in the mouth caused by, e.g., ice crusts or coarse feed, and possibly also associated with the eruption of new teeth in young animals (Tryland et al. 2019a). The incubation period in reindeer is 3–5 days (Tryland et al. 2013). There is no specific treatment, but supportive therapy and antibiotic treatment against secondary bacterial infections may be indicated. The humoral immune response is thought to be short-lived in reindeer, as in sheep (months) (Damon 2007).

An early sign of contagious ecthyma in reindeer is that an animal is not able to chew and has accumulations of food and regurgitated matter in and around their mouth, causing a foul smell. Upon inspection, there may be one or more proliferative lesions in the oral mucosa (gingiva and palatum) (Fig. 2a), often with secondary bacterial infections (Tryland et al. 2019a). During a disease outbreak, typical contagious ecthyma lesions are often easily visible and recognizable in some of the animals (Fig. 2b).

In semidomesticated reindeer, ORFV is presumably transferred from sheep or goats, through direct contact or indirectly via shared pastures containing scabs from infected animals or via shared transport vehicles, feeding troughs, or other equipment. Contagious ecthyma may be seen associated with supplementary feeding of reindeer, sometimes in combination with other diseases, and the disease may affect many animals in a herd (Tryland et al. 2019b; Tikkanen et al. 2004).

One report has been published on the finding of pathology and parapoxvirus-specific DNA, with sequences indicating ORFV, in lesions of a caribou (*R. t. granti*) from Admiralty Bay, Northern Alaska, USA. The animal had cutaneous lesions on the upper lip and on the plantar side of the hooves and on the coronary bands (Tryland et al. 2018). The prevalence of such infections in wild populations of *Rangifer* is, however, unknown.

3 Parapoxvirus Infections in Muskoxen

As an integral component of the Arctic fauna, muskoxen have a circumpolar distribution, with uncertainties counting about 170,000 animals and comprising 55 endemic or translocated populations, of which six are in decline due to climate conditions, anthropogenic changes, and diseases (Cuyler et al. 2020). Parapoxvirus infections have been diagnosed in captive muskoxen, at the research animal stations at the University of Fairbanks (Alaska, USA; Dieterich et al. 1981) and UiT The Arctic University of Norway (Tromsø, Norway; Mathiesen et al. 1985), as well as in captive muskoxen in a zoo in Minnesota (USA; Guo et al. 2004). Genetical studies of some of the causative virus isolates have indicated ORFV (Guo et al. 2004; Moens et al. 1990), and a possible source of infection may have been the use of pastures and infrastructure previously used for small ruminants, in particular sheep. Serological investigations in Alaska revealed parapoxvirus antibodies in 10 of 45 muskoxen tested at Nunivak Island, Alaska (Zarnke et al. 1983). There are no reports of parapoxvirus infections in the muskoxen populations in Greenland, which has been the source of several translocations to other regions (Cuyler et al. 2020).

In Dovre National Park (Norway), a muskox population that during normal conditions is kept at around 200 animals (winter) is based on the import of 21 calves and yearlings from Eastern Greenland (1947–1953). Contagious ecthyma was diagnosed in one calf in 1987 and one calf in 1994. A regular disease outbreak occurred in 2004, when a total of 16 animals were euthanized due to severe symptoms of contagious ecthyma, and PCR (*B2L* gene) and sequencing (326 nt) revealed 100% homology with ORFV, suggesting transmission from sheep on shared pastures (Vikøren et al. 2008). Similar disease outbreaks occurred in 2012 and 2016, and the disease was diagnosed in a few calves during the winter of 2018. The disease has been particularly severe in calves and young individuals, and diseased individuals are euthanized.

In Victoria Island (Canada), contagious ecthyma was recently diagnosed in an adult male muskox in 2014, also having abscesses from which the bacterium *Brucella suis* biovar 4 was isolated and having severe pathology from lungworm infections with *Umingmakstrongylus pallikuukensis* and *Varestrongylus eleguneniensis* (Tomaselli et al. 2016). Also, in this case, the PCR and sequencing resembled the isolate from Minnesota Zoo, indicating ORFV.

Contagious ecthyma in muskoxen has been characterized by typical proliferative lesions in the oral mucosa (gingiva and tongue) and around the mouth and nostrils. In contrast to the appearance in semidomesticated reindeer, the lesions, multifocal or

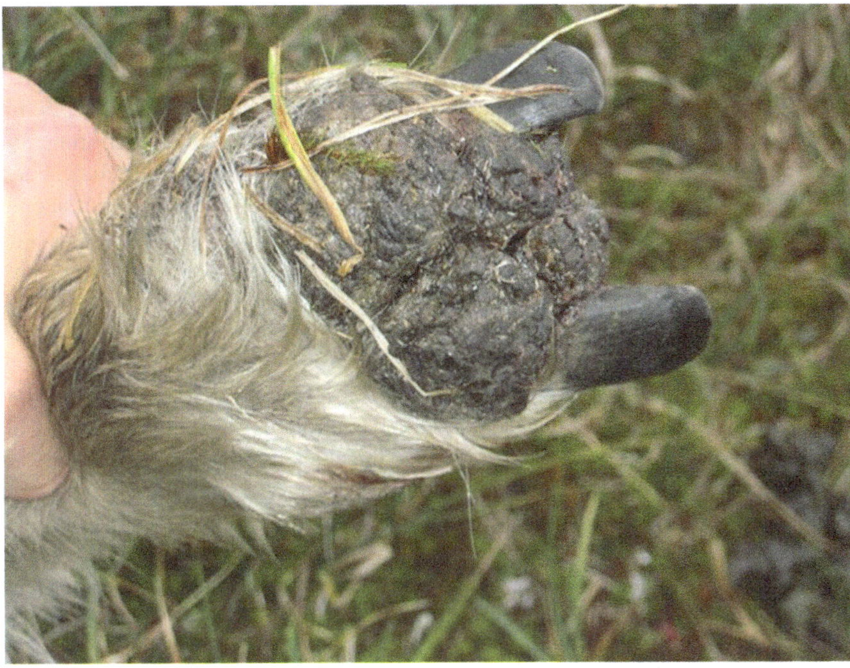

Fig. 3 Contagious ecthyma in a muskox calf at Dovre, Norway, 2004. Muskoxen are often affected by proliferative lesions around the muzzle and in the oral mucosa, but also on distal parts of the limbs, as shown here in interdigital space (Photo: Tord Bretten)

coalescing, have also been found in the skin around the eyes and on the distal parts of the legs and in the interdigital cleft, often covered with thick, black crusts (Fig. 3).

4 Parapoxvirus Infections in Other Northern Wildlife Ruminants

A retrospective study, summarizing and verifying pathological findings and initial parapoxvirus identification in necropsied wildlife species in Alaska, revealed Parapoxvirus-specific DNA in pathological lesions characteristic of contagious ecthyma from mountain goats (*Oreamnos americanus*, $n = 8$), Dall's sheep (*Ovis dalli dalli*, $n = 3$), muskox (*Ovibos moschatus*, $n = 3$), and caribou (*Rangifer tarandus granti*, $n = 1$). Similar results were obtained when investigating tissues from a fibroma-like structure from a Sitka black-tailed deer (*Odocoileus hemionus sitkensis*, $n = 1$). Sequencing of PCR amplicons (*B2L*, *GIF*, *vIL-10*, and *ATI* genes) from all the cases indicated ORFV (Tryland et al. 2018). Interestingly, the virus sequences obtained from Dall's sheep formed a separate cluster comparable to ORFV from domestic sheep, whereas sequences from the other species were almost identical to each other.

Since parapoxvirus infections cause proliferative lesions, in the literature often called "wart-like" or "papillomas," there is confusion with regard to which infectious agent is involved. In contrast to contagious ecthyma lesions, papillomas (i.e., warts) are caused by papillomaviruses (family *Papillomaviridae*). Papillomaviruses are usually very species-specific, the disease affects mostly single and often immunocompromised individuals, and the lesions develop slowly. Even if their clinical and epidemiological appearance may be easily distinguished, a clinical diagnosis should be verified by virological investigations.

5 Parapoxvirus Infections in Seals

Parapoxviruses that are infecting seals are different from ORFV and other parapoxviruses that are infecting terrestrial hosts (Nollens et al. 2006; Costa et al. 2021). The diagnosis is often only based on clinical signs, and the causative viruses are often not well characterized, but rather just called "sealpox" viruses. Seal parapoxvirus, which is a tentative member of the genus *Parapoxvirus* (i.e., not yet accepted by the International Committee on Taxonomy of Viruses, ICTV), is morphologically similar to other parapoxviruses in shape, size, and superficial structures of the protein tubules. These structures together make such virus particles easily recognizable in the electron microscope when the sample is prepared by the negative staining method, using phosphotungstic acid. A recent study compared parapoxviruses from 11 pinniped cases of five different species in rehabilitation centers on the west and east coast of the USA. In addition to distinguishing them from parapoxvirus isolates from terrestrial mammals, the comparison revealed that they formed a separate cluster of the genus *Parapoxvirus*, supporting the suggestion of including seal parapoxvirus as a species. Further, it revealed that animals of the same species but originating from geographically separated places (i.e., west and east coast of the USA) had genetically distant parapoxviruses (Costa et al. 2021).

Parapoxvirus infections in seals were reported for the first time in California sea lions (Wilson et al. 1969) but have since been identified from a wide range of seal species representing all the three families that constitute the suborder Pinnipedia, e.g., Phocidae ("true seals"), Otariidae ("eared seals"), and Odobenidae (walrus) (Tryland 2011; Scotter et al. 2019). Most of these reports are from the northern hemisphere, with the exceptions of a finding in South American sea lions (*Otaria flavescens*) in Peru (Wilson and Poglayen-Neuwall 1971), a case report on a Weddell seal (*Leptonychotes weddellii*) in the Weddel Sea, Antarctica (Tryland et al. 2005), and the recent detection of parapoxvirus-specific DNA in swab samples from two of 29 investigated Antarctic fur seal (*Arctocephalus gazella*) pups from Bouvet Island (Núñez-Egido et al. 2020).

Clinical signs are usually observed as skin lesions, 1.0 to 2.5 cm in diameter, as a single process or in some cases as a generalized infection with multiple single or coalescing lesions. Skin lesions are often seen on the neck, chest, flippers, and perineum. However, as with other parapoxviruses (e.g., ORFV), seal parapoxvirus may cause similar proliferative lesions also in the oral mucosa, as seen during an

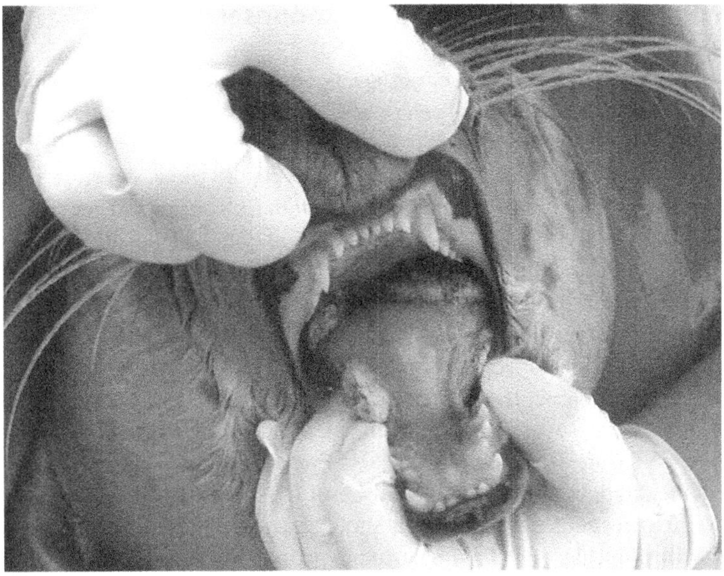

Fig. 4 Parapoxvirus infection (seal parapoxvirus, often called "sealpox") in a harbor seal in a rehabilitation center in Germany. Large proliferative lesions are seen on the basis and tip of the tongue. Reprinted from Müller, G. et al., Vet Pathol., 40, 445, 2003. With permission from the American College of Veterinary Pathologists

outbreak among harbor seals from the German North Sea after they had been taken into captivity (Fig. 4). Parapoxvirus infections in seals are most common in young animals in captivity and often appear and spread among wild individuals a few weeks after they have been taken into the facility (Simpson et al. 1994; Müller et al. 2003; Tryland 2011). This may indicate that the appearance of clinical disease is associated with stress and changes in the environment of the seals. The virus may be transmitted via water and through direct contact between mother and pup and on haul-out sites.

The terms "sealpox" and "poxlike lesions" are often used in clinical reports, but it is clear from clinical cases in sea lions for which the causative agent has been identified that also poxviruses with sequence homology with orthopoxviruses may cause lesions that are indistinguishable from those caused by seal parapoxvirus (Burek et al. 2005). Further, mixed infections with both parapoxvirus and orthopoxvirus have been reported in grey seals (*Halichoerus grypus*) (Osterhaus et al. 1990). Sealpox may thus be useful to describe the clinical signs, whereas virological identification is necessary to address the etiology and epidemiology of the disease.

In the wild, it is not likely that parapoxvirus infections have high morbidity and mortality, but the virus spreads more easily in pools and facilities, especially among young individuals that also experience stress and new environments.

6 Parapoxvirus Infections as Zoonoses

Parapoxvirus infections represent an occupational hazard for people handling affected animals, such as farmers, veterinarians, hunters, game managers, and animal handlers in parks, zoos, and rehabilitation facilities. Lesions occur mainly on fingers and hands (Fig. 5a, b) and usually start as a single papule which develops into a large vesicle or bulla, sometimes being hemorrhagic with a black appearance. Such lesions are usually painful and can be associated with local lymphadenopathy and fever, but there is no specific medical treatment and the lesion heals spontaneously within weeks (Palatsi et al. 1993). Such infections are probably most common in sheep farmers that may have been exposed before and know what it is and thus maybe do not seek medical assistance. This may contribute to parapoxvirus infections in humans that are often neglected and underreported. However, such infections may also become more severe and generalized, especially in immunocompromised persons (de Vicq de Cumptich et al. 2015), and it is listed among diseases that people in the Arctic region should be aware of (Anonymous 2011).

People in direct contact with animals should be aware of the disease and be able to recognize the clinical signs in animals and in people. They should also keep in mind that transmission may occur from animals with no clinical signs of disease, as demonstrated in Finland when people handling slaughtered and apparently healthy animals during the first major outbreak of contagious ecthyma in Finland (Palatsi et al. 1993). Infection can be avoided by the use of gloves and personal hygiene (e.g., hand washing, a separate set of field clothes).

(a) (b)

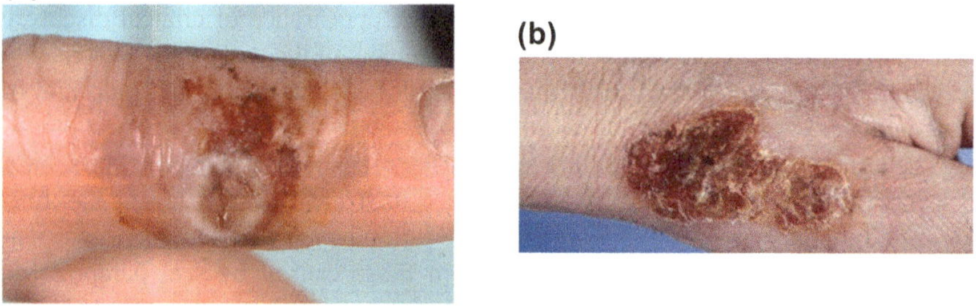

Fig. 5 Parapoxviruses are zoonotic. During the outbreak among semi-domesticated reindeer in Finland during the winter 1992–1993, about 400 reindeer died. Several human cases appeared, most of them with typical lesions on fingers (**a**) and one with a larger lesion on the hand (**b**) (Photo: Riitta Liisa Palatsi)

7 Concluding Remarks

ORFV is distributed worldwide with sheep and goats, but evidence suggests that parapoxviruses of different kinds are present in a wide variety of terrestrial and marine mammal species, sometimes affecting single animals without causing disease outbreaks or as subclinical infections with no suspicion of the virus being present (Tryland et al. 2018; Palatsi et al. 1993). The finding of parapoxvirus infections in a wide variety of animal species that are harvested in arctic regions, such as caribou, reindeer, muskoxen, Dall's sheep, mountain goats, and seals, underscores the importance of wildlife health surveillance to maintain health and food security in the region, including subsistence-oriented communities (Tomaselli et al. 2016).

References

Anonymous (2011) Food safety for first nations people of Canada: a manual for healthy practices. https://www.gov.mb.ca/inr/pdf/pubs/nhfi_food_safety_for_first_nations_people_of_canada.pdf. Accessed 30 May 2021

Burek KA, Beckmen K, Gelatt T et al (2005) Poxvirus infection of Steller Sea lions (*Eumetopias jubatus*) in Alaska. J Wildl Dis 41(4):745–752. https://doi.org/10.7589/0090-3558-41.4.745

Büttner M, von Einem C, McInnes C et al (1995) Clinical findings and diagnosis of a severe parapoxvirus epidemic in Finnish reindeer. *Tierarztl Prax* 23(6):614–618. (German)

Costa H, Klein J, Breines EM et al (2021) A comparison of parapoxviruses in north American pinnipeds. Front Vet Sci. https://doi.org/10.3389/fvets.2021.653094

Cuyler C, Rowell J, Adamczewski J et al (2020) Muskox status, recent variation, and uncertain future. Ambio 49:805–819. https://doi.org/10.1007/s13280-019-01205-x

Damon I (2007) Poxviruses. In: Knipe DM, Howley PM (eds) Fields virology, 5th edn. Lippincott, Williams & Wilkins, London, pp 2947–2975

de Vicq de Cumptich M, Snoeck R, Sass U et al (2015) Orf nodules and immunosuppression: a case report and review of therapeutics. Rev Med Brux 36:439–443. (in French)

Dieterich RA, Spencer GR, Burger D et al (1981) Contagious ecthyma in Alaskan musk-oxen and Dall sheep. J Am Vet Med Assoc 179:1140–1143

Guo J, Rasmussen J, Wünschmann A et al (2004) Genetic characterization of orf viruses isolated from various ruminant species of a zoo. Vet Microbiol 99:81–92

Hautaniemi M, Vaccari F, Scagliarini A et al (2011) Analysis of deletion within the reindeer pseudocowpoxvirus genome. Virus Res 160:326–332

Kitching RP (2004) Orf. In: Coetzer JAW, Tustin RC (eds) Infectious diseases of livestock, 2nd edn. Oxford University Press, Oxford, pp 1282–1286

Mathiesen SD, Jørgensen T, Traavik T et al (1985) On contagious echtyma and its treatment in muskoxen (*Ovibos moschatus*). Acta Vet Scand 26:120–126

Moens U, Wold I, Mathiesen SD et al (1990) Parapoxvirus papillomatosis in the muskoxen (*Ovibos moschatus*): genetical difference between the virus causing new outbreak in a vaccinated herd, the vaccine virus and a local orf virus. Acta Vet Scand 31:17–25

Müller G, Gröters S, Siebert U et al (2003) Parapoxvirus infection in harbor seals (*Phoca vitulina*) from the German North Sea. Vet Pathol 40:445–454. https://doi.org/10.1354/vp.40-4-445

Nollens HH, Gulland FM, Jacobson ER et al (2006) Parapoxviruses of seals and sea lions make up a distinct subclade within the genus Parapoxvirus. Virology 349:316–324. https://doi.org/10.1016/j.virol.2006.01.020

Nordkvist M (1973) Munvårtsjuka - en ny rensjukdom? Rennäringsnytt 8-9:6–8

Núñez-Egido S, Lowther A, Nymo IH et al (2020) Pathogen surveillance in Southern Ocean pinnipeds. Polar Res 39:3841. https://doi.org/10.33265/polar.v39.3841

Osterhaus AD, Broeders HW, Visser IK et al (1990) Isolation of an orthopoxvirus from pox-like lesions of a grey seal (*Halichoerus grypus*). Vet Rec 127:91–92

Palatsi R, Oksanen A, Sormunen R et al (1993) The first Orf virus epidemic diagnosed in man and reindeer in 1992-1993 in Finland. Duodecim 109:1945–1950. (In Finnish)

Scotter SE, Tryland M, Nymo IH et al (2019) Contaminants in Atlantic walruses in Svalbard part 1: relationships between exposure, diet and pathogen prevalence. Environ Pollut 244:9–18. https://doi.org/10.1016/j.envpol.2018.10.001

Simpson VR, Stuart NC, Stack MJ et al (1994) Parapox infection in grey seals (*Halichoerus grypus*) in Cornwall. Vet Rec 134:292–296. https://doi.org/10.1136/vr.134.12.292

Skinner MA, Buller RM, Damon IK et al (2012) Poxviridae. In: King AMQ, Adams MJ, Carstens EB, Lefkowitz EJ (eds) Virus taxonomy: classification and nomenclature of viruses: ninth report of the international committee on taxonomy of viruses. Elsevier Academic Press, San Diego, pp 291–309

Tikkanen MK, McInnes CJ, Mercer AA et al (2004) Recent isolates of parapoxvirus of Finnish reindeer (*Rangifer tarandus tarandus*) are closely related to bovine pseudocowpox virus. J Gen Virol 85:1413–1418

Tomaselli M, Dalton C, Duignan PJ et al (2016) Contagious ecthyma, rangiferine brucellosis, and lungworm infection in a muskox (*Ovibos moschatus*) from the Canadian Arctic, 2014. J Wildl Dis 52(3):719–724

Tryland M (2011) Seal parapoxvirus. In: Liu D (ed) Molecular detection of human viral pathogens. CRC Press, Boca Raton, FL, pp 1029–1037

Tryland M, Josefsen TD, Oksanen A et al (2001) Parapoxvirus infection in Norwegian semi-domesticated reindeer (*Rangifer tarandus tarandus*). Vet Rec 149:394–395

Tryland M, Klein J, Nordøy E et al (2005) Isolation and partial characterization of a parapoxvirus isolated from a skin lesion of a Weddell seal. Virus Res 108:83–87

Tryland M, Klein J, Berger T et al (2013) Experimental parapoxvirus infection (contagious ecthyma) in semi-domesticated reindeer (*Rangifer tarandus tarandus*). Vet Microbiol 162: 499–506. https://doi.org/10.1016/j.vetmic.2012.10.039

Tryland M, Beckmen KB, Burek-Huntington KA et al (2018) Orf virus infection in Alaskan mountain goats, Dall's sheep, muskoxen, caribou and Sitka black-tailed deer. Acta Vet Scand 60(1):12. https://doi.org/10.1186/s13028-018-0366-8

Tryland M, Das Neves CG, Klein J et al (2019a) Viral infections and diseases. In: Tryland M, Kutz SJ (eds) Reindeer and Caribou – health and disease. CRC Press, Boca Raton, FL, pp 273–303

Tryland M, Nymo IH, Sánchez Romano J et al (2019b) Infectious disease outbreak associated with supplementary feeding of semi-domesticated reindeer. Front Vet Sci 6:126. https://doi.org/10.3389/fvets.2019.00126

Ueda N, Inder MK, Wise LM et al (2007) Parapoxvirus of red deer in New Zealand encodes a variant of viral vascular endothelial growth factor. Virus Res 124:50–58

Vikøren T, Haugum M, Schulze J et al (2008) A severe outbreak of contagious ecthyma (orf) in a free-ranging musk ox (*Ovibos moschatus*) population in Norway. Vet Microbiol 127:10–20

Wilson TM, Poglayen-Neuwall I (1971) Pox in South American sea lions (*Otaria byronia*). Can J Comp Med 35:174

Wilson TM, Cheville NF, Karstad L (1969) Seal pox. Bull Wildl Dis Assoc 5:412

Zarnke RL, Dieterich RA, Neiland KA et al (1983) Serologic and experimental investigations of contagious ecthyma in Alaska. J Wildl Dis 19:170–174

Part IV

Harvesting the Arctic: Potential Health Threats for Arctic People

Hunting with Lead Ammunition: A One Health Perspective

Jon M. Arnemo, Boris Fuchs, Christian Sonne, and Sigbjørn Stokke

1 Introduction

Lead (Pb) from spent hunting ammunition is a serious health risk for humans, wildlife, and ecosystems and is a global One Health issue (Arnemo et al. 2016, Fig. 1). Humans are exposed to lead from ammunition by consumption of meat from game harvested with lead shot or lead-based bullets. Sources of exposure in scavenging wildlife includes slaughter remains and non-retrieved carcasses from game shot with lead-based ammunition. Predators are exposed through prey wounded by lead shot or lead-based bullets. Waterfowl and terrestrial birds pick up lead shot in shallow waters or from the ground or vegetation.

Lead-based bullets or lead shot are widely used for both small and large game hunting. Stokke et al. (2017) reported that 90% of the bullets used to harvest moose (*Alces alces*) in Finland, Sweden, and Norway were lead-based. In Norway, 88% of

J. M. Arnemo (✉)
Department of Forestry and Wildlife Management, Inland Norway University of Applied Sciences, Koppang, Norway

Department of Wildlife, Fish and Environmental Studies, Swedish University of Agricultural Sciences, Umeå, Sweden
e-mail: jon.arnemo@inn.no

B. Fuchs
Department of Forestry and Wildlife Management, Inland Norway University of Applied Sciences, Koppang, Norway
e-mail: boris.fuchs@inn.no

C. Sonne
Department of Ecoscience, Aarhus University, Roskilde, Denmark
e-mail: cs@bios.au.dk

S. Stokke
Norwegian Institute for Nature Research, Trondheim, Norway
e-mail: sigbjorn.stokke@nina.no

M. Tryland (ed.), *Arctic One Health*, https://doi.org/10.1007/978-3-030-87853-5_21

Fig. 1 The cycle of lead from spent hunting ammunition in an ecosystem perspective. (Illustration by Diogo Guerra)

ptarmigan (*Lagopus* sp.) hunters changed to lead shot after the total ban was lifted in 2015 (Arnemo et al. 2019a). In countries with a ban on the use of lead shot for hunting over wetlands, compliance is low, and enforcement is illusory. Several studies from the United Kingdom revealed that 70–90% of waterfowl were harvested with lead shot (e.g. Cromie et al. 2015; Pain et al. 2010). In a recent survey, Widemo (2021) showed that the proportion of Swedish hunters using lead shot over wetlands was 25.3% and 40.9% among those with and without a hunters' exam, respectively.

This chapter covers wound ballistics and lead toxicology and gives an overview of lead exposure from hunting ammunition in humans and wildlife with emphasis on the Arctic region.

2 Wound Ballistics

Wound ballistics describes the interactions between projectiles and live tissues (Kneubuehl et al. 2011). A basic understanding of this subject is fundamental for assessment of lead residues from ammunition in harvested game and killing efficiency of hunting bullets and shot.

2.1 Rifle Bullets

Most rifle bullets used for large game hunting are designed to expand upon impact (Stokke et al. 2018). There are two main types of expanding hunting bullets, heterogenous and homogenous. Heterogenous ("lead") bullets consist of a lead core. These types of bullets are covered with a copper alloy jacket except for the tip where the lead core is exposed. Homogenous ("non-lead") bullets consist of solid copper or a copper-based alloy with a hollow point at the tip. Expansion, which occurs within 0.1 ms after impact, is characterized by a mushroom-like frontal increase in the cross-sectional area of the bullet. Lead bullets expand due to the force acting on the exposed lead tip upon impact with tissue. The drag forces generated by the stagnation pressure exceed the yield limit for lead, which then behaves like an incompressible fluid. Pressure disperses within the floating lead and acts on the jacket from inside of the bullet, causing it to burst (Kneubuehl et al. 2011). Deformation and fragmentation persist as long as the stagnation pressure exceeds the yield limit for lead. Copper bullets expand in a similar manner if the cavity at the tip of the bullet is large enough for viscous pressure to enter. Copper bullets, however, deform significantly faster than lead bullets (Kneubuehl et al. 2011).

Bullet penetration is characterized by a cavity created by tissue impelled radially in relation to the velocity vector as momentum is imparted from the projectile to soft tissue, which undergoes elastic deformation due to stretching and compression (Kneubuehl et al. 2011; Stefanopoulos et al. 2014). This creates a temporary empty space behind the bullet, termed the temporary wound cavity. The vacuum and the elastic energy imparted to the tissue rapidly forces the displaced tissue to recoil towards its initial position, thus generating a brief oscillation (Fackler 2001; Harvey et al. 1946; Kneubuehl et al. 2011). The residual wound cavity, which is filled with blood, damaged tissue, and contaminants sucked in from the outside, is termed the permanent wound cavity (Fackler 1988; Janzon et al. 1997). The extravasation zone is the transition zone between the permanent wound cavity and surrounding tissue where crushing and laceration by direct contact with the bullet have not occurred. Hemorrhage in the extravasation zone results from the distention of the temporal cavity, inflicting damage to blood vessels as a result of overstretching and shearing effects (Kneubuehl et al. 2011; Stefanopoulos et al. 2014). Due to a proportional relationship between the kinetic energy of the penetrating bullet and the expansion of the temporary cavity, the potential energy stored in the tissue equals the work done to create the maximum expansion. The potential for this energy to cause wounding depends on four factors: (1) the magnitude of the stored energy in the tissue, (2) the ability of the tissue to sustain strain, (3) the size of the organ structure, and (4) the anatomical constraints to tissue movements (MacPherson 1994). The affected tissue ruptures and causes permanent wounding if the energy stored in tissue exceeds the elastic limit of the tissue. Tissue elasticity is therefore an important factor as it impairs the extent of permanent damage caused by a penetrating bullet.

Elastic tissues like muscle, skin, blood vessels, and lungs can absorb energy generated by a penetrating bullet and tend to recoil towards the bullet trajectory (Fackler 1988; MacPherson 1994; Karger 2008). Other organs, such as liver, kidney, and brain, may disrupt from penetrating projectiles because they are less resilient or constrained from movements (Caudell 2013; Roberts 1988; Stefanopoulos et al. 2014). The size of the organ or body is important because there will be a lower size limit whereby the temporary cavity will stretch all tissues beyond the elastic limit of the organ or body, causing it to rupture. For organs or bodies larger than this critical size, tissue damage primarily occurs by crushing, tearing, and stress (MacPherson 1994). The dimension of the permanent cavitation therefore tends to be approximately unchanged regardless of the body size (Stokke et al. 2018). As a result, time from bullet impact to exsanguination will increase with body size for mammals dispatched with rifle hunting bullets.

Expanding, high-velocity lead bullets fragment upon impact in game animals (e.g. Cruz-Martinez et al. 2015; Hunt et al. 2009; Knott et al. 2010; Menozzi et al. 2019; Stokke et al. 2017; Trinogga et al. 2019). Stokke et al. (2017) found that lead bullets retrieved from harvested moose lost on average 2.8 g of lead. The difference in lead loss between so-called bonded lead bullets and traditional lead-core bullets was marginal, 2.6 versus 3.0 g of lead, respectively. X-ray studies and necropsies of game animals shot with lead bullets have shown that hundreds of small fragments are broadly distributed along the wound channel and can be found as far as 45 cm from the bullet trajectory in the carcass (Fig. 2). Most fragments, however, are smaller than the resolution of an x-ray image and too small to be seen by the naked eye or detected while chewing. Kollander et al. (2017) found 27–50 million nanoparticles (range 40–750 nm) per gram of meat from a roe deer (*Capreolus capreolus*) and a wild boar (*Sus scrofa*) shot with one lead bullet each. Nanoparticles are so small that they can cross intact cell membranes.

Most non-lead bullets have a non-fragmenting design to ensure proper penetration and expansion upon impact. Non-lead bullets do, however, fragment to some degree, although much less than lead bullets. The fragments are fewer, larger and closer to the wound channel (Trinogga et al. 2019). Apparently, fragmentation of non-lead bullets varies between ammunition brands. Menozzi et al. (2019) radiographed two wild boars shot with non-lead bullets (Federal Vital Shock 270 Winchester 8.4 g) and did not find any signs of fragmentation. Irschik et al. (2013) x-rayed 46 wild ungulates of four different species shot with non-lead bullets, either Barnes TSX or Styria Arms Aero (calibers and mass of bullets not reported). For Barnes TSX, a fragment was found only in 1 of 34 carcasses, whereas fragments were found in all 12 carcasses from animals shot with Styria Arms Aero. A similar

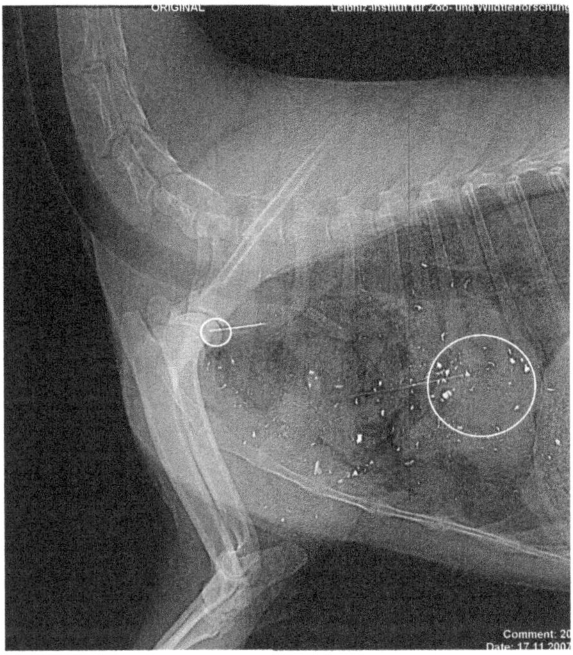

Fig. 2 X-ray image of a roe deer (*Capreolus capreolus*) shot with a lead-based hunting bullet (RWS H-Mantle 30-06 Springfield 11.7 g). Lead fragments are visible as white, radio-dense spots. Small white circle: bullet entry. Large white circle: bullet exit. (Courtesy of Anna Trinogga/ Oliver Krone). (Reproduced with permission)

low rate of fragmentation of Barnes TSX (308 Winchester 10 g) was reported by Grund et al. (2010), who found an average of two fragments per bullet fired into sheep carcasses. Stokke et al. (2017) examined bullets retrieved from shot moose and found that the average metal loss from non-lead bullets ($n = 181$) was 0.5 g compared to 2.8 g for lead bullets ($n = 1147$).

Some non-lead bullets are designed to break into three or four petals that will continue as separate projectiles after impact, whereas the remaining larger shank ensures proper penetration. The bullet subunits will generate their own wound channels, likely increasing the total amount of tissue trauma and hemorrhage and possibly shortening the time to death and the flight distance after fatal wounding. In addition to the main subunits, this type of bullet also leave more small fragments in tisues than bullets with a non-fragmenting design (Trinogga et al. 2019). Compared to lead fragments, however, these petals are larger and much harder and have sharp edges that may harm the gastrointestinal tract of scavengers and humans (Hampton et al. 2021). Non-lead bullets that are neither expanding nor fragmenting have also been designed for hunting (Gremse et al. 2014). These bullets are instable and will tumble during their trajectory in an animal, creating a larger wound channel than expanding bullets of the same caliber.

We are not aware of any scientific assessment of the efficiency of fragmenting non-lead bullets used for recreational or subsistence hunting. In a recent study, Hampton et al. (2021) compared fragmenting lead and non-lead bullets for aerial shooting ("culling") of wild pigs. They reported that the two bullet types were similarly effective but that lead bullets displayed a higher degree of fragmentation, indicating that fragamenting lead bullets may be an important contributor to lead exposure in scavenging wildlife.

Concerns have been raised over the efficiency of non-lead bullets (Caudell et al. 2012), including the perception of limited supply, higher costs, inferior killing efficiency, and correspondingly longer flight distances and higher wounding rates compared to traditional lead-based bullets (Thomas et al. 2016). Trinogga et al. (2013), using computed tomography to examine 34 carcasses of five different wild ungulate species, found no difference in wound channel diameters for lead and non-lead hunting bullets. In an extensive study that included 1234 roe deer and 825 wild boars, Martin et al. (2017) found no difference in flight distances between non-lead and lead ammunition. Several other studies comparing killing efficiency between non-lead and lead hunting bullets showed no significant differences (Kanstrup et al. 2016; Knott et al. 2009; McCann et al. 2016; Stokke et al. 2019).

2.2 Shotgun Shot

Wound ballistics of shot differs from rifle bullets in several ways. Shot are spherical, and although they may deform or fragment upon impact, they do not expand. One shotgun shell may contain 200–400 pellets, and thus, game animals are usually hit by several shot from the same shell. Similar to rifle bullets, however, the killing efficiency of shot depends on the physical trauma inflicted on vital organs and large blood vessels from individual pellets (Stokke et al. 2018). There is no so-called "shock effect", neither from shot nor from bullets. Instant death only occurs if vital parts of the brain are traumatized. Most game animals harvested with shot or bullets die from fatal hemorrhage secondary to tissue trauma. The concept of "shock effect" is a long-lived myth that has been perpetuated in the popular literature.

Contrary to a widespread belief among hunters, lead shot fragment upon impact in game animals (e.g. Andreotti and Borghesi 2013; Andreotti et al. 2016; Johansen et al. 2004; Mann et al. 1994; Pain et al. 2010; Scheuhammer et al. 1998). Pain et al. (2010) examined 121 birds of six different species harvested with lead shot and found lead fragments in 76% of the birds (Fig. 3). They stated that "the majority of fragments found were very tiny (i.e. less than about a tenth of a shot in size) and both too small and too scattered to be detected or removed by a consumer". Andreotti and coworkers x-rayed 196 European starlings (*Sturnus vulgaris*) (Andreotti and Borghesi 2013) and 59 Eurasian woodcocks (*Scolopax rusticola*) (Andreotti et al. 2016) harvested with lead shot and found ammunition residues in 60.2% and 96.6%, respectively, of the the birds.

Deformation, fragmentation, and/or shattering of non-lead shot depends on their chemical composition (Mann et al. 1994). Steel shot neither deform nor fragment. Bismuth shot may fragment and shatter upon impact with bone but do not fragment in soft tissue. Tungsten shot may fragment but do not shatter upon impact with bone and do not fragment in soft tissue. Wound ballistics of shot may change due to the addition of other elements, either as components of the shot alloy (e.g. antimony) or as coating of the shot (e.g. copper, nickel or zinc).

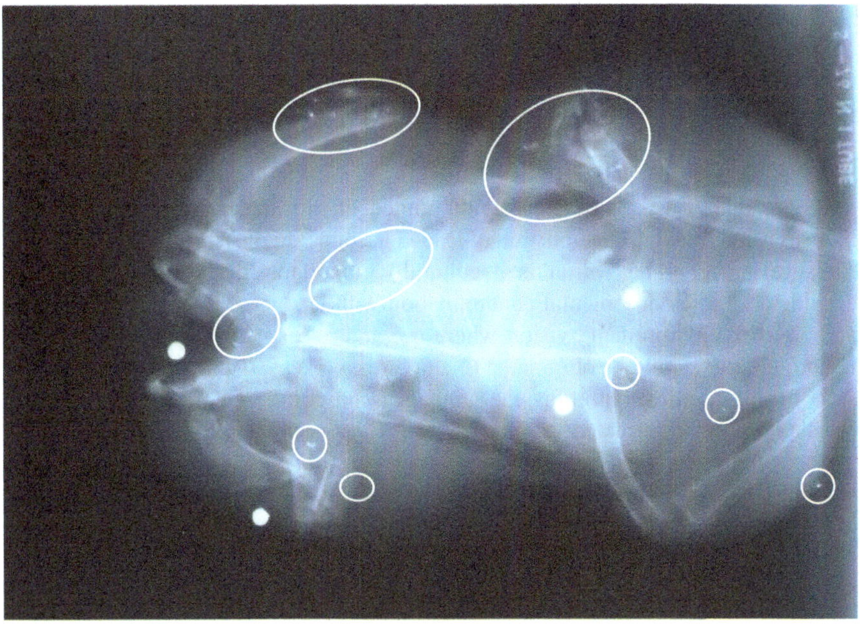

Fig. 3 X-ray image of a harvested wood pigeon (*Columba palumbus*) with lead shot (white dots) and lead shot fragments (inside white circles). (Reproduced from Pain et al. 2010)

Concerns have been raised over the killing efficiency of non-lead shot. In a well-designed, double-blinded study, Pierce et al. (2014) showed that steel shot and lead shot were equally effective for harvesting mourning doves (*Zenaida macroura*) under typical hunting conditions. The apparent widespread scepticism among hunters to non-lead shot alternatives is not based on scientific evidence.

3 Lead in the Environment

The 2500 years of lead production, use and release by humans have contributed to a global contamination with the element and have increased the environmental burden of lead in northern Europe to 1000 times the natural background concentrations (Renberg et al. 2001; Settle and Patterson 1980). Thus, environmental lead exposure of humans and wildlife is now entirely due to anthropogenic sources. In many high-income countries, the environmental deposition of lead has been significantly reduced during the past few decades, mainly due to a decrease in atmospheric deposition, improved waste management, and banning of leaded gasoline. In Sweden and Norway, lead levels in moss were reduced by 96% and 90%, respectively, from the mid 1970s to 2015 (Danielsson and Karlsson 2016; Miljøstatus 2019).

Hunting ammunition, however, remains a significant source of lead exposure in humans and wildlife. The worldwide consumption of lead for manufacturing of ammunition is 120,000 tons/year (ILZSG 2019). Although this only constitutes 1% of the global use of lead, the use of lead ammunition is now considered the greatest source of lead deliberately discharged into the environment in both the USA and the EU (Bellinger et al. 2013a; ECHA 2018). This may be even more pronounced in remote and less urbanized areas with limited historical deposition of lead from industrial activities and combustion of leaded gasoline. In the Arctic regions, hunting for subsistence or recreation is a culturally important lifestyle that contributes considerably to environmental contamination with lead. Sweden and Norway have a "non-toxic environment" vision where use and release of dangerous substances, including lead, should be eliminated (Naturvårdssverket 2019; Miljøstatus 2020). In spite of this political ambition, the annual discharge of lead from ammunition in Sweden has been estimated to 580–709 tons (Naturvårdsverket 2006). In Norway, 58 tons of lead shot were used in 2017, which constituted 67% of the total release of lead into the environment that year (Miljøstatus 2019).

4 General Aspects of Lead Exposure in Humans

Lead is a non-essential element that is toxic to all physiological systems even at very low levels of exposure. The toxicity of lead at the subcellular and molecular levels is complex (Gorkhali et al. 2016). Lead interferes with a variety of enzymes and creates reactive radicals that damage cell structures, including membranes and DNA. The developing brain is the organ most sensitive to lead exposure. Neurotoxic effects of lead include reduced neuronal growth, reduced numbers of neurons, disruption of synaptic activity, and interference with neurotransmission and neuronal signalling pathways.

Lead has been ranked as the worst toxic threat to human health (Pure Earth 2015). Although there is no safe level of lead exposure in humans (WHO 2018), health authorities have issued thresholds of blood lead levels (BLL) for potential negative health effects. The Centers for Disease Control and Prevention (CDC) in the USA now uses a BLL of 35 μg/L, termed "blood lead reference value", to identify children in the highest 2.5% of BLLs and to identify children who require case management (CDC 2021). This reference value, previously called "blood lead level of concern", was reduced from 100 μg/L to 50 μg/L in 2012. The European Food Safety Authority (EFSA) has set benchmark dose levels (BMDL) for developmental neurotoxicity ($BMDL_{01}$: 1% or 1 point reduction of IQ in children), chronic kidney disease ($BMDL_{10}$: 10% increase in prevalence of chronic kidney disease in adults), and increased systolic blood pressure ($BMDL_{01}$: 1% increase in systolic blood pressure in adults) (EFSA 2013). These BLL threshold values are 12, 15, and 36 μg/L, respectively.

Especially at risk for lead toxicity is the developing central nervous system in both fetuses and children. Lead is transferred over the placenta to fetuses in umbilical cord blood and to infants in breast milk. Lead exposure in early life is known to reduce IQ, cognitive abilities, and academic achievement and to cause idiopathic intellectual disability and mild to moderate mental retardation (CDC 2021; Landrigan et al. 2018). These effects are permanent and cannot be treated or reversed. On a global scale, one in three children, approximately 800 millions, has BLLs above the CDC reference value (UNICEF and Pure Earth 2020). The vast majority of these children live in low-income countries. This is in stark contrast to the situation in many high-income countries where the BLLs of children has been significantly reduced during the past 40 years. Monitoring of children in southern Sweden showed that their mean BLL dropped by nearly 90% from 1978 (64 μg/L) to 2019 (7 μg/L) (Naturvårdsverket 2020). The most important contribution to this decrease is most likely the gradual reduction of lead in gasoline followed by a ban of leaded gasoline in 1994.

Lead toxicity in adults includes elevated risk of cardiovasular and chronic kidney disease (EFSA 2013). Low-level environmental lead exposure has been termed a "silent killer" because the negative effects largely went undetected (Nawrot and Staessen 2006). A recent study showed that 400,000 annual deaths from cardiovascular diseases in the USA are caused by lead exposure (Lanphear et al. 2018). Increased rates of mortality were seen at BLLs below the CDC reference value and at BLLs commonly found in people with moderate intakes of meat from game harvested with lead-based ammunition.

In humans, the pharmacokinetics of lead follows a three-compartment model, with different half-lives of elimination from blood, soft tissues, and bones (EFSA 2013; Rabinowitz et al. 1973). In adults, 94% of body lead is stored in bones where Pb^{2+} substitutes Ca^{2+}. The half-life of lead in bones is extremely long, 10–30 years. Thus, the total body burden of lead may continue to increase even at very low levels of exposure. Only 2% of body lead is found in the blood, and 99% of this lead is bound to red blood cells. The half-life of lead in blood is 4–5 weeks. The BLL of an individual mainly reflects recent exposure but may also be highly influenced by the current metabolic status of that person. Bone-to-blood mobilization of lead occurs during pregnancy, lactation, fasting, menopause, reduced kidney function, osteoporosis, and fractures and may bring the BLL above toxic thresholds. Lead is mainly excreted in the bile and urine, with some lead being reabsorbed from the intestinal tract where conjugated lead is broken down by enzymatic activity. Small amounts of lead are eliminated in deciduous teeth, hair, nails, and sweat. In pregnant or lactating women, transfer of lead to the fetus and milk may constitute significant routes of elimination.

5 Lead Exposure from Ammunition in Humans

The main pathway of lead exposure from ammunition is the ingestion of shot or fragments of shot or bullets in meat from game harvested with lead-based ammunition. Although the bioavailability of metallic lead from ammunition is lower than that of organic forms of lead, there is no doubt that lead in ammunition fragments is dissolved at the low pH found in the stomach (shown by laboratory experiments, Livsmedelsverket 2014a) and that lead from ammunition residues in game meat is absorbed into the blood, causing an elevated BLL (shown by experiments in pigs, Hunt et al. 2009). In general, infants and older children absorb lead in the diet at a much higher rate than do adults (Mushak 1998). Studies on infants show that lead uptake is in the order of 50%, whereas adults typically absorb 10–15% of the lead in the diet. Hunters may also be exposed to lead from shooting or handloading of lead-based ammunition, due to inhalation of lead fumes/dust or ingestion from contaminated hands (University of Washington 2019).

Globally, millions of people eat wild game. Signs of lead exposure are typically non-specific and include headache, abdominal pain, constipation, learning difficulties, loss of appetite and weight, increased blood pressure, joint and muscle pain, mood disorders, and difficulties with memory or concentration. Thus, diagnosis of lead poisoning may be difficult to differentiate from other conditions. Parry and Buenz (2020) suggested to ask patients with such subtle symptoms the simple question: "Do you eat wild meat shot with lead?"

5.1 Lead Residues in Game Meat

Numerous studies have shown that meat from game harvested with lead bullets or shot contain lead residues (see review by Green and Pain (2019) for a European perspective). Here, we cover selected reports mainly from the Arctic region.

All lead shot and bullets fragment to some degree, depending on several factors such as velocity of the projectile, bullet design, point of impact in the target animal, and the type of tissue penetrated. The size of the fragments varies from large pieces to miniscule nanoparticles. The health risks of lead residues ingestion are likely more dependent on the total surface of the fragments rather than the total mass of lead ingested (Green and Pain 2019; Livsmedelsverket 2014a). Lead is easily dissolved at low pH, such as in the stomach, and more lead becomes bioavailable when the surface area of the lead fragments increases (Livsmedelsverket 2014a). Also, recipes based on marination and cooking of the meat in wine or vinegar will dissolve lead and increase its bioaccessibility, potentially contaminating the entire product (Mateo et al. 2007, 2011).

The European Commission (EC) has set a maximum allowable level (ML) of lead at 0.10 and 0.50 mg/kg wet weight for meat and offal, respectively, from bovines, sheep, pigs, and poultry (Thomas et al. 2020). Except for Russia, no such regulations exist for edible parts of wild game, neither in the EU nor in any other region or country. In Russia, the ML for lead in meat from all types of domestic and game animals is 0.50 mg/kg wet weight (Chief State Sanitary Doctor of the Russian Federation 2011).

Lindboe et al. (2012) measured the lead concentration in 52 batches of minced (ground) meat from moose harvested with lead-based bullets in Norway. The samples were intended for private consumption and were provided by hunters. Lead above the limit of quantification (0.03 mg/kg) was detected in 81% of the samples. The mean lead concentration was 5.6 mg/kg, which is 56 times the ML set by the EC for meat from domestic animals. The authors also analyzed a subsample of five packages of minced moose meat purchased at stores in Oslo, Norway, and found a mean lead concentration of 1.5 mg/kg.

Based on the study by Lindboe et al. (2012), a risk assessment of lead exposure from cervid meat in Norwegian consumers was carried out (VKM 2013). This risk assessment was followed by dressing recommendations from the Norwegian Food Safety Authority. In large game harvested with lead bullets, tissues visibly traumatized by the projectile plus an additional 10 cm radius of tissues around the wound channel should be removed and discarded. One year after these recommendations were issued, the Norwegian Food Safety Authority carried out a follow-up study on lead in moose meat (Eurofins 2015). In 150 samples of minced moose meat purchased from 27 different stores in southern Norway, lead concentrations above the limit of quantification (0.01 mg/kg) were detected in 100% of the samples and 73% of them were above the ML set by the EC for meat from domestic animals. The mean lead concentration was 1.79 mg/kg, which is 18 times the ML set by the EC.

A similar study on lead residues in minced moose meat was carried out in Sweden (Livsmedelsverket 2014a). A total of 54 packages intended for human consumption were collected nationwide—48 were provided by hunting teams and six were bought in five different stores. Before chemical analysis of the the lead concentration, all samples were x-rayed, and visible lead fragments (resolution $\geq$0.1 mm) were found in 35% of the samples. In 54% of the samples, the lead concentration was above the limit of quantification (0.02 mg/kg), and in 33% of the samples, the lead content was above the ML set by the EC for meat from domestic animals. Analysis of samples from animals not harvested with lead-based ammunition, using a different laboratory technique, showed that the "background" lead concentration in moose muscle was $\leq$0.002 mg/kg.

Recently, the Swedish Food Agency (Livsmedelsverket 2020b) reported a nationwide, follow-up assessment of lead in 100 minced meat samples (50 from moose and 50 from wild boars) purchased at 47 different game slaughterhouses. The results showed that 21% of the samples contained traces of lead (0.015–0.1 mg/kg) and that 15% of the samples had lead levels above the ML set by the EC. Samples from 45% of the slaughterhouses all contained lead residues, whereas no traces of lead were found in samples from the other 55% of the slaughterhouses.

In a study from Canada, McAuley et al. (2018) measured lead concentrations in the pectoral muscles from 13 grouse (*Bonasa umbellus* and *Falcipennis canadensis*) harvested with lead ammunition. Twelve birds were shot with a caliber 0.22 rifle and one with a shotgun. Prior to analysis, all visible bullet fragments and shot were removed. Muscle samples were categorized as having visible signs of bullet/shot impact or not. The mean lead concentration was 0.968 mg/kg in impacted muscle compared to 0.013 mg/kg in non-impacted muscle.

Several studies show that edible parts of small game harvested with lead shot may contain large amounts of lead, even after the removal of visible shot and fragments (e.g. Johansen et al. 2004; Livsmedelsverket 2014a; McAuley et al. 2018; Scheuhammer et al. 1998). Scheuhammer et al. (1998) carried out a nationwide study in Canada in order to determine the frequency of elevated lead concentrations

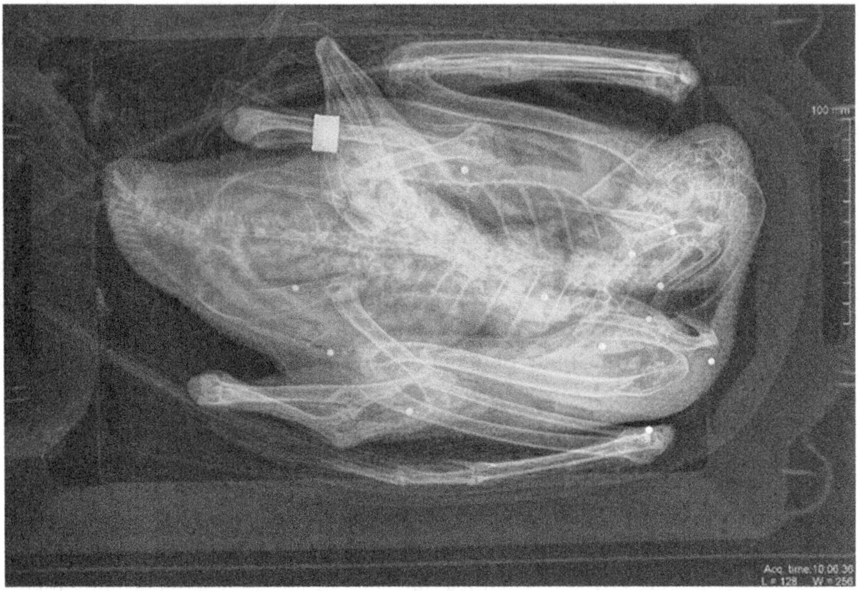

Fig. 4 X-ray image of a live pink-footed goose (*Anser brachyrhynchus*) captured for tagging. White dots ($n = 12$) are shotgun shot of unknown type. (Courtesy of Department of Bioscience, Aarhus University, Denmark) (Reproduced with permission)

in pectoral muscles of hunter-killed birds. They collected nearly 4000 individual birds of 44 different species from 125 sites across the country. The birds were harvested using shotguns and lead shot. Muscle samples were pooled according to species and location. Of 827 pools, 92 (11%) had a lead concentration above 0.5 mg/kg. The mean lead concentration for these 92 pools was 12 mg/kg, which is 120 times the ML set by the EC for meat from poultry. Johansen et al. (2004) analyzed the lead content in meat from the two most important bird species in the diet in Greenland, thick-billed murre (*Uria lomvia*) and common eider (*Somateria mollissima*). In the latter species, the mean lead concentration in meat from 25 birds harvested with lead shot was 6.1 mg/kg, which was 44 times higher than for 25 common eiders that drowned in fishing nets, even though 11 of the drowned birds had 1–3 embedded lead shot from previous wounding. The Swedish Food Agency analyzed the lead content in one European hare (*Lepus europaeus*) and one roe deer killed with lead shot (Livsmedelsverket 2014a). After the removal of traumatized and bloody tissues, the mean meat lead concentration in various parts (number of samples not given) ranged from 0.02 to 5.5 mg/kg in the hare and 0.1 to 21 mg/kg in the roe deer.

Apparently healthy small game may have shot from previous hunting attempts embedded in their tissues (Fig. 4). The proportion of animals with embedded shot is referred to as wounding or "crippling" rate. In a study carried out in northern Norway, 43 mountain hares (*Lepus timidus*) were harvested with rifle bullets and x-rayed using computerized tomography (Larsen and Nybakk 2011). At least 16% of the carcasses contained embedded shot. Falk et al. (2006) conducted an x-ray study on wounding rates in 993 common eiders and king eiders (*Somateria spectabilis*) in Greenland and found that on average 22% of common eiders carried embedded shot. The prevalence of embedded shot was age-dependent, varying from 13% in immatures to 29% in adults. Hicklin and Barrow (2004) used fluoroscopy to examine 1624 live waterfowl of four different species in order to detect embedded shot. They found that 25% of the birds carried embedded shot. These reports show that in areas with a high hunting pressure, where the use of lead shot is allowed, a very high proportion of apparently healthy small game will contain lead residues from previous hunting attempts. Thus, consumers may unknowingly be exposed to lead, even if the game eventually was harvested with lead-free ammunition.

5.2 Lead Pellet Ingestion in Game Meat Consumers

Although lead shot commonly used for hunting are relatively large (2.5–3.5 mm diameter), they are frequently ingested when people eat meat from game hunted with shotguns. Most shot probably pass through the intestinal tract to be eliminated over time. Several studies, however, show that shot may be retained in the appendix. During routine radiographical examination, Reddy (1985) identified 62 patients with retained lead shot in their appendices. The number of shot per patient varied from one to more than 200. The patients were native to northern Newfoundland and Labrador, Canada, and included First Nation and settler populations. The author stated that "eating of wild game was the source of lead shot". In another Canadian study, Tsuji and Nieboer (1997) examined 132 archived radiographs for evidence of retainted lead shot in First Nation Cree of the western James Bay and northern Ontario regions. Lead shot were found in 15% of the randomly selected cases, either in the appendix (7%) or intraluminally in other parts of the gastrointestinal tract (8%). The authors assumed that consumption of wild game meat with embedded lead shot was the source of the pellets. Carey (1977) reported two cases of acute, perforated and gangrenous appendicitis in First Nation Inuit who hunted and ate game regurlarly. During surgery, approximately 100 shot was removed from a 56-year-old female and more than 500 pellets from a 57-year-old male. In another case report, Cox and Pesola (2005) presented a radiograph of a 73-year-old Inuit woman with the appendix literally full of lead shot (Fig. 5). According to the authors, retained shot in the appendix are commonly seen in Alaskan natives, and the source is the consumption of meat from hunted waterfowl.

5.3 Blood Lead Level (BLL) in Game Meat Consumers

Consumption of meat from game harvested with lead-based ammunition poses a health risk to humans (see review by Green and Pain 2019). Even a moderate intake of meat from game shot with lead-based ammunition may cause an elevated BLL.

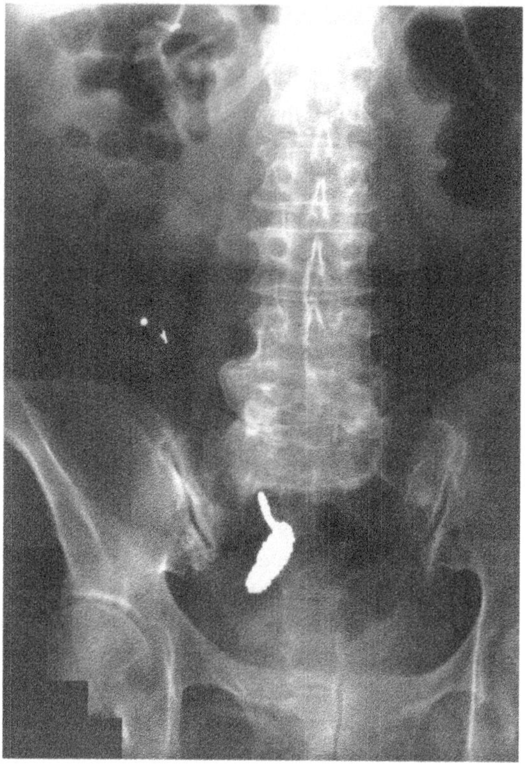

Fig. 5 X-ray image of a 73-year-old Inuit woman with the appendix completely full of lead shot, with the contour of the appendix easily visualized. An additional lead shot (round white dot) from a recent meal is seen in the intestinal tract. (©Massachusetts Medical Society). (Reproduced with permission from Cox and Pesola 2005)

Due to the long half-life of body lead in humans, low-level exposure may increase the total body burden of lead over time and eventually bring the BLL above thresholds for documented toxic effects. Here, we review relevant studies on the BLLs of game-consuming people in the Arctic regions.

The Arctic Monitoring and Assessment Programme (AMAP) has reported levels and trends of contaminants in humans in the eight Arctic countries (AMAP 2015). In the late 1990s, hotspots for lead exposure, attributed to traditional food consumption, were identified in Greenland and Canada. In general, BLLs have shown a decreasing trend, which most likely is due to reduced lead emissions and a partial ban on lead shot in the circumpolar region. Interestingly, another contributing factor is probably a gradual change from a traditional diet largely based on hunted game meat to processed food bought in stores, especially in the the younger population. Although the reported BLLs were relatively low (<40 μg/L), children and women of reproductive age in Russian and Canadian subpopulations were identified as groups at risk.

Due to the effects on the developing central nervous system, fetuses and infants are especially at risk for lead toxicity. Studies from Canada have shown that lead levels in maternal blood, breast milk, umbilical cord blood, and infant blood were correlated and that these levels were also correlated with maternal traditional game

Fig. 6 Lead from spent hunting ammunition is poisoning consumers of game meat and scavenging wildlife. (Illustration by Juliana D. Spahr, SciVisuals.com)

meat intake (Hanning et al. 2003, Fig. 6). Two studies on Greenland Inuits showed that frequent consumption of seabirds harvested with lead shot caused very high BLLs. In the first study (Bjerregaard et al. 2004) conducted in 1993–1994, the mean BLLs of people eating seabird meat several times a week or daily were 110 and 170 μg/L, respectively. The highest level exceeds the EFSA thresholds for increased systolic blood pressure and chronic kidney disease by factors 5 and 11, respectively. In 2003–2004, Johansen et al. (2006) carried out a similar study and found that the mean BLL of people eating seabirds regularly was 128 μg/L compared to 15 μg/L in participants reporting no consumption of seabirds.

Several reports from Norway and Sweden have shown a positive association between BLLs and consumption of game meat (Birgisdottir et al. 2013; Bjermo et al. 2013; Livsmedelsverket 2014b; Meltzer et al. 2013; VKM 2013; Wennberg et al.

2017). Other hunting-related factors positively correlated with the BLL were shooting and hand-loading of ammunition (Livsmedelsverket 2014b; Meltzer et al. 2013). Although the mean or median BLLs in game consumers were relatively low (<30 µg/L), they were considerably higher than in non-game consumers or the general population and exceeded the EFSA thresholds for developmental neurotoxicity and chronic kidney disease.

Canadian studies over the past three decades have shown that elevated BLLs in First Nation people were associated with a traditional diet based on meat from game harvested with lead-based ammunition (e.g. Dallaire et al. 2003; Dewailly et al. 2001, 2007; Fontaine et al. 2008; Hanning et al. 2003; Lévesque et al. 2003; Tsuji et al. 2008c). A significant risk of elevated BLLs has also been attributed to the use of firearms, lead shot, and lead bullets (Liberda et al. 2018). Although time trends showed a significant decrease in lead exposure (Dallaire et al. 2003; Fontaine et al. 2008; Lévesque et al. 2003), likely due to a change in dietary habits, a partial ban on lead shot and a public intervention to reduce the use of lead shot, elevated levels of lead exposure are still found in some communities. Two studies showed that the the continuous use of lead ammunition for hunting posed a risk for elevated BLLs. Tsuji et al. (2008c) reported that 18% of 196 First Nation people of northern Ontario examined in 1999–2000 had BLLs ≥100 µg/L, the CDC level of concern at that time. In another study, Liberda et al. (2018) found that in one Cree community of Quebec the mean BLL was 87 µg/L, which is almost six times higher than the EFSA threshold for chronic kidney disease.

In a study carried out on the Russian Kola Penninsula, very high BLLs were found in children (5–14 years) living in a remote and isolated village with Sami, Komi and Nenets populations (Odland et al. 1999). The mean BLL was 83 µg/L, which is seven times higher than the EFSA threshold for developmental neurotoxicity, and 36% of the children had a BLL ≥100 µg/L, the CDC level of concern at the time of the study. The authors concluded that substantial contributions to the elevated BLLs came from dietary sources, especially meat from game harvested with lead ammunition.

In a recent study from Alaska, Bressler et al. (2019) reported that 80 (1.6%) of 4962 children (<6 years) tested 2011–2015 had a BLL of 50 µg/L or more. Of those, 54 (68%) had a BLL of 50–99 µg/L, 13 (16%) had a BLL of 100–149 µg/L, and 13 (16%) had a BLL of 150–450 µg/L. In addition to the 80 BLLs of 50 µg/L or more found during 2011–2015, 74 more were identified during 2016–2018, yielding a total of 154 BLLs of 50 µg/L or more during 2011–2018. Possible exposure sources were identified based on interviews with the families of 74 children with BLLs ≥50 µg/L. Except for the parents' occupation, consumption of game or subsistence meat hunted with lead-based ammunition was listed as the most frequent possible cause of lead exposure and elevated BLL.

Although foreign bodies in the appendix is quite rare in the general population, lead shot may be retained in the appendix in people who frequently consume meat from game harvested with shotguns (Klingler et al. 1998; Reddy 1985; Tsuji and Nieboer 1997). Even a small number of retained shot may increase the BLL. In a study from Denmark, rarely cited in current literature, Madsen et al. (1988) measured the BLL of seven patients with 1–2 retained lead shot in the appendix and 14 sex- and age-matched controls without lead shot in the appendix (presence or abscence of lead shot was determined by x-ray). The mean BLL in the patients with retained lead shot was 114 μg/L, which was almost twice the mean BLL in the controls. The authors concluded, based on contemporary understanding of lead toxicity, that "none of the seven patients with lead shot in the appendix had blood lead levels approaching the toxic level". The mean BLL in the patients, however, was 3 and 7.6 times higher than the EFSA thresholds for increased systolic blood pressure and chronic kidney disease, respectively.

All studies cited above were based on epidemiological or other indirect data to show an association between elevated BLLs and consumption of meat from game harvested with lead-based ammunition, shooting/hunting with lead shot or lead bullets, hand-loading of lead-based ammunition and ingestion of lead pellets. The element lead (Pb) consists of four stable isotopes, ^{204}Pb, ^{206}Pb, ^{207}Pb, and ^{208}Pb, varying in their ratios depending on the source of lead. Although the current recycling of lead is a complicating factor, specific lead isotope ratios have the potential as signatures ("fingerprints") for tracing lead in blood or tissue samples. Four studies in Canadian First Nation people used this lead isotope approach to identify ammunition as a major source of lead exposure (Fillion et al. 2014; Lévesque et al. 2003; Tsuji et al. 2008a, b).

6 Lead Exposure in Dogs Fed Game Meat

Dogs may be exposed to lead residues if they are fed trimmings or offal from game harvested with lead-based ammunition (Fig. 1). The Norwegian Scientific Committee for Food Safety carried out a risk assessment of lead exposure from cervid meat in hunting dogs and concluded that "the risk for chronic health effects in dogs repeatedly fed cervid meat or offal originating from trimmings from the wound channel after use of lead-based ammunition can be considered as high and of concern" (VKM 2013). In a follow-up study, Høgåsen et al. (2016) identified data gaps and pointed out the need for more research to assess the consequences of feeding dogs game meat that may contain lead residues. They concluded that "trimmings close to the wound channel should be made inaccessible to dogs, as well as to other domestic or wild animals". We are aware of only one published study on actual lead exposure in dogs fed meat and offal from game hunted with lead-based ammunition. Fernández et al. (2021) found elevated lead concentrations in blood and hair from 31 dogs consuming meat and offal from culled wild boars and axis deer (*Axis axis*) in a national park in Argentina. The mean BLL was 189.1 μg/L and 19% of the dogs had a BLL above 400 μg/L, which is considered a marker for lead poisoning in dogs (Høgåsen et al. 2016).

7 Lead Exposure in Wildlife

Lead poisoning of wild birds from spent hunting ammunition has been recognized for almost 150 years. Several letters to the magazine The Field in the 1870s and 1880s reported poisoning and deaths in pheasants from ingested lead shot (e.g. Calvert 1876). Since then, lead intoxication linked to ammunition sources has been documented in a large number of avian species, especially waterfowl, terrestrial game birds, raptors, and scavengers (see reviews by Pain et al. 2019 and Monclús et al. 2020). Lead toxicity from spent ammunition is a threat to avian species across the globe and on all continets (Lohr et al. 2020). Protection of the iconic California condor (*Gymnogyps californianus*) from lead posioning (Church et al. 2006) was a main driver behind the State of California banning all use of lead-based ammunition for hunting from 1 July 2019.

There are multiple routes of lead exposure from spent ammunition in birds. Waterfowl and terrestrial game birds pick up shot in shallow waters and from the ground, mistaking them for grit or food. Predatory birds are exposed from shot embedded in tissues of wounded prey, whereas scavengers are poisoned by shot or bullet fragments in carcasses or slaughter remains (Pain et al. 2019). Mörner and Petersson (1999) reported lead posioning of woodpeckers (*Picus canus* and *Dendrocopus leucotos*) in Sweden. Although the authors were uanable to determine the source of lead, they hypothesized that woodpeckers, during search for insects, could be exposed from lead pellets shot into trees by hunters.

There are very few published studies on lead exposure in scavenging or predatory mammals (Rodríguez-Jorquera et al. 2017; Lazarus et al. 2017). Rogers et al. (2012) investigated whether large carnivores in the greater Yellowstone ecosystem, USA, were exposed to lead. They reported that brown bears (*Ursus arctos*, $n = 82$) had a mean (range) BLL of 55 (11–186) μg/L, whereas the corresponding values in American black bears (*U. americanus*, $n = 35$) were 19 (5–69) μg/L. In a recent study from Croatia and Poland, Lazarus et al. (2020) found that brown bears ($n = 26$) had a mean (range) BLL of 61 (28–139) μg/L. None of these studies identified possible sources of lead exposure in the bears.

Lead exposure in Arctic biota was reported by the Arctic Monitoring and Assessment Programme nearly 20 years ago (AMAP 2002). Although there had been a significant reduction in the deposition of atmospheric lead in the Arctic regions, levels in wildlife had not measurably declined. The report pointed out that lead shot from hunting remained a problem and recommended a complete ban on the use of lead shot in the Arctic. Additionally, the report highlighted the need for effect studies in priority species experiencing heavy metal exposure of concern.

In Sweden and Finland, lead exposure from spent ammunition is a significant morbidity and mortality factor in white-tailed sea eagles (*Haliaeetus albicilla*) (Axelsson 2009; Helander et al. 2009; Isomursu et al. 2018) and golden eagles (*Aquila chrysaetos*) (Axelsson 2009; Ecke et al. 2017). Ecke et al. (2017) showed that BLLs in Swedish golden eagles were correlated with progress of the moose hunting season and that sublethal lead exposure may impair flight performance and thus increase the risk of mortality from other causes. Isomursu et al. (2018) found the highest frequency of lead poisoning in Finnish white-tailed sea eagles during the main hunting season. A study from Canada (Legagneux et al. 2014) revealed that BLLs in common ravens (*Corvus corax*) increased over time during the moose-hunting season in an area with high hunting success, matching the number of rifle-shot moose. The main source of lead in scavenging birds is most likely bullet fragments in gut piles, offal, and carcasses from animals hunted with lead-based ammunition (Helander et al. 2009; Isomursu et al. 2018). Stokke et al. (2017) estimated that the annual amount of lead deposited in slaughter remains left in nature from moose hunting in Finland, Sweden, and Norway constitutes >100,000 lethal doses for eagles. Also, lead shot in wounded prey pose a risk to predators. There are numerous reports showing that a high percentage of small game has embedded lead shot in their tissues, e.g. as much as 35% of common eiders in heavily hunted coastal areas of Greenland (Falk et al. (2006), at least 16% of mountain hares in northern Norway (Larsen and Nybakk 2011) and on average 25% of waterfowl in Canada's Atlantic provinces and eastern subarctic region (Hicklin and Barrow 2004).

In an ongoing study in south-central Scandinavia, Fuchs et al. (2021) measured the BLL in 136 samples from 99 brown bears (1–25 years), including 22 females with a total of 45 dependent cubs. Additionally, the lead concentration was measured in milk samples from 15 lactating females. The mean (range) BLL was 96 (36–221) μg/L. Lactating females had higher BLLs than non-lactating (younger) females. The BLLs in dependent cubs were correlated with their mother's BLL, which in turn was correlated with the milk lead concentration. Thus, bear cubs will start accumulating lead from the first day of their lives and will continue to absorb lead from the milk as long as they suckle. In Scandinavia, two of three brown bear litters separate from their mothers at 16 months of age; the rest separate 1 year later. In humans, high blood lead levels are positively correlated with the duration of breastfeeding (Lozoff et al. 2009). Although critical levels have not been established in brown bears, the authors hypothesized that life-long lead exposure may have adverse effects on both individual and population levels. The high BLLs found in brown bears contrast the general reduction in environmental lead contamination over the past decades in Scandinavia. The sources and routes of lead exposure in the brown bears have not been identified.

Bakka (2020) compared lead exposure in brown bears in Scandinavia and three national parks in Alaska, USA, and found that the mean BLL in Scandinavia was 2, 5, and 13 times higher than the mean BLLs in Gates of the Arctic, Lake Clark, and Katmai, respectively. These results indicate that differences in diet and environmental lead sources have a huge impact on the BLLs in this species.

Thresholds for toxic effects of lead have largely not been established in wildlife. The challenge of establishing critical blood or tissue concentrations is due to the lack of data on cause-effect relationships, especially in Arctic species (AMAP 2002). Valid thresholds for toxic effects are most likely species dependent and translation of critical levels across species is unreliable at best. Buekers et al. (2009) reviewed 80 toxicity studies of which 19 experimental and 6 field studies on mammals and birds were selected in order to establish a critical blood lead concentration in wildlife. The authors reported that the highest "no observed effect concentrations"

(NOECs) varied 50-fold among species. They then used the fifth percentile of species NOEC values with data from 15 different species to calculate a critical blood lead concentration in mammals and birds, which turned out to be 180 and 710 μg/L, respectively. One problem with this approach is that it is mainly based on effects on easily observable variables, such as growth and reproduction. Subtle changes in acute senses, mental capacity, organ function, and behavior are unlikely to be detected in wild animals. In a study on golden eagles, however, Ecke et al. (2017) found that negative effects on flight performance occurred at a BLL of 250 μg/L, which is only 35% of the avian critical blood lead concentration suggested by Buekers et al. (2009). In this context, the population-wide lead exposure in Scandinavian brown bears reported by Fuchs et al. (2021) is interesting. The mean BLL, 96 μg/L, is only half of the mammalian critical blood lead concentration calculated by Buekers et al. (2009) and should pose no adverse health effect, at least not on growth and reproduction. It is, however, approximatey three, six, and eight times higher than the thresholds for increased systolic blood pressure, chronic kidney disease, and developmental neurotoxicity, respectively, in humans (EFSA 2013). Obviously, species-specific toxicity studies are highly needed in wildlife.

8 How to Avoid Lead Exposure from Game Meat

There is an obvious and simple way to avoid lead exposure from hunting ammunition and that is to use lead-free alternatives. Non-toxic ammunition is now available in most calibers and guns commonly used for both small and large game hunting. The price and efficiency of lead-free bullets and shot are comparable to lead-based ammunition (e.g. Trinogga et al. 2013; Kanstrup and Thomas 2020). Due to tradition, conservatism, and ignorance, however, an overwhelming majority of hunters are still using lead-based ammunition. Buenz (2016) indicated a contradictory attitude among hunters: "People generally reject the idea of injecting toxic substances into food, except when it involves hunting wild game". Based on risk assessments, health authorities have therefore issued dietary and dressing recommendations for game harvested with lead bullets and shot.

The Swedish Food Agency recommends that in game harvested with lead bullets, tissues traumatized by the projectile and at least an additional 10 cm radius of tissues around the wound channel should be removed and should not be used as food for humans or animals (Livsmedelsverket 2020a). Figure 7 shows the carcass of a fallow deer (*Dama dama*) calf harvested with lead ammunition and dressed according to the Swedish recommendations. In game harvested with lead shot, visible lead pellets and traumatized tissues should be removed. If these dressing practices cannot be verified, children <7 years, pregnant women, and women who plan pregnancy within 3 months should not eat products such as minced meat and sausages, whereas other people, including breastfeeding women, should not eat such products more than once a month.

The Norwegian Food Safety Authority has issued similar recommendations (Mattilsynet 2019). In large game harvested with lead ammunition, tissues

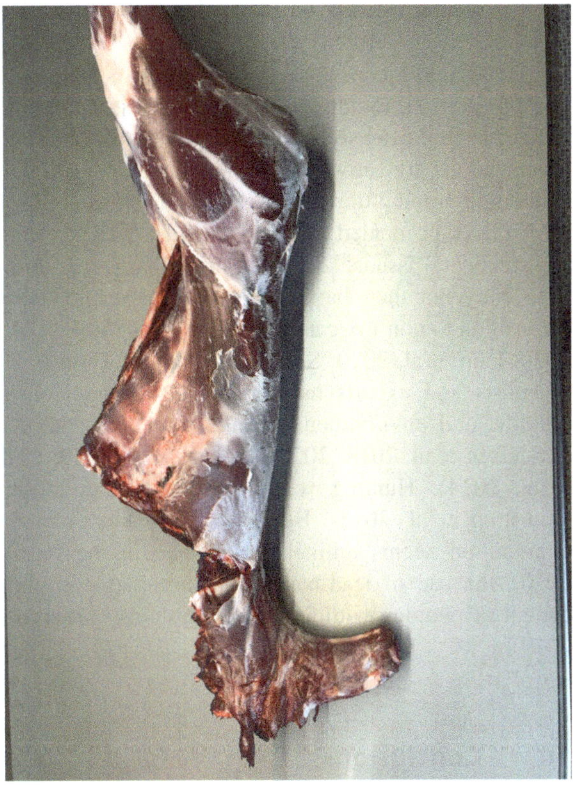

Fig. 7 Carcass of a fallow deer (*Dama dama*) calf harvested with a lead bullet (6.5 × 55 Lapua Mega 10.1 g) and dressed according to Swedish recommendations (Livsmedelsverket 2020a). Both front limbs were removed (approximately 2 kg), and net weight after dressing was ca. 12 kg. (Courtesy of Niels Kanstrup) (Reproduced with permission)

traumatized by the bullet and at least an additional 10 cm radius of tissues around the wound channel should be removed and discarded. People should not eat small game harvested with lead shot more than once a month, and visible lead pellets and traumatized tissues should be removed. Women who plan pregnancy, are pregnant, or are breastfeeding and children <7 years should not eat game meat that can possibly contain lead.

In Finland, a less specific recommendation has been issued. The Finnish Food Authority (2020) simply states that "people who eat game shot with lead bullets should avoid not only the gunshot wound but also an area of at least 10 cm of the normal-looking meat surrounding it".

In Norway, it has been estimated that due to the dressing recommendations, 2 kg of meat will be discarded from each animal shot with lead bullets (NTB 2016). The annual harvest of large game in Norway is approximately 110,000 animals (moose, red deer (*Cervus elaphus*), reindeer (*Rangifer tarandus tarandus*) and roe deer). Assuming that 90% of the hunters use lead-based ammunition (Stokke et al. 2017), the discarded meat amounts to 200 metric tons per year, which should neither be fed to pets nor left in nature.

9 Lead-Based Ammunition: Science Versus Politics

In spite of a 99% consensus among scientists that lead from spent hunting ammunition poses a risk to human, wildlife, and environmental health, this issue is highly politicized and extensively debated worldwide (Arnemo et al. 2016). Major lobbyist organizations are actively opposing any attempt to ban, or even restrict, the use of lead-based ammunition (Kanstrup and Thomas 2020). Extensive scientific evidence is disputed or denied, using strategies well known from discussions on other environmental issues, such as climate change and pesticides (Arnemo et al. 2016). In recent years, there have been numerous science-based proposals and iniatives for a transition to non-toxic ammunition (e.g. Cromie et al. 2019; Kanstrup and Thomas 2020; Pain et al. 2020; Schulz et al. 2019; Thomas et al. 2019; Treu et al. 2020). Scientists and experts have published several consensus statements on human, wildlife, and environmental health risks from the use of lead-based ammunition (Andersen et al. 2018, 2020; Arnemo et al. 2020; Bellinger et al. 2013b; Bernhoft et al. 2014). Hunting with lead-based ammunition is simply not sustainable (Kanstrup et al. 2018). Because a voluntary, major switch to lead-free hunting ammunition seems unlikely to happen in the foreseeable future (Schulz et al. 2020), the use of lead-based ammunition for hunting should be banned, to the benefit of people, wildlife, and ecosystems (Arnemo et al. 2019a, b; Sonne et al. 2019).

10 Conclusion

Hunting with lead-based ammunition poses a significant health risk to humans, wildlife, and ecosystems and is thus a One Health issue in the Arctic. In humans, even a moderate intake of meat from game harvested with lead-based ammunition may bring the BLL above thresholds for documented toxic effects. Improved dressing practices will reduce the amount of lead in edible parts, but all lead residues can never be removed. The use of lead-based ammunition for hunting will certainly also poison wildlife and contaminate ecosystems. A sustainable solution to this One Health problem requires national and international science-based legislative actions. In essence, however, it depends on hunters' choice of ammunition and is thus an ethical issue that can easily be solved at an individual level.

References

AMAP (2002) Heavy metals in the Arctic. https://www.amap.no/documents/doc/amap-assessment-2002-heavy-metals-in-the-arctic/97

AMAP (2015) Human health in the Arctic. https://www.amap.no/documents/doc/amap-assessment-2015-human-health-in-the-arctic/1346

Andersen O, Andreotti A, Arnemo JM et al (2018) European scientists' open letter on the risks of lead ammunition. http://www.europeanscientists.eu/open-letter-2018/

Andersen O, Arnemo JM, Bellinger DC et al (2020) European scientists' open letter on the risks of lead ammunition. http://www.europeanscientists.eu/open-letter-2020/
Andreotti A, Borghesi F (2013) Embedded lead shot in European starlings *Sturnus vulgaris*: an underestimated hazard for humans and birds of prey. Eur J Wildl Res 59:705–712. https://doi.org/10.1007/s10344-013-0725-8
Andreotti A, Borghesi F, Aradis A (2016) Lead ammunition residues in the meat of hunted woodcock: a potential health risk to consumers. Ital J Anim Sci 15:22–29. https://doi.org/10.1080/1828051X.2016.1142360
Arnemo JM, Andersen O, Stokke S et al (2016) Health and environmental risks from lead-based ammunition: science versus socio-politics. EcoHealth 13:618–622. https://doi.org/10.1007/s10393-016-1177-x
Arnemo JM, Andersen O, Stokke S et al (2019a) Norske jegere forgifter økosystemet [in Norwegian: Norwegian hunters are poisoning the ecosystem]. Norsk Vet Tidsskr 131:94–98
Arnemo JM, Cromie R, Fox AD et al (2019b) Transition to lead-free ammunition benefits all. Ambio 48:1097–1098. https://doi.org/10.1007/s13280-019-01221-x
Arnemo JM, Brainerd SM, Gremse C et al (2020) Alternatives to lead ammunition: sorting fact from fiction. A fact sheet for non-hunting decision makers from scientists and others who are hunters with extensive experience in this issue. https://europeanhuntingexperts.org/
Axelsson J (2009) Bly från ammunition som förgiftningsrisk hos rovfåglar - en kunskapsöversikt [in Swedish: Lead from ammuntion as risk of poisoning in raptors – a review]. Viltforum #1/2009. Svenska Jägareförbundet, Nyköping
Bakka MH (2020) Blood concentrations of lead (Pb), mercury (Hg) and cadmium (Cd) in Scandinavian and Alaskan brown bears (*Ursus arctos*). MSc thesis. Inland Norway University of Applied Sciences, Evenstad. https://brage.inn.no/inn-xmlui/handle/11250/2660562
Bellinger DC, Burger J, Cade TJ et al (2013a) Health risks from lead-based ammunition in the environment. Environ Health Persp 121:A178–A179. https://doi.org/10.1289/ehp.1306945
Bellinger DC, Bradman A, Burger J et al (2013b) Health risks from lead-based ammunition in the environment - a consensus statement of scientists. https://escholarship.org/uc/item/6dq3h64x
Bernhoft A, Cromie R, Devineau O et al (2014) Wildlife and human health risks from lead-based ammunition in Europe: a consensus statement by scientists. http://www.zoo.cam.ac.uk/leadammunitionstatement/
Birgisdottir B, Knutsen H, Haugen M et al (2013) Essential and toxic element concentrations in blood and urine and their associations with diet: results from a Norwegian population study including high-consumers of seafood and game. Sci Total Environ 463:836–844. https://doi.org/10.1016/j.scitotenv.2013.06.078
Bjermo H, Sand S, Nälsén C et al (2013) Lead, mercury, and cadmium in blood and their relation to diet among Swedish adults. Food Chem Toxicol 57:161–169. https://doi.org/10.1016/j.fct.2013.03.024
Bjerregaard P, Johansen P, Mulvad G et al (2004) Lead sources in human diet in Greenland. Environ Health Persp 112:1496–1498. https://doi.org/10.1289/ehp.7083
Bressler JM, Yoder S, Cooper S et al (2019) Blood lead surveillance and exposure sources among Alaska children. J Public Health Man 25:S71–S75. https://doi.org/10.1097/PHH.0000000000000877
Buekers J, Steen E, Redeker ES (2009) Lead toxicity to wildlife: derivation of a critical blood concentration for wildlife monitoring based on literature data. Sci Total Environ 407:3431–3438. https://doi.org/10.1016/j.scitotenv.2009.01.044
Buenz EJ (2016) Lead exposure through eating wild game. Am J Med 129:457–458. https://doi.org/10.1016/j.amjmed.2015.12.022
Calvert HS (1876) Pheasants poisoned by swallowing shot. Field 47:189
Carey LS (1977) Lead shot appendicitis in northern native people. J Can Assoc Radiol 28:171–174
Caudell JN (2013) Review of wound ballistic research and its applicability to wildlife management. Wildl Soc Bull 37:824–831. https://doi.org/10.1002/wsb.311

Caudell JN, Stopak SR, Wolf PC (2012) Lead-free, high-powered rifle bullets and their applicability in wildlife management. Hum Wildl Interact 6:105–111. https://doi.org/10.26077/qajj-wf35

CDC (2021) Blood lead levels in children. https://www.cdc.gov/nceh/lead/acclpp/blood_lead_levels.htm

Chief State Sanitary Doctor of the Russian Federation (2011) SanPiN 2.3.2.1078-01 Hygienic requirements for safety and nutritional value of food products [in Russian]. http://docs.cntd.ru/document/901806306

Church M, Gwiazda R, Risebrough RW et al (2006) Ammunition is the principal source of lead accumulated by California condors re-introduced to the wild. Environ Sci Technol 40:6143–6150. https://doi.org/10.1021/es060765s

Cox WM, Pesola GR (2005) Buckshot ingestion. N Engl J Med 353:e23. https://doi.org/10.1056/NEJMicm041154

Cromie R, Newth J, Reeves J et al (2015) The sociological and political aspects of reducing lead poisoning from ammunition in the UK: why the transition to non-toxic ammunition is so difficult. In: Delahay RJ, Spray CJ (eds) Proceedings of the Oxford Lead Symposium. Lead ammunition: understanding and minimising the risks to human and environmental health. University of Oxford, Oxford, pp 104–124. http://oxfordleadsymposium.info/proceedings/

Cromie R, Newth J, Strong E (2019) Transition to non-toxic ammunition: making change happen. Ambio 48:1079–1096. https://doi.org/10.1007/s13280-019-01204-y

Cruz-Martinez L, Grund MD, Redig PT (2015) Quantative assessment of bullet fragments in viscera of sheep carcasses as surrogates for white-tailed deer. Hum Wildl Interact 9:211–218. https://doi.org/10.26077/rxm7-x083

Dallaire F, Dewailly É, Muckle G (2003) Time trends of persistent organic pollutants and heavy metals in mbilical cord blood of Inuit infants born in Nunavik (Québec, Canada) between 1994 and 2001. Environ Health Persp 111:1660–1664. https://doi.org/10.1289/ehp.6269

Danielsson H, Karlsson GP (2016) Metaller i mossa [in Swedish: metals in mosses]. Rapport C 204. IVL Svenska Miljöinstitutet, Stockholm. https://www.ivl.se/download/18.76c6e08e1573302315f20e/1474381195136/C204.pdf

Dewailly É, Ayotte P, Bruneau S et al (2001) Exposure of the Inuit population of Nunavik (Arctic Québec) to lead and mercury. Arch Environ Health 56:350–357. https://doi.org/10.1080/00039890109604467

Dewailly É, Ayotte P, Pereg D et al (2007) Exposure to environmental contaminants in Nunavik: metals. Gouvernement du Québec. https://www.inspq.qc.ca/pdf/publications/661_esi_contaminants.pdf

ECHA (2018) A review of the available information on lead in shot used in terrestrial environments, in ammunition and in fishing tackle. ECHA, Helsinki. https://echa.europa.eu/documents/10162/13641/lead_ammunition_investigation_report_en.pdf/efdc0ae4-c7be-ee71-48a3-bb8abe20374a

Ecke F, Singh NJ, Arnemo JM et al (2017) Sub-lethal lead exposure changes movement behavior in free-ranging golden eagles. Environ Sci Technol 51:5729–5736. https://doi.org/10.1021/acs.est.6b06024

EFSA (2013) Scientific opinion on lead in food. https://efsa.onlinelibrary.wiley.com/doi/epdf/10.2903/j.efsa.2010.1570

Eurofins (2015) Lead in minced beef from Norwegian hunted game, elg (*Alces alces*). https://www.mattilsynet.no/mat_og_vann/jakt_og_viltkjott/rapport_lead_in_minced_beef_from_norwegian_hunted_game_elg.20383/binary/Rapport:%20Lead%20in%20minced%20beef%20from%20Norwegian%20hunted%20game%20Elg

Fackler ML (1988) Wound ballistics: a review of common misconceptions. AFTE J 21:25–29

Fackler ML (2001) Wound profiles. Wound Ballistic Rev 5:25–38

Falk K, Merkel F, Kampp K et al (2006) Embedded lead shot and infliction rates in common eiders *Somateria mollissima* and king eiders *S. spectabilis* wintering in Southwest Greenland. Wildl Biol 12:257–265. https://doi.org/10.2981/0909-6396(2006)12[257:ELSAIR]2.0.CO;2

Fernández V, Caselli A, Tammone A et al (2021) Lead exposure in dogs fed game meat and offal from culled invasive species in El Palmar National Park, Argentina. Environ Sci Pollut Res. 28(33):45486–45495

Fillion M, Blais JM, Yumvihoze E et al (2014) Identification of environmental sources of lead exposure in Nunavut (Canada) using stable isotope analyses. Environ Int 71:63–73. https://doi.org/10.1016/j.envint.2014.06.004

Finnish Food Authority (2020) Consumer possibilities to reduce lead intake. https://www.ruokavirasto.fi/en/companies/food-sector/production/common-requirements-for-composition/contaminants/metals-in-foods/lyijy/consumer-possibilities-to-reduce-lead-intake/

Fontaine J, Dewailly É, Benedetti J-L et al (2008) Re-evaluation of blood mercury, lead and cadmium concentrations in the Inuit population of Nunavik (Québec): a cross-sectional study. Environ Health 7:25. https://doi.org/10.1186/1476-069X-7-25

Fuchs B, Thiel A, Zedrosser A et al (2021) High blood lead concentrations in Scandinavian brown bears (*Ursus arctos*). Environ Pollut, 289:117595. https://doi.org/10.1016/j.envpol.2021.117595

Gorkhali R, Huang K, Kirberger M et al (2016) Defining potential roles of Pb^{2+} in neurotoxicity from a calciomics approach. Metallomics 8:563–578. https://doi.org/10.1039/c6mt00038j

Green RE, Pain DJ (2019) Risks to human health from ammunition-derived lead in Europe. Ambio 48:954–968. https://doi.org/10.1007/s13280-019-01194-x

Gremse F, Krone O, Thamm M et al (2014) Performance of lead-free versus lead-based hunting ammunition in ballistic soap. PLoS One 9:e102015. https://doi.org/10.1371/journal.pone.0102015

Grund MD, Cornicelli L, Carslon LT et al (2010) Bullet fragmentation and lead deposition in white-tailed deer and domestic sheep. Hum Wildl Interact 4:257–265. https://doi.org/10.26077/72ts-6n20

Hampton JO, Eccles G, Hunt R et al (2021) A comparison of fragmenting lead-based and lead-free bullets for aerial shooting of wild pigs. PLoS One 16:e0247785. https://doi.org/10.1371/journal.pone.0247785

Hanning RM, Sandhu R, MacMillan A et al (2003) Impact on blood Pb levels of maternal and early infant feeding practices of First Nation Cree in the Mushkegowuk Territory of northern Ontario, Canada. J Environ Monitor 5:241–245. https://doi.org/10.1039/B208220A

Harvey EN, Whiteeley AH, Grundfest H et al (1946) Piezoelectric crystal measurements of pressure changes in the abdomen of deeply anaesthetized animals during the passage of HV missiles. Mil Surg 98:509–528

Helander B, Axelsson J, Borg H et al (2009) Ingestion of lead from ammunition and lead concentrations in white-tailed sea eagles (*Haliaeetus albicilla*) in Sweden. Sci Tot Environ 407:5555–5563. https://doi.org/10.1016/j.scitotenv.2009.07.027

Hicklin P, Barrow B (2004) The incidence of embedded shot in waterfowl in Atlantic Canada and Hudson Strait. Waterbirds 27:41–45. https://doi.org/10.1675/1524-4695(2004)027[0041:TIOESI]2.0.CO;2

Høgåsen HR, Ørnsrud R, Knutsen HK et al (2016) Lead intoxication in dogs: risk assessment of feeding dogs trimmings of lead-shot game. BMC Vet Res 12:152. https://doi.org/10.1186/s12917-016-0771-z

Hunt WG, Watson RT, Oaks JL et al (2009) Lead bullet fragments in venison from rifle-killed deer: potential for human dietary exposure. PLoS One 4:e5330. https://doi.org/10.1371/journal.pone.0005330

ILZSG (2019) The world lead factbook. http://www.ilzsg.org/pages/1181/document.aspx?page=7&ff_aa_document_type=B&from=2

Irschik I, Bauer F, Sager M et al (2013) Copper residues in meat from wild artiodactyls hunted with two types of rifle bullets manufactured from copper. Eur J Wildl Res 59:129–136. https://doi.org/10.1007/s10344-012-0656-9

Isomursu M, Koivusaari J, Stjernberg T et al (2018) Lead poisoning and other human-related factors cause significant mortality in white-tailed eagles. Ambio 47:858–868. https://doi.org/10.1007/s13280-018-1052-9

Janzon B, Hull JB, Ryan JM (1997) Projectile, material interactions: soft tissue and bone. In: Cooper GJ, Dudley HAF, Gann DS et al (eds) Scientific foundations of trauma. Butterworth Heinemann, Oxford, pp 37–52

Johansen P, Asmund G, Riget F (2004) High human exposure to lead through consumption of birds hunted with lead shot. Environ Pollut 127:125–129. https://doi.org/10.1016/S0269-7491(03)00255-0

Johansen P, Pedersen HS, Asmund G et al (2006) Lead shot from hunting as a source of lead in human blood. Environ Pollut 142:93–97. https://doi.org/10.1016/j.envpol.2005.09.015

Kanstrup N, Thomas VG (2020) Transitioning to lead-free ammunition use in hunting: socio-economic and regulatory considerations for the European Union and other jurisdictions. Environ Sci Eur 32:91. https://doi.org/10.1186/s12302-020-00368-9

Kanstrup N, Balsby TJS, Thomas VG (2016) Efficacy of non-lead ammunition for hunting in Denmark. Eur J Wildl Res. 62:333–340. https://doi.org/10.1007/s10344-016-1006-0

Kanstrup N, Swift J, Stroud DA et al (2018) Hunting with lead ammunition is not sustainable: European perspectives. Ambio 47:846–857. https://doi.org/10.1007/s13280-018-1042-y

Karger B (2008) Forensic ballistics. Forensic Pathol Rev 5:139–172

Klingler PJ, Seelig MH, DeVault KR et al (1998) Ingested foreign bodies within the appendix: a 100-year review of the literature. Digest Dis 16:308–314. https://doi.org/10.1159/000016880

Kneubuehl BP, Coupland RM, Rothschild MA et al (2011) Wound ballistics, basics and applications. Springer, Berlin

Knott J, Gilbert J, Green RE et al (2009) Comparison of the lethality of lead and copper bullets in deer control operations to reduce incidental lead poisoning; field trials in England and Scotland. Conserv Evid 6:71–78

Knott J, Gilbert J, Knott J et al (2010) Implications for wildlife and humans of dietary exposure to lead from fragments of lead rifle bullets in deer shot in the UK. Sci Tot Environ 409:95–99. https://doi.org/10.1016/j.scitotenv.2010.08.053

Kollander B, Widemo F, Ågren E et al (2017) Detection of lead nanoparticles in game meat by single particle ICP-MS following use of lead-containing bullets. Anal Bioanal Chem 409:1877–1885. https://doi.org/10.1007/s00216-016-0132-6

Landrigan PJ, Fuller R, Acosta NJR et al (2018) The Lancet Commission on pollution and health. Lancet 391:462–512. https://doi.org/10.1016/S0140-6736(17)32345-0

Lanphear BP, Rauch S, Auinger P et al (2018) Low-level lead exposure and mortality in US adults: a population-based cohort study. Lancet Public Health 3:e177–e184. https://doi.org/10.1016/S2468-2667(18)30025-2

Larsen S, Nybakk D (2011) Riflejakt som jaktform på hare (*Lepus timidus*) og omfanget av skadeskytning med haglegevær på hare (*Lepus timidus*) [in Norwegian: hunting hares with rifle and the extent of wounding with shotgun]. Bachelor thesis, Hedmark University College, Evenstad

Lazarus M, Sekovanić A, Orct T et al (2017) Apex predatory mammals as bioindicator species in environmental monitoring of elements in Dinaric Alps (Croatia). Environ Sci Pollut Res 24:23977–23991. https://doi.org/10.1007/s11356-017-0008-0

Lazarus M, Orcta T, Sergiel A et al (2020) Metal(loid) exposure assessment and biomarker responses in captive and free-ranging European brown bear (*Ursus arctos*). Environ Res 183:109166. https://doi.org/10.1016/j.envres.2020.109166

Legagneux P, Suffice P, Messier JS et al (2014) High risk of lead contamination for scavengers in an area with high moose hunting success. PLoS One 9:e111546. https://doi.org/10.1371/journal.pone.0111546

Lévesque B, Duchesne J-F, Gariépy C et al (2003) Monitoring of umbilical cord blood lead levels and sources assessment among the Inuit. Occup Environ Med 60:693–695. https://doi.org/10.1136/oem.60.9.693

Liberda EN, Tsuji LJS, Martin ID et al (2018) Source identification of human exposure to lead in nine Cree Nations from Quebec, Canada (Eeyou Istchee territory). Environ Res 161:409–417. https://doi.org/10.1016/j.envres.2017.11.023

Lindboe M, Henrichsen EN, Høgåsen HR et al (2012) Lead concentration in meat from lead-killed moose and predicted human exposure using Monte Carlo simulation. Food Addit Contam Part A 29:1052–1057. https://doi.org/10.1080/19440049.2012.680201

Livsmedelsverket (2014a) Bly i viltkött. Del 1 – ammunitionsrester och kemisk analys [in Swedish: Lead in game meat. Part 1 – ammunition residues and chemical analysis]. https://www.livsmedelsverket.se/globalassets/publikationsdatabas/rapporter/2014/bly-i-viltkott%2D%2D-del-1-ammunitionsrester-och-kemisk-analys.pdf

Livsmedelsverket (2014b) Bly i viltkött. Del 2 – halter av bly i i blod hos jägarfamiljer [in Swedish: Lead in game meat. Part 2 - blood lead levels in hunter families]. https://www.livsmedelsverket.se/globalassets/publikationsdatabas/rapporter/2014/bly-i-viltkott-del-2%2D%2D-halter-i-bly-hos-jagarfamiljer.pdf

Livsmedelsverket (2020a) Jakt [in Swedish: Hunting]. https://www.livsmedelsverket.se/produktion-handel%2D%2Dkontroll/produktion-av-livsmedel/primarproduktion/jakt

Livsmedelsverket (2020b) Ammunitionsbly i viltkött [in Swedish: Ammunition lead in game meat]. L-2020 nr 15. https://www.livsmedelsverket.se/sok?q=ammunitionsbly

Lohr MT, Hampton JO, Cherriman S et al (2020) Completing a worldwide picture: preliminary evidence of lead exposure in a scavenging bird from mainland Australia. Sci Tot Environ 715: 135913. https://doi.org/10.1016/j.scitotenv.2019.135913

Lozoff B, Jimenez E, Wolf AW et al (2009) Higher infant blood lead levels with longer duration of breastfeeding. J Pediatr 155:663–667. https://doi.org/10.1016/j.jpeds.2009.04.032

MacPherson D (1994) Bullet penetration—modelling the dynamics and incapacitation resulting from wound trauma, 2nd edn. Ballistic Publications, El Segundo

Madsen HHT, Skjødt T, Jørgensen PJ et al (1988) Blood lead levels in patients with lead shot in the appendix. Acta Radiol 29:745–746

Mann MJ, Aspinoza EO, Ralston RM et al (1994) Shot pellets: an overview. AFTE J 26:224–241

Martin A, Gremse C, Selhorst T et al (2017) Hunting of roe deer and wild boar in Germany: is non-lead ammunition suitable for hunting? PLoS One 12:e0185029. https://doi.org/10.1371/journal.pone.0185029

Mateo R, Rodriguez-de la Cruz M, Vidal D et al (2007) Transfer of lead from shot pellets to game meat during cooking. Sci Tot Environ 372:480–485. https://doi.org/10.1016/j.scitotenv.2006.10.022

Mateo R, Baos AR, Vidal D, Camarero PR et al (2011) Bioaccessibility of Pb from ammunition in game meat is affected by cooking treatment. PLoS One 6:e15892. https://doi.org/10.1371/journal.pone.0015892

Mattilsynet (2019) Begrens inntaket av viltkjøtt med blyrester [in Norwegian: Limit the intake of game meat with lead residues]. https://www.matportalen.no/matvaregrupper/tema/fjorfe_og_kjott/begrens_inntaket_av_viltkjott_med_blyrester-1

McAuley C, Ng C, McFarland C et al (2018) Lead exposure through consumption of small game harvested using lead-based ammunition and the corresponding health risks to First Nations in Alberta. Cogent Environ Sci 4:1557316. https://doi.org/10.1080/23311843.2018.1557316

McCann BE, Whitworth W, Newman RA (2016) Efficacy of non-lead ammunition for culling elk at Theodore Roosevelt National Park. Hum Wildl Interact 10:268–282. https://doi.org/10.26077/8gma-q214

Meltzer HM, Dahl H, Brantsæter AL, Birgisdottir BE et al (2013) Consumption of lead-shot cervid meat and blood lead concentrations in a group of adult Norwegians. Environ Res 127:29–39. https://doi.org/10.1016/j.envres.2013.08.007

Menozzi A, Menotta S, Fedrizzi G et al (2019) Lead and copper in hunted wild boars and radiographic evaluation of bullet fragmentation between ammunitions. Food Addit Contam Part B 12:182–190. https://doi.org/10.1080/19393210.2019.1588389

Miljøstatus (2019) Nivåene av bly i miljøet reduseres [in Norwegian: Environmental lead levels are reduced]. https://miljostatus.miljodirektoratet.no/tema/miljogifter/prioriterte-miljogifter/bly-og-blyforbindelser/

Miljøstatus (2020) Den norske prioritetslista [in Norwegian: The Norwegian priority list]. https://miljostatus.miljodirektoratet.no/tema/miljogifter/prioriterte-miljogifter/om-prioriterte-miljogifter/

Monclús L, Shore RF, Krone O (2020) Lead contamination in raptors in Europe: a systematic review and meta-analysis. Sci Tot Environ 748:141437. https://doi.org/10.1016/j.scitotenv.2020.141437

Mörner T, Petersson L (1999) Lead poisoning in woodpeckers in Sweden. J Wildl Dis 35:763–765. https://doi.org/10.7589/0090-3558-35.4.763

Mushak P (1998) Uses and limits of empirical data in measuring and modeling human lead exposure. Environ Health Persp 106:1467–1484. https://doi.org/10.1289/ehp.98106s61467

Naturvårdssverket (2019) A non-toxic environment. http://www.swedishepa.se/Environmental-objectives-and-cooperation/Swedens-environmental-objectives/The-national-environmental-objectives/A-Non-Toxic-Environment/

Naturvårdsverket (2006) Konsekvenser av förbud mot bly i ammunition [in Swedish: consequences of banning lead in ammunition]. Naturvårdsverket, Stockholm, Sweden. https://www.naturvardsverket.se/Documents/publikationer/620-5627-1.pdf?pid=3239

Naturvårdsverket (2020) Blyhalter i blod hos barn [in Swedish: Blood lead levels in children]. http://www.naturvardsverket.se/Sa-mar-miljon/Statistik-A-O/Blyhalter-i-blod-hos-barn/

Nawrot TS, Staessen JA (2006) Low-level environmental exposure to lead unmasked as silent killer. Circulation 114:1347–1349. https://doi.org/10.1161/CIRCULATIONAHA.106.650440

NTB (2016) Nye regler: 200 tonn viltkjøtt med bly må kastes [in Norwegian: New regulations: 200 tons of game meat have to be discarded]. https://www.opdalingen.no/nyheter/2016/09/26/Nye-regler-200-tonn-viltkj%C3%B8tt-med-bly-m%C3%A5-kastes-13527508.ece

Odland JO, Perminova I, Romanova N et al (1999) Elevated blood lead concentrations in children living in isolated communities of the Kola Peninsula, Russia. Ecosyst Health 5:76–81. https://doi.org/10.1046/j.1526-0992.1999.09912.x

Pain DJ, Cromie RL, Newth J et al (2010) Potential hazard to human health from exposure to fragments of lead bullets and shot in the tissues of game animals. PLoS One 5:e10315. https://doi.org/10.1371/journal.pone.0010315

Pain DJ, Mateo R, Green RE (2019) Effects of lead from ammunition on birds and other wildlife: a review and update. Ambio 48:935–953. https://doi.org/10.1007/s13280-019-01159-0

Pain D, Swift J, Green R et al (2020) The tide is turning for lead ammunition. Brit Birds 113:110–118. https://britishbirds.co.uk/article/the-tide-is-turning-for-lead-ammunition/

Parry G, Buenz EJ (2020) Including eating lead-shot meat in the differential diagnosis of non-specific symptoms. Int Med J 50:1293–1294. https://doi.org/10.1111/imj.15024

Pierce BL, Roster TA, Frisbie MC et al (2014) A comparison of lead and steel shot loads for harvesting mourning doves. Wildl Soc Bull 39:103–115. https://doi.org/10.1002/wsb.504

Pure Earth (2015) World's worst pollution problems. http://worstpolluted.org/docs/WWP15.pdf

Rabinowitz MB, Wetherill GW, Kopple JD (1973) Lead metabolism in the normal human: stable isotope studies. Science 182:725–727. https://doi.org/10.1126/science.182.4113.725

Reddy ER (1985) Retained lead shot in the appendix. J Can Assoc Radiol 36:47–48

Renberg I, Bindler R, Bränvall MJ (2001) Using the historical atmospheric lead-deposition record as a chronological marker in sediment deposits in Europe. Holocene 11:511–516. https://doi.org/10.1191/095968301680223468

Roberts GK (1988) The wounding effect of 5.56 mm/.223 law enforcement general purpose shoulder fired carbines compared with 12 GA. Shotguns and pistol caliber weapons using 10% ordnance gelatine as a tissue simulant. Wound Ballistics Rev 3:16–28

Rodríguez-Jorquera IA, Vitale N, Garner L et al (2017) Contamination of the upper class: occurrence and effects of chemical pollutants in terrestrial top predators. Curr Pollut Rep 3:206–219. https://doi.org/10.1007/s40726-017-0061-9

Rogers TA, Bedrosian B, Graham J et al (2012) Lead exposure in large carnivores in the Greater Yellowstone ecosystem. J Wildl Manag 76:575–582. https://doi.org/10.1002/jwmg.277
Scheuhammer AM, Perrault JA, Routhier E et al (1998) Elevated lead concentrations in edible parts of game birds harvested with lead shot. Environ Pollut 102:251–257. https://doi.org/10.1016/S0269-7491(98)00083-9
Schulz JH, Stanis SJW, Webb EB et al (2019) Communication strategies for reducing lead poisoning in wildlife and human health risks. Wildl Soc Bull 43:131–140. https://doi.org/10.1002/wsb.955
Schulz JH, Stanis SJW, Hall DM et al (2020) Until it's a regulation it's not my fight: complexities of a voluntary nonlead hunting ammunition program. J Environ Manag 276:111438. https://doi.org/10.1016/j.jenvman.2020.111438
Settle DM, Patterson CC (1980) Lead in albacore: guide to lead pollution in Americans. Science 207:1167–1176. https://doi.org/10.1126/science.6986654
Sonne C, Alstrup AKO, Ok YS et al (2019) Time to ban lead hunting ammunition. Science 366: 961–962. https://doi.org/10.1126/science.aaz8150
Stefanopoulos PK, Filippakis K, Soupiou OT et al (2014) Wound ballistics of firearm-related injuries - part 1: missile characteristics and mechanisms of soft tissue wounding. Int J Oral Max Surg 43:1445–1458. https://doi.org/10.1016/j.ijom.2014.07.013
Stokke S, Brainerd S, Arnemo JM (2017) Metal deposition of copper and lead in moose harvested in Fennoscandia. Wildl Soc Bull 41:98–106. https://doi.org/10.1002/wsb.731
Stokke S, Arnemo JM, Brainerd S et al (2018) Defining animal welfare standards in hunting: body mass determines thresholds for incapacitation time and flight distance. Sci Rep 8:13786. https://doi.org/10.1038/s41598-018-32102-0
Stokke S, Arnemo JM, Brainerd S (2019) Unleaded hunting: are copper bullets and lead-based bullets equally effective for killing big game? Ambio 48:1044–1055. https://doi.org/10.1007/s13280-019-01171-4
Thomas VG, Gremse C, Kanstrup N (2016) Non-lead rifle hunting ammunition: issues of availability and performance in Europe. Eur J Wildl Res 62:633–641. https://doi.org/10.1007/s10344-016-1044-7
Thomas VG, Kanstrup N, Fox AD (2019) The transition to non-lead sporting ammunition and fishing weights: review of progress and barriers to implementation. Ambio 48:925–934. https://doi.org/10.1007/s13280-018-1132-x
Thomas VG, Pain DJ, Kanstrup N et al (2020) Setting maximum levels for lead in game meat in EC regulations: an adjunct to replacement of lead ammunition. Ambio 49:2026–2037. https://link.springer.com/article/10.1007%2Fs13280-020-01336-6
Treu G, Drost W, Stock F (2020) An evaluation of the proposal to regulate lead in hunting ammunition through the European Union's REACH regulation. Environ Sci Eur 32:68. https://doi.org/10.1186/s12302-020-00345-2
Trinogga A, Fritsch G, Hofer H (2013) Are lead-free hunting rifle bullets as effective at killing wildlife as conventional lead bullets? A comparison based on wound size and morphology. Sci Tot Environ 443:226–232. https://doi.org/10.1016/j.scitotenv.2012.10.084
Trinogga AL, Courtiol A, Krone O (2019) Fragmentation of lead-free and lead-based hunting rifle bullets under real life hunting conditions in Germany. Ambio 48:1056–1064. https://doi.org/10.1007/s13280-019-01168-z
Tsuji LJS, Nieboer E (1997) Lead pellets ingestion in First Nation Cree of the western James Bay Region and northern Ontario, Canada: implications for a nontoxic shot alternative. Ecosyst Health 3:54–61. https://doi.org/10.1111/j.1526-0992.1997.00706.pp.x
Tsuji LJS, Wainman BC, Martin ID et al (2008a) The identification of lead ammunition as a source of lead exposure in First Nations: the use of lead isotope ratios. Sci Tot Environ 393:291–298. https://doi.org/10.1016/j.scitotenv.2008.01.022
Tsuji LJS, Wainman BC, Martin ID et al (2008b) Lead shot contribution to blood lead of First Nations people: the use of lead isotopes to identify the source of exposure. Sci Tot Environ 405: 180–185. https://doi.org/10.1016/j.scitotenv.2008.06.048

Tsuji LJS, Wainman BC, Martin ID et al (2008c) Elevated blood-lead levels in First Nation people of Northern Ontario Canada: policy implications. Bull Environ Contam Toxicol 80:14–18. https://doi.org/10.1007/s00128-007-9281-9

UNICEF, Pure Earth (2020) The toxic truth: children's exposure to lead pollution undermines a generation of future potential. https://www.unicef.org/reports/toxic-truth-childrens-exposure-to-lead-pollution-2020

University of Washington (2019) Metallic lead safety. https://www.ehs.washington.edu/system/files/resources/metallic-lead-safety.pdf

VKM (2013) Risk assessment of lead exposure from cervid meat in Norwegian consumers and in hunting dogs. https://www.matportalen.no/uonskedestoffer_i_mat/tema/miljogifter/article54010.ece/BINARY/VKM%E2%80%99s%20vurdering%20av%20risiko%20knyttet%20til%20konsum%20av%20hjortevilt%20felt%20med%20blyholdig%20ammunisjon

Wennberg M, Lundh T, Sommar JN et al (2017) Time trends and exposure determinants of lead and cadmium in the adult population of northern Sweden 1990–2014. Environ Res 189:111–117. https://doi.org/10.1016/j.envres.2017.07.029

WHO (2018) Lead poisoning and health. http://www.who.int/en/news-room/fact-sheets/detail/lead-poisoning-and-health

Widemo F (2021) Shooting habits and habitats - effects of education and legislation on the phasing out of lead shot. Environ Sci Pol 118:56–62. https://doi.org/10.1016/j.envsci.2021.01.010

Traditional Conservation Methods and Food Habits in the Arctic

Raphaela Stimmelmayr and Gay Sheffield

1 Traditional Foods and Indigenous People of the Arctic

> We all have to go by seasons for fishing and trapping. I can't say I'll catch herring later when I need them. By then there will be nothing, as the fish and game don't wait for our time.
> Frank Cockney (N-1992-253:1-44-28) as found in Hart and Amos 2004

Traditional aboriginal foods encompass the spectrum of Arctic animals (ocean, land, sky) and plants. These foods are an integral part of cultural identity, community, collective identity, and cultural stability and a key component of nutritional and spiritual well-being within traditional food systems (Fig. 1). Across northern communities, traditional seasonal rounds of hunting, trapping, fishing, egging, picking, collecting, etc., associated food handling practices, and the use of animals and plants (traditional customary food practices) generally mirror the prevailing environmental conditions, seasonal availability, and accessibility of these essential resources, while being flexible enough to adapt to natural variations in resource abundance. Given the intricate complex human biological–physical coupling of resources and users, it is not surprising that "no two societies shared exactly the same annual cycle of subsistence activities or dependence on specific resources. Subsistence strategies at the societal and local family level were aimed at effective use of virtually all local plant and animal resources" (Burch 1980 cited in Wolfe 2004). Within the larger migratory and strongly seasonal context of all animal and plant resources in the North, traditional customary practices of food processing

R. Stimmelmayr (✉)
Department of Wildlife Management, North Slope Borough, Utqiaġvik, AK, USA

Institute of Arctic Biology, University of Alaska Fairbanks, Fairbanks, AK, USA
e-mail: raphaela.stimmelmayr@north-slope.org

G. Sheffield
Alaska Sea Grant, Marine Advisory Program, University of Alaska Fairbanks, Nome, AK, USA
e-mail: ggsheffield@alaska.edu

M. Tryland (ed.), *Arctic One Health*, https://doi.org/10.1007/978-3-030-87853-5_22

Fig. 1 A delicious nutritious meal of Arctic cuisine is the end result of traditional wildlife health assessment protocols as well as traditional processing, preservation, and ageing, Bering Strait region. Photo: G. Sheffield

including food preparation, safe storage, and preservation methods are at the center of subsistence strategies to ensure aboriginal food security by making seasonally available surplus foods obtainable year-round. Across the Arctic, these culinary methods are based on traditional Arctic food wisdom, but like subsistence strategies, they are also shaped by environmental–societal–individual influences and preferences at the societal and local family level. Therefore, traditional food preferences, cultural uses, and customary practices of traditional Arctic food processing and storing practices at the societal and local family level have a distinctly local flavor by being similar—but not the same. To this day, Arctic food processing, in all of its nuances of skills and knowledge, is a task mostly performed by women (Starks 2002, 2007, 2011). This chapter provides an introduction to the guiding principles of traditional Arctic food systems, with a special focus on the Arctic subsistence food chain (hunting–food processing–storing–sharing) from a wildlife health, food safety, and food science perspective. Arctic traditional food systems are living testaments to the ingenuity of Arctic people's traditional ecological knowledge and food wisdom and remain deeply steeped in cultural identity, prevailing local conditions, and history, so any single text and examples given are useful in showcasing a specific aspect but will ultimately fall short in being representative of the cultural diversity of indigenous people's traditional food systems in the Arctic. Regardless, inherent to northern people's societies is a clear understanding of the connection between the health of animals, the health of the people who depend on them, and the environment they both inhabit. This understanding provides a common platform akin to the One Health perspective (Zinsstag et al. 2011) to discuss aspects of resilience and potential vulnerability of transitioning Arctic traditional food systems to zoonotic diseases in the context of a rapidly changing Arctic.

2 Harvesting and Principles of Wildlife Health Assessment

> We travelled a little way (and) we saw two[..] seals. we had nothing..." I'll go to the first seal. If I can't (kill it) I'll try the other if it doesn't go down. I went to the other and got the seal...it was a big seal but it looked like it was sick, skinny, with no blubber. We didn't eat out of the seal, we were afraid to eat it. We just gave our dogs a little bit of blubber. Jimi Wolki (N-1992-007:0043B) as found in Hart and Amos 2004

Circumpolar indigenous people have long-term traditions in harvesting wildlife that reach back to ancestral times. Hunting is not just procurement of food but is deeply steeped in complex spiritual relationships with the animals and the environments (i.e., Puiguitkaat 1978; Bennett and Rowley 2004; Kochneva 2007; Gadamus and Raymond-Yakoubian 2015). The focus of the following sections will be on the pragmatic aspects of hunting as it is beyond the scope of this chapter to discuss the complex spiritual and cosmological aspects of northern people's relations with the animals, their land, and oceans.

Central to their respective hunting culture, traditional customary practices are at the core of traditional ante- and post-mortem examinations of hunted animals. Understanding normal species-specific animal behavior within and outside the hunting context gives insight about the general health that ultimately guides the selection and exclusion of an animal from becoming food for the individual, the family, and the community. The eloquence of animals in showing behavioral and physiological cues of health (i.e., alertness, responsiveness, breathing rate, fight and flight behavior, movement, speed, body condition, pelt condition) or signs of sickness (i.e., separating from the herd, diving shorter than usual, being in the wrong habitat, abnormal locomotion, approachability, aggressiveness, discharge, etc.) is understood by the keen observation skills and in-depth ecological knowledge that come from the rigorous study by the experienced and well-taught hunters. Taking these health and sickness cues from live animals (ante-mortem) is part of the hunter's animal health assessment. After the kill, catch assessments are designed to provide a thorough health assessment and, by default, a public health judgement on food safety, similar to western meat hygiene inspections. The identification of sick animals is critical to the health of the individual, family, and extended community food chain. The knowledge of those who process the catch, from the hunters, to processors, to meal preparers, makes this decision in equal parts using in-depth traditional knowledge of the ecology and behavior of the hunted animal, as well as the shared generational knowledge about animal anatomy and health, for example, Inuvialuit marine resources (Hart and Amos 2004), bowhead whale (*Balaena mysticetus*) (Brewster 1997; Hay 2000; Albert 2001; Noongwook et al. 2007), beluga whale (*Delphinapterus leucas*) (Waugh et al. 2018; Ostertag et al. 2018), ice-associated seals and Pacific walrus (*Odobenus rosmarus*) (Gadamus and Raymond-Yakoubian 2015), polar bear (*Ursus maritimus*) (Armitage, 2015), caribou (*Rangifer tarandus*) (Benson 2019), and fish (Jones 2006; Brewster et al. 2008). Traditional food wisdom and its precautionary customary practices generally do not depend on detailed information about the nature of the sickness (i.e., infections, toxins). Although the cause of disease and death of wildlife are topics of great interest and discussions among community members, it is a robust mechanistic public health understanding regarding animals that determines if foods should not be eaten. Interestingly this is often how wildlife health information is presented in a western/scientific approach and in traditional medical and veterinary school training. Co-developing recommendations for effective wildlife health and disease information sharing in the context of climate change and emergence of zoonotic diseases in the Arctic are needed to bridge this gap. Several guiding principles are inherent to most scientific types of knowledge acquisition and inquiries, namely, observations, experimentation, detailed oral documentation of outcomes, and repetitions of various experimental designs (contexts), in order to be able to predict outcomes.

Therefore, observed abnormalities should trigger precautionary measures ranging from discarding only the abnormal part to the entire animal. Abnormalities that do not pass traditional guidelines include but are not limited to unusual differences in structure and texture (i.e., lumps and bumps, pus pockets, bruises, swollen joints, weakened shear resistance of tissue and hides, old and fresh wounds), odor, abnormal discharge from natural openings and joints, abnormal coloration, and uncommon parasites. Commonly, animals of concern are left where they were hunted (i.e., sea ice, ocean, land) in order to avoid contamination of additional (healthy) harvested animals in close proximity, hunting equipment, and/or transportation means (i.e., boats, sleds, all-terrain vehicles). This is a counter to what is often recommended in a scientific context in order to identify the exact cause of the abnormality. In cases where abnormalities are discovered during a more detailed processing in the community, expert consultation with experienced hunters, food preparers, and elders is often essential for ad hoc decision making about hunter-concern animals. In recent years, northern communities use social media, such as Facebook groups and regional observer networks, to share hunter-concern cases and successful hunts. Sharing information on hunter-concern cases and/or observations is critical to alerting federal and/or tribal wildlife management agencies to emerging northern wildlife diseases and pollution events that have implications for regional food safety and food security (i.e., Kwan 2004; Kutz et al. 2015; Bodenstein et al. 2015; Iverson et al. 2016; Sformo et al. 2017; Henri et al. 2018; Stimmelmayr et al. 2018, 2021; Tryland et al. 2018; Sheffield and Stimmelmayr 2020).

Skinning, butchering, preparing, and other handling of hunted animals pose the risk of a skin injury and thus potential exposure to zoonotic microbes and parasites. Rare exposures during butchering activities to selected microbial agents (e.g.,

Francisella tularensis (muskrat, hare), *Mycoplasma* spp. (pinnipeds), *Brucella suis serovar IV* (caribou/reindeer), *Parapoxvirus* (mountain goat, Dall's sheep)) have been reported from Alaska and Canada (Kennedy 1988; Wobeser et al. 2009; Castrodale 2009, Castrodale 2011; Hansen et al. 2011; Westley et al. 2016; Carr cited in Tryland et al. 2018). Interestingly, the finding of *Francisella*-specific antibodies (tularemia) in people did not correlate with case reporting of tularemia-like illness from the northern villages studied, suggesting infection with a less virulent type favoring asymptomatic disease or resistance to the pathogen (Hansen et al. 2011). Tularemia is among the group of "climate-sensitive" pathogens (Parkinson et al. 2014; Omazic et al. 2019). Several cases of naturally acquired rabies antibodies in Inupiaq–Inuit Arctic fox trappers have also been reported from Alaska and Canada (Orr 1988 cited in Follmann et al. 1994). Skinning and hide processing are also known risk factors for human cutaneous anthrax infection (Mwakapeje et al. 2018). For Arctic regions with endemic anthrax (Revich and Podolnaya 2011; Hueffer et al. 2020), carcass handling and the use of animal hides (e.g., reindeer) for clothing, bedding, etc., are regionally relevant risk factors. Localized morbidity and one human mortality were reported during the recent emergence of anthrax among Siberian native reindeer herders (Timofeev et al. 2019). Most harvesting of wolves as well as Arctic and red foxes occurs during winter and spring months, coinciding with prime pelt condition. When skinning wild canids, hunters may be exposed to shed parasite eggs or larvae (e.g., *Echinococcus* spp.) clinging onto the fur around the anus and tail base. General risk reduction measures include good hand hygiene, wearing gloves, and keeping cutting tools clean. Given that sled dogs are also suitable hosts, important local risk reduction measures for human infection control include but are not limited to regular deworming, not feeding raw infected parts from caribou or moose, good hand hygiene, and village waste management of dog feces (Rausch 2003).

Although infection from skinning and other handling of hunted animals is a possibility, wearing protective gear such as disposable gloves during handling is mostly a personal preference among communities. Generally, any skin breaks or cuts are immediately attended to, by cleaning wounds with soap and applying a disinfectant solution and protective covering. This is especially important during butchering seals in order to avoid exposure to the bacteria that causes "seal finger." In cases of minor cuts and wounds occurring on the hands, local and regional guidance from elders and experienced food processors discourage further participation in community butchering activities of marine mammals (whales, pinnipeds) until the wound heals (Taqulik Hepa pers. comm.). "Seal finger" is a well-documented clinical entity, and during commercial sealing, "seal finger" was frequently diagnosed in seal fishermen in Canada and Norway (Tryland 2000; White and Jewer 2009). The bacterial etiology remains unclear with a variety of isolates cultured over the years (*Staphylococcus* sp., *Erysipelothrix* spp., *Mycoplasma* sp.). Although "seal finger" is mostly a clinically localized infection of the finger joint(s), disseminated seal finger mycoplasmosis, associated with a novel *Mycoplasma* sp., was diagnosed in a seal hunter from the Bering Strait region of Alaska (Westley et al. 2016). Skin infections (i.e., suppurative lesions, trauma with and without sepsis) among whaling crews

Fig. 2 Processing a landed bowhead whale on the sea ice in the Bering Strait region. Photo: G. Sheffield

were also reported during commercial whaling (Case 1948; Cockrill 1960). Working on a "dauhvaler" (an unclaimed dead whale found floating) or "fender whales" (whales used as ship fenders during processing) was a known risk factor for contracting "blubber finger."

In the process of "cutting up," evisceration of animals brings with it the possibility of carcasses and organs becoming tainted due to contact with the surrounding physical environment. Keeping hunting camps and butchering areas clean and choosing unsoiled sea ice, beaches, and tundra for processing are part of a successful traditional hunt (Fig. 2). On the North Slope, Alaska (i.e., Utqiaġvik, Point Hope), cardboard, wood pallets, and plywood sheets are used as clean butchering surfaces when cutting up seals in the community (Fig. 3). Soil reservoirs of emerging pathogens of concern that form environmentally stable spores, such as *Coxiella burnetii* (Duncan et al. 2013, 2014) and *Clostridium botulinum* (Fagan et al. 2011), are regionally present in the Arctic, and there is limited evidence to support potential regional public health exposure risks from environmental contamination of harvested food during field dressing or by inhalation. For example, inhalation of aerosolized *Coxiella burnetii* from nearby northern fur seal rookeries is thought to be the primary exposure route to community members of St. Paul, Alaska, rather than ingestion from northern fur seal meat (Kersh et al. 2020; Duncan et al. 2014). Interestingly, no locally acquired clinical cases of Q fever in people and/or various

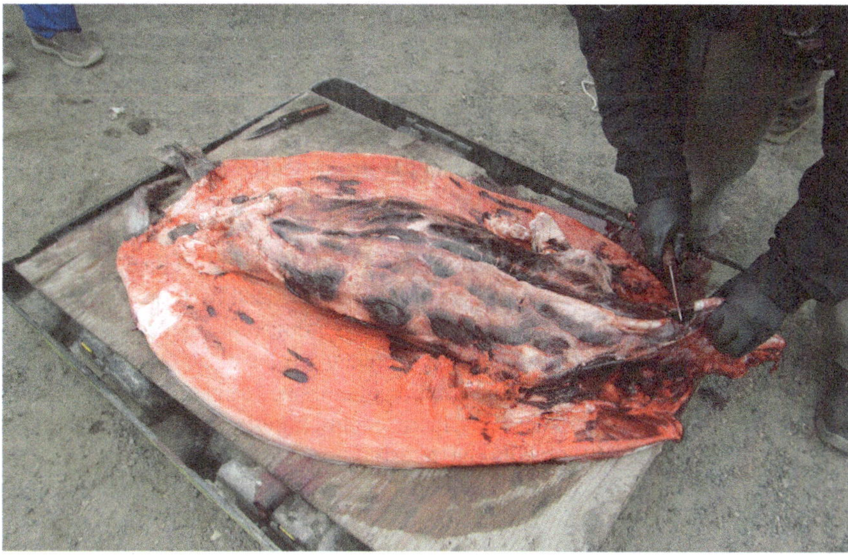

Fig. 3 Cutting up a harvested ringed seal, Utqiaġvik, Alaska. Photo: R. Stimmelmayr

wildlife species (ruminants, canids, ursids) have been reported in Alaska. In Greenland, however, a single locally acquired clinical case of Q fever was reported—with sled dogs and seals implicated as a possible source of infection (Koch et al. 2010). In Kotzebue, Alaska, and Nunavik, eastern Canada, *C. botulinum type E* contamination of seal meat and blubber from butchering on the beach is thought to contribute to botulism risk during the fermentation of marine mammal food products (Miller 1975; LeClair et al. 2017).

A clean evisceration, that is with no contact with bile, stomach contents, feces, or urine, is a critical step during butchering, especially as pathogenic bacteria are normally found in the digestive tract of healthy animals (i.e., Harper et al. 2002; Glad et al. 2010a, b; Tryland et al. 2014; Daoust and Stacey 2014; Sanders et al. 2015). Though regional butchering practices exist, all practices focus on the avoidance of spilling these animal liquids onto edible meat and organs. For example, the removal of viscera in landed bowhead whales at Utqiaġvik, Alaska, occurs largely as one unit, whereby a rope is tied to the larynx, and the entire viscera and entrails are pulled backward and away from the animal (Stimmelmayr et al. 2017). Concurrent cutting of serosal organ attachments and the diaphragmatic muscle facilitates the backward pulling process. The same approach is taken with pinnipeds, whereby the trachea is used as a handle to pull the viscera backward, and the pelvic symphysis is split at the end to facilitate the complete removal of the entire gastrointestinal tract and urinary bladder. Both techniques limit the risk for soiling meat and organs with bile, stomach content, feces, and urine during field dressing. In case of meat contamination due to accidental puncturing of the gastrointestinal tract, the tainted meat is usually trimmed generously or discarded in entirety. Carcasses of inadvertently gut shot animals with stomach contents spilled into the body cavity are often discarded. Urine spillage during butchering is uncommon, as hunted animals often voided their urine. However, accidental exposure of broken skin to urine and/or urine-stained pelt from an animal infected with *Leptospira* sp. bacteria could pose an infection risk during butchering and skinning. *Leptospira* sp. can affect a variety of Arctic wildlife, ranging from seals and walruses to other carnivores and ruminants. Interestingly, among Inuit from Northern Quebec with a seropositivity around 5.9%, no risk factor for *Leptospira* sp. infection was identified (Messier et al. 2012). *Leptospira* sp. is among the group of "climate-sensitive" pathogens (Parkinson et al. 2014). Apart from gut-associated bacteria, many important Arctic marine mammals are also host to zoonotic protozoal agents such as *Cryptosporidium* sp. and *Giardia* sp. (for review Miller et al. 2018). Although waterborne transmission through consumption of untreated drinking water is thought to be a primary exposure route in northern communities (i.e., Pardhan-Ali et al. 2013; Masina et al. 2019), foodborne transmission through the consumption of marine mammal intestines used in traditional northern dishes, for example, *inaluaq* (boiled small intestines from the bowhead whale), boiled ringed seal intestines, *qulivrat* (braided ringed seal intestines), and *qiaq* (chopped/minced bearded seal intestines that are prepared on a slab of bearded seal blubber) (Carroll 1976; Lisa Adams pers. comm.), has been proposed as a potential source for protozoal infections (Jenkins et al. 2013). Although not explicitly tested, traditional customary practices place a great emphasis on thoroughly emptying intestines by repeatedly flushing the inside using copious amounts of water. This practice offers some form of risk reduction measure from gut-associated microbes and parasites. Most recently, the hypothesized risk of marine biotoxin exposure (i.e., domoic acid, saxitoxin) from eating traditional foods of marine mammals, such as intestines, clams from walrus stomachs, meat and blubber, etc., has been a concern among Alaskan coastal communities. Both marine biotoxins have been detected in feces from harvested Alaskan marine mammals—statewide (Lefebvre et al. 2016). Whether customary cleaning with the traditionally practiced removal of the inner intestinal lining (mucosa) has the potential to reduce exposure risk to domoic acid and saxitoxin from consumption of marine mammal (e.g., seal, walrus, and bowhead whales), the intestines have not yet been explicitly tested (Bering Strait: Walruses and Saxitoxin 2017). Most recently, these toxins have also been detected in Alaskan seabirds (Van Hemert et al. 2020). Given that seabirds are cherished seasonal resources and consumed in various ways including (in some cases) whole de-feathered birds contains a potential exposure risk that should be considered.

3 Use of Beachcast Marine Mammals and Invertebrates

> Now some of the young men had gone to Kitigaaryuit for supplies, like tobacco and matches, and had never shown up again. Their names were Situgatkak, Akauyaluk. the young men had also taken their wives along and just before getting to Kitigaaryuit, they had eaten

bowhead whale meat which had gone bad and they died of it from poisoning Joe Nasogaluak (N-1992-253:3rd Series, Tape 6) as found in Hart and Amos 2004

The use of dead beachcast (stranded) or drifting whales, walruses, or other marine resources as food is a long-standing practice, but it comes with widely acknowledged risks. Post-mortem decomposition of meat and organs in marine mammals of good physical condition is accelerated due to the highly insulative properties of blubber, which slows post-mortem carcass cooling. The post-mortem body temperature of large whales left at sea for 5–6 days remained within $\pm 1°$ of 35 °C (Case 1948). Under these circumstances, whale meat is rapidly colonized by mixed bacteria flora (commonly sporulating and non-sporulating gram-positive bacilli, gram-negative bacilli, fecal *Streptococcus* sp., rarely *Staphylococcus albus*) originating from the gastrointestinal tract (Case 1948). Dried cultures of bacteria isolated from decomposed whale muscle included a mixed flora of clostridia, *Cl. oedematiens* (*novyi*) and *Cl. chauvoei* (Case 1948). Rare cases of gas gangrene were reported among Antarctic commercial whaling crews (Case 1948). While the whale is afloat with decomposing gasses, the outer blubber and skin layer continues to cool through heat loss to the aquatic environment and undergoes a more limited decomposition (ageing) process. Inuvialuit elders make reference to the concept that the outer blubber of whales that is in close (below the epidermis) contact with water is salvageable (see Hart and Amos 2004). Specific traditional guidelines based on blubber appearance, color, smell, oiliness, and overall evaluation of beachcast whales are in place to assess if muktuk (blubber with skin) can still be utilized from such a whale. Although natural mortality of whales can be caused by disease and toxins, a variety of other non-infectious causes of death including old age, predation (e.g., orcas, polar bears, and humans), ship strikes, and entanglements in marine debris contribute to seasonal availability of floating and beachcast marine mammals (i.e., Fay et al. 1977; Kalxdorff 1997; Hay 2000; George et al. 2017). Among bowhead and beluga hunters, it is a customary practice to wait for a few days until the internal gas from decomposition floats a carcass and then salvage "struck and lost" animals and harvest muktuk (the epidermis and blubber layer) and baleen (bowhead whale) only. "Struck and lost" hereby refers to animals that were hit (struck) by a weapon (rifle, grenade, harpoon) and were not landed for various hunt-related reasons (NAMMCO 2017). Blubber that has "aged" during the decomposition process from these whales has its place among other cherished aged foods. Despite community precautions, isolated historical community outbreaks of *Salmonella*-associated food poisoning have occurred from consuming blubber and meat from beachcast dead whales (*beluga whale*) (Boggild 1969) (unknown whale species, Bender et al. 1972). Though not in the Arctic, historic cases of sickness from consumption of meat from moribund whales (fin and sei whales) have also been reported from Japan. Isolates from whale muscle included *Salmonella* and various *Clostridia* (*novyi*, *septicum*) (unknown whale, Kawamura et al. 1927; *bottlenose whale*, Nakaya 1950). Given the traditional practice of food sharing and/or bartering, providing for those in the community that cannot hunt (i.e., Kishigami 2009; Braund 2018), food poisoning outbreaks, although rare occasions, will often involve large numbers of community members. Personal and past community experiences, in combination with traditional stories and place names, likely remain important cultural education tools among northern communities to maintain community health vigilance on the uses of beachcast marine animals.

Unrelated to subsistence hunts, there is a limited body of literature on occupational human exposure (i.e., veterinarians, wildlife biologists, and marine mammal trainers) to various marine zoonotic agents (viral, bacterial, fungal) from handling dead and sick marine mammals (see Geraci and Ridgway 1991; Hunt et al. 2008; Waltzek et al. 2011; Tryland et al. 2014; Tryland 2018). With the exception of "seal finger"–associated pathogens, most marine zoonotic agents cause some form of disease in marine mammals as well (see Waltzek et al. 2011; Hunt et al. 2008). Given the precautionary nature of traditional health examination of hunted animals, obvious signs of sickness would preclude further handling of a carcass. Sick and wounded animals encountered during hunts are generally euthanized and left in the field even at the risk of opposing mandatory regulatory salvage obligations.

Northern subsistence lifestyle and customary practices have not been exempted from economic hardship. Monetary and climatic challenges of regional modern subsistence, such as the increasing cost of fuel and hunting equipment, could result in individuals and communities being forced to choose between no eating or utilizing animals that do not pass traditional customary safety practices. During historical food shortages and recently during the 2011–2016 Alaskan Northern Pinniped unusual mortality event (https://www.fisheries.noaa.gov/alaska/marine-life-distress/diseased-ice-seals-and-unusual-mortality-events) where hundreds of seals were affected, some seals that appeared sick were fed to dogs but excluded from the human food chain (Hart and Amos 2004; Sheffield unpubl. data).

Coastal storms during the summer and fall regularly deposit marine invertebrates onto northern shores, and these beachcast ocean foods are enjoyed as seasonally available food resources (see Kuhnlein and Humphries 2017; Rapinski et al. 2018). Although consumption of shellfish in more temperate regions has commonly brought the risk for paralytic shellfish poisoning (PSP), few outbreaks related to PSP from consumption of seafood have been reported in the North (Orlova et al. 2007; Fortune 1975 cited in Himelbloom et al. 1996). With sea surface temperatures (SST) of the Bering–Chukchi–Beaufort seas (BCB) increasing, risks for harmful algal blooms (HABs) in the western and northern Arctic are a concern (Anderson et al. 2018). Hence, exposure to PSP from eating subsistence-harvested seafoods, gathered from the beach or from the stomach of harvested marine mammals (i.e. clams from walrus stomach or fish from ringed seal stomachs) may pose an emergent risk for northern communities (Bering Strait: Walruses and Saxitoxin 2017; Alexandrium algae, saxitoxin, and clams: Bering Strait and Chukchi Sea 2018–2019 2020) (Fig. 4). The most recent Alaskan human fatality due to paralytic shellfish poisoning linked to consumption of locally gathered blue mussels and marine snails occurred during 2020 on Unalaska Island (https://content.govdelivery.com/bulletins/gd/AKDHSS-295e317). In Alaska, the Southeast Alaska Tribal Ocean Research (SEATOR) is currently the only local tribally owned laboratory for testing of subsistence-harvested shellfish. Studies to systematically monitor

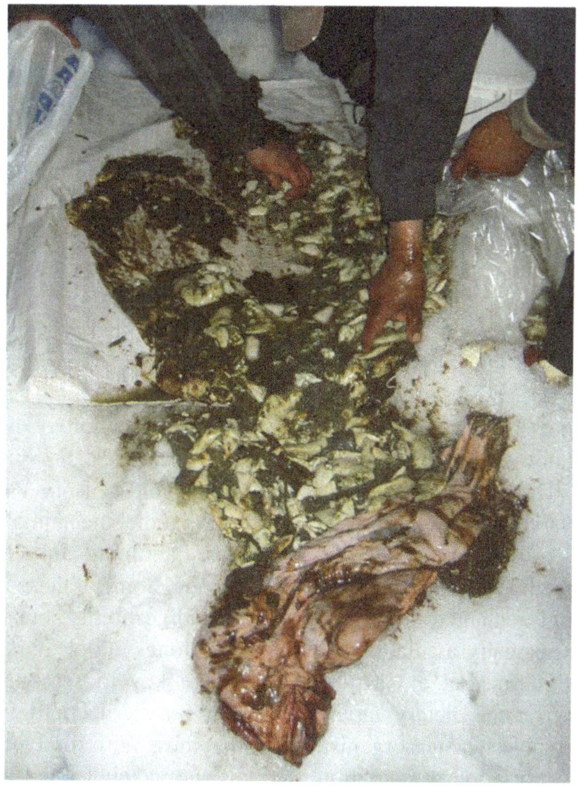

Fig. 4 Harvesting clams from the stomach content of a walrus in the Bering Strait region. Photo: G. Sheffield

the bioaccumulation of marine biotoxins in marine mammal and seabird tissues are lacking; however, preliminary findings from Alaskan marine biotoxin monitoring studies suggest limited bioaccumulation in liver tissue of seabirds (Van Hemert et al. 2020) and the meat/blubber of Pacific walrus (Lefevbre and Sheffield unpubl. data).

4 Traditional Food Preservation and Arctic Cuisine

> When people gathered together they told stories about the people before them. To hear them tell these stories one would think our ancestors were rich people never going hungry. Every bit of food they caught was stored away and nothing wasted. Even the fish they caught every bit of it they saved. They were scaled and dried into dry fish. The liver, eggs, and gizzards were all cleaned, dried and stored in bags of fish oil. This was very delicious and one could not stop eating once he started. How wealthy our people used to live in those days. Amos Tuma (N-1992-253: Life History # 8) as found in Hart and Amos 2004

Regional insights into the variety of wildlife resources used and their seasonal importance for northern communities can be gleaned from harvesting surveys, dietary recall studies, ethnographic accounts, and traditional ecological knowledge studies (i.e., Wolfe 2004; Council of Canadian Academies 2014; Kuhnlein and

Humphries 2017). Fish, marine invertebrate, animal meats (i.e., muscle, blubber, organ meat, bone marrow, brain, eye, skin, mammary glands, and blood), oils/animal fats, plants, and berries are central to northern food systems. Food safety, next to food quality (i.e., taste, texture, color, smell, nutrient content), is a universal need and prerogative to regional Arctic cuisines (Food Safety Network 2009). Arctic food preparation methods encompass universal cooking methods such as boiling, roasting, grilling, baking, and frying and include northern food preparations such as raw (fresh/frozen), frozen (raw/aged), dried/smoked (aged) and aged/rendered (fermented) foods. Furthermore, Arctic traditional cuisine often uses combinations of preservation methods to control food safety, shelf life, sensory, and nutritive food properties (i.e., water activity reduction [drying] with cold storage/freezing; fermentation [ageing] combined with cold storage; limiting oxidation by storing various fresh and dried foods in seal oil combined with cold storage). Multi-targeted food preservation approaches, called hurdle technology in modern food science, are known to improve bactericidal or bacteriostatic outcomes (Singh and Shalini 2016). Food properties (i.e., nutritive and sensory value such as lean versus fat, variable tastes to meat, fat pending what diet the animal utilized), in concert with the seasonal harvest environment, shape the timing and selection of which preservation methods are appropriate for which food. Some of these food preparation methods not only enhance flavors, edibility, and digestibility, but they extend the shelf life of seasonally available resources (i.e., Jones 2006; Starks 2002, 2007, 2011;Yamin-Pasternak et al. 2014) and thus are key to seasonal food security, while also providing dietary variety. Foods are not sterile, and given time, they will spoil due to present microorganisms and enzymes. With food processing that uses traditional Arctic food preservation methods (i.e., ageing, rendering oil, drying), community members, especially elder knowledge holders, possess acute awareness that faulty processing methods can lead to failure of preservation with outcomes ranging from food deterioration and spoilage to food becoming extremely hazardous and/or lethal to ingest. Therefore, like other cuisines, food safety of Arctic traditional food systems is influenced by (1) the skill and knowledge of who is preparing the food items, (2) inherent food properties per se (nutritive, physical, physicochemical, sensory, microbial-death-growth-related properties, health properties, etc.), (3) food preservation method used, (4) environmental conditions such as climate, geography, etc., and (5) microbiological contamination (i.e., protozoa, parasites, zoonotic pathogens, and biotoxins). Elder guidance of best food processing practices and the education of younger generations are essential components of northern oral history–based societies. The passing of the keepers of orally transmitted knowledge has actively contributed to the loss of Arctic food wisdom on how to most effectively process and store country foods safely (survival knowledge) among northern communities, thereby contributing to food insecurity. Many northern communities struggle with food insecurity (i.e., Inuit Circumpolar Council 2012, Wakegijig et al. 2013; Council of Canadian Academies, 2014, Inuit Circumpolar Council-Alaska 2015; Sowa 2015; Kenny et al. 2018; Walch et al. 2018).

5 Preservation by Heat Treatment

> Mom would always boil the fresh smelt and we drank the broth. We ate them with fresh sura (willow leaves) and oil. Hannah Stalker as found in Jones 2006

Across the North, variations of food heating technologies were used, including stone boiling (indirect heating of water with hot rocks using clay pottery, baskets, and animal stomachs as cooking containers), spit-roasting food, cooking meat in caribou/mountain goat stomach, using hot springs, using hot sand to cook eggs, using pit ovens, roasting on a flat stone, stone baking, etc. An ancient method of preservation, heating food is effective in controlling enzymes and heat-sensitive microbes and parasites in foods by inactivation (Cebrián et al. 2017). Ethnographic accounts attest traditional "boiled" foods were, in fact, parboiled (Harry and Frink 2009; Starks 2002, 2007). Parboiling, although less effective in inactivating microbes, protects the natural vitamin content in foods (Fig. 5). Many Arctic animal meats, including fish, birds, and land- and marine-based mammals, are excellent sources for vitamins and other micronutrients apart from being good sources of protein and long-chain n-3 fatty acids (see *Canada*: Northwest Territories Traditional Foods Fact Sheets 2002, Nunavut Nutrition Fact Sheets 2005; *Alaska:* ANTH Traditional Food guide 2015; *Greenland:* Pars et al. 2001; Anderson 2005; Deutch

Fig. 5 Bowhead whale blubber and skin (maktak) cut and ready to be boiled for uunaalik during fall whaling, Utqiaġvik, Alaska. Photo: R. Stimmelmayr

et al. 2007). Parboiling and simmering of dried and frozen foods among indigenous northern cooking coexists with prolonged boiling for making meat stews, bird soup, etc. (Starks 2002). For example, fresh kidneys from hunted bowhead whales are traditionally boiled so that the center of the serving piece remains pink (Barrow Whaling Captains' Wife Association pers. comm.). Although not as effective at destroying pathogens, parboiling has emerged as a practical and effective community method for determining kidney worm infection (*Crassicauda* sp.) in hunted bowhead whales. As the kidneys are being warmed, kidney worms will start migrating out of the kidney (Barrow Whaling Captains' Wife Association pers. comm.). An increased anisakid larval migration from the gut to the flesh due to cold smoking temperatures and brining salinity has been reported (Hauck 1977). Kidney worm infection is an emerging parasitic disease in the Bering–Chukchi–Beaufort seas (BCB) population of bowhead whales (Stimmelmayr et al. 2021) and has direct influence on cultural uses of bowhead whale kidney. Heavily parasitized kidneys with many worms, granulomas and fibrous nodules throughout are generally discarded and do not enter the community food chain (Barrow Whaling Captains' Wife Association pers. comm.).

Undercooking meat and organs poses an infection risk from a range of zoonotic parasites (e.g., *Trichinella* spp., *Toxoplasma gondii*, and anisakid nematode larvae). Undercooked meats of walruses and polar bears are well-documented reservoirs for *Trichinella nativa* and have commonly been the cause of trichinosis outbreaks of in Arctic communities (i.e., Margolis et al. 1979; Serhir et al. 2001;Møller et al. 2005, 2010; Fearey and Provo 2015; Springer et al. 2017; Uspensky et al. 2019; Martinez-Levasseur et al. 2020). In Greenland, historic *Trichinella* outbreaks were also linked to sled dog consumption (Thorborg et al. 1948). Although not an everyday fare, ethnographic accounts (Rausch et al. 1956; McManus-Fry et al. 2018) and Yupik-Eskimo traditional sayings (Mary Pete, pers. comm) on "1, 2, 3 dog trips" (number of sled dogs that had been killed and consumed) attest to this old northern practice during food shortage. Interestingly, dog meat consumption continues to play a role in trichinosis outbreaks in China (Wang et al. 2006). Independent of the animal species involved, human trichinosis, although infrequent and treatable, remains an important northern disease. Surprisingly, few regional public health measures for the control of trichinosis are in place in northern communities (e.g., *Nunavik Trichinellosis Prevention Program*, Canada (Larrat et al. 2012; Ducrocq et al. 2020); general game meat inspection in Greenland (Møller et al. 2010); *Trichinella* sp. surveillance of subsistence foods, North Slope, Alaska (Seymour et al. 2014; http://www.north-slope.org/departments/wildlife-management trichinosis).

Antibodies to and occasionally muscle tissue-dwelling cysts of *Toxoplasma gondii* have been detected in a variety of northern wildlife species (for review Miller et al. 2018; Jenkins et al. 2013). The role of marine and terrestrial wildlife in human infection remains inconclusive for northern communities (Peterson et al. 1974; Messier et al. 2009; Sharma et al. 2018; Reiling and Dixon 2019). A pilot study using northern village sled dogs fed traditional foods (i.e., caribou, fish, seals, beluga whales) as environmental sentinels for *Toxoplasma gondii* confirmed antibodies in the majority of dogs that were tested (Stimmelmayr et al. 2019). In addition to foodborne transmission from uncooked Arctic foods (raw, frozen, dried, aged), environmental exposure through subsistence harvest activities (i.e., skinning, butchering, and food processing) and contaminated aquatic reservoirs likely plays a role as well (for review Jenkins et al. 2013). The use of traditional water sources (untreated) including gathering ice or snow melt, water from brooks, tundra ponds, lakes, and rainwater collection is still practiced in many northern communities. The taste of water gathered from the air, land, and ocean is commonly judged superior to tap and/or imported bottled water (e.g., Marino et al. 2009; Goldhar et al. 2013). Good hand hygiene practices during food production and processing are considered important risk reduction measures. *Toxoplasma gondii* tissue cysts and oocysts are heat-labile, and cooking is effective in inactivating the organism (Franssen et al. 2019).

6 Preservation by Drying

> We dry smelt whole by stringing a few dozen at a time on a willow stick. We cut last years' length of willow and poke the big end though their gills. Then tie the small end of the stick around the large end to form a hoop that we hang up to dry and the drying racks or anywhere. We turn and move them around each days so they dry faster and more evenly. They get a little stinky too, which only makes them better (Jones 2006).

Preservation by dehydration is a traditional northern food processing technique that rests on the principle of reducing water activity (content) through various thermal drying processes (i.e., sun and air drying or hot/cold smoking). Customarily, food to be dried (e.g., fish, meat, intestines) undergoes much preparation before the process begins. Across the North, culture-specific ethnographic accounts showcase a plethora of regionally specific food-cutting techniques (i.e., slicing, cross-cutting, and half cutting) and various natural and engineered drying structures (i.e., drying racks, stringing, wood sticks, natural surfaces, and smoke houses) that are used to maximize drying of foods under prevailing local conditions (Fig. 6). From a food quality perspective, the nutrient content of most dried animal foods becomes concentrated through water reduction. However, as a food preservation method, drying is generally less effective for inactivation of food-associated enzymes, microbes, and parasites (see Beuchat et al. 2013; Kuhnlein and Soueida 1992). Several studies have characterized the microbiological status of traditional dried foods (*Greenland*: caribou, muskox, warble larvae, fish, Hauptmann et al. 2020a, b; *Canada*: salmon, Allen et al. 2012). Findings suggested that drying methods were effective in controlling microbial growth. Microbial content of traditionally smoked salmon (*Alaska*: lightly brined and air dried) after a 4-day drying period indicated an increased microbial growth, including *Staphylococcus aureus* (Himelbloom 1998). Initial microbial growth was driven by the slow onset of water activity loss. The recontamination of dried foods from handling was the major reason for the increased presence of *Staphylococcus aureus*, the latter being absent in fresh salmon. Thus,

Fig. 6 Processing and preservation of meat by drying, Bering Strait region. Photo: G. Sheffield

good hand hygiene is an effective risk reduction measure for avoiding microbial recontamination.

Botulism outbreaks have been linked to the consumption of dried fish (Botulism Alaska review 2017). Besides pH and salt, water activity is an important growth-controlling factor for *C. botulinum* and other microorganisms. Generally, water activity below 0.94 inhibits bacteria growth and the germination of bacterial spores (Zottola et al. 1983). Half-smoked fish, a fish that is brined and then smoked in a smokehouse for only one day, has a higher water content than smoked fish and may thus pose a risk, especially if kept at temperatures above freezing (Zottola et al. 1983). Freezing, through crystallization of water, further reduces water activity in foods. Interestingly, fully smoked salmon samples from the latter Alaskan study that were analyzed with mouse bioassay for the presence *C. botulinum* toxin were negative despite microbial analysis indicating the presence of anaerobic spores (i.e., *C. botulinum*, *perfringens*, *sporogenes*). Only after enrichment, two samples demonstrated positive reactions with the mouse bioassay (Zottola et al. 1983). Findings from this study suggest that spores were not destroyed by the traditional drying techniques, and incomplete salting or storing at >3.3° could pose a risk from further anaerobic growth.

As discussed earlier, foodborne transmission of *Toxoplasma gondii* from uncooked Arctic foods (i.e., raw, frozen, dried, and aged) is thought to play an important role (Jenkins et al. 2013). Dried seal meat (*nikku*, air-dried meat) prepared from grey seal meat experimentally spiked with *Toxoplasma gondii* oocysts (swine origin) and stored at 4°C and bioassayed in cats at 49, 95, and 140 days post preparation was not infective for cats (Forbes et al. 2009). Very limited data are available with respect to the effectiveness of methods of drying meat in killing parasitic nematode larvae (i.e., *Trichinella* sp., *Anisakis* sp., and *Diphyllobothrium* larvae) (Franssen et al. 2019). For example, cold smoking does not impact the survival of *Anisakis* larvae in herring and salmon (Hauck 1977; Gardiner 1990). Combining drying with the addition of brine (salt solution) appears to be relatively more effective for inactivating various parasites (Franssen et al. 2019). Salt content (%) determined in traditionally smoked and half smoked salmon from Alaska ranged between 1.6 and 2.0 (Zottola et al. 1983). Given the traditional practice to store some dried foods in meat pits to keep cool during the summer and frozen during the winter (Starks 2002), the latter cold treatment may further enhance bacterial inactivation and inactivation of freeze-intolerant parasites, but this hypothesis has not been tested. Starks (2011) further suggests that during fall fishing, large amounts of fish would be piled for drying, and under ambient conditions, the outside mound would freeze while the core would age, adding another layer to the complex food preparation of drying fish. The use of hurdle technology in modern food science (multi-targeted food preservation approach) generally improves bactericidal or bacteriostatic outcomes (Singh and Shalini 2016).

Effective traditional drying depends on appropriate environmental conditions (i.e., temperature, wind, humidity, and insect activity) (Starks 2011). Because ongoing climate change will likely impact all of these parameters, the process of traditionally drying foods is becoming more difficult, and experimentation with novel drying methods (i.e., using regular ovens to dry beluga meat (Kristi Frankeson pers. comm.) and the use of fans for increased air movement for fish drying) are just some of the mitigation measures that have been attempted in some communities (Helen Aderman pers. comm). In many instances, the prevailing microclimate may not only limit when one is able to hunt, but it may also limit food processing (i.e., Waugh et al. 2018; Nancarrow and Chan 2010). In summer 2019, multiple cases of scombroid poisoning from consuming salmon were reported in Alaska (Frick et al. 2019). Scombroid poisoning is caused by fish that have been inadequately cooled after capture resulting in bacterial overgrowth that converts histidine present in fish tissue to histamine. Histamine (scombroid toxin) is not destroyed by either cooking, freezing, drying, curing, smoking, or canning. It is unclear what led to the spike in Alaskan cases, as previously (2015–2018) only five cases of scombroid poisoning were reported. Although speculative, higher than normal summer temperatures in 2019 that was Alaska's hottest year on record could have played a role.

7 Preservation by Cold Treatment

> During the whaling season he would dig a pit into which he put whale heads [beluga] and tails and flippers until the pit was full. Now he covered this up with logs, side by side. These whale heads remained in there until just before freeze-up. By the time the flippers and whale

heads and tails had aged. They were taken out. Then he got some grass or hay stacked up and bailed all the water and oil out of the pit till it was dry. Then he got some nice thick willows which were clean, and put them in the bottom of the pit as flooring. Then he got the grass and hay and wiped..the sides of the pit clean of all blubber. Now he got the heads and tails and flippers. These were washed clean, and then put back into the pit. The logs were put back and the cracks chinked with grass, and mud was put over this so water and snow would not go into it. He kept two or three heads out, handy for winter use. What had gone into the pit was for emergency use. Frank Cockney (N-1992-253:1-44-01) as found in Hart and Amos 2004

Chilling and, ultimately, freezing using traditional Arctic methods (e.g., meat pit, fish pit, ice houses, and ice cellars) and modern freezers are widely used practices aimed to preserve Arctic foods and their properties (color, flavor, and texture) beyond the harvest season. Freezing, in contrast to heat preservation, mostly inhibits enzyme activity, slows down microbial growth of foods, and is effective in killing freeze-intolerant parasites. The ability of freezing to kill microbes depends on cold tolerance of microbes and the duration of frozen storage time. Recurring failures of modern freezers, as well as a prolonged time between thawing foods and consumption, will result in recovery of microbial growth leading to food spoilage. Infectious organisms present in the meats and organs reflect pathogens harbored by the animal (carrier) and any contamination that occurred during harvesting, butchering, and processing.

Traditional northern freezer technology is based on excavating a compartment in the upper permafrost (Hart and Amos 2004; Starks 2011; Peterson and Wendler 2011). In general, the deeper the pit, the colder the average temperature at depth. Although temperature profiles of traditional freezer technology (i.e., ice cellars, ice houses, and pits) have not been sufficiently characterized, functioning ice cellars in Utqiaġvik and Kaktovik, Alaska, at a location below the active permafrost layer maintain fairly constant sub-freezing temperatures (−2 °C to 12 °C) that vary seasonally (winter versus summer) (Peterson and Wendler 2011; Nyland et al. 2017). Storing meat at temperatures below −18 °C is recommended by the U.S. Department of Agriculture. Ice cellars have been used for thousands of years, but climate change and loss of or unfamiliarity with traditional knowledge on how to construct and maintain traditional ice cellars have led to occasional degradation or failure in northern Alaska (Osterkamp and Romanovsky 1999; Barrow Whaling Captains Association pers. comm.; Nyland et al. 2017). Increased rates of food spoilage caused by saline water intrusion and subsequent mold growth on ice cellar walls have been reported due to temperature failure during a time of climatic warming (Brubaker et al. 2009). The use of large-sized electric powered walk-in freezers by bowhead whaling communities in the North Slope region of Alaska can be viewed as a temporary fix, until innovative mitigation such as prototypes of hybrid ice cellar designs using thermal siphon technology with traditional Inuit ice cellar engineering designs is ready for production (Peterson and Wendler 2011). The taste of traditional ocean-based foods such as meat and muktuk (whale skin/blubber) frozen by traditional ice cellar method (slow chilling, which allows for ageing to occur) is preferred to the taste of traditional foods frozen by modern freezers (rapid chilling and freezing) (George Kaleak Sr. cited in D'Oro 2019; Nyland et al. 2017). It is likely that meat processes, such as cold shortening caused by rapid cooling of meat, play a role in the noted taste differences (Kadim and Mahgoub 2002). Lastly, permafrost-driven chilling and freezing technology of foods is uniquely northern, and the 2005 discovery of whale blubber that had been processed for consumption and subsequently frozen 1050 years prior (±70 years) was remarkable (George et al. 2008). Although the whale blubber was no longer considered edible, its existence underlines the unique preservation ability of permafrost-based freezing technology.

Frozen foods (meat, fish, bird) are staples of Arctic cuisine, and organoleptic assessment (i.e., taste, sound, color, smell, touch) is an important culinary element of eating Arctic frozen foods (Starks 2002, 2007). Changes in color, smell, consistency, and taste are valuable indicators for food integrity and general food safety (i.e., food spoilage and foods that have gone beyond shelf life). Trimming surfaces of frozen meat and/or blubber and discarding these cuts prior to serving is a customary practice. Norwegian whale meat hygiene studies (Tryland et al. 2014) provide evidence that this food practice is an effective risk reduction measure with respect to surface-associated bacterial flora. The pathogenic significance of *Clostridia* and *Streptococci* isolated from fresh whale meat and organ meat (liver) obtained during commercial whaling was deemed "no greater than other types of meat" (Cockrill 1960). The microbiological status of raw/frozen meat/blubber (post-harvest/butchering) from Arctic animals originating from subsistence hunts has not been characterized.

As discussed earlier, foods that make it into the traditional food preservation methodology of freezing have already undergone customary food quality assessment. Parasites can be considered a normal finding in wild animals, and customary assessment of animals for visible parasites is a step in traditional food safety assessment. In general, known parasites (e.g., worms) within the gastrointestinal tract are expected and deemed normal. For example, a substantial gastrointestinal parasite burden in bearded seals of physical good condition (i.e., thick blubber layer with seal oil running off freely) can be a useful health proxy, indicating that the animal is healthy and was eating well (Roberta Leavitt pers. comm.). Although unrelated to parasites, a detailed attention is also paid to the composition of stomach contents from seals and walruses as they can inform an experienced and well-taught consumer about diet changes in the marine mammals and, by extension, about the food web at large. Stomach examination occurs after visceral organs have been removed from the animal and generally away from the carcass to avoid accidental tainting of meat. Traditional customary attitudes toward parasites and parasitic cysts associated with other organs such as the liver, kidney, body cavities, and meat are conservative and very precautionary. The discovery of these anomalies often results in discarding the organ, parts of meat, or the entire animal. Very small to microscopic meat-dwelling parasites (i.e., *Trichinella* sp., *Sarcocystis* sp., *Toxoplasma gondii*, *Anisakis* sp., and *Diphyllobothrium* larvae) pose detection problems to traditional food systems, especially if animals do not show behavioral or clinical disease symptoms associated with being infected by the parasite. Under these circumstances, the safety of frozen foods will depend largely on the cold tolerance of the parasite. For example, among helminths, freezing survival is high for *Trichinella nativa* in traditionally frozen seal walrus meat (Kapel et al. 2003; Leclair et al. 2004), while freezing survival is low to zero for *anisakid* (fish roundworm) and *Diphyllobothrium* larvae (fish tapeworm) (Franssen et al. 2019). Interestingly, human infection from either fish parasite is rarely reported in northern communities despite high fish consumption (Jenkins et al. 2013). Cultural consumption preferences for raw-frozen, dried, or aged fish versus raw fish may partially explain the overall low prevalence. The distinction between raw and raw-frozen Arctic foods is very important, and Starks (2007) speculated that early ethnographic accounts of raw meat eaten by Arctic peoples missed the subtlety that these foods were probably *quaq* (raw-frozen foods). Meat-dwelling *Toxoplasma gondii* cysts are also sensitive to low temperature and freezing, albeit inactivation versus killing of tissue cysts depends on freezing temperature and duration (Alizadeh et al. 2018; Franssen et al. 2019).

8 Preservation by Ageing

> Beluga should be aged in a cool place that lets air in. If you are not sure how to prepare igunaq [aged meat] or other country foods, ask an elder who has this knowledge Department of Health and Social Services, Government of Nunavut, 2005

Ageing food is a very old food preparation method that improves tenderness and flavors (Kadim and Mahgoub 2002). Inuit aged food generally does not include the addition of preservatives or salt and thus differs from other ageing-based food science technologies. However, whole animal ageing (e.g., seabirds and fish) probably introduces a suite of unknown components that serve similar purposes. By definition, fermentation is an enzyme-driven breakdown of tissues with degradation of carbohydrates to lactic acid, thereby lowering the pH of the aged food. Although the carbohydrate content of traditional meats, organs, and whole animals used in ageing has not been characterized, it is generally thought that these foods contain few fermentable carbohydrates. Given these differences, it has been argued that aged traditional foods are not true fermented foods, but that they rather reflect controlled meat putrefaction processes (see Yamin-Pasternak et al. 2014). Under carbohydrate-limited conditions, ageing-related reductions of pH in aged foods are expected to be minimal. Low pH (acidic foods) limits growth of foodborne pathogens, especially *C. botulinum.* Few studies have characterized pH in traditionally aged foods, such as fresh salmon eggs (pH 4.2–5.1), freshly fermented salmon eggs (pH 5.9; Hauschild and Gauvreau 1985), and *mikiyuk* (aged bowhead whale meat, Inupiaq) (pH 4.58; Zottola et al. 1983). Based on the findings of the latter study, *mikiyuk* (alternatively *mikigaq*) reached a pH that is outside the pH range (6.5–7.5) for optimum growth of *C. botulinum,* and subsequent mice bioassay tests were unable to detect *C. botulinum* toxin (Zottola et al. 1983). *Migikak*, a cherished aged dish in the North Slope region of Alaska, is made from blubber with skin, tongue meat, and blood from bowhead whales. It is "aged" for about 2–4 weeks above ground in an open container that is

Fig. 7 Seal oil with dried seal meat, Utqiaġvik, Alaska. R. Stimmelmayr

kept in a cool shaded place, turned/stirred by hand, and only covered lightly to allow for generous air movement (Sarah Skin, Utqiaġvik, Alaska cited in NSB DWM traditional food recipes). A similar preparation approach (using open containers, keeping cool and shaded) is taken while rendering seal blubber to make seal oil. The latter is among the most important condiments and base ingredients for many Arctic dishes (Starks 2002, 2007, 2011). Immersing foods in seal oil (i.e., bird eggs, dried fish, dried whale and seal meat strips, and plants) for storage purposes is widely practiced in some regions (see Hart and Amos 2004; Starks 2002, 2007, 2011; Dudarev et al. 2019; Gay Sheffield unpubl. data) (Fig. 7). Oil treatment with commercially available, plant-based oils (i.e., olive oil) of meats in western food science is known to preserve and alter flavors and prevent oxidation that limits color changes as well as most microbial growth (although mold and yeast can still flourish) (Moon et al. 2012). Lastly, direct sunlight alters food properties quickly, and the customary practice to keep things in cool, shaded places and/or underground helps to control photodegradation of nutritive food properties (i.e., vitamin loss, oxidative rancidity, color changes) and therefore maintains high food quality. The majority of aged Inuit meat dishes (i.e., fish, bird, whale, walrus, and seal) are commonly stored in the ground. Some meats are sealed in seal pokes (boned out seal skins free of meat) (Starks 2002, 2007, 2011; Tigullaraq 2008), marine mammal stomach "balloons" (consisting of the waterproof membrane prepared from the stomachs of whales, walruses, and seals) (Hart and Amos 2004), and wrapped "meat rolls"

(Yamin-Pasternak et al. 2014). A detailed knowledge and precise memory of sensory properties of aged food are needed to properly age the foods and decide when the product is ready to be eaten versus when it has spoiled. Underground storage time for aged foods varies by region but can be up to a year. During the preparation of *igunaq* (aged meat), gas production occurs during the liquefaction processes, and the liquefied blubber soaks into the meat and through the seal skin poke (*igunaq* Forbes et al. 2003; unknown aged dish Starks 2011). The aroma of *igunaq* is described as strong but does not smell of putrefaction (Forbes et al. 2003). Many botulism outbreaks in Canada and Alaska (Todd 1988; Fagan et al. 2011; LeClair et al. 2013; Alaska DHSS Epibulletin 2019) have been linked to improperly fermented (aged) meats (often whales, seals, and fish). Botulism cases in humans linked to seal oil use are seen less frequently, but this may be a reporting artifact with seal oil as a primary source being overlooked as seal oil, a common condiment with dried and aged foods (Funk 1997). A recent collaborative science-to-action approach between the Maniilaq Association in Kotzebue, Alaska, and academic researchers was successful in characterizing the food science of regional seal oil rendering processes. This collaborative effort led to the development of a state-approved food processing plan to minimize food safety risks from seal oil. As of 2021, the Maniilaq Association has received authorization to add seal oil to the menu at the regional Maniilaq Association elder care facility (http://kotz.org/2021/01/28/it-brings-back-memories-maniilaq-cleared-to-make-and-serve-long-prohibited-seal-oil-to-elders/).

Today, at least in the Bering Strait region of Alaska, traditional recipes for aged foods are generally shared within family kinship units, but not with strangers (Gay Sheffield unpubl. data). It is unclear if this reflects ongoing concerns about the safety associated with "outside" people learning traditional food ageing preparation in a transitioning Arctic climate, or if it has always been that traditional customary practices of Arctic food preparation needed to stay local as they reflect place-based needs (preferences), seasonal animal resource availability, and specific place-based environmental microclimate processing conditions and substrates (i.e., temperature, humidity, direct sunlight, and/or the presence of insects) (Starks 2002, 2007). Where and when aged foods (i.e., *mikigaq*, seal oil, and *igunaq*) are traded and shared in the communities, as well as the knowledge of who has prepared the food, is a critical part of the local qualitative food safety assessment; local evidence-based trust is part of the decision process to enjoy these traditional dishes. Lastly, the transition from traditional methods (*tried and true*) to incorporating modern-day tools and equipment has added risks when ageing foods. For example, a higher frequency of botulism in Alaska has strong epidemiological correlates to the use of tightly sealable (non-breathable) containers, such as plastic totes/plastic bags, when "ageing" fish, fish eggs, whale meat, and walrus meat (Botulism in Alaska- November 2017). Other likely correlates include the loss of the transmission of knowledge to younger generations; processing food in modern houses that do not have good air circulation; ageing foods above ground versus underground due to relocation of people away from coastal villages to more semi-urban settings (i.e., regional hubs) and/or interior climates, as well as separation from elders transmission of traditional food preparation knowledge; ageing meat of animals that are novel due to range expansion; traditional gender role switching in food processing; and lastly, not following other traditionally customary guidelines (see Todd 1988; Shaffer et al. 1990; Starks 2002; Yamin-Pasternak et al. 2014; Gay Sheffield unpubl. data).

The resilience of a food preparation method can be crudely gauged by its effectiveness to inactivate foodborne pathogens and toxins. A few studies have explored whether traditionally aged food products can inactivate important zoonotic parasites such as *Toxoplasma gondii* and *Trichinella nativa*. For example, seal *igunaq* (aged seal dish) prepared from experimentally spiked grey seal meat with *Toxoplasma gondii* oocysts (swine origin), stored at 4 °C ,and bioassayed in cats at 49, 95, and 140 days post preparation were not infective for cats (Forbes et al. 2009). Raw aged sausages prepared from experimentally spiked grey seal meat with *Toxoplasma gondii* oocysts (swine origin), stored at −20 °C, and bioassayed in cats at 49, 95, and 140 days post preparation were not infective for cats (Forbes et al. 2009). *Toxoplasma gondii* tissue cysts have a high pH tolerance with infectivity not reduced at a lower pH (pH 5 to 6 compared to pH 7) with exposure for 24–26 days at 4 °C but are freeze intolerant (Franssen et al. 2019). It remains to be seen if Arctic strains of *Toxoplasma gondii* are more freeze tolerant. Three traditional marine mammal dishes, including seal *igunaq*, *nikku*, and raw and undercooked sausages prepared from experimentally spiked grey seal meat with *T. nativa*, were bioassayed in cats and mice over a 317-day period (Forbes et al. 2003). Under controlled laboratory conditions, infective larvae survived for a minimum of 5 months (153 days post preparation) in these aged foods; however, larvae count and infectivity declined throughout the experimental period. Given these initial findings, consumption of aged seal food (*igunaq* and *nikku*) and raw sausages made from seal carries a food safety risk. However, as suggested by the authors (Forbes et al. 2003), more studies are needed to clarify the effectiveness of ageing combined with freezing on *Trichinella* inactivation, since subsamples of seal *igunaq* removed at 219 days post preparation and then stored frozen at −20 °C for 89 to 95 days prior to being bioassayed in cats and mice did not contain infective larvae. Preliminary investigation on the infectivity of *Trichinella* sp. larvae in traditional preparations of aged walrus meat (*igunaq*) reported similar findings, stating that "none of the 4–5 and 10-month-old *igunaq* preparations contained infective *T. nativa* larvae as measured by bioassays using mice and guinea pigs at inoculation doses ranging from 6 to 500 larvae" (Leclair et al. 2004). The authors concluded that "the degradation process that occurred in the field can be sufficient to either kill *Trichinella* larvae or render them non-infective for mice and guinea pigs" but emphasized that further studies are needed (Leclair et al. 2004).

9 Post Food Processing Quality Measures

> … we had to go overland to where we had our fish cached. We had a stage after arriving there, Niulummaaluk and I saw two polar bears. They were eating fish from our stage. Looking in our sleighs we found that Niulummaaluk had forgotten to take his rifle. The bears

> had just finished eating. They had also spotted one whale stomach which was filled with blubber. Once in a while the younger cub, which was about a year old, would stand up on his hind legs and the mother was stretched out on the ice, for now they were full. Joe Nasogaluak (ISDP n.d.: Life Story, November 21 to April 1922, 1-1-15) as found in as found in Hart and Amos 2004

The construction of seasonal and long-term food caches strengthens short-term and long-term food security in a highly dynamic yet predictable environment. However, despite robust engineering, meat and fish pits enforced with logs, rocks, and other natural vegetation are not always able to prevent pilfering by polar bears, brown bears, wolverines, dogs, and foxes or be resistant to coastal erosion and flooding. From a food science packaging perspective, traditionally used materials included animal hides, seal pokes, and marine mammal stomach "balloons." The use of such items to contain foods remains an ideal packaging being regionally sourced, low cost, reusable, biodegradable, and non-toxic. Robust packaging from available materials (i.e., fish bundles, meat roll packet, pokes, and marine mammal stomach balloons) allowed for a safe spill-proof transport and kept out most pests such as insects and rodents.

10 The Future of Arctic Traditional Food Systems

Like other indigenous traditional food systems (Kuhnlein et al. 2009), Arctic food wisdom is characterized by inherent resilience and vulnerability. Under the Arctic Anthropocene, the complex nature of climate change shaping the new Arctic is widely affecting terrestrial and marine habitats, animal ecophysiology, and both wildlife and human health as we currently understand it. Successful adaptation to these complex food security and food safety challenges will require novel "science-to-action" collaborative research that is shaped by Arctic indigenous people's perception of food security while maintaining cultural identity—concurrent with adapting and mitigating to ongoing environmental and ecological transitions. Actionable One Health research in the Arctic should initiate in-depth research on Arctic wild foods cuisine. This research should characterize community-based food science technology while integrating those agencies/entities tasked with wildlife health monitoring and public health research. Research on Arctic cuisine and food science technology is critical to generating evidence-based community guidelines and risk communication strategies under circumstances where climate change–driven hypothesized pathogen impacts on Arctic food safety seem logical but have not yet been explicitly and rigorously tested by scientific methods. Evidence-based approaches are essential due to highly complex and dynamic socio-ecological contexts of subsistence harvesting, in which food (in)securities have emerged as a major threat for many northern communities. One epistemology cannot define a research action framework—as a lack of scientific data in the Arctic does not mean there is a lack of knowledge (Berkes 1999; Stimmelmayr 2003). Rather, it must be co-produced by indigenous and western science. The outcomes from such a co-produced One Health/Eco-health research effort will equally benefit the northern communities, essential wildlife resources, and the ecosystems at large.

References

Alaska: ANTH Traditional Food guide (2015) Traditional food guide for the Alaska native people second edition, Publisher Alaska Native Tribal Health Consortium, 168. https://www.amazon.com/Traditional-Alaska-Native-People-Second/dp/0692392165. ISBN-10: 0692392165 ISBN-1 978–0692392164

Alaska DHSS Epibulletin (2019 2017) list. https://dhss.alaska.gov/dph/Epi/id/SiteAssets/Pages/Botulism/Monograph.pdf

Albert TF (2001) The influence of Harry Brower, Sr., an Iñupiaq Eskimo hunter, on the bowhead whale research program conducted at the UIC-NARL Facility by the North Slope Borough. In: Norton DW (ed) Fifty more years below zero: tributes and meditations for the Naval Arctic Research Laboratory's first half century at Barrow, Alaska. University of Alaska Press, Fairbanks, AK, and Arctic Institute of North America, Calgary, Alberta and Fairbanks, AK, pp 265–278

Alexandrium algae, saxitoxin, and clams: Bering Strait and Chukchi Sea 2018–2019. 2020 UAF Alaska Sea Grant publication, MAB-75, Nome Alaska. Available from https://seagrant.uaf.edu/bookstore/pubs/MAB-75.html

Alizadeh A, Jazaeri S, Shemshadi B, Hashempour-Baltork F, Sarlak Z, Pilevar Z, Hosseini H (2018) A review on inactivation methods of Toxoplasma gondii in foods. Pathog Glob Health 112(6):306–319

Allen KJ, Chen XM, Mesak LR, Kitts DD (2012) Antimicrobial activity of salmon extracts derived from traditional First Nations smoke processing. J Food Prot 75(10):1878–1882

Anderson SM (2005) Vitamins and minerals in traditional Greenland diet. NERI National Environmental Research Institute, Denmark technical report no 528. Available from https://www2.dmu.dk/1_viden/2_publikationer/3_fagrapporter/rapporter/fr528.pdf

Anderson DM, Richlen ML, Lefebvre KA (2018) Harmful algal blooms in the Arctic [in Arctic Report Card 2018]. Available from https://www.arctic.noaa.gov/Report-Card

Armitage P (2015) Inuvialuit and Nanuq: a polar bear traditional knowledge study. Joint Secretariat, Canada. Available from https://wmacns.ca/resources/inuvialuit-and-nanuq-polar-bear-traditional-knowledge-study/

Bender TR, Jones TS, DeWitt WE, Kaplan GJ, Saslow AR, Nevius SE, Clark PS, Gangarosa EJ (1972) Salmonellosis associated with whale meat in an Eskimo community. Serologic and bacteriologic methods as adjuncts to an epidemiologic investigation. Am J Epidemiol 96(2):153–160

Bennett J, Rowley S (2004) Uqalurait: an oral history of Nunavut. McGill-Queen's University Press, Montreal and Kingston

Benson K (2019) Gwich'in knowledge of porcupine caribou: state of current knowledge and gaps assessment. Department of Cultural Heritage, Gwich'in Tribal Council. Available from https://www.pcmb.ca/PDF/EIS/GTC%20current%20knowledge%20and%20gaps%20assessment.pdf

Bering Strait: Walruses and Saxitoxin (2017) UAF Alaska Sea Grant publication, MAB-74, Nome Alaska. Available from https://seagrant.uaf.edu/bookstore/pubs/MAB-74.html

Berkes F (1999) Sacred ecology: traditional ecological knowledge and resource management. Taylor & Francis, Philadelphia

Beuchat LR, Komitopoulou E, Beckers H, Betts RP, Bourdichon F, Fanning S, Joosten HM, TerKuile BH (2013) Low-water activity foods: increased concern as vehicles of foodborne pathogens. J Food Prot 76(1):150–172

Bodenstein B, Beckmen K, Sheffield G, Kuletz K, Van Hemert C, Berlowski B, Shearn-Bochsler V (2015) Avian cholera causes marine bird mortality in the Bering Sea of Alaska. J Wildl Dis 51(4):934–937

Boggild J (1969) Hygienic problems in Greenland. Arch Environ Health 18(1):138–143

Botulism in Alaska- November 2017 State of Alaska Epi Bulletin publication. Available from http://dhss.alaska.gov/dph/Epi/id/siteassets/Pages/botulism/monograph.pdf

Braund SR (2018) Description of Alaskan Bowhead whale Subsistence Sharing practices. Available from http://www.north-slope.org/assets/images/uploads/Braund_AEWC16_Bowhead_Sharing_Report_5-23-18.pdf

Brewster K (1997) Native contributions to arctic science at Barrow, Alaska. Arctic 50(3):277–288

Brewster K, George C, Aiken M, Brower A, Itta M, Itta N, Lou Leavitt M, Leavitt O, Matumeak W (2008) Iñupiat knowledge of selected subsistence fish near Barrow, Alaska. Report prepared for the Bureau of Land Management through an assistance agreement to the North Slope Borough Department of Wildlife Management. Available from http://www.north-slope.org/assets/images/uploads/Brewster_Fish%20TEK%20paper.pdf

Brubaker M, Bell J, Rolin A (2009) Climate change effects on traditional Inupiaq food cellars. Center for Climate and Health Bulletin 1:1–7. Available from https://data.globalchange.gov/report/aknativetribalhealth-cchbulletin-1-2009

Canada: Northwest Territories Traditional Foods Fact Sheets (2002) https://pubs.aina.ucalgary.ca//health/62512.pdf

Carroll GM (1976) Utilization of the bowhead whale. Mar Fish Rev 38(8):18–21

Case RAM (1948) A study of the incidence of disease in a whaling expedition to the Antarctic pelagic whaling grounds, 1946-7. Br Soc Med 2:1–17

Castrodale L (2009) Two cases of tularemia – interior Alaska, Bulletin No. 20 August 26, 2009. Available from http://epibulletins.dhss.alaska.gov/Bulletin/DisplayClassificationBulletins/130

Castrodale L (2011) Human and animal brucellosis Alaska bulletin no. 31 December 28, 2011. Available from http://epibulletins.dhss.alaska.gov/Bulletin/DisplayClassificationBulletins/96

Cebrián G, Condón S, Mañas P (2017) Physiology of the inactivation of vegetative bacteria by thermal treatments: mode of action, influence of environmental factors and inactivation kinetics. Foods 6(12):107

Cockrill WR (1960) Pathology of Cetacea. A veterinary study on whales. Part I and II. Br Vet J 116: 133–189

Council of Canadian Academies (2014) Aboriginal food security in northern Canada: an assessment of the state of knowledge, Ottawa, ON. The Expert Panel on the State of Knowledge of Food Security in Northern Canada, Council of Canadian Academies. Available from https://cca-reports.ca/wp-content/uploads/2018/10/foodsecurity_fullreporten.pdf

Daoust PY, Stacey Z (2014) Harvesting seal products of high quality for human consumption. DFO Cn Sci Adv Sec Res doc 2014/009.v +34p. Available from https://waves-vagues.dfo-mpo.gc.ca/Library/360812.pdf

D'Oro (2019) News paper article. https://www.adn.com/alaska-news/rural-alaska/2019/11/25/failing-permafrost-cellars-signal-change-in-alaska-whaling-towns/

Deutch B, Dyerberg J, Pedersen HS, Aschlund E, Hansen JC (2007) Traditional and modern Greenlandic food - dietary composition, nutrients and contaminants. Sci Total Environ 384(1–3):106–119

Ducrocq J, Proulx JF, Simard M, Lévesque B, Iqaluk M, Elijassiapik L, Ningiuk E, Perkins P, Jacques S, Lemire M (2020) The unique contribution of a local response group in the field investigation and management of a trichinellosis outbreak in Nunavik (Québec, Canada). Can J Public Health 111(1):31–39

Dudarev AA, Yamin-Pasternak S, Pasternak I, Chupakhin VS (2019) Traditional diet and environmental contaminants in coastal Chukotka IV: recommended intake criteria. Int J Environ Res Public Health 16(5):696

Duncan C, Savage K, Williams M, Dickerson B, Kondas AV, Fitzpatrick KA, Guerrero JL, Spraker T, Kersh GJ (2013) Multiple strains of Coxiella burnetii are present in the environment of St. Paul Island, Alaska. Transbound Emerg Dis 60(4):345–350

Duncan C, Dickerson B, Pabilonia K, Miller A, Gelatt T (2014) Prevalence of Coxiella burnetii and Brucella spp. in tissues from subsistence-harvested northern fur seals (*Callorhinus ursinus*) of St. Paul Island, Alaska. Acta Vet Scand 56(1):67

Fagan RP, McLaughlin JB, Castrodale LJ, Gessner BD, Jenkerson SA, Funk EA, Hennessy TW, Middaugh JP, Butler JC (2011) Endemic foodborne botulism among Alaska Native persons—Alaska, 1947–2007. Clin Infect Dis 52(5):585–592

Fay F, Dieterich RA, Shults LM (1977) Morbidity and mortality of marine mammals Bering Sea. Final report Outer Continental Shelf Environmental Assessment Program 29. Available from http://www.adfg.alaska.gov/static/home/library/pdfs/wildlife/federal_aid/77_fay_dieterich_shults_morbidity_mortality_marine.pdf

Fearey D, Provo G (2015) Trichinellosis cases – Alaska, 2005-2014. Alaska Bulletin No 6, March 18, 2015. Available from http://epibulletins.dhss.alaska.gov/Bulletin/DisplayClassificationBulletins/47

Follmann EH, Ritter DG, Beller M (1994) Survey of fox trappers in northern Alaska for rabies antibody. Epidemiol Infect 113(1):137–141

Forbes LB, Measures L, Gajadhar A, Kapel C (2003) Infectivity of Trichinella nativa in traditional northern (country) foods prepared with meat from experimentally infected seals. J Food Prot 66 (10):1857–1863. https://doi.org/10.4315/0362-028x-66.10.1857. PMID: 14572224

Forbes LB, Measures L, Gajadhar A (2009) Infectivity of toxoplasma gondii in northern traditional (country) foods prepared with meat from experimentally infected seals. J Food Prot 72(8):1756–1760. https://doi.org/10.4315/0362-028x-72.8.1756. PMID: 19722415

Franssen F, Gerard C, Cozma-Petrut A, Vieira-Pinto M, Režek AJambrak A, Rowan N, Paulsen P, Rozycki M, Tysnes K, Rodriguez-Lazaro D, Robertson L (2019) Inactivation of parasite transmission stages: efficacy of treatments on food of animal origin. Trends Food Sci Technol 83:114–128

Frick A, Mooring E, Helfrich K (2019) Seven patients diagnosed with Scombroid Poisoning-Alaska Summer 2019. State of Alaska DHSS Epibulletin 21, October 15. Available from http://www.epi.alaska.gov/bulletins/docs/b2019_21.pdf

FSN (Food Safety Network) (2009) Safe preparation and storage of aboriginal traditional/country foods: a review. National Collaborating Centre for Environmental Health, Guelph, ON. Available from https://ncceh.ca/sites/default/files/Aboriginal_Foods_Mar_2009.pdf

Funk B (1997) Botulism caused by seal oil- 1997 DHSS Epi Bulletin No. 24 May 22, 1997. Available from http://epibulletins.dhss.alaska.gov/Document/Display?DocumentId=1693

Gadamus L, Raymond-Yakoubian J (2015) A Bering Strait indigenous framework for resource management: respectful seal and walrus. Hunting Arctic Anthro 52(2):87–101

Gardiner MA (1990) Survival of Anisakis in Cold Smoked Salmon. Can Inst Food Sci Technol J 23(2–3):143–144

George JC, Wetzel D, O'Hara TM, Robertson K, Dehn L, Leduc R, Reynolds J (2008) An analysis of ancient bowhead whale *Mangtak* from Gambell Alaska: what can it tell us? Paper SC/60/E2 presented to the International Whaling Commission Scientific Committee, 11 pp

George JC, Sheffield G, Reed DJ, Tudor B, Stimmelmayr R, Person BT, Sformo T, Suydam R (2017) Frequency of injuries from line entanglements, killer whales, and ship strikes on Bering-Chukchi-Beaufort seas bowhead whales. Arctic 70(1):37–46

Geraci JR, Ridgway SH (1991) On disease transmission between cetaceans and humans. Mar Mamm Sci 7(2):191–194

Glad T, Kristiansen VF, Nielsen KM, Brusetti L, Wright AD, Sundset MA (2010a) Ecological characterization of the colonic microbiota in arctic and sub-arctic seals. Microb Ecol 60(2): 320–330

Glad T, Bernhardsen P, Nielsen KM, Brusetti L, Andersen M, Aars J, Sundset MA (2010b) Bacterial diversity in faeces from polar bear (Ursus maritimus) in Arctic Svalbard. BMC Microbiol 10:10
Goldhar C, Bell T, Wolf J (2013) Rethinking existing approaches to water security in remote communities: an analysis of two drinking water systems in Nunatsiavut, Labrador, Canada. Water Alternatives 6(3):462–486
Hansen CM, Vogler AJ, Keim P, Wagner DM, Hueffer K (2011) Tularemia in Alaska, 1938–2010. Acta Vet Scand 53:61
Harry K, Frink L (2009) The Arctic cooking pot: why was it adopted? Am Anthropol 111(3):330–343. http://www.jstor.org/stable/40300845
Harper CG, Feng Y, Xu S, Taylor NS, Kinsel M, Dewhirst FE, Paster BJ, Greenwell M, Levine G, Rogers A, Fox JG (2002) Helicobacter cetorum sp. nov., a urease-positive helicobacter species isolated from dolphins and whales. J Clin Microbiol 40(12):4536–4543
Hart EJ, Amos B (2004) Learning about marine resources and their use through Inuvialuit Oral history. Inuvialuit cultural resource Centre. Available from https://waves-vagues.dfo-mpo.gc.ca/Library/279627.pdf
Hauck AK (1977) Occurrence and survival of the Larval Nematode Anisakis Sp. in the Flesh of Fresh, Frozen, Brined, and Smoked Pacific Herring, Clupea Harengus Pallasi. J Parasitol 63(3): 515–519
Hauptmann AL, Paulová P, Hansen LH, Sicheritz-Pontén T, Mulvad G, Nielsen DS (2020a) Microbiota in foods from Inuit traditional hunting. PLoS One 15(1):e0227819
Hauptmann AL, Paulová P, Castro-Mejía JL, Hansen LH, Sicheritz-Pontén T, Mulvad G, Nielsen DS (2020b) The microbial composition of dried fish prepared according to Greenlandic Inuit traditions and industrial counterparts. Food Microbiol 85:103305
Hauschild AH, Gauvreau L (1985) Food-borne botulism in Canada, 1971-84. CMAJ 133(11): 1141–1146
Hay K (2000) Final report of the Inuit Bowhead Knowledge Study, Nunavut, Canada. Nunavut Wildlife Management Board, 2000. Available at: https://www.nwmb.com/en/publications/bowhead-knowledge-study/1819-bowhead-knowledge-study-eng/file
Henri DA, Jean-Gagnon F, Gilchrist HG (2018) Using Inuit traditional ecological knowledge for detecting and monitoring avian cholera among common eiders in the eastern Canadian Arctic. Ecol Soc 23(1):22
Himelbloom BH (1998) Primer on food-borne pathogens for subsistence food handlers. Int J Circumpolar Health 57(Suppl 1):228–234
Himelbloom BH, Crapo CA, Pfutzenreuter RC (1996) Microbial quality of an Alaska native smoked Salmon process. J Food Prot 59(1):56–58
Hueffer K, Drown D, Romanovsky V, Hennessy T (2020) Factors contributing to anthrax outbreaks in the circumpolar north. EcoHealth 17(1):174–180
Hunt TD, Ziccardi MH, Gulland FMD, Yochem PK, Hird DW, Rowles T, Mazet JAK (2008) Health risks for marine mammal workers. Dis Aquat Org 81:81–92
Inuit Circumpolar Council (2012) Food security across the arctic. Available from https://www.inuitcircumpolar.com/project/food-security-across-the-arctic/
Inuit Circumpolar Council-Alaska (2015) Alaskan Inuit food security conceptual framework: how to assess the Arctic from an Inuit perspective: summary report and recommendations report. Anchorage, AK. Available from https://iccalaska.org/wp-icc/wp-content/uploads/2016/05/Food-Security-Full-Technical-Report.pdf
Iverson SA, Forbes MR, Simard M, Soos C, Gilchrist HG (2016) Avian cholera emergence in Arctic-nesting northern common eiders: using community-based, participatory surveillance to delineate disease outbreak patterns and predict transmission risk. Ecol Soc 21(4)
Jenkins EJ, Castrodale LJ, de Rosemond SJ, Dixon BR, Elmore SA, Gesy KM, Hoberg EP, Polley L, Schurer JM, Simard M, Thompson RC (2013) Tradition and transition: parasitic zoonoses of people and animals in Alaska, northern Canada, and Greenland. Adv Parasitol 82: 33–204

Jones A (2006) Iqaluich Nibieaqtuat, fish that we eat. United States fish and wildlife service, Office of Subsistence Management, fisheries resource monitoring program, final report no. FIS02-023, Anchorage, Alaska. Available from https://www.fws.gov/uploadedFiles/Region_7/NWRS/Zone_2/Selawik/PDF/02-023_Iqualich_Niginaqtuat,_Fish_that_we_eat_final.pdf

Kadim IT, Mahgoub O (2002) Post harvest handling of red meat. In: Rahman MS (ed) Handbook of food preservation, 2nd edn. CRC Press, Boca Raton, FL, pp 173–202

Kalxdorff SB (1997) Collection of local knowledge regarding polar bear habitat use in Alaska. Anchorage, Alaska: Marine Mammals Management, U.S. Fish and Wildlife Service Region 7, U.S. Department of the Interior

Kapel CM, Measures L, Møller LN, Forbes L, Gajadhar A (2003) Experimental Trichinella infection in seals. Int J Parasitol 33(13):1463–1470

Kawamura Y, Sannomiya Y, Akechi T (1927) Notes upon a certain anaerobe isolated from whale muscle. J Jpn Soc Vet Sci 6(2):191–213

Kennedy (1988) Tularemia Alaska Bulletin No 25, December, 1988. Available from http://epibulletins.dhss.alaska.gov/Bulletin/DisplayClassificationBulletins/130

Kenny TA, Fillion M, Simpkin S, Wesche SD, Chan HM (2018) Caribou (Rangifer tarandus) and Inuit nutrition security in Canada. EcoHealth 15(3):590–607

Kersh GJ, Fitzpatrick K, Pletnikoff K, Brubaker M, Bruce M, Parkinson A (2020) Prevalence of serum antibodies to Coxiella burnetii in Alaska native persons from the Pribilof Islands. Zoonoses Public Health 67(1):89–92

Kishigami N (2009) Contemporary Inuit food sharing and hunter support program of Nunavik. Senri Ethnol Stud 53:171–192

Koch A, Svendsen CB, Christensen JJ, Bundgaard H, Vindfeld L, Christiansen CB, Kemp M, Villumsen S (2010) Q fever in Greenland. Emerg Infect Dis 16(3):511–513

Kochneva S (2007) Polar bear in material and spiritual culture of the native peoples of Chukotka. Chukotka Association of Traditional Marine Mammal Hunters, Anadyr

Kuhnlein HV, Humphries MM (2017) Traditional animal foods of indigenous peoples of northern North America. Centre for Indigenous Peoples' Nutrition and Environment, McGill University, Montreal. Available from http://traditionalanimalfoods.org/

Kuhnlein HV, Soueida R (1992) Use and nutrient composition of traditional Baffin Inuit foods. J Food Comp Anal 5:112–126

Kuhnlein HV, Erasmus B, Spigelski D (2009) Indigenous peoples' food systems: the many dimensions of culture, diversity and environment for nutrition and health. Food and Agriculture Organization and Centre for Indigenous Peoples' Nutrition and Environment (CINE), Rome, Italy

Kutz S, Bollinger T, Branigan M, Checkley S, Davison T, Dumond M, Elkin B, Forde T, Hutchins W, Niptanatiak A, Orsel K (2015) Erysipelothrix rhusiopathiae associated with recent widespread muskox mortalities in the Canadian Arctic. Can Vet J 56(6):560–563

Kwan M (2004) Avian cholera outbreak in Ivujivik. Makvik Mag 70:1–32

Larrat S, Simard M, Lair S, Bélanger D, Proulx JF (2012) From science to action and from action to science: the Nunavik Trichinellosis prevention program. Int J Circumpolar Health 71:18595

Leclair D, Forbes LB, Suppa S, Proulx JF, Gajadhar AA (2004) A preliminary investigation on the infectivity of Trichinella larvae in traditional preparations of walrus meat. Parasitol Res 93(6): 507–509

Leclair D, Fung J, Isaac-Renton JL, Proulx JF, May-Hadford J, Ellis A, Ashton E, Bekal S, Farber JM, Blanchfield B, Austin JW (2013) Foodborne botulism in Canada, 1985-2005. Emerg Infect Dis 19(6):961–968. https://doi.org/10.3201/eid1906.120873. PMID: 23735780; PMCID: PMC3713816

Leclair D, Farber JM, Pagotto F, Suppa S, Doidge B, Austin JW (2017) Tracking sources of Clostridium botulinum type E contamination in seal meat. Int J Circumpolar Health 76(1): 1380994

Lefebvre KA, Quakenbush L, Frame E, Huntington KB, Sheffield G, Stimmelmayr R, Bryan A, Kendrick P, Ziel H, Goldstein T, Snyder JA, Gelatt T, Gulland F, Dickerson B, Gill V (2016)

Prevalence of algal toxins in Alaskan marine mammals foraging in a changing arctic and subarctic environment. Harmful Algae 55:13–24

Margolis HS, Middaugh JP, Burgess RD (1979) Arctic trichinosis: two Alaskan outbreaks from walrus meat. J Infect Dis 139:102–105

Marino E, White D, Schweitzer P, Chambers M, Wisniewski J (2009) Drinking water in northwestern Alaska: using or not using centralized water Systems in two rural communities. Arctic 62(1): 75–82

Martinez-Levasseur LM, Simard M, Furgal CM, Burness G, Bertrand P, Suppa S, Avard E, Lemire M (2020) Towards a better understanding of the benefits and risks of country food consumption using the case of walruses in Nunavik (Northern Quebec, Canada). Sci Total Environ 719: 137307

Masina S, Shirley J, Allen J, Sargeant JM, Guy RA, Wallis PM, Weese JS, Cunsolo A, Bunce A, Harper SL (2019) Weather, environmental conditions, and waterborne giardia and cryptosporidium in Iqaluit, Nunavut. J Water Health 17(1):84–97

McManus-Fry E, Knecht R, Dobney K, Richards MP, Britton K (2018) Dog-human dietary relationships in Yup'ik western Alaska: the stable isotope and zooarchaeological evidence from pre-contact Nunalleq. J Archaeol Sci Rep 17:964–972

Messier V, Lévesque B, Proulx JF, Rochette L, Libman MD, Ward BJ, Serhir B, Couillard M, Ogden NH, Dewailly E, Hubert B, Déry S, Barthe C, Murphy D, Dixon B (2009) Seroprevalence of toxoplasma gondii among Nunavik Inuit (Canada). Zoonoses Public Health 56(4): 188–197

Messier V, Lévesque B, Proulx JF, Rochette L, Serhir B, Couillard M, Ward BJ, Libman MD, Dewailly E, Déry S (2012) Seroprevalence of seven zoonotic infections in Nunavik, Quebec (Canada). Zoonoses Public Health 59(2):107–117

Miller LG (1975) Observations on the distribution and ecology of Clostridium botulinum type E in Alaska. Can J Microbiol 21(6):920–926

Miller MA, Shapiro K, Murray M, Haulena MJ, Raverty S (2018) Protozoan parasites of marine mammals. In: Gulland FMD, Dierauf LA, Whitman KL (eds) Handbook of marine mammal medicine, 3rd edn. CRC Press, Boca Raton, FL, pp 425–470

Møller LN, Petersen E, Kapel CM, Melbye M, Koch A (2005) Outbreak of trichinellosis associated with consumption of game meat in West Greenland. Vet Parasitol 132(1–2):131–136

Møller LN, Koch A, Petersen E, Hjuler T, Kapel CM, Andersen A, Melbye M (2010) Trichinella infection in a hunting community in East Greenland. Epidemiol Infect 138(9):1252–1256

Moon SS, Jo C, Ahn DU, Kang SN, Kim YT, Kim IS (2012). Meat products manufactured with olive oil. In: Olive-oil constituents, quality, health properties and bioconversions, pp 421–436. Retrieved from http://lib.dr.iastate.edu/abe_eng_pubs/83

Mwakapeje ER, Høgset S, Softic A, Mghamba J, Nonga HE, Mdegela RH, Skjerve E (2018) Risk factors for human cutaneous anthrax outbreaks in the hotspot districts of Northern Tanzania: an unmatched case-control study. R Soc Open Sci 5(9):180479

Nakaya R (1950) Salmonella enteritidis in a whale. Jpn Med J (Natl Inst Health Jpn) 3(5):279–280

NAMMCO (2017) Information sheet on struck and lost. Available from https://nammco.no/wp-content/uploads/2018/08/information-sheet-on-struck-and-lost-rev.pdf

Nancarrow TL, Chan L (2010) Observations of environmental changes and potential dietary impacts in two communities in Nunavut, Canada. Rural Remote Health 10:1370. www.rrh.org.au/journal/article/1370

Noongwook G, the Native Village of Gambell, the Native Village of Savoonga, Huntington HP, George JC (2007) Traditional knowledge of the bowhead whale (Balaena mysticetus) around St. Lawrence Island, Alaska. Arctic 60(1):47–54

Nunavut Nutrition Fact Sheets (2005). https://livehealthy.gov.nu.ca/sites/default/files/resource_attachments/EN_WEB_itf%2D%2Dnutrition-fact-sheet-series.pdf

Nyland KE, Klene AE, Brown J, Shiklomanov NI, Nelson FE, Streletskiy DA, Yoshikawa K (2017) Traditional Iñupiat Ice Cellars (SIĠḶUAQ) in Barrow, Alaska: characteristics, temperature monitoring, and distribution. Geogr Rev 107(1):143–158

Puiguitkaat (pwē-weet-kaht): the 1978 Elder's Conference ed. Gary Kean, Translated by Kisautaq-Leona Okakok, North Slope Borough, Commission on History and Culture, 1981

Omazic A, Bylund H, Boqvist S, Högberg A, Björkman C, Tryland M, Evengård B, Koch A, Berggren C, Malogolovkin A, Kolbasov D, Pavelko N, Thierfelder T, Albihn A (2019) Identifying climate-sensitive infectious diseases in animals and humans in northern regions. Acta Vet Scand 61(1):53

Orlova TY, Selina MS, Lilly EL, Kulis DM, Anderson DM (2007) Morphogenetic and toxin composition variability of *Alexandrium tamarense* (Dinophyceae) from the east coast of Russia. Phycologia 46:534–548

Osterkamp TE, Romanovsky VE (1999) Evidence for warming and thawing of discontinuous permafrost in Alaska. Permafr Periglac Process 10:17–37

Ostertag SK, Loseto LL, Snow K, Lam J, Hynes K, Gillman DV (2018) "That's how we know they're healthy": the inclusion of traditional ecological knowledge in beluga health monitoring in the Inuvialuit settlement region. Arctic Sci 4:292–320

Pardhan-Ali A, Wilson J, Edge VL, Furgal C, Reid-Smith R, Santos M, McEwen SA (2013) Community-level risk factors for notifiable gastrointestinal illness in the Northwest Territories, Canada, 1991-2008. BMC Public Health 13:63

Parkinson AJ, Evengard B, Semenza JC, Ogden N, Børresen ML, Berner J, Brubaker M, Sjöstedt A, Evander M, Hondula DM, Menne B, Pshenichnaya N, Gounder P, Larose T, Revich B, Hueffer K, Albihn A (2014) Climate change and infectious diseases in the Arctic: establishment of a circumpolar working group. Int J Circumpolar Health. 73:25163

Pars T, Osler M, Bjerregaard P (2001) Contemporary use of traditional and imported food among Greenlandic Inuit. Arctic 54(1):22–31

Peterson RA, Wendler K (2011) Preservation of traditional ice cellars in permafrost. Available from http://acep.uaf.edu/media/260236/IceCellarsModelingFinalReport.pdf

Peterson DR, Cooney MK, Beasley RP (1974) Prevalence of antibody to toxoplasma among Alaska natives: relation to exposure to Felidae. J Infect Dis 130(6):557–563

Rapinski M, Cuerrier A, Harris C, Elders of Ivujivik, Elders of Kangiqsujuaq, Lemire M (2018) Inuit perception of marine organisms: from folk classification to food harvest. J Ethnobiol 38(3): 333–355

Rausch RL (2003) Cystic echinococcosis in the Arctic and sub-Arctic. Parasitology 127(Suppl): S73–S85

Rausch RL, Babero BB, Rausch VR Schiller E (1956) Studies on the helminth Fauna of Alaska. XXVII. The Occurrence of Larvae of Trichinella spiralis in Alaskan Mammals Faculty Publications from the Harold W. Manter Laboratory of Parasitology, 519

Reiling SJ, Dixon BR (2019) Toxoplasma gondii: how an Amazonian parasite became an Inuit health issue. Can Commun Dis Rep 45(7–8):183–190

Revich BA, Podolnaya MA (2011) Thawing of permafrost may disturb historic cattle burial grounds in East Siberia. Glob Health Action 4

Sanders JG, Beichman AC, Roman J, Scott JJ, Emerson D, McCarthy JJ, Girguis PR (2015) Baleen whales host a unique gut microbiome with similarities to both carnivores and herbivores. Nat Commun 6:8285

Serhir B, MacLean JD, Healey S, Segal B, Forbes L (2001) Outbreak of trichinellosis associated with arctic walruses in northern Canada, 1999. Can Commun Dis Rep. 27(4):31–36

Seymour J, Horstmann-Dehn L, Rosa C, Lopez JA (2014) Occurrence and genotypic analysis of Trichinella species in Alaska marine-associated mammals of the Bering and Chukchi seas. Vet Parasitol 200(1–2):153–164

Sformo TL, Adams B, Seigle JC, Jayde A, Ferguson M, Jason M, Purcelle K, Stimmelmayr R, Welchc JH, Ellis LM, Leppi JC, George JC (2017) Observations and first reports of saprolegniosis in Aanaakłiq, broad whitefish (Coregonus nasus), from the Colville River near Nuiqsut, Alaska. Polar Sci 14:78–82

Shaffer N, Wainwright RB, Middaugh JP, Tauxe RV (1990) Botulism among Alaska natives. The role of changing food preparation and consumption practices. West J Med 153(4):390–393

Sharma R, Loseto LL, Ostertag SK, Tomaselli M, Bredtmann CM, Crill C, Rodríguez-Pinacho C, Schultz D, Jung D, Shrestha K, Jindal P, Jenkins EJ (2018) Qualitative risk assessment of impact of toxoplasma gondii on health of beluga whales, Delphinapterus leucas, from the eastern Beaufort Sea, Northwest Territories. Arctic Sci 4:321–337

Sheffield G, Stimmelmayr R (2020) Lessons learned: moving forward marine wildlife response, health investigations, and research in Western and Northern Coastal Alaska during unparalleled ecosystem transition. In: Smith B (ed) 2020 Alaska Marine Science Symposium Abstracts & Poster Presentations. Available from https://pwssc.org/wp-content/uploads/2020/01/2020_AMSS_Abstract_Book.pdf

Singh S, Shalini R (2016) Effect of hurdle technology in food preservation: a review. Crit Rev Food Sci Nutr 56(4):641–649

Sowa F (2015) Kalaalimernit: the Greenlandic taste for local foods in a globalized world. Polar Record 51(3):290–300

Springer YPS, Casillas K, Helfrich D, Mocan M, Smith G, Arriaga L, Mixson L, Castrodale L, McLaughlin J (2017) Two outbreaks of Trichinellosis linked to consumption of walrus meat – Alaska, 2016–2017. MMWR Morb Mortal Wkly Rep 66(26):692–696

Starks ZS (2002) Alaska's vanishing Arctic cuisine. Gastronomica 2(1):30–40

Starks ZS (2007) Arctic foodways and contemporary cuisine. Gastronomica 7(1):41–49

Starks ZS (2011) Drying and fermenting in the Arctic dictating women's role in Alaskan' Inupiat culture in cured, fermented and smoked foods. In: Proceedings of the Oxford Symposium on Food and Cookery 2010. Prospect Books, UK

Stimmelmayr R (2003) Traditional ecological knowledge and remote sensing imagery: a new millennium initiative for power sharing in the north. University of Arctic. In: Kohn G (ed) Circumpolar connections, Proceedings of the 8th Circumpolar Cooperation Conference, November 7–10, 2003. Whitehorse, Yukon, pp 67–71

Stimmelmayr R, Rotstein D, Seguel M, Gottdenker N (2017) Hepatic lipomas and Myelolipomas in subsistence harvested bowhead whales (Balaena mysticetus), Barrow, Alaska: a case review 1980-2016. Dis Aquat Organ 127(1):71–74

Stimmelmayr R, Ylitalo GM, Sheffield G, Beckmen KB, Burek-Huntington KA, Metcalf V, Rowles T (2018) Oil fouling in three subsistence-harvested ringed (Phoca hispida) and spotted seals (Phoca largha) from the Bering Strait region, Alaska: Polycyclic aromatic hydrocarbon bile and tissue levels and pathological findings. Mar Pollut Bull 130:311–323

Stimmelmayr R, Coburn S, Neakok F, Valencia R (2019) Toxoplasma gondii seroprevalence in village sled dogs from the North Slope, Alaska. In: Smith B (ed) 2019 Alaska Marine Science Symposium, Anchorage, Alaska

Stimmelmayr R, Rotstein D, Sheffield G, Brower HK, George JC (2021) Diseases and parasites. The bowhead whale.:471–98. Epub 2020 Sep 25

Thorborg NB, Tulinius S, Roth H (1948) Trichinosis in Greenland. Acta Pathol Microbiol Scand 35:778–794

Tigullaraq E (2008) Igunaq. Available from https://www.nmto.ca/sites/default/files/igunaq.pdf

Timofeev V, Bahtejeva I, Mironova R, Titareva G, Lev I, Christiany D, Borzilov A, Bogun A, Vergnaud G (2019 May 22) Insights from Bacillus anthracis strains isolated from permafrost in the tundra zone of Russia. PLoS One 14(5):e0209140

Todd ECD (1988) Botulism in native peoples: an economic study. J Food Prot 51(7):581–587

Tryland M (2000) Zoonoses of arctic marine mammals. Infect Dis Rev 2(2):55–64

Tryland M (2018) Zoonosis and public health. In: Gulland FMD, Dierauf LA, Whitman KL (eds) CRC handbook of marine mammal medicine, 3rd edn. CRC Press, Boca Raton, FL, pp 47–61

Tryland M, Nesbakken T, Robertson L, Grahek-Ogden D, Lunestad BT (2014) Human pathogens in marine mammal meat – a northern perspective. Zoonoses Public Health 61(6):377–394

Tryland M, Beckmen KB, Burek-Huntington KA, Breines EM, Klein J (2018) Orf virus infection in Alaskan mountain goats, Dall's sheep, muskoxen, caribou and Sitka black-tailed deer. Acta Vet Scand. 60(1):12

Uspensky A, Bukina L, Odoevskaya I, Movsesyan S, Voronin M (2019) The epidemiology of trichinellosis in the Arctic territories of a far Eastern District of the Russian Federation. J Helminthol 93(1):42–49

Van Hemert C, Schoen SK, Litaker RW, Smith MM, Arimitsu ML, Piatt JF, Holland WC, Ransom Hardison D, Pearce JM (2020) Algal toxins in Alaskan seabirds: evaluating the role of saxitoxin and domoic acid in a large-scale die-off of common Murres. Harmful Algae 92:101730

Wakegijig J, Osborne G, Statham S, Issaluk MD (2013) Collaborating toward improving food security in Nunavut. Int J Circumpolar Health 72

Walch A, Loring P, Johnson R, Tholl M, Bersamin A (2018) A scoping review of traditional food security in Alaska. Int J Circumpolar Health 77(1):1419678

Waltzek TB, Cortés-Hinojosa G, Wellehan JF Jr, Gray GC (2011) Marine mammal zoonoses: a review of disease manifestations. Zoonoses Public Health 59(8):521–535

Wang ZQ, Cui J, Xu BL (2006) The epidemiology of human trichinellosis in China during 2000-2003. Acta Tropica 97(3):247–251

Waugh D, Pearce T, Ostertag SK, Pokiak V, Collings P, Loseto LL (2018) Inuvialuit traditional ecological knowledge of beluga whale (*Delphinapterus leucas*) under changing climatic conditions in Tuktoyaktuk, NT. Arctic Sci 2018(4):242–258

Westley BP, Horazdovsky RD, Michaels DL, Brown DR (2016) Identification of a novel mycoplasma species in a patient with septic arthritis of the hip and seal finger. Clin Inf Dis 62(4): 491–493

White CP, Jewer DD (2009) Seal finger: a case report and review of the literature. Can J Plast Surg 17:133–135

Wobeser G, Campbell GD, Dallaire A, McBurney S (2009) Tularemia, plague, yersiniosis, and Tyzzer's disease in wild rodents and lagomorphs in Canada: a review. Can Vet J 50(12): 1251–1256

Wolfe RJ (2004) Local tradition and subsistence: a synopsis of 25 years of research by the state of Alaska technical report 284. Alaska Department of Fish and Game, Subsistence Division, 89 p. Available from http://www.adfg.alaska.gov/techpap/tp284.pdf

Yamin-Pasternak S, Kliskey A, Alessa L, Pasternak I, Schweitzer P (2014) The rotten renaissance in the Bering Strait: loving, loathing, and washing the smell of foods with a (re)acquired taste. Curr Anthropol 55(5):619–646

Zinsstag J, Schelling E, Waltner-Toews D, Tanner M (2011) From "one medicine" to "one health" and systemic approaches to health and well-being. Prev Vet Med 101(3–4):148–156. https://doi.org/10.1016/j.prevetmed.2010.07.003. Epub 2010 Sep 15. PMID: 20832879; PMCID: PMC3145159

Zottola EA, Wagner MK, Zoltai PT (1983) Composition and microbiology of some native Alaskan preserved foods. J Food Prot 46(5):441–443

Part V

Working with Arctic Communities

Wildlife Health Surveillance in the Arctic

Sylvia L. Checkley, Matilde Tomaselli, and Nigel Caulkett

1 Introduction

1.1 Health Surveillance

Surveillance has been defined by the World Organization for Animal Health (OIE) as "the systematic ongoing collection, collation, and analysis of information related to animal health and the timely dissemination of information so that action can be taken" (OIE 2019). This definition highlights the key components of surveillance, which are detection of diseases and pathogens, analysis of data, and response to the information that includes communication and reporting to stakeholders and disease investigation and/or implementation of disease control measures when indicated. Surveillance is often described as "Information for Action" (Orenstein and Bernier 1990). Health surveillance could also be called disease surveillance, but using "health surveillance" clarifies the relationship with health and allows for a broader definition of health than just the presence or absence of disease. Biosurveillance is a focus of surveillance that emphasizes characterization of disease outbreaks with a focus on the infectious processes, in addition to characteristics of the main definition focusing on detection and response (Wagner 2006). Monitoring is another word that may refer to a surveillance system, or it may refer to "Ongoing efforts directed at assessing the health and disease status of a given population" (Salman 2003). It does not necessarily include a response. Therefore, due to varying definitions, it is important to clarify meaning when discussing monitoring and surveillance.

S. L. Checkley (✉) · N. Caulkett
Faculty of Veterinary Medicine, University of Calgary, Calgary, AB, Canada
e-mail: slcheckl@ucalgary.ca

M. Tomaselli
Faculty of Veterinary Medicine, University of Calgary, Calgary, AB, Canada

Polar Knowledge Canada, Canadian High Arctic Research Station, Cambridge Bay, NU, Canada

M. Tryland (ed.), *Arctic One Health*, https://doi.org/10.1007/978-3-030-87853-5_23

The World Organization for Animal Health (OIE) is responsible for overseeing animal disease at the global level:

> *"We ensure transparency in the global animal disease situation, collect, analyse and disseminate veterinary scientific information, encourage international solidarity in the control of animal diseases, safeguard world trade by publishing health standards for international trade in animals and animal products, improve the legal framework and resources of national Veterinary Services, to provide a better guarantee of food of animal origin and to promote animal welfare through a science-based approach."* https://www.oie.int/en/what-we-do/

The OIE has listed diseases that are of global concern for which member countries have to report information about the number of cases in their country (OIE 2019). Each country therefore has to collect information on these diseases and infections, using various types of surveillance. Local governments, academia, and non-governmental organizations also carry out surveillance.

There are several main purposes for surveillance systems: early warning (early detection of disease), measuring levels of disease along with changing trends, and demonstrating freedom from disease. Early detection of disease outbreaks is critical so that rapid response is launched to investigate and control disease outbreaks, changes in trends or presentation of disease, and new, emerging diseases. Surveillance is also important to understand what diseases are present in a population; with wildlife populations, this provides information, which might otherwise be missing, to support conservation and other decision-making. Demonstrating freedom from disease within livestock populations is important for countries as regulations for international trade in animal products are based on preventing transmission of disease between countries (OIE 2019). Adequate surveillance must be in place for the country to "prove" the country doesn't have a specific disease. Sometimes wildlife surveillance might be used to support these claims, depending on the disease epidemiology within a specific country. Surveillance is also important to generate data to inform decision-making, risk assessments, emergency management, and exploratory modelling of pathogen transmission and/or to direct and evaluate interventions. Another important function of surveillance systems is to provide a framework from which to launch targeted epidemiological research.

Surveillance systems are often described by the method of data collection, which varies related to many factors including the purpose, logistics, and population of interest. For example, sentinel site or sentinel unit surveillance refers to repeated sampling of healthy test negative animals or humans, possibly chosen to represent a specific geographic area, in order to identify new disease or seroconversion representing disease exposure (OIE 2019). This type of surveillance is also important when trying to prove freedom from a disease or alternately to identify disease incursion in a specific area. An example of this is West Nile virus surveillance, using either captive or free-ranging birds in sentinel sites chosen as probable areas for early virus circulation, for the purpose of early warning (Komar 2002). Conversely, in syndromic surveillance, a real-time analysis of cases with a specific set of clinical signs is used for early detection of an outbreak instead of waiting for confirmatory clinical and/or laboratory testing (CDC 2021; Berezowski et al. 2015). For example, coughing, heavy breathing, and/or nasal discharge might be used to represent a respiratory syndrome. Case definitions are typically utilized in syndromic surveillance, and they can be more or less specific depending on the need for increased sensitivity or specificity in the system. Although more typical of public health surveillance, other surrogate measures of disease can also be used in syndromic surveillance; for example, increased over-the-counter medicine sales might be an early indication of an infectious disease outbreak. Syndromic approaches are also often applied in participatory surveillance together with event-based or One Health approaches (Smolinski et al. 2017). Participatory surveillance strongly relies on participatory epidemiology methods, characterized by interviews with techniques like mapping, diagrams, calendars, and scoring, to gather surveillance information from experts, knowledge holders from a community, or a specific group of people (OIE 2018, 2019). Participatory surveillance systems exist across many countries and are a promising approach to improve veterinary surveillance in the Arctic (Tomaselli et al. 2018b).

Health data sourced from abattoirs and diagnostic labs can also be used for surveillance for humans and animals as can diagnostic results from environmental samples like those submitted for drinking water quality testing. Risk-based approaches to sampling are also used, usually to improve the likelihood of detection of a rare disease using the most cost-effective method. This is the basis for surveillance systems that aim to identify transmissible spongiform encephalopathies within high-risk sub-populations of animals showing clinical signs. Data for surveillance of wildlife can also be collected through surveys and sampling, clinical disease surveillance, laboratory data, and specimen archives. Ideally, other data sources of interest would be available to provide robust analysis and interpretation of surveillance information. These might include domestic animal data and public health data, which could be assessed together along with relevant environmental information to provide the most robust analysis of the data.

Surveillance can also be described as scanning or targeted, sometimes called passive and active surveillance, respectively (OIE 2015). Scanning surveillance is opportunistic, relying on samples collected for other reasons such as routine laboratory submissions or sample archives. Targeted surveillance surveillance is focussed on a specific question often within a specific subgroup requiring specific methodology to be implemented to collect the data. Some surveillance systems combine methods from both approaches.

There are other methods that are intrinsically linked to health surveillance such as population estimation. The estimates of population size, density, and range and the proportion of young inform management activities and trigger investigation when populations are not flourishing as expected. Population decreases may be multifactorial, but emerging disease or changing disease patterns is likely involved (Cuyler et al. 2020). Health surveillance might also involve tagging animals to follow them forward through time or using tracking collars to follow their movements for potential recapture (Clausen et al. 1984; Alasaad et al. 2013). This may be done in association with relocation efforts. These methods require capture and immobilization of the animals using pharmaceutical agents. There are inherent risks associated with capture and immobilization, and mortalities can occur. In addition, collars and field operations are expensive so these techniques are only used in specialized circumstances. Generally, when an animal is immobilized for these purposes, biological samples (blood, feces) and physiological measurements are also collected to make the most of the opportunity.

Immobilization of Muskoxen, as an Example

In recent years, there have been many advances in technology that have enabled surveillance with a more "hands off" approach to the animal. Unfortunately, there are still situations that require the animal to be captured and immobilized for tissue sampling, marking, or placement of tracking devices. Fortunately, there have been many advances in capture and supportive care that have resulted in decreased morbidity and mortality.

Immobilization of muskoxen can be challenging; their thick hair coat can make it difficult to identify anatomical landmarks for dart placement, and the dart must be delivered with sufficient velocity to penetrate the coat. Muskoxen are prone to hyperthermia, regurgitation, and hypoxemia. Captive animals should be fasted prior to anesthesia to decrease the risk of regurgitation. Supplemental inspired oxygen is recommended to prevent and treat hypoxemia. This can be administered by nasal insufflation. Chase times should be kept to a minimum to avoid hyperthermia, and immobilization should be avoided on warm days whenever possible.

Muskoxen are best maintained in sternal recumbency to reduce ventilation/perfusion mismatching. Jugular venous access can be challenging in large males; the cephalic vein and the lateral saphenous vein are alternate routes for blood sampling or catheterization.

Calm, captive muskoxen can be immobilized with 0.05–0.06 mg/kg of medetomidine, combined with 1.5 mg/kg of ketamine. This combination is suitable for minor procedures, and medetomidine can be antagonized with atipamezole at five times the medetomidine dose to speed recovery.

Muskoxen have been immobilized with a combination of 0.05 mg/kg etorphine and 0.15 mg/kg xylazine. This combination was described for capture of 133 captive muskoxen (Blix et al. 2011). A dose of 0.015 mg/kg of etorphine combined with 0.1 mg/kg of xylazine has been described for immobilization of captive muskoxen (Lian et al. 2017). The major side effect of this combination was hypoxemia. Although hypoxemia was successfully treated with supplemental inspired oxygen, the authors noted that hypercarbia

(continued)

became more pronounced during treatment with supplemental oxygen, and careful tiration of oxygen should be considered to prevent severe hypoventilation. Etorphine was antagonized with 20 mg naltrexone/1 mg of etorphine. Xylazine was antagonized with 1 mg atipamezole/10 mg of xylazine. The BAM combination (butorphanol, azaperone, medetomidine) is effective in muskoxen; the author has used this combination at a dose of 0.075 mg/kg butorphanol, 0.075 mg/kg azaperone, and 0.1 mg/kg medetomidine via dart administration in a captive muskoxen. Further research will need to be performed with the BAM combination to characterize the efficacy, safety, and dose requirements for captive and wild muskoxen.

1.2 One Health Approaches to Surveillance

One Health has been described as "a transdisciplinary approach to address issues that emanate from the interfaces of animals, humans and their environment" (One Health Research Group 2020). A natural application of the One Health approach is animal health surveillance that can be defined as the "collaborative on-going systematic collection and analysis of data from multiple domains at local, national and global levels to detect health related events and produce information which leads to action aimed at attaining optimal health for people, animals and environment" (Nwafor-Okoli et al. 2014).

Health surveillance systems are designed to answer different questions so the implementation methods of the surveillance system and the type of information generated vary between types of surveillance systems. A recent survey of the global biosurveillance community aimed to better understand the application, perceptions, and motivations surrounding the practice of One Health surveillance; over 50% of participants had implemented aspects of One Health surveillance, while over 85% saw benefits to the One Health approach to surveillance (Berezowski et al. 2019). Findings were not affected by domain (animal health, public health, or environmental health) nor country income status, and all respondents had challenges with cross domain data access (Berezowski et al. 2019). In addition, survey respondents from low-income and middle-income countries were more motivated to make improvements in their surveillance systems and had different priorities than those from high-income countries (Berezowski et al. 2019).

One Health approaches are needed to robustly address many issues and challenges at the interface between humans, wildlife, and the environment, which do not involve zoonotic disease transmission, such as food security, species conservation, and species biodiversity. However, it is prudent to remember that over 60% of current human pathogens and 75% of emerging and re-emerging human pathogens are zoonotic or have an animal origin (Stallknecht 2007). Zoonotic pathogens include, for example, *Bacillus anthracis* (causing the disease anthrax),

Brucella suis biovar 4 (causing brucellosis), *Echinococcus granulosus* (causing cystic echinococcosis), *Trichinella nativa* (causing trichinosis), orf virus (causing contagious ecthyma), and *Francisella tularensis* (causing tularemia).

1.3 Wildlife Health Surveillance

Wildlife health surveillance, like other types of health surveillance, is important for early detection of outbreaks or changes in disease prevalence so an early response can protect wildlife health, domestic animal health, and human health. It may refer to surveillance for clinical disease in populations of wildlife or for pathogens or other markers of health in samples collected from wildlife, such as glucocorticoid levels, as a biomarker of physiological stress (Di Francesco et al. 2021; OIE 2015). Wildlife health surveillance also informs activities related to species conservation, reservoirs of zoonotic and livestock diseases, and food security (Watsa 2020; Grogan et al. 2014; Holmes et al. 2019). Food safety is also important with respect to zoonotic pathogen transmission through hunting, preparation, and ingestion of country (traditional) foods (Kuisma et al. 2019). This is linked to food habits and food conservation and preparation methods (see Chap. "Traditional Conservation Methods and Food Habits in the Arctic"). Wildlife can also potentially function as sentinels and provide biological information on individual-level and population-level health that could serve as an early warning for climate change informing public health decision-making (Stephen and Duncan 2017). A role for wildlife health surveillance of diseases that threaten biodiversity is also proposed (Grogan et al. 2014). This type of surveillance would focus on identifying emerging diseases and new presentations of existing diseases that have the potential to impact wildlife populations as has been experienced with white-nose syndrome in bats, caused by *Pseudogymnoascus destructans*, and chytridiomycosis in amphibians, caused by *Batrachochytrium dendrobatidis* (Grogan et al. 2014).

Key challenges specific to wildlife health surveillance exist (OIE 2015; Stallknecht 2007). The close association of humans, livestock, and companion animals facilitates the early identification of disease in animals, facilitating prompt treatment. This type of close observation of individuals and populations is lacking with wildlife surveillance and is often combined with remote geographic locations where even mass die-offs of animals could go undetected. In addition, diagnostic tests are often not validated for use in wildlife. There is a lack of existing surveillance infrastructure from which other programs can be launched, a lack of representative sampling, and a variety of potential biases. When working with wild species, the total number of animals is typically unknown, so there is no denominator to make accurate rate and prevalence calculations.

A strength of wildlife health surveillance is the broad group of stakeholders that are often involved. Stakeholders in wildlife surveillance often include representatives from government (possibly at different levels), non-governmental agencies, academia, community representatives, the broader public, and organizations that represent and protect the interests of indigenous people from the area. However, a large stakeholder platform can also be problematic especially when the roles of the different groups are not well defined and it becomes unclear who leads the surveillance activities. In Canada, arctic wildlife are co-managed through a formal co-management process involving the management board, Inuit organizations, and government at local, regional, and territorial levels (NWMB 2021). This brings together numerous people with multiple ideas and approaches, potentially provides more funding options, and often allows collaborative sample collection and access to wildlife in otherwise inaccessible areas and times of the year. This strength can also sometimes be a challenge as stakeholders may have different goals and perspectives. The group must use team-based approaches and make decisions together about leadership in order to build consensus, work together, and realize benefits.

1.4 Wildlife Health Surveillance in the Arctic

There is a need for health surveillance of wildlife in the Arctic. Disease investigation is a critical response arm for surveillance when unusual or otherwise unexpected morbidity or mortality is identified. Numerous disease outbreaks across the Arctic have drawn attention to the need for health surveillance to allow for possible prevention of outbreaks or at least rapid response to outbreak situations including outbreak investigation. Sporadic anthrax outbreaks in bison in northern Canada over the past 60 years led to annual surveillance flights over the vast area, diagnostic testing of carcasses found dead during the high-risk season, and an emergency response plan to outbreaks that aims to minimize public health risk and reduce environmental contamination with anthrax spores (Nishi et al. 2007).

Fatal disease outbreaks in muskoxen have occurred in Norway and Canada over the last two decades. The deaths in Norway were attributed to *Mycoplasma ovipneumoniae* and those in Canada to septicemia caused by *Erysipelothrix rhusiopathiae* (Handeland et al. 2020; Kutz et al. 2015). Previous mass die-offs of muskoxen on the remote Banks Island, Northwest Territories, Canada, were attributed to yersiniosis (Blake et al. 1991). These diseases were not previously well described in disease outbreaks in wild muskoxen, and the epidemiology was unknown. These recent outbreaks led to further work to better understand each outbreak. A serosurvey of muskox blood samples archived from 1976 to 2017 from various locations in northern Canada and Alaska found evidence of *E. rhusiopathiae* exposure in the earliest samples with increased seroprevalence associated with large-scale mortality events in the same area (Mavrot et al. 2020). Whole genome sequencing provided valuable insight into the epidemiology and transmission dynamics of *E. rhusiopathiae* (Forde et al. 2016). Creation of sample archives is critical to provide a broader picture when diseases appear to emerge; these samples can be tested later to provide a temporal baseline and allow for assessment of seasonal and annual patterns of diseases, as well as potential changes in these patterns (Kutz et al. 2015; Handeland et al. 2014; Tomaselli et al. 2019).

Further investigation into these outbreaks has underlined the fact that disease in many arctic species is multifactorial, and there are many potential components that could be related to the expression of clinical disease. This can be thought of in terms of the epidemiologic triad, with factors associated with each component:

- *Host animal component:* age, species, innate resistance, prior pathogen exposure, behavioral traits, reproductive status, genetics, co-infections
- *Pathogen component:* dose, virulence, environmental hardiness, infectivity
- *Environment component:* animal density, animal movements, environmental conditions, season, nutrition, climate/weather, human activities, air traffic, habitat changes

To better represent these complex arctic systems, we can add further complexity to this simple epidemiologic triad accounting for multiple interfacing triads to representing multiple pathogens or strains of pathogens, multiple animal hosts and reservoir species, and multiple environments. When considering the system as a whole, the simple and schematic triad of causation often becomes a more complex web (Wobeser 2007).

2 Examples of Arctic Wildlife Surveillance Programs

Numerous surveillance systems have previously or are currently collecting data and samples that will be used to optimize health for various wildlife species. Over the years, disease surveys have been carried out assessing exposure to key pathogens such as *Salmonella* spp. in moose (*Alces alces*), reindeer (*Rangifer tarandus*), and harp seals (*Pagophilus groenlandicus*) (Aschfalk et al. 2002, 2003; Aschfalk and Thórisson 2004; Andersen-Ranberg et al. 2018). In Canada, disease surveys have been centered around commercial or community harvests of arctic ungulates; there is also a long-running program looking at transboundary contaminants in the arctic ecosystem (GOC 2021).

Semi-domesticated reindeer herds in Norway and Sweden commonly deal with infectious keratoconjunctivitis in Eurasian reindeer (*Rangifer t. tarandus*), a multi-factorial disease, involving cervid herpesvirus 2, varying bacteria species, and environmental factors (Tryland et al. 2017). The herders have the opportunity to observe the animals more closely than other wild populations. They sometimes corral the animals for supplemental feeding causing additional stress and increased opportunity for disease transmission between animals. A survey was used to gain information about this and other diseases as well as management practices, ultimately providing information to herders to improve disease prevention and treatment (Tryland et al. 2016).

The Northern Contaminants Program (NCP) is an interesting example of One Health surveillance in the Arctic in many ways (GOC 2021). Since 1991, this program has used a combination of research and surveillance with the objective to decrease or eliminate contaminants, such as persistent organic pollutants (POPs) and mercury, in the arctic environment and in country (traditional) foods. It also assesses public health risks associated with contaminants in these foods, in particular bioaccumulation of contaminants over time and cumulative effects on people. Information from this program has informed numerous international agreements concerning regulations surrounding these contaminants. This is a government program with formal partnerships between federal and territorial government departments and representatives of northern indigenous people's organizations. Although atmospheric monitoring for contaminants is a component of the surveillance, the NCP has prioritized human health and the environment, including animals in terrestrial, freshwater, and marine ecosystems. In particular, collaboration with hunters is key to the long-term surveillance of contaminants in ringed seals (*Pusa hispida*), polar bears (*Ursus maritimus*), caribou (*Rangifer tarandus*), beluga whales (*Delphinapterus leucas*), and various types of fish and seabird eggs. To assess human health, human dietary surveys are important to estimate contaminant exposure and public health. Country (traditional) food preferences vary in different communities so this must be assessed locally (GOC 2021; Tomaselli et al. 2018a). Once results are available, NCP partners come together to provide robust interpretation of the findings (GOC 2021). Finally, this program also provides training and capacity building to partnered northern communities, including educational opportunities for community members and within local schools.

Hunters and researchers from the government, academia, and/or non-governmental organizations often work together to collect samples for wildlife surveillance. This is an efficient way to access samples for testing and has the advantage of building relationships between researchers and community members. For example, researchers worked with hunters during narwhal (*Monodon monoceros*) harvests in Nunavut, Canada, and performed full necropsies on the harvested animals. This collaborative project informed a baseline for narwhal health for use in ongoing community-based monitoring (Black et al. 2018).

Early muskox health surveillance initiatives commenced in the community of Sachs Harbour, Northwest Territories, and in Cambridge Bay, Nunavut, Canada, in the 1980s with abattoir sampling during hunter-led muskox hunts, processing larger numbers of animals in local, sometimes portable, abattoirs, with meat inspection, to allow sale of products within the Territory first and then to southern locales (Wu et al. 2011; Gunn et al. 1991; Tomaselli et al. 2018a). The initial goal was to create information about the health of the local muskox populations specifically their exposure to known diseases and then also to comply with federal inspection standards to allow for meat export. In Cambridge Bay, abattoir testing was temporally associated with a decreased population size of muskoxen in the local area; however, as the surveillance system was set up, it gave no clear warning that something was going on with the local muskox population. Despite the inability of the system to detect changes in real time, the creation of sample archives from the commercial harvests proved important; these samples have later been tested to provide temporal baselines of diseases of concern, including assessments of their annual patterns and changes in these patterns (Kutz et al. 2015; Tomaselli et al. 2019; Mavrot et al. 2020).

Fig. 1 Representation from design to field use of the sample collection kit used in hunter-based sampling in the context of the participatory muskox health surveillance program implemented in Cambridge Bay, Nunavut, Canada. Pre-packaged and lightweight kits are designed together with hunters and include samples and information collected easily in harsh field conditions; once collected, kits remain relatively light and contained allowing for easy transportation; once samples are analyzed, results are reported back to the community

While commercial harvesting of muskoxen has been halted or significantly reduced, hunter-based sampling methods have been implemented to allow continuity of surveillance activities in declining muskox populations. Researchers worked with subsistence hunters to understand barriers and opportunities for field sampling associated with regular harvesting activities, which resulted in designing lightweight, prepackaged sampling kits for the collection of a standard suite of samples and data from harvested muskoxen and caribou as shown in Fig. 1 (Tomaselli 2018; Tomaselli and Curry 2019). Sample collection techniques have also been modified to enable ease of blood collection from caribou and muskoxen by hunters in the field immediately after harvest using Nobuto filter paper strips (Curry et al. 2011; Carlsson et al. 2019; Tomaselli and Curry 2019).

Surveillance of trichinosis is another important example of wildlife surveillance in the Arctic that contributes to public health protection. *Trichinella nativa* is a zoonotic parasite causing chronic disease in people characterized by swelling, muscle pain, and/or gastrointestinal symptoms (see Chap. "Trichinella spp. in the North"). It is found in a number of arctic species that are part of the traditional northern diet, and as such, it is a food safety concern (Forbes 2000; Uspensky et al. 2019; Seymour et al. 2014). Surveillance of walrus meat for *Trichinella nativa* is important in arctic communities to prevent human trichinosis (Proulx et al. 2002). As with other arctic-adapted parasites, *Trichinella nativa* is resistant to freezing temperatures and only inactivated if meat is well cooked, reaching a core temperature of not less than 71 °C (159.8 °F) for at least 1 min (Gottstein et al. 2009). Walrus is often eaten raw or fermented. Although *T. nativa* is not common in walrus meat, it is associated with large outbreaks of trichinosis in people as the meat is often shared through established family and community social networks (Proulx et al. 2002). In the Eastern Canadian Arctic, hunters collect a tongue sample during butchering that is sent away for diagnostic testing and the meat is held until results are received, which typically occurs within days (Proulx et al. 2002). This program demonstrates characteristics of One Health surveillance with broad stakeholders, community engagement, a public health education component, and the ability to implement outbreak investigations as needed.

While wildlife surveillance programs are critical in the Arctic due to the strong interconnectedness between wildlife and human health, unexpected outcomes can hinder their effectiveness. For example, hunters' participation may decrease in surveillance programs for zoonotic diseases of harvested wildlife due to the possible stigma associated with "catching the wrong animal" and potentially sharing diseased meat with family, friends, and the community. This phenomenon may be greater when the prevalence of zoonotic diseases increases in wildlife and, therefore, when the surveillance program is needed the most. A rigorous evaluation of the surveillance with the continuous engagement of stakeholders is critical to avoid pitfalls and negative outcomes.

3 Case Study Illustrating a One Health Approach to Wildlife Surveillance in the Arctic: The Participatory Muskox Health Surveillance Program

A project was designed to explore the integration of local and indigenous knowledge and western science in a community-based participatory muskox health surveillance system in the community of Cambridge Bay, Nunavut, Canada (Tomaselli 2018). The purpose of the surveillance was to assess the health status of the local muskox population while developing a system capable of early detection of new diseases and changes in disease patterns. The surveillance was designed to include multiple components that included local/traditional knowledge systems and a more conventional method of surveillance based on diagnostics of samples accessed through different means (i.e., newly developed hunter-based collections, targeted field disease investigations, sample archives) as demonstrated in Fig. 2. Interviews of community members were the foundation of the system as they allowed the collaborative research group to understand the local contexts, design conventional surveillance components that were locally adapted thus effective (i.e., hunter-based sample

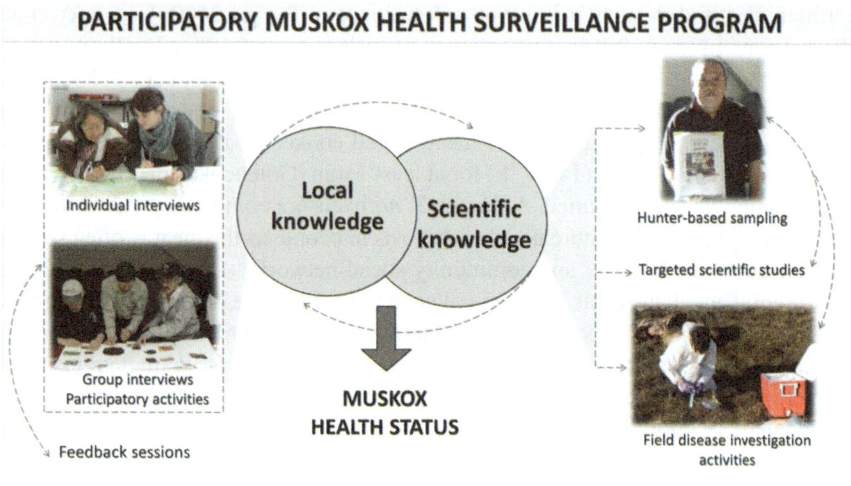

Fig. 2 Schematic representation of the participatory muskox health surveillance program implemented in the community of Cambridge Bay, Nunavut, Canada. The surveillance components developed draw from different knowledge systems, local and scientific knowledge, and, working in synergy, allow for a more accurate and reliable muskox health status assessment than single surveillance components and knowledge systems would have achieved in isolation. Dashed lines represent how the different components relate and influence each other within the system. Figure modified from Tomaselli and Curry (2019)

collections, Fig. 1), gather missing health data on muskoxen and caribou, and interpret sample results within context.

Robust qualitative research design was implemented for the interview component of the surveillance, which was characterized by a combination of individual, group, and follow-up interviews. Syndromic surveillance approaches and participatory epidemiology techniques were applied to gather quantitative data on wildlife health with spatial and temporal resolutions, including population trends, demographics, body condition status, relative prevalence of diseases, causes of mortality, and disease outbreaks. For example, community members provided observations that characterized large declines in muskox and caribou populations and possible reasons for this, including emerging and re-emerging diseases. Interviews were also essential to identify large mortality outbreaks that were underdetected by the standard passive surveillance (Tomaselli et al. 2018b).

Zoonoses such as orf virus infection (parapoxvirus) and rangiferine brucellosis (*Brucella suis* biovar 4) were identified through both syndromic approaches and conventional diagnostics (Tomaselli et al. 2016, 2018b). Samples obtained through various means where then used in targeted scientific studies that, working in synergy with the interview data, helped further characterize spatial and temporal trends of disease. For example, *Brucella suis* biovar 4 was detected in the muskox population under surveillance, and its increasing trend was temporally associated with the local muskox decline (Tomaselli et al. 2019).

Overall, the inclusion in the surveillance of local knowledge holders with their perspectives and rich understanding of local wildlife facilitated the collection of important data and significantly enriched the interpretation of results. Interview data utilized in synergy with conventional sample diagnostics have provided rich information of more value for early detection of new diseases or changing disease trends. Bridging and valuing multiple ways of knowing have many benefits overall, and working together with multiple stakeholders facilitates the application of surveillance interventions (Kutz and Tomaselli 2019). Participatory approaches to wildlife surveillance, such as the one described here, hold a great potential in the Arctic to improve surveillance capacity and interventions for both wildlife management and public health.

4 Conclusions

A One Health approach to health surveillance emphasizing the interaction of humans, animals, and the environment is clearly a critical component in effective and efficient health surveillance of arctic wildlife. It is clear that muskox surveillance in arctic communities is an approach that fully fits within this paradigm. One Health approaches to any complex problems are not without challenges, and the same is true for One Health disease surveillance. Challenges include logistical issues, such as poor performance of electronic data collection devices during field work in the Arctic and aggregation of data on different spatial scales leading to misalignment in spatial analysis, barriers to data sharing related to questions about confidentiality and mandate, and questions about who will lead the initiative. Trust builds as issues are worked through together; inputs from all stakeholders are valued allowing shared and more robust data interpretation, information creation, and knowledge transfer and translation. Within existing health surveillance initiatives that could be considered to have a One Health approach, human health is usually the goal even when the surveillance system has a significant wildlife health surveillance outcome in early warning, species conservation, or other goals. Despite these challenges, wildlife heath surveillance in the Arctic clearly benefits from and easily demonstrates a One Health approach, and is expected to be the approach of future surveillance initiatives.

References

Alasaad S, Granados JE, Fandos P, Cano-Manuel FJ, Soriguer RC, Pérez JM (2013) The use of radio-collars for monitoring wildlife diseases: a case study from Iberian ibex affected by Sarcoptes scabiei in Sierra Nevada, Spain. Parasit Vectors 6:242

Andersen-Ranberg EU, Barnes CJ, Rasmussen I, Salgado-Flores A, Grøndahl C, Mosbacher JB, Hansen AJ, Sundset MA, Schmidt NM, Sonne C (2018) A comparative study on the Faecal bacterial community and potential zoonotic bacteria of muskoxen (Ovibos moschatus) in Northeast Greenland, Northwest Greenland and Norway. Microorganisms 6

Aschfalk A, Thórisson SG (2004) Seroprevalence of Salmonella spp. in wild reindeer (Rangifer tarandus tarandus) in Iceland. Vet Res Commun 28:191–195

Aschfalk A, Folkow L, Rud H, Denzin N (2002) Apparent seroprevalence of salmonella spp. in harp seals in the Greenland Sea as determined by enzyme-linked immunosorbent assay. Vet Res Commun 26:523–530

Aschfalk A, Hundertmark KJ, Bendiksen HR, Arnemo JM, Denzin N (2003) Serosurvey for antibodies against Salmonella species in free-ranging moose (Alces alces) from Norway. Berl Munch Tierarztl Wochenschr 116:417–420

Berezowski J, Byra C, Brockhoff E, Hurnik D, Klopfenstein C, Kloeze H, Bergeron L, Charbonneau G, Cardinal F, Jamal I, Herntier T (2015) Surveillance to manage disease on Canadian Swine Farms. Online J Public Health Informatics 7:e8

Berezowski J, Akkina J, Del Rio Vilas VJ, Devore K, Dorea FC, Dupuy C, Maxwell MJ, Singh VV, Vial F, Contadini FM, Streichert LC (2019) One health surveillance: perceived benefits and workforce motivations. Rev Sci Tech 38:251–260

Black SR, Simonee J, Tomaselli M, Mablick N, Duignan PJ (2018) Community Health Monitoring of Narwhal (Monodon monoceros): Necropsy Findings 2013-2017. In: ArcticNet annual scientific meeting, 2018 Ottawa, Ontario, Canada, p 20

Blake JE, Mclean BD, Gunn A (1991) Yersiniosis in free-ranging muskoxen on Banks Island, Northwest Territories, Canada. J Wildl Dis 27:527–533

Blix AS, Lian H, Ness J (2011) Immobilization of muskoxen (Ovibos moschatus) with etorphine and xylazine. Acta Vet Scand 53:42

Carlsson AM, Curry P, Elkin B, Russell D, Veitch A, Branigan M, Campbell M, Croft B, Cuyler C, Côté SD, Leclerc LM, Tryland M, Nymo IH, Kutz SJ (2019) Multi-pathogen serological survey of migratory caribou herds: a snapshot in time. PLoS One 14:e0219838–e0219838

CDC, Center for Disease Control and Prevention (2021) What is Syndromic Surveillance? [Online]. Available from https://www.cdc.gov/nssp/overview.html. Accessed 27 April 2021

Clausen B, Hjort P, Strandgaard H, Soerensen PL (1984) Immobilization and tagging of muskoxen (Ovibos moschatus) in Jameson Land, northeastern Greenland. J Wildl Dis 20:141–145

Curry PS, Elkin BT, Campbell M, Nielsen K, Hutchins W, Ribble C, Kutz SJ (2011) Filter-paper blood samples for ELISA detection of Brucella antibodies in caribou. J Wildl Dis 47:12–20

Cuyler C, Rowell J, Adamczewski J, Anderson M, Blake J, Bretten T, Brodeur V, Campbell M, Checkley SL, Cluff HD, Côté SD, Davison T, Dumond M, Ford B, Gruzdev A, Gunn A, Jones P, Kutz S, Leclerc LM, Mallory C, Mavrot F, Mosbacher JB, Okhlopkov IM, Reynolds P, Schmidt NM, Sipko T, Suitor M, Tomaselli M, Ytrehus B (2020) Muskox status, recent variation, and uncertain future. Ambio 49:805–819

Di Francesco J, Mastromonaco GF, Rowell JE, Blake J, Checkley SL, Kutz S (2021) Fecal glucocorticoid metabolites reflect hypothalamic-pituitary-adrenal axis activity in muskoxen (Ovibos moschatus). PLoS One 16:e0249281

Forbes LB (2000) The occurrence and ecology of Trichinella in marine mammals. Vet Parasitol 93:321–334

Forde T, Biek R, Zadoks R, Workentine ML, De Buck J, Kutz S, Opriessnig T, Trewby H, Van Der Meer F, Orsel K (2016) Genomic analysis of the multi-host pathogen Erysipelothrix rhusiopathiae reveals extensive recombination as well as the existence of three generalist clades with wide geographic distribution. BMC Genomics 17:461–461

GOC, Government of Canada (2021) Northern Contaminants Program [Online]. Available from http://www.science.gc.ca/eic/site/063.nsf/eng/h_7A463DBA.html. Accessed 27 April 2021

Gottstein B, Pozio E, Nöckler K (2009) Epidemiology, diagnosis, treatment, and control of trichinellosis. Clin Microbiol Rev 22:127–145

Grogan LF, Berger L, Rose K, Grillo V, Cashins SD, Skerratt LF (2014) Surveillance for emerging biodiversity diseases of wildlife. PLoS Pathog 10:e1004015

Gunn A, Shank C, Mclean B (1991) The history, status and management of Muskoxen on Banks Island. Arctic 44:188–195

Handeland K, Tengs T, Kokotovic B, Vikøren T, Ayling RD, Bergsjø B, Sigurðardóttir OG, Bretten T (2014) Mycoplasma ovipneumoniae: a primary cause of severe pneumonia epizootics in the Norwegian Muskox (Ovibos moschatus) population. PLoS One 9:e106116

Handeland K, Madslien K, Bretten T, Røtvei I, Våge J, Tengs T (2020) Mycoplasma conjunctivae-associated Keratoconjunctivitis in Norwegian Muskox (Ovibos moschatus). J Wildl Dis 56:489–491

Holmes JP, Duff JP, Barlow A, Everest D, Man C, Smith F, Twomey F (2019) 20 years of national wildlife disease surveillance. Vet Rec 184:520–521

Komar N (2002) West Nile virus surveillance using sentinel birds. Ann N Y Acad Sci 951:58–73

Kuisma E, Olson SH, Cameron KN, Reed PE, Karesh WB, Ondzie AI, Akongo M-J, Kaba SD, Fischer RJ, Seifert SN, Muñoz-Fontela C, Becker-Ziaja B, Escudero-Pérez B, Goma-Nkoua C, Munster VJ, Mombouli J-V (2019) Long-term wildlife mortality surveillance in northern Congo: a model for the detection of Ebola virus disease epizootics. Philos Trans R Soc B Biol Sci 374:20180339

Kutz S, Tomaselli M (2019) "Two-eyed seeing" supports wildlife health. Science (AAAS) 364:1135–1137

Kutz S, Bollinger T, Branigan M, Checkley S, Davison T, Dumond M, Elkin B, Forde T, Hutchins W, Niptanatiak A, Orsel K (2015) Erysipelothrix rhusiopathiae associated with recent widespread muskox mortalities in the Canadian Arctic. Can Vet J 56:560–563

Lian M, Björck S, Arnemo JM, Esteruelas NF, Angel M, Minsaas SC, Jones KL, Alina LE (2017) Severe hypoxemia in muskoxen (Ovibos moschatus) immobilized with Etorphine and Xylazine corrected with supplemental nasal oxygen. J Wildl Dis 53:356–360

Mavrot F, Orsel K, Hutchins W, Adams LG, Beckmen K, Blake JE, Checkley SL, Davison T, Di Francesco J, Elkin B, Leclerc LM, Schneider A, Tomaselli M, Kutz SJ (2020) Novel insights into serodiagnosis and epidemiology of Erysipelothrix rhusiopathiae, a newly recognized pathogen in muskoxen (Ovibos moschatus). PLoS One 15:e0231724

Nishi JS, Ellsworth TR, Lee N, Dewar D, Elkin BT, Dragon DC (2007) Northwest Territories. An outbreak of anthrax (Bacillus anthracis) in free-roaming bison in the Northwest Territories, June-July 2006. Can Vet J 48:37–38

Nwafor-Okoli C, Berezowski J, Jamal I, Reformat M, Checkley S (2014) International Conference for Animal Health Surveillance. In: International Conference for Animal Health Surveillance, 2014 La Havana, Cuba

NWMB, Nunavut Wildlife Management Board (2021) Nunavut Wildlife Management Board [Online]. Available from https://www.nwmb.com/en/. Accessed 27 April 2021

OIE, World Organisation for Animal Health (2015) Guidelines for wildlife disease surveillance: an overview. France, Paris

OIE, World Organisation for Animal Health (2018) Manual 5 surveillance and epidemiology. OIE, Paris

OIE, World Organisation for Animal Health (2019) Terrestrial animal health code. OIE, Paris

One Health Research Group (2020) One Health Research Group [Online]. Available from https://vet.ucalgary.ca/research/research-groups/one-health-research-group/home. Accessed 18 April 2021

Orenstein WA, Bernier RH (1990) Surveillance: information for action. Pediatr Clin N Am 37:709–734

Proulx J-F, Maclean JD, Gyorkos TW, Leclair D, Richter A-K, Serhir B, Forbes L, Gajadhar AA (2002) Novel prevention program for Trichinellosis in Inuit communities. Clin Infect Dis 34:1508–1514

Salman MD (2003) Surveillance and monitoring systems for animal health programs and disease surveys. Animal Disease Surveillance and Survey Systems

Seymour J, Horstmann-Dehn L, Rosa C, Lopez JA (2014) Occurrence and genotypic analysis of Trichinella species in Alaska marine-associated mammals of the Bering and Chukchi seas. Vet Parasitol 200:153–164

Smolinski MS, Crawley AW, Olsen JM, Jayaraman T, Libel M (2017) Participatory disease surveillance: engaging communities directly in reporting, monitoring, and responding to health threats. JMIR Public Health Surveill 3:e62–e62

Stallknecht DE (2007) Impediments to wildlife disease surveillance, research, and diagnostics. In: Childs JE, MacKenzie JS, Richt JA (eds) Wildlife and emerging zoonotic diseases: the biology, circumstances and consequences of cross-species transmission. Springer, Berlin

Stephen C, Duncan C (2017) Can wildlife surveillance contribute to public health preparedness for climate change? A Canadian perspective. Clim Chang 141:259–271

Tomaselli M, Dalton C, Duignan PJ, Kutz S, van der Meer F, Kafle P, Surujballi O, Turcotte C, Checkley S (2016) Contagious Ecthyma dermatitis as a portal of entry for Erysipelothrix rhusiopathiae in muskoxen (*Ovibos moschatus*) of the Canadian Arctic. J Wildl Dis 52(3): 719–724

Tomaselli M (2018) Improved wildlife health and disease surveillance through the combined use of local knowledge and scientific knowledge. PhD, University of Calgary

Tomaselli M, Curry P (2019) Wildlife health and disease surveillance. In: Cork SC, Halliwell R (eds) The veterinary laboratory & field manual, 3rd edn. 5M Publishing, Sheffield, UK

Tomaselli M, Gerlach SC, Kutz SJ, Checkley SL, Iqaluktutiaq TCO (2018a) Iqaluktutiaq voices: local perspectives about the importance of muskoxen, contemporary and traditional use and practices + Supplementary Appendices S1–S5 (see article tools). Arctic 71

Tomaselli M, Kutz S, Gerlach C, Checkley S (2018b) Local knowledge to enhance wildlife population health surveillance: conserving muskoxen and caribou in the Canadian Arctic. Biol Conserv 217:337–348

Tomaselli M, Elkin B, Kutz S, Harms NJ, Nymo HI, Davison T, Leclerc LM, Branigan M, Dumond M, Tryland M, Checkley S (2019) A transdisciplinary approach to Brucella in muskoxen of the Western Canadian Arctic 1989-2016. EcoHealth 16:488–501

Tryland M, Stubsjøen SM, Agren E, Johansen B, Kielland C (2016) Herding conditions related to infectious keratoconjunctivitis in semi-domesticated reindeer: a questionnaire-based survey among reindeer herders. Acta Vet Scand 58:22

Tryland M, Romano JS, Marcin N, Nymo IH, Josefsen TD, Sørensen KK, Mørk T (2017) Cervid herpesvirus 2 and not Moraxella bovoculi caused keratoconjunctivitis in experimentally inoculated semi-domesticated Eurasian tundra reindeer. Acta Vet Scand 59:23

Uspensky A, Bukina L, Odoevskaya I, Movsesyan S, Voronin M (2019) The epidemiology of trichinellosis in the Arctic territories of a far Eastern District of the Russian Federation. J Helminthol 93:42–49

Wagner MM (2006) Chapter 1 - Introduction. In: Wagner MM, Moore AW, Aryel RM (eds) Handbook of biosurveillance. Academic Press, Burlington

Watsa M (2020) Rigorous wildlife disease surveillance. Science 369:145–147

Wobeser GA (2007) Special problems in working with free-living animals. In: Czeschlik D (ed) Disease in wild animals - investigation and management, 2nd edn. Springer, New York

Wu J, Checkley SL, Dumond M, Kutz S (2011) 2010 muskox health survey: Victoria Island. In: GON, Government of Nunavut

Dogs and People: Providing Veterinary Services to Remote Arctic Communities

Tessa Baker, Laurie Meythaler-Mullins, Arleigh Reynolds, and Susan Kutz

1 Introduction

The relationship between dogs and humans in the Circumpolar North is an ancient and symbiotic accord. Recent archeological evidence from Eastern Siberia has revealed the earliest known evidence anywhere of dogs being bred for a specific purpose (Pitulko and Kasparov 2017; Ameen et al. 2019). This research indicates that these dogs, which were used for polar bear hunting, were also bred to pull sleds, suggesting that dogs and people have shared a symbiotic relationship in the North for over 15,000 years.

Nomadic life centered around taking care of and working with dogs. In the morning, fires were built to cook for the dogs and people, and then the work of the day, hunting, trapping, and hauling, was performed cooperatively with dogs, and in the evening, they were fed again and often shared a campsite. Human and dog lives literally depended upon each other as they worked together to obtain food and travel safely over often hazardous and unstable frozen terrain. Even with advances in modern technology, there is not a machine made that can sense a polar bear in a

T. Baker (✉) · S. Kutz
Department of Ecosystem and Public Health, University of Calgary Faculty of Veterinary Medicine, Calgary, AB, Canada
e-mail: tessa.baker1@ucalgary.ca; skutz@ucalgary.ca

L. Meythaler-Mullins
College of Veterinary Medicine and Biomedical Sciences, Colorado State University, Fort Collins, CO, USA

Department of Veterinary Medicine, University of Alaska Fairbanks, Fairbanks, AK, USA
e-mail: Laurie.Meythaler-Mullins@colostate.edu

A. Reynolds
Department of Veterinary Medicine, University of Alaska Fairbanks, Fairbanks, AK, USA
e-mail: ajreynolds@alaska.edu

M. Tryland (ed.), *Arctic One Health*, https://doi.org/10.1007/978-3-030-87853-5_24

blizzard or unsafe ice on a river or coast as well as an experienced team of sled dogs (Coppenger 1977).

For millennia, this relationship remained unchanged, with dogs helping with hunting, draft work, protection and providing warmth for people, whom in turn provided food and shelter. In northern Europe, the advent of reindeer herding emerged as another activity shared between dogs and humans (Salmi et al. 2021). In Alaska, northern Canada, and Greenland, sled dogs remained the most common means of winter transportation, for Indigenous and settlers alike, until after the middle of the twentieth century. One of the most famous examples of this is the 1925 relay of dog teams that transported diphtheria antiserum to stem an epidemic in the gold rush town of Nome, Alaska. The 650-mile route that often took mail teams 2–4 weeks to cover was traversed in less than 6 days under conditions that saw air temperatures hit −70 °F (−57 °C) and wind chills to −90 °F (−68 °C). This heroic saga is detailed in *The Cruelest Miles* by Gay and Laney Salisbury (Salisbury 2003).

The second half of the twentieth century brought rapid social and economic change to the Circumpolar North. These changes included the settling of nomadic cultures into permanent communities (Bonesteel 2006), a shift from subsistence-based toward cash economies and the colonial implementation of western education and language (Krauss 1980). This forced settlement resulted in improved access to health care and other government supported services but also imposed changes that have resulted in loss of language, culture, self-efficacy, and self-determination. For Alaskan First Nations, the net result has been a tragic increase in substance abuse, domestic violence, and suicide, all of which were extremely rare conditions before 1950 (Berman 2014).

In many ways, the holistic health of Arctic communities has been closely reflected by the health of the dogs with which they live. Practices involved in forced settlement were disruptive to the traditional human–dog relationship. In the Canadian North, government policies around safety and animal control in the 1950–1970s saw the Royal Canadian Mounted Police destroy hundreds to thousands of sled dogs kept by Inuit in settlements, limiting Inuit access to the land and increasing their dependence on cash and services within settlements (QTC 2013). At the same time that colonial and assimilatory changes were being imposed upon northern Indigenous people, the arrival of the snow machine was associated with a dramatic decrease in the dependence upon and, therefore, the number of sled dogs in these newly formed communities. The introduction of the snow machine strengthened the reliance on a cash economy and further decreased the practice of traditional activities. The irony of this situation is that people now had to work more to support the purchase, maintenance, and fuel for these machines that left them less time to participate in traditional subsistence activities (Hueffer and Murphy 2018).

A decrease in the need for dogs and the concentration of people into settled communities has also been associated with changes in the way people keep and relate to dogs. In many northern North American communities, these changes have resulted in health disparities for the dogs and the humans alike. Unwanted feral dog populations in villages pose significant health threats to the community. For example, the Yukon-Kuskokwim Delta of Alaska is an area about the size of the state of

Louisiana that lies 400 miles from the nearest road system, contains about 50 villages, and is severely underserved in access to veterinary care. This area has seven to nine times the annual national average of per capita of reported dog bites upon children (Bjork et al. 2013; Castrodale 2007). This number is likely to be underestimated as, in order to be reported, a bite has to be severe enough to warrant travel to a regional clinic for treatment. Overall, Alaska Indigenous children are hospitalized from dog bites more than anyone else in the Indian Health Service system (Bjork et al. 2013). Rabies is enzootic in the fox population in this area of Alaska and elsewhere across the circumarctic and is occasionally transmitted to susceptible animals, such as unvaccinated dogs (Castrodale 2013). This results in several people, bitten by dogs, being treated for rabies exposure each year from the Yukon-Kuskokwim Delta. This is a common concern elsewhere, with similar dog control and rabies transmission concerns in northern Canada. Other zoonotic health concerns include parasitic infections such as echinococcosis (Salb et al. 2008). Furthermore, dogs can be a source of parasites to wildlife populations as well (Salb et al. 2008).

This relatively recent and rapid change in the relationship between people and dogs in northern communities has had a far-reaching negative impact upon both human and canine health. While the risks to human and canine physical health are easy to document, the impact on human mental and behavioral health may be more difficult to measure but just as impactful. When a foundationally favorable aspect of a culture and knowledge structure becomes threatened, this challenges belief systems and can result in severe and negative impacts on mental and behavioral health and well-being (Hueffer and Murphy 2018). Community-based One Health approaches have demonstrated success in restoring the human–canine relationship to its original mutually beneficial state and resolving physical and mental/behavioral health issues for the communities involved. The Frank Attla Youth and Sled Dog Care Program (FAYSDCP), the Alaska Care and Husbandry Instruction for Lifelong Learning (ACHILL) (see Box 1), and the Sahtu Northern Community Health Program in the NWT, Canada (Baker et al. 2018a, 2020), are examples of community-based One Health programs that have helped holistically improve the health of dogs and humans in remote northern communities and serve as examples of what can be done when community members, educators, and veterinarians work together to address the challenges described above.

Box 1 Frank Attla Youth and Sled Dog Care Program

The Frank Attla Youth and Sled Dog Care Program (FAYSDCP) was founded by the late George Attla Jr. in memory of his son. When George Attla retired from an illustrious career as a professional dog musher, he returned with his young son Frank to his home community of Huslia, Alaska, a small village on the Koyukuk River that lies over 200 km from the nearest road. He moved home so Frank could have access to traditional ways of knowing and his

(continued)

Box 1 (continued)

extended family. When Frank died unexpectedly from an asthma attack, George decided to do something for his community in honor of his son's memory. As a child, George had contracted tuberculosis and spent 9 years in a Native hospital almost 1500 km from home until antibiotics had been developed that would cure his infection. His left knee and ankle were so badly damaged by the disease that they had to be fused so he could walk. He returned a young man who had missed much of his cultural training, lost much of his language, and felt out of place. He found solace and self-esteem through his relationship with sled dogs and later from the acknowledgment he received for his skill in racing with them. Years later, he described his relationship with sled dogs as literally "saving his life." When George retired to Huslia, he was concerned with the state of the youth in the community and the disappearance of dog teams due to the advent of snow machines. Many students did not finish high school, and some struggled with substance abuse and suicidal ideation. Remembering how dogs had helped him through his own troubled youth, George went to the Tribal Council and the school board and asked if he could start a program in memory of his late son that would teach sled dog husbandry in the schools and bring students into the homes and kennels of elders where they could learn how to care for, train, and race the dogs and in the process learn their own history.

The FAYSDCP has been transformative not only for the students but for the community as a whole. Everyone has a job and everyone works together with a common goal. Young children gather firewood so that older students can build fires and cook fish for the dogs. The students learn how to catch, cut, and dry fish for their dogs. They learn how to build dog houses, cut and dry grass for bedding, feed the dogs, and train them. Elders mentor older students who in turn mentor younger students (Fig. 1). In daily caring for their dogs, these students, many of whom were struggling, began to find purpose. No matter how cold or how dark it was outside, they had to get up, go out, and take care of their dogs. The dogs were always happy to see them, and as many students said, they felt the dogs understood them when no one else did. George often said, "Dogs accept you for who you are," and for these students, that turned out to be a powerfully healing influence.

Classroom assignments required students to research mushing in their community, and they discovered that their valley was a "cradle of champions," and each of them had ancestors who were famous champions. This gave them pride in their heritage and in themselves when their accomplishments were acknowledged by community members. The community put in trails, held races, and then bake sales so the students could travel to Fairbanks to race in and succeed at the highest level. Each student's success was seen as a success for the whole community. George told every student that they were a

(continued)

Box 1 (continued)
champion, and they began to believe in this and in themselves. The FAYSCP has been so successful in Huslia that it became the inspiration for a large educational grant (ACHILL), which currently supports similar programs in 17 rural Alaskan communities. Sled dogs have traditionally been a part of the culture of Northern communities spanning from Russia, across North America, and as far as Greenland. And so similar programs may benefit other communities across this region. One of the great benefits of this program is how the revitalized practice of working with dogs has become a platform for the transfer of traditional knowledge and cultural skills. The dogs serve as a bridge between modern and traditional worlds where elders and youth can find common ground and interests.

Fig. 1 Older Frank Attla Youth and Sled Dog Care Program student mentoring a younger student. Photo Credit: Kathy Turco

This chapter describes many ways in which healthy dog populations support overall community health, with a focus on Alaska and northern Canada. As veterinarians, we are well trained in promoting canine health, and in doing so, we can also promote physical and mental health of pet owners and communities. We also value the power of the human–animal bond. Programs highlighted in this chapter show how this relationship can be healing for whole communities and why it is so important that underserved regions have access to veterinary care so these programs and the communities they serve can thrive.

2 Determinants of Health for Dogs

The health of dogs in remote communities can reflect the overall health and well-being of the community in which they live. Factors that affect the health of people and communities are often called determinants of health. These factors include the social and economic circumstances, the physical environment, individual behaviors, and genetics with which people live (WHO 2017). These factors can also be applied to animal health (Baker et al. 2018a; Card et al. 2018). Determinants of dog health in the Arctic include the physical environment in which they live (e.g., exposure to climate and wildlife), biology (e.g., dog breeds that have evolved in the region are better suited to the local conditions than many of those brought in from the south), behavior and animal social factors (e.g., roaming, playing, and guarding behaviors and interactions with other dogs), human social factors (e.g., husbandry choices, cultural and social factors that influence their value, community norms), and access to veterinary care (e.g., remote Arctic communities often do not have access to such services) (Baker et al. 2018a; Card et al. 2018).

3 Veterinary Services in Remote Arctic Communities

Veterinary services are often inaccessible, unavailable, and unaffordable to residents of remote northern communities (Brook et al. 2010). Many northern North American communities are accessible by winter road for only a few weeks of the year or by air, boat, or snow machine. Thus, veterinary services are limited to those residents who can afford to send their animal to larger city centers, at a cost of hundreds of dollars, a prospect unaffordable to the majority of dog owners. Residents of many small, remote communities across northern North America are relatively poor compared to those in larger urban centers, with lower household incomes, more families living in core housing need, and higher rates of unemployment (GNWT 2018; GovYK 2016; StateOfAlaska 2020). These communities are too small to sustainably support local veterinary practitioners and the costs associated with travel to the communities make it economically untenable for the veterinary practices to provide on-site services. These economic barriers combined with the high cost of living (often 1.5–2 times that of major city centers) further limit community access to veterinary services.

Acceptability of veterinary services is also a critical consideration. Some northern communities have never had access to veterinary care and may not be familiar with, or value, the services that veterinarians can provide. Different cultural perspectives and beliefs around dogs and previous experiences and history (e.g., colonization, residential school) will influence the individual and community-wide acceptability of services. Acknowledging this history and these perspectives and employing a culturally sensitive and respectful approach are essential for the veterinary team when introducing and offering services.

Access to veterinary services is important not only to animal health and welfare but also to community wellness. Left unchecked, dog populations can grow quickly and lead to safety concerns, dog bites, disease transmission (including zoonoses), and noise and waste complaints. The benefits of access to preventative veterinary services such as vaccinations (e.g., rabies), anti-parasitic drugs, and sterilization surgeries on animal health and welfare are clear: a more stable population of healthier dogs that are less likely to act as vectors of zoonotic diseases and less likely to roam or fight over female dogs in heat. Less appreciated is how services can also impact public health and community well-being. A considerable emotional trauma can occur when communities have to make difficult decisions on how to handle problem dogs or dog overpopulation. In some instances, the emotional stress caused by the culling of loose dogs has led to teachers leaving the communities (Brook et al. 2010), and high turnover rates of teachers are a major challenge in northern communities with negative consequences for educational continuity and youth learning (Kaden et al. 2016).

The mechanisms for provision of veterinary services in remote northern regions vary. Services range from fully subsidized to cost-shared initiatives and are generally delivered by local, regional, national, or even international charity groups or academic institutions as a component of experiential learning initiatives. Patterns of delivery also vary, with some programs occurring repeatedly in the same communities while others serve communities once. There are few examples of sustainable delivery that do not require some degree of government, charity, or academic institutional support. Two programs involving different models of academic–community partnerships are explored in more detail in Boxes 2 and 3.

Box 2 Northern Community Health Rotation, Sathu Settlement Area, Northwest Territories (NWT), Canada

Where Does It Occur? The Sahtu Settlement Area is a Dene and Metis Land Claim Area in the NWT, Canada, with five small communities (142–818 people in 2018), accessible only by air or winter road (Brook et al. 2010). See red shaded area on the map in Fig. 2.

Who Does It? The Northern Community Health Rotation (NCHR) is a final year veterinary student clinical rotation organized by the University of Calgary Faculty of Veterinary Medicine (UCVM). The program was initiated in 2008.

(continued)

Box 2 (continued)

Four final year veterinary students participate in the clinics each year, benefiting from a unique service–learning opportunity in a remote and culturally different setting. Extensive pre-trip preparation on the local context and provision of veterinary services in remote regions is done to prepare the team. The team works with the government of the NWT, community liaisons, and the local schools to provide clinics that are built on a One Health model to broadly benefit community health.

Who Does It Serve? The NCHR provides annual preventive veterinary services to the communities of the Sahtu Settlement Area. This program was developed in collaboration with the communities after a need assessment identified a desire and demand for veterinary services in the region (Brook et al. 2010). Local youth are involved in the clinics, and outreach sessions are done in the classrooms to raise awareness of health sciences education and animal health and husbandry.

When Does It Occur? Every February when the winter roads to each of the community are open.

What Services Are Provided? The program provides wellness examinations, core canine and feline vaccinations, deworming medications, and surgical sterilization. The clinics are hosted in schools and community halls and have a strong youth outreach and engagement components (Fig. 3).

How Is It Funded? The program is funded through a combination of university funding, local community government, corporate, and charity funding and material donations, and in-kind support from the communities for accommodation, food, and clinic space.

What Has It Accomplished? Prior to the start of the program, few dogs were vaccinated against rabies and other canine viruses or dewormed against parasitic infections, including zoonotic worms, and less than one quarter of dogs had been sterilized (with community members commenting there were too many dogs in their communities) (Brook et al. 2010). A program assessment conducted after a decade of services, through household questionnaires, community dog censuses, and a chart review, found that as veterinary service uptake increased, there were improvements in dog population health and welfare indicators, changes in the husbandry of dogs, and differences in community concerns about and relationships with dogs since the start of the program (Baker et al. 2020). Significant improvements in dog vaccination and deworming rates led to a better overall dog health and a reduced public health risk of rabies and zoonotic parasites from dogs (Baker et al. 2020). An increase in both the proportion of dogs described as sterilized and the median age of dogs suggests a better control of dog population size and increased stability and fewer puppies and more longer living dogs in the communities compared to in 2008, both of which benefit dog health and public health and well-being

(continued)

Box 2 (continued)
(Baker et al. 2020). Community members felt there were no longer too many dogs in their communities and attitudes toward dogs were generally more positive across communities. Interviewees perceived improvements in how dogs were cared for, less fear of dogs, fewer problem dogs in the communities, and more access to options, other than culling, for residents who could no longer care for their animal. In addition, most residents described their dog as a companion and a part of their family. However, some concerns remained, including the ongoing occurrence of roaming dogs and how these dogs were dealt with by the community (Baker et al. 2020). Notably, the positive findings documented were contingent on the long-term commitment of the program.

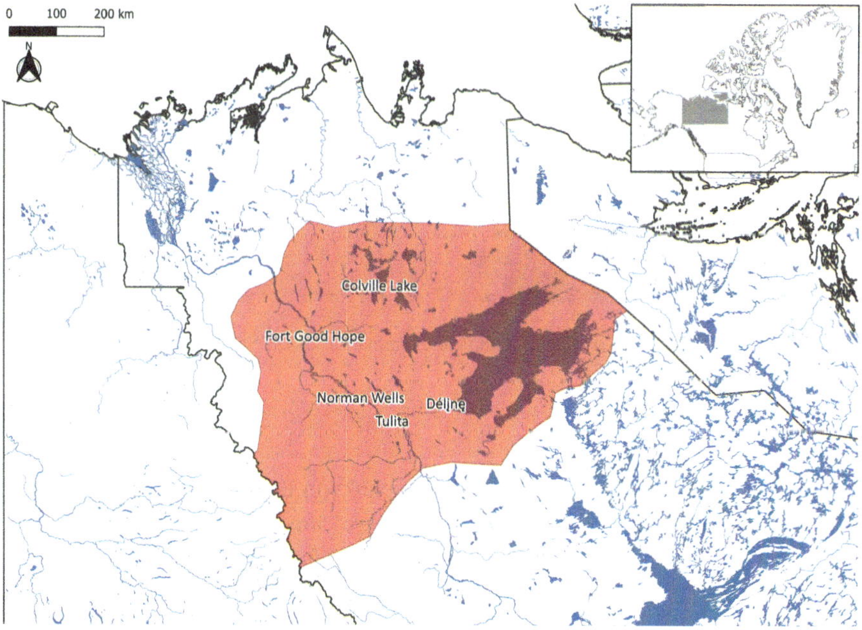

Fig. 2 Map showing the location of the Sahtu Settlement Area, Northwest Territories, Canada

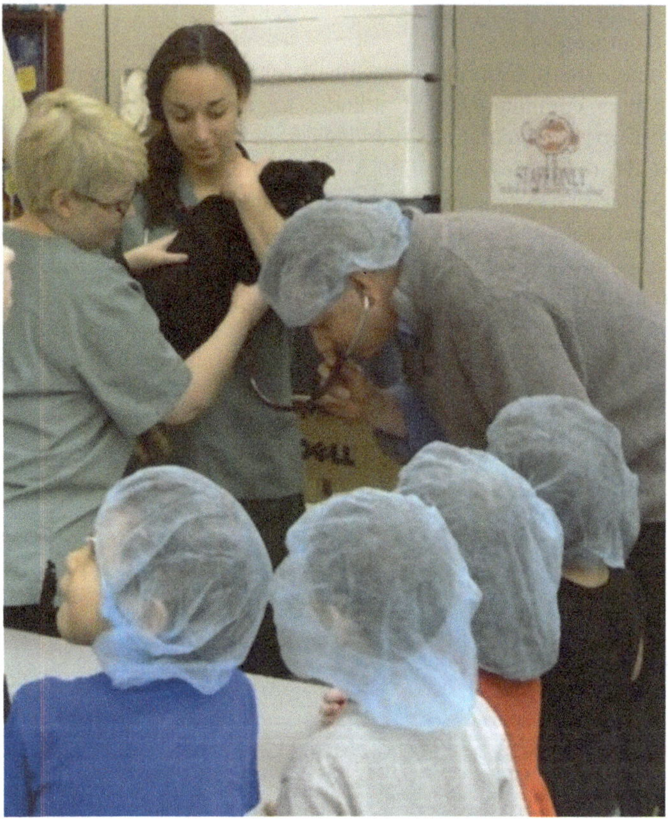

Fig. 3 Elementary students watching a clinical examination in the Sahtu Settlement Area. Photo Credit: Susan Kutz

Box 3 Hub Outreach Project, Yukon-Kuskokwin Delta, Alaska, USA

Where Does It Occur? This rural area of Southwest Alaska is comprised of 58 federally recognized tribal communities, ranging in size from 25 people to 1000 people, with the exception of the larger regional service and transportation hub of Bethel. See map highlighting this area in Fig. 4.

When Does It Occur? Year round. These communities are completely off the road system, only accessible by small plane, boats in the summer, or snow machines in the winter.

What Services Are Provided? The program provides wellness examinations, core canine and feline vaccinations, deworming medications, and surgical sterilization.

How Is It Funded? The program is funded through a combination of university funding, local community government, corporate, and charity

(continued)

Box 3 (continued)
funding and material donations, and in-kind support from the communities for accommodation, food, and clinic space.

What Has It Accomplished? The Hub Outreach Project (HOP) access to veterinary care model was designed with input from the regional health care provider, the Yukon-Kuskokwim Health Corporation, following their human health care model. With the design of a hub-and-spoke model, strategically identified communities are travelled from the "hub" of Bethel, and veterinary field clinics are set up in the smaller outlying "spoke" communities. Each of these "spoke" communities are accessible to 5–10 other smaller outlying communities. This brings a high-quality preventative veterinary care directly to those in the Southwest Alaska who previously lacked access due to transportation, socioeconomic, and geographic barriers. Surveys are completed that address the real needs that exist in these communities and provide a metric for demonstrating improvement.

The costs associated with travel and treatment for rabies exposure, and the mental and behavioral health impacts associated with dogs are significant. In remote underserved areas, the most economically feasible solution to these issues is to bring the clinic to the pet owners rather than have them travel individually to distant centers where veterinary services are available. For this region, having the HOP model in place far outweighs the cost of the consequences that would ensue in its absence. It is a framework for areas lacking access to veterinary care, which helps control zoonotic disease, as well as negative human health impacts resulting from dog overpopulation. By improving the lives of animals, the lives of people are improved, not only physically but also through well-being and mental health.

A One Health model for program sustainability was developed for the Hub Outreach Project that highlights four goals to engage stakeholders in the development of a holistic and sustainable veterinary program that demonstrates impacts to animal, human, and environmental health through stakeholder participation and knowledge sharing (Fig. 5).

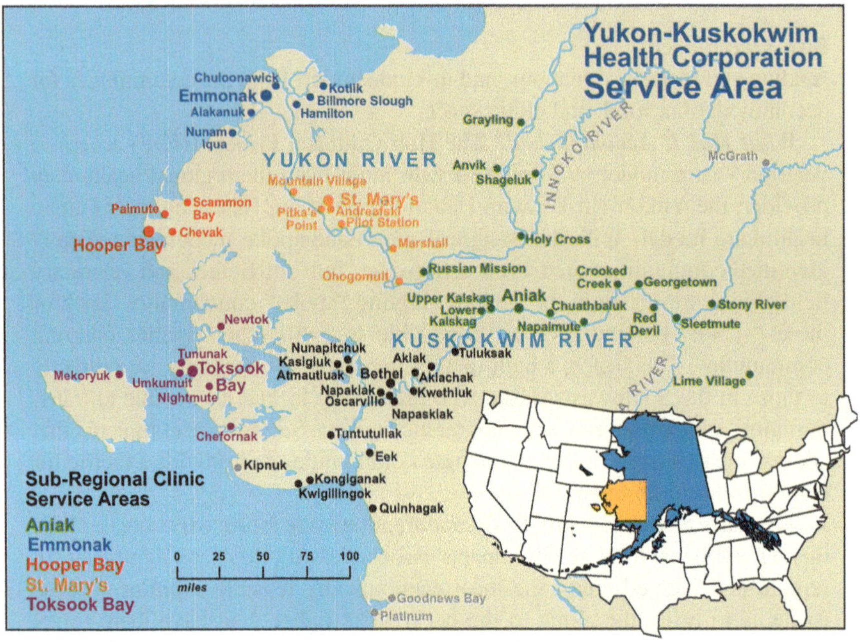

Fig. 4 Map of Alaska showing the Yukon-Kuskokwin Delta

4 Key Principles for Successful Subsidized Veterinary Programs

Considering the broad impact that healthy dogs can have on a community, a One Health approach to veterinary service delivery is key to program success and sustainability. There are several overarching principles to consider when developing subsidized veterinary programs to ensure the long-term success for animals, people, and the environment.

4.1 Has a Need for Veterinary Services Been Identified by the Community as a Priority?

Using a One Health model, a crucial first step is to understand the community needs and plan for service provision relevant to each specific community context. Input and direction from community elders, local government, schools, and health providers can guide this. Taking the time to understand local needs will ensure the team and services provided are trusted and welcome. For example, the community of Aniak, Alaska, requested a pet first aid class, and the local school in Nunapitchuk, Alaska, requested education on rabies and the threat to human and animal health.

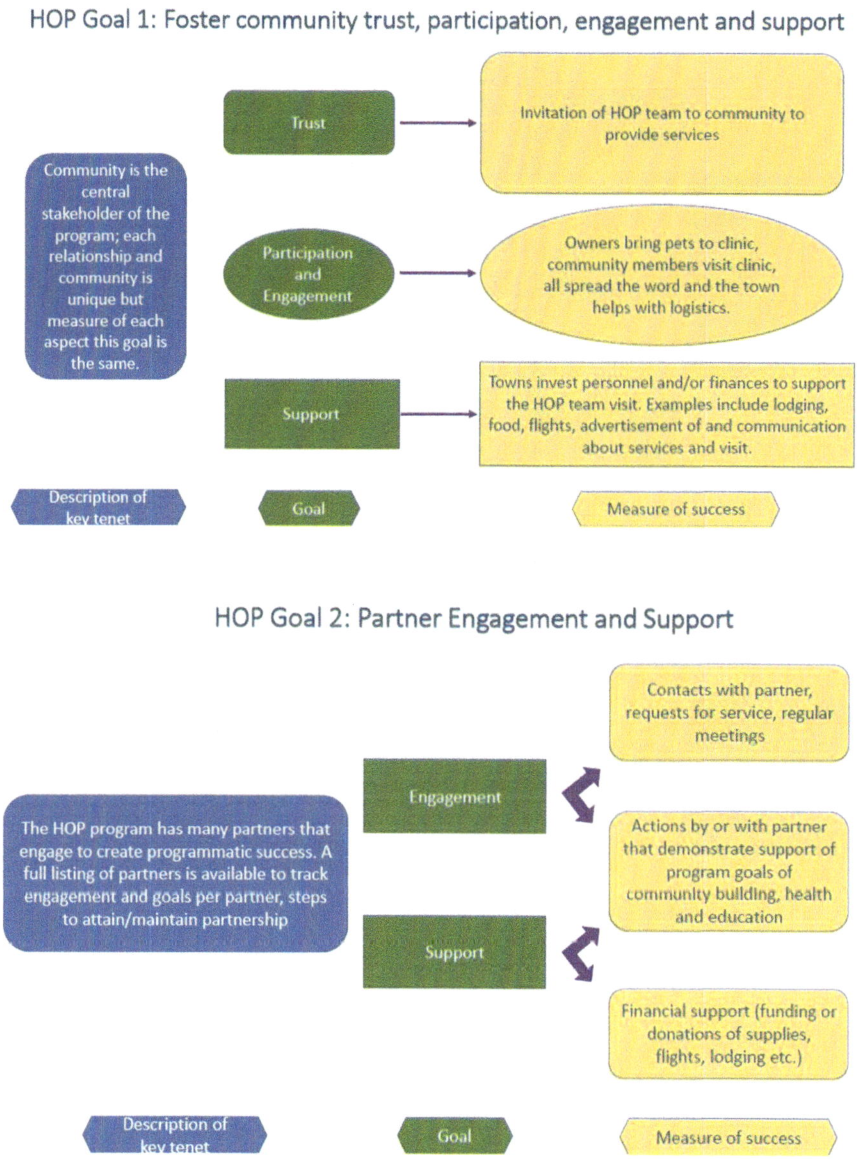

Fig. 5 The Hub Outreach Project's One Health model for sustainability

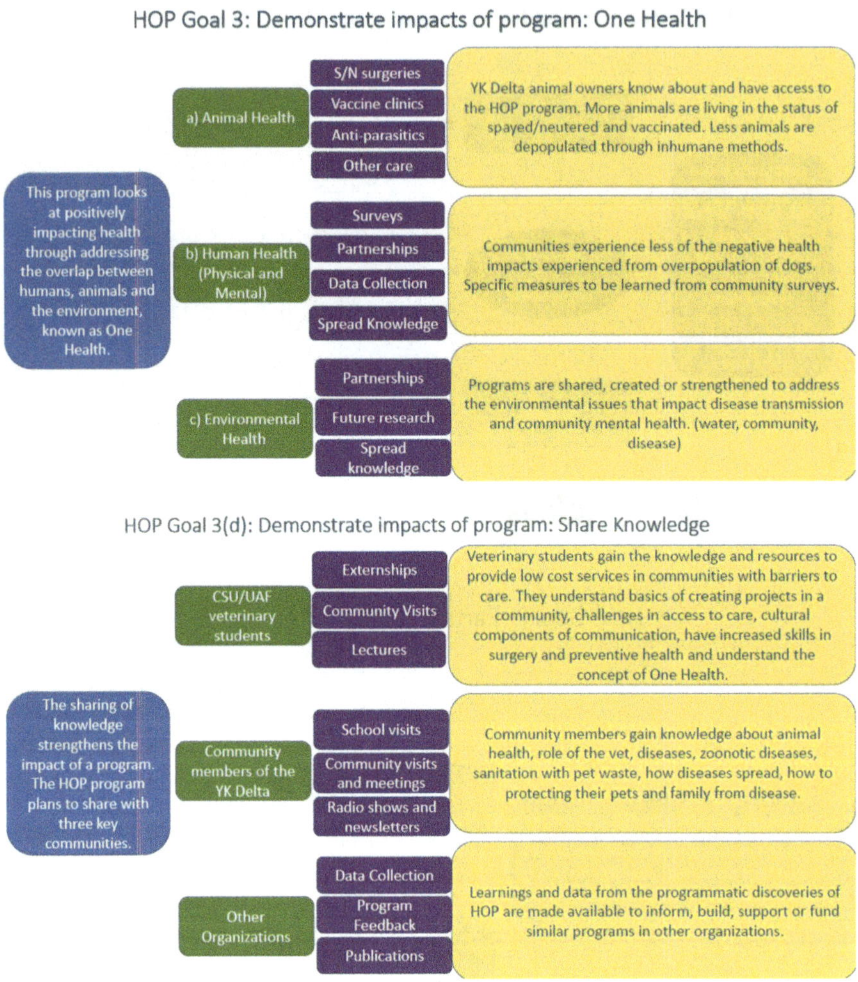

Fig. 5 (continued)

Both community requests were met by offering an evening pet first aid class in Aniak and lecturing to a class during the school day in Nunapitchuk.

4.2 Spend Time Building Collaborative and Trusting Relationships with Communities

It is clear that relationship building is essential for effective provision of veterinary services in northern communities. These relationships are established far before the first mobile veterinary clinic and, ideally, lead to long-term community partnerships that facilitate local trust and buy-in, enhance program uptake, and ensure sustainability. Community partners can include the local governments, schools, businesses, and individual community members. Partnership support can be provided in the form of food and accommodation, a location for a clinic venue, administrative support, logistical and financial support, and liaising with the community. For example, the Hub Outreach Project (HOP) works with each community's tribal council to discuss when the veterinary clinic visits will be the most beneficial and to consider when access to the community would be most practical for the mobile veterinarian. The tribal councils also help secure space for the mobile clinic, transportation upon arrival in the community, and advertising of HOP services. HOP also engages with each community's Village Safety Public Officer, who oversees the handling of unowned dogs in these communities.

In the Sahtu, community liaisons, which can include dog owners, teachers, government employees, or others, act as champions for the program and play an essential role in program implementation. These liaisons are integral for distributing program information, booking appointments, and liaising between the clinic team and the community in the interim months.

4.3 Secure a Long-Term Commitment to Providing Services and Develop a Sustainability Plan

Delivery models vary from providing the services in one large community and serving anyone who comes from the surrounding communities (i.e., hub and spoke model) to traveling to and setting up in each community. Funding also varies with support received from charities, academic institutions, and different levels of government. Services are often offered as subsidized services where clients do not pay, but donations are encouraged. Regardless of the delivery or funding model, a long-term commitment and a sustainability plan are required as they are important to maintaining continuity of services and community support for the program.

The positive impacts of the program on both animal health and welfare and community health and well-being can take several years to become evident, especially if animal turnover is initially high. Thus, the continued commitment to communities to return at least annually is important to effect changes in dog population health and welfare and community concerns about dogs. A sustainability plan to ensure the continuity of the services that communities come to rely on is important for programs funded and delivered with significant charity and academic institutional support. The long-term vision should include transferring to more support from local and regional partners and decreasing the reliance on external charities. This is also important to supporting community capacity development. For example, the Hub Outreach Project in Alaska hopes to demonstrate to governments that the cost of providing services is less than paying for the consequences of not having the program (e.g., providing post-exposure rabies prophylaxis or wound care to a dog bite victim), leading to the incorporation of the program into local health services' budgets and planning.

4.4 Have an Interim Plan in Place for the Time When the Clinics Are Not in the Community

For the communities that are provided with veterinary services annually or on occasion, dog health is often improved through decreased infectious disease occurrence, improved body condition, reduced fecundity, and longer survival. However, between veterinary visits, community members are left without on-site veterinary services to respond to emergency needs or ongoing medical problems. While local animal health advocates may be able to respond to minor needs, engaging vets in "nearby" urban centers can improve continuity of care. These practices may be the first point of contact that could see animals in the interim months if the owners can afford it and also may be able to provide remote support in a telemedicine-type framework.

Emergency situations, such as if a dog is hit by a car, can be incredibly difficult for community members to deal with. They may consult with veterinarians at urban veterinary clinics or humane societies, but the difficulty of getting the animal to the clinic for care remains. The use of telemedicine and teletriage are options in these emergency situations. As a long-term plan, engaging local youth to pursue careers in the veterinary field, like as a registered animal health technologist, could provide a knowledgeable individual in the community to provide first aid and consult with a regional center vet to determine the best course of action. Community-based animal health workers (CAHWs) trained to vaccinate, provide treatment, and perform disease surveillance under the guidance of a veterinarian through teleconsultation provide essential care to animals in remote and underserved areas of some countries (AMRRIC 2019), particularly developing countries (Catley et al. 2004). This model of care, while difficult to implement under the current professional regulatory restrictions in some Arctic jurisdictions, deserves further exploration.

4.5 Perform Regular Program Evaluations to Assess Value and Progress

Evaluations of long-running subsidized veterinary programs for their value to stakeholders are rare, particularly in Indigenous communities (Baker et al. 2018b). Evaluations provide important information about program uptake, reach, and barriers, vaccination and sterilization rates, and acceptance of services such as surgical sterilization. Assessments also provide information on community perceptions of critical changes, such as overall dog welfare, the numbers of roaming or packing dogs, and the sense of safety in the community. Information from regular evaluations can be used to further tailor a program to community needs and reduce barriers to care and to better reach public, animal, and community health goals.

4.6 Be Flexible and Adaptable

While a careful and detailed planning for mobile clinics is important, flexibility, creativity and adaptability are essential for success. It is not unusual for community working locations to have unreliable power sources and limited availability or no access to running water. Unplanned community events, such as a death of a community member, can take priority over, and conflict with clinic dates and times. Of utmost importance in these situations is respecting the local customs and taking guidance from community partners on how to proceed.

5 Practical Considerations for Provision of Subsidized Veterinary Services

Despite the remoteness of the clinics, a commitment to providing the best standard of care for that animal and client within the socioeconomic and cultural context in which the services are being provided is key to any subsidized veterinary clinic. The services should be delivered professionally, with a high regard for both the quality of care provided to the patient and the interactions with the client. Furthermore, the highest standard of clinic equipment and protocols should be used that can be reliably, safely, and affordably delivered in the community setting. Financial, logistical, and practical considerations mean that ancillary diagnostic testing may be limited, and the veterinary team must rely on a thorough history and physical examination in order to develop a differential diagnosis list. Treatment plans also must take into account and work within the client's financial and logistical limitations, such as the inability to do regular follow-up testing.

There are many practical considerations to successfully operating a subsidized veterinary clinic in remote Arctic communities.

5.1 Veterinary License Regulations

Veterinarians and technicians participating in the clinics are often licensed professionals from other regions who are volunteering their time to provide services. Jurisdictions in the Arctic have their own regulations around professional licensing that must be respected. It is important to contact the professional licensing body in the state or territory of interest well in advance of the clinics to ensure appropriate licensure.

5.2 Animal Health Care Needs

5.2.1 Diseases

Key viral diseases of concern in northern and Arctic regions include canine distemper virus (CDV), canine parvovirus (CPV), and rabies virus. A combination vaccination (e.g., DA2PP or DHPPi) provides protection against CDV, CPV, and canine adeno- and parainfluenza viruses. Rabies virus vaccination is provided through another vaccine. CDV and CPV cycle through wild canid species (e.g., Arctic fox, red fox, and wolves) and can pass to domesticated dogs through contact with respiratory secretions (CDV) and diarrhea (CPV) (Beineke et al. 2015; Steinel et al. 2001). Rabies virus (Arctic strain) is endemic in fox populations in some parts of the Arctic and spillover events can occur with domesticated dogs (Filejski 2016). This poses an increased risk to people in communities that then come into contact with the dogs. Vaccination against all these diseases is very effective to reduce the risk of disease in dogs and people (i.e., rabies).

Dogs and cats are host to a variety of parasites in the Arctic. For dogs, the most common parasites include (i) the tapeworms *Diphyllobothrium* spp., *Echinococcus* spp., and *Taenia* spp., (ii) roundworms and *Toxocara canis* spp., *Toxascaris leonina*, and *Uncinaria*, (iii) the fluke *Alaria*, and (iv) protozoa *Giardia duodenalis*, *Cryptosporidium* spp., *Sarcocystis* spp., *Neospora caninum*, and *Cystoisospora* spp. (Salb et al. 2008; Jenkins et al. 2013). Many of these parasites have complex, predator–prey lifecycles, and dogs can acquire them through diets of country foods such as fish (*Diphyllobothrium*) and wild game (*Echinococcus canadensis* or *granulosus*, *Taenia*, *Sarcocystis*, *Neospora*), by ingestion of small mammals (*Toxocara*, *Toxascaris*, *Cystoisospora*, *Echinococcus multilocularis*) or amphibians (*Alaria*), or through direct transmission from mother to fetus/pup (*Toxocara*). Several parasites are also acquired through environmental contamination (*Giardia*, *Cystoisospora*, *Cryptosporidium*, *Toxascaris*, *Uncinaria*, *Toxocara*). Several other parasites, including *Eucoleus*, *Strongyloides*, eggs of the family Physalopteridae, lungworms (*Oslerus*, *Crenosoma*, and *Filaroides*), and various trematode and protozoal species have also been reported in Arctic fox, wolves, or dogs in boreal to arctic regions and may also be expected (Kapel and Nansen 1996; Elmore et al. 2013; Mech and Luigi 2010). For cats, *Taenia* spp., *Toxocara cati*, and *Toxoplasma gondii* are likely the most common parasites. Unlike in more temperate and tropical parts, ectoparasites do not tend to be a major issue for most northern dogs or cats, although juvenile demodicosis and fleas are occasionally observed, and lice is a known problem in wild canids in the north (Woldstad et al. 2014).

Some parasites infecting dogs and cats in the Arctic are important zoonoses, perhaps the most important and common of which are *Echinococcus* spp. Dogs acquire *Echinococcus* spp. through ingestion of hydatid or multilocular cysts in the lungs, liver, or abdomen of ungulate or rodent species, respectively. Adult parasites develop in the canid small intestine and produce eggs that are shed in the feces. These eggs are environmentally resistant and are immediately infective to people, manifesting as cysts in the liver, lungs, and occasionally elsewhere in the body. *Echinococcus multilocularis* is the most pathogenic of these parasites in people (Jenkins et al. 2013). Other zoonotic parasites to which people may be exposed through eggs/cysts excreted in dog feces and subsequently available in the environment, including water systems, include *Toxocara*, *Cryptosporidium*, and *Giardia*.

In cats, *Toxoplasma* is the most important zoonotic parasites of concern. *Toxoplasma* is an increasing concern in the Arctic, with seroprevalence in people ranging from 8% to 60% (Reiling and Dixon 2019) and detection, primarily through serology, in several subsistence wildlife species (Reiling and Dixon 2019; Jenkins et al. 2013; Kutz et al. 2012). Although there is little to no current evidence of domestic cats in the Arctic being a source of exposure for people, an increased broader concern about the parasite in Arctic wildlife requires that veterinarians are aware of the parasite and can appropriately communicate risks to owners. Prevention of human exposure to zoonotic parasites of dog or cat origin is best accomplished through good hygiene practices and reducing the burden of parasites in peridomestic animals.

Parasite control for dogs in the Arctic should take a multi-pronged approach that addresses both exposure risks and pharmaceutical treatment options. Appropriate hygiene should be encouraged, especially for dogs that are maintained outside in enclosures or on chains where accumulation of feces can lead to substantial environmental contamination. For most parasites transmitted through food sources (i.e., predator–prey lifecycle), freezing can destroy the infective stages from the intermediate (prey) hosts. Raw fish is a common food for many traditional dogs but is also a risk for infection with *Diphyllobothrium.* To mitigate this risk, freezing below −4 °F (−20 °C) for 7 days or below −31 °F (−35 °C) for 15 h is recommended (see Chap. "Zoonotic Marine Helminths: Anisakid Nematodes and Diphyllobothriid Cestodes") (FDA 2020; Wharton and Aalders 2002). Freezing of *Echinococcus* cysts is thought to kill the parasites (−112 °F (−80 °C) for 2 days (Krauss et al. 2005; CFSPH 2011)), but the general recommendation is to not feed organ meat (lungs or livers are the most common location for hydatid cysts) to dogs. However, exposure to *E. multilocularis* for dogs that roam also occurs through hunting wild rodents.

Selection of anti-parasitic drugs should be done with the goal of reducing the burden of the most common and important parasites. For general preventive treatment, a single-dose, broad-spectrum anthelmintic, with efficacy against *Echinococcus*, is the best choice, taking into account housing environment and whether there are other dogs on the premise that should be treated as well. Appropriate protocols for puppies and lactating bitches should be developed using established guidelines such as those outlined in the Canadian Parasitology Expert Panel Guidelines for the Management of Parasites in Dogs and Cats produced by the Canadian Parasitology Expert Panel and updated regularly (Conboy et al. 2019). This is particularly important for the zoonotic parasite *Toxocara canis*, where the burden in puppies in the first few months of life can be high, continued transmission can occur from the mother to pups through milk, and environmental contamination with resistant eggs can build up. An individualized risk assessment should be performed for each dog, with recommendations for follow-up deworming taking into account the dog's lifestyle (source of food, housing, etc.), accessibility and affordability of appropriate parasiticides, and the owners' means and needs.

5.2.2 Nutrition and Husbandry

The climate in the winter months in the Arctic is cold and harsh. Many dogs continue to be kept outdoors as has been done since relationships with dogs in the region began. Some dogs are provided with dog houses or other forms of shelter against the elements. An important component of thriving through the winter is access to foods that are high in fats. Traditionally, dogs in the region were fed fish, wild game, and table scraps. In some regions, relationships with dogs, and the type of dogs being kept, have changed from that of a working relationship with working breeds to relationships that are more companion based with a variety of imported breeds. These changes along with social, economic, and cultural shifts have resulted in more dogs being fed store-bought commercial dog foods. Feeding this type of food may make it challenging for owners to provide sufficient calories to maintain a healthy body condition for dogs that continue to live outside through the winter months. Furthermore, the moisture content found in traditional diets is important when water left in a bowl would quickly freeze. A consideration for adding moisture to commercial dry dog food diets must be made. Discussions around diet and body condition of dogs and appropriate shelters suitable to the local context and resource limitations are important conversations between veterinary professionals and clients.

5.2.3 Euthanasia

Access to humane euthanasia becomes more complicated as traditional techniques of shooting a dog to euthanize them if they are seriously injured, sick, or old are less palatable if dogs are considered family members. The ability to provide a peaceful euthanasia for a pet is an important act that a veterinarian can provide when they are available. In some communities where organizations provide services on an annual basis, community members will often wait until the veterinarians are in their community to euthanize their pet. This can result in them having to wait longer than they would otherwise be comfortable with or to make the decision sooner than they want to, just to be able to access the service in their community.

When euthanasia with injectable barbiturates, which provides a quick and painless death, is performed by a veterinarian, the ability to safely dispose of the animal's body is often an added difficulty in small, remote communities, particularly in the winter months. The accumulation of barbiturates in the body after death makes the body a risk to any animal who may scavenge on it through secondary poisoning. As such, any animal euthanized in this way should be either completely cremated, ideally through incineration, or buried at least 4 feet deep. These requirements are difficult to achieve when euthanasia is performed in the winter months since most communities do not have an incinerator but can be accomplished by educating owners about how to build a pyre and can even become a type of funeral for the pet.

5.3 Providing Preventive Veterinary Services in Remote Communities

Subsidized veterinary clinics typically provide wellness examinations (Fig. 6), core vaccinations, anti-parasitic treatment, and sterilization surgery for a few days, once or a few times per year. This makes in-person follow-up after surgical procedures and booster doses of vaccine or regular deworming treatments difficult. To address

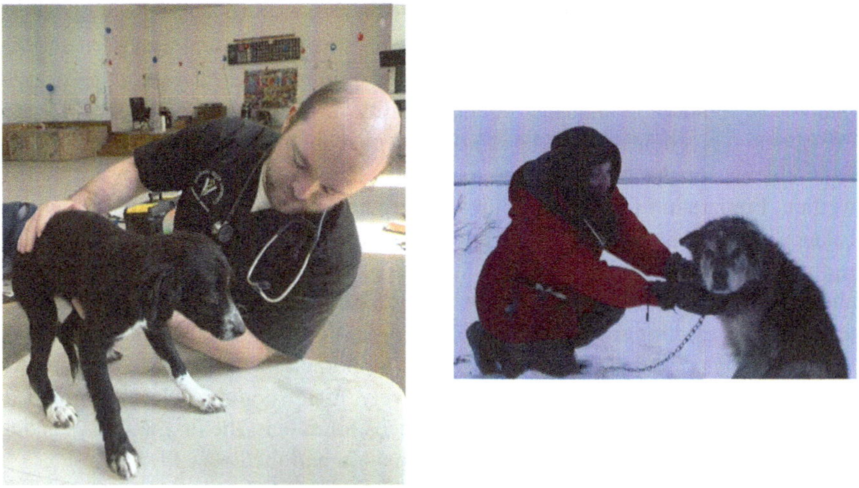

Fig. 6 Completing a wellness examination on a puppy in the Yukon-Kuskokwin Delta, Alaska (left), and completing a wellness examination on a dog living outside in the Sahtu Settlement Area, NWT (right). Photo Credit: Laurie Meythaler-Mullins (left) and Tessa Baker (right)

these challenges, veterinary service providers and clients need to work together to adapt medical and surgical protocols and procedures so that they are most appropriate in the given context (e.g., animal age/sex/species, client's situation, geographic location, and availability of other supports).

During the wellness examination, the veterinarian discusses with the owner the animal's history, diet, and housing situation and any concerns that the owner may have. If the animal is deemed to be healthy, core vaccinations are given and an anti-parasitic medication is chosen based on the risk of parasites from the history and diet. With the limited equipment available to mobile veterinary clinics, tests to confirm the presence and type of parasites in feces are not routinely done. Fecal testing is recommended, when possible, to clearly dictate anti-parasitic treatment. While services are only provided to people who bring their animal to the clinic, the recommendation from the World Health Organization (WHO) and the World Organization for Animal Health (OIE) to prevent the spread of rabies virus (canine variant) to people is to vaccinate at least 70% of the dog population (WHO and OIE 2016). In the Arctic, achieving this level can provide herd immunity in the dog population to prevent spillover of rabies from wild canids to people through domesticated dogs.

Surgical sterilization is offered as a means of population control. A female dog can have up to 2 litters of puppies per year with as many as 12 puppies per litter. Depending on her size, a female dog will begin cycling around 6 months of age. As a result, many remote communities have issues with dog overpopulation. Dogs in these communities can form packs that are dangerous to people, result in dog bites and noise complaints. A sterilized dog is often more docile and less dominant, wanders less, and requires fewer calories (Root Kustritz 2012).

The type of sterilization procedure performed (ovariohysterectomy versus ovariectomy, flank versus midline approach), pre-operative preparation, and post-surgical care requirements will depend on community setting, time of year, dog lifestyle (indoor or outdoor), and the client's ability to provide post-operative care. Pre-operatively, clients should be clearly informed about the minimal post-operative requirements for the dog to ensure that a plan can be developed to ensure animal welfare. Preparation for surgery should take into account how much hair to clip to balance both the maintenance of aseptic technique and the protection of exposed skin from frostbite or flies, depending on the time of year. Minimizing the length of the abdominal incision and the length of the procedure through experienced surgical technique or selecting a different procedure (e.g., choosing to do an ovariectomy in a young healthy prepubertal dog) is particularly important if animals have a higher risk or minimal post-operative care. Post-operatively, there should be clear communication with the client about the care the dog needs for at least the first few days and to develop a plan that works for their living situation and the dog's lifestyle.

An alternative to surgical sterilization is the hormonal implant, deslorelin, which can stop a female dog from cycling for approximately 1 year. This procedure eliminates the invasiveness, equipment, and personnel intensiveness of surgery. While a promising technique, it is off-label use (the implant was originally developed to suppress fertility in male dogs), and there is variation in how female dogs react. For example, depending on where a dog is in its cycle, the implant may induce an estrus cycle prior to being effective (Lucas 2014; Marino et al. 2014). For communities where the turnover of dogs is high due to a high mortality rate (predation, vehicle accidents, or community culling events) or the cost or logistics of a full surgical team are prohibitive, the hormonal implant is an option that could help curb reproduction and build community support until surgical sterilization becomes feasible and acceptable.

The intermittent or annual nature of service provision requires adapting standard vaccination and deworming protocols that are quite intensive until about 16 weeks of age (Ford et al. 2017). Vaccination protocols require a full series of shots as puppies, again at 1 year old, and then most vaccines are labelled for every 3 years (Ford et al. 2017). Because of the variable survival of dogs, difficulty tracking individuals that change ownership, the infrequent or unpredictable frequency of clinics, and the high risk of spillover of rabies and other diseases from wild canids, annual vaccination against CDV, CPV, and rabies virus (the key viral diseases of concern in the region) (e.g., DA2PP) may be considered. Training of the owner or other willing community members on how to properly store, prepare, and inject the vaccine subcutaneously is an option for vaccinating against these viruses to ensure that puppies are given boosters at the appropriate intervals. This approach may not be broadly applicable for rabies due to laws that limit delivery of rabies vaccinations by veterinarians only. However, due to the remoteness of Arctic communities and the inaccessibility of veterinary care, both Canada and the USA have an exemption to allow for a trained rabies lay vaccinator to administer rabies vaccines (provided by the territorial or state government) in their community in the absence of a veterinarian (Middaugh and

Ritter 1982; GNWT 1996). Unfortunately, a consistent rabies lay vaccinator is not always available in each community.

5.4 Remote Veterinary Clinic Planning

Detailed planning for preventive veterinary clinics in remote Arctic communities is essential as veterinary supplies are not available locally and shipment of supplies from major centers may take several weeks. Travel to communities is often by plane, boat, or snow machine (Fig. 7), as is the case in the Yukon-Kuskokwim Delta, Alaska, or on winter roads (Fig. 8), as is the case in the Sahtu Settlement Area, NWT. This requires that equipment is carefully chosen and packed to withstand freezing temperatures and rough transport conditions, and that weight and size are minimized.

Flexibility with veterinary procedures is also important. Unreliable power sources and no access to running water are situations that may be encountered, requiring pre-planning for alternate methods to anesthetize and safely perform surgical procedures. The Hub Outreach Project in Alaska uses an injectable anesthetic protocol and brings pre-autoclaved surgery packs for surgery in the communities they serve. This eliminates the need to travel with an anesthetic machine and an autoclave. A portable oxygen concentrator has been used by the Sahtu program to provide inhalational anesthesia. In this case, a small oxygen tank is carried as a

Fig. 7 Arriving by small plane and continuing travel via snow machine to Nunapitchuk, Alaska. Photo Credit: Laurie Meythaler-Mullins

Fig. 8 Convoy of three trucks used to transport veterinary clinic equipment and personnel between the communities in the Sahtu Settlement Area, NWT. Photo Credit: Susan Kutz (left) and NCHR 2020 veterinary student participant (right)

backup in case of a power outage, and the team is prepared to use injectable anesthesia only if required. Small oxygen tanks are important regardless of anesthetic protocol, for ensuring patient safety and standards of care. Contributions by communities, such as tables, blankets for patient recovery, paper towels, and garbage bags, can cut down on packing. A typical clinic packing list is outlined in Table 1 and can be modified based on clinic location and needs.

Prior to arrival, every community must secure a location for the veterinary clinic. These may include the local community center, curling arena/sports complex, government lab, or school. In addition to each community, providing tables at the veterinary clinic location, wooden blocks, or other materials should be available to raise the tables to a proper surgical height, to save the surgeons' backs during long working days.

Upon arrival at the community mobile clinic location, the veterinary team evaluates the space and sets up designated areas as best as possible. These include Animal Intake, Examination/Appointment area, Anesthesia & Surgery Prep, Surgery, and Recovery/Discharge. Intake, anesthesia, surgery, and post-operative protocols, any required paperwork (examination forms, discharge instructions, etc.), and consistent client communication protocols should be prepared in advance. In general, patients present for wellness examinations, vaccinations, anti-parasitic treatments and sterilization surgeries, although other needs can be addressed. Physical examinations, vaccinations, and anti-parasitic treatments are completed in appointment area, and those that will be undergoing surgery are then moved to the anesthesia area where the pre-anesthetic assessment is completed. Sturdy but foldable kennels (necessary for traveling) are a great asset to hold the animals pre- and post-surgery, as kennels are often scarce in most communities. When surgical patients are able to walk on their own, they are discharged. Upon discharge, owners are given oral and written and pictorial discharge instructions that clearly outline what care an animal requires after surgery and what to watch for that might be of

Table 1 A typical remote veterinary clinic packing list

Animal intake and appointments	Anesthesia and surgery prep	Surgery	Emergency box	Other
Intake and consent forms	Injectable anesthetics and spare O_2 tank	Surgical caps and masks	Endotracheal tubes	First aid kit
Stethoscopes	Or: Anesthetic machine and gas	Disposable surgery gowns	IV fluids and lines	Euthanasia solution
Clipboards and pens	Oxygen concentrator	Sterile gloves	Catheters	Surgical instrument cleaning solution
Scale	Spare O_2 tank	Disposable surgery drapes	Emergency drugs	Dental kit
Tape	Premedicant drugs	Battery-operated head lamps with batteries	Handheld manual resuscitator	Bandaging materials
Non-sterile gloves	Medicine dosage charts	Autoclaved surgical packs		Surface cleaning solutions
Non-sterile gauze	Battery-operated SpO2 monitor with extra batteries	V-tray for patient placement		Otoscope
Sharps container	Doppler	Surgical blades and suture material		Ophthalmoscope
Syringes and needles	Ophthalmic lubricating solution	Formulary		Blood collection tubes
Lint rollers	Surgical prep scrubs and alcohol	Tissue adhesive		Cleaning supplies
Nail trimmers	Hydrogen peroxide	Tattoo gun and ink		Duct tape
Vaccinations (DA2PP, rabies)		Or: Microchip installer, microchips, and reader		Tool box
Anti-parasitic treatments		Hand sanitizer		± Portable autoclave

(continued)

Table 1 (continued)

Animal intake and appointments	Anesthesia and surgery prep	Surgery	Emergency box	Other
Battery-operated clippers, extra batteries		Or: Surgical hand scrub		
± Portable kennels for surgical recovery		Pain medications		
Animal handling equipment (e.g., leashes, collars, and muzzles and cat gloves)				
Medications (e.g., ear and eye medications, antibiotics, anti-inflammatories)				

concern. Language and literacy can be a barrier, so attention should be paid to ensuring that discharge instructions are understood. Translation of the discharge instructions into the local language may be appropriate in many cases.

6 Conclusions

While improved access to veterinary care will not solve all health disparities in remote northern communities, it may help improve well-being for people and animals in many ways. When provided respectfully, consistently, and over the long term, veterinary services can be well accepted by communities and result in a broad vaccination and anti-parasitic coverage and a more stable local dog population through increased sterilization. These changes can reduce the risk of rabies in dogs as spillover events from wildlife and, therefore, human exposure from dogs, reduce zoonotic parasite risk and environmental contamination from local dogs, and lead to fewer negative dog encounters as the dog population stabilizes. A more stable dog population with fewer roaming dogs can also improve the sense of safety in the community, making residents feel more comfortable getting outside for activities. Furthermore, a more stable dog population can change how the community deals with loose and roaming dogs (i.e., fewer culls), which can have positive implications for mental health.

These positive changes, however, can only occur with continued commitment to provide services over the long term. In addition, provision of veterinary services in remote communities can be resource intensive. Therefore, evaluations of programs to understand if they are meeting the stated program and stakeholder goals are necessary to sustain funding support, the commitment of personnel, and community buy-in. Engaging community members on multiple levels ensures program sustainability as residents participate in the planning and implementation of clinics, fundraising, and as program liaisons.

Lack of access to veterinary services is common in remote communities across the Circumpolar North, and the benefits of provision of these services on dog health and welfare and community health and well-being are clear. While a one-size-fits-all approach will likely not be successful when implementing programs in communities requesting services, important considerations include first assessing community needs and priorities, approaching program delivery with a no-pressure approach to build trust and understanding of the services being offered, and regular evaluations that involve the community to ensure that goals and expectations are being met. In this way, a sustainable community-specific dog management plan is developed with community buy-in and support.

References

Ameen C, Feuerborn TR, Brown SK et al (2019) Specialized sledge dogs accompanied Inuit dispersal across the north American Arctic. Proc Biol Sci 286:20191929

AMRRIC (2019) Delivering animal health & management programs in remote indigenous communities: an environmental health practitioner's guide

Baker T, Flaig J, Shillingford M et al (2018a) Ice road vets: perspectives on the role of veterinarians in northern community health. Can Vet J 59:668–672

Baker T, Kutz S, Toews L et al (2018b) Are we adequately evaluating subsidized veterinary services? A scoping review. Prev Vet Med 157:59–69

Baker T, Rock M, Brook R et al (2020) Indigenous community perspectives on dogs in northern Canada after 10 years of veterinary services indicated improved animal and human welfare. Prev Vet Med 181:105061

Beineke A, Baumgärtner W, Wohlsein P (2015) Cross-species transmission of canine distemper virus—an update. One Health 1:49–59

Berman M (2014) Suicide among young Alaska native men: community risk factors and alcohol control. Am J Public Health 104(Suppl 3):S329–S335

Bjork A, Holman RC, Callinan LS et al (2013) Dog bite injuries among American Indian and Alaska native children. J Pediatr 162:1270–1275

Bonesteel S (2006) Canada's relationship with Inuit: a history of policy and program development. Indian and Northern Affairs Canada, Toronto

Brook RK, Kutz SJ, Millins C et al (2010) Evaluation and delivery of domestic animal health services in remote communities in the Northwest Territories: a case study of status and needs. Can Vet J 51:1115–1122

Card C, Epp T, Lem M (2018) Exploring the social determinants of animal health. J Vet Med Educ 45:437–447

Castrodale L (2007) Hospitalizations resulting from dog bite injuries – Alaska, 1991-2002. Int J Circumpolar Health 66:320–327

Castrodale L (2013) Alaska rabies post-exposure prophylaxis: 2008–2012 data summary and policy change. State of Alaska Epidemioogy Bulletin

Catley A, Leyland T, Mariner JC et al (2004) Para-veterinary professionals and the development of quality, self-sustaining community-based services. Revue Scientifique et Technique-OIE 23: 225–252

CFSPH (2011) The Center for Food Security and Public Health (CFSPH): Echinococcosis. Available from http://www.cfsph.iastate.edu/Factsheets/pdfs/echinococcosis.pdf

Conboy G, Prada CF, Gilleard J, Jenkins E, Langelier K, Lee A, Peregrine A, Stevenson S, Stevenson B (2019). Canadian parasitology expert panel guidelines for the management of parasites in dogs and cats
Coppenger L (1977) The World of Sked dogs: from Siberia to sport racing. Howell Book House, New York, pp 19–30
Elmore SA, Lalonde LF, Samelius G et al (2013) Endoparasites in the feces of arctic foxes in a terrestrial ecosystem in Canada. Int J Parasitol Parasites Wildl 2:90–96
FDA (2020) Fish and fishery products hazards and controls guidance Chapter 5: Parasites. In: USFAD (ed.) Administration. USFAD, Washington
Filejski C (2016) The changing face of rabies in Canada. Canada Commun Dis Rep 42:118–120
Ford RB, Larson LJ, Schultz RD, Welborn LV (2017) 2017 AAHA Canine vaccination guidelines
GNWT (1996) Government of the Northwest Territories: rabies control in the Northwest Territories: community-based "lay vaccinator" program
GNWT (2018) Summary of NWT community statistics. Available from https://www.statsnwt.ca/community-data/index.html. Accessed 18 April 2018
GovYK (2016) Government of Yukon Socio-Economic Web Portal [Online]. Available from http://sewp.gov.yk.ca/home. Accessed 15 December 2020
Hueffer K, Murphy M (2018) Rabies in Alaska, from the past to an uncertain future. Int J Circumpolar Health 77:1475185
Jenkins EJ, Castrodale LJ, de Rosemond SJC et al (2013) Chapter two - tradition and transition: parasitic Zoonoses of people and animals in Alaska, northern Canada, and Greenland. In: Rollinson D (ed) Advances in parasitology. Academic Press, New York
Kaden U, Patterson PP, Healy J, Adams BL (2016) Stemming the revolving door: teacher retention and attrition in Arctic Alaska schools. Int Rev Educ 3:129–147
Kapel CM, Nansen P (1996) Gastrointestinal helminths of Arctic foxes (Alopex lagopus) from different bioclimatological regions in Greenland. J Parasitol 82:17–24
Krauss H, Weber A, Appel M, Enders B, Isenberg HD, Schiefer HG, Slenczka W, von Graevenitz A, Zahner H (2005) Parasitic zoonoses. ASM Press, Washington, DC
Krauss ME (1980) Alaska native languages [microform]: past, present, and future. Alaska Native Language Center Research Papers No. 4/Michael E. Krauss, Distributed by ERIC Clearinghouse, Washington, DC
Kutz SJ, Ducrocq J, Verocai GG et al (2012) Chapter 2 - Parasites in ungulates of Arctic North America and Greenland: a view of contemporary diversity, ecology, and impact in a world under change. In: Rollinson D, Hay SI (eds) Advances in parasitology. Academic Press, New York
Lucas X (2014) Clinical use of Deslorelin (GnRH agonist) in companion animals: a review. Reprod Domest Anim 49:64–71
Marino G, Rizzo S, Quartuccio M et al (2014) Deslorelin implants in pre-pubertal female dogs: short- and long-term effects on the genital tract. Reprod Domest Anim 49:297–301
Mech, LD, Luigi, B (2010) Wolves. In: Mech, LD & Luigi, B (eds.) Author index. University of Chicago Press, Chicago
Middaugh J, Ritter D (1982) A comprehensive rabies control program in Alaska. Am J Public Health 72:384–386
Pitulko VV, Kasparov AK (2017) Archaeological dogs from the early Holocene Zhokhov site in the eastern Siberian Arctic. J Archaeol Sci Rep 13:491–515
QTC (2013) Qimmiliriniq: Inuit Sled Dogs in Qikiqtaaluk. Qikiqtani Truth Commission Thematic Reports and Special Studies 1950–1975. Qikiqtani Inuit Association, Iqaluit, Nunavut
Reiling SJ, Dixon BR (2019) Toxoplasma gondii: how an Amazonian parasite became an Inuit health issue. Can Commun Dis Rep 45:183–190
Root Kustritz MV (2012) Effects of surgical sterilization on canine and feline health and on society. Reprod Domest Anim 47(Suppl 4):214–222
Salb AL, Barkema HW, Elkin BT et al (2008) Dogs as sources and sentinels of parasites in humans and wildlife, northern Canada. Emerg Infect Dis 14:60–63

Salisbury G (2003) The cruelest miles: the heroic story of dogs and men in a race against an epidemic. W.W. Norton & Co, New York

Salmi A-K, van den Berg M, Niinimäki S, Pelletier M (2021) Earliest archaeological evidence for domesticated reindeer economy among the Sámi of northeastern Fennoscandia AD 1300 onwards. J Anth Arch 62:01303

StateOfAlaska (2020) Department of Labor & Workforce Development: Research and Analysis [Online]. Available from https://live.laborstats.alaska.gov/. Accessed 15 December 2020

Steinel A, Parrish CR, Bloom ME et al (2001) Parvovirus infections in wild carnivores. J Wildl Dis 37:594–607

Wharton DA, Aalders O (2002) The response of Anisakis larvae to freezing. J Helminthol 76:363–368

WHO (2017) Determinants of health [Online]. Available from https://www.who.int/news-room/q-a-detail/determinants-of-health. Accessed 23 December 2020

WHO & OIE (2016) Global elimination of dog-mediated human rabies report of the Rabies Global Conference, Geneva, Switzerland, 10–11 December 2015 [Online]. Available from http://www.who.int/rabies/resources/who_htm_ntd_nzd_2016.02/en/. Accessed 21 April 2017

Woldstad TM, Dullen KN, Hundertmark KJ et al (2014) Restricted evaluation of Trichodectes canis (Phthiraptera: Trichodectidae) detection methods in Alaska gray wolves. Int J Parasitol Parasites Wildl 3:239–241

Semi-Domesticated Reindeer, Health, and Animal Welfare

Morten Tryland

1 Introduction

Reindeer and caribou (*Rangifer tarandus*), often called just *Rangifer*, are distributed throughout the northern Holarctic, a zoogeographical region comprised of the two continents that once were connected by the Bering land bridge and have closely related faunas. The *Rangifer* species belongs to the Cervidae (deer) family of ruminants.

The evolutionary history of *Rangifer* has been highly influenced by glacial and interglacial conditions (Lorenzen et al. 2011; Yannic et al. 2014). At the time of the last glacial maximum (25000–19,000 years before present), *Rangifer* existed south of the ice cover in North America, in Eurasia, and in Beringia, encompassing most of Siberia and the Bering land bridge including parts of Yukon and Alaska (Yannic et al. 2014; Røed et al. 2014). Due to genetic diversity, this species has been able to adapt to local environments and to evolve, thus forming the different *Rangifer* ecotypes and subspecies present today.

The use of two vernacular names, Caribou for wild animals in North America and Reindeer for wild and semi-domesticated animals in Eurasia, has contributed to confusion regarding the number of subspecies. However, it is clear that some major ecotypes exist and that these are adapted to their environmental conditions, namely, the Arctic or High Arctic type (e.g., *R. t. platyrhynchus* and *R. t. pearyi*), the mountain type (e.g., *R. t. caribou*), the migratory barren ground or tundra type (e.g., *R. t. granti*, *R. t. tarandus*), and the boreal forest or woodland type (e.g., *R. t. fennicus*) (Røed et al. 2019).

M. Tryland (✉)
Department of Arctic and Marine Biology, UiT The Arctic University of Tromsø, Tromsø, Norway

Department of Forestry and Wildlife Management, Inland Norway University of Applied Sciences, Koppang, Norway
e-mail: morten.tryland@inn.no

M. Tryland (ed.), *Arctic One Health*, https://doi.org/10.1007/978-3-030-87853-5_25

For tens of thousands of years, *Rangifer* has been an important resource for people in the northern regions of the world, and this animal has had a major impact on human migrations and settlements, both in Eurasia and Northern America. The majority of *Rangifer* in North America are wild and, in total, are approximately 2–3,000,000 animals (Røed et al. 2019). However, these populations have been declining the past decades (Vors and Boyce 2009; Solveig and Boyce 2009). Most of the 3–4,000,000 wild and semi-domesticated reindeer in Eurasia belong to the Eurasian tundra reindeer subspecies, *Rangifer tarandus tarandus* (Røed et al. 2019), which is also the basis for the semi-domesticated reindeer and the Sámi reindeer herding in Fennoscandia, that is, Norway, Sweden, and Finland, as well as the Kola Peninsula and Karelia in Russia. Genetic characterization of the different subspecies has suggested that this subspecies originated from a glacial refugium in southern Europe and not from the Beringia refugium (Røed et al. 2019). When the ice retreated approximately 10–12,000 years ago, reindeer increased in their northern distribution. To illustrate, there is evidence of the existence of reindeer in Denmark dated approximately 14,700 years ago (Aaris-Sørensen et al. 2007), whereas in Norway, the oldest remnants of reindeer are estimated to be approximately 14,000 years old (Lie 1986).

2 Semi-Domesticated Reindeer and Reindeer Herding

Archeological evidence, as found in Combe Grenal and Vergisson caves in France, have suggested that reindeer was hunted at least 45,000 years ago. Reindeer have constituted the main game species from many peoples in different geographical regions, as indicated by remains of bones, tools, hunting pits, and fences for wild reindeer hunting (Nieminen 2019). The close tie between reindeer and man has also been reflected in the traditions, culture, and religion of many peoples. Although exactly when domestication of reindeer occurred is disputed, several theories exist. Domesticated animals were used as decoys when hunting wild reindeer and later as draught and milking animals. Later, people held larger numbers of reindeer as semi-domesticated herds. The domestication of reindeer has taken place already some thousand years ago, probably originating at different times and in different geographical regions (Nieminen 2019; Istomin and Dwyer 2010; Bjørklund 2013).

Reindeer represent not only food but also fur and warm clothes and other valuable materials, and its importance is thereby reflected among many peoples and their cultures. Although reindeer herding originated in Eurasia, it is today practiced in 10 countries and by about 30 different ethnic and arctic peoples, engaging about 100,000 people in different kinds of reindeer husbandry (Nieminen 2019). Reindeer herding has its core distribution in northern parts of Russia (from west to east) as well as in Fennoscandia. In addition, there are smaller reindeer herding units in Alaska (USA), China, Mongolia, and Greenland. Russia is hosting about 1.6 million semi-domesticated reindeer, which is approximately 70% of the semi-domesticated reindeer in the world.

It has been assumed that a nuclear family was able to herd and manage a herd size up to 30–40 animals, and up to the nineteenth century, herd size in the Sámi reindeer herding was usually below 100 animals (Bjørklund 2013). Reindeer herding was conducted in combination with fishing, hunting, and trading (Hansen 2008) and was organized in Sámi herding siida consisting of several households sharing the work with the herd (Bjørklund 2013). The tundra type of reindeer herding, characterized by long seasonal migrations of reindeer, became the dominant type of herding in Norway and Sweden and the northern part of Finland. However, a coastal type (local seasonal migrations; Norway) and a taiga type (small scale in forested areas; Finland and Sweden) also exist (Riseth et al. 2019). Reindeer herding was originally performed on a small scale only producing meat and milk for subsistence. However, during the first half of the twentieth century, a transition to a more extensive reindeer herding occurred and included larger herds and herders establishing permanent settlements in houses. During the second half of the twentieth century, increased motorization, such as the use of snowmobiles and all-terrain vehicles for herding, helicopter for gathering animals, and trucks and barges for transport, took place. The reindeer herding switched from production for subsistence to production for profit.

Reindeer herding in Fennoscandia today is based on the semi-domesticated reindeer. Every semi-domesticated reindeer is owned and is identified (tagged) by the owner. Although these animals are not entirely wild, the degree of domestication varies, and they are seldom referred to as tame (Skarin and Åhman 2014). In 2019–2020 (after slaughter, before calving) about 214,000 semi-domesticated reindeer existed in Norway, 240,000 in Sweden and about 195,000 in Finland (Anonymous 2020) (Fig. 1).

3 Historical Reindeer Epizootics

Early historical sources from the eighteenth and nineteenth century from Norway and Sweden have documented that infectious diseases could be devastating to both reindeer herds and the small vulnerable communities that were dependent on them for subsistence. Knowledge of infectious diseases, what caused them and how they spread, was scarce, as were the ideas of prophylaxis and treatment. From some of the sources and seen in the light of the knowledge and methods we have today for characterizing diseases and infectious agents, it is not always evident which infectious disease they were reporting.

One example is the "reindeer pest." This disease was reported from Jokkmokk, Sweden, in 1896. It was highly contagious, had a high mortality, and caused acute or peracute disease. Reindeer were observed with an unstable walk and lagged behind the herd. Animals were found dead with foul-smelling mucopurulent discharge from the mouth and nose and having dark, thick blood with gas bubbles. Further, subcutaneous gas bubbles and enlarged liver and spleen were observed (Nordkvist 1960). Several of the old epizootics are called "reindeer pest," and a clostridial bacterium (*Clostridium septicum*) has been suggested as the causative agent. However, the documentation is rather scarce, and it is likely that different diseases and

Fig. 1 A. Semi-domesticated reindeer are gathered in roundups only a few times a year. In the working fence, herders can inspect animals closely and conduct antiparasitic treatment. (Photo: Morten Tryland)

infectious agents were classified as pests among reindeer in these old reports (Josefsen et al. 2019).

On the other hand, different clinical appearances caused by the same infectious agent could sometimes be described as different diseases and epizootics. The bacterium *Fusobacterium necrophorum* is one example. This bacterium is found in the alimentary tract of many ruminant species, including reindeer (Aagnes et al. 1995). Through abrasions in the skin, it may cause severe infections and swellings, usually on the distal parts of the feet of reindeer. This disease is called *slubbo* in the Sámi language, meaning "club," and could be highly transmissible. Slubbo is not currently reported in the semi-domesticated reindeer herds but was recently reported as outbreaks among wild reindeer (Handeland et al. 2010).

Anders Eira, as referred to in Qvigstad's report on reindeer diseases (Qvigstad 1941), explained:

> If a reindeer has slubbo and is licking the lesions, it will get ruodno and dies (Translated from Swedish, from Qvigstad 1941)

While slubbo today is called digital necrobacillosis, or sometimes "foot rot," the terms "ruodno" and "ruodnovihki" are called alimentary necrobacillosis. The latter is caused by the same pathogen but instead causes lesions in the mouth and digestive tract. Thus, when necrobacillosis was identified in the mouth, the disease could also be called "njalbmevihki," meaning disease in the mouth. However, it seems like the digital form, often associated with keeping reindeer in corrals for milking, represented the form of the disease that caused the largest and most severe disease outbreaks.

Anthrax, caused by the bacterium *Bacillus anthracis*, was described by Nils Gissler in 1759:

> Ten years ago, a disease appeared among the reindeer in Jukkasjärvi, that was called redock-tauta or redock-maine by the Sámi people. The disease spread to all the herds in Torne Lappmark in Northern Sweden. The disease could kill two-thirds of the herds. It caused poverty and within four years, more than 100 persons had left the village, most of them migrating to Norway to live from the fisheries (Extracted sentences translated from Swedish, from Qvigstad 1941).

This description also indicates how important reindeer herding was for these northern Sámi communities. The disease described is similar to what is observed during anthrax outbreaks today (see chapter "Anthrax in the North").

Infectious keratoconjunctivitis (IKC), called *Čalbmivihki* in the Sámi language and meaning eye disease, is a transmissible eye infection that can cause uni- or bilateral blindness. It has been shown that the reindeer alphaherpesvirus (Cervid herpesvirus 2, CvHV2) can cause the disease (Tryland et al. 2009, 2017) by producing mucosal lesions in the eyes, which may rapidly be colonized by pathogenic or opportunistic bacteria. Bacteriological cultivation has revealed a variety of bacteria but no firm conclusion with regard to being causative agents of this disease (Sánchez Romano et al. 2018).

Professor Arvid M. Bergman (1872–1923) reported some serious outbreaks among reindeer herds in Sweden during the fall in 1909 and in the spring and fall 1910. Clinical signs started with increased lacrimation and then mucopurulent secretions in the medial eye angle, followed by corneal edema and a whitish appearance of the cornea—the disease is also called "kvitøye," which means white eye (Norwegian). The disease can heal itself from this stage or progress to panophthalmitis, a stage involving the several structures of the inner eye and the orbit, leading to permanent blindness. Bacteriological cultivation revealed a variety of bacteria and no firm conclusion with regard to the causative agent was drawn. Bergman reported from two herds in Arjeplog, Sweden:

> In one of the herds, about 90% of the calves were affected, 20% of them in both eyes, and about 40% of all the calves were lost. Only 10% of all the calves of the year remained healthy. In the other herd, 58 calves were affected. The disease spread to other herds, and the herders tried to manage the outbreak by slaughtering the sick animals and dispersing the herd on the mountain pasture. The disease appeared in calves only and was spread from calf to calf, but in 1910 also in older reindeer. (Extracted sentences translated from Swedish; Bergman 1912.)

4 Health and Diseases Today

In contrast to the historical reindeer epizootics discussed above, which could affect hundreds and sometimes thousands of animals and spread to many herds, disease outbreaks among reindeer today seldom affect so many animals. One exception, however, is anthrax. In 2016 on the Yamal Peninsula in Russia, anthrax caused the death of about 2000 reindeer and led to the culling of about 200,000 animals (Hueffer et al. 2020) (see chapter "Anthrax in the North" for further details).

4.1 The Loss of Reindeer

During the reindeer herding year 2019–2020, the annual loss of reindeer in Norway was reported to be 38% for calves and 10% for adult animals (Tveraa and Kulseng 2020). Starvation, predation by large carnivores, extreme weather events, and accidents are identified as the main causes of reindeer loss (Nybakk et al. 2002; Tveraa et al. 2014).

Of the annual reported losses, reindeer herders report the majority of loss (96% in calves and 86% in adults) from predation (Anonymous 2020). Semi-domesticated reindeer are found on about 40% of the land areas of Norway, Sweden, and Finland, and predator type and density vary largely between regions. In Norway and Sweden, the wolverine (*Gulo gulo*) and the Eurasian lynx (*Lynx lynx*) are regarded as the most important predators of reindeer, and conversely, reindeer represent the most important prey for these predators in many areas. In contrast, the numbers of brown bears (*Ursus arctos*) and wolfs (*Canis lupus*) are scarce; brown bears are present in some reindeer herding districts and may be predating on reindeer calves, whereas the wolfs for the most part are distributed in regions with little or no reindeer herding. However, eagles, in particular the white-tailed eagle (*Haliaeetus albicilla*) but also the golden eagle (*Aquila chrysaetos*) and the red fox (*Vulpes vulpes*), are quite numerous and widely distributed and may prey on reindeer, in particular calves during their first weeks of life (Nieminen et al. 2011; Skonhoft et al. 2017).

Except for the red fox, these predator species and populations are for the most part protected in Fennoscandia, a condition which has contributed to higher predator populations and thus more interactions with the reindeer herding (Hobbs et al. 2012). To be compensated for reindeer loss due to predators (protected under law), herders must collect proof that a dead reindeer has in fact been killed by a predator. Because predators often hide the remnants of their prey and because these remnants also quickly attract scavengers, such documentation is often difficult.

Although recent research on the Norwegian herds of semi-domesticated reindeer confirms that many reindeer are taken by predators, it also emphasizes the importance of the weight of the animals. Light-weighted females may die, abort their calves, or lose their calves during the first weeks or months of life, whereas heavier females are more likely to survive and produce a fit calf that can survive the winter (Tveraa et al. 2014). However, these mechanisms are complicated and may vary geographically and over the years, and the role of predators and pasture resources on the loss of reindeer are disputed (Tveraa et al. 2014; Skonhoft et al. 2017; Eilertsen and Riseth 2021).

From the annual statistics for Norway (Anonymous 2020), only 2% of the calves and 5% of adult reindeer were lost due to unknown causes. Many different factors and complex interactions may be hidden in these figures. Although the epizootics described in historical sources do not take place anymore, health and diseases, beyond what can be explained by predators and starvation, are definitely involved. However, in the same way as for confirming the loss of animals due to predators, it is challenging to document losses due to health and diseases in reindeer. Veterinarians and management bodies are used to become involved and take action when large disease outbreaks appear, particularly those that can cause severe loss and spread to other herds or to livestock and humans. These diseases are therefore also diagnosed and appear in the statistics. However, there is little focus on conditions and infections that have less obvious impacts, such as those causing reduced fitness or poor reproductivity. If these subtle conditions affect a large number of animals, they may actually have much more impact on the population and the herding industry over time than the regular and identified disease outbreaks.

4.2 Loss of Pastureland, Climate Change, and Reindeer Feeding

One of the most critical challenges for reindeer herding today is fragmentation and loss of pastureland, which with few exceptions is a non-reversible process (see chapter "Loss of Untouched Land"). Modern forest management eliminates lichen resources that are an essential feed for reindeer and has thus impacted reindeer herding, especially in Sweden and Finland (Kivinen et al. 2010, 2012; Horstkotte et al. 2020). Furthermore, the establishment of roads, cabins, windmills, and mining is increasing, which impacts reindeer habitat use and migration (Skarin et al. 2015; Vors and Boyce 2009; Riseth and Johansen 2019). Another major challenge, today and in future generations, is the climatic changes, which are expected to impact the Arctic and northern regions to a larger extent than other regions of the world (see chapter "Climate Change in Northern Regions"). Higher winter temperatures and more frequent temperature fluctuations combined with increased precipitation as rain during winter will create ice-covered winter pastures for reindeer (Moen 2008; Forbes et al. 2016; Box et al. 2019). These changes in climate during winter may also cause lakes and rivers not to freeze over and thus affect reindeer migration routes in spring and autumn. The forecasted future changes from climate change have contributed to stress and anxiety among reindeer herders, some fearing that they represent the last generation practicing reindeer herding (Furberg et al. 2011).

Starvation may be an important cause of loss of reindeer in certain regions, especially in snow-rich winters and in winters with unfavorable weather conditions and ice-locked pastures. Feeding of reindeer was traditionally conducted when animals were held for draught and milking or if the herd was corralled over a number of days. However, during the last decades, knowledge on how to successfully feed reindeer has increased, along with the availability of suitable reindeer feed (Åhman et al. 2019).

Fig. 2 Supplementary feeding has become more common. Pelleted feed designed for reindeer can be dispersed directly on the snow. (Photo: Morten Tryland)

Dry hay is rarely used as reindeer feed today, but silage bales and pelleted feed, the latter adapted to the digestive system of reindeer and their needs, are common, as well as different species of lichens, when available. In 2020, approximately 3600 tons of reindeer pelleted feed were sold in Norway (Felleskjøpet, Norway) increasing from less than 1000 tons before 2016 (Fig. 2).

This indicates that feeding has become an important measure to avoid starving during challenging winter conditions. Feeding during late winter and early spring may be especially relevant for pregnant females at a time when energy reserves are diminishing (Turunen and Vuojala-Magga 2014; Horstkotte et al. 2020). Feeding is also practiced in some herds that are corralled during calving, to prevent or restrict heavy predator losses of newborn calves. Feeding may also be used to keep the reindeer herd more gathered and controlled over a period of time.

Feeding reindeer has undoubtedly saved many animals from starvation and emaciation. With the right knowledge and experience, suitable feed, and a good feeding regime, reindeer may be fed full rations over longer periods of time. This strategy is increasingly being used as a supplement to natural pasture resources (Horstkotte et al. 2020). In addition, economic compensation for feeding is offered to herders when reindeer pastures are exploited for other purposes, such as constructions of windmill parks, roads, and other infrastructure.

However, feeding may also have other consequences. Feeding animals over time may increase their tameness. While this makes them easier to handle, they may also start to approach snowmobiles and other motorized vehicles as well as humans as they associate all these things with food. This can make them less alert and could increase conflicts with people, dogs, and settlements and increase the number of animals hit by car or train. Long-term feeding may also change their foraging behavior. In Finland, a high degree of permanent feeding has been adopted (Turunen and Vuojala-Magga 2014). An animal that early in life has learned that food is always easily available may keep waiting for the feed to appear instead of searching and digging in the snow for natural forage. In addition, the survival strategies of large ungulates involve a following behavior (Fisher et al. 2002), whereby the offspring follow the mother, often during the whole first year of life, and during this time, they learn foraging techniques and migration patterns. Post-weaning care is important for reindeer calf survival during their first winter when food sources are patchy and covered with snow (Holand et al. 2012). In contrast to reindeer bulls, female reindeer keep their antlers throughout winter and defend feeding craters for themselves and their calves. In a natural foraging system, the calf will learn from its mother how to find and defend forage. Long-term feeding may thus have severe effects on the calves' future foraging performance, and feeding may also change or even eliminate the reindeer' spatial foraging and migration patterns.

Herders across Fennoscandia also fear that an increased use of supplementary feeding will erode traditional and experience-based knowledge, regarding for instance how to use and protect winter pastures in the best way (Horstkotte et al. 2020). Herders have across generations gained knowledge about the interactions and dynamics between the landscape, animals, and people. With less time spent in the landscape and more time devoted to feeding the herd, there may be less time to focus on herding culture and knowledge, reducing the coming generation's ability to use these pastoral landscapes.

4.3 Digestive Disorders Associated with Feeding

The *Rangifer* subspecies has adapted to many different types of forage throughout their distribution including grasses, graminoids, lichens, mosses, willows, and shrubs (Åhman and White 2019). However, when fed with supplemental feed, the forage may be very different from natural forage, and the availability and intake may be restricted to only a few short periods during the day, which may impact the rumen microbiota and the digestion. The animal needs time to adapt to a new diet and "forage strategy" to handle the transition. Therefore, when trying to save starved reindeer by feeding, it is necessary to start feeding before the body condition is too poor to allow time for the animal and the rumen microbiota to adapt to and utilize the nutrients in the feed.

Sometimes, poor quality feed or quick changes may cause health problems associated with dysfunction of the digestive system. For instance, diarrhea is one condition in reindeer that is almost exclusively associated with feeding, mostly through the use of reindeer pelleted feed, but also to some extent when feeding with ensilage (Åhman et al. 2019).

Ruminal acidosis, which may also cause diarrhea, is a result of high intake, and sometimes too rapid increase, of easily digestible carbohydrates so that the rumen microbiota is not able to adapt fast enough. This condition favors the growth of lactobacilli (*Lactobacillus* spp.). A rumen microbiota dominated by these bacteria causes rumen digestion to stop because the content has become too acidic, dropping from a normal pH of 6–7 to a pH of 4–5. Acid is absorbed from the rumen and may exceed the buffering capacity of the blood, leading to acidosis with a poor prognosis (Åhman et al. 2019).

When animals are fed with fiber-rich hay or silage, this food material can accumulate in the rumen and stop or slow the digestion process. Still, the appetite of the reindeer remains, and the animal continues to fill the rumen with fodder. Digestion is, however, poor and will not be able to provide the animal with energy and nutrients. The animal enters a negative energy balance and may face emaciation if the feed is not changed. This condition illustrates that reindeer are sensitive to rough forage with a high fiber content (Åhman et al. 2019).

A condition called "wet belly" has also been observed associated with feeding. The reindeer fur, typically in the axilla and lower parts of thorax and abdomen and on the legs, becomes wet. The reason for wet belly is unknown, and it has not been associated with any specific types of feed (Åhman et al. 2019).

4.4 Some Infectious Diseases in Semi-Domesticated Reindeer

Feeding reindeer full rations during longer periods of time creates increased animal-to-animal contact and challenging hygienic conditions, which may facilitate pathogen transmission and disease outbreaks (Tryland et al. 2001, 2019a; Sànchez Romano et al. 2019).

Contagious ecthyma (CE) is a disease caused by orf virus (ORFV), which is a zoonotic parapoxvirus within the family *Poxviridae*. The virus has a worldwide distribution, affecting primarily sheep and goats, but also many wild ruminant species and people. In reindeer, ORFV causes the disease contagious ecthyma (CE). In Norway, CE was first documented in Hattfjelldal in 1999, among reindeer corralled in a fence previously used for sheep, and several animals died (Tryland et al. 2001). Whereas there have been a few outbreaks in Sweden (Tryland et al. 2019b), severe outbreaks have been reported in Finnish reindeer herds since the winter of 1992–1993, when about 2800 animals were affected and 400 died (Büttner et al. 1995). In Finland, pseudocowpox virus, a close relative to ORFV, has also caused contagious ecthyma in reindeer (Tikkanen et al. 2004). While ORFV may only produce a single or only a few proliferative nodules that may not significantly affect the animal, these nodules may also develop on the gingiva and the palate, making it difficult to feed and causing the animal to die from starvation. ORFV lesions in the mucosa may also become the port of entry for other pathogens, such as *Fusobacterium necrophorum*.

The bacterium *Fusobacterium necrophorum* is a part of the normal reindeer rumen microbiota (Aagnes et al. 1995). The bacterium is thus normally present in the environment but needs a lesion or abrasion of the skin or mucosal membrane to be able to start an infection (Josefsen et al. 2019). As mentioned above, the digital form ("slubbo") was associated with reindeer in corrals but disappeared when the practice of milking reindeer stopped around the 1950s. In contrast, disease outbreaks with the alimentary form ("njalbmevihki") seem to be increasing, as reported from Sweden among animals that have been corralled for supplementary feeding (Tryland et al. 2019b). As with parapoxvirus infections and contagious ecthyma, severe outbreaks of alimentary necrobacillosis may be associated with secondary infections and death or euthanasia. Oral necrobacillosis is regarded as one of the most common diseases in corralled reindeer in Finland (Laaksonen 2019).

Outbreaks of the transmissible eye infections (IKC) discussed above still appear in Fennoscandia today and appear in a similar way as described from the 1909 outbreak in Sweden (Sánchez Romano et al. 2018, 2019). The virus CvHV2 is enzootic in the Fennoscandian reindeer herds. As with other herpesviruses, CvHV2 may establish latency, a dormant stage of the virus. During latency, the virus does not produce any clinical signs but can be reactivated upon stress (i.e., immunosuppression) and other stimuli. When active, the virus causes clinical lesions very similar to the herpes simplex virus, which causes recurrent cold sores in humans. When the reindeer herpes virus is spread to immunologically naïve calves, it replicates in the cell lining of cornea and conjunctiva in the eyes and can be followed by secondary bacterial infections of many kinds (e.g., *Moraxella bovoculi*, *Chlamydia pecorum*) (Sànchez Romano et al. 2019). In fact, IKC outbreaks are most often seen in young animals (calves) that have been gathered, transported, and placed in a new environment. As these calves are assumed to be exposed to the virus for the first time, it can take 8–10 days to mount an antibody response (Das Neves et al. 2009). In the meantime, the disease is developing in one or both eyes and sometimes progressing to involve several inner structures of the eye (e.g., panophthalmitis) and permanent blindness. Infectious keratoconjunctivitis is a painful disease that represents an animal welfare issue and an economical loss for the herder. Treatment consists of cleaning the eye and the surrounding area with sterile saline and the use of an antibiotic eye ointment against bacterial infections.

The bacterium *Pasteurella multocida* is an opportunistic pathogen, present in the upper respiratory tract (e.g., tonsils) of many animal species. It sometimes causes pasteurellosis, a disease which is characterized by septicemia and severe pneumonia. Outbreaks of pasteurellosis causing the death of more than 1000 reindeer have been reported (Josefsen et al. 2019), outbreaks being associated with stress factors such as insect harassment, heat, and drought (Nordkvist and Karlsson 1962; Dwyer and Istomin 2006). The disease has also been associated with the presence of lungworms (*Dictyocaulus* sp.) and the throat bot fly (*Cephenemyia trompe*). Pasteurellosis is most severe in calves and animals in poor condition (Josefsen et al. 2019).

4.5 Parasites in Semi-Domesticated Reindeer

More than 45 species of helminth, protozoal, and arthropod parasites have been reported in wild and semi-domesticated reindeer (Kutz et al. 2019). Parasites may impact reproduction, survival, behavior, and overall population dynamics of a host population. In a natural population, parasites tend to be unevenly distributed with few animals hosting the majority of the parasites. Most of the animals will experience few to no effects; however, individuals with a high parasite burden likely experience reduced fitness and even elimination from the population, particularly when parasite effects are combined with other factors such as predation, stress, and starvation. In semi-domesticated reindeer herds, animals may be sheltered from many of these challenges, particularly as seasonal roundups and physical handling of the animals enables antiparasitic treatment. Consistent parasite treatment over time may contribute to parasites becoming more evenly distributed in the population as compared to a wild population (Kutz et al. 2019), which will reduce the loss of heavily infected individuals but may instead create a herd with constant, subtle effects from parasitism.

4.5.1 Ectoparasites

Ectoparasites are parasites that live on or in the skin of their host. The most eye-catching reindeer parasite in the Sámi reindeer herding is the larval stage of the warble fly or reindeer bot fly, *Hypoderma tarandi* (Diptera, Oestridae), as it grows big under the skin and prepares to leave the host in the spring (Fig. 3).

At that time, the larvae are about 2.5–3 cm long, weigh approximately 1.5 grams each, and may count from a few to many hundreds, thus representing a considerable

Fig. 3 (**a**) The warble fly (*Hypoderma tarandi*) lays eggs in the haircoat of reindeer during summer. Upon hatching, the larvae penetrate the skin and live under the skin during the winter. (**b**, **c**) In late May–early June, the larvae, now about 2.5–3 cm long, make a hole in the skin, which they use to leave their host. (Photo: **a** Arne C. Nilssen, **b**, **c** Morten Tryland)

biomass. At this stage, the larvae have established a breathing hole through the skin and start the process of exiting the host. The yellowish, soft larvae drop to the ground and become dark brown pupae, which then develop into warble flies in July. The animals are free of larvae during only a short period of time from late May to July. A new generation of bot flies deposit their eggs on the haircoat of the reindeer in July–August. After these eggs hatch, the larvae penetrate the skin and establish themselves subcutaneously over the back and sides of the animal where they overwinter.

The throat bot fly, or nose bot fly, *Cephenemyia trompe*, is also a very common finding in semi-domesticated reindeer. Both the warble fly and the throat bot fly appear very similar to each other as adults (15–18 mm long) and resemble bumblebees with a gold and black haircoat. The life cycles are also similar, but instead of depositing eggs in the haircoat like the warble fly, the female throat bot fly sprays live larvae on the nose and mouth of the reindeer. The larvae migrate into the nasal passages and live in the nasal sinuses over the winter before migrating to the pharynx and attaching with hooks. Although larvae of different sizes may be found in the same animal, most larvae grow to around 4 cm in length by the late spring. In the period late April to June–July, they are ejected from the host. This process is often associated with transient coughing and sneezing that may be confused with infections in the upper respiratory tract or pneumonia. The larvae bury themselves in the soil, become pupa, and develop to flies after 4–5 weeks depending on the temperature (Kutz et al. 2019).

These two types of bot flies represent a significant cost to reindeer herding. Fly harassment stresses the reindeer and causes continuous restlessness and fleeing behavior. This has a negative impact on grazing and resting time and thus the animals' ability to build winter reserves. In high numbers, the larvae themselves create a considerable energy and nutrition demand on the host. In addition, the breathing holes of the warble fly larvae may cost the reindeer heat loss, and the open holes are also at risk of infection.

The deer ked, *Lipoptena cervi* (Diptera, Hippoboscidae), has expanded its geographic range in Fennoscandia and is establishing itself as a new ectoparasite in semi-domesticated reindeer and other cervid species (Välimäki et al. 2010; Madslien et al. 2012). Adult flies develop from pupae on the ground in late summer and autumn. When they find a new host, the flies detach their wings to live in the haircoat of the new host through the winter and to feed on blood and interstitial fluids (Kutz et al. 2019). The fly is viviparous and deposits larvae that, after becoming pupae, drop to the ground to develop further into adult flies. In the southern reindeer herding regions of Finland, the deer ked has become a common finding, sometimes numbering many hundreds on one animal. Similar to the effects in moose, hair loss, heat loss, and energy costs are likely effects of heavy infestations (Madslien et al. 2011).

Several species of ticks are associated with different *Rangifer* subspecies. In Fennoscandia, new tick species such as the taiga tick (*Ixodes persulcatus*), are emerging and are expanding their distribution ranges (Kjær et al. 2019), although *Ixodes ricinus* remains as the most common tick in this region. Having a distribution mostly in southern and coastal regions, *I. ricinus* has not been associated with reindeer to a large extent historically. However, the distribution of this tick species is changing. The increased presence in northern regions and from the coast to inland ecosystems are likely due to higher winter temperatures and increased deciduous woodland, which favor the survival of the ticks (Medlock et al. 2013; Jaenson et al. 2012; Hvidsten et al. 2015; van Oort et al. 2020). Thus, *I. ricinus* has also emerged in the semi-domesticated reindeer herds in some areas (Tryland et al. 2019c). *Ixodes ricinus* has been identified as the likely vector of two bacterial infections, which affect both humans and animals, namely, *Borrelia burgdorferi* s.l., which causes borreliosis (Lyme's disease) and *Anaplasma phagocytophilum*, which causes anaplasmosis (tick-borne fever). These ticks may also cause these and other diseases in reindeer (Stuen et al. 2013; Tryland et al. 2019c).

4.5.2 Endoparasites

Endoparasites live inside their host, and reindeer can be a host for many different species. Endoparasites are associated with most organ systems of reindeer, such as the respiratory tract (nematodes and cestodes), the gastrointestinal tract (nematodes, trematodes, and protozoa), the peritoneal cavity (nematodes and cestodes) and liver (trematodes and cestodes), the heart and skeletal muscles (protozoa and cestodes), the central nervous system (CNS, nematodes and protozoa), and other tissues (Kutz et al. 2019). However, it is beyond the scope of this chapter to present or discuss all of these parasites in detail. Because reindeer herding is for the most part based on the utilization of natural pastures and because reindeer are usually gathered only a few times during the year, the observation and diagnosis of clinical aspects of infections, including parasites, are challenging. Heavy parasite burdens can be observed in some individuals. This is presumably a result of high infection pressure, lack of immunity (particularly calves), poor nutritional status and body condition, and/or other factors making the animal more prone to suffer from parasites. Based on the diversity of parasite species that are commonly found in reindeer, it is obvious that these parasites have evolved as integrated parts of the life of their host. It has, in fact, been shown that a high richness in parasite diversity is associated with a reduced disease risk (Morand 2015). Furthermore, parasite richness has been positively correlated with host density (Waller 2005).

An endoparasite that has received increased interest due to climatic changes is the filaroid nematode *Setaria tundra*. The adult worms (males 35 mm long, females 67 mm long) inhabit the abdominal cavity of reindeer. Clinical signs include peritonitis, ascites, fibrin deposits and adhesions, and findings of live worms at slaughter. Animals with a high intensity of infection may have reduced body condition. These findings may result in carcasses being condemned for human consumption. The adult worms produce microfilaria that are picked up by blood-sucking mosquitos (*Aedes* spp. and to some extent *Anopheles* spp.). Although the development of larvae to the third developmental molt stage (L3 larvae, infective stage) is halted at 14 °C, it only takes 2 weeks to develop at 21 °C (Laaksonen et al. 2009). Thus, warmer summers due to climate change may promote mosquito populations and the development and transmission of this parasite and its clinical impact in reindeer (Laaksonen et al. 2009).

Another parasite linked to climate change is the nematode *Elaphostrongylus rangiferi* (brain worm, meningeal worm). This nematode has an indirect life cycle with snails and slugs as its intermediate host. When reindeer ingest the intermediate host through grazing, they also ingest the infective stage (L3 larvae). The larvae migrate from the gastrointestinal system of the reindeer to the CNS where they develop to adult threadlike brownish worms, 3–7 cm long, on the subdural and epidural surfaces. During their development here, the larvae may cause inflammation, which generates neurological signs, such as weakness, poor coordination, hindlimb paresis, and ataxia. The animals may not be able to stand or may sit like a dog and then rise up as if normal again but with relapsing poor locomotion. Animals may also stand with a bent neck, fixed to one side, and exhibit circling behavior. Lying animals may also display a kicking and circling action, leaving characteristic traces on the ground or in the snow. Healing may occur but may take months, and affected animals are often lost or euthanized (Kutz et al. 2019). The larval stages outside the reindeer are temperature dependent, and cooler temperatures (below 8–10 °C) halt or slow development (Josefsen et al. 2019). Climatic conditions thus influence the survival of the larvae, their development within the intermediation host and the survival of the gastropod host itself. Therefore, warm summers may be followed by an increased disease prevalence the following winter (Davidson et al. 2020). *Elaphostrongylus rangiferi* larvae may also cause considerable pathology in the lungs (Davidson et al. 2020). Even when such changes are compensated for and produce no obvious clinical signs, they may affect respiration and represent a complicating factor during chemical immobilization procedures (Tryland et al. 2021) or other types of stressful situations for the animals.

The impact of endoparasites may also be affected by increased feeding of semi-domesticated reindeer (Davidson et al. 2020). Thus, climate change, which is often mitigated by supplementary feeding, may indirectly impact the parasite burden of the animals.

4.5.3 Antiparasitic Treatment

Antiparasitic treatment will always be a cost-benefit calculation and should take into consideration not only economical aspects but also animal welfare aspects, such as the cost for the animal to live with a high burden of parasites. Herders need to consider several factors when evaluating whether or not to treat including the practicality of the treatment regimen in relation to administration method (e.g., oral/pour-on), frequency, and financial gain in relation to expected production increase. Expected financial gain may also vary depending on if the herder chooses to treat all animals or chooses to focus on selected groups of the herd, such as the calves, which are usually more vulnerable to parasite burdens. Many different antiparasitic drugs have been experimentally tested in reindeer, but currently, the macrocyclic lactones Ivermectin and Doramectin are the only anthelmintics licensed for use in reindeer in Fennoscandia. Although these drugs are only labelled to use against bot fly larvae, these drugs may also have effect on many other parasites. Still, like other macrocyclic lactones, these drugs have poor blood–brain barrier penetration and, therefore, likely have little effect on brain worm larvae in the CNS

(McKellar and Gokbulut 2012). Nevertheless, Ivermectin treatment has been shown to reduce the brain worm burden in the lungs as compared to untreated control animals (Nordkvist et al. 1983). Antiparasitic treatments are usually conducted by trained herders during reindeer roundups. Because of the extensive growth of warble and throat bot larvae in late winter and spring, treatment should be conducted in autumn or early winter to avoid a large biomass of dead and decomposing larvae under the skin and in the throat in spring.

4.6 Diseases of Zoonotic Concern

Reindeer herding, handling, and the consumption of reindeer meat and products are not associated with a high risk of zoonotic infections in Fennoscandia. The zoosanitary situation in livestock and wildlife in this region is generally favorable, and reindeer do not under normal circumstances act as the main host or main reservoir of human pathogens. However, there are still a number of pathogens that can be transmitted from reindeer to humans given the right circumstances.

Contagious ecthyma, caused by orf virus (ORFV, genus *Parapoxvirus*), is a known zoonotic infection with a worldwide distribution in sheep and goats, which can also infect a wide range of wildlife species. During the severe outbreak among reindeer in Finland in 1992–1993, more than ten human cases were reported (Büttner et al. 1995), which demonstrates the zoonotic nature of this pathogen. ORFV causes local skin infections on fingers and hands, which are painful, but heals spontaneously within a few weeks (see chapter "Parapoxvirus Infections in Northern Species and Populations" for more details). Parapoxvirus-specific DNA has been documented in reindeer found dead on the mountain pastures and from nose swab samples from live animals in Finnmark, Norway, despite the fact that this disease has hitherto never been reported in reindeer from this region. This indicates an ever-present zoonotic risk when handling reindeer, especially with contact with the mouth and saliva, even from animals with no clinical signs of contagious ecthyma.

Toxoplasma gondii is an obligate intracellular protozoan with a worldwide distribution and the causative agent for the disease toxoplasmosis. Felids are the only known definitive host, in which the parasite can undertake sexual reproduction. However, the parasite may infect virtually all warm-blooded animals. People are infected by consuming raw or undercooked meat containing tissue cysts or by ingesting oocysts through contaminated water or poorly washed vegetables or by exposure to cat (felid) feces. Most infections in humans are asymptomatic, but if infected for the first time during pregnancy, toxoplasmosis may cause abortion, stillbirth, or malformations. It has been shown that *T. gondii* may cause encephalitis and placentitis in reindeer, leading to stillbirth (Dubey et al. 2002), and experimentally also fatal enteritis (Oksanen et al. 1996). Antibodies have been detected in wild cervids in Norway, including wild reindeer (Vikøren et al. 2004).

The tape worm *Echinococcus granulosus* uses dogs and wolves as its definitive hosts but may produce tissue cysts (metacestodes) in other animals, such as in reindeer and humans. In western Finnmark, Norway, cysts were found in about

10% of the animals in the 1950s. Parasite treatment of dogs and an increased awareness to not feed raw reindeer meat or organs to dogs during slaughter have resulted in a decline, and cysts have not been found during meat control of reindeer in Norway and Sweden since the 1990s (Kutz et al. 2019). Humans are considered an accidental parasite host, and in humans, the disease is called cystic echinococcosis (hydatid disease). Cysts may develop in the lungs and other organs after the ingestion of eggs from dog feces. These cysts can reach a size of 20 cm and can be fatal to the host, particularly without treatment (Eckert and Deplazes 2004).

The warble fly (*H. tarandi*) may deposit its eggs on humans, causing human myiasis. A total of 39 human cases was reported in Norway in 2011–2016 (Landehag et al. 2017). Clinical manifestations included migratory dermal swellings of the face and head, periorbital edema, and regional swelling of the lymph nodes (lymphadenopathy). Ivermectin treatment is recommended to avoid larval invasion of the eye (ophthalmomyiasis) but has not always been successful.

The oral form of necrobacillosis (*F. necrophorum*) is increasingly diagnosed in semi-domesticated reindeer in Fennoscandia. This bacterium may also be a human pathogen, causing abscesses and throat infections or life-threatening systemic infections (Brazier 2006), but there are no reports of this disease in humans that have been associated with disease outbreaks in reindeer.

Chlamydia pecorum (family *Chlamydiaceae*) is an obligate intracellular Gram-negative bacterium that was recently isolated from reindeer with infectious kerato-conjunctivitis (IKC) in Sweden (Sànchez Romano et al. 2019). Other chlamydia species, that is, *Ch. abortus* and *Ch. psittaci*, are known zoonotic pathogens, but little is known about the zoonotic potential of *Ch. pecorum* (Longbottom and Coulter 2003).

Brucellosis (*Brucella suis* biovar 4) and anthrax (*Bacillus anthracis*) are two zoonoses that can be associated with *Rangifer*, and both are discussed in detail in other chapters.

Chronic wasting disease (CWD) is a transmissible spongiform encephalopathy (TSE) affecting cervids. It is caused by a prion, which is a misfolded prion protein that is highly transmissible and resistant in the environment. Since 2016, CWD has been diagnosed in wild reindeer in Norway, and there is a risk that it may spread to adjacent herds of semi-domesticated reindeer (EFSA 2017). The zoonotic potential of CWD remains uncertain, but available data suggest that the risk of transmission of CWD to humans is low but not negligible (Osterholm et al. 2019).

5 Addressing Health and Diseases in Reindeer Herding

About 95% of the semi-domesticated reindeer in Norway are owned by Sámi people. As with other indigenous peoples, there have been historical conflicts when outsiders (in this instance, Norwegians) colonized the land and thereafter. Closing of country borders in the north of Fennoscandia for border-crossing reindeer husbandry (approximately 1850–1920) hindered the traditional migration routes for the semi-nomadic reindeer herding, and from the start of Norwegian colonization, the Sámi people were continuously exposed to "Norwegianization" with regard to religion, language, and the way of living. The semi-domesticated reindeer is privately owned, and as with other livestock, it is the duty of the animal owner to maintain healthy animals and good animal welfare. However, reindeer herding is not a large industry, and in contrast to the production systems for, for example, pigs, poultry, and farmed fish, there are no organized health services for reindeer. Reindeer health is thus the responsibility of the herder, with support from local veterinarians. Even though most, if not all, Sámi people today speak Norwegian, historical conflicts from the "Norwegianization" periods may still to some extent be relevant today, as some reindeer herders are, for various reasons, reluctant to involve a veterinarian or to be involved in specific research projects addressing reindeer health. Reindeer herders may also evaluate the skills of veterinarians as insufficient to successfully treat disease conditions in reindeer, as veterinarians receive poor education in reindeer health and often lack experience with reindeer practice.

Traditionally, reindeer are marked by owner. Marks are performed by cutting a specific pattern, unique for each person who has the right to own reindeer, in the ear cartilage of the animals. This is still practiced for most of the herds, but some herders also use plastic tags or are testing the use of electronic tags (chips) that are injected under the skin, usually in addition to the ear tagging. With the two latter techniques, it is possible to identify reindeer by individual, which may enable a more systematic overview of the herd. From a veterinary point of view, individual marking also facilitates records for medical treatment, which is useful if an animal has been treated and then subsequently is sent to slaughter.

Reindeer health is, in general, a challenge for today's veterinarians (Fig. 4). With the exception of the antiparasitic drug Ivermectin, there are no drugs registered for use in reindeer today (2021), and dosages and treatment regimens of any drugs used must be improvised based on comparative aspects and experience with other species. Reindeer will always represent a very small part of the clinical work of a veterinarian, and building experience with the species may take many years. Furthermore, reindeer herding, reindeer as a production animal, and its health and diseases presently occupy a very little part, if any, of the veterinary education in the Fennoscandian countries. This results in new generations of veterinarians with not only restricted knowledge but also restricted interest and understanding of reindeer diseases, Sámi reindeer herding, reindeer population dynamics, and the reindeer needs and welfare demands.

Today, reindeer meat ranks highly in people's attitude and experience due to several factors. Reindeer meat has a characteristic reindeer taste due to the animal's natural foraging and comes from free-ranging animals harvested from mountain pastures not exploited by other means. People also view reindeer meat as free of pathogens and drugs and therefore healthy and safe to eat. However, it cannot be taken for granted that these positive aspects of reindeer production will hold true in the future. With the increased practice of feeding, traditional herding is changing more and more toward farming. Free-range grazing is becoming restricted to corrals and fenced areas, and consumption of natural foods is instead switched to ensilage and pelleted feedstuff. Finally, with this change of husbandry comes the emergence

Fig. 4 A reindeer was discovered with an eye infection during a roundup, and the animal is examined, sampled, and treated by two veterinarians. (Photo: Morten Tryland)

of different health and disease challenges and possibly increased medical treatment of the animals.

References

Aagnes TH, Sørmo W, Mathiesen SD (1995) Ruminal microbial digestion in free-living, in captive lichen-fed, and in starved reindeer (*Rangifer tarandus tarandus*) in winter. Appl Environ Microbiol 61:583–591

Aaris-Sørensen K, Muhldorff R, Petersen EB (2007) The Scandinavian reindeer (*Rangifer tarandus* L.) after the last glacial maximum: time, seasonality and human exploitation. J Archaeol Sci 34: 914–923

Åhman B, White RG (2019) Rangifer diet and nutritional needs. In: Tryland M, Kutz S (eds) Reindeer and Caribou - health and disease, 1st edn. CRC Press, Boca Raton, FL, pp 107–134. https://doi.org/10.1201/9780429489617-3

Åhman B, Finstad GL, Josefsen TD (2019) Feeding and associated health problems. In: Tryland M, Kutz S (eds) Reindeer and Caribou - health and disease, 1st edn. CRC Press, Boca Raton, FL, pp 135–156

Anonymous (2020) Ressursregnskap for Reindriftsnæringen. Norwegian Reindeer Husbandry Authority, Alta, Norway. Annual reports. https://www.landbruksdirektoratet.no/no/reindriften/for-siidaandeler/publikasjoner?index=0

Bergman A (1912) Contagious keratitis in reindeer. Skand Vet 2:145–177

Bjørklund I (2013) Domestication, reindeer husbandry and the development of Sámi pastoralism. Acta Borealia 30:174–189. https://doi.org/10.1080/08003831.2013.847676

Box JE, Colgan WT, Christensen TR et al (2019) Key indicators of Arctic climate change: 1971–2017. Environ Res Lett 14(4):045010

Brazier JS (2006) Human infections with fusobacterium necrophorum. Anaerobe 12(4):165–172. https://doi.org/10.1016/j.anaerobe.2005.11.003

Büttner M, von Einem C, McInnes C, Oksanen A (1995) Clinical findings and diagnosis of a severe parapoxvirus epidemic in Finnish reindeer. Tierarztl Prax 23(6):614–618

Das Neves CG, Mørk T, Godfroid T, Sørensen KK, Breines E, Hareide E, Thiry J, Rimstad E, Thiry E, Tryland M (2009) Experimental infection of reindeer with Cervid herpesvirus 2. Clin Vaccine Immunol 16(12):1758–1765

Davidson R, Mørk T, Holmgren KE, Oksanen A (2020) Infection with brainworm (*Elaphostrongylus rangiferi*) in reindeer (*Rangifer tarandus* ssp.) in Fennoscandia. Acta Vet Scand 62:24

Dubey JP, Lewis B, Beam K, Abbitt B (2002) Transplacental toxoplasmosis in a reindeer (*Rangifer tarandus*) fetus. Vet Parasitol 110:131–135. https://doi.org/10.1016/s0304-4017(02)00320-5

Dwyer MJ, Istomin KV (2006) Mobility and technology: understanding the vulnerability of two groups of nomadic pastoralists to reindeer losses. Nomadic Peoples 10:142–165

Eckert J, Deplazes P (2004) Biological, epidemiological, and clinical aspects of echinococcosis, a zoonosis of increasing concern. Clin Microbiol Rev 17(1):107–135. https://doi.org/10.1128/cmr.17.1.107-135.2004

EFSA Panel on Biological Hazards (BIOHAZ), Ricci A, Allende A, Bolton D, Chemaly M, Davies R, Fernández Escámez PS, Gironés R, Herman L, Koutsoumanis K, Lindqvist R, Nørrung B, Robertson L, Sanaa M, Skandamis P, Snary E, Speybroeck N, Ter Kuile B, Threlfall J, Wahlström H, Benestad S, Gavier-Widen D, Miller MW, Ru G, Telling GC, Tryland M, Ortiz Pelaez A, Simmons M (2017) Chronic wasting disease (CWD) in cervids. EFSA J 15(1):e04667. https://doi.org/10.2903/j.efsa.2017.4667

Eilertsen SM, Riseth JA (2021) Rovviltforvaltning og reindriftsnæringen-Hvordan kan todelt målsetting i samsvar med internasjonale konvensjoner praktiseres? I: Flemsæter Frode og Bjørn Egil Flø (red.) Utmark i endring-frå beitemark til rekreativ koloni. Cappelen Damm Akademisk

Fisher DO, Blomberg SP, Owens IPF (2002) Convergent maternal care strategies in ungulates and macropods. Evolution 56:167–176

Forbes BC, Kumpula T, Meschtyb N et al (2016) Sea ice, rain-on-snow and tundra reindeer nomadism in Arctic Russia. Biol Lett 12:20160466

Furberg M, Evengård B, Nilsson M (2011) Facing the limit of resilience: perceptions of climate change among reindeer herding Sami in Sweden. Glob Health Action 4. https://doi.org/10.3402/gha.v4i0.8417

Handeland K, Boye M, Bergsjø B, Bondal H, Isaksen K, Agerholm JS (2010) Digital necrobacillosis in Norwegian wild tundra reindeer (*Rangifer tarandus tarandus*). J Comp Pathol 143(1):29–38. https://doi.org/10.1016/j.jcpa.2009.12.018

Hansen LI (2008) Spesialisert reindrift eller kombinasjonsnæring? Reinholdet i Sør-Troms på 1600- og 1700-tallet. In: Andersen O (ed) Fra villreinjakt til reindrift/Gåddebivdos boatsojsujttuj. Tjálaráddjo/Skriftserie 1/2005, pp 165–183

Hobbs NT, Andrén H, Persson J, Aronsson M, Chapron G (2012) Native predators reduce harvest of reindeer by Sámi pastoralists. Ecol Appl 22:1640–1654

Holand O, Weladji RB, Mysterud A, Roed K, Reimers E, Nieminen M (2012) Induced orphaning reveals post-weaning maternal care in reindeer. Eur J Wildl Res 58:589–596

Horstkotte T, Lépy É, Risvoll C (2020) Supplementary feeding of semi-domesticated reindeer in Fennoscandia: challenges and opportunities. Report, Umeå University

Hueffer K, Drown D, Romanovsky V, Hennessy T (2020) Factors contributing to anthrax outbreaks in the circumpolar north. EcoHealth 17:174–180

Hvidsten D, Stordal F, Lager M, Rognerud B, Kristiansen BE, Matussek A, Gray J, Stuen S (2015) *Borrelia burgdorferi sensu lato*-infected *Ixodes ricinus* collected from vegetation near the Arctic circle. Ticks Tick Borne Dis 6(6):768–773. https://doi.org/10.1016/j.ttbdis.2015.07.002

Istomin KV, Dwyer MJ (2010) Dynamic mutual adaptation: human-animal interaction in reindeer herding pastoralism. Hum Ecol 38:613–623

Jaenson TG, Jaenson DG, Eisen L et al (2012) Changes in the geographical distribution and abundance of the tick *Ixodes ricinus* during the past 30 years in Sweden. Parasit Vectors 5(1):8

Josefsen TD, Mørk T, Nymo IH (2019) Bacterial infections and diseases. In: Tryland M, Kutz SJ (eds) Reindeer and Caribou. Health and disease. CRC Press, Boca Raton, FL, pp 237–271

Kivinen S, Moen J, Berg A, Eriksson Å (2010) Effects of modern forest management on winter grazing resources for reindeer in Sweden. AMBIO J Hum Environ 39:269–278

Kivinen S, Berg A, Moen J, Östlund L, Olofsson J (2012) Forest fragmentation and landscape transformation in a reindeer husbandry area in Sweden. Environ Manag 49:295–304

Kjær LJ, Soleng A, Edgar KS, Lindstedt HEH, Paulsen KM, Andreassen ÅK, Korslund L, Kjelland V, Slettan A, Stuen S, Kjellander P, Christensson M, Teräväinen M, Baum A, Isbrand A, Jensen LM, Klitgaard K, Bødker R (2019) A large-scale screening for the taiga tick, *Ixodes persulcatus*, and the meadow tick, *Dermacentor reticulatus*, in southern Scandinavia, 2016. Parsit Vectors 12:338. https://doi.org/10.1186/s13071-019-3596-3

Kutz S, Laaksonen S, Åsbakk K, Nilssen AC (2019) Parasitic infections and diseases. In: Tryland M, Kutz SJ (eds) Reindeer and Caribou – health and disease. CRC Press, Boca Raton, FL, pp 177–235

Laaksonen S (2019) Assessment and treatment of reindeer diseases. In: Tryland M, Kutz SJ (eds) Reindeer and Caribou – health and disease. CRC Press, Boca Raton, FL, pp 383–444

Laaksonen S, Solismaa M, Kortet R, Kuusela J, Oksanen A (2009) Vectors and transmission dynamics for *Setaria tundra* (Filarioidea; Onchocercidae), a parasite of reindeer in Finland. Parasit Vectors 2:3

Landehag J, Skogen A, Åsbakk K, Kan B (2017) Human myiasis caused by the reindeer warble fly, *Hypoderma tarandi*, case series from Norway, 2011 to 2016. Euro Surveill 22(29):30576. https://doi.org/10.2807/1560-7917.ES.2017.22.29.30576

Lie RW (1986) Animal bones from the late Weichselian in Norway. Fauna Norv Ser A 7:41–46

Longbottom D, Coulter LJ (2003) Animal chlamydiosis and zoonotic implications. J Comp Pathol 128:217–244. https://doi.org/10.1053/jcpa.2002.0629

Lorenzen ED, Nogués-Bravo D, Orlando L et al (2011) Species-specific responses of late quaternary megafauna to climate and humans. Nature 479:359–365

Madslien K, Ytrehus B, Vikoren T et al (2011) Hair-loss epizootic in moose (*Alces alces*) associated with massive deer ked (*Lipoptena cervi*) infestation. J Wildl Dis 47(4):893–906

Madslien K, Ytrehus B, Viljugrein H, Solberg EJ, Bråten KR, Mysterud A (2012) Factors affecting deer ked (Lipoptena cervi) prevalence and infestation intensity in moose (Alces alces) in Norway. Parasit Vectors 5:251. https://doi.org/10.1186/1756-3305-5-251

McKellar QA, Gokbulut C (2012) Pharmacokinetic features of the antiparasitic macrocyclic lactones. Curr Pharm Biotechnol 13:888–911

Medlock JM, Hansford KM, Bormane A, Derdakova M, Estrada-Peña A, George JC, Golovljova I, Jaenson TG, Jensen JK, Jensen PM, Kazimirova M, Oteo JA, Papa A, Pfister K, Plantard O, Randolph SE, Rizzoli A, Santos-Silva MM, Sprong H, Vial L, Hendrickx G, Zeller H, Van Bortel W (2013) Driving forces for changes in geographical distribution of *Ixodes ricinus* ticks in Europe. Parasit Vectors 6:1. https://doi.org/10.1186/1756-3305-6-1

Moen J (2008) Climate change: effects on the ecological basis for reindeer husbandry in Sweden. AMBIO J Hum Environ 37:304–311

Morand S (2015) (macro-) Evolutionary ecology of parasite diversity: from determinants of parasite species richness to host diversification. Int J Parasitol Parasit Wildl 4:80–87

Nieminen M (2019) Reindeer and man: from hunting to domestication. In: Tryland M, Kutz S (eds) Reindeer and Caribou - health and disease, 1st edn. CRC Press, Boca Raton, FL, pp 13–23. https://doi.org/10.1201/9780429489617-3

Nieminen M, Norberg H, Maijala V (2011) Mortality and survival of semi-domesticated reindeer (*Rangifer tarandus tarandus* L.) calves in northern Finland. Rangifer 31:71–84. https://doi.org/10.7557/2.31.1.2029

Nordkvist M (1960) Renens sjukdomar: kort oversikt. Stockholm 1960. (Lappväsendet – Renforskningen. Småskrift, 4) (In Swedish)

Nordkvist M, Karlsson KA (1962) Epizootisk förlöpande infektion med *Pasteurella multocida* hos ren. (Pasteurella epizootic in reindeer). Nord Vet Med 14:1–15. (In Swedish)

Nordkvist MR, Rehbinder C, Christensson D (1983) A comparative study on the efficacy of four anthelmintics on some important reindeer parasites. Rangifer 3:19–38

Nybakk K, Kjelvik O, Kvam T et al (2002) Mortality of semi-domestic reindeer Rangifer tarandus in Central Norway. Wildl Biol 8:63–68

Oksanen A, Gustafsson K, Lundén A, Dubey JP, Thulliez P, Uggla A (1996) Experimental *Toxoplasma gondii* infection leading to fatal enteritis in reindeer (*Rangifer tarandus*). J Parsitol 82:843–845

Osterholm MT, Anderson CJ, Zabel MD, Scheftel JM, Moore KA, Appleby BS (2019) Chronic wasting disease in Cervids: implications for prion transmission to humans and other animal species. mBio 10(4):e01091. https://doi.org/10.1128/mBio.01091-19

Qvigstad J (1941) Den tamme rens sykdommer. Tromsø museums Årshefter 59(1):1–56. (printed 1941; in Swedish)

Riseth JÅ, Johansen B (2019) Inngrepsanalyse for reindrifta i Troms. Rapport 23/2018. Norut, Tromsø

Riseth JÅ, Tømmervik H, Forbes BC (2019) Sustainable and resilient reindeer herding. In: Tryland M, Kutz SJ (eds) Reindeer and Caribou. Health and disease. CRC Press, Boca Raton, FL, pp 23–43

Røed KH, Bjørnstad G, Flagstad Ø et al (2014) Ancient DNA reveals prehistoric habitat fragmentation and recent domestic introgression into native wild reindeer. Conserv Genet 14:1137–1149

Røed KH, Côté SD, Yannic G (2019) *Rangifer tarandus* – classification and genetic variation. In: Tryland M, Kutz SJ (eds) Reindeer and Caribou – health and disease. CRC Press, Boca Raton, FL, pp 1–13

Sánchez Romano J, Mørk T, Laaksonen S, Ågren E, Nymo IH, Sunde M, Tryland M (2018) Infectious keratoconjunctivitis in semi-domesticated Eurasian tundra reindeer (*Rangifer tarandus tarandus*): microbiological study of clinically affected and unaffected animals with special reference to cervid herpesvirus 2. BMC Vet Res 14(1):15. https://doi.org/10.1186/s12917-018-1338-y

Sànchez Romano J, Leijon M, Hagström Å, Jinnerot T, Rockström UK, Tryland M (2019) *Chlamydia pecorum* associated with an outbreak of infectious Keratoconjunctivitis in semi-domesticated reindeer in Sweden. Front Vet Med 6:14. https://doi.org/10.3389/fvets.2019.00014

Skarin A, Åhman B (2014) Do human activity and infrastructure disturb domesticated reindeer? The need for the reindeer's perspective. Polar Biol 37:1041–1054

Skarin A, Nellemann C, Rönnegård L et al (2015) Wind farm construction impacts reindeer migration and movement corridors. Landsc Ecol 30:1527–1540

Skonhoft A, Johannesen AB, Olaussen JO (2017) On the tragedy of the commons: when predation and livestock loss may improve the economic lot of herders. Ambio 46:644–654. https://doi.org/10.1007/s13280-017-0910-1

Solveig LV, Boyce MS (2009) Global declines of caribou and reindeer. Glob Chang Biol 15:2626–2633

Stuen S, Granquist EG, Silaghi C (2013) *Anaplasma phagocytophilum* – a widespread multi-host pathogen with highly adaptive strategies. Front Cell Infect Microbiol 3(31). https://doi.org/10.3389/fcimb.2013.00031

Tikkanen MK, McInnes CJ, Mercer AA, Büttner M, Tuimala J, Hirvelä-Koski V, Neuvonen E, Huovilainen A (2004) Recent isolates of parapoxvirus of Finnish reindeer (*Rangifer tarandus tarandus*) are closely related to bovine pseudocowpox virus. J Gen Virol 85:1413–1418

Tryland M, Josefsen TD, Oksanen A, Aschfalk A (2001) Contagious ecthyma in Norwegian semidomesticated reindeer (*Rangifer tarandus tarandus*). Vet Rec 149:394–395

Tryland M, Das Neves CG, Sunde M, Mørk T (2009) Cervid herpesvirus 2, the primary agent in an outbreak of infectious keratoconjunctivitis in semi-domesticated reindeer. J Clin Microbiol 47(11):3707–3713

Tryland M, Romano JS, Marcin N, Nymo IH, Josefsen TD, Sørensen KK, Mørk T (2017) Cervid herpesvirus 2 and not *Moraxella bovoculi* caused keratoconjunctivitis in experimentally inoculated semi-domesticated Eurasian tundra reindeer. Acta Vet Scand 59(1):23. https://doi.org/10.1186/s13028-017-0291-2

Tryland M, Das Neves CG, Klein J, Mørk T, Hautaniemi M, Wensman J (2019a) Viral infections and diseases. In: Tryland M, Kutz SJ (eds) Reindeer and Caribou – health and disease. CRC Press, Boca Raton, FL, pp 273–303

Tryland M, Nymo IH, Sánchez Romano J, Mørk T, Klein J, Rockström U (2019b) Infectious disease outbreak associated with supplementary feeding of semi-domesticated reindeer. Front Vet Sci 6:126. https://doi.org/10.3389/fvets.2019.00126

Tryland M, Ravolainen V, Pedersen ÅØ (2019c) Climate change. Potential impacts on pasture resources, health and diseases of reindeer and caribou. In: Tryland M, Kutz SJ (eds) Reindeer and Caribou – health and disease. CRC Press, Boca Raton, FL, pp 493–514

Tryland M, Josefsen TD, Sánchez Romano J, Marcin N, Mørk T, Arnemo JM (2021) Case report: severe morbidity and mortality in semi-domesticated eurasian tundra reindeer (*Rangifer tarandus tarandus*) immobilized with medetomidine-ketamine. Front Vet Sci. https://doi.org/10.3389/fvets.2021.606323

Turunen M, Vuojala-Magga T (2014) Past and present winter feeding of reindeer in Finland: herders' adaptive learning of feeding practices. Arctic 67:173–188. https://doi.org/10.14430/arctic4385

Tveraa T, Kulseng A (2020) Ressursregnskap for reindriftsnæringen. Reindriftsåret 1. april – 31. mars 2020. Report 43 (in Norwegian)

Tveraa T, Stien A, Brøseth H et al (2014) The role of predation and food limitation on claims for compensation, reindeer demography and population dynamics. J Appl Ecol 51:1264–1272

Välimäki P, Madlsien K, Malmstedn J, Härkönen L, Härkönen S, Kaitala A, Kortet R, Laaksonen S, Mehl R, Redford L, Ylönen H, Ytrehus B (2010) Fennoscandian distribution of an important parasite of cervids, the deer ked (*Lipoptena cervi*), revisited. Parasitol Res 107(1):117–125. https://doi.org/10.1007/s00436-010-1845-7

van Oort BEH, Hovelsrud GK, Risvoll C, Mohr CW, Jore S (2020) A mini-review of *Ixodes* ticks climate sensitive infection dispersion risk in the Nordic region. Int J Environ Res Public Health 17(15):5387. https://doi.org/10.3390/ijerph17155387

Vikøren T, Tharaldsen J, Fredriksen B, Handeland K (2004) Prevalence of *toxoplasma gondii* antibodies in wild red deer, roe deer, moose, and reindeer from Norway. Vet Parasitol 120:159–169

Vors LS, Boyce MS (2009) Global declines of caribou and reindeer. Glob Chang Biol 15:2626–2633

Waller PJ (2005) Domestication of ruminant livestock and the impact of nematode parasites: possible implications for the reindeer industry. Rangifer 25(1):39–50

Yannic G, Pellissier L, Ortego J et al (2014) Genetic diversity in caribou linked to past and future climate change. Nat Clim Chang 4:132–137

The manufacturer's authorised representative in the EU is Springer Nature Customer Service Centre GmbH, Europaplatz 3, 69115 Heidelberg, Germany. If you have any concerns regarding our products, please contact ProductSafety@springernature.com

Printed and bound by CPI Group (UK) Ltd, Croydon, CR0 4YY
22/09/2026
02220954-0013